Handbook of Research on Mobile Multimedia

Ismail Khalil Ibrahim
Johannes Kepler University Linz, Austria

Idea Group
REFERENCE

IDEA GROUP REFERENCE
Hershey · London · Melbourne · Singapore

Acquisitions Editor: Michelle Potter
Development Editor: Kristin Roth
Senior Managing Editor: Jennifer Neidig
Managing Editor: Sara Reed
Copy Editor: Larissa Vinci
Typesetter: Sharon Berger
Cover Design: Lisa Tosheff
Printed at: Yurchak Printing Inc.

Published in the United States of America by
Idea Group Reference (an imprint of Idea Group Inc.)
701 E. Chocolate Avenue, Suite 200
Hershey PA 17033
Tel: 717-533-8845
Fax: 717-533-8661
E-mail: cust@idea-group.com
Web site: http://www.idea-group-ref.com

and in the United Kingdom by
Idea Group Reference (an imprint of Idea Group Inc.)
3 Henrietta Street
Covent Garden
London WC2E 8LU
Tel: 44 20 7240 0856
Fax: 44 20 7379 0609
Web site: http://www.eurospanonline.com

Library of Congress Cataloging-in-Publication Data

Handbook of research on mobile multimedia / Ismail Khalil Ibrahim, editor.
p. cm.
Summary: "This handbook provides insight into the field of mobile multimedia and associated applications and services"--Provided by publisher.
Includes bibliographical references and index.
ISBN 1-59140-866-0 (hardcover) -- ISBN 1-59140-868-7 (ebook)
1. Mobile communication systems. 2. Wireless communication systems. 3. Multimedia systems. 4. Mobile computing. I. Ibrahim, Ismail Khalil.
TK6570.M6H27 2006
384.3'3--dc22

2006000378

British Cataloguing in Publication Data
A Cataloguing in Publication record for this book is available from the British Library.

Editorial Advisory Board

List of Contributors

Ahmad, Ashraf M. A. / *National Chiao Tung University, Taiwan* 357
Alesanco Iglesias, Álvaro / *University of Zaragoza, Spain* 521
Angelides, Marios C. / *Brunel University, UK* 1
Blechar, Jennifer / *University of Oslo, Norway* 119
Breiteneder, Christian / *Vienna University of Technology, Austria* 383
Bressan, Stéphane / *National University of Singapore, Singapore* 103
Canalda, Philippe / *University of Franche-Comté, France* 491
Chang, Li-Pin / *National Chiao-Tung University, Taiwan* 191
Charlet, Damien / *INRIA-Rocquencourt (ARLES Project), France* 491
Chatonnay, Pascal / *University of Franche-Comté, France* 491
Constantiou, Ioanna D. / *Copenhagen Business School, Denmark* 119
Costa, Patrícia Dockhorn / *Centre for Telematics and Information Technology, University of Twente, The Netherlands* 456
Damsgaard, Jan / *Copenhagen Business School, Denmark* 119
Derballa, Volker / *Universität Augsburg, Germany* 11
Djoudi, Mahieddine / *Université de Poitiers, France* 368
Doolan, Daniel C. / *University College Cork, Ireland* 399
Downes, Barry / *Telecommunications Software & Systems Group (TSSG) and Waterford Institute of Technology (WIT), Ireland* 555
Dustdar, Schahram / *Vienna University of Technology, Austria* 414
Feki, Mohamed Ali / *Handicom Lab, INT/GET, France* 440
Fernández Navajas, Julián / *University of Zaragoza, Spain* 521
Fouliras, Panayotis / *University of Macedonia, Thessaloniki, Greece* 38
García Moros, José / *University of Zaragoza, Spain* 521
Georgiadis, Christos K. / *University of Macedonia, Greece* 266
Giroux, Sylvain / *Université de Sherbrooke, Canada* 544
Gruber, Franz / *RISC Software GmbH, Austria* 507
Hadjiefthymiades, Stathes / *University of Athens, Greece* 139
Häkkilä, Jonna / *Nokia Multimedia, Finland* 326
Hämäläinen, Timo / *University of Jyväskylä, Finland* 179
Harous, Saad / *University of Sharjah, UAE* 368
Hartmann, Werner / *FAW Software Engineering GmbH, Austria* 507
Hernández Ramos, Carolina / *University of Zaragoza, Spain* 521
Istepanian, Robert S. H. / *Kingston University, UK* 521
Jørstad, Ivar / *Norwegian University of Science and Technology, Norway* 414
Kalnis, Panagiotis / *National University of Singapore, Singapore* 103
King, Ross / *Research Studio Digital Memory Engineering, Austria* 232
Klas, Wolfgang / *University of Vienna, Austria* 232

Kostakos, Vassilis / *University of Bath, UK* 71
Kouadri Mostéfaoui, Ghita / *University of Fribourg, Switzerland* 251
Koubaa, Hend / *Norwegian University of Science and Technology (NTNU), Norway* 165
Kronsteiner, Reinhard / *Johannes Kepler University, Austria* 86
Lahti, Janne / *VTT Technical Research Centre of Finland, Finland* 340
Lassabe, Frédéric / *University of Franche-Comté, France* 491
Ledermann, Florian / *Vienna University of Technology, Austria* 383
Lim, Say Ying / *Monash University, Australia* 49
Mahdi, Abdulhussain E. / *University of Limerick, Ireland* 210
Mäntyjärvi, Jani / *VTT Technical Centre of Finland, Finland* 326
Mokhtari, Mounir / *Handicom Lab, INT/GET, France* 440
Moreau, Jean-François / *Université de Sherbrooke, Canada* 544
Nösekabel, Holger / *University of Passau, Germany* 430
O'Neill, Eamonn / *University of Bath, UK* 71
Palola, Marko / *VTT Technical Research Centre of Finland, Finland* 340
Pang, Ai-Chun / *National Taiwan University, Taiwan* 191
Peltola, Johannes / *VTT Technical Research Centre of Finland, Finland* 340
Pfeifer, Tom / *Telecommunications Software & Systems Group (TSSG) and Waterford Institute of Technology (WIT), Ireland* 555
Picovici, Dorel / *University of Limerick, Ireland* 210
Pigot, Hélène / *Université de Sherbrooke, Canada* 544
Pires, Luís Ferreira / *Centre for Telematics and Information Technology, University of Twente, The Netherlands* 456
Pousttchi, Key / *Universität Augsburg, Germany* 11
Priggouris, Ioannis / *University of Athens, Greece* 139
Puttonen, Jani / *University of Jyväskylä, Finland* 179
Röckelein, Wolfgang / *EMPRISE Consulting Düseldorf, Germany* 430
Ruiz Mas, José / *University of Zaragoza, Spain* 521
Savary, Jean-Pierre / *Division R&D CRD, France* 544
Schizas, Christos N. / *University of Cyprus, Cyprus* 1
Sinderen, Marten van / *Centre for Telematics and Information Technology, University of Twente, The Netherlands* 456
Sofokleous, Anastasis A. / *Brunel University, UK* 1
Spies, François / *University of Franche-Comté, France* 491
Srinivasan, Bala / *Monash University, Australia* 49
Stary, Chris / *University of Linz, Austria* 291
Stormer, Henrik / *University of Fribourg, Switzerland* 278
Sulander, Miska / *University of Jyväskylä, Finland* 179
Susilo, Willy / *University of Wollongong, Australia* 534
Tabirca, Sabin / *University College Cork, Ireland* 399
Taniar, David / *Monash University, Australia* 49
Tok, Wee Hyong / *National University of Singapore, Singapore* 103
Turowski, Klaus / *Universität Augsburg, Germany* 11
Valdovinos Bardají, Antonio / *University of Zaragoza, Spain* 521
Van Thanh, Do / *Telenor R & D and Norwegian University of Science and Technology, Norway* 414
Viinikainen, Ari / *University of Jyväskylä, Finland* 179
Vildjiounaite, Elena / *VTT Technical Research Centre of Finland, Finland* 340
Viruete Navarro, Eduardo Antonio / *University of Zaragoza, Spain* 521

Wagner, Roland / *Johannes Kepler University Linz, Austria* 507
Wang, Zhou / *Fraunhofer Integrated Publication and Information Systems Institute (IPSI), Germany* 165
Weippl, Edgar R. / *Vienna University of Technology, Austria* 22
Welzl, Michael / *University of Innsbruck, Austria* 129
Westermann, Utz / *VTT Technical Research Centre of Finland, Finland* 340
Williams, M. Howard / *Heriot-Watt University, UK* 311
Win, Khin Than / *University of Wollongong, Australia* 534
Yang, Laurence T. / *St. Francis Xavier University, Canada* 399
Yang, Yuping / *Heriot-Watt University, UK* 311
Yu, Zhiwen / *Northwestern Polytechnical University, China* 476
Zehetmayer, Robert / *University of Vienna, Austria* 232
Zervas, Evangelos / *TEI-Athens, Greece* 139
Zhang, Daqing / *Institute for Infocomm Research, Singapore* 476
Zheng, Baihua / *Singapore Management University, Singapore* 103

Table of Contents

Foreword xx

Preface xxii

Section I
Basic Concepts

Chapter I
Mobile Computing: Technology Challenges, Constraints, and Standards / *Anastasis A. Sofokleous, Marios C. Angelides, and Christos N. Schizas* 1

Chapter II
Business Model Typology for Mobile Commerce / *Volker Derballa, Key Pousttchi, and Klaus Turowski* 11

Chapter III
Security and Trust in Mobile Multimedia / *Edgar R. Weippl* 22

Chapter IV
Data Dissemination in Mobile Environments / *Panayotis Fouliras* 38

Chapter V
A Taxonomy of Database Operations on Mobile Devices / *Say Ying Lim, David Taniar, and Bala Srinivasan* 49

Chapter VI
Interacting with Mobile and Pervasive Computer Systems / *Vassilis Kostakos and Eamonn O'Neill* 71

Chapter VII
Engineering Mobile Group Decision Support / *Reinhard Kronsteiner* 86

Chapter VIII
Spatial Data on the Move / *Wee Hyong Tok, Stéphane Bressan, Panagiotis Kalnis, and Baihua Zheng* 103

Chapter IX
Key Attributes and the Use of Advanced Mobile Services: Lessons Learned from a Field Study / *Jennifer Blechar, Ioanna D. Constantiou, and Jan Damsgaard* 119

Section II
Standards and Protocols

Chapter X
New Internet Protocols for Multimedia Transmission / *Michael Welzl* 129

Chapter XI
Location-Based Network Resource Management / *Ioannis Priggouris, Evangelos Zervas, and Stathes Hadjiefthymiades* 139

Chapter XII
Discovering Multimedia Services and Contents in Mobile Environments / *Zhou Wang and Hend Koubaa* 165

Chapter XIII
A Fast Handover Method for Real-Time Multimedia Services / *Jani Puttonen, Ari Viinikainen, Miska Sulander, and Timo Hämäläinen* 179

Chapter XIV
Real-Time Multimedia Delivery for All-IP Mobile Networks / *Li-Pin Chang and Ai-Chun Pang* 191

Chapter XV
Perceptual Voice Quality Measurement – Can You Hear Me Loud and Clear? / *Abdulhussain E. Mahdi and Dorel Picovici* 210

Chapter XVI
Modular Implementation of an Ontology-Driven Multimedia Content Delivery Application for Mobile Networks / *Robert Zehetmayer, Wolfgang Klas, and Ross King* 232

Chapter XVII
Software Engineering for Mobile Multimedia: A Roadmap / *Ghita Kouadri Mostéfaoui* 251

Section III
Multimedia Information

Chapter XVIII
Adaptation and Personalization of User Interface and Content / *Christos K. Georgiadis* 266

Chapter XIX
Adapting Web Sites for Mobile Devices – A Comparison of Different Approaches / *Henrik Stormer* 278

Chapter XX
Ensuring Task Conformance and Adaptability of Polymorph Multimedia Systems / *Chris Stary* 291

Chapter XXI
Personalized Redirection of Communication and Data / *Yuping Yang and M. Howard Williams* 311

Chapter XXII
Situated Multimedia for Mobile Communications / *Jonna Häkkilä and Jani Mäntyjärvi* 326

Chapter XXIII
Context-Aware Mobile Capture and Sharing of Video Clips / *Janne Lahti, Utz Westermann, Marko Palola, Johannes Peltola, and Elena Vildjiounaite* 340

Chapter XXIV
Content-Based Video Streaming Approaches and Challenges / *Ashraf M. A. Ahmad* 357

Chapter XXV
Portable MP3 Players for Oral Comprehension of a Foreign Language / *Mahieddine Djoudi and Saad Harous* 368

Chapter XXVI
Towards a Taxonomy of Display Styles for Ubiquitous Multimedia / *Florian Ledermann and Christian Breiteneder* 383

Chapter XXVII
Mobile Fractal Generation / *Daniel C. Doolan, Sabin Tabirca, and Laurence T. Yang* 399

Section IV
Applications and Services

Chapter XXVIII
Mobile Multimedia Collaborative Services / *Do Van Thanh, Ivar Jørstad, and Schahram Dustdar* 414

Chapter XXIX
V-Card: Mobile Multimedia for Mobile Marketing / *Holger Nösekabel and Wolfgang Röckelein* 430

Chapter XXX
Context Awareness for Pervasive Assistive Environment / *Mohamed Ali Feki and Mounir Mokhtari* 440

Chapter XXXI
Architectural Support for Mobile Context-Aware Applications / *Patrícia Dockhorn Costa, Luís Ferreira Pires, and Marten van Sinderen* 456

Chapter XXXII
Middleware Support for Context-Aware Ubiquitous Multimedia Services / *Zhiwen Yu and Daqing Zhang* 476

Chapter XXXIII
Mobility Prediction for Multimedia Services / *Damien Charlet, Frédéric Lassabe, Philippe Canalda, Pascal Chatonnay, and François Spies* 491

Chapter XXXIV
Distribution Patterns for Mobile Internet Applications / *Roland Wagner, Franz Gruber, and Werner Hartmann* 507

Chapter XXXV
Design of an Enhanced 3G-Based Mobile Healthcare System / *José Ruiz Mas, Eduardo Antonio Viruete Navarro, Carolina Hernández Ramos, Álvaro Alesanco Iglesias, Julián Fernández Navajas, Antonio Valdovinos Bardají, Robert S. H. Istepanian, and José García Moros* 521

Chapter XXXVI
Securing Mobile Data Computing in Healthcare / *Willy Susilo and Khin Than Win* 534

Chapter XXXVII
Distributed Mobile Services and Interfaces for People Suffering from Cognitive Deficits / *Sylvain Giroux, Hélène Pigot, Jean-François Moreau, and Jean-Pierre Savary* 544

Chapter XXXVIII
Mobile Magazines / *Tom Pfeifer and Barry Downes* 555

About the Authors 573

Detailed Table of Contents

Foreword xx

Preface xxii

Section I
Basic Concepts

Mobile multimedia is the set of standards and protocols for the exchange of multimedia information over wireless networks. It enables information systems to process and transmit multimedia data to provide end users with access to data, no matter where the data is stored or where the user happens to be. Section I consists of nine chapters to introduce the readers to the basic ideas behind mobile multimedia and provides the business and technical drivers, which initiated the mobile multimedia revolution.

Chapter I
Mobile Computing: Technology Challenges, Constraints, and Standards / *Anastasis A. Sofokleous, Marios C. Angelides, and Christos N. Schizas* 1

Ubiquitous and mobile computing has made any information, any device, any network, any time, anywhere an everyday reality. This chapter discusses the main research and development in mobile technology and standards that make ubiquity a reality: from wireless middleware client profiling to m-commerce services.

Chapter II
Business Model Typology for Mobile Commerce / *Volker Derballa, Key Pousttchi, and Klaus Turowski* 11

Mobile technology enables enterprises to introduce new business models by applying new forms of organization or offering new products and services. In this chapter, a business model typology is introduced where the building blocks in the form of generic business model types are identified and used to create concrete business models.

Chapter III
Security and Trust in Mobile Multimedia / *Edgar R. Weippl* 22

Mobile multimedia applications are becoming increasingly popular because today's cell phones and PDAs often include digital cameras and can also record audio. It is a challenge to accommodate existing techniques for protecting multimedia content on the limited hardware and software basis provided by mobile devices. This chapter provides a comprehensive overview of mobile multimedia security.

Chapter IV
Data Dissemination in Mobile Environments / *Panayotis Fouliras* 38

Data dissemination in mobile environments represents the cornerstone of network-based services. This chapter outlines the existing proposals and the related issues employing a simple but concise methodology.

Chapter V
A Taxonomy of Database Operations on Mobile Devices / *Say Ying Lim, David Taniar, and Bala Srinivasan* 49

Database operations on mobile devices represent a critical research issue. This chapter presents an extensive study of database operations on mobile devices, which provides an understanding and directions for processing data locally on mobile devices.

Chapter VI
Interacting with Mobile and Pervasive Computer Systems / *Vassilis Kostakos and Eamonn O'Neill* 71

Human-computer interaction presents an exciting and timely research direction in mobile multimedia. This chapter introduces novel interaction techniques aiming at improving the way users interact with mobile and pervasive systems. Three broad categories: stroke interaction, kinesthetic interaction, and text entry are presented.

Chapter VII
Engineering Mobile Group Decision Support / *Reinhard Kronsteiner* 86

Group decision support in mobile environments is one of the promising research directions in mobile multimedia. In this chapter, mobile decision support systems were categorized based on the complexity of the decision problem space and group composition. This categorization leads to a set of requirements that are used for designing and implementing a collaborative decision support system.

Chapter VIII
Spatial Data on The Move / *Wee Hyong Tok, Stéphane Bressan, Panagiotis Kalnis, and Baihua Zheng* 103

Advances in mobile devices and wireless networking infrastructure have created a plethora of location-based services where users need to pose queries to remote servers. This chapter identifies the issues and challenges of processing spatial data on the move and presents insights on the state-of-the-art spatial query processing techniques.

Chapter IX
Key Attributes and the Use of Advanced Mobile Services: Lessons Learned from a Field Study / *Jennifer Blechar, Ioanna D. Constantiou, and Jan Damsgaard* 119

This chapter investigates key attributes deemed to provide indications of the behavior of consumers in the m-services market. It illustrates the manner in which users' perceptions related to the key attributes of service quality, content-device fit and personalization were adversely affected by trial of the m-services offered.

Section II
Standards and Protocols

The key feature of mobile multimedia is to combine the Internet, telephones, and broadcast media into a single device. Section II, which consists of eight chapters, explains the enabling technologies for mobile multimedia with respect to communication networking protocols and standards.

Chapter X
New Internet Protocols for Multimedia Transmission / *Michael Welzl* 129

This chapter introduces three new IETF transport layer protocols in support of multimedia data transmission and discusses their usage. In addition, the chapter concludes with an overview of the DCCP protocol for the transmission of real-time multimedia data streams.

Chapter XI
Location-Based Network Resource Management / *Ioannis Priggouris, Evangelos Zervas, and Stathes Hadjiefthymiades* 139

Extensive research on mobile multimedia communications concentrates on how to provide mobile users with at least similar multimedia services as those available to fixed hosts. This chapter aims to provide a general introduction to the emerging research area of mobile communications where the user's location is exploited to optimally manage both the capacity of the network and the offered quality of service.

Chapter XII
Discovering Multimedia Services and Contents in Mobile Environments / *Zhou Wang and Hend Koubaa* 165

Accessing multimedia services from portable devices in nomadic environments is of increasing interest for mobile users. Service discovery mechanisms help mobile users to freely and efficiently locate multimedia services they want. This chapter provides an introduction to the state-of-the-art in service discovery, architectures, technologies, emerging industry standards and advances in the research world. The chapter also describes in great depth the approaches for content location in mobile ad-hoc networks.

Chapter XIII
A Fast Handover Method for Real-Time Multimedia Services / *Jani Puttonen, Ari Viinikainen, Miska Sulander, and Timo Hämäläinen* 179

Mobile IPv6 has been standardized for mobility management in the IPv6 networks. In this chapter, a fast handover method called flow-based fast handover for Mobile IPv6 (FFHMIPv6) is introduced and its performance is compared to other fundamental handover methods.

Chapter XIV
Real-Time Multimedia Delivery for All-IP Mobile Networks / *Li-Pin Chang and Ai-Chun Pang* 191

The introduction of mobile/wireless systems such as 3G and WLAN has driven the Internet into new markets to support mobile users. This chapter focuses on QoS support for multimedia streaming and the dynamic session management for VoIP applications. An efficient multimedia broadcasting/multicasting approach is introduced to provide different levels of QoS, and a dynamic session refreshing approach for the management of disconnected VoIP sessions is proposed.

Chapter XV
Perceptual Voice Quality Measurement – Can You Hear Me Loud and Clear? / *Abdulhussain E. Mahdi and Dorel Picovici* 210

For telecommunication systems, voice communication quality is the most visible and important aspects to QoS, and the ability to monitor and design for this quality should be a top priority. This chapter examines some of the technological issues related to voice quality measurement and describes their various classes.

Chapter XVI
Modular Implementation of an Ontology-Driven Multimedia Content Delivery Application for Mobile Networks / *Robert Zehetmayer, Wolfgang Klas, and Ross King* 232

Mobile multimedia applications provide users with only limited means to define what information they wish to receive. However, users would prefer to receive content that reflect specific personal interests. This chapter presents a prototype multimedia application that demonstrates personalized content delivery using the multimedia messaging service (MMS) protocol.

Chapter XVII
Software Engineering for Mobile Multimedia: A Roadmap / *Ghita Kouadri Mostéfaoui* 251

Research on mobile multimedia mainly focuses on improving wireless protocols in order to improve the quality of service. This chapter argues that software engineering perspective should be investigated in more depth in order to boost the mobile multimedia industry.

Section III
Multimedia Information

Multimedia information as combined information presented by various media types (text, pictures, graphics, sounds, animations, videos) enriches the quality of the information and represents the reality as adequately as possible. Section III contains ten chapters and is dedicated to how information can be exchanged over wireless networks whether it is voice, text, or multimedia information.

Chapter XVIII
Adaptation and Personalization of User Interface and Content / *Christos K. Georgiadis* 266

This chapter is concerned with the building of an adaptive multimedia system that can customize the representation of multimedia content to the specific needs of a user. A personalization perspective is deployed to classify the multimedia interface elements and to analyze their influence on the effectiveness of mobile applications.

Chapter XIX
Adapting Web Sites for Mobile Devices – A Comparison of Different Approaches / *Henrik Stormer* 278

Currently, almost all Web sites are designed for stationary computers and cannot be shown directly on mobile devices due to small display size, delicate data input facilities, and smaller bandwidth. This chapter compares different server side solutions to adapt Web sites for mobile devices.

Chapter XX
Ensuring Task Conformance and Adaptability of Polymorph Multimedia Systems / *Chris Stary* 291

The characteristics of mobile multimedia interaction are captured through accommodating multiple styles and devices at a generic layer of abstraction in an interaction model. This model is related to text representations in terms of work tasks, user roles and preferences, and problem-domain data at an implementation-independent layer. This chapter shows how specifications of mobile multimedia applications can be checked against usability principles very early in software development through an analytical approach.

Chapter XXI
Personalized Redirection of Communication and Data / *Yuping Yang and M. Howard Williams* 311

The vision of mobile multimedia lies in a universal system that can deliver information and communications at any time and place and in any form. Personalized redirection is concerned with providing the user with appropriate control over what communication is delivered and where, depending on his/her context and nature of communication and data. This chapter provides an understanding of what is meant by personalized redirection through a set of scenarios.

Chapter XXII
Situated Multimedia for Mobile Communications / *Jonna Häkkilä and Jani Mäntyjärvi* 326

Situated mobile multimedia has been enabled by technological developments in recent years, including mobile phone integrated cameras, audio-video players, and multimedia editing tools, as well as improved sensing technologies and data transfer formats. This chapter presents the state of the art in situated mobile multimedia, identifies the existing developments trends, and builds a roadmap for future directions.

Chapter XXIII
Context-Aware Mobile Capture and Sharing of Video Clips / *Janne Lahti, Utz Westermann, Marko Palola, Johannes Peltola, and Elena Vildjiounaite* 340

The current research in video management has been neglecting the increased attractiveness of using camera-equipped mobile phones for the production of short home video clips. This chapter presents MobiCon, a mobile, context aware home video production tools that allows users to capture video clips with their camera phones, to semi automatically create MPEG-7 conformant annotations, to upload both clips and annotations to the users' video collections, and to share these clips with friends using OMA DRM.

Chapter XXIV
Content-Based Video Streaming Approaches and Challenges / *Ashraf M. A. Ahmad* 357

Video streaming poses significant technical challenges in quality of service guarantee and efficient resource management in mobile multimedia. This chapter investigates current approaches and their related challenges of content-based video streaming under various network resource requirements.

Chapter XXV
Portable MP3 Players for Oral Comprehension of a Foreign Language / *Mahieddine Djoudi and Saad Harous* 368

Portable MP3 players can be adopted as a useful tool for teaching/learning of languages. This chapter proposes a method for using portable MP3 players for oral comprehension of a foreign language in a diversified population.

Chapter XXVI
Towards a Taxonomy of Display Styles for Ubiquitous Multimedia / *Florian Ledermann and Christian Breiteneder* 383

Classification of display styles for ubiquitous multimedia is essential for the construction of future multimedia systems that are capable of automatically generating complex yet legible graphical responses from an underlying abstract information space such as a semantic network. In this chapter, a domain independent taxonomy of sign functions, rooted in an analysis of physical signs found in public space, is presented.

Chapter XXVII
Mobile Fractal Generation / *Daniel C. Doolan, Sabin Tabirca, and Laurence T. Yang*............ 399

In the past years, there have been few applications developed to generate fractal images on mobile phones. This chapter discusses three possible methodologies for visualizing images on mobile devices. These methodologies include: the generation of an image on a phone, the use of a server to generate the image, and the use of a network of phones to distribute the processing task.

Section IV
Applications and Services

The explosive growth of the Internet and the rising popularity of mobile devices have created a dynamic business environment where a wide range of mobile multimedia applications and services, such as mobile working place, mobile entertainment, mobile information retrieval, and context-based services are emerging everyday. Section IV with its eleven chapters will clarify in a simple and self-implemented way how to implement basic applications for mobile multimedia services.

Chapter XXVIII
Mobile Multimedia Collaborative Services / *Do Van Thanh, Ivar Jørstad, and Schahram Dustdar*.. 414

Mobile multimedia collaborative services allow people, teams, and organizations, to collaborate in a dynamic, flexible, and efficient manner. This chapter studies different collaboration forms in mobile multimedia by reviewing existing collaborative services and describing the service-oriented architecture platform supporting mobile multimedia collaborative services.

Chapter XXIX
V-Card: Mobile Multimedia for Mobile Marketing / *Holger Nösekabel and Wolfgang Röckelein* 430

V-card is a service to create personalized multimedia messages. This chapter presents the use of mobile multimedia for marketing services by introducing the V-card technical infrastructure, related projects, a field test evaluation as well as the social and legal issues emerging from mobile marketing.

Chapter XXX
Context Awareness for Pervasive Assistive Environment / *Mohamed Ali Feki and Mounir Mokhtari* 440

This chapter describes a model-based method for environment design in the field of smart homes dedicated to people with disabilities. This model introduces two constraints in a context-aware environment: the control of different types of assistive devices (environmental control system) and the presence of the user with disabilities (user profile).

Chapter XXXI
Architectural Support for Mobile Context-Aware Applications / *Patrícia Dockhorn Costa, Luís Ferreira Pires, and Marten van Sinderen* 456

Context awareness has emerged as an important and desirable research discipline in distributed mobile systems, since it benefits from the changes in the user's context to dynamically tailor services based on the user's current situation and needs. This chapter presents the design of a flexible infrastructure to support the development of mobile context-aware applications.

Chapter XXXII
Middleware Support for Context-Aware Ubiquitous Multimedia Services / *Zhiwen Yu and Daqing Zhang* 476

In order to facilitate the development and proliferation of multimedia services in ubiquitous environments, a context-aware multimedia middleware is essential. This chapter discusses the middleware support issues for context-aware multimedia services. The design and implementation of a context-aware multimedia middleware called CMM is presented.

Chapter XXXIII
Mobility Prediction for Multimedia Services / *Damien Charlet, Frédéric Lassabe, Philippe Canalda, Pascal Chatonnay, and François Spies* 491

Advances in technology have enabled a broad and outbreaking solutions for new mobile multimedia applications and services. It is necessary to predict adaptation behavior which not only addresses the mobile usage or the infrastructure availability but also the service quality especially the continuity of services.

Chapter XXXIV
Distribution Patterns for Mobile Internet Applications / *Roland Wagner, Franz Gruber, and Werner Hartmann* 507

Developing applications for mobile multimedia is a challenging task due to the limitation of mobile devices such as small memory, limited bandwidth, and the probability of connection losses. This chapter analyses application distribution patterns for their applicability for the mobile environment and the IP multimedia subsystem which is part of the current specification of 3G mobile network is introduced.

Chapter XXXV
Design of an Enhanced 3G-Based Mobile Healthcare System / *José Ruiz Mas, Eduardo Antonio Viruete Navarro, Carolina Hernández Ramos, Álvaro Alesanco Iglesias, Julián Fernández Navajas, Antonio Valdovinos Bardají, Robert S. H. Istepanian, and José García Moros* 521

This chapter describes the design and use of an enhanced mobile healthcare multi-collaborative system operating over a 3G mobile network. The system provides real-time and other non-real-time transmission of medical data using the most appropriate codecs.

Chapter XXXVI
Securing Mobile Data Computing in Healthcare / *Willy Susilo and Khin Than Win* 534

Access to mobile data and messages is essential in healthcare environment as patients and healthcare providers are mobile by providing easy availability of data at the point of care. In the chapter, the need of mobile devices in healthcare, usage of these devices, underlying technology and applications, securing mobile data communication are outlined and studied through different security models and case examples.

Chapter XXXVII
Distributed Mobile Services and Interfaces for People Suffering from Cognitive Deficits /
Sylvain Giroux, Hélène Pigot, Jean-François Moreau, and Jean-Pierre Savary 544

This chapter presents a mobile device that is designed to offer several services to enhance autonomy, security, and communication among the cognitively impaired people and their caregivers. These services include a simplified reminder, an assistance request service, and an ecological information gathering service.

Chapter XXXVIII
Mobile Magazines / *Tom Pfeifer and Barry Downes* 555

Mobile magazines are magazines over mobile computing and communication platforms providing valuable, current multimedia content. This chapter introduces the m-Mag eco-system as the next generation mobile publishing service. Using Parlay/OSA as an open approach, the m-Mag platform can be integrated into an operator's network using standardized APIs and is portable across different operator networks.

About the Authors 573

Foreword

Recent years have witnessed a sustained growth of interest in mobile computing and communications. Indicators are the rapidly increasing penetration of the cellular phone market in Europe, or the mobile computing market growing nearly twice as fast as the desktop market. In addition, technological advancements have significantly enhanced the usability of mobile communication and computer devices. From the first CT1 cordless telephones to today's Iridium mobile phones and laptops/PDAs with wireless Internet connection, mobile tools and utilities have made the life of many people at work and at home much easier and more comfortable. As a result, mobility and wireless connectivity are expected to play a dominant role in the future in all branches of economy. This is also motivated by the large number of potential users (a U.S. study reports of one in six workers spending at least 20% of their time away from their primary workplace, similar trends are observed in Europe). The addition of mobility to data communications systems has not only the potential to put the vision of "being always on" into practice, but has also enabled new generation of services (e.g., location-based services).

Mobile applications are based on a computational paradigm, which is quite different from the traditional model, in which programs are executed on a stationary single computer. In mobile computing, processes may migrate (with users) according to the tasks they perform, providing the user with his or her particular work environment wherever he or she is. To accomplish this goal of ubiquitous access, key requirements are platform independence but also automatic adaptation of applications to: (1) the processing capabilities that the current execution platform is able to offer and (2) the connectivity that is currently provided by the network. Mobile services and applications differ with respect to the quality of service delivered (in terms of reliability and performance) and the degree of mobility they support, ranging from stationary, to walking, to even faster movements in cars, trains, or airplanes. A particular challenge is imposed by (interactive) multimedia applications, which are characterized by high QoS demands. New methods and techniques for characterizing the workload and for QoS modeling are needed to adequately capture the characteristics of mobile commerce applications and services.

A fundamental necessity for mobile information delivery is to understand the behavior and needs of the users (i.e., of the people). Recent research issues include efficient mechanisms for the prediction of user behavior (e.g., location of users in cellular systems) in order to allow for proactive management of the underlying networks. Besides this quantitative evaluation user behavior can also be studied from a quantitative point of view (how well is the user able to do her or his job, what is the level of user satisfaction, etc.) to provide information to other services, which can adapt accordingly. This kind of adaptation may for example, include changes in the user interface, but also chances in the type of information transmitted to the user.

From a telecommunications infrastructure point of view, the key enabling technology for mobility are wireless networks and mobile computing/communication devices, including smart phones, PDAs, or (Ultra)portables. Wireless technologies are deployed in global and wide area networks (GSM, GPRS, and future UMTS, wireless broadband networks, GEO and LEO satellite systems) in local area networks (WLAN, mobile IP), but also in even smaller regional units such as a campus or a room (Bluetooth). Research on wireless networking technologies is mainly driven by the quality of service requirements of distributed (multimedia) applications with respect to the availability of bandwidth as well as performance, reliability, and security of access.

Being provocative, one might state that the situation that the application developers are facing nowadays in mobile computing is similar to the early days of mainframe computing. Comparatively "dumb" clients with restricted graphical capabilities are connected to remote servers over limited bandwidth. Although significant improvements have been achieved increasing the capabilities of networks and devices, there will always be a plethora of networks and devices, and the challenge is to provide a seamlessly integrated access as well as adaptability to devices in application development making utmost use of the available resources.

I am delighted to write the Foreword to this handbook, as its scope, content, and coverage provides a descriptive, analytical, and comprehensive assessment of factors, trends, and issues in the ever-changing field of mobile multimedia. This authoritative research-based publication also offers in-depth explanations of mobile solutions and their specific applications areas, as well as an overview of the future outlook for mobile multimedia.

I am pleased to be able to recommend this timely reference source to readers, be they it researchers looking for future directions to pursue when examining issues in the field or practitioners interested in applying pioneering concepts in practical situations and looking for the perfect tool.

Gabriele Kotsis
President of the Austrian Computer Society, Austria
September 2005

Preface

The demand for mobile access to data no matter where the data is stored and where the user happens to be, in addition to the explosive growth of the Internet and the rising popularity of mobile devices, are among the factors that have created a dynamic business environment, where companies are competing to provide customers access to information resources and services any time, any where.

Advances in wireless networking, specifically the development of the IEEE 802.11 protocol family and the rapid deployment and growth of GSM (and GPRS), have enabled a broad spectrum of novel and out breaking solutions for new applications and services. Voice services are no longer sufficient to satisfy customers' business and personal requirements. More and more people and companies are demanding for mobile access to multimedia services.

Mobile multimedia seems to be the next mass market in mobile communications following the success of GSM and SMS. It enables the industry to create products and services to better meet the consumer needs. However, an innovation in itself does not guarantee a success; it is necessary to be able to predict the new technology adaptation behaviour and to try to fulfil customer needs rather than to wait for a demand pattern to surface.

It is beyond all expectations that mobile multimedia will create significant added values for customers by providing mobile access to Internet-based, multimedia services, video conferencing, and streaming. Mobile multimedia is one of the mainstream systems for the next generation mobile communications, featuring large voice capacity, multimedia applications, and high-speed mobile data services. As for the technology, the trend in the radio frequency area is to shift from narrowband to wideband with a family of standards tailored to a variety of application needs. Many enabling technologies including WCDMA, software-defined radio, intelligent antennas, and digital processing devices are greatly improving the spectral efficiency of third generation systems. In the mobile network area, the trend is to move from traditional circuit-switched systems to packet-switched programmable networks that integrate both voice and packet services, and eventually evolve towards an all-IP network.

While for the information explosion, the addition of mobility to data communications systems has enabled new generation of services not meaningful in a fixed network (that is, positioning-based services. However, the development of mobile multimedia services has only started, and, in the future, we will see new application areas opening up.

Research in mobile multimedia is typically focused on bridging the gap between the high resource demands of multimedia applications and the limited bandwidth and capabilities offered by state-of-the art networking technologies and mobile devices.

OVERVIEW OF MOBILE MULTIMEDIA

Mobile multimedia can be defined as a set of protocols and standards for multimedia information exchange over wireless networks. It enables information systems to process and transmit multimedia data to provide end users with services from various areas, such as mobile working place, mobile entertainment, mobile information retrieval and context-based services.

Multimedia information as combined information presented by more than one media type (text [+pictures] [+graphics] [+sounds] [+animations] [+videos]) enriches the quality of the information and is a way to represent reality as adequately as possible. Multimedia allows users to enhance their understanding of the provided information and increases the potential of person to person and person to system communication.

Mobility as one of the key drivers of mobile multimedia can be decomposed into:

- **User mobility:** The user is forced to move from one location to location during fulfilling his activities. For the user, the access to information and computing resources is necessary regardless of his actual position (e.g., terminal services, VPNs to company-intern information systems).
- **Device mobility:** User activities require a device to fulfill his needs regardless of the location in a mobile environment (e.g., PDAs, notebooks, cell phones, etc.).
- **Service mobility:** The service itself is mobile and can be used in different systems and can be moved seamlessly among those systems (e.g., mobile agents).

The special requirements coming along with the mobility of users, devices, and services, and, specifically the requirements of multimedia as traffic type bring the need of new paradigms in software-engineering and system-development but also in non-technical issues such as the emergence of new business models and concerns about privacy, security, or digital inclusion to name a few.

The key feature of mobile multimedia is around the idea of reaching customers and partners, regardless of their location and delivering multimedia content to the right place at the right time. Key drivers of this technology are, on the one hand technical, and on the other, business drivers.

Evolutions in technology pushed the penetration of the mobile multimedia market and made services in this field feasible. The miniaturization of devices and the coverage of radio networks are the key technical drivers in the field of mobile multimedia.

- **Miniaturization:** The first mobile phones had brick-like dimensions. Their limited battery capacity and transmission range restricted their usage in mobile environments. Actual mobile devices with multiple features fit into cases with minimal dimensions and can be (and are) carried by the user in every situation.
- **Radio networks:** Today's technology allows radio networks of every size for every application scenario. Nowadays, public wireless wide area networks cover the bulk of areas especially in congested areas. They enable (most of the time) adequate quality of service. They allow location-independent service provision and virtual private network access.
- **Market evolution:** The market for mobile devices changed in the last years. Ten years ago, the devices have not been really mobile (short-time battery operation, heavy and large devices), but therefore, they have been expensive and affordable just for high-class business people. Shrinking devices and falling operation- (network-) costs made mobile devices to a mass-consumer-good available and affordable for everyone. The result is a dramatically subscriber growth, and therefore, a new increasing market for mobile multimedia services.
- **Service evolution:** The permanent increasing market brought more and more sophisticated services, starting in the field of telecommunication from poor quality speech-communication to real-time video conferencing. Meanwhile, mobile multimedia services provide rich media content and intelligent context-based services.

The value chain of mobile multimedia services describes the players involved in the business with mobile multimedia. Every service in the field of mobile multimedia requires that their output and service fees must be divided to them considering interdependencies in the complete service life cycle.

- **Network operators:** They provide end-users with the infrastructure to access services mobile via wireless networks (e.g., via GSM/GPRS/UMTS).
- **Content provider:** Content provider and aggregators license content and prepare it for end-users. They collect information and services to provide customers with convenient service collection adapted for mobile use.
- **Fixed Internet Company:** Those companies create the multimedia content. Usually, they provide it already via the fixed Internet but are not specialized on mobile service provisioning. They handle the computing infrastructure and content creation.
- **App developers and device manufacturers:** They deliver hardware and software for mobile multimedia services and are not involved with any type of content creation and delivering.

WHO SHOULD READ THIS HANDBOOK

This handbook provides:

- An insight into the field of mobile multimedia and associated technologies;
- The background for understanding those emerging applications and services;
- Major advantages and disadvantages of individual technologies and the problems that must be overcome;
- An outlook in the future of mobile multimedia.

The handbook is intended for people interested in mobile multimedia at all levels. The primary audience of this book includes students, developers, engineers, innovators, research strategists, and IT-managers who are looking for the big picture of how to integrate and deliver mobile multimedia products and services.

While the handbook can be used as a textbook, system developers and technology innovators can also use it, which gives the book a competitive advantage over existing publications.

WHAT MAKES THIS HANDBOOK DIFFERENT?

Despite the fact that mobile multimedia is the next generation information revolution and the cash cow that presents an opportunity and a challenge for most people and businesses. This book is intended to clarify the hype, which surrounds the concept of mobile multimedia through introducing the idea in a clear and understandable way. This book has a strong focus on mobile solutions, addressing specific application areas. It gives an overview of the key future trends on mobile multimedia including UMTS focusing on mobile applications as well as on future technologies. It also serves as a forum for discussions on economic, political as well as strategic aspects of mobile communications and aims to bring together user groups with operators, manufacturers, service providers, content providers and developers from different sectors like business, health care, public administration and regional development agencies, as well as to developers, telecommunication, and infrastructure operators,...etc.

ORGANIZATION OF THIS HANDBOOK

Mobile multimedia is defined as a set of protocols and standards for multimedia information exchange over wireless networks. Therefore, the book will be organized into four sections. The introduction section, which consists of nine chapters, introduces the readers to the basic ideas behind mobile multimedia and provides the business and technical drivers, which initiated the mobile multimedia revolution. Section II, which consists of eight chapters, explains the enabling technologies for mobile multimedia with respect to communication networking protocols and standards. Section III contains 10 chapters and is dedicated to how information can be exchanged over wireless networks whether it is voice, text, or multimedia information. Section IV with its eleven chapters will clarify, in a simple a self-implemented way, how to implement basic applications for mobile multimedia services.

A CLOSING REMARK

This handbook has been compiled from extensive work done by the contributing authors, who are researchers and industry professionals in this area and who particularly have expertise in the topic area addressed in their respective chapters. We hope readers will benefit from the works presented in this handbook.

Ismail Khalil Ibrahim
September 2005

Acknowledgments

The editor would like to acknowledge the help of all involved in the collation and review process of the handbook, without whose support the project could not have been satisfactorily completed. A special thanks goes to Idea Group Inc. Special thanks goes to Mehdi Khosrow-Pour, Jan Travers, Kristin Roth, Renée Davies, Amanda Phillips, and Dorsey Howard, whose contributions throughout the whole process from initial idea to final publication have been invaluable. I would like to express my sincere thanks to the advisory board and my employer Johannes Kepler University Linz and my colleagues at the Institute of Telecooperation for supporting this project. In closing, I wish to thank all of the authors for their insights and excellent contributions to this handbook, in addition to all those who assisted in the review process.

Ismail Khalil Ibrahim
Johannes Kepler University Linz, Austria

Section I

Basic Concepts

Mobile multimedia is the set of standards and protocols for the exchange of multimedia information over wireless networks. It enables information systems to process and transmit multimedia data to provide end users with access to data, no matter where the data is stored or where the user happens to be. Section I consists of nine chapters to introduce the readers to the basic ideas behind mobile multimedia and provides the business and technical drivers, which initiated the mobile multimedia revolution.

Chapter I
Mobile Computing:
Technology Challenges, Constraints, and Standards

Anastasis A. Sofokleous
Brunel University, UK

Marios C. Angelides
Brunel University, UK

Christos N. Schizas
University of Cyprus, Cyprus

ABSTRACT

Mobile communications and computing has changed forever the way people communicate and interact and it has made "any information, any device, any network, anytime, anywhere" an everyday reality which we all take for granted. This chapter discusses the main research and development in the mobile technology and standards that made ubiquity a reality: from wireless middleware to wireless client profiling to m-commerce services.

INTRODUCTION

What motivates the ordinary household to embark on mobile computing is the availability of low-cost, lightweight, portable "Internet" computers. What fuels this further are protocols and standards developed specially, or modified, to enable mobile devices to work pervasively: "any information, any device, any network, anytime, anywhere" and hence to support mobile applications especially m-commerce. Mobile devices are usually being utilized based on the location and mobile users' profile, and therefore content has to be provided and most of the times to be adapted in a suitable format. Although mobile devices' constrains vary (e.g., data transfer speed, performance, memory capabilities, display resolution, etc), researchers and practitioners taking advantage of new technologies and standards, are trying to overcome every limitation and constraint.

This chapter presents an overview of mobile computing and discusses its current limitations. In addition, it presents research and development work currently carried out in the area of technology and standards, and emphases the effect industry has on mobile computing. Furthermore, this chapter aims to provide a complete picture of mobile computing challenges in terms of payment, commerce, middleware and services in m-commerce. The proceeding chapter presents the most popular technologies and standards implemented for mobile devices whilst the chapters thereafter discuss wireless middleware, the importance of client profiling for wireless devices. The final chapter concludes with discussion of challenges and trends.

WIRELESS TECHNOLOGIES AND STANDARDS

Currently, the focus is on wireless technologies and standards, such as in the area of network connectivity, communication protocols, standards and device characteristics (e.g., computing performance, memory, and presentation). A lot of technologies are being proposed and investigated by researchers and practitioners, some of which have been incorporated in industrial wireless products whose aim is to dominate the next generation market (Figure 1).

Among the most known communication standards and wireless deployments are the GSM, TDMA, FDMA, TDMA, CDMA, GPRS, SMS, MMS, HSCSD, Bluetooth, IEEE 802.11, etc. **GSM** (global system for mobile communications) is a 2G digital wireless standard, which is the most widely used digital mobile phone system. GSM uses the three classical multiple access processes, space division multiple access (**SDMA**), frequency division multiple access (**FDMA**), and time division multiple access (**TDMA**) in parallel and simultaneously (Heine & Sagkob, 2003). **CDMA** (code division multiple access), which is also a second generation (2G) wireless standard, works by some means different than the previous wireless. It can be distinguished in the way information is transmitted over the air, since it uses unique coding for each call or data session, which allows a mobile device to distinguish other transmissions on the same frequency. Therefore this technology allows every wireless device in the same area to utilize the same channel of spectrum, and at the same time to sort out the calls by encoding each one uniquely. **GPRS** (General Packet Radio Service) is a packet-switched service that allows data communications (with data rates significantly faster than a GSM — 53.6kbps for downloading data) to be sent and received over the existing global system for mobile (GSM) communications network. The introduction of **EDGE** (enhanced data rates for GSM evolution) enhances the connection bandwidth over the GSM network. It is a 3G technology that enables the provision of advanced mobile services (e.g., the downloading of video and music files, the high-speed color Internet access and e-mail) anywhere and anytime.

The **SMS** (short message service) is a technology that allows sending and receiving text messages to and from mobile telephones. Although the very first text message was sent in December 1992, commercially, SMS was launched in 1995. The rapid evolution of SMS is evident, since by 2002, over a billion text messages were being exchanged globally per day and by 2003, that figure had jumped to almost 17 billion. One reason mobile phone carriers continue to push text messaging is that they derive up to 20% of their annual revenues from SMS

Figure 1. Wireless technologies and standards

Wireless Technologies and Standards	**Application Development and Deployment**
	wireless application protocol (WAP); use of HTTP; i-mode; wireless middleware; compression technologies; IP telephony, SMS, MMS
	Personal Area Networks and Local-Area Networks
	Infrared; Bluetooth; IEEE 802.11; IEEE 802.11a; IEEE 802.11b; HiperLAN; HomeRF; Unlicensed National Information Infrastructure (UNII); security standards; quality-of-service mechanisms; public broadband access
	Digital Cellular and PCS
	cellular digital packet data (CDPD); global system for mobile communications (GSM), code division multiple access (CDMA), time division multiple access (TDMA); general packet radio service (GPRS); enhanced data rates for GSM evolution (EDGE); high speed circuit switched data (HSCSD)
	Third Generation Cellular
	International Mobile Telephone (IMT) 2000; 3G standards; wideband CDMA (WCDMA); Universal Mobile Telephone System (UMTS); CDMA 2000 (1X, 1XEV); voice over IP; quality-of-service mechanisms; all-IP core networks

service (Johnson, 2005). **MMS** (multimedia messaging service) is the descendant service of SMS, a store and forward messaging service that allows mobile subscribers to exchange multimedia messages with other mobile subscribers. **HSCSD** (high speed circuit switched data) is an enhancement of data services (circuit switched data — CSD) of all current GSM networks enabling higher rates by using multiple channels. It allows access to non-voice services at speeds 3-times faster. For example, it enables wireless devices to send and receive data at a speed of up to 28.8 kbps (some networks support up to 43.2 kbps). **Bluetooth** is a technology that provides short-range radio links between devices. When Bluetooth-enabled devices come into range with one another, they automatically detect each other and establish a network connection for exchanging files or using each other's services.

Most of the preveriously discussed standards and technologies pushed the evolution of e-commerce for mobile devices (m-commerce). Mobile commerce is referring to all forms of e-commerce that takes place when a consumer makes an online purchase using any mobile device (WAP phone, wireless handheld, etc). M-commerce is discussed in the following section.

M-COMMERCE

M-commerce is rapidly becoming the new defacto standard for buying goods and services. However, it appears that like e-commerce, it also requires a number of security mechanisms for mobile transactions, middleware for content retrieval and adaptation using user, standards and methods for retrieving and managing device, user and network characteristics so as to be used during mobile commerce interaction (Figure 2). M-commerce is expected to exceed wired e-commerce as the method of preference for digital commerce transactions, since it is already being used by a number of common services and applications, such as financial services (e.g., mobile banking), telecommunications, retail and service, and information services (e.g., delivery of financial news and traffic updates).

Figure 2. M-commerce

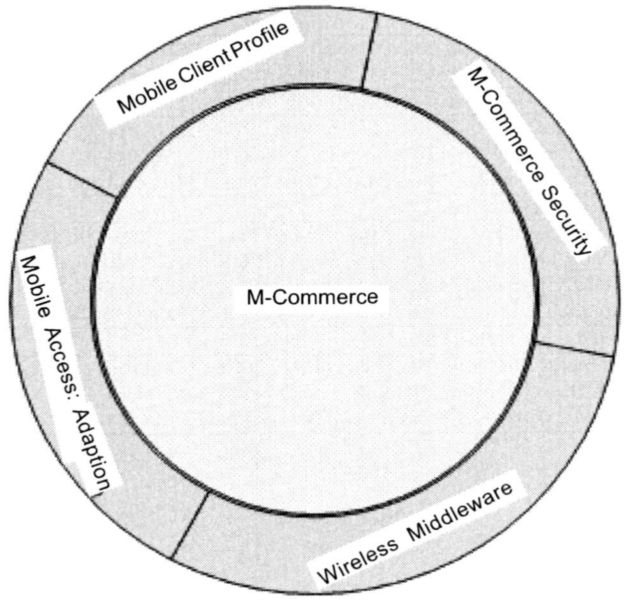

Mobile security (m-security) and mobile payment (m-payment) are essential to mobile commerce and the mobile world. Consumers and merchants have benefited from the virtual payments, which information technology has enabled. Due to the extensive use of mobile devices nowadays, a number of payment methods have been deployed which allows the payment of services/goods from any mobile device. The success of mobile payments is contingent on the same factors that have fuelled the growth of traditional non-cash payments: security, interoperability, privacy, global acceptance, and ease of use. Existing mobile payment applications are categorized based on the payment settlement methods which they implement: prepaid (using smart cards or digital wallet), instant paid (direct debiting or off-line payments), and post paid (credit card or telephone bill) (Seema & Chang-Tien, 2004). Developers deploying applications using mobile payments must consider security, interoperability, and usability requirements. A secure mobile application has to allow an issuer to identify a user, authenticate a transaction and prevent unauthorized parties from obtaining any information on a transaction. Interoperability guarantees completion of a transaction between different mobile devices or distribution of a transaction across devices and usability ensures user-friendliness and multi-users. M-commerce security and other essential treads are discussed in the following section.

M-COMMERCE TREADS

Mobile computing applications may be classified into three categories: client-server, client-proxy-server, and peer-to-peer depending on the interaction model. Each transaction, especially for m-commerce usually requires the involvement of mobile security, wireless middleware, mobile access adaptation, and mobile client profile.

M-Commerce Security

While m-commerce may be used anywhere and on the move, security threats are on the increase because personal information has to been delivered to a number of mobile workers engaged in online activities outside the secure perimeter of a corporate area and so access or use of private and personal data by unauthorized persons is easy. A number of methods and standards have been developed for the purpose of increasing the security model and being used also for mobile applications and services such as simple usernames and passwords, special single use passwords from electronic tokens, cryptographic keys and certificates from public key infrastructures (PKI). Additionally, developers are using authentication mechanisms to determine what data and applications the user can access (after login authorization). These mechanisms, often called policies or directories, are handled by databases that authenticate users and determine their permissions to access specific data simultaneously. However, the current mobile business (m-business) environment runs over the TCP/IPv4 protocol stack which poses serious security level threats with respect to user authentication, integrity and confidentiality. In a mobile environment, it is necessary to have identification and non-repudiation and service availability, mostly a concern for Internet and or Application service providers. For these purposes, carriers (telecomm operators and access providers), services, application providers and users demand end-to-end security as far as possible (Leonidou et al., 2003), (Tsaoussidis & Matta, 2002).

Although m-business services and applications such as iMode, Hand-held Device Markup Language (HDML) and wireless access protocol (WAP) are used daily for securing and encrypting the transfer of data between different type of end systems, however this kind of technologies cannot provide applicable security layers to secure transactions such as user PIN-protected digital signatures. Therefore, consumers cannot acknowledge that indeed their transactions are automatically generated and transmitted secured by their mobile devices. Many security concerns exist in Internet2 and IPv6, such as the denial-of-service attack. New technologies and standards provide adequate mechanisms and allow developers to implement security controls for mobile devices that do afford a reasonable level of protection in each of the four main problem areas: virus attacks, data storage, synchronization, and security.

Wireless Middleware

Desktop applications (applications that have been developed for the wired Internet) cannot be directly used by mobile devices since some of the regular assumptions made in building Internet applications, such as presence of high bandwidth disconnection-free network connections, resource-rich machines and computation platforms, are not valid in mobile environments (Avancha, Chakraborty, Perich, & Joshi, 2003). Content delivery and transformation of applications to wireless devices without rewriting the application can be facilitated by wireless middleware. Additionally, a middleware framework can support multiple wireless device types and provide continuous access to content or services (Sofokleous, Mavromoustakos, Andreou, Papadopoulos, & Samaras, 2004). The main functionality of wireless middleware is the data transformation shaping a bridge from one programming language to another, and in a number of circumstances is the manipulation of content in order to suit different device specifications. Wireless middleware components can detect and store device characteristics in a database and later optimize the wireless data output according to device attributes by using various data-compression algorithms such as Huffman coding, dynamic Huffman coding, arithmetic coding, and Lempel-Ziv coding. Data compression algorithms serve to minimize the amount of data being sent over wireless links, thus improving overall performance on a handheld device. Additionally, they ensure end-to-end security from handheld devices to application servers and finally they perform message storage and forwarding should the user get disconnected from the network. They provide operation support by offering utilities and tools to allow MIS personnel to manage and troubleshoot wireless devices. Choosing the right wireless middleware depends on the following key factors: platform language, platform support and security, middleware integration with other products, synchronization, scalability, convergence, adaptability, and fault tolerance (Vichr & Malhotra, 2001).

Mobile Access Adaptation

The combination of diversity in user preferences and device characteristics with the many different services that are everyday deployed requires the extensive adaptation of content. The network topology and physical connections between hosts in the network must constantly be recomputed and application software must adapt its behavior continuously in response to this changing context (Julien, Roman, & Huang, 2003) either when server-usage is light or if users pay for the privilege (Ghinea & Angelides, 2004).

The developed architecture of m-commerce communications exploits user perceptual tolerance to varying QoS in order to optimize network bandwidth and data sizing. This will provide quality of service (QoS) impacts upon the success of m-commerce applications without doubt, as it plays a pivotal role in attracting and retaining customers. As the content adaptation and in general the mobile access personalization concept is budding, central role plays the utilization of the mobile client profile, which is analyzed in the next section.

Mobile Client Profile

Profile management aims to provide content that match user needs and interests. This can be achieved by gathering all the required information for user's preferences and user's device in (e.g., display resolution, content format and type, supported codec, performance, and memory, etc.). The particular data may be used for determining the content and the presentation that best fit the user's expectations and the device capabilities (Chang & Vetro, 2005). The information may be combined with the location of the user and the action context of the user at the time of the request (Agostini, Bettini, Cesa-Bianchi, Maggiorini, & Riboni, 2003).

Different entities are assembled from different logical locations to create a complete user profile (e.g., the personal data is provided by the user, whereas the information about the user's current location is usually provided by the network operator). Using the profile, service providers may search and retrieve information for a user. However, several problems and methods for holdback the privacy of data are raised, as mobile devices allow the control of personal identifying information (Srivastava, 2004). Specifically, there is a growing ability to trace and cross-reference a person's activities via his various digitally assisted transactions. The resulting picture might provide insight into his medical condition, buying habits, or particular demographic situation. In addition various location-transmission devices allow the location and movement tracking of someone (Ling, 2004). And that is the main reason people are instantly concerned for location privacy generated by location tracking services.

CURRENT CHALLENGES OF MOBILE COMPUTING AND FUTURE TRENDS

Mobile devices suffer from several constraints calling for immediate development of a variety of mechanisms in order to be able to accommodate high quality, user-friendly, and ubiquitous access to information based on the needs and preferences of mobile users. The latter is required as the demand of new mobile services and applications based on a local and personal profile of the mobile is significantly increasing in the last decades. Current mobile devices exhibit several constraints such as limited screen space (screens cannot be made physically bigger as the devices must fit into hand or pocket to enable portability) (Brewster & Cryer, 1999), unfriendly user interfaces, limited resources (memory, processing power, energy power, tracking), variable connectivity performance and reliability, constantly changing environment, and low security mechanisms.

The relationship between mobility, portability, human ergonomics, and cost is intriguing. As the mobility refers to the ability to move or be moved easily, portability refers to the ability to move user data along with the users. The use of traditional hard-drive and keyboard designs in mobile devices is impossible as a portable device has to be small and lightweight. The greatest assets of mobile devices are the small size, its inherent portability, and easy access to information (Newcomb, Pashley, & Stasko, 2003). Although mobile devices were initially been used for calendar and contact management, wireless connectivity has led to new uses such as user location tracking on-the-move. The ability to change locations while connected to the internet increases the volatility of some information.

Mobile phones are sold better than PCs these days but the idea that the PC is going away and probably it is going to be replaced by mobile phones is definitely incorrect if not a myth. Mobile devices cannot serve the same purposes as personal computers. It is almost impossible to imagine PCs replaced by mobiles, especially for raw interactivity with the user, flexibility of purpose, richness of display, and in-depth experience (the same was said for video recorders). For instance, writing a book on a mobile phone or designing complicated spreadsheets on a PDA is very time-consuming

Figure 3. Areas of mobility evolution

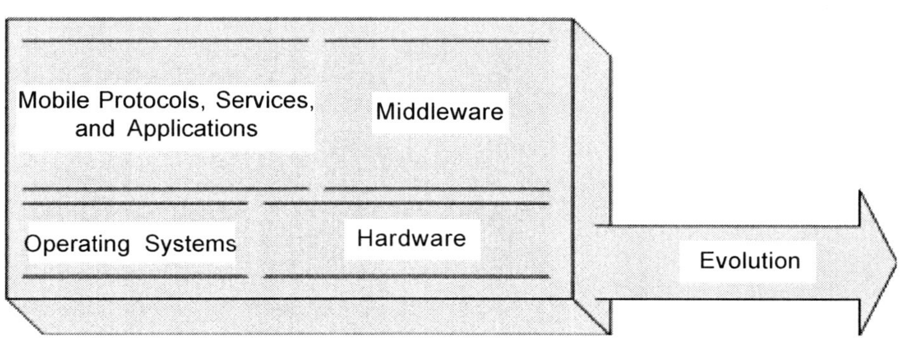

and difficult (Salmre, 2005). Mobile computing has changed the business and consumer perception and there is no doubt that it has already exceeded most expectations. The evolution of mobility is being achieved by the architectures and protocols standards, management, services and applications, mobile operating systems (Angelides, 2004).

Although applications in the area of mobile computing and m-commerce are restricted by the available hardware and software resources, more than a few applications, such transactional applications (financial services/banking and home shopping, instant messages, stock quotes, sale details, client information, and locations-based services) have already showed potential for expansion making the mobile computing environment capable of changing the daily lifestyle.

CONCLUDING DISCUSSION

This chapter presents the concept of mobile computation, its standards and underlying technologies, and continues by discussing the basic trends of m-commerce. As it is anticipated, information will be more important if it is provided based on user's preferences and location and that can be borne out since new mobile services and applications maintain and deal with location and profile management. Security for mobile devices and wireless communication still continue to need further investigation and consideration especially during the design steps of mobile frameworks. Although m-commerce and e-commerce are both concerned with trading of goods and services over the Web, however m-commerce explores opportunities from a different perspective as business transactions conducted while on the move. Having many requirements and many devices to support, developers have to adapt the content in order to fit on a user screen and at the same time consider network requirements (bandwidth, packet loss rate, etc.) and device characteristics (resolution, supported content, performance, and memory, etc.).

REFERENCES

Agostini, A., Bettini, C., Cesa-Bianchi, N., Maggiorini, D., & Riboni, D. (2003). *Integrated profile management for mobile computing.* Workshop on Artificial Intelligence, Information Access, and Mobile Computing—IJCAI 2003, Acapulco, Mexico.

Angelides, C. M. (2004). Mobile multimedia and communications and m-commerce. *Multimedia Tools and Applications, 22*(2), 107-108.

Avancha, S., Chakraborty, D., Perich, F., & Joshi, A. (2003). *Data and services for mobile computing. Handbook of Internet computing.* Baton Rouge, FL: CRC Press.

Brewster, A. S., & Cryer, P. G. (1999). Maximizing screen-space on mobile computing devices. *Proceedings of ACM SIGCHI Conference on Human Factors in Computing Systems* (pp. 224-225). Pittsburgh; New York.

Chang, S. F., & Vetro, A. (2005). Video adaptation: Concepts, technologies, and open issues. *Proceedings of the IEEE, 93*(1), 148-158.

Dahleberg, T., & Tuunainen, V. (2001). Mobile payments: The trust perspective. *International Workshop on Seamless Mobility.* Sollentuna.

Ghinea, G., & Angelides, C. M. (2004). A user perspective of quality of service in m-commerce. *Multimedia Tools and Applications, 22*(2), 187-206.

Heine, G., & Sagkob, H. (2003). *GPRS: Gateway to third generation mobile networks.* Norwood, MA: Artech House.

Johnson, F. (2005) *Global mobile connecting without walls. Wires or borders*. Berkeley, CA: Peachpit Press.

Julien, C., Roman, G., & Huang, Q. (2003). *Declarative and dynamic context specification supporting mobile computing in ad hoc networks* (Tech. Rep. No. WUCSE-03-13). St. Louis, Missouri: Washington University, CS Department.

Juniper Research. (2004). *The big micropayment opportunity*. White paper. Retrieved September 24, 2004, from http://industries.bnet.com/abstract.aspx?scid=2552&docid=121277

Leonidou, C., Andreou, S. A., Sofokleous, A., Chrysostomou, C., Mavromoustakos, S., Pitsillides, A., Samaras, G., & Schizas, C. (2003). A security tunnel for conducting mobile business over the TCP protocol. *2nd International Conference on Mobile Business* (pp. 219-227). Vienna, Austria.

Ling, R. (2004). *The mobile connection: The cell phone's impact on society*. San Francisco: Morgan Kaufmann.

Nambiar, S. & Lu, C.-T. (2005). M-payment solutions and m-commerce fraud management. In W.-C. Hu, C.-w. Lee, & W. Kou (Eds.), *Advances in security and payment methods for mobile commerce* (pp. 192-213). Hershey, PA: Idea Group Publishing.

Newcomb, E., Pashley, T., & Stasko, J. (2003). Mobile computing in the retail arena. *ACM Proceedings of the Conference on Human Factors in Computing Systems* (pp. 337-344). Florida, USA.

Salmre, I. (2005). *Writing mobile code essential software engineering for building mobile application*. Hagerstown, MD: Addison Wesley Professional.

Sofokleous, A., Mavromoustakos, S., Andreou, A. S., Papadopoulos, A. G., & Samaras, G. (2004). Jinius-link: A distributed architecture for mobile services based on localization and personalization. *IADIS International Conference*, Portugal, Lisbon.

Srivastava, L. (2004). Social and human consideration for a mobile world. *ITU/MIC Workshop on Shaping the Future Mobile Information Society*, Seoul, Korea.

Tsaoussidis, V. & Matta, I. (2002). Open issues on TCP for mobile computing. *Journal of Wireless Communications and Mobile Computing*, *2*(1), 3-20.

Vichr, R., & Malhotra, V. (2001). *Middleware smoothes the bumpy road to wireless integration*. An IBM article retrieved August 11, 2004, from http://www-106.ibm.com/developerworks/library/wi-midarch/index.html

KEY TERMS

EDGE: EDGE (enhanced data rates for GSM evolution) is a 3G technology, which enables the provision of advanced mobile services and enhances the connection bandwidth over the GSM network.

GPRS: GPRS (General Packet Radio Service) is a packet-switched service that allows data communications (with data rates significantly faster than a GSM—53.6kbps for downloading data) to be sent and received over the existing global system for mobile (GSM) communications network.

GSM: GSM (global system for mobile communications) is a 2G digital wireless standard and is the most widely used digital mobile phone system.

GSM Multiple Access Processes: GSM use space division multiple access (SDMA), frequency division multiple access (FDMA), and time division multiple access (TDMA) in parallel and simultaneously.

M-Business: Mobile business means using any mobile device to make business practice more efficient, easier and profitable.

M-Commerce: Mobile commerce is the transactions of goods and services through wireless handheld devices such as cellular telephone and personal digital assistants (PDAs).

MMS: MMS (multimedia messaging service) is a store and forward messaging service, which allows mobile subscribers to exchange multimedia messages with other mobile subscribers.

Mobile Computing: Mobile computing encompasses a number of technologies and devices, such as wireless LANs, notebook computers, cell and smart phones, tablet PCs, and PDAs helping the organization of our life, the communication with coworkers or friends, or the accomplishment of our job more efficiently.

M-Payment: Mobile payment is defined as the process of two parties exchanging financial value using a mobile device in return for goods or services.

M-Security: Mobile security is the technologies and method used for securing the wireless communication between the mobile device and the other point of communication such as other mobile client or pc.

Chapter II
Business Model Typology for Mobile Commerce

Volker Derballa
Universität Augsburg, Germany

Key Pousttchi
Universität Augsburg, Germany

Klaus Turowski
Universität Augsburg, Germany

ABSTRACT

Mobile technology enables enterprises to invent new business models by applying new forms of organization or offering new products and services. In order to assess these new business models, there is a need for a methodology that allows classifying mobile commerce business models according to their typical characteristics. For that purpose a business model typology is introduced. Doing so, building blocks in the form of generic business model types are identified, which can be combined to create concrete business models. The business model typology presented is conceptualized as generic as possible to be generally applicable, even to business models that are not known today.

INTRODUCTION

Having seen failures like WAP, the hype that was predominant for the area of mobile commerce (MC) up until the year 2001 has gone. About one year ago however, this negative trend has begun to change again. Based on more realistic expectations, the mobile access and use of data, applications and services is considered important by an increasing number of users. This trend becomes obvious in the light of the remarkable success of mobile communication devices. Substantial growth rates are expected in the next years, not only in the area of B2C but also for B2E and B2B. Along with that development go new challenges for the operators of mobile services resulting in reassessed validations and alterations of existing business models and the creation of new business models. In order to estimate the economic success of particular business models, a thorough analysis of those models is necessary. There is a need for an evaluation methodology in order to assess existing and future business models based on modern information and communication technologies. Technological capabilities have to be identified as well as benefits that users and producers of electronic offers can achieve when using them.

The work presented here is part of comprehensive research on mobile commerce (Turowski & Pousttchi, 2003). Closely related is a methodology for the qualitative assessment of electronic and mobile business models (Bazijanec, Pousttchi, & Turowski, 2004). In that work, the focus is on the added value for which the customers is ready to pay. The theory of informational added values is extended by the definition of technology-specific properties that are advantageous when using them to build up business models or other solutions based on information and communication techniques. As mobile communication techniques extend Internet technologies and add some more characteristics that can be considered as additional benefits, a own class of technology-specific added values is defined and named mobile added values (MAV), which are the cause of informational added values. These added values based on mobility of mobile devices are then used to assess mobile business models.

In order to be able to qualitatively assess mobile business models, those business models need to be unambiguously identified. For that purpose, we introduce in this chapter a business model typology. Further, the business model typology presented here is conceptualized as generic as possible, in order to be robust and be generally applicable — even to business models that are not known today. In the following we are building the foundation for the discussion of the business model typology by defining our view of MC. After that, alternative business model typologies are presented and distinguished from our approach, which is introduced in the subsequent section. The proposed approach is then used on an existing MC business model. The chapter ends with a conclusion and implications for further research.

BACKGROUND AND RELATED WORK

Mobile Commerce: A Definition

Before addressing the business model typology for MC, our understanding of MC needs to be defined. If one does agree with the Global Mobile Commerce Forum, mobile commerce can be defined as "the delivery of electronic commerce capabilities directly into the consumer's device, anywhere, anytime via wireless networks." Although this is no precise definition yet, the underlying idea becomes clear. Mobile commerce is considered a specific characteristic of electronic commerce and as such comprises specific attributes, as for example the utilization of wireless communication and mobile devices. Thus, mobile commerce can be defined as every form of business transaction in which the participants use mobile electronic communication techniques in connection with mobile devices for initiation, agreement or the provision of services. The concept mobile electronic communication techniques is used for different forms of wireless communication. That includes foremost cellular radio, but also technologies like wireless LAN, Bluetooth or infrared communication. We use the term mobile devices for information and communication devices that have been developed for mobile use. Thus, the category of mobile devices encompasses a wide spectrum of appliances. Although the laptop is often included in the definition of mobile devices, we have reservations to include it here without precincts due to its special characteristics: It can be moved easily, but it is usually not used during that process. For that reason we argue that the laptop can only be seen to some extend as amobile device.

Related Work

Every business model has to prove that it is able to generate a benefit for the customers. This is especially true for businesses that offer their products or services in the area of EC and MC. Since the beginning of Internet business in the mid 1990s, models have been developed that tried to explain advantages that arose from electronic offers. An extensive overview of approaches can be found in (Pateli & Giaglis, 2002). At first, models were rather a collection of the few business models that had already proven to be able to generate a revenue stream (Fedewa, 1996; Schlachter, 1995; Timmers, 1998). Later approaches extended these collections to a comprehensive taxonomy of business models observable on the web (Rappa, 2004; Tapscott, Lowi, & Ticoll, 2000). Only Timmers (1998) provided a first classification of eleven business models along two dimensions: innovation and functional integration. Due to many different aspects that have to be considered when comparing business models, some authors introduced taxonomies with different views on Internet business. This provides an overall picture of a firm doing Internet business (Osterwalder, 2002), where the views are discussed separately (Afuah & Tucci, 2001; Bartelt & Lamersdorf, 2000; Hamel, 2000; Rayport & Jaworski, 2001; Wirtz & Kleineicken, 2000). Views are for example commerce strategy, organizational structure or business process. The two most important views that can be found in every approach are value proposition and revenue. A comparison of views proposed in different approaches can be found in (Schwickert, 2004). While the view revenue describes the rather short-term monetary aspect of a business model the value proposition characterizes the type of business that is the basis of any revenue stream. To describe this value proposition authors decomposed business models into their atomic elements (Mahadevan, 2000). These elements represent offered services or products. Models that follow this approach are for example (Afuah & Tucci, 2001) and (Wirtz & Kleineicken, 2000). Another approach that already focuses on generated value can be found in (Mahadevan, 2000). There, four so-called value streams are identified: virtual communities, reduction of transaction costs, gainful exploitation of information asymmetry, and a value added marketing process.

In this work however, we are pursuing another approach: The evaluation of real business models showed that some few business model types recur. These basic business model types have been used for building up more complex business models. They can be classified according to the type of product or service offered. A categorization based on this criterion is highly extensible and thus very generic (Turowski & Pousttchi, 2003). Unlike the classifications of electronic offers introduced previously, this approach can also be applied to mobile business models that use for example location-based services to provide a user context. In the following sections, we are describing this business model typology in detail.

BUSINESS MODEL TYPOLOGY

Business Idea

Starting point for every value creation process is a product or business idea. An instance of a

Figure 1. Business idea and business model

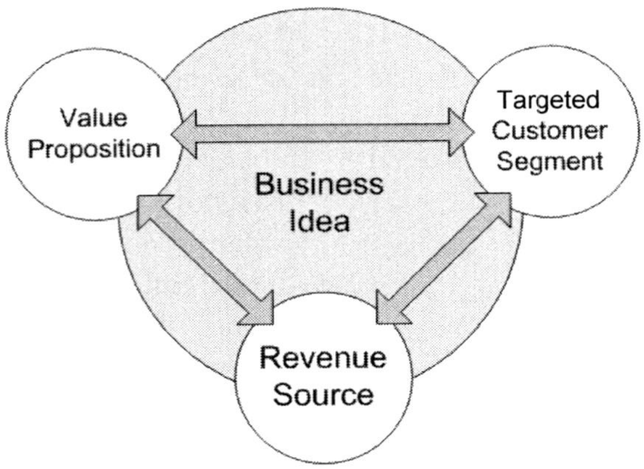

business idea is the offer to participate in auctions or conduct auctions — using any mobile device without tempo-spatial restrictions. Precondition for the economic, organisational, and technical implementation and assessment of that idea is its transparent specification. That abstracting specification of a business idea's functionality is called business model. It foremost includes an answer to the question: Why has this idea the potential to be successful? The following aspects have to be considered for that purpose:

- Value proposition (which value can be created)
- Targeted customer segment (which customers can and should be addressed)
- Revenue source (who, how much and in which manner will pay for the offer)

Figure 1 shows the interrelationship between those concepts. It needs to be assessed how the business idea can be implemented regarding organisational, technical, legal, and investment-related issues. Further, it has to be verified whether the combination of value proposition, targeted customer segment and revenue source that is considered optimal for the business model fits the particular company's competitive strategy. Let's assume an enterprise is pursuing a cost leader strategy using offers based on SMS, it is unclear whether the enterprise can be successful with premium-SMS.

It needs to be pointed out that different business models can exist for every single business idea. Coming back to the example of offering auctions without tempo-spatial restrictions, revenues can be generated in different ways with one business model recurring to revenues generated by advertisements and the other recurring to revenues generated by fees.

Revenue Models

The instance introduced previously used the mode of revenue generation in order to distinguish business models. In this case, the revenue model is defined as the part of the business model describing the revenue sources, their volume and their distribution. In general, revenues can be generated by using the following revenue sources:

- Direct revenues from the user of a MC-offer

Figure 2. Revenue sources in MC (based on Wirtz & Kleineicken, 2000)

	Direct	Indirect
Transaction based	• Transaction revenues • Event-based billing • Time-based billing	• Commissions
Transaction independent	• Set-up fees • Subscription fees	• Advertising • Trading user profiles

Revenue source

- Indirect revenues, in respect to the user of the MC-offer (i.e., revenues generated by 3rd parties); and
- Indirect revenues, in respect to the MC-offer (i.e., in the context of a non-EC offer).

Further, revenues can be distinguished according to their underlying mode in transaction-based and transaction-independent. The resulting revenue matrix is depicted in Figure 2.

Direct transaction-based revenues can include event-based billing (e.g., for file download) or time-based billing (e.g., for the participation in a blind-date game). Direct transaction-independent revenues are generated as set-up fees, (e.g., to cover administrative costs for the first-time registration to a friend finder service) or subscription fees (e.g., for streaming audio offers).

The different revenue modes as well as the individual revenue sources are not necessarily mutually excluding. Rather, the provider is able to decide which aspects of the revenue matrix he wants to refer to. In the context of MC-offers, revenues are generated that are considered (relating to the user) indirect revenues. That refers to payments of third parties, which in turn can be transaction-based or transaction-independent. Transaction-based revenues (e.g., as commissions) accrue if, for example, restaurants or hotels pay a certain amount to the operator of mobile tourist guide for guiding a customer to their locality. Transaction-independent revenues are generated by advertisements or trading user profiles. Especially the latter revenue source should not be neglected, as the operator of a MC-offer possesses considerable possibilities for the generation of user profiles due to the inherent characteristics of context sensitivity and identifying functions (compared to the ordinary EC-vendor). Revenues that are not generated by the actual MC-offer are a further specificity of indirect revenues. This includes MC-offers pertaining to customer retention, effecting on other business activities (e.g., free SMS-information on a soccer team leading to an improvement in merchandising sales).

MC-Business Models

In the first step, the specificity of the value offered is evaluated. Is the service exclusively based on the exchange of digitally encoded data or is a significant not digital part existent (i.e., a good needs to be manufactured or a service is accomplished that demands some kind of manipulation conducted on a physically existing object)? *Not digital* services can be subdivided into *tangible* and *intangible* services. Whereas *tangible* services need to have a significant physical component, this classification assumes the following: The category of *intangible* services only includes services that demand manipulation conducted on a physically existing object.

Services that can be created through the exchange of digitally encoded data are subdivided into *action* and *information*. The category *information* focuses on the provision of data (e.g., multi-media contents in the area of entertainment or the supply of information). Opposed to that, the category *action* includes processing, manipulating, transforming, extracting, or arranging of data.

On the lowermost level, building blocks for business models are created through the further subdivision according to the value offered. For that purpose, a distinction is made between the concrete business models that can include one or more business model types and those business model types as such. These act as building blocks that can constitute concrete business models.

The business model type *classical goods* is included in all concrete business models aiming at the vending of tangible goods (e.g., CDs or flowers, i.e., goods that are manufactured as industrial products or created as agricultural produce). Those goods can include some digital components (e.g., cars, washing machines). However, decision criterion in that case is the fact that a significant part of the good is of physical nature and requires the physical transfer from one owner to the other.

Figure 3. Categorization of basic business model types

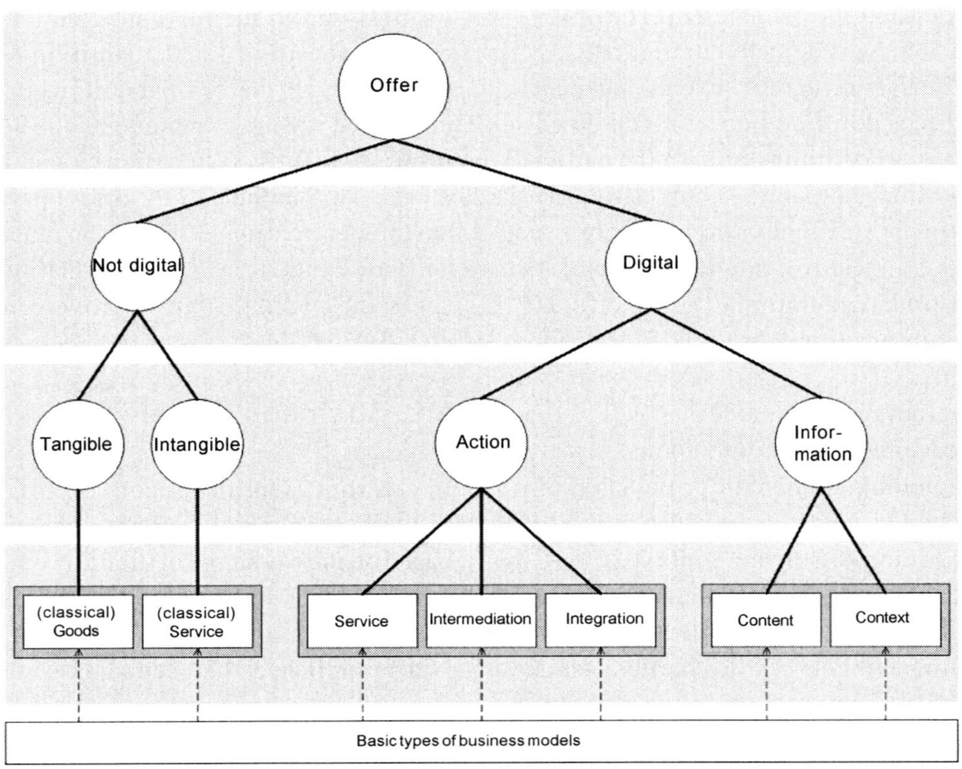

Concrete business models include the business model type *classical service* if some manipulation activities have to be conducted on a physical object. That comprises e.g. vacation trips and maintenance activities.

The basic business model type *service* comprises concrete business models, if they comprise an original service that is considered by the customer as such and requires some action based on digitally encoded data as described previously, without having intermediation characteristics (c.f., basic business model type *intermediation*). Such services, e.g., route planning or mobile brokerage are discrete services and can be combined to new services through bundling. A typical offer that belongs to the business model type service is mobile banking. Further, it might be required (e.g., in order to enable mobile payment or ensure particular security goals (data confidentiality) to add further services, which require some kind of action, as described previously. As the emphasis is on the original service, these services can be considered as supporting factors. Depending on the circumstances, they might be seen as an original service. Due to that, those supporting services will not be attributed to a basic business model type. Rather, those services are assigned to the business model type service.

A concrete business model includes the business model type *intermediation* if it aims at the execution of classifying, systemising, searching, selecting, or interceding actions. The following offers are included:

- Typical search engines/offers (e.g., www.a1.net);
- Offers for detecting and interacting with other consumers demanding similar products;
- Offers for detecting and interacting with persons having similar interests;
- Offers for the intermediation of consumers and suppliers;
- Any kind of intermediation or brokering action, especially the execution of online auctions; and
- In general the operations of platforms (portals), which advance, simplify or enable the interaction of the aforementioned economic entities.

Taking all together, the focus is on matching of appropriate parings (i.e., the initiation of a transaction). Nevertheless, some offers provide more functionality by for example supporting the agreement process as well (e.g., the hotel finder and reservation service (wap.hotelkatalog.de)): This service lets the user search for hotels, make room reservations, and cancel reservations. All the relevant data is shown and hotel rooms can be booked, cancelled, or reserved. The user is contacted using e-mail, telephone, fax, or mail. Revenues are generated indirectly and transaction-independent, as the user agrees to obtain advertisements from third parties.

The basic business model type *integration* comprises concrete business models aiming at the combination of (original) services in order to create a bundle of services. The individual services might be a product of concrete business models that in turn can be combined to create new offers. Further, the fact that services have been combined is not necessarily transparent for the consumer. This can even lead to user individual offers where the user does not even know about the combination of different offers. For example, an offer could be an insurance bundle specifically adjusted to a customer's needs. The individual products may come from different insurance companies. On the other hand, it is possible to present this combination to the consumer as the result of a customization process (custom-made service bundle).

The basic business model type *content* can be identified in every concrete business model that generates and offers digitally encoded multi-media content in the areas of entertainment, education, arts, culture sport etc. Additionally, this type comprises games. Wetter-Online (pda.wetteronline.de) can be considered a typical example for that business model type. The user can access free weather information using a PDA. The information offered includes forecasts, actual weather data, and holiday weather. The PDA-version of this service generates no revenues, as it is used as promotion for a similar EC-offer, which in turn is ad sponsored.

A concrete business model comprises the basic business model type *context* if information describing the context (i.e., situation, position, environments, needs, requirements etc.) of a user is utilised or provided. For example, every business model building on location-based services comprises or utilises typical services of the basic business model type context. This is also termed context-sensitivity. A multiplicity of further applications is realised in connection with the utilisation of sensor technology integrated in or directly connected to the mobile device. An instance is the offer of Vitaphone (www.vitaphone.de). It makes it possible to permanently monitor the cardiovascular system of endangered patients. In case of an emergency, prompt assistance can be provided. Using a specially developed mobile phone, biological signals, biochemical parameters, and the users' position are transmitted to the Vitaphone service centre. Additionally to the aforementioned sensors, the mobile phone has GPS functionality and a special emergency button to establish quick contact with the service centre.

Figure 4 depicts the classification of that business model using the systematics introduced previously. It shows that vita phone's

Figure 4. Classification of Vitaphone's business model

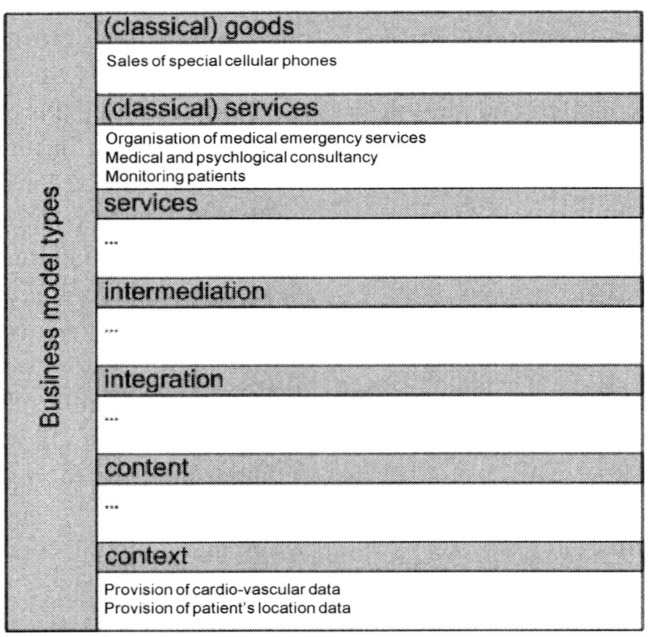

Figure 5. Vitaphone's revenue model

	MC-Business		Non-MC-Business
	Direct	Indirect	
Transaction based	• Communication with the service centre	...	Sales of special cellular phones
Transaction independent	• Subscription fee	...	

business model uses mainly the building blocks from the area of *classical service*. Those services are supplemented with additional building blocks from the area of *context*. This leads to the weakening of the essential requirement — physical proximity of patient and medical practioner — at least what the medical monitoring is concerned. This creates several added values for the patient, which will lead to the willingness to accept that offer.

Analysing the offer of Vitaphone in more detail leads to the conclusion that the current offer is only a first step. The offer results indeed in increased freedom of movement, but requires active participation of the patient. He has to operate the monitoring process and actively transmit the generated data to the service centre. To round of the analysis of Vitaphone's business model, the revenue model is presented in Figure 5.

Non MC-relevant revues are generated by selling special cellular phones. Further, direct MC revenues are generated by subscription fees (with or without the utilisation of the service centre) and transmission fees (for data generated and telephone calls using the emergency button).

CONCLUSION

This chapter presents an approach to classify mobile business models by introducing a generic mobile business model typology. The aim was to create a typology that is as generic as possible, in order to be robust and applicable for business models that do not exist today. The specific characteristics of MC make it appropriate to classify the business models according to the mode of the service offered. Doing so, building blocks in the form of business model types can be identified. Those business model types then can be combined to create concrete business model. The resulting tree of building blocks for MC business models differentiates digital and not digital services. Not digital services can be subdivided into the business model types classical goods for tangible services and classical service for intangible services. Digital services are divided into the category action with the business model types service, intermediation, integration and the category information with the business model types content and context.

Although the typology is generic and is based on the analysis of a very large number of actual business models, further research is necessary to validate this claim for new business models from time to time.

REFERENCES

Afuah, A., & Tucci, C. (2001). *Internet business models and strategies*. Boston: McGraw Hill.

Bartelt, A., & Lamersdorf, W. (2000). *Geschäftsmodelle des Electronic Commerce: Modell-bildung und Klassifikation*. Paper presented at the Verbundtagung Wirtschaftsinformatik.

Bazijanec, B., Pousttchi, K., & Turowski, K. (2004). *An approach for assessment of electronic offers*. Paper presented at the FORTE 2004, Toledo.

Fedewa, C. S. (1996). *Business models for Internetpreneurs*. Retrieved from http://www.gen.com/iess/articles/art4.html

Hamel, G. (2000). *Leading the revolution*. Boston: Harvard Business School Press.

Mahadevan, B. (2000). Business models for Internet based e-commerce: An anatomy. *California Management Review, 42*(4), 55-69.

Osterwalder, A. (2002). *An e-business model ontology for the creation of new management software tools and IS requirement engineering*. CAiSE 2002 Doctoral Consortium, Toronto.

Pateli, A., & Giaglis, G. M. (2002). *A domain area report on business models*. Athens, Greece: Athens University of Economics and Business.

Rappa, M. (2004). *Managing the digital enterprise — Business models on the Web*. Retrieved June 14, 2004, from http://digitalenterprise.org/models/models.html

Rayport, J. F., & Jaworski, B. J. (2001). *E-Commerce*. New York: McGraw Hill/Irwin.

Schlachter, E. (1995). *Generating revenues from Web sites*. Retrieved from http://boardwatch.internet.com/mag/95/jul/bwm39

Schwickert, A. C. (2004). *Geschäftsmodelle im electronic business—Bestandsaufnahme und relativierung*. Gießen: Professur BWL-Wirtschaftsinformatik, Justus-Liebig-Universität.

Tapscott, D., Lowi, A., & Ticoll, D. (2000). *Digital capital—Harnessing the power of business Webs*. Boston.

Timmers, P. (1998). Business models for electronic markets. *Electronic Markets, 8*, 3-8.

Turowski, K., & Pousttchi, K. (2003). *Mobile Commerce—Grundlagen und Techniken*. Heidelberg: Springer Verlag.

Wirtz, B., & Kleineicken, A. (2000). Geschäftsmodelltypen im Internet. *WiSt, 29*(11), 628-636.

KEY TERMS

Business Model: Business model is defined as the abstracting description of the functionality of a business idea, focusing on the value proposition, customer segmentation, and revenue source.

Business Model Types: Building blocks for the creation of concrete business models.

Electronic Commerce: Every form of business transaction in which the participants use electronic communication techniques for initiation, agreement or the provision of services.

Mobile Commerce: Every form of business transaction in which the participants use mobile electronic communication techniques in connection with mobile devices for initiation, agreement or the provision of services.

Revenue Model: The part of the business model describing the revenue sources, their volume and their distribution.

Chapter III
Security and Trust in Mobile Multimedia

Edgar R. Weippl
Vienna University of Technology, Austria

ABSTRACT

While security in general is increasingly well addressed, both mobile security and multimedia security are still areas of research undergoing major changes. Mobile security is characterized by small devices that, for instance, make it difficult to enter long passwords and that cannot perform complex cryptographic operations due to power constraints. Multimedia security has focused on digital rights management and watermarks; as we all know, there are yet no good solutions to prevent illegal copying of audio and video files.

INTRODUCTION TO SECURITY

Traditionally, there are at least three fundamentally different areas of security illustrated in Figure 1 (Olovsson, 1992): Hardware security, information security and organizational security. A forth area, that is outside the scope of this chapter, are legal aspects of security.

Hardware security encompasses all aspects of physical security and emanation. Compromising emanation refers to unintentional signals that, if intercepted and analyzed, would disclose the information transmitted, received, handled, or otherwise processed by telecommunications or automated systems equipment (NIS, 1992).

Information security includes computer security and communication security. Computer security deals with the prevention and detection of unauthorized actions by users of a computer system (Gollmann, 1999). Communication security encompasses measures and controls taken to deny unauthorized persons access to information derived from telecommunications and ensure the authenticity of such telecommunications (NIS, 1992).

Organizational or administration security is highly relevant even though people tend to neglect it in favor of fancy technical solutions. Both personnel security and operation security pertain to this aspect of security.

Figure 1. Categorization of areas in security

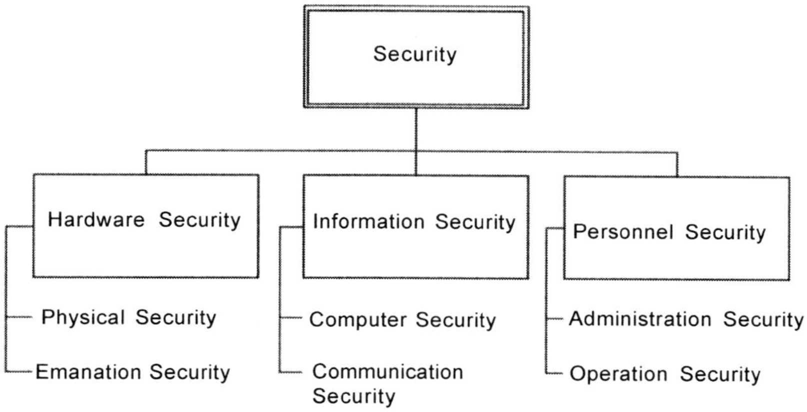

Systematic Categorization of Requirements

Whether a system is "secure" or not, merely depends on the definition of the requirements. As nothing can ever be absolutely secure, the definition of an appropriate security policy based on the requirements is the first essential step to implement security.

Security requirements can generally be defined in terms of four basic requirements: secrecy, integrity, availability, and non-repudiation. All other requirements that we perceive can be traced back to one of these four requirements. The forth requirement, non-repudiation, could also be seen as a special case of integrity, (i.e., the integrity of log data recording who has accessed which object).

Secrecy

The perhaps most well known security requirement is secrecy. It means that users may obtain access only to those objects for which they have received authorization, and will not get access to information they must not see.

The security policies guaranteeing secrecy are implemented by means of access control.

Integrity

The integrity of the data and programs is just as important as secrecy but in daily life it is frequently neglected. Integrity means that only authorized people are permitted to modify data (or programs). Secrecy of data is closely connected to the integrity of programs of operating systems. If the integrity of the operating system is violated, then the reference monitor may not work properly any more. The reference monitor is a mechanism which insures that only authorized people are able to conduct operations. It is obvious that secrecy of information cannot be guaranteed any longer if this mechanism is not working. For this reason it is important to protect the integrity of operating systems just as properly as the secrecy of information.

The security policy guaranteeing integrity is implemented by means of access control as like previously discussed.

Availability

It is through the Internet that many users have become aware that availability is one of the major security requirements for computer systems. Productivity decreases dramatically if network-based applications are not or only limitedly available.

There are no effective mechanisms for the prevention of denial-of-service, which is the opposite of availability. However, through permanent monitoring of applications and network connections it can be recognized when a denial-of-service occurs. At this point one can either switch to a backup system or take other appropriate measures.

Non-Repudiation

The fourth important security requirement is that users are not able to deny (plausibly) to have carried out operations. Let us assume that a teacher deletes his or her students' exam results. In this case, it should be possible to trace back who deleted documents and the traceing records must be so reliable that one can believe them. Auditing is the mechanism used to implement this requirement. All the security requirements are central requirements discussed in this section for computer security as well as network security.

Mechanisms

In this subsection we will elaborate on the mechanisms that are used to implement the aforementioned requirements (secrecy, integrity, availability, and non-repudiation).

Authentication

Authentication means proving that a person is the one he/she claims to be. A simple example illustrates what authentication is about. If a user logs on to the system, he or she will usually enter a name for identification purposes. The name *identifies* but does not *authenticate* the user since any other person can enter the same name as well. To prove his or her identity beyond all doubt, the user must enter a password that is known exclusively to him or her. After this proof the user is not just identified but (his or her identity) also authenticated.

Just as in many other areas, the most widely spread solutions for authentication are not necessarily the most secure ones. Security and simplicity of use frequently conflict each other. One must take into consideration that what is secure in theory may not mean secure in practice because it is not user-friendly; thus prompting users to circumvent the mechanisms. For example, in theory it is more secure to use long and frequently changed passwords. Obviously, many users will avoid these mechanisms effectively by writing down their passwords and possibly sticking post-its on their computers.

A number of approaches for authentication can be distinguished:

- What you know (e.g., password);
- What you do (e.g., signature);
- What you are (e.g., biometric methods such as face identification or fingerprints); and
- What you have (e.g., key or identity card).

Access Control

Access control is used to limit access (reading or writing operations) to specific objects (e.g., files) only to those who are authorized.

Access control can only work with a reliable authentication. Only if the user's identity can be established reliably, is it possible to check the access rights. Access control can take place on different levels. Everyone who has ever worked with a networked computer will know the rights supported by common operating systems such as read access, write access, and execution rights.

Irrespective of the form of access control (DAC (Samarati, 1996), RBAC (Sandhu & Coyne, 1996), or MAC (Bell & Padula, 1996)), each access can be described in terms of a triplet (S,O,Op). S stands for the subject that is about to conduct an operation (Op) on an object (O). A specific mechanism of the operating system (often referred to as reference monitor) then checks whether or not the access is to be permitted.

In database systems, access restrictions can usually be defined on a finer level of granularity compared to operating systems. Various mechanisms make it possible to grant access authorizations not only at the level of relations (tables) but on each tuple (data record).

Closely linked to access control is auditing, which means that various operations such as successful and unsuccessful logon attempts can be recorded in order to trace back. It is possible to specify for each object which operations by whom should be recorded. Clearly, the integrity of the resulting log files is of utmost importance. No one should be able to modify (i.e., forge) and only system administrations should be able to delete them.

Cryptography

Cryptography has a long tradition. Humans have probably encrypted and decrypted communication contents since the early days. The so called Caesar encryption is a classical method, which Caesar is said to have used to send messages to his generals. The method is extremely simple: Every letter of the alphabet shifts for a certain distance denoted by k. K — meaning key — is a number between 1 and 25.

Although Caesar's code has obvious weaknesses, it clearly shows that sender as well as receiver must know the same secret. This secret is the key (k) and hence the method is a so-called secret key algorithm.

This contrasts with the public key method, where the encryption keys and the decryption keys are not the same. There are mathematical methods, which make it possible to generate the keys in such a way that the decryption keys (private keys) cannot be deduced from the encryption keys (public keys). The public key methods are also applicable for making digital signatures.

Cryptography alone is no solution to a security problem in most cases. Cryptography usually solves problems of communication security. However, it creates new problems in the form of key management which belongs to the field of computer security.

As long as cryptography has existed, people have been trying to break the cipher by means of cryptanalysis.

MOBILE SECURITY

The security risks particular to mobile devices result from their inherent properties. Mobile devices are personal, portable, have limited resources and are used to connect to various networks which are usually not trustworthy. In addition, mobile devices are usually connected to wireless networks that are often easier to compromise than their wired counterparts.

Their portability makes mobile devices subject to loss or theft. If a mobile device has been stolen or lost, unauthorized individuals are likely to gain direct access to the data stored on the device's resources.

Another, completely different risk are trojans devices which means that a stolen device is copied and a Trojan device is returned to the user. Thus attackers are able to access the recordings of all the actions performed by the user.

Unfortunately, the current practice when addressing resource limitations is to ignore well-known security concepts. For instance, to empower WML scripts, implementations lack the established sandbox model; thus downloaded scripts can access all local resources without restriction (Ghosh & Swaminatha, 2001).

Communication Security

Security cannot be confined to a device itself. Mobile devices are mostly used to communicate and thus securing this process is a first step.

The following security threats are not particular to mobile devices, but with wireless communication technologies certain new aspects arise (Mahan), (Eckert, 2000).

- **Denial-of-Service (DoS)** occurs when an adversary causes a system or a network to become unavailable to legitimate users or causes services to be interrupted or delayed. A wireless DoS attack could be the scenario, where an external signal jams the wireless channel. Up to now there is little that can be done to keep a serious adversary from mounting a DoS attack. A possible solution is to keep external persons away from the signal coverage, but this is rarely realizable.
- **Interception** has more than one meaning. An external user can masquerade himself as a legitimate user and therefore receive internal or confidential data. Also the data stream itself can be intercepted and decrypted for the purpose of disclosing private information. Therefore some form of strong encryption as well as authentication is necessary to protect the signals coverage area.
- **Manipulation** means that data can be modified on a system or during transmission. An example would be the insertion of a Trojan horse or a virus on a computer device. Protection of access to the network and its attached systems is one means of avoiding manipulation.
- **Masquerading** refers to claiming to be an authorized user while actually being a malicious external source. Strong authentication is required to avoid masquerade attacks.
- **Repudiation** is when a user denies having performed an action on the network. Strong authentication, integrity assurance methods and digital signatures can minimize this security threat.

Wireless LAN (IEEE 802.11)

Wireless LAN (WLAN) specifies two security services; the authentication and the privacy service. These services are mostly handled by the wired equivalency privacy (WEP). WEP is based on the RC4 encryption algorithm developed by Ron Rivest at MIT for RSA data security. RC4 is a strong encryption algorithm used in many commercial products. The key management, needed for the en/decryption is not standardized in WLAN but two key-lengths have come up: 40bit keys for export controlled applications and 128bit keys for strong encryption in domestic applications. Some papers on the weaknesses of the WEP standard have been published by Borisov, Goldberg, and Wagner (n.d.), but Kelly (2001) from the 802.11 standardization committee responded in the following way: WEP was not intended to give more protection than a physically protected (i.e., wired) LAN. So WEP is not a complete security solution and additional security mechanisms like end-to-end encryption, virtual private networks (VPNs) etc. need to be provided.

Bluetooth

In the Bluetooth Generic Access Profile (GAP, see Bluetooth Specification), the basis on which all other profiles are based, three security modes are defined:

- **Security Mode 1:** Non-secure.
- **Security Mode 2:** Service level enforced security.
- **Security Mode 3:** Link level enforced security.

In security mode 1, a device will not initiate any security—this is the non-secure mode. In security mode 2, the Bluetooth device initiates security procedures after the channel is established (at the higher layers), while in security mode 3, the Bluetooth device initiates security procedures before the channel is established (at the lower layers). At the same time two possibilities exist for the device's access to services: "trusted device" and "untrusted device." Trusted devices have unrestricted access to all services. Untrusted devices do not have fixed relationships and their access to services is limited. For services, 3 security levels are defined: services that require authorization and authentication, services that require authentication only and services that are open to all devices. These levels of access are obviously based on the results of the security mechanisms themselves. Thus we will concentrate on the two areas where the security mechanisms are implemented: the service level and the link level. Details on how security is handled on these levels can be found in (Daid).

Although Bluetooth design has focused on security, it still suffers from vulnerabilities. Vainio (Vainio) and Sutherland (Sutherland) present various risks.

Infrared

The standard infrared communication protocol does not include any security-related mechanisms. The standardization committee, the Infrared Data Association justifies this with the limited spatial range and with the required line-of-sight connection. To the best of our knowledge there has been no research on eavesdropping on infrared connections.

3.1.4 GSM, GPRS, UMTS

The security of digital wireless wide-area networks (WAN) depends on the protocols used. Details on GSM, GPRS, HSCSD, etc. can be found in (Gruber & Wolfmaier, 2001).

According to Walke (2000) and Hansmann and Nicklous (2001), it is required to identify user first and foremost to enable billing. Secondly, the transmitted data must be protected for privacy reasons. Since GSM and GPRS are the most widely used standards, we will focus on these standards.

In today's mobile phones a unique device ID can be used to identify the phone regardless of the SIM card used. A second unique ID is assigned to the SIM card. The SIM card is assigned a telephone number and, in addition, can usually store 16-32 KByte of data such as short message service (SMS) or phone numbers.

When a mobile phone tries to connect to the operator, the two unique IDs are transmitted. Based on these Ids, a decision is taken whether to allow the device to connect to the network.

1. **White-listed:** Access is allowed.
2. **Gray-listed:** Access is allowed, but mobile device remains under observation.
3. **Black-listed:** Access is not allowed (e.g., mobile device has been reported stolen).

The next step is to authenticate the user. Each subscriber is issued a unique security key and a security algorithm. Both are stored in the operator's system and in the mobile device. When accessing the network for the first time, the security system of the network sends a random number to the mobile device. The mobile device encrypts this random number with its security key and algorithm and returns it to the network. Subsequently, the security system of the network performs the same calculations and finally compares the result to the number transmitted by the mobile device. If both numbers match, the authentication process is completed successfully. Since random numbers are sent each time, replay attacks are not possible. In addition, the secret keys are never transmitted over the network.

Cryptography is not only used by the authentication process but the transmission of data is encrypted, too. Once a connection is established, a random session key is generated. Based on this session key and a security algorithm, a security key is generated. Using this security key and yet another security algorithm, all transmitted data are encrypted. Each connection is encrypted with a different session key.

Even if this concepts seems secure, there are various vulnerabilities as discussed, for instance, by Pesonen (1999).

Computer Security

According to Gollmann (1999) "computer security deals with the prevention and detection of unauthorized actions by users of a computer system."

Physical Protection

Mobile devices can be stolen easily because of their small form factor. Thus, anti-theft devices such as steel cables and holsters can be used to secure the devices.

Authentication

Authentication on the mobile device establishes the identity of the user to the particular mobile device, which then can act on behalf of that user. Most of the available mobile devices do not support any other authentication mechanisms than passwords or PINs. Some offer fingerprint sensors but they are not widely used and reported to be not very reliable.

Some products are already available that provide personal digital assistants (PDAs) with enhanced authentication features. For instance, PDASecure (for PalmOS) and Sign-On (both for PalmOS and WinCE), support password-based encryption for data. PINPrint (both for PalmOS and WinCE) provides fingerprint authentication.

OneTouchPass[1] offers an image-based way of authentication. When the device is switched on, an image is diplayed. The user authenticates himself by tapping on the previously specified places in the picture. The level of security offered by this program is similar to passwords; however, since the process of authentication is faster, more people are likely to use it. Hence, overall security may be improved.

Access Control

Based on the authenticated identity, the mobile device should further restrict access to its resources. Even though a PDA is a device that it typically used by only one person—hence the name *personal* digital assistant—access to files and other resources should still be restricted according to a policy for access control. In some cases, users may share devices or allow coworkers to access certain entries (business vs. personal). Most of the mobile devices do not provide any access control at all. For PalmOS, some products (e.g., Enforcer, Restrictor) are available that provide profiles limiting access to specific data.

On-Device Encryption

Authentication and access control may not suffice to protect highly sensitive corporate or private data stored on a mobile device. A common attack is to circumvent the access control mechanisms provided by the device by resetting the password or updating the operating system. It thus makes sense to encrypt sensitive data. Several products that offer various encryption algorithms are available, including JawzDataGator or MemoSafe for PalmOS. CryptoGrapher encrypts data stored on flash cards. On WinCE PocketLock encrypts documents, seNTry 2020 encrypts entire volumes, folders and also single files.

Anti-Virus Software

Installing anti-virus software is a standard security procedure for all corporate and most private computers and laptops. For mobile devices, special anti-virus packages are available. We expect that in the future more malicious software will be distributed that specifically targets handheld devices. However, just as anti-virus software developers generally keep up with new virus developments within hours, we expect similar success for anti-virus software for mobile devices. Examples for currently available software are InoculateIT and Palm Scanner for PalmOS; VirusScan and Anti-Virus for WinCE.

Application Security

It is not sufficient to protect the mobile device itself and the wireless communication protocols. In addition, precautions are also required on an application level. Applications should be designed in a way that authentication, authorization, access-control, and encryption mechanisms are supported. Standard technologies like SSL should be used as default settings.

MULTIMEDIA SECURITY

According to Memon and Wong (1998), today's copyright laws may be inadequate for digital data. He identifies four major application scenarios for multimedia security.

- **Ownership assertion:** The author can later prove that he really is the author.
- **Fingerprinting:** identifies each copy uniquely for each user. If unauthorized copies are found, one can determine who was the last rightful user. One can infer that he either willingly or unwillingly handed the content on to others illegally.
- **Authentication and integrity verification:** Necessary when digital content is used in medical applications and for legal purposes.
- **Usage control:** Mechanisms allow, for instance, to make copies of the original disk but not a copy of the copy.

These four requirements can again be systematically analyzed by looking at the basic requirements integrity, secrecy and availability.

Integrity and Authenticity

A possible definition of the integrity of multimedia content is to prove that the content's origin is in fact the alleged source (authenticity).

For example, a video or a still image may be used in court or for an insurance claim. Establishing both the authenticity (source) and the integrity (original content) of such clips is of paramount importance.

Why is this a new problem? When analog media (i.e., exposable film) were used there was always an original that could be faked only with a lot of additional effort.

Authenticity and integrity are also required in the context of electronic commerce (i.e., the buyer requires that the content has not been altered after leaving the certified producer's premises). Thus, authenticity is the answer to two distinct user requirements: (1) electronic evidence and (2) certified product.

Digital Signatures

The authenticity of traditional original sources can be established using various physical clues such as negatives (its age, material defects, etc). With the rise of digital multimedia data there is no longer an original because the content is a combination of bytes which can only be authenticated by non-physical clues. One option, which is referred to as *blind authentication*, is to examine the characteristics of the content and hope to be able to detect any forged parts by discontinuities in the content. Another option which is technically easier to implement is the use of digital signatures (Diffie & Hellman, 1976). However, the management effort required for a working public-key infrastructure should not be underestimated.

A method to verify whether a video clip has been forged is the *trustworthy camera* proposed by Friedman (1993). Using a chip inside the camera, the captured multimedia data can be signed. Since it is more difficult to manipulate hardware, a video clip signed by a trustworthy camera can usually be trusted.

Watermarking

The fundamental difference to other security measures is that watermarks primarily protect the copyright (copyright protection) and do not prevent copying (copy protection).

When watermarking graphics, information invisible to the viewer is hidden in the picture. The hidden information pertains to the original author, identifying, for instance, his name and address. The changes caused by embedding information are so marginal that they are not or only hardly perceptible.

Figure 2. This image is the most famous test image for watermarking

Embedding of Digital Watermarks

The information to be embedded is not uniformly distributed across the picture. That is to say, in large areas of one color, in which modifications would be immediately recognized, there is less information than as in patterned areas. In Figure 2, the area of the woman's hair and her plume would be ideal locations to hide information.

This image is the most famous test image for watermarking. The original copyright holder is Playboy (Nov 1971); researchers (illegally) used the image in their publications. Since it was so widely distributed Playboy eventually waived its rights and placed the image in the public domain.

A frequently used procedure (Figure 3) is that the hidden message can be seen as signal and the picture, in which the message is to be embedded, as interfering signal.

Detecting Digital Watermarks

To every picture, regardless of whether or not it contains a watermark, a detector is applicable, which searches the picture for watermarks. Depending on the detector used, it can be established whether a specific watermark has been embedded or whether it was taken from a multitude of watermarks, and if, which one. According to the sensitivity value for detection, the rate of false positive and false negative detection processes varies.

Robustness

An important quality characteristic of watermarks is their robustness when the image is being changed. Typical manipulations include changes in the resolution, cutting out details of the image, and application of different filters.

Well-known tests include Stirmarks[2], Checkmarks[3] (also contains Stirmarks), and Optimark[4].

Products

Digimarc[5] markets software that enables watermarks to be embedded in graphics. A distinctive code will be created for authors if they subscribe to Digimarc at MarcCenter. This ID can then be linked with personal information including name or e-mail address.

Most watermarks are based on random patterns, which are hidden in the brightness component of the image. Good watermarks are relatively robust and detectable even after printing and rescanning.

Digimarc have developed another interesting system[6], which can hide a URL in an image. Its primary aim is not so much copy protection

Figure 3. A signal is added to the original image

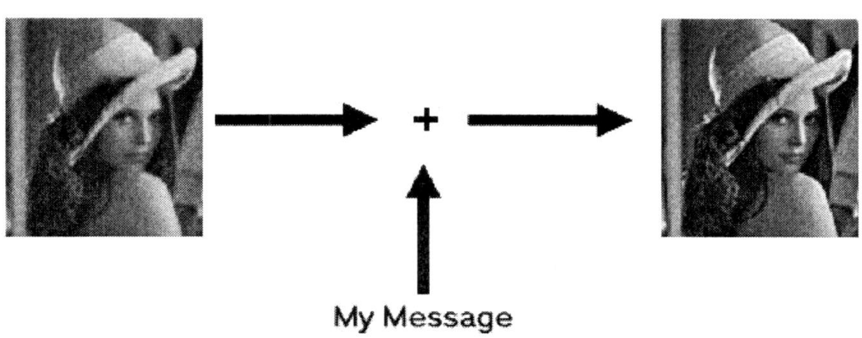

but rather the possibility to open a particular URL quickly in case a printout is held in the Web-camera.

MediaSec Technologies[7] Ltd. specializes in marketing watermarking software and in consulting services concerning media security. MediaSec sales the commercial version of SysCoP[8] watermarking technology. MediaTrust combines watermarks with digital signatures.

A good survey about watermarking is provided by Watermarkingworld[9]. Peter Meerwald wrote a diploma thesis[10] on this topic at the University of Salzburg.

Secrecy

Multimedia can be used in a very effective way to keep data secret. Steganography is about hiding data inside images, videos or audio recordings. Similar to watermarks, additional information is embedded so that the human observer does not or can only hardly notice it. However, the requirements are different compared to watermarks.

By definition, *visible* watermarks are not steganography because they are not hidden. The primary difference is the user's intention. Digital watermarks are used to store *additional* information inseparably with the multimedia object. Steganography, however, attempts to *conceal* information. The multimedia object is merely used as a cover in which the message is concealed. Steganography can be effectively combined with cryptography. First, the message is encrypted and then it is hidden in a multimedia object. This procedure is especially useful when one needs to hide the fact that encrypted information is transmitted (e.g., in countries that outlaw the use of cryptography, or if governments or employers consider all encrypted communication to be suspicious).

Watermarks are expected to be robust whereas the most important characteristic of steganographic marks is that they are difficult to detect — even with tools.

There are two kinds of compression for multimedia data: lossless and lossy. Clearly, both methods compress multimedia data but the resulting image differs. As the name indicates, lossless compression compresses the image without any changes. Thus, the original image can be reconstructed with all bytes being identical. Any information that is hidden in the image can be extracted without modification. Typical image formats for lossless compression are GIF, BMP and PNG.

Lossy compression changes the bytes of the image is a way that the human observer sees little difference but that it can be better compressed. That said, it is evident that the hidden message is changed, too, making extraction more difficult or even impossible. JPEG is among the most common lossy compression algorithms.

For steganography it is therefore preferred when the original information remains intact. Lossless compression are used when saving images saved as GIF (graphic interchange format) and 8-bit BMP (a Microsoft Windows and OS/2 bitmap file).

There are various programs available that implement steganography. Johnson and Jajodia (1998) provide an excellent overview of available solutions. The author also maintains a Web site[11] with various links to tools, research papers, and books.

Availability

Availability becomes especially important for streaming data. Even brief (less than 1 sec) interruptions of service will be noticed. Standards such as MPEG4 (Koenen, 2000) address this issue by using buffers. For a data stream from a specific source a minimum buffer may be specified.

Using this buffer, real time information can still be displayed even if the channel's current capacity is exceeded or transmission errors occur. Clearly, it is essential that the employed algorithms allow for a quick recovery from such errors. Most compression algorithms transfer a complete image only every few seconds and only updates in-between. Good algorithms allow to recalculate those in-between pictures not only in forward direction but also backwards. This improves *error resilience*.

For the aforementioned error resilience to work efficiently, good *error concealment* is also required. Error concealment refers to the ability to quickly locate the position of the erroneous data as accurately as possible.

Even if the network transmitting the data provides sufficient bandwidth, data-intensive multimedia content such as streaming video requires also unprecedented server performance. A few dozen requests may suffice to overload a server's disk array unless special measures (such as tremendous amounts of main memory) are taken.

Digital Rights Management

Digital rights management (DRM) is one of the greatest challenges for content producers in the digital age. In the past, the obstacle of non-authorized use of the content was much more difficult to overcome because the content was always bound to some physical product such as a book. However, the ease of producing digital copies without a loss of quality can lead to breaches of the copyright law. Typically, DRM addresses content integrity and availability.

In the past, DRM was concentrating on encryption to prevent the use of unauthorized copies. Today, DRM comprises the description, labeling, trading, protection, and monitoring of all forms of content. DRM is the "digital management of rights" and not the "management of digital rights." That is to say, DRM can also include the management of rights in non-digital media (e.g., print-outs).

It is essential for future DRM systems that they will be used starting with the initial creation of the content. This is the only way that the protection can comprise the whole process of development and increasing value of intellectual property. Meta-information is used to specify the information (e.g., author and type of permitted use). In order to enable the use and reuse, all meta-information must be inextricably connected to the content. Despite some basic approaches to such systems (e.g., digital watermarking), there are still no wide-spread systems today.

There is a collection of numerous links on the Web site[12] of Internet Engineering Task Force concerning the topic of intellectual property.

MOBILE MULTIMEDIA SECURITY

In this section we combine the knowledge presented in the previous sections. Clearly, mobile multimedia security comprises general security aspects. Since mobile devices are used, issues of mobile security are relevant; in particular methods and algorithms in the context of multimedia security will be applied.

We discuss the influence of mobile hardware and software designed for operating systems such as PalmOS or WindowsCE on multimedia security.

Hardware

Mobile devices are small and portable. Even though the processing power has increased in the past, they are not only a lot slower than any desktop PC but also suffer from a limited power supply. Although it is theoretically possible to have a personal device perform complex calculations when it is not used otherwise, this background processing very quickly drains the batteries.

Recently, mobile devices are often combined with (low resolution) digital cameras. Today's top models include cameras with a resolution of up to one mega pixel. Compared to "single purpose" digital cameras the images' quality is clearly inferior.

Lower image quality makes it harder to use physical clues in the image to establish its authenticity. However, smaller images can be processed quicker for digital signatures. By first calculating a secure hash value — a not too power-consuming operation — and secondly signing this hash value, a trustworthy camera can be implemented. Since both the camera and the processing unit are built into one hardware device that also has unique hardware IDs, tampering with the device is rendered more difficult.

Additionally, images of lower quality are more suitable for steganographic purposes. Since the images already contain various artifacts caused by poor lenses and low quality CCD chips, additional changes introduced by the steganographic algorithms cannot be seen as easily compared to high-quality digital images.

The same considerations apply to audio content. Generally the quality both of recording and playback of audio data is lower on mobile devices. Hence, it is again easier to hide information (either steganography or watermarks).

Software Limits

The advantage that mobile devices offer is that the operating system can be specifically tailored to the hardware. As previously mentioned, the integrity of the operating system is a prerequisite for all data related to security such as access control. Only if all operations accessing a resource, pass through the reference monitor, access control can work reliably. Mobile devices offer the opportunity to store the most basic kernel functions in read-only memory which clearly makes it difficult to change them. However, the last years have shown that device vendors usually need to update the operating systems quite frequently so that a pure ROM-based operating system will no longer be available.

New Combinations

Mobile devices often contain multiple devices that can be combined to improve multimedia security. For instance, a very trustworthy camera can be implemented using a GPS module and wireless communication. The built-in camera creates an image that can immediately be digitally signed even before it is stored to the device's filesystem.

Using the time and position signals of GPS, precise location information can be appended and a message digest (hash value) computed. This value is subsequently sent to a trusted third party (cell service provider) via wireless communication. The provider can verify the approximate location because the geographical location of the receiving cell is known. The message digest has a small fingerprint and can thus be stored easily.

This approach allows to establish not only the authenticity of the image itself but also its context, ie. the time and location where it was taken.

SUMMARY

This chapter provides a comprehensive overview of mobile multimedia security. Since nothing can be totally secure, security heavily depends on the requirements in a specific application domain.

All security requirements can be traced back to one of the four basic requirements:

- Secrecy (also known as confidentiality)
- Integrity
- Availability
- Non-repudiation

When looking at security in mobile computing, we distinguish between communication security and computer security. Communication security focuses on securing the communication between devices, whereas computer security refers to securing data on the device. Since mobile device rarely use wire-bound communication, we have elaborated on wireless standards (Bluetooth, WLAN, GPRS) and their implications on security requirements.

Multimedia security has received a lot of attention in mass media because of file sharing systems that are used to share music in MP3 format. However, even long before this hype, many researchers worked on watermarking techniques to embed copyright information in digital works such as images, audio and video. Digital rights management (DRM) works primarily based on embedded copyright information to allow or prevent copying and distribution of content. Even though research shows theoretical solutions how DRM could work, there is currently little incentive for hardware and software manufacturers to implement such a system. Most users will always choose a platform restricting them as little as possible.

Mobile multimedia applications are becoming increasingly popular because today's cell phones and PDAs often include digital cameras and can also record audio. It is a challenge to accommodate existing techniques for protecting multimedia content on the limited hardware and software basis provided by mobile devices. The importance of adequate protection of content on mobile devices will increase simply because such devices will become even more widespread. Since in near future, most of the data stored on mobile devices will undoubtedly be multimedia content, we can be certain that *mobile multimedia security* will be a focus of security research.

REFERENCES

Bell, D., & Padula, L. L. (1996). Mitre technical report 2547 (secure computer system). *Journal of Computer Security*, *4*(2), 239-263.

Borisov, N., Goldberg, I., & Wagner, D. (2001). Intercepting mobile communications: The insecurity of 802.11. *Proceedings of the 7th Annual International Conference on Mobile Computing and Networking* (pp. 180-189), Rome, Italy. New York: ACM Press. Retrieved August 1, 2003, from citeseer.ist.psu.edu/borisov01intercepting.html

Daid, M. (2000). *Bluetooth security, parts 1, 2, and 3*. Retrieved August 1, 2003, from http://www.palowireless.com/bluearticle/cc1_security1.asp and http://www.palowireless.com/bluearticle/cc1_security2.asp http://www.palowireless.com/bluearticle/cc1_security3.asp

Diffie, W., & Hellman, M. (1976). New directions in cryptography. *IEEE Transactions Information Theory, IT22*(6), 644-654.

Eckert, C. (2000). Mobile devices in ebusiness — new opportunities and new risks. *Proceedings Fachtagung Sicherheit in Informations Systemen (SIS)*, Zurich, Switzerland.

Friedman, G. (1993). The trustworthy digital camera: Restoring credibility to the photographic image. *IEEE Transactions Consumer Electronics*, *39*(4), 905-910.

Ghosh, K., & Swaminatha, T. (2001). Software security and privacy risks in mobile e-commerce. *Communications of the ACM, 44*(2), 51-57.

Gollmann, D. (1999). *Computer security*. West Sussex, UK: John Wiley & Sons.

Gruber, F., & Wolfmaier, K. (2001). *State of the art in wireless communication* (Tech. Rep. No. Scch-tr-0171). Hagenberg, Austria: Software Competence Center Hagenberg.

Hansmann, M., & Nicklous, S. (2001). *Pervasive computing-handbook*. Böbling, Germany: Springer Verlag.

Johnson, F., & Jajodia, S. (1998). Steganography: Seeing the unseen. *IEEE Computer*, *31*(2), 26-34.

Kerry, S. J. (2001). *Chair of ieee 802.11 responds to wep security flaws*. Retrieved from http://slashdot.org/it/01/02/15/1745204.shtml

Koenen, R. (2000). *Overview of the mpeg-4 standard* (Tech. Rep. No. jtc1/sc29/wg11 n3536). International Organisation for Standardization ISO/IEC JTC1/SC29/WG11, Dpt. Of Computer Science and Engineering.

Kwok, S. H. (2003). Watermark-based copyright protection system security. *Communications of the ACM*, *46*(10), 98-101. Retrieved from http://doi.acm.org/10.1145/944217.944219

Mahan, R. E. (2001). *Security in wireless networks*. Sans Institute. Retrieved August 1, 2003, from http://rr.sans.org/wireless/wireless_net3.php

Memon, N., & Wong, P. W. (1998). Protecting digital media content. *Communications of the ACM*, *41*(7), 35-43.

NIS. (1992). *National information systems security (infosec) glossary* (NSTISSI No. 4009 4009). NIS, Computer Science Department, Fanstord, California. Federal Standard 1037C.

Olovsson, T. (1992). *A structured approach to computer security* (Tech. Rep. No. 122 122). Gothenburg, Sweden: Chalmers University of Technology, Department of Computer Engineering. Retrieved from http://www.securityfocus.com/library/661

Pesonen, L. (1999). *Gsm interception*. Technical report, Helsinki University of Technology, Dpt. Of Computer Science and Engineering.

Samarati, R. S. R. P. (1996). Authentication, access control, and audit. *ACM Computing Surveys*, *28*(1), 241-243.

Sandhu, R., & Coyne, E. (1996). Role-based access control models. *IEEE Computer*, *29*(2), 38-47.

Sutherland, E.. (n.d.). *Bluetooth security: An oxymoren*? Retrieved August 1, 2003, from http://www.mcommercetimes.com/Technology/41

Vainio, J. (2000). *Bluetooth security*. Retrieved August 1, 2003, from http://www.niksula.cs.hut.fi/~jiitv/bluesec.html

Walke. (2000). *Mobilfunknetze und ihre Protokolle*, volume 1. B. G. Teubner Verlag, Stuttgart.

KEY TERMS

Availability: Refers to the state that a system can perform the specified service. Denial-of-Service (DoS) attacks target a system's availability.

Authentication: Means proving that a person is the one he/she claims to be.

Integrity: Only authorized people are permitted to modify data.

Non-Repudiation: Users are not able to deny (plausibly) to have carried out operations.

Secrecy: Users may obtain access only to those objects for which they have received authorization, and will not get access to information they must not see.

Security: Encompasses secrecy (aka, confidentiality), integrity, and availability. Nonrepudiation is a composite requirement that can be traced back to integrity.

Watermarking: Refers to the process of hiding information in graphics. In some cases visible watermarks are used (such as on paper currency) so that people can detect the presence of a mark without special equipment.

ENDNOTES

1 http://www.onetouchpass.comhttp://www.onetouchpass.com

2 http://www.watermarkingworld.org/stirmark/stirmark.htmlhttp://www.watermarkingworld.org/stirmark/stirmark.html

3 http://www.watermarkingworld.org/checkmark/checkmark.htmlhttp://www.wate[4]markingworld.org/checkmark/checkmark.html

5 http://www.watermarkingworld.org/optimark/index.htmlhttp://www.watermarkingworld.org/optimark/index.html

6 http://www.digimarc.com/mediabridgehttp://www.digimarc.com/mediabridge

7 http://www.mediasec.de/http://www.mediasec.de/

8 http://www.mediasec.de/html/de/products\s\do5(s)ervices/syscop.htmhttp://www.mediasec.de/html/de/products_services/syscop.htm

9 http://www.watermarkingworld.org/http://www.watermarkingworld.org/

10 http://www.cosy.sbg.ac.at/ pmeerw/Watermarking/MasterThesis/http://www.cosy.sbg.ac.at/ pmeerw/Watermarking/MasterThesis/

11 http://www.jjtc.com/Steganography/http://www.jjtc.com/Steganography/

12 http://www.ietf.org/ipr.htmlhttp://www.ietf.org/ipr.html

Chapter IV
Data Dissemination in Mobile Environments

Panayotis Fouliras
University of Macedonia, Greece

ABSTRACT

Data dissemination today represents one of the cornerstones of network-based services and even more so for mobile environments. This becomes more important for large volumes of multimedia data such as video, which have the additional constraints of speedy, accurate, and isochronous delivery often to thousands of clients. In this chapter, we focus on video streaming with emphasis on the mobile environment, first outlining the related issues and then the most important of the existing proposals employing a simple but concise classification. New trends are included such as overlay and p2p network-based methods. The advantages and disadvantages for each proposal are also presented so that the reader can better appreciate their relative value.

INTRODUCTION

A well-established fact throughout history is that many social endeavors require dissemination of information to a large audience in a fast, reliable, and cost-effective way. For example, mass education could not have been possible without paper and typography. Therefore, the main factors for the success of any data dissemination effort are supporting technology and low cost.

The rapid evolution of computers and networks has allowed the creation of the Internet with a myriad of services, all based on rapid and low cost data dissemination. During recent years, we have witnessed a similar revolution in mobile devices, both in relation to their processing power as well as their respective network infrastructure. Typical representatives of such networks are the 802.11x for LANs and GSM for WANs.

In this context, it is not surprising that the main effort has been focusing on the dissemination of multimedia content–especially audio and video, since the popularity of such services is high, with RTP the de-facto protocol for multimedia data transfer on the Internet. Although both audio and video have strict requirements in terms of packet jitter (the variability of packet delays within the same packet stream), video additionally requires significant amount of bandwidth due to its data size. Moreover, a typical user requires multimedia to be played in real-time, (i.e., shortly after his request, instead of waiting for the complete file to be downloaded; this is commonly referred to as *multimedia streaming*.

In most cases, it is assumed that the item in demand is already stored at some server(s) from where the clients may request it. Nevertheless, if the item is popular and the client population very large, additional methods must be devised in order to avoid a possible drain of available resources. Simple additional services such as fast forward (FF) and rewind (RW) are difficult to support, let alone interactive video. Moreover, the case of asymmetric links (different upstream and downstream bandwidth) can introduce more problems. Also, if the item on demand is not previously stored but represents an ongoing event, many of the proposed techniques are not feasible.

In the case of mobile networks, the situation is further aggravated, since the probability of packet loss is higher and the variation in device capabilities is larger than in the case of desktop computers. Furthermore, ad-hoc networks are introduced, where it is straightforward to follow the bazaar model, under which a client may enter a wall mart and receive or even exchange videos in real time from other clients, such as specially targeted promotions, based on its profile. Such a model complicates the problem even further.

In this chapter, we are focusing on video streaming, since video is the most popular and demanding multimedia data type (Sripanidkulchai, Ganjam, Maggs, & Zhang, 2004). In the following sections, we are identifying the key issues, present metrics to measure the efficiency of some of the most important proposals and perform a comparative evaluation in order to provide an adequate guide to the appropriate solutions.

ISSUES

As stated earlier, streaming popular multimedia content with large size such as video has been a challenging problem, since a large client population demands the same item to be delivered and played out within a short period of time. This period should be smaller that the time t_w a client would be willing to wait after it made its request. Typically there are on average a constant number of requests over a long time period, which suggests that a single broadcast should suffice for each batch of requests. However, the capabilities of all entities involved (server, clients, and network) are finite and often of varying degree (e.g., effective available network and client bandwidth). Hence the issues and challenges involved can be summarized as follows:

- What should the broadcasting schedule of the server be so that the maximum number of clients' requests is satisfied without having them wait more than t_w?
- How can overall network bandwidth be minimized?
- How can the network infrastructure be minimally affected?

- How can the clients assist if at all?
- What are the security considerations?

In the case of mobile networks, the mobile devices are the clients; the rest of the network typically is static, leading to a mixed, hybrid result. Nevertheless, there are exceptions to this rule, such as the ad hoc networks. Hence, for mobile clients there are some additional issues:

- Mobile clients may leave or appear to leave a session due to higher probability of packet loss. How does such a system recover from this situation?
- How can redirection (or *handoff*) take place without any disruption in play out quality?
- How can the bazaar model be accommodated?

BACKGROUND

In general, without prior knowledge on how the data is provided by the server, a client has to send a request to the server. The server then either directly delivers the data (on demand service) or replies with the broadcast channel access information (e.g., channel identifier, estimated access time, etc.). In the latter case, if the mobile client decides so, it monitors the broadcast channels (Hu, Lee, & Lee, 1998). In both cases, there have been many proposals, many of which are also suitable for mobile clients. Nevertheless, many proposals regarding mobile networks are not suitable for the multimedia dissemination. For example, Coda is a file replication system, Bayou a database replication system and Roam a slightly more scalable general file replication system (Ratner, Reiher, & Popek, 2004), all of which do not assume strict temporal requirements.

The basic elements which comprise a dissemination system are the server(s), the clients, and the intermittent network. Depending on which of these is the focus, the various proposals can be classified into two broad categories: Proposals regarding the server organization and its broadcast schedule, and those regarding modifications in the intermittent network or client model of computation and communication.

Proposals According to Server Organization and Broadcasting Schedule

Let us first examine the various proposals in terms of the server(s) organization and broadcasting schedule. These can be classified in two broad classes, namely *push*-based scheduling (or *proactive*) and *pull*-based scheduling (or *reactive*). Under the first class, the clients continuously monitor the broadcast process from the server and retrieve the required data without explicit requests, whereas under the second class the clients make explicit requests which are used by the server to make a schedule which satisfies them. Typically, a hybrid combination of the two is employed with push-based scheduling for popular and pull-based scheduling for less popular items (Guo, Das, & Pinotti, 2001).

Proposals for Popular Videos

For the case of pushed-based scheduling broadcasting schedules of the so-called *periodic broadcasting* type are usually employed: The server organizes each item in segments of appropriate size, which it broadcasts periodically. Interested clients simply start downloading from the beginning of the first segment and play it out immediately. The clients must be able to preload some segments of the item and be capable of downlink bandwidth higher than that for a single video stream. Obviously this scheme works for popular videos, assuming there is adequate bandwidth at the server in relation to the amount and size of items broadcasted.

Pyramid broadcasting (PB) (Viswanathan & Imielinski, 1995) has been the first proposal in this category. Here, each client is capable of downloading from up to two channels simultaneously. The video is segmented in s segments of increasing size, so that $s_{i+1}=\alpha\cdot s_i$, where $\alpha=\frac{B}{MK}$ and B is the total server bandwidth expressed in terms of the minimum bandwidth b_{min} required to play out a single item, M the total number of videos and K the total number of virtual server channels. Each channel broadcasts a separate segment of the same video periodically, at a speed higher than b_{min}. Thus, with M=4, K=4 and B=32, we have α=2, which means that each successive segment is twice the size of the previous one. Each segment is broadcasted continuously from a dedicated channel as depicted in Figure 1. In our example, each server channel has bandwidth $B'=B/K=8\cdot b_{min}$, which means that the clients must have a download bandwidth of $16\cdot b_{min}$.

If D is the duration of the video, then the waiting time of a client is at most $M\cdot s_1/B'$. With $D = 120$ and $K = M = 4$, we have $M\cdot s_1/B' = 4\cdot 8/8 = 4$ time units. Each segment from the first channel requires 1 time unit to be downloaded, but has a play out time of 8 units. Consider the case that a client requests video 1 at the time indicated by the thick vertical arrow. Here the first three segments to be downloaded are indicated by small grey rectangles. By the time the client has played out half of the first segment from channel 1 it will start downloading the second segment from channel 2 and so on. The obvious drawback of this scheme is that it requires a very large download bandwidth at the client as well as a large buffer to store the preloaded segments (as high as 70% of the video).

In order to address these problems, other methods have been proposed, such as *permutation-based pyramid broadcasting* (PPB) (Aggarwal, Wolf, & Yu, 1996) and *skyscraper broadcasting* (SB) (Hua & Sheu, 1997). Under PPB each of the K channels is multiplexed into P subchannels with P times lower rate, where the client may alternate the selection of subchannel during download. However, the

Figure 1. Example of pyramid broadcasting with 4 videos and 4 channels

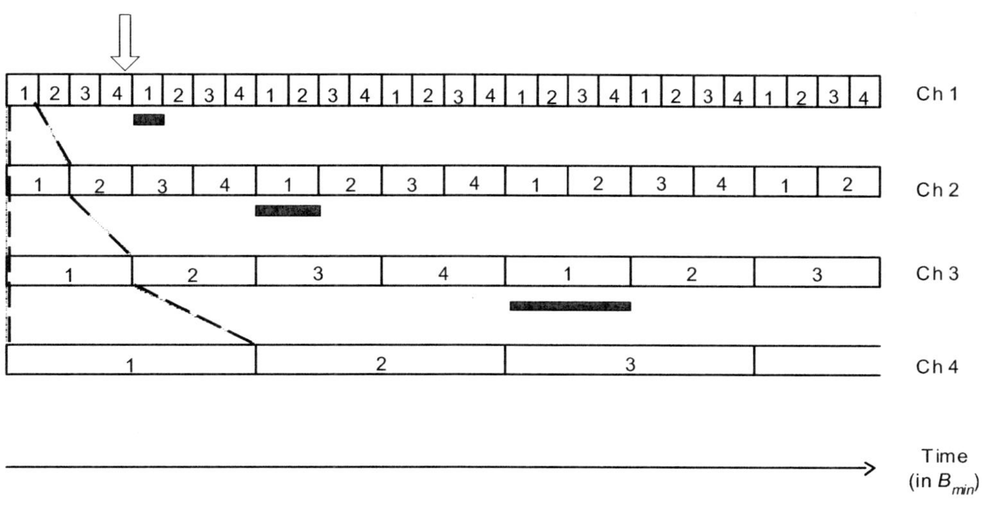

buffer requirements are still high (about 50% of the video) and synchronization is difficult. Under SB, two channels are used for downloading, but with a rate equal to the playing rate B_{min}. Relative segment sizes are 1, 2, 2, 5, 5, 12, 12, 25, 25,... W, where W the width of the skyscraper. This leads to much lower demand on the client, but is inefficient in terms of server bandwidth. The latter goal is achieved by *fast broadcasting* (FB) (Juhn & Tseng, 1998) which divides the video into segments of geometric series, with K channels of B_{min} bandwidth, but where the clients download from *all* K channels.

Yet another important variation is *harmonic broadcasting* (HB) (Juhn & Tseng, 1997) which divides the video in segments of equal size and broadcasts them on K successive channels of bandwidth B_{min}/i, where $i = 1,\ldots K$. The client downloads from all channels as soon as the first segment has started downloading. The client download bandwidth is thus equal to the server's and the buffer requirements low (about 37% of the total video). However, the timing requirements may not be met, which is a serious drawback. Other variations exist that solve this problem with the same requirements (Paris, Carter, & Long, 1998) or are hybrid versions of the schemes discussed so far, with approximately the same cost in resources as well as efficiency.

Proposals for Less Popular Videos or Varying Request Pattern

In the case of less popular videos or of a varying request pattern pulled-based or reactive methods are more appropriate. More specifically, the server gathers clients' requests within a specific time interval $t_{in} < t_w$. In the simplest case all requests are for the beginning of the same video, although they may be for different videos or for different parts of the same video (e.g., after a FF or RW). For each group (*batch*) of similar requests a new broadcast is scheduled by reserving a separate server channel, (*batching*). With a video duration t_D a maximum of $\lceil t_D/t_{in} \rceil$ server channels are required for a single video assuming multicast.

The most important proposals for *static* multicast batching are: *first-come-first-served* (FCFS) where the oldest batch is served first, *maximum-queue-length-first* (MQLF) where the batch containing the largest amount of requests is served first, reducing average system throughput by being unfair and *maximum-factor-queue-length* (MFQL) where the batch containing the largest amount of requests for some video weighted by the factor $1/\sqrt{f_i}$ is selected, where f_i is the access frequency of the particular video. In this way the popular videos are not always favored (Hua, Tantaoui, & Tavanapong, 2004).

A common drawback of the proposals above is that client requests which miss a particular video broadcasting schedule cannot hope for a reasonably quick service time, in a relatively busy server. Hence, *dynamic* multicast proposals have emerged, which allow the existing multicast tree for the same video to be extended in order to include late requests. The most notable proposals are *patching*, *bandwidth skimming,* and *chaining*.

Patching (Hua, Cai, & Sheu, 1998) and its variations allow a late client to join an existing multicast stream and buffer it, while simultaneously the missing portion is delivered by the server via a separate patching stream. The latter is of short duration, thus quickly releasing the bandwidth used by the server. Should the clients arrive towards the end of the normal stream broadcast, a new normal broadcast is scheduled instead of a patch one. In more recent variations it is also possible to have double patching, where a patching stream is created on top of a previous patching stream, but requires more bandwidth on both the client(s) and the server and synchronization is more difficult to achieve.

The main idea in Bandwidth Skimming (Eager, Vernon, & Zahorjan, 2000) is for clients to download a multicast stream, while reserving a small portion of their download bandwidth (*skim*) in order to listen to the closest active stream other than theirs. In this way, hierarchical merging of the various streams is possible to achieve. It has been shown that it is better than patching in terms of server bandwidth utilization, though more complex to implement.

Chaining (Sheu, Hua, & Tavanapong, 1997) on the other hand is essentially a pipeline of clients, operating in a peer-to-peer scheme, where the server is at the root of the pipeline. New clients are added at the bottom of the tree, receiving the first portion of the requested video. If an appropriate pipeline does not exist, a new one is created by having the server feed the new clients directly. This scheme reduces the server bandwidth and is scalable, but it requires a collaborative environment and implementation is a challenge, especially for clients who are in the middle of a pipeline and suddenly lose network connection or simply decide to withdraw. It also requires substantial upload bandwidth to exist at the clients, so it is not generally suitable for asymmetric connections.

Proposals According to Network and Client Organization

Proxies and Content Distribution Networks

Proxies have been used for decades for delivering all sorts of data and especially on the Web, with considerable success. Hence there have been proposals for their use for multimedia dissemination. Actually, some of the p2p proposals discussed later represent a form of proxies, since they cache part of the data they receive for use by their peers. A more general form of this approach, however, involves dedicated proxies strategically placed so that they are more effective.

Wang, Sen, Adler, and Towsley, (2004) base their proposal on the principle of prefix proxy cache allocation in order to reduce the aggregate network bandwidth cost and startup delays at the clients. Although they report substantial savings in transmission cost, this is based on the assumption that all clients request a video from its beginning.

A more comprehensive study based on Akamai's streaming network appears in (Sripanidkulchai, Ganjam, Maggs, & Zhang, 2004). The latter is a static overlay composed of edge nodes located close to the clients and intermediate nodes that take streams from the original content publisher and split and replicate them to the edge nodes. This scheme effectively constitutes a content distribution network (CDN), used not only for multimedia, but other traffic as well. It is reported that under several techniques and assumptions tested, application end-point architectures have enough resources, inherent stability and can support large-scale groups. Hence, such proposals (including p2p) are promising for real-world applications. Client buffers and uplink bandwidth can contribute significantly if it is possible to use them.

Multicast Overlay Networks

Most of the proposals so far work for multicast broadcasts. This suggests that the network infrastructure supports IP multicasting completely. Unfortunately, most routers in the Internet do not support multicast routing. As the experience from MBone (multicast backbone) (Kurose, & Ross, 2004) shows, an overlay virtual network interconnecting "islands" of multicasting-capable routers must be established over the existing Internet using the rest of the routers as end-points of "tunnels." Nevertheless, since IP multicasting is still a best effort service and therefore unsuitable for multimedia streaming, appropriate reservation of resources at the participating routers is necessary. The signaling protocol of choice is RSVP under which potential receivers signal their intention to join the multicast tree. This is a de-facto part of the Intserv mechanism proposed by IETF. However, this solution does not scale well. A similar proposal but with better scaling is DiffServ which has still to be deployed in numbers (Kurose, & Ross, 2004).

A more recent trend is to create an overlay multicast network at the application layer, using unicast transmissions. Although worse than pure multicast in theory, it has been an active area of research due to its relative simplicity, scalability and the complete absence of necessity for modifications at the network level. Thus, the complexity is now placed at the end points, (i.e., the participating clients and server(s)) and the popular point-to-point (p2p) computation model can be employed in most cases. Asymmetric connections must still include uplink connections of adequate bandwidth in order to support the p2p principle.

Variations include P2Cast (Guo, Suh, Kurose, & Towsley, 2003) which essentially is patching in the p2p environment: Late clients receive the patch stream(s) from old clients, by having two download streams, namely the normal and the patch stream. Any failure of the parent involves the source (the initial server), which makes the whole mechanism vulnerable and prone to bottlenecks.

ZigZag (Tran, Hua, & Do, 2003) creates a logical hierarchy of clusters of peers, with each member at a bounded distance from each other and one of them the cluster leader. The name of this technique emanates from the fact that the leader of each cluster forwards data only to peers in different clusters from its own. An example is shown in Figure 2, where there are 16 peers, organized in clusters of four at level 0. One peer from each cluster is the cluster leader or *head* (additionally depicted for clarity) at level 1. The main advantages of ZigZag are the small height of the multicast tree and the amount of data and control traffic at the server. However, leader failures can cause significant disruption, since both data and control traffic pass through a leader.

LEMP (Fouliras, Xanthos, Tsantalis., & Manitsaris, 2004) is a another variation which

Figure 2. ZigZag: Example multicast tree of peers (3 layers, 4 peers per cluster)

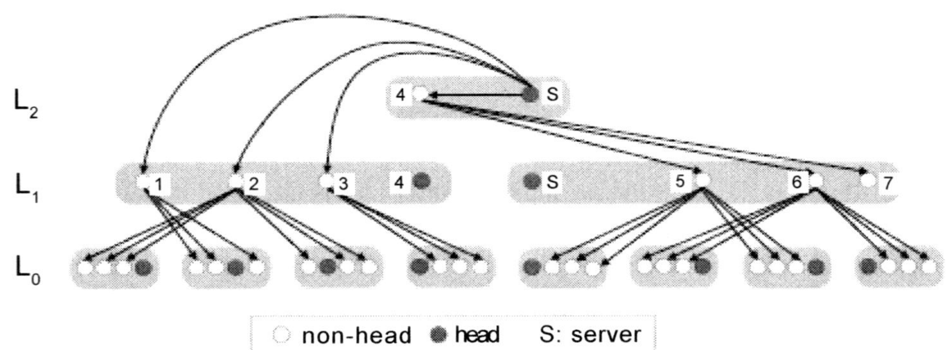

forms a simple overlay multicast tree with an upper bound on the number of peers receiving data from their parent. However, each level of the multicast tree forms a virtual cluster where one peer is the local representative (LR) and another peer is its backup, both initially selected by the server. Most of the control traffic remains at the same level between the LR and the rest of the peers. Should the LR fail, the backup takes its place, selecting a new backup. All new clients are assigned by the server to an additional level under the most recent or form a new level under the server with a separate broadcast. Furthermore, special care has been made for the case of frequent disconnections and reconnections, typical for mobile environments; peers require a single downlink channel at play rate and varying, but bounded uplink channels. This scheme has better response to failures and shorter trees than ZigZag, but for very populous levels there can be some bottleneck for the light control traffic at the LR.

Other Proposals

Most of the existing proposals have been designed without taking into consideration the issues specific to mobile networks. Therefore, there has recently been considerable interest for research in this area. Most of the proposed solutions, however, are simple variations of the proposals presented already. This is natural, since the network infrastructure is typically static and only clients are mobile. The main exception to this rule comes from ad hoc networks.

Add hoc networks are more likely to show packet loss, due to the unpredictable behavior of all or most of the participant nodes. For this reason there has been considerable research effort to address this particular problem, mostly by resorting to multipath routing, since connectivity is less likely to be broken along multiple paths. For example, (Zhu, Han, & Girod, 2004) elaborate on this scheme, by proposing a suitable objective function which determines the appropriate rate allocation among multiple routes. In this way congestion is also avoided considerably, providing better results at the receiver. Also (Wei, & Zakhor, 2004) propose a multipath extension to an existing on-demand source routing protocol (DSR), where the packet carries the end-to-end information in its header and a route discovery process is initiated in case of problems and (Wu, & Huang, 2004) for the case of heterogeneous wireless networks.

All these schemes work reasonably well for small networks, but their scalability is questionable, since they have been tested for small size networks.

COMPARATIVE EVALUATION

We assume that the play out duration t_D of the item on demand is in general longer than at least an order of magnitude compared to t_w. Furthermore, we assume that the arrival of client requests is a Poisson distribution and that the popularity of items stored at the server follows the Zipf distribution. These assumptions are in line with those appearing in most of the proposals.

In order to evaluate the various proposals we need to define appropriate metrics. More specifically:

- Item access time: this should be smaller than t_w as detailed previously
- The bandwidth required at the server as a function of client requests
- The download and upload bandwidth required at a client expressed in units of the minimum bandwidth b_{min} for playing out a single item
- The minimum buffer size required at a client

- The maximum delay during redirection, if at all; obviously this should not exceed the remainder in the client's buffer
- The overall network bandwidth requirements
- Network infrastructure modification; obviously minimal modification is preferable
- Interactive capabilities

Examining the proposals for popular videos presented earlier, we note that they are unsuitable for mobile environments, either because they require a large client buffer, large bandwidth for downloads or very strict and complex synchronization. Furthermore, they were designed for popular videos with a static request pattern, where clients always request videos from their beginning.

On the other hand, patching, bandwidth skimming are better equipped to address these problems, but unless multicasting is supported, may overwhelm the server. Chaining was designed for multicasting, but uses the p2p computation model, lowering server load and bandwidth.

Nevertheless, unicast-based schemes are better in practice for both wired and mobile networks as stated earlier. Although several proposals exist, Zigzag and LEMP are better suited for mobile environments, since they have the advantages of chaining, but are designed having taken into consideration the existence of a significant probability of peer failures, as well as the case of ad hoc networks and are scalable. Their main disadvantage is that they require a collaborative environment and considerable client upload bandwidth capability, which is not always the case for asymmetric mobile networks. Furthermore, they reduce server bandwidth load, but not the load of the overall network.

The remaining proposals either assume a radical reorganization of the network infrastructure (CDN) or are not proven to be scalable.

CONCLUSION AND FUTURE TRENDS

The research conducted by IETF for quality of service (QoS) in IP-based mobile networks and QoS policy control is of particular importance. Such research is directly applicable to the dissemination of multimedia data, since the temporal requirement may lead to an early decision for packet control, providing better network bandwidth utilization. The new requirements of policy control in mobile networks are set by the user's home network operator, depending upon a profile created for the user. Thus, certain sessions may not be allowed to be initiated under certain circumstances (Zheng, & Greis, 2004).

In this sense, most mobile networks will continue being hybrid in nature for the foreseeable future, since this scheme offers better control for administrative and charging reasons, as well as higher effective throughput and connectivity to the Internet. Therefore, proposals based on some form of CDN are better suited for commercial providers. Nevertheless, from a purely technical point of view, the p2p computation model is better suited for the mobile environment, with low server bandwidth requirements, providing failure tolerance and, most important, inherently supporting ad hoc networks and interactive multimedia.

REFERENCES

Aggarwal, C., Wolf, J., & Yu, P. (1996). A permutation based pyramid broadcasting scheme for video on-demand systems. *IEEE International Conference on Multimedia*

Computing and Systems (ICMCS '96), (pp. 118-126), Hiroshima, Japan.

Eager, D., Vernon, M., & Zahorjan, J. (2000). Bandwidth skimming: A technique for cost-effective video-on-demand. *Proceedings of IS&T/SPIE Conference on Multimedia Computing and Networking (MMCN 2000)* (pp. 206-215).

Fouliras, P., Xanthos, S., Tsantalis, N., & Manitsaris, A. (2004). LEMP: Lightweight efficient multicast protocol for video on demand. *ACM Symposium on Applied Computing (SAC'04)* (pp. 1226-1231), Nicosia, Cyprus.

Guo, Y., Das, S., & Pinotti, M. (2001). A new hybrid broadcast scheduling algorithm for asymmetric communication systems: Push and pull data based on optimal cut-off point. *Mobile Computing and Communications Review (MC2R)*, *5*(3), 39-54. ACM.

Guo, Y., Suh, K., Kurose, J., & Towsley, D. (2003). A peer-to-peer on-demand streaming service and its performance evaluation. *IEEE International Conference on Multimedia Expo (ICME '03)* (pp. 649-652).

Hu, Q., Lee, D., & Lee, W. (1998). Optimal channel allocation for data dissemination in mobile computing environments. *International Conference on Distributed Computing Systems* (pp. 480-487).

Hua, K., Tantaoui, M., & Tavanapong, W. (2004). Video delivery technologies for large-scale deployment of multimedia applications. *Proceedings of the IEEE*, *92*(9), 1439-1451.

Hua, K., & Sheu, S. (1997). Skyscraper broadcasting: A new broadcasting scheme for metropolitan video-on-demand systems. *ACM Special Interest Group on Data Communication (SIGCOMM '97)* (pp. 89-100), Sophia, Antipolis, France.

Hua, K., Cai, Y. & Sheu, S. (1998). Patching: A multicast technique for true video-on-demand services. *ACM Multimedia '98* (pp. 191-200), Bristol, UK.

Juhn, L., & Tseng, L. (1997). Harmonic broadcasting for video-on-demand service. *IEEE Transactions on Broadcasting*, *43*(3), 268-271.

Juhn, L., & Tseng, L. (1998). Fast data broadcasting and receiving scheme for popular video service. *IEEE Transactions on Broadcasting*, *44*(1), 100-105.

Kurose, J., & Ross, K. (2004). *Computer networking: A top-down approach featuring the Internet* (3rd ed.). Salford, UK: Addison Wesley; Pearson Education.

Paris, J., Carter, S., & Long, D. (1998). A low bandwidth broadcasting protocol for video on demand. *IEEE International Conference on Computer Communications and Networks (IC3N'98)* (pp. 690-697).

Ratner, D., Reiher, P., & Popek, G. (2004). Roam: A scalable replication system for mobility. *Mobile Networks and Applications, 9,* 537-544). Kluwer Academic Publishers.

Sheu, S., Hua, K., & Tavanapong, W. (1997). Chaining: A generalized batching technique for video-on-demand systems. *Proceedings of the IEEE ICMCS'97* (pp. 110-117).

Sripanidkulchai, K., Ganjam, A., Maggs, B., & Zhang, H. (2004). The feasibility of supporting large-scale live streaming applications with dynamic application end-points. *ACM Special Interest Group on Data Communication (SIGCOMM'04)* (pp. 107-120), Portland, OR.

Tran, D., Hua, K., & Do, T. (2003). Zigzag: An efficient peer-to-peer scheme for media streaming. *Proceedings of IEEE Infocom* (pp. 1283-1293).

Viswanathan, S., & Imielinski, T. (1995). Pyramid broadcasting for video-on-demand service. *Proceedings of the SPIE Multimedia Computing and Networking Conference* (pp. 66-77).

Wang, B., Sen, S., Adler, M., & Towsley, D. (2004). Optimal proxy cache allocation for efficient streaming media distribution. *IEEE Transaction on Multimedia*, *6*(2), 366-374.

Wei, W., & Zakhor, A. (2004). Robust multipath source routing protocol (RMPSR) for video communication over wireless ad hoc networks. *International Conference on Multimedia and Expo (ICME)* (pp. 27-30).

Wu, E., & Huang, Y. (2004). Dynamic adaptive routing for a heterogeneous wireless network. *Mobile Networks and Applications*, *9*, 219-233.

Zheng, H., & Greis, M. (2004). Ongoing research on QoS policy control schemes in mobile networks. *Mobile Networks and Applications*, *9*, 235-241. Kluwer Academic Publishers.

Zhu, X., Han, S., & Girod, B. (2004). Congestion-aware rate allocation for multipath video streaming over ad hoc wireless networks. *IEEE International Conference on Image Processing (ICIP-04)*.

KEY TERMS

CDN: Content distribution network is a network where the ISP has placed proxies in strategically selected points, so that the bandwidth used and response time to clients' requests is minimized.

Overlay Network: A virtual network built over a physical network, where the participants communicate with a special protocol, transparent to the non-participants.

QoS: A notion stating that transmission quality and service availability can be measured, improved, and, to some extent, guaranteed in advance. QoS is of particular concern for the continuous transmission of multimedia information and declares the ability of a network to deliver traffic with minimum delay and maximum availability.

Streaming: The scheme under which clients start playing out the multimedia immediately or shortly after they have received the first portion without waiting for the transmission to be completed.

Chapter V
A Taxonomy of Database Operations on Mobile Devices

Say Ying Lim
Monash University, Australia

David Taniar
Monash University, Australia

Bala Srinivasan
Monash University, Australia

ABSTRACT

In this chapter, we present an extensive study of database operations on mobile devices which provides an understanding and direction for processing data locally on mobile devices. Generally, it is not efficient to download everything from the remote databases and display on a small screen. Also in a mobile environment, where users move when issuing queries to the servers, location has become a crucial aspect. Our taxonomy of database operations on mobile devices mainly consists of on-mobile join operations and on-mobile location dependent operations. For the on-mobile join operation, we include pre- and post-processing whereas for on-mobile location dependent operations, we focus on set operations arise from location-dependent queries.

INTRODUCTION

In these days, mobile technology has been increasingly in demand and is widely used to allow people to be connected wirelessly without having to worry about the distance barrier (Myers, 2003; Kapp, 2002). Mobile technologies can be seen as new resources for accomplishing various everyday activities that are carried out on the move. The direction of the mobile technology industry is beginning to emerge as more mobile users have been evolved. The emergence of this new technology provides the ability for users to access information anytime, anywhere (Lee, Zhu, & Hu, 2005; Seydim, Dunham, & Kumar, 2001). Quick and easy access of information at anytime anywhere is now becoming more and more popular.

People have tremendous capabilities for utilizing mobile devices in various innovative ways for various purposes. Mobile devices are capable to process and retrieve data from multiple remote databases (Lo, Mamoulis, Cheung, Ho, & Kalnis, 2003; Malladi & Davis, 2002). This allows mobile users who wish to collect data from different remote databases by sending queries to the servers and then be able to process the multiple information gathered from these sources locally on the mobile devices (Mamoulis, Kalnis, Bakiras, Li, 2003; Ozakar, Morvan, & Hameurlain, 2005). By processing the data locally, mobile users would have more control on to what they actually want as the final results of the query. They can therefore choose to query information from different servers and join them to be process locally according to their requirements. Also, by being able to obtain specific information over several different sites would help bring optimum results to mobile users queries. This is because different sites may give different insights on a particular thing and with this different insights being join together the return would be more complete. Also processing that is done locally would helps reduce communication cost which is cost of sending the query to and from the servers (Lee & Chen, 2002; Lo et al, 2003).

Example 1: A Japanese tourist while traveling to Malaysia wants to know the available vegetarian restaurants in Malaysia. He looks for restaurants recommended by both the Malaysian Tourist Office and Malaysian Vegetarian Community. First, using his wireless PDA, he would download information broadcast from the Malaysian Tourist Office. Then, he would download the information provided by the second organization mentioned previously. Once he obtains the two lists from the two information providers, he may perform an operation on his mobile device that joins the contents from the two relations that may not be collaborative to each other. This illustrates the importance of assembling information obtained from various non-collaborative sources in a mobile device.

This chapter proposes a framework of the various kinds of join queries for mobile devices for the benefits of the mobile users that may want to retrieve information from several different non-collaborative sites. Our query taxonomy concentrates on various database operations, including not only join, but as well as location-dependent information processing, which are performed on mobile devices.

The main difference between this chapter and other mobile query processing papers is that the query processing proposed here is carried out locally on mobile devices, and not in the server. Our approach is whereby the mobile users gather information from multiple servers and process them locally on a mobile device. This study is important, not only due to the need for local processing, but also due to reducing communication costs as well as giving the mobile users more control on what information they want to assemble. Frequent disconnections and low bandwidth also play a major motivation to our work which focuses on local processing.

The rest of this chapter is organized as follows. In the next section, we will briefly explain the background knowledge of mobile database technology, related work, as well as the issues and constraints imposed by mobile devices. We will then present a taxonomy of various database operations on mobile devices, including join operation in the client-side and describes how location-dependent affects information gathering processing scheme on mobile devices. Last but not least, we will discuss the future trend which includes the potential applications for database processing on mobile devices.

PRELIMINARIES

As the preliminary of our work, we will briefly discuss the general background of mobile database environment which includes some basic knowledge behind a mobile environment. Next, we will discuss related work of mobile query processing done by other researchers. Lastly, we will also cover the issues and complexity of local mobile database operations.

Mobile Database Environment: A Background

Mobile devices are defined as electronic equipments which operate without cables for the purposes of communication, data processing, and exchange, which can be carried by its user and which can receive, send, or transmit information anywhere, anytime due to its mobility and portability (Myers, 2003). In particular, mobile devices include mobile phones, personal digital assistants (PDA), laptops that can be connected to network and mixes of these such as PDA-mobile phones that add mobile phone to the functionality of a PDA. This chapter is concerned with devices categorized as PDA-mobile phones or PDAs.

Generally, mobile users with their mobile devices and servers that store data are involved in a typical mobile environment (Lee, Zhu, & Hu, 2005; Madria, Bhargava, Pitoura, & Kumar, 2000; Wolfson, 2002). Each of these mobile users communicates with a single server or multiple servers that may or may not be collaborative with one another. However, communication between mobile users and servers are required in order to carry out any transaction and information retrieval. Basically, the servers are more or less static and do not move, whereas the mobile users can move from one

Figure 1. A mobile database environment

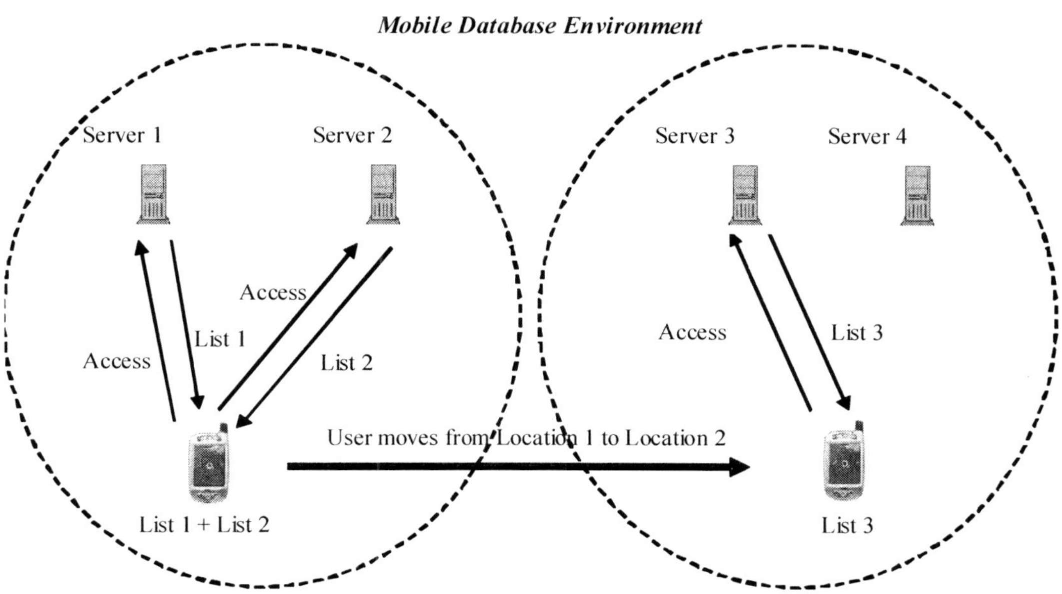

place to another and are therefore dynamic. Nevertheless, mobile users have to be within specific region to be able to received signal in order to connect to the servers (Goh & Taniar, 2005; Jayaputera & Taniar, 2005). Figure 1 illustrates a scenario of a mobile database environment.

It can be seen from Figure 1 that mobile user 1 when within a specific location is able to access servers 1 and 2. By downloading from both servers, the data will be stored in the mobile device which can be manipulated later locally. And if mobile user 1 moves to a different location, the server to access maybe the same but the list downloaded would be different since this mobile client is located in a different location now. The user might also be able to access to a different server that is not available in his pervious location before he moves.

Due to the dynamic nature of this mobile environment, mobile devices face several limitations (Paulson, 2003; Trivedi, Dharmaraja, & Ma, 2002). These include limited processing capacity as well as storage capacity. Moreover, limited bandwidth is an issue because this wireless bandwidth is smaller compared with the fixed network. This leads to poor connection and frequent disconnection. Another major issue would be the small display which causes limitations in the visualizations. Therefore, it is important to comprehensively study how database operations may be carried out locally on mobile devices.

Mobile Query Processing: Related Work

As a result of the desire to process queries between servers that might not be collaborative, traditional join query techniques might not be applicable (Lo et al, 2003). Recent related work done by others in the field of mobile database queries includes processing query via *server strategy*, *on-air strategy* and *client strategy*(Waluyo, Srinivasan, & Taniar, 2005b). Figure 2 gives an illustration of the three strategies of query processing on a mobile environment.

In general, the *server strategy* is referring to mobile users sending a query to the server for processing and then the results are returned to the user (Seydim, Dunham, & Kumar, 2001; Waluyo, Srinivasan, & Taniar, 2005b). Issues, such as location-dependent, take into account since different location will be accessing different servers, and subsequently it relates to the processing by the server and the return of the results based on the new location of the mobile user (Jayaputera & Taniar, 2005). Our approach differs from this strategy in the sense that we focus on how to process the already downloaded data on a mobile device and ma-

Figure 2. Mobile query processing strategies

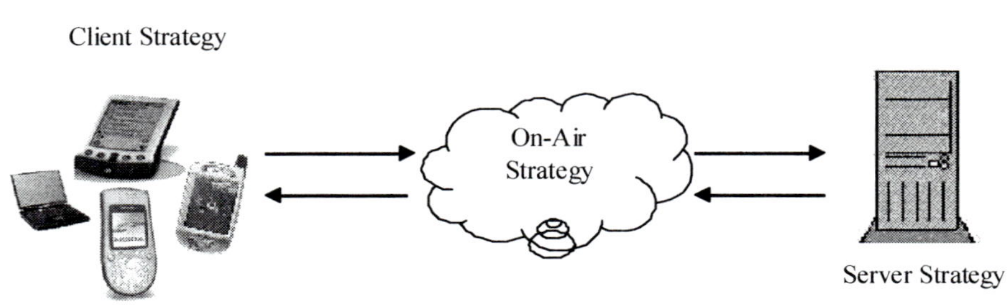

nipulate the data locally to return satisfactorily results taken into account the limitations of mobile devices.

As for the *on-air strategy* which is also known as the *broadcasting strategy* is basically the server broadcasts data to the air and mobile users tune into a channel to download the necessary data (Tran, Hua, & Jiang, 2001; Triantafillou, Harpantidou, & Paterakis, 2001). This broadcasting technique broadcasts a set of database items to the air to a large number of mobile users over a single channel or multiple channels (Huang & Chen, 2003; Prabhakara, Hua, & Jiang, 2000; Waluyo, Srinivasan, & Taniar, 2005a, 2005c). This strategy greatly deals with problem of channel distortion and fault transmission. With the set of data on the air, mobile users can tune into one or more channel to get the data. This, subsequently, improves query performance. This also differs from our approach in the sense that our focus is not how the mobile users download the data in terms of whether it is downloaded from data on the air or whether downloaded from data in the server, but rather how we process the downloaded data locally on mobile devices.

The *client strategy* is whereby the mobile user downloads multiple lists of data from the server and processes them locally on their mobile device (Lo et al, 2003; Ozakar, Morvan, & Hameurlein, 2005). This strategy deals with processing locally on the mobile devices itself, such as when data are downloaded from remote databases and need to be process to return a join result. Downloading both non-collaborative relations entirely may not be a good method due to the limitations of mobile devices which have limited memory space to hold large volume of data and small display which limits the visualization (Lo et al, 2003). Thus efficient space management of output contents has to be taken into account. In addition, this strategy also relates to maintaining cached data in the local storage, since efficient cache management is critical in mobile query processing (Cao, 2003; Elmagarmid, Jing, Helal, & Lee, 2003; Xu, Hu, Lee, & Lee, 2004; Zheng, Xu, & Lee, 2002). This approach is similar to our work in terms of processing data that are downloaded from remote databases locally and readily for further processing.

The related work intends to concentrate on using different strategies, such as via server or on air to download data and how to perform join queries locally on mobile devices taking into account the mobile devices limitations. However our approach focus on using a combination of various possible join queries that is to be carried out locally to attend to the major issues such as the limited memory and limited screen space of mobile devices. We also incorporate the location-dependent aspects in the local processing.

Issues and Complexity of Local Mobile Database Operations

Our database wireless environment consists of PDAs (personal digital assistant), wireless network connections, and changing user environment (e.g., car, street, building site). This arises some issues and complexity of the mobile operations. And also secondly, the limited screen space is another constraint. If the results of the join are too long, then it is cumbersome to be shown on the small mobile device screen. The visualization is thus limited by the small screen of the mobile devices. Figure 3 shows an illustration of how join results are displayed on a PDA.

Processors may also be overloading with time consuming joins especially those that involve thousands of records from many different servers, and completion time will be expected to be longer.

Another issue to be taken account is by having a complex join that involves large amount of data, the consequences would lead to in-

Figure 3. Join display on a PDA

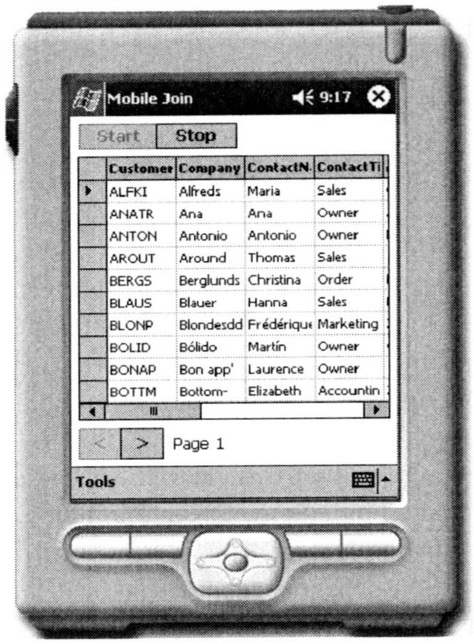

crease communication cost. One must keep in mind that using mobile devices, our aim is to minimize the communication cost with is the cost to ship query and results from database site to requested site.

The previous limitations such as small displays, low bandwidth, low processor power, and operating memory are dramatically limiting the quality of obtaining more resourceful information. The problem of keeping mobile users on the satisfactory level becomes a big challenge. Due to the previously mentioned hardware limitations and changing user environment, the limitations must be drastically overcome and adapted to the mobile environment capabilities. As a result, it is extremely important to study comprehensive database operations that are performed on mobile devices taking into account all the issues and complexities. By minimizing and overcoming these limitations it can further help to boost the number of mobile users in the near future.

TAXONOMY OF DATABASE OPERATIONS ON MOBILE DEVICES

This chapter proposes a taxonomy of database operations on mobile devices. These operations give flexibility to mobile users in retrieving information from remote databases and processing them locally on their mobile devices. This is important because users may want to have more control over the lists of data that are downloaded from multiple servers. They may be interested in only a selection of specific information that can only be derived by processing the data that are obtained from different servers, and this processing should be done locally when all the data have been downloaded from the respective servers. As a result, one of the reasons for presenting the taxonomy of database operations on mobile results is because there is a need to process data locally based on user requests. And since it is quite a complex task that requires more processing from the mobile device itself, it is important to study and further investigate. It also indicates some implications of the various choices one may make when making a query.

We classify database operations on mobile devices into two main groups: (*1*) *on-mobile join processing*, and (*2*) *on-mobile location-dependent information processing.*

On-Mobile Join Processing

It is basically a process of combining data from one relation with another relation. In a mobile environment, joins are used to bring together information from two or more different information that is stored in non-collaborative servers or remote databases. It joins multiple data from different servers into a single output to be displayed on the mobile device. In on-mobile join, due to a small visualization screen, mobile users who are joining information from various servers normally require some pre- and post-processing.

Consider Example 1 presented earlier. It shows how a join operation is needed to be performed on a mobile device as the mobile user downloads information from two different sources which are not collaborative between each other and wants to assemble information through a join operation on his mobile device. This example illustrates a simple on-mobile join case.

On-Mobile Location-Dependent Information Processing

The emerging growth of the use of intelligent mobile devices (e.g., mobile phones and PDAs) opens up a whole new world of possibilities which includes delivering information to the mobile devices that are customized and tailored according to their current location. The intention is to take into account location dependent factors which allow mobile users to query information without facing location problem. Data that are downloaded from different location would be different and there is a need to bring together these data according to user request who may want to synchronize the data that are downloaded from different location to be consolidated into a single output.

Example 2: A property investor while driving his car downloads a list of nearby apartments for sale from a real-estate agent. As he moves, he downloads the requested information again from the same real-estate agent. Because his position has changed since he first enquires, the two lists of apartments for sale would be different due to the relative location when this investor was inquiring the information. Based on these two lists, the investor would probably like to perform an operation on his mobile device to show only those apartments exist in the latest list, and not in the first list. This kind of list operation is often known as a "difference" or "minus" or "exclude" operation, and this is incurred due to information which is location-dependent and is very much relevant in a mobile environment.

Each of the previous classifications will be further explained into more detail in the succeeding sections.

ON-MOBILE JOIN OPERATIONS

Joins are used in queries to explain how different tables are related (Mamoulis, Kalnis, & Bakiras, 2003; Ozakar, Morvan, & Hameurlain, 2005). In a mobile environment, joins are useful especially when you want to bring together information from two or more different information that is stored in non-collaborative servers. Basically, it is an operation that provides access to data from two tables at the same time from different remote databases. This relational computing feature consolidates multiple data from different servers for use in a single output on the mobile devices.

Based on the limitations of mobile devices which are the limited mount of memory and small screen space, it is important to take into account the output results to ensure that it is not too large. And furthermore, sometimes user may want to join items together from different databases but they do not want to see everything. They may only want to see certain related information that satisfies their criteria. Due to this user's demand, a join alone is not sufficient because it does not limit the conditions based on user's requirements. The idea of this is basically to ensure mobile users has the ability to reduce the query results with maximum return of satisfaction because with the pre

Figure 4. On-mobile join taxonomy

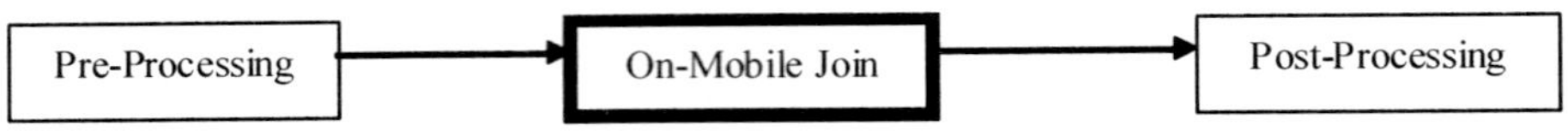

and post-processing, the output results will greatly reduce base on the user's requirements without having to sacrifice any possible wanted information. There will also be more potential of data manipulation that a mobile user can perform.

Therefore we will need to combine a pre-processing which is executed before mobile join and/or a post-processing which is executed after the mobile join. Figure 4 shows an illustration of the combination of pre and post-processing with the mobile join.

Join Operations

Generally, there are various kinds of joins available (Elmasri & Navathe, 2003). However, when using joins in a mobile environment, we would like to particularly focus on two types of joins which is equi-join and anti-join. Whenever there are two relations from different servers that wanted to be joined together into a single relation, this is known as equi or simple join. What it actually does is basically combining data from relation one with data from relation two.

Referring to Example 1 presented earlier, which shows an equi-join, which joins the relations from the first server (i.e., Malaysian Tourist Office) with the second server (i.e., Malaysian Vegetarian Community) to have a more complete output based on user requirements. The contents of the two relations which are hosted by the two different servers that is needed to be joined can be seen on Figure 5.

An *anti-join* is a form of join with reverse logic (Elmasri & Navathe, 2003). Instead of returning rows when there is a match (according to the join predicate) between the left and right side, an anti-join returns those rows from the left side of the predicate for which there is no match on the right. However one of the limitations of using anti-join is that the columns involved in the anti-join must both have not null constraints (Kifer, Bernstein, & Lewis, 2006).

Example 3: A tourist who visits Australia uses his mobile device to issue a query on current local events held in Australia. There is a server holds all types of events happened all year in 2005. The tourist may want to know if a particular event is a remake in the past years and is only interested in non-remake events. So if the list obtained from Current Local Events list matches with events in Past Events list, then he will not be interested and hence it is not needed to display as output on his mobile device.

Example 3 shows an example of the opposite of an equi-join. The tourist only wants to collect information that is not matched with the previous list. In other words, when you get the match, then you do not want it.

Nevertheless, if join is done alone, it may raise issues and complexity especially when applying to a mobile device that has a limited memory capacity and a limited screen space. Therefore, in a mobile device environment, it is likely that we impose pre and post-processing to make on-mobile join more efficient and cost effective.

Figure 5. An equi-join between two relations

Name	**Address**	**Category**	**Rating**
Restaurant A	Address 1	Chinese	Excellent
Restaurant B	Address 2	Vietnamese	Satisfactory
Restaurant C	Address 3	Thai	Excellent
Restaurant D	Address 4	Thai	Satisfactory
--------	--------	-------	---------

Server 1 : Malaysian Tourist Office

Name	**Address**
Restaurant A	Address 1
Restaurant F	Address 6
Restaurant X	Address 24
Restaurant G	Address 7
---------	-----------

Server 2 : Malaysian Vegetarian Community

Pre-Processing Operations

Pre-processing is an operation that is being carried out before the actual join between two or more relations in the non-collaborative servers are carried out (in this context, we then also call it a *pre-join* operation). The importance of the existence of pre-processing in a mobile environment is because mobile users might not be interested in all the data from the server that he wants to download from. The mobile users may only be interested in a selection of specific data from one of the server and another selection of data from another server. Therefore, pre-processing is needed to get the specific selection from each of the servers before being downloaded into the mobile device to be further processed. This also leads to reducing communication cost since less data is needed to download from each server and also helps to discard unwanted data from being downloaded into the mobile devices.

Filtering is a well-known operation of pre-processing. It is similar to the selection operation in relational algebra (Elmasri & Navathe, 2003). Filter is best applied before join because it will helps reduce size of the relations before join between relations occurs. Basically it is being used when the user only needs selective rows of items so that only those requested are being process to be joined. This is extremely handy for use in a mobile environment because this helps to limits the number of rows being process which in return helps to reduce the communication cost since the data being process has been reduced. Filtering can be done in several different ways. Figure 6 shows illustration of pre-processing whereby two lists of data from two different servers that are filtered by the respective server before they are downloaded into the mobile device.

Example 4: A student is in the city centre and wants to know which of the bookshops in the city centre sell networking books. So using his mobile device, he looks for the books recommended by two of the nearest bookshops based on his current location which are called bookshop1 and bookshop2. The student's query

Figure 6. Filtering

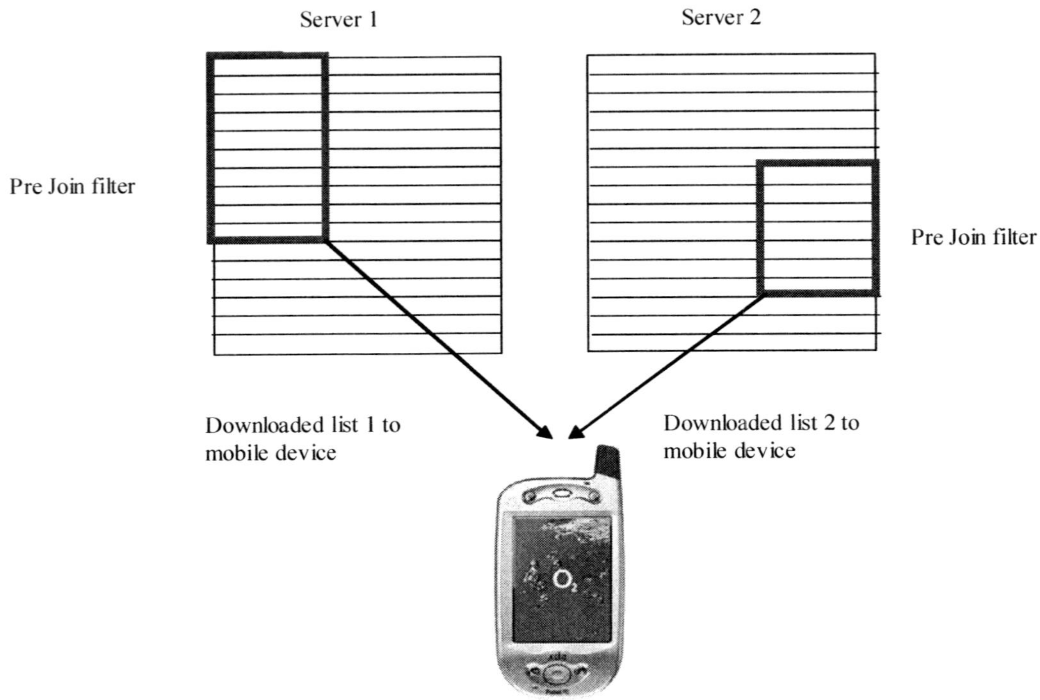

would first scans through all the books and filters out only those that he is interested which in this case is networking books, and then joins together the relation from both bookshop1 and bookshop2.

Filtering one particular type of item can be expressed as in terms of a table of books titles. In this case, the user may be only interested in networking book, so filter comes in to ensure only networking books are being processed.

Filtering a selection group of items can be expressed as in terms having a large list of data and you want to select out only those that are base on the list which contains a specific amount of data, such as top 10 list and so on.

Example 5: A customer is interested in buying a notebook during his visit to a computer fair. However, he is only interested in the top 10 best selling based in Japan and he wants to know the specifications of the notebook from the top 10 list. And because he is in a computer fair in Singapore, so he uses his mobile device to make a query to get the ten notebooks from the top 10 Japan list and then joins with the respective vendors to get the details of the specifications. This type of filter gets the top ten records, instead of a specific one like in the previous example.

From Examples 4 and 5, we use pre-processing because the first list of data has to be filtered first before joining to get the matching with the second list of data.

Post-Processing Operations

Post-processing is an operation that is being carried out after the actual join (in this context, we then also call it a *post-join* operation). It is when the successive rows output from one step which is the pre-processing and then join with the other relation are then fed into the next step that is a post-join. The importance of the existence of post-processing in a mobile environment is because after mobile joins are carried out which combines lists from several remote databases, the results maybe too large and may contain some data that are neither needed nor interested by the users. So with post-processing comes into operation, the results of the output can further be reduced and manipulated in a way that it shows the results in which the user is interested. Therefore, post-processing operation is important because it is the final step that is being taken to produce the users the outputs that meets their requirements.

In general, there is a range of different post-processing operations that is available. However, in this chapter, we would like to focus only on aggregation, sorting, and projection that are to be used in a mobile environment.

Aggregation

Aggregation is a process of grouping distinct data (Taniar, Jiang, Liu, & Leung, 2002). The aggregated data set has a smaller number of data elements than the input data set which therefore helps reduce the output results to meet the limitation of the mobile device of smaller memory capacity. This also appears to be one of the ways for speeding up query performance due to facts are summed up for selected dimensions from the original fact table. The resulting aggregate table will have fewer rows, thus making queries that can use them go faster. Positioning, count, and calculations are commonly used to implement the aggregation concepts.

Positioning aggregation gives the return of a particular position or ranking after joins are completed (Tan, Taniar, & Lu, 2004). Fundamentally, after joining required information from several remote databases, the user may want to know a particular location of a point base on the new joined list of data. Positioning can be relevant and useful in a mobile environment especially when a mobile user who has two lists of data on hand and wants to know the position of a particular item in the list base on the previous list of data.

Example 6: A music fan who attends the Annual Grammy Award event is interested in knowing what the ranking of the songs that won the best romantic song in the top 100 best songs list. So using his mobile device, he first gets that particular song he is interested in and then joins with the top 100 best songs list to get the position of that romantic song that won the best award.

From Example 6, it shows an example of post-processing, because getting the position of the song that has won a Grammy Award from the top 100 best songs list can only be obtained after the join between the two lists is performed.

Count aggregation is an aggregate function which returns the number of rows of a query or some part of a query (Elmasri & Navathe, 2003). Count can be used to return a single count of the rows a query selects, or the rows for each group in a query. This is relevant for a mobile environment especially when a mobile user, for instance, is interested in knowing the number of petrol kiosks in his nearby location.

Example 7: Referring to Example 6 on the Grammy Award Event, in this example the mobile user wants to know the number of awards previously won which is obtained from the idol biography server who is a current winner in the Grammy Award. So using his mobile device, he first gets the name of his idol he is interested in and then joins with the idol biography server site to get the number of awards previously won and return the number of count of all awards he/she has won.

From Example 7, the post-processing shows that the return of the specific numeric value which is the count of the previously won awards, is also only obtainable after the join between the two lists to the final value.

Calculation aggregation is a process of mathematical or logical methods and problem solving that involves numbers (Elmasri & Navathe, 2003). This is relevant for a mobile environment especially when a mobile user who is on the road wants to calculate distance or an exact amount of the two geographical coordinates between two different lists of data.

Example 8: A tourist who was stranded in the city and wants to get home but do not know which public transport and where to take them. He wants to know which is the nearest available transportation and how far it is from its current standing position. He only wants the nearest available with its timetable. So using his mobile device, he gets a list of all surrounding transportation available but narrows down based on the shortest distance calculated by kilometers and then joins both relations together so that both the timetable information and the map getting there for that transportation are available. As a result of looking for the shortest distance, calculations are needed in order to get the numeric value.

From Example 8, post-processing is carried after joining two different lists from different sources and if the user wants to make calculation on specific thing such as the distance, it can only be calculated when the query joins together with the type of transportation selected with the other list which shows the tourist current coordinate location.

Sorting

Sorting is another type of post-processing operation, which sorts the query results (Taniar & Rahayu, 2002). It can help user to minimize looking at the unwanted output. Therefore, mobile users might use sorting techniques after performing the mobile join to sort the data possibly based on the importance of user desire. This means that the more important or most close related to user desire conditions would be listed at the top in a descending order. This makes it more convenient for the mobile user to choose what they would like to see first since the more important items have been placed on top. Another possible reason for using this technique is because the mobile device screen is small and the screen itself it might not cover everything on a single page. So by sorting the data then the user can save time looking further at other pages since the user can probably have found what he wants at the top of the list.

Example 9: By referring to previous Example 1 on vegetarian restaurants, the mobile user is only interested in high rating vegetarian restaurants. So in this case, sorting comes into consideration because there is no point to list vegetarian restaurants that is low ratings since the tourist is not interested at all.

From Example 9, sorting is classified as post-processing because it is done when you have got the final list that has been joined. Sorting basically reorders the list in terms of user preference.

Projection

Projection is defined as the list of attributes, which a user wants to display as a result of the execution of the query (Elmasri & Navathe, 2003). One of the main reasons that projection is important in a mobile environment is because of the limitation of mobile device which has small screen that may not be able to display all the results of the data at once. Hence, with projection, those more irrelevant data without ignoring user requirements will be further discarded and so less number of items would be produced and displayed on the limited screen space of a mobile device.

Example 10: By referring to previous Example 5 regarding enquiring the top 10 note-

Figure 7. Ratio between PDA screen and join results

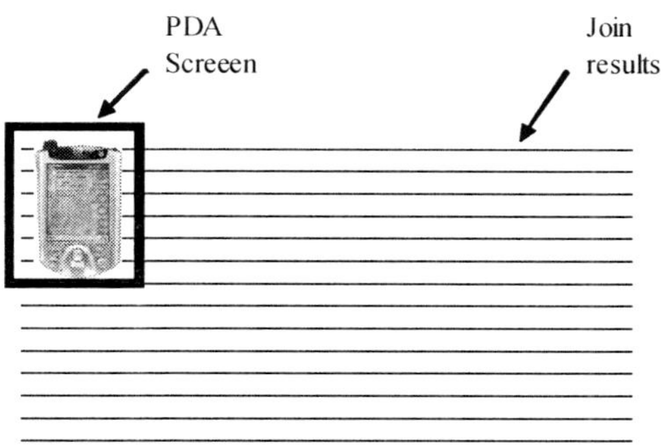

books, the user may only want to know which of the top 10 notebooks in Japan that has DVD-RW. Generally, the top 10 list only contains names of the notebook and may not show the specification. Hence in order to see the specification, it can only be obtained by making another query to a second list which contains detail of the specification.

From Example 10, projection is a sub class of post-processing in the sense that the user only wants specific information after the join which get every details of the other specifications.

Figure 7 shows an illustration of how aggregation, projection, and sorting are important in a mobile device after performing a typical join which has returned a large amount of data. As can be seen, the screen of a mobile device is too small and may affect the viewing results of a typical join situation which has produced too many join results.

ON-MOBILE LOCATION-DEPENDENT OPERATIONS

Location-dependent processing is of interest in a number of applications, especially those that involves geographical information systems (Cai & Hua, 2002; Cheverst, Davies, Mitchell, 2000; Jung, You, Lee, & Kim, 2002; Tsalgatidou, Veijalainen, Markkula, Katasonov, & Hadjiefthymiades, 2003). An example query might be "to find the nearest petrol kiosk" or "find the three nearest vegetarian restaurants" queries that are issued from mobile users. As the mobile users move around, the query results may change and would therefore depend on the location of the issuer. This means that if a user sends a query and then changes his/her location, the answer of that query has to be based on the location of the user issuing the query (Seydim, Dunham, & Kumar, 2001; Waluyo, Srinivasan, & Taniar, 2005a).

Figure 8 shows a general illustration of how general mobile location dependent processing is carried out in a typical mobile environment (Jayaputera & Taniar, 2005). The query is first transmitted from a mobile user to the small base station which will send it to the master station to get the required downloaded list and sent back. Then as the user moves from point *A* to point *B* the query will be transmitted to a different small base station that is within the current location of the user. Then again, this query is send to the master station to get relevant data to be downloaded or update if the data already exist in the mobile device and sent back.

In order to provide powerful functions in a mobile environment, we have to let mobile users to query information without facing the location problem. This involves data acquirement and manipulation from multiple lists over remote databases (Liberatore, 2002). We will explain the type of operations that can be carried out to synchronize different lists that a mobile user downloads due to his moving position to a new location. Hence, the list the mobile user downloaded is actually location dependent which depends on where is his current location and will change if he/she moves. Since this operation is performed locally on a mobile

Figure 8. A typical location-dependent query

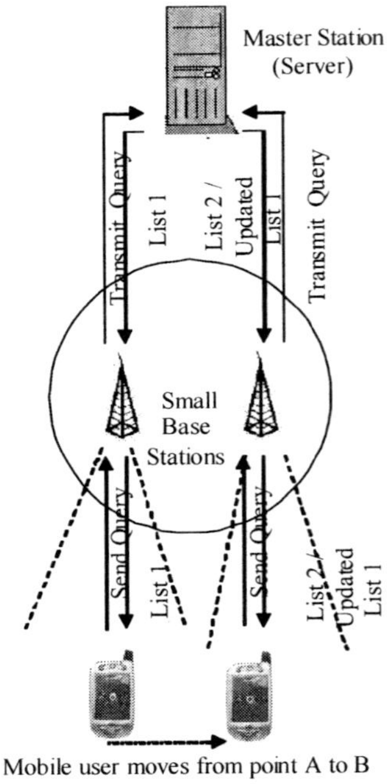

device, we call it "on-mobile location-dependent operations."

On-mobile location dependent operations have been becoming a growing trend due to the constant behavior of mobile users who move around. In this section, we look at examples of location dependent processing utilizing traditional set operations commonly used in relational algebra and other set operations. It involves the circumstances when mobile users are in the situation where they download a list when in a certain location and then they move around and download another list in their new current location. Or another circumstance might be mobile user might already have a list in his mobile device but moves and require to download the same list again but from different location. In any case, there is a need to synchronize these lists that has been downloaded from different location.

Figure 9 shows an example of how location dependent play a role when a mobile user who is on the highway going from location *A* to location *B* and wants to find the nearest available petrol kiosk. First, the mobile user establishes contact with server located at location *A* and downloads the first list which contains petrol kiosk around location *A*. As he moves and comes nearby to new location *B* he downloads another new list and this time the list is different from the previously downloaded because the location has been changed and therefore only contains petrol kiosk around location *B*. These two lists represent possible solutions for the mobile user. Through a local list processing, it can determine by comparing both the

Figure 9. On-mobile location-dependent operations

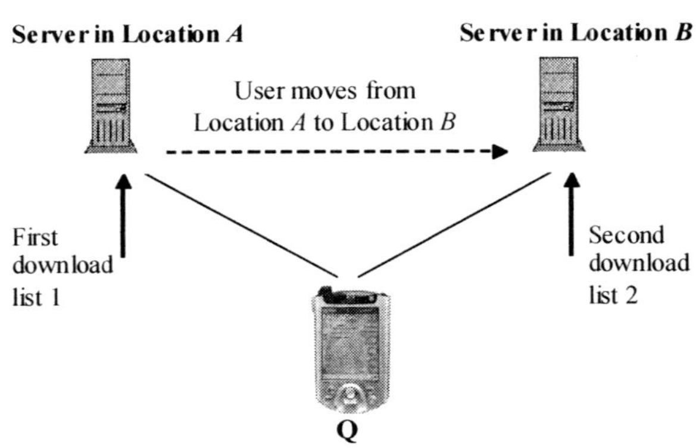

lists, which is indeed its nearest gas station based on current location.

Traditional Relational Algebra Set Operations

In a mobile environment, mobile users would possibly face a situation when he/she is required to download a list of data from one location and then download again another list of data from the same source but from different location. So, the relevance of using set operations to on-mobile location dependent processing is that both involve more than one relation. Due to the possible situation that mobile users face concerning downloading different list of data from similar source but different location, the needs of processing the two lists of data into a single list is highly desirable, particularly in this mobile environment.

Therefore, relational algebra set operations can be used for list processing on mobile devices which involves processing the data that are obtained from the same source but different locations. Different types of traditional relational algebra set operations that can be used include union, intersection and difference (Elmasri & Navathe, 2003).

Union Set Operation

Union operation combines the results of two or more independent queries into a single output (Kifer, Bernstein, & Lewis, 2006). By default, no duplicate records are returned when you use a union operation. Given that the union operation discards duplicate record, this type of set operation is therefore handy when processing user query that requires only distinct results that are obtained by combining two similar kinds of lists. For instance, when a mobile user needs to download data from the same source but different location, and wishes to get only distinct results. This operation can help bring together all possible output downloaded from same source but different location into a single output list of result.

However, the limitation is that the mobile user that access queries in a union operation must ensure the relations are union compatible. For achieving union compatible in mobile environment, a user must ensure the lists are downloaded from the same source. This means that the user may download from one source and then moves to a new location and download again but from the same source. Then only the user can perform a union operation on the mobile device. However the contents may be different between the two lists of data downloaded from different location although the same source. This is because in a location dependent processing when the user moves to a new location, the data downloaded is different from the data downloaded in the previous location.

Nevertheless, if both lists are too large then using union operation by itself may not be substantial. This brings in post-processing operation. Post-processing are processing that are further executed after a typical on-mobile join operation is being carried out.

Example 11: A tourist currently visiting Melbourne wants to know places of interest and downloads a list of interesting places in Melbourne from tourist attraction site and stores in his mobile device. Then he visits Sydney and again downloads another list of interesting places from tourist attraction site but this time it shows places in Sydney. He wants to perform a join that shows only the places regardless of the states but in terms of the types of places such as whether it is a historical building, zoo, religions centre and so on.

Example 11 demonstrates a union operation whereby the query combine all data from the first relation which contains places in Melbourne together with places in Sydney that are downloaded from similar source but the list are different because they are in different location. And since they are similar source, the number of fields is basically the same and so union operator is relevant. In this example, the results of the union operation are further post-processed to do the grouping based on type of places.

Intersection Set Operation

Given collections R1 and R2, the set of elements that is contained in both R1 and R2 are basically called *intersection*. It only returns results that appear in both R1 and R2. The intersection set operation is handy in a mobile environment when the user would like to know only information that has common attribute that exist in both relations that he/she has downloaded when moving from one place to another. An intersection of two lists basically gives the information that appears in both lists (Elmasri & Navathe, 2003). However, a post-processing operation might be highly desirable if the current output result is too large. With the post-processing, it can further reduce the final results by manipulating the multiple list of data in a way that shows only results in which the user is interested.

Example 12: A group of student in Location *A* wants to know where is the nearest McDonalds and using the mobile device they downloaded a list of McDonalds locations which shows all available McDonalds in surrounding location. As they travel further until they arrive in Location *B*, they download another McDonalds lists again and realize the list is somewhat different since they have move from *A* to *B*. Therefore based on these two lists, the student wants to display only those McDonalds that provide drive through service regardless of whether it is in *A* or *B*.

Example 12 demonstrates an intersection operation because what the students are interested is based on both the downloaded lists as well as they want to know which McDonalds has the common field of providing drive through service. The drive-through service can also be thought as part of the post-processing.

Difference Set Operation

Difference set operation is also sometimes known as minus or excludes operation (Elmasri & Navathe, 2003). Given collections R1 and R2, the set of elements that is contained in R1 and not in R2 or vice versa is called difference. Therefore, the output results return only results that appear in R1 that does not appear in R2. The difference set operation may come into benefit especially when the mobile user would like to find certain information that is unique and only appears in one relation and not both from the downloaded list of data, and in the context of location-dependent the information requested must come from one location only.

Example 13: A student wants to know what movie is currently showing in a shopping complex that houses a number of cinemas. He downloads a list when he is at the complex. Then he goes to another shopping complex and wants to know the movies currently showing there. So now the new list is downloaded which contain movies in his new location. The student then wants to know which movies are only showing in this current location and not shown in the previous location.

Example 13 demonstrates a difference in operation because having two different lists downloaded from the two shopping complex, the student only wants the query to return movies that show in either one of the cinemas only and not both.

Other Set Operations

Besides the traditional relational algebra set operations, there are different types of set operations that maybe applicable for location dependent processing on mobile devices. An example of this is a list comparison operation that maybe useful in local mobile device processing between two list of data that is downloaded from the same source. Mobile users are often on the move — moving from one place to another. However, they may typically send query to similar source in different locations. With the implementation of comparison operation in the mobile device, a mobile user can now obtain a view side by side and weight against each other between the two lists of data that is downloaded from similar source but different location. This is useful when mobile user want to compare between the two different lists together.

Example 14: In the city market, a user has downloaded a list of current vegetables prices and keeps then in her mobile device. Then she went to a countryside market and downloaded another list of vegetables prices. With these two lists, she wants to make a comparison and show which vegetables type is cheaper in which market.

From Example 14, it is known that the first list which contains the city price list has been downloaded and kept in the mobile device locally. And then the user further downloads a new list when she is in the country which contains a different list of prices. With these two different lists on hand that contain common items, the mobile user wants her mobile device to locally process these two lists by making a comparison result and then show which of the two list has cheaper price for the respectively vegetables items.

FUTURE TRENDS

Database operations on mobile devices are indeed a potential area for further investigation, because accessing and downloading multiple data anywhere and anytime from multiple remote databases and process them locally through mobile devices is becoming an important emerging element for mobile users who want to have more control over the final output. Also, location dependent processing has becoming more important in playing a role on operations on mobile devices (Goh, & Taniar, 2005; Kubach & Rothernel, 2001; Lee, Xu, Zheng, & Lee, 2002; Ren & Dunham, 2000). The future remains positive but there are some issues need to be addressed. Hence, this section discusses some future trend of database operations on mobile devices in terms of various perspectives, including query processing perspective, user application perspective, technological perspective, as well as security and privacy perspective. Each of the perspectives gives different view of the future work in the area of mobile database processing and applications.

Query Processing Perspective

From the query processing perspective, the most important element is to help reduce the communication cost, which occurs due to data transfer between to and from the servers and mobile devices (Xu, Zheng, Lee, & Lee, 2003). These also includes are location dependent processing, future processing that takes into consideration various screen types and storage capacity.

The need for collecting information from multiple remote databases and processing locally becomes apparent especially when mobile users collect information from several non-collaborative remote databases. Therefore, it is of great magnitude to investigating the optimization of database processing on mobile devices, because it helps addresses issue of communication cost. It would also be of a great interest to be able to work on optimizing processing of the database operations to make the processing more efficient and cost effective.

For location dependent processing, whenever mobile users move from one location to another location, the downloaded data would be different even though the query is direct to similar source. And because of this, whenever the downloaded data differ as the users move to a new location, the database server must be intelligent enough to inform that existing list contains different information and prompt if user wants to download a new list.

There are various types of mobile devices available in the market today. Some of them may have bigger screen and some of them may have smaller screen. Therefore, in the future the processing must be able to be personalized or to be adopted to any screen types or sizes. The same goes for storage space. Some mobile phones may have just built in limited memory, whereas PDAs may allow expansion of storage capacity through the use of storage card. So, future intelligent query processing must be able to adapt to any storage requirement such as when downloading list of data to limited build in memory, the data size is reduced to a different format that can adapt to the storage requirement. As we notice, one of the major limitations of mobile devices is the limited storage capacity. Thus, filtering possible irrelevant data from mobile users before being downloaded would most likely help the storage limitation in terms of having irrelevant data automatically filtered out before being downloaded into a mobile device. This also helps in increasing the speed of returning downloaded list of data to the mobile devices.

User Application Perspective

User application perspective looks at the type of future applications that may be developed taking into account the current limitations of mobile devices and its environment processing capabilities. This includes developing future applications taking into account location dependent technology, communication bandwidth, and different capabilities of mobile devices.

There are numerous opportunities for future development of applications especially those that incorporate the need for extensive location dependent processing (Goh & Taniar, 2005). In this case, we would like to explain an example of a particular application that uses location dependent technology. Essentially, there is a need for constant monitoring movement of people because it may be useful in locating missing persons. Therefore, operators are required to provide police with information allowing them to locate an individual's mobile device in order to retrieve the persons that were reported as missing. This can be made possible by inserting tracking software according to user agreement (Wolfson, 2002).

Although, communication bandwidth is still relatively small at the moment, but as more and more demand towards the use of mobile devices, there has been a trend in 3G communication to provide a wider bandwidth (Kapp, 2002; Lee, Leong, & Si, 2002; Myers & Beigl, 2003). This makes it available for mobile users to be able to do more things with their mobile devices such as downloading video and so on. Therefore, future applications can make use of a faster bandwidth and query processing can be easier.

Despite the fact that processing capabilities of mobile devices varies such as small mobile phone which does not have processing capabilities to PDAs which has bigger memory and processors, and so, future applications must be able to distinguish these and program applications that has the option of whether it is to be loaded into mobile phones or PDAs.

Technological Perspective

Technological perspective looks at how technology plays a role for future development of better and more powerful mobile devices. This may includes producing mobile devices that are capable to handle massive amount of data and devices that are able to have combined voice and data capabilities (Myers & Beigl, 2003).

Another case from a technological point of view is that when operationally active, mobile users will often handle large amount of data in real time which may cause overload processing. Hence, this requires hardware that is capable of processing these data with minimum usage of processing power. The processing power required increases as the number of servers and data downloaded by the user increases. Therefore, strategies would be to further develop hardware that capable to process faster.

There are some users who prefer to listen than reading from a mobile device especially the user is driving from point *A* to *B* and is querying directions. This is practical since the screen display of a mobile device is so small and it may require constant scrolling up down and left right to get see the map from one point to another point on the mobile device. It would be proficient if there is a convergence towards voice and data combination whereby the mobile device are voice enabled in the sense that as the user drives the mobile device read out the direction to the user.

Security and Privacy Perspective

Security and privacy perspective arises due to more and more mobile users from all over the world accessing data from remote servers wirelessly through an open mobile environment. As a result, mobile users are often vulnerable to issues such as possible interference from others in this open network. This exists largely due to the need for protecting human rights by allowing them to remain anonymous, and allowing the user to be able to do things freely with minimal interference from others. Therefore, security and privacy issue remain important factors (Lee et al, 2002).

Hence, it is important to have the option for enabling the user to remain anonymous and unknown of their choice and behavior unless required by legal system. This also includes higher security levels whenever accessing the open network wirelessly. This issue could potentially be addressed by means of privacy preserving methods, such as user personal information are carefully being protected and when the user are connected to the network, identify the user with a nickname rather than the real name.

CONCLUSION

In this chapter, we have presented a comprehensive taxonomy of database operations on mobile devices. The decision of choosing the right usage of operations to minimize results without neglecting user requirements is essential especially when processing queries locally on mobile devices from multiple list of remote database by taking into account considerations of the issues and complexity of mobile operations. And, this chapter also covers issues on location-dependent queries processing in mobile database environment. As the wireless and mobile communication of mobile users has increased, location has become a very important constraint. Lists of data from different locations would be different and there is a need to bring together these data according to user requirements who may want need these two separate lists of data to be synchronized into a single output.

REFERENCES

Cai, Y., & Hua, K. A. (2002). An adaptive query management technique for real-time monitoring of spatial regions in mobile database systems. *Proceedings of the 21st IEEE International Conference on Performance, Computing, and Communications* (pp. 259-266).

Cao, G. (2003). A scalable low-latency cache invalidation strategy for mobile environments. *IEEE Transactions on Knowledge and Data Engineering, 15*(5), 1251-1265.

Cheverst, K., Davies, N., Mitchell, K., & Friday, A. (2000). Experiences of developing and deploying a context-aware tourist guide. *Proceedings of the 6th Annual International Conference on Mobile Computing and Networking* (pp. 20-31).

Elmargamid, A., Jing, J., Helal, A., & Lee, C. (2003). Scalable cache invalidation algorithms for mobile data access. *IEEE Transactions on Knowledge and Data Engineering, 15*(6), 1498-1511.

Elmasri, R., & Navathe, S. B. (2003). *Fundamentals of database systems* (4th ed.). Reading, MA: Addison Wesley.

Goh, J., & Taniar, D. (2005, Jan-Mar). Mining parallel pattern from mobile users. *International Journal of Business Data Communications and Networking, 1*(1), 50-76.

Huang, J. L., & Chen, M. S. (2003) Broadcast program generation for unordered queries with data replication. *Proceedings of the 8th ACM Symposium on Applied Computing* (pp. 866-870).

Jayaputera, J., & Taniar, D. (2005). Data retrieval for location-dependent query in a multicell wireless environment. *Mobile Information Systems, IOS Press, 1*(2), 91-108.

Jung, II, D., You, Y. H., Lee, J. J., & Kim, K. (2002). Broadcasting and caching policies for location-dependent queries in urban areas. *Proceedings of the 2nd International Workshop on Mobile Commerce* (pp. 54-59).

Kapp, S. (2002). 802.11: Leaving the wire behind. *IEEE Internet Computing, 6*(1).

Kifer, M., Bernstein, A., & Lewis, P. M. (2006). *Database systems: An application-oriented approach* (2nd ed.). Addison Wesley.

Kubach, U., & Rothermel, K. (2001). A map-based hoarding mechanism for location- dependent information. *Proceedings of the 2nd International Conference on Mobile Data Management* (pp. 145-157).

Lee, C. H., & Chen, M. S. (2002). Processing distributed mobile queries with interleaved remote mobile joins. *IEEE Tran. on Computers, 51*(10), 1182-1195.

Lee, D. K., Xu, J., Zheng, B., & Lee, W. C. (2002, July-Sept.). Data management in location-dependent information services. *IEEE Pervasive Computing, 2*(3), 65-72.

Lee, D. K., Zhu, M., & Hu, H. (2005). When location-based services meet databases. *Mobile Information Systems, 1*(2), 81-90.

Lee, K. C. K., Leong, H. V., & Si, A. (2002). Semantic data access in an asymmetric mobile environment. *Proceedings of the 3rd Mobile Data Management* (pp. 94-101).

Liberatore, V. (2002). Multicast scheduling for list requests. *Proceedings of IEEE INFOCOM Conference* (pp. 1129-1137).

Lo, E., Mamoulis, N., Cheung, D. W., Ho, W. S., & Kalnis, P. (2003). Processing ad-hoc joins on mobile devices. *Database and Expert Systems Applications, Lecure Notes in Computer Science, 3180,* 611-621.

Madria, S. K., Bhargava, B., Pitoura, E., & Kumar, V. (2000). Data organisation for location-dependent queries in mobile computing. *Proceedings of ADBIS-DASFAA* (pp. 142-156).

Malladi, R., & Davis, K. C. (2002). Applying multiple query optimization in mobile databases.

Proceedings of the 36th Hawaii International Conference on System Sciences (pp. 294-303).

Mamoulis, N., Kalnis, P., Bakiras, S., & Li, X. (2003). Optimization of spatial joins on mobile devices. *Proceedings of the SSTD.*

Myers, B. A., & Beigl, M. (2003). Handheld computing. *IEEE Computer Magazine, 36*(9), 27-29.

Ozakar, B., Morvan, F., & Hameurlain, A. (2005). Mobile join operators for restricted sources. *Mobile Information Systems, 1*(3).

Paulson, L. D. (2003). Will fuel cells replace batteries in mobile devices? *IEEE Computer Magazine, 36*(11), 10-12.

Prabhakara, K., Hua, K. A., & Jiang, N. (2000). Multi-level multi-channel air cache designs for broadcasting in a mobile environment. *Proceedings of the IEEE International Conference on Data Engineering (ICDE'00)* (pp. 167-176).

Ren, Q., & Dunham, M. H. (1999). Using clustering for effective management of a semantic cache in mobile computing. *Proceedings of the ACM International Workshop on Data Engineering for Wireless and Mobile Access* (pp. 94-101).

Ren, Q., & Dunham, M. H. (2000). Using semantic caching to manage location-dependent data in mobile computing. *Proceedings of the 6th International Conference on Mobile Computing and Networking* (pp. 210-221). 2000.

Seydim, A. Y., Dunham, M. H., & Kumar, V. (2001). Location-dependent query processing. *Proceedings of the 2nd International Workshop on Data Engineering on Mobile and Wireless Access (MobiDE'01)* (pp. 47-53).

Tan, R. B. N., Taniar, D., & Lu, G. J. (2004, Sept.). A taxonomy for data cube query. *International Journal of Computers and Their Applications, 11*(3), 171-185.

Taniar, D., & Rahayu, J. W. (2002). Parallel database sorting. *Information Sciences, 146*(1-4), 171-219.

Taniar, D., Jiang, Y., Liu, K. H., & Leung, C. H. C. (2002). Parallel aggregate-join query processing. *Informatica: An International Journal of Computing and Informatics, 26*(3), 321-332.

Tran, D. A., Hua, K. A., & Jiang, N. (2001). A generalized design for broadcasting on multiple physical-channel air-cache. *Proceedings of the ACM SIGAPP Symposium on Applied Computing (SAC'01)* (pp. 387-392).

Triantafillou, P., Harpantidou, R., & Paterakis, M. (2001). High performance data broadcasting: A comprehensive systems perspective. *Proceedings of the 2nd International Conference on Mobile Data Management (MDM 2001)* (pp. 79-90).

Trivedi, K. S., Dharmaraja, S., & Ma, X. (2002). Analytic modelling of handoffs in wireless cellular networks. *Information Sciences, 148*(1-4), 155-166.

Tsalgatidou, A., Veijalainen, J., Markkula, J., Katasonov, A., & Hadjiefthymiades, S. (2003). Mobile e-commerce and location-based services: Technology and requirements. *Proceedings of the 9th Scandinavian Research Conference on Geographical Information Services* (pp. 1-14).

Waluyo, A. B., Srinivasan, B., & Taniar, D. (2005a). Indexing schemes for multi channel data broadcasting in mobile databases. *International Journal of Wireless and Mobile Computing*. To appear Mar/Apr.

Waluyo, A. B., Srinivasan, B., & Taniar, D. (2005b, Mar.). Research on location-dependent queries in mobile databases. *International Journal of Computer Systems Science & Engineering, 20*(3), 77-93.

Waluyo, A. B., Srinivasan, B., & Taniar, D. (2005c). Global indexing scheme for location-dependent queries in multi-channels broadcast environment. *Proceedings of the 19th IEEE International Conference on Advanced Information Networking and Applications, Volume 1, AINA 2005,* (pp. 1011-1016).

Wolfson, O. (2002). Moving objects information management: The database challenge. *Proceedings of the 5th Workshop on Next Generation Information Technology and Systems (NGITS)* (pp. 75-89).

Xu, J., Hu, Q., Lee, W. C., & Lee, D. L. (2004). Performance evaluation of an optimal cache replacement policy for wireless data dissemination. *IEEE Transaction on Knowledge and Data Engineering (TKDE), 16*(1), 125-139.

Xu, J., Zheng, B., Lee, W. C., & Lee, D. L. (2003). Energy efficient index for querying location-dependent data in mobile broadcast environments. *Proceedings of the 19th IEEE International Conference on Data Engineering (ICDE '03)* (pp. 239-250).

Zheng, B., Xu, J., Lee, D. L. (2002). Cache invalidation and replacement strategies for location-dependent data in mobile environments. *IEEE Transactions on Computers, 51*(10), 1141-1153.

KEY TERMS

Location-Dependent Information Processing: Information processing whereby the information requested is based on the current location of the user.

Mobile Database: Databases which are available for access by users using a wireless media through a wireless medium.

Mobile Query Processing: Join processing carried out in a mobile device.

On-Mobile Location-Dependent Information Processing: Location-dependent information processing carried out in a mobile device.

Post-Join: Database operations which are performed after the join operations are completed. These operations are normally carried out to further filter the information obtained from the join.

Pre-Join: Database operations which are carried out before the actual join operations are performed. A pre-join operation is commonly done to reduce the number of records being processed in the join.

Chapter VI
Interacting with Mobile and Pervasive Computer Systems

Vassilis Kostakos
University of Bath, UK

Eamonn O'Neill
University of Bath, UK

ABSTRACT

In this chapter, we present existing and ongoing research within the Human-Computer Interaction group at the University of Bath into the development of novel interaction techniques. With our research, we aim to improve the way in which users interact with mobile and pervasive systems. More specifically, we present work in three broad categories of interaction: stroke interaction, kinaesthetic interaction, and text entry. Finally, we describe some of our currently ongoing work as well as planned future work.

INTRODUCTION

One of the most exciting developments in current human-computer interaction research is the shift in focus from computing on the desktop to computing in the wider world. Computational power and the interfaces to that power are moving rapidly into our streets, our vehicles, our buildings, and our pockets. The combination of mobile/wearable computing and pervasive/ubiquitous computing is generating great expectations.

We face, however, many challenges in designing human interaction with mobile and pervasive technologies. In particular, the input and output devices and methods of using them that work (at least some of the time!) with deskbound computers are often inappropriate for interaction on the street.

Physically shrinking everything including the input and output devices does not create a usable mobile computer. Instead, we need radical changes in our interaction techniques, comparable to the sea change in the 1980s from command line to graphical user interfaces. As with that development, the breakthrough we need in interaction techniques will most likely come not from relatively minor adjustments to existing interface hardware and software but from a less predictable mixture of inspiration and experimentation. For example, Brewster and colleagues have investigated overcoming the limitations of tiny screens on mobile devices by utilising sound and gesture to augment or to replace conventional mobile device interfaces (Brewster, 2002; Brewster, Lumsden, Bell, Hall, & Tasker, 2003).

In this chapter, we present existing and ongoing research within the Human-Computer Interaction group at the University of Bath into the development of novel interaction techniques. With our research, we aim to improve the way in which users interact with mobile and pervasive systems. More specifically, we present work in three broad categories of interaction:

- Stroke interaction
- Kinaesthetic interaction
- Text entry

Finally, we describe some of our currently ongoing work as well as planned future work. Before we discuss our research, we present some existing work in the areas mentioned previously.

RELATED WORK

One of the first applications to implement stroke recognition was Sutherland's sketchpad (1963). Strokes-based interaction involves the recognition of pre-defined movement patterns of an input device (typically mouse or touch screen). The idea of mouse strokes as gestures dates back to the 1970s and pie menus (Callahan, Hopkins, Weiser, & Shneiderman, 1998). Since then, numerous applications have used similar techniques for allowing users to perform complex actions using an input device. For instance, design programs like (Zhao, 1993) allow users to perform actions on objects by performing mouse or pen strokes on the object. Recently, Web browsing applications, like Opera[1] and Mozilla Firefox,[2] have incorporated similar capabilities. There are numerous open source projects which involve the development of stroke recognition, including Mozilla, Libstroke,[3] X Scribble,[4] and WayV.[5]

Furthermore, a number of pervasive systems have been developed to date, and most have been designed for, and deployed in, specific physical locations and social situations (Harrison & Dourish, 1996) such as smart homes and living rooms, cars, labs, and offices. As each project was faced with the challenges of its own particular situation, new technologies and interaction techniques were developed, or new ways of combining existing ones. This has led to a number of technological developments, such as tracking via sensing equipment and ultra sound (Hightower & Borriello, 2001), or even motion and object tracking using cameras (Brumitt & Shafer, 2001). Furthermore, various input and output technologies have been developed including speech, gesture, tactile feedback, and kinaesthetic input (Rekimoto, 2001). Additionally, environmental parameters have been used with the help of environmental sensors, and toolkits have been developed towards this end (Dey, Abowd, & Salber, 2001). Another strand of research has focused on historical data analysis, which is not directly related to pervasive systems but has found practical applications in this area. Finally, many attempts have been made to provide an interface to these systems using tangible interfaces (Rekimoto, Ullmer, & Oba, 2001), or a metaphoric relationship between atoms and bits (Ishii & Ullmer, 1997).

Some projects have incorporated a wide range of such technologies into one system. For instance, Microsoft's EasyLiving project (Brumitt, Meyers, Krumm, Kern, & Shafer,

2000; Brumitt & Shafer, 2001) utilized smart card readers, video camera tracking, and voice input/output in order to set up a home with a pervasive computing environment. In this environment, users would be able to interact with each other, as well as have casual access to digital devices and resources.

Additionally, text entry on small devices has taken a number of different approaches. One approach is to recognise normal handwriting on the device screen, which will allow users to enter text naturally. The Microsoft PocketPC[6] operating system, for instance, supports this feature. Another approach aiming to minimise the required screen space is the Graffitti[7] system used by Palm PDAs, which allows users to enter text one character at a time. Text entry happens on a specific part of the screen, therefore only a small area is required for text entry. An extension of this approach is provided by Boukreev,[8] who has implemented stroke recognition using neural networks. This approach allows for a system that learns from user input, thus becoming more accurate. A third approach is to display a virtual keyboard on-screen, and allow the users to enter text using a stylus.

The work we report in the section *Stroke Interaction* presents a technique for recognising input strokes which can be used successfully on devices with very low processing capabilities and very limited space for the input area (i.e., small touch-screens). The technique is based on the user's denoting a direction rather than an actual shape and has the twin benefits of computational efficiency and a very small input area requirement. We have demonstrated the technique with mouse input on a desktop computer, stylus, and touch-screen input on a wearable computer and hand movement input using real-time video capture.

Furthermore, the work on kinaesthetic user input we present in the section *Kinaesthetic Interaction* provides valuable insight into different application domains. The first prototype we present gives real-time feedback to athletes performing weight lifting exercises. Although a number of commercial software packages are available to help athletes with their training programme, most of them are designed to be used after the exercises have been carried out and the data collected. Our system, on the other hand provides instant feedback, both visual and audio, in order to improve the accuracy and timing of the athletes. The second prototype we present is a mixed reality game. We present a pilot study we carried out with three different version of our game, effectively comparing traditional mouse input with abstract, token-based kinaesthetic input and mixed-reality kinaesthetic input.

Finally, the text-entry prototypes we present in the section *Text Entry* provide novel ways of entering text in small and embedded devices. An additional design constraint has been the assumption that the users will be attending to other tasks simultaneously (such as driving a car) and that they will only be able to use one hand to carry out text entry. The two prototypes we present address this issue in two distinct ways. The first prototype utilises only three hardware buttons, similar to the traditional buttons used in car stereos. Our second prototype makes the best use of a small touch screen and utilises the users' peripheral vision and awareness in order to enhance users' performance. By maximising the size of buttons on the screen, users are given a larger target to aim for, as well as a larger target to notice with their peripheral vision.

STROKE INTERACTION

In our recent work (Kostakos & O'Neill, 2003) we have developed a technique for recognising input strokes. This technique can be used successfully on a wide range of devices right across this scale. Previously, we have demonstrated the technique with mouse input on a desktop computer, stylus, and touch screen input on a wearable computer and hand movement input using real-time video capture. We have termed our technique directional stroke recognition (DSR). As its name implies, it uses strokes as a means of accepting input and commands from the user. In this section we give a brief synopsis of how our technique works and in which situations it can be utilised. A fuller description of the technique is available in (Kostakos & O'Neill, 2003).

The technique is based exclusively on the direction of strokes and discards other characteristics such as the position of a stroke or the relative positions of many strokes. The algorithm is given an ordered set of coordinates (x, y) that describes the path of the performed stroke. These coordinates may be generated in a number of different ways, including conventional pointing devices such as mice and touch screens, but also smart cards, smart rings, and visual object tracking. The coordinates are then translated into a "signature" which is a symbolic representation of the stroke. For instance, an L-shaped stroke could have a signature of "South, East." This signature can then be looked up against a table of pre-defined commands, much as a mouse button double-click has a different result in different contexts. An advantage of using only the direction of the strokes is that a complex stroke may be broken down into a series of simpler strokes that can be performed in situations with very limited input space (Figure 1).

Figure 1. The recognition algorithm allows a signature to be accessed via different strokes

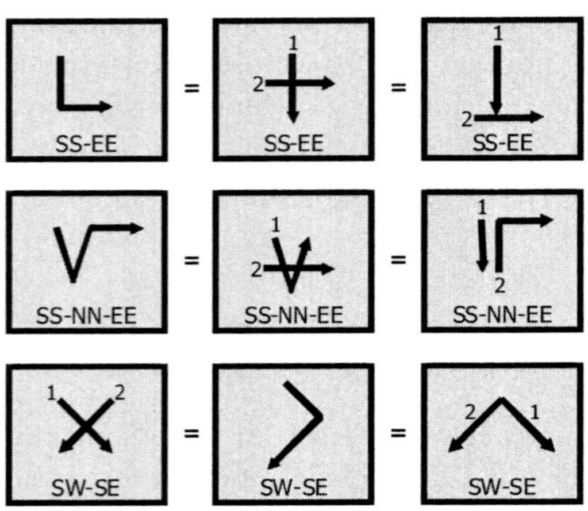

The flexibility of our method allows switching between input devices and methods with no need to learn a new interaction technique. For example, someone may at one moment wish to interact with their PDA using a common set of gestures and in the next moment move seamlessly to interacting with a wall display using the same set of gestures. At one moment, the PDA provides the interaction area on which the gestures are made using a stylus; in the next moment, the PDA itself becomes the "stylus" as is it waved in the air during the interaction with the wall display. Any object or device that can provide a meaningful way of generating coordinates and directions can provide input to the gesture recognition algorithm (Figure 2).

Some important characteristics of this technique include the ability for users to choose the scale and nature of the interaction space they create (Kostakos, 2005; Kostakos & O'Neill, 2005), thus influencing the privacy of their interaction and others' awareness of it. In addition, the physical manifestation of our interaction technique can be tailored according to the situation's requirements. As a result, the technique also allows for easy access, literally just walking up to a system and using it, with no need for special equipment on the part of the users. This makes the technique very suitable

Figure 2. Using various techniques with the stroke recognition engine

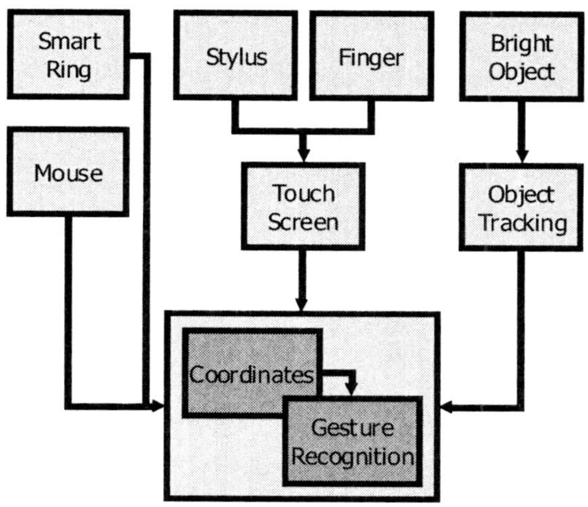

for use in domains such as the hospital A&E department's waiting area.

The directional stroke recognition technique is flexible enough to accommodate a range of technologies (and their physical forms) yet provide the same functionality wherever used. Thus, issues concerning physical form may be addressed independently. In contrast, standard GUI-based interaction techniques are closely tied to physical form: mouse, keyboard, and monitor. The technique we have described goes a long way towards the separation of the physical form and interaction technique.

As a proof of principle, we implemented a real-time object tracking technique that we then used along with our stroke recognition algorithm as an input technique. For our prototype, we implemented an algorithm that performs real-time object tracking on live input from a Web camera (shown in Figure 3). The user can select a specific object by sampling its colour, and the algorithm tracks this object in order to generate a series of coordinates that describe the position of the object on the screen, or to be precise, the position of the object relative to the camera's view. We then pass these generated coordinates to our stroke recognition algorithm, which proceeds with the recognition of the strokes. Due to the characteristics of our stroke recognition method, the coordinates may be supplied at any rate. So long as this rate is kept steady, the stroke recognition is very successful. Thus, despite the fact that our object-tracking algorithm is not optimal, it still provides us with a useful prototype.

Experimental Evaluation

Our concerns to test the usability of interaction techniques in the absence of visual displays led us to develop a prototype system for providing information to A&E patients through a combination of gesture input and audio output. We used our DSR technique for the gesture input and speech synthesis for the audio output. We ran an experimental evaluation of this prototype system. The main question addressed by the evaluation was: if we move away from the standard desktop GUI paradigm and its focus on the visual display, do we decrease usability by losing the major benefit that the GUI brought (i.e., being able to see the currently available functionality and how to invoke it)?

The experiment itself (screenshots shown in Figure 4) is extensively reported in (O'Neill, Kaenampornpan, Kostakos, Warr, & Woodgate, 2005). The results of our evaluation may be interpreted as good news for those developers of multimodal interaction who want to mitigate our reliance on the increasingly unsuitable visual displays of small mobile and wearable devices and ubiquitous systems. We found no significant evidence that usability suffered in the absence of one of the major benefits of the GUI paradigm: a visual display of available services and how to access them. However,

Figure 3. Our prototype system for object tracking used with DSR

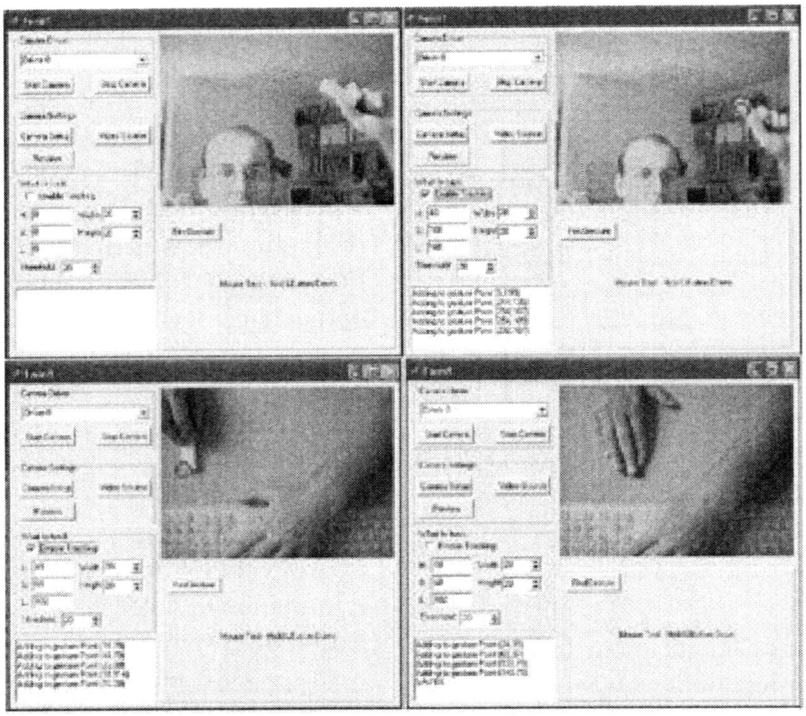

A control object is identified by clicking on it (top left), and then this object is tracked across the image to generate coordinates (top right). The same object can be tracked in different setups (bottom left). By obscuring the object (bottom right) the stroke recognition algorithm is initiated.

Figure 4. Our experimental setup shown on the left and a sample stroke as entered by a user shown on the right

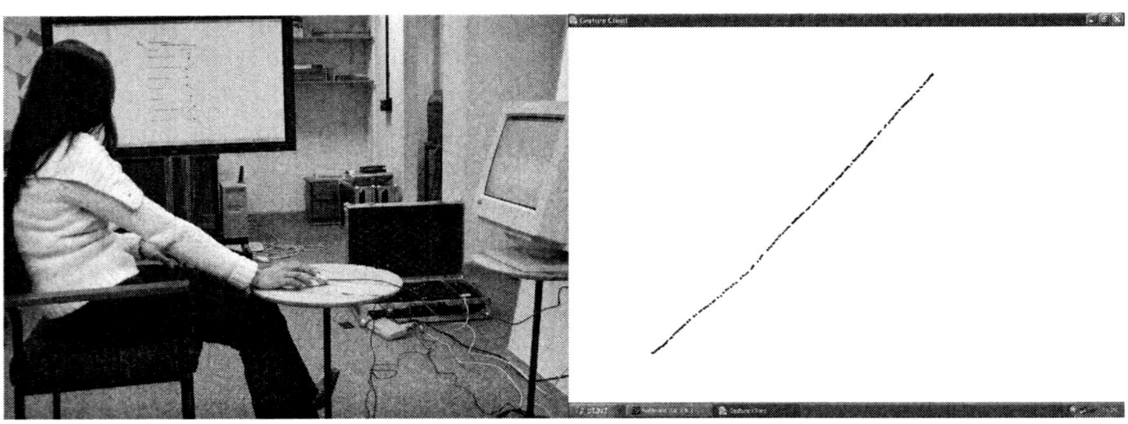

we must sound a note of caution. Our study suggests that *with particular constraints*, the effects of losing the cognitive support provided by a standard GUI visual display are mitigated. These constraints include having a small set of available functions, a small set of simple input gestures in a memorable pattern (e.g., the points of the compass), a tightly constrained user context, and semantically very distinct functions.

Our initial concern remains for the development of non-visual interaction techniques for *general* use in a mobile and pervasive computing world. Our DSR technique for gestural input can handle arbitrarily complex gestures comprised of multiple strokes. There is no requirement for it to be confined to simple single strokes to compass points. Its potential for much richer syntax (similar to a type of alphabet) coincides with the requirement for much richer semantics in general purpose mobile devices.

KINAESTHETIC INTERACTION

Another focus of our research is on developing interaction techniques that utilise implicit user input. More specifically, the prototypes we describe here utilise kinaesthetic user input as a means of interaction. The two prototypes were developed by undergraduate students at the University of Bath and utilise motion-tracking technology (XSens MT9 XBus system[9] with Bluetooth) to sense user movements. The first prototype we describe is a training assistant for weight lifting and provides real-time feedback to athletes about their posture and timing. The second prototype described here is a game application which turns a Tablet PC into a mixed-reality maze game in which players must navigate a virtual ball through a trapped maze by means of tilting the Tablet PC.

Weight Lifting Trainer

For our first prototype we utilised our motion sensors to build an interactive weight lifting trainer application. Our system is designed to be used by athletes whilst they are actually performing an exercise. The system gives feedback as to how well the exercise is being performed (i.e., if the user has the correct posture and timing). The prototype system is shown in Figure 5.

To use the system, users need to attach the motion sensors to specific parts of the body. The system itself provided guidance on how to do this (top left image in Figure 5). The sensors we used are self-powered and communicate via Bluetooth with a laptop or desktop computer. Therefore, the athlete only has some wiring from each individual sensor to a hub. The hub is placed on the athlete's lower back. This allows users for complete freedom of movement in relation to the computer.

Once the user selects an exercise to be performed, the system loads the hard-coded set of data for the "correct" way of carrying out the exercise. This data was produced by recording a professional athlete carrying out the exercise. The skeleton image on the left provided indications for the main stages of an exercise (such as "Lift", "Hold," "Drop"). The right stick-man diagram (top right image in Figure 5) demonstrates the correct posture and timing for performing the exercise, whilst the stick-man to its left represents the user's actual position. There is also a bar meter on the right which describes the degree of match between optimal and actual position and timing. All these diagrams were updated in real-time and in reaction to user movement. Furthermore, the system provided speech feedback with predetermined cues in order to help the users with the exercise.

Figure 5. The weight lifting trainer prototype

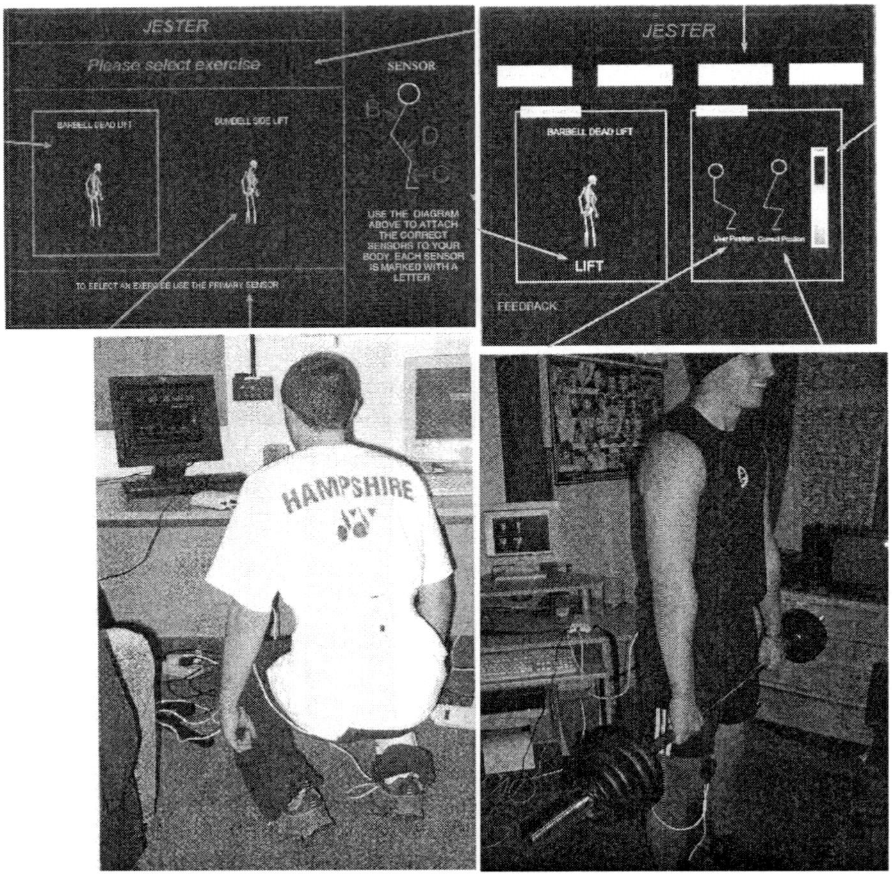

The two images at the top show screenshots of the system. The two images below were taken during our evaluation session.

To evaluate this prototype we carried out an initial cooperative evaluation (Wright & Monk, 1991) with five participants (bottom left and bottom right in Figure 5). Our evaluation revealed that users found it difficult to strap on the sensors, due to the ineffective strapping mechanism we provided. Additionally, we discovered that the sensors didn't always stay in exactly the same positions. Both of these problems can be addressed by providing a more secure strapping mechanism and smaller motion sensors. These problems, however, caused some users to believe that the system was not functioning properly. The users thought that the bar meter feedback was useful and easy to understand. Some of the users found that the skeleton didn't help them. Finally, some users found the voice annoying, while others found that the voice helped them to keep up with the exercise. Most users, however, agreed that more motivational comments (such as the comments that a real life trainer makes) would have been appropriate.

Tilt the Maze

With this prototype we explored the use of motion sensors in a mixed-reality game of tilt the maze. Utilising motion sensors we build three different versions of the game. The objective was to navigate a ball through a maze by tilting the maze in different directions. This tilt was achieved though the use of:

- A mouse connected to a typical desktop PC. The maze was displayed on a typical desktop monitor
- A lightweight board fitted with motion sensors. The maze was displayed on a large plasma screen
- A Tablet PC fitted with motion sensors. The maze was displayed on the Tablet PC itself so that tilting the tablet would appear to be tilting the virtual maze itself

We carried out a pilot study to compare performance and user preference for all three conditions. During this study we collected qualitative data in the form of questionnaires, as well as quantitative date by recording the number of

Figure 6. Tilt the maze

At the top, we see the system being used by means of a paper cardboard acting as a control token. At the bottom left, we see the condition with the PC and mouse, and at the bottom right, we see the condition with a Tablet PC acting both as a screen and a control token.

aborts, errors, and time to completion. The three experimental conditions are shown in Figure 6.

Each participant was given the chance to try all systems. The order in which each participant tried each of the systems was determined at random. The interaction technique of using motion sensors to move the board was well received by the participants. This was not only shown in the high numbers of participants which "preferred" the Tablet PC (78%) but also in the very low number of participants who "preferred" the standard and most commonly used interaction technique of a mouse (3%). This was also comparatively low to the percentage of people who preferred the plasma screen (19%), which also used the motion sensors to tilt the board. The questionnaires showed that participants found the Tablet PC the least difficult, then the plasma screen and found the mouse the hardest way of interacting with the system. Using the Tablet PC participants took on average 79 seconds using than the plasma screen 91 seconds, and with the mouse 154 seconds. The mouse on average took almost twice as long as the Tablet PC to complete. The number of aborted games was also least on the Tablet PC (1) and most by the mouse (9), while the plasma screen had four aborts. It should be noted however, that the average number of errors made was greatest on the mouse (160), but the plasma screen seemed to produce on average less errors (94) than the Tablet PC (104), although the difference was relatively small.

These results show that on average the participants liked using the Tablet PC the most, made slightly more errors on it than on the plasma screen but finished in a faster time. The lab experiment has given some confirmation that the novel interaction technique of using motion detectors to manipulate a maze (and hopefully an indication that similar tasks will behave in a similar manner) was received well and that it outclassed the most common interaction technique of using a mouse.

TEXT ENTRY

In our earlier work on gestural interaction we noted that the DSR may be utilised to communicate complex strokes, essentially acting as a kind of alphabet with eight distinct tokens. Although this allows for complex interactions, it does not address the perennial issue of text entry in mobile and pervasive systems. In this section we describe two prototype systems for text entry in embedded devices. These prototypes were developed by undergraduate students at the University of Bath. The first prototype makes use of two keys and a dial to enter text. The second prototype allows for text entry on a small size touch screen. Both prototypes address the entry of text on embedded devices. The application domain for both prototypes were designed is embedded digital music players. We designed these systems so that users can interact with them using only one had and situations were the users have to attend to other tasks simultaneously (such as driving a car).

Key and Dial Text Entry

The first prototype we present allows for text entry on an embedded digital music player. We envision this system to be used in cars, an application domain in which traditionally all interaction takes place via a minimum number of hardware keys. One of the main purposes of this approach is to minimise the cognitive load on drivers who are concurrently interacting with the steering controls as well as the music player.

In Figure 7 we can see our first prototype. The top of the figure is a mock-up of the actual

Figure 7. Our mock-up prototype for text entry

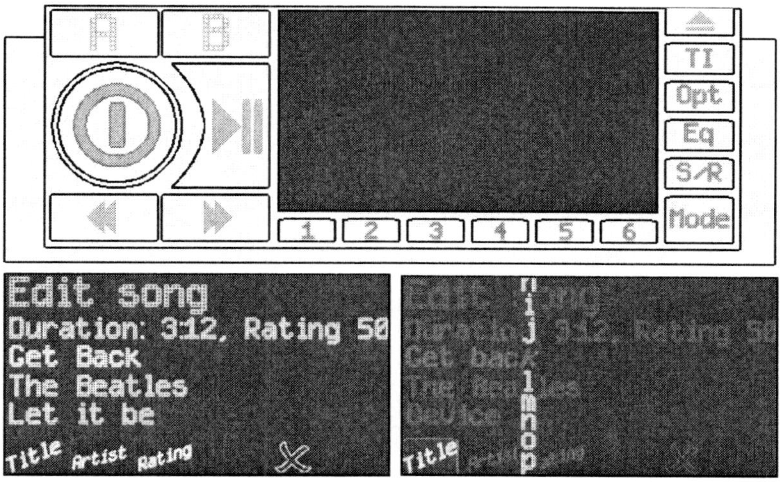

The circular dial on the left is used to select a letter from the alphabet. The left/right arrows below the dial are used to shift the edited character in the word.

hardware façade that would be visible in a car. The main aspects of this façade we focus on is the circular dial on the left, the left/right arrows below it, as well as the grey area which denotes a simple LCD screen. At the bottom of Figure 7 we see the screenshots of our functional prototype's screen. Bottom left depicts normal operation, while bottom right depicts edit mode.

When the user enters text edit mode, the system greys everything on the screen except the current line of text being edited. In Figure 7, the text being edited is the title of a song called "Get back." Text entry with this system takes place as follows. The user uses the left/right buttons to select the character they wish to change. The character to be changed is placed in the middle of a column of characters making up the alphabet. For example, in the bottom right part of Figure 7 we can see that character "k" is about to be changed. To actually change the character, the user turns the dial clockwise or anti-clockwise, which has the effect of scrolling up and down the column of characters. When the user has selected the desired character, they can move on to the next character in the word using the left/right buttons.

We have carried out an initial set of cooperative evaluation sessions with 10 participants. The evaluation itself was carried out on the whole spectrum of the prototype's functionality, which included playing music tracks from a database, adding/deleting tracks and tuning to radio stations. We received very positive feedback in relation to text entry interaction. Some users were able to pick up the interaction technique without any prompt or instructions from us. A few users, on the other hand asked for instruction on how text entry worked. Generally, however, towards the end of the evaluation sessions all users felt happy and comfortable with entering text using the dial and keys.

Text Entry on Small Touch Screens

The second prototype we have developed and evaluated utilises small-sized touch screens for text entry. Once more, this prototype was developed for text entry in environments where the users are distracted or must be focused on various tasks. For this prototype we wished to take advantage of user's peripheral vision and awareness. For this reason, the prototype utilises the whole of the touch screen for text entry. This enables users to aim for bigger targets on the screen while entering text. Furthermore, this prototype was designed to allow for single-handed interaction. The prototype is shown in Figure 8.

To enable text entry, the system brings up a keyboard screen, shown in the bottom left in Figure 8. This design closely resembles the layout of text used in traditional phones and mobile phones. At this stage, the background functionality of the system has been disabled. When a user presses a button, a new screen is displayed with four options from which the user may choose (bottom right in Figure 8). Notice that the user can only enter text, and no other functionality is accessible. This decision was made in order to accommodate for clumsy targeting resulting in the use of a finger, instead of a stylus, to touch the screen.

We evaluated this prototype by carrying out six cooperative evaluation sessions. The initial phase of our evaluation was used to gauge the skill level of the user. The co-operative evaluation was then carried out following a brief introduction to the system. During the evaluation, breakdowns and critical incidents were noted either via user prompting or by the evaluator noticing user problems. After the evaluation was complete, the user was queried on these breakdowns and instances. A brief qualitative questionnaire was given followed by a longer quantitative questionnaire. These gave us both feedback on user opinions, and suggestions about the overall system.

Figure 8. A second mock-up prototype for text entry

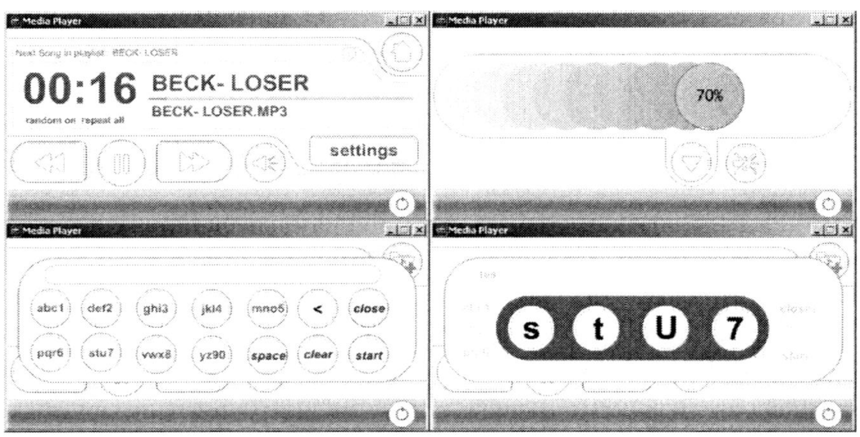

The prototype's main playing screen is shown in the top left. The volume control screen is shown in the top right. The keyboard screen is shown in the bottom left. Once a key is pressed, the four options come up, as shown in the bottom right.

According to our questionnaire data, users found the text entry functionality quite intuitive. Specifically, on a scale of 0 (very difficult) to 9 (very easy), the text entry functionality was rated 8 on average. Based on the qualitative data collected, we believe that the design employed, that of the simulation of a mobile phone keyboard, worked well and was highly intuitive.

ONGOING AND FUTURE WORK

In our research, we are currently exploring new ways of interacting with big and small displays. One of the systems we are currently developing is used for exploring high-resolution images on small displays. This system, shown in Figure 9, provides an overview of the image, and then proceeds to zoom into hot spots, or areas of interest within the image. The feedback area at the top provides information about the progress of the task (progress bar), the current zoom level (circle), and the location of the next hot spot to be shown (arrow).

Another research strand we are currently exploring is the use of both large screen and small screen devices in situations were public and private information is to be shared between groups of people. We are exploring the use of small-screen devices as a private portal, and are developing interaction techniques for controlling where and how public and private information is displayed. Our overall aim is to develop interaction techniques that match our theoretical work on the design of pervasive systems (Kostakos, 2005), the presentation and delivery of public and private information (O'Neill, Woodgate, & Kostakos, 2004), and making use of physical and interaction spaces for delivering such information (Kostakos & O'Neill, 2005).

ACKNOWLEDGMENTS

We wish to thank Andy Warr and Manatsawee (Jay) Kaenampornpan for their contribution and assistance. We are also very grateful to Adrian Merville-Tugg, Avri Bilovich, Christos Bechlivanidis, Colin Paxton, David Taylor, Gareth Roberts, Hemal Patel, Ian Saunders, Ieuan Pearn, James Wynn, Jason Lovell, John

Figure 9. Our image explorer provides an overview of the image to be explored, and then proceeds to zoom into specific areas of interest within the image

Quesnell, Jon Bailyes, Jonathan Mason, Ka Tang, Mark Bryant, Mary Estall, Nick Brunwin, Nick Wells, Richard Pearcy, and Simon Jones for developing the prototypes presented in sections *Kinaesthetic Interaction* and *Text Entry*. Special thanks to John Collomosse for his assistance in the development of the image explorer application.

REFERENCES

Brewster, S. A. (2002). Overcoming the lack of screen space on mobile computers. *Personal and Ubiquitous Computing, 6*(3), 188-205.

Brewster, S. A., Lumsden, J., Bell, M., Hall, M., & Tasker, S. (2003). Multi-modal "eyes free" interaction techniques for wearable devices. In G. Cockton & P. Korhonen (Eds.), *Proceedings of CHI'03 Conference on Human Factors in Computing Systems, CHI Letters, 5*(1), p. 473-80.

Brumitt, B., Meyers, B., Krumm, J., Kern, A., & Shafer, S. (2000). EasyLiving: Technologies for intelligent environments. *Lecture Notes in Computer Science,* (pp. 12-29).

Brumitt, B., & Shafer, S. (2001). Better living through geometry. *Personal and Ubiquitous Computing, 2001, 5*(1), 42-45.

Callahan, J., Hopkins, D., Weiser, M., & Shneiderman, B. (1998). An empirical comparison of pie vs. linear menus. In M. E. Atwood, C. M. Karat, A. Lund, J. Coutaz, & J. Karat (Eds.), *Proceedings of the CHI'98 Conference on Human Factors in Computing Systems* (pp. 95-100). ACM Press.

Dey, A. K., Abowd, G. D., & Salber, D. (2001). A conceptual framework and a toolkit for supporting the rapid prototyping of context-aware applications. *Human Computer Interaction, 2001, 16*(2/4), 97-166.

Harrison, S., & Dourish, P. (1996). Re-placing space: The roles of place and space in collaborative systems. In *Proceedings of the 1996 ACM Conference on Computer Supported Cooperative Work* (pp. 67-76). ACM Press.

Hightower, J., & Borriello, G. (2001). Location systems for ubiquitous computing. *Computer, 2001, 34*(8), 57-66.

Ishii, H., & Ullmer, B. (1997). Tangible bits: Towards seamless interfaces between people, bits, and atoms. In *Proceedings of the SIGCHI Conference on Human factors in Computing Systems (CHI '97)* (pp. 234-241). New York: ACM Press.

Kostakos, V. (2005). *A design framework for pervasive computing* (Tech. Rep. No. CSBU-2005-02). PhD Dissertation in Technical Report Series ISSN 1740-9497. University of Bath: Department of Computer Science.

Kostakos, V., & O'Neill, E. (2003, September). A directional stroke recognition technique for mobile interaction in a pervasive computing world, people and computers XVII. In *Proceedings of HCI 2003: Designing for Society,* Bath (pp. 197-206).

Kostakos, V., & O'Neill, E. (2005, February 9-11). A space oriented approach to designing pervasive systems. In *Proceedings of the 3rd Uk-UbiNet Workshop*, University of Bath, UK.

O'Neill, E., Kaenampornpan, M., Kostakos, V., Warr, A., & Woodgate, D. (2005). Can we do without GUIs? Gesture and speech interaction with a patient information system. *Personal and Ubiquitous Computing,* 1617-4917.

O'Neill, E., Woodgate, D., & Kostakos, V. (2004, August) Easing the wait in the emergency room: Building a theory of public information systems. In *Proceedings of the ACM Designing Interactive Systems 2004,* Boston (pp. 17-25).

Rekimoto, J. (2001). GestureWrist and GesturePad: Unobtrusive wearable interaction devices. In *Wearable Computers,* (pp. 21-30). Zurich, Switzerland: IEEE.

Rekimoto, J., Ullmer, B., & Oba, H. (2001). DataTiles: A modular platform for mixed physical and graphical interactions. In *Proceedings of the SIGCHI Conference on Human Factors in Computing Systems,* (pp. 269-276). ACM Press.

Sutherland, I. (1963). Sketchpad: A man-machine graphical communication system. In *Proceedings of the Spring Joint Computer Conference* (pp. 329-346). IFIP.

Wright, P. C., & Monk A. F. (1991). A cost-effective evaluation method for use by designers. *International Journal of Man-machine Studies*, *35*(6), 891-912.

Zhao, R. (1993). Incremental recognition in gesture-based and syntax-directed diagram editors. In S. Ashlund, K. Mullet, A. Henderson, E. Hollnagel, & T. White (Eds.), *Proceedings of INTERCHI'93* (pp. 96-100). ACM Press/ IOS Press.

KEY TERMS

Cooperative Evaluation: The process by which a computer system developer observes users using the system. The purpose of this process is for the developer to identify problems with the system.

Gesture Interaction: Interacting with a computer using movements (not restricted to strokes) performed by a token object.

Kinaestetic Interaction: Interacting with a computer via body movement (i.e., hand, arm, leg movement).

Pilot Study: An initial, small-scale evaluation of a system.

Stroke Interaction: Interacting with a computer using strokes. To perform the strokes a user needs a token object, such as the mouse, their hand, or a tennis ball.

Strokes: Straight lines of movement.

Text Entry: Entering alphanumeric characters into a computer system.

ENDNOTES

1 See http://www.opera.com
2 See http://www.mozilla.org/
3 See http://www.etla.net/libstroke/libstroke.pdf
4 See http://www.handhelds.org/projects/xscribble.html
5 http://www.stressbunny.com/wayv/
6 See http://www.pocketpc.com
7 See http://www.palm.com
8 See http://www.generation5.org/aisolutions/gestureapp.shtml
9 See http://www.xsens.com

Chapter VII
Engineering Mobile Group Decision Support

Reinhard Kronsteiner
Johannes Kepler University, Austria

ABSTRACT

This chapter investigates the potential of mobile multimedia for group decisions. Decision support systems can be categorized based on the complexity of the decision problem space and group composition. The combination of the dimensions of the problem space and group compositions in mobile environments in terms of time, spatial distribution, and interaction will result in a set of requirements that need to be addressed in different phases of decision process. Mobility analysis of group decision processes leads to the development of appropriate mobile group decision support tools. In this chapter, we explore the different requirements for designing and implementing a collaborative decision support systems.

INTRODUCTION

Mobile multimedia has become an essential part in our daily life and accompanies many work processes (Gruhn & Koehler 2004, Pinelle, Dyck, & Gutwin, 2003b). Mobile technologies are now indispensable for communication and personal information management. Their combination with wireless communication networks allows the usage in various business relevant activities (such as group decisions). This chapter investigates in the potential of mobile multimedia for group decisions. It builds upon the characteristics of group decision support with respect to mobile decision participants. Mobility analysis of group decision processes leads to the development of appropriate mobile group decision support tools. Research in-group decision support mainly focuses on the support of communication processes in-group decision scenarios. Research in mobile computing concentrates on technological achievements, on mobile networking and ubiquitous penetration of everyday processes with mobile technologies. This chapter concentrates on the facilities of mobile multimedia for group decision processes based on structured process analysis of group decisions with respect to mobile decision participants. The following section defines the theoretical foundation of group decisions in order to agree on an exemplary group decision process. Following this, a taxonomy for the complexity of group decision is presented as the foundation for requirements of mobile group decision support systems. The chapter closes by outlining the implications for the design of mobile group decision support systems.

GROUP DECISION THEORY

The ongoing research in this field focuses on group decisions as communication processes, in which a set of more than two people need to reach a mutual result, need to answer a question or to solve a problem. A group decision occurs as the result of interpersonal communication (the exchange of information) among a group's members, and aims at detecting and structuring a problem, generating alternative solutions to the problem, and evaluating these solutions (DeSanctis & Gallupe, 1987).

The aim of decision support tools is the minimization of decision effort with satisfactional decision quality. Following Janis and Mann (1979), decision makers, within their information process capabilities, canvas a wide range of alternative courses of action. Surveying the full range of objectives to be fulfilled and the values implicated by the choice, they carefully weigh the costs and risks of consequences. Decision makers undertake an intense search for new information or for expert judgment that is relevant to further evaluation of the alternatives. Furthermore, a decision maker needs to be aware of decision constraints (money, time, norms, etc.), must respect actors and their needs affected by the course of action, and lastly has to document decision for further post decision process evaluation and argumentation.

Vigilant information processing and a high degree of selectivity ought to save the decision maker from unproductive confusion, unnecessary delays, and waste of resources in a fruitless quest for an elusive, faultless alternative.

Nowadays technology can assist decision makers not only in selective information retrieval and algorithmic methodology in the judgment of alternatives. They can also direct the decision makers in a process-oriented walkthrough of decisions to avoid post-decisional regret.

PROCESS-ORIENTED VIEW ON DECISIONS

In order to support human actions as efficiently as possible with information technology, a formal process needs to be identified. Examples of decision process-models are given by Simon (1960) and Dix (1994).

According the decision process model of Herbert A. Simon (1960), the group decision process consists of the following phases and sub processes that are interdependent (illustrated in Figure 1):

- **Pre-decision phase:** Selection of the decision topic/domain, Forming of the group (introduction of the decision participants)
- **Intelligence phase:** Collection of information regarding the problem (in-/outside the group), Collection of alternatives
- **Design phase:** Organization of information, Declaration of each participant's position regarding the decision topic, Discussion of the topic and various alternatives based on existing information, Col-

Figure 1. Decision process by Simon

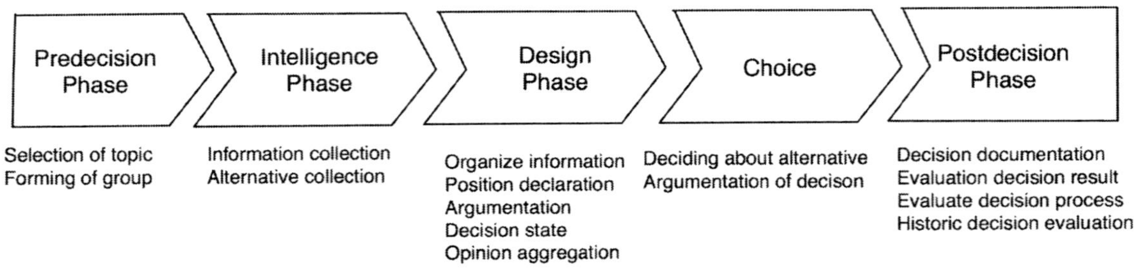

lection and communication of the actual opinion (decision state), aggregation of individual opinions to a group opinion/Identify majority

- **Choice:** Deciding on an alternative, Discussion of decision
- **Post-decision phase:** Documenting the decision, Evaluating the result, Evaluating the decision process, Historic decision evaluation

According to the taxonomy formulated by Dix (1994), as shown in Figure 2, these process phases can take place under various circumstances. Decisions can be distributed spatially, temporally, or in a combination of the two. In the case of spatially distributed decision groups, the participants involved in the process benefit from the communication facilities in mobile technology (wireless wide area networks or ad-hoc networks).

Considering asynchronous decision scenarios, personal direct communication between the decision participants needs to be respected as well as the communication via shared documents and databases (shared artifacts) that represent the group knowledge.

Artifacts shared in groups can be of static and dynamic nature. Static artifacts are introduced by one or more group members and do

Figure 2. Decisions under the view of groupware taxonomy

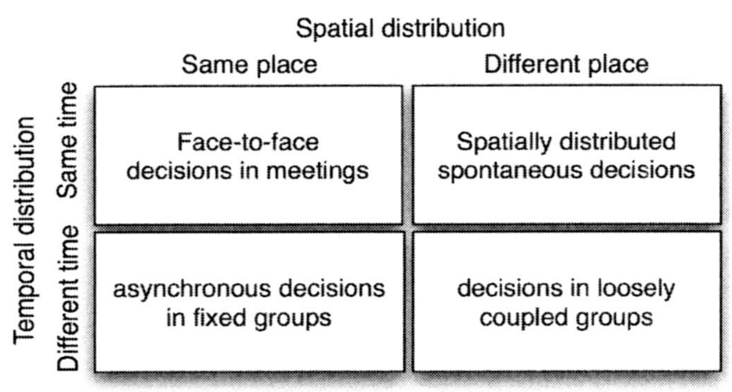

not change for the duration of their presence in the system. Dynamic artifacts are explicitly or implicitly modified by the group. An example of explicit manipulation by one or more group members is the manipulation of shared documents). Implicit manipulation of artifacts can be found at artifacts containing aggregated information concerning the actual work process or the actual state of the group.

DIMENSIONS OF GROUP DECISIONS

For the support of group decisions in mobile scenarios the dimensions of group decisions must be respected. The specific category in each dimension affects the complexity of the system, and therefore influences the need for supporting systems (shown in Figure 3). Each category can be distinguished between the dimensions of problem-space, group-composition, and decision distribution.

Problem Space of the Decision

The decision's problem space is related to the number of possible different courses of action (i.e., the number of alternatives) and their dynamicity. The first category in this dimension is the one of unidimensional problem spaces. In a unidimensional problem space, the decision handles a course of action by posing the simple question of Yes or No to it (do it or leave it). An example for such a unidimensional group decision is a simple vote regarding agreement or disagreement on a specific course of action (such as the decision of whether or not to accept a new group member). The second category (bidimensional decisions) handles the decision over more than one course of action and orders the various alternatives. This includes a ranking method for comparing courses of action. An example is the election of political parties, where a group of decision participants (elective inhabitants) choose competing political parties and derive a specific order. The third category in the dimension problem space is multidimensional decision. Compared to uni- and bidimensional decisions, here the set of available courses of actions is not fixed at the outset of the decision process. In such cases, in-group communication (discussing the arguments of decision participants) increases the available alternatives. Examples for this are creative group processes, where the course of action is generated through grouping-group communication during the decision process. With increasing intricacy of the problem-space, the complexity of the decision process increases,

Figure 3. Dimensions of group decisions TIPFEHLER

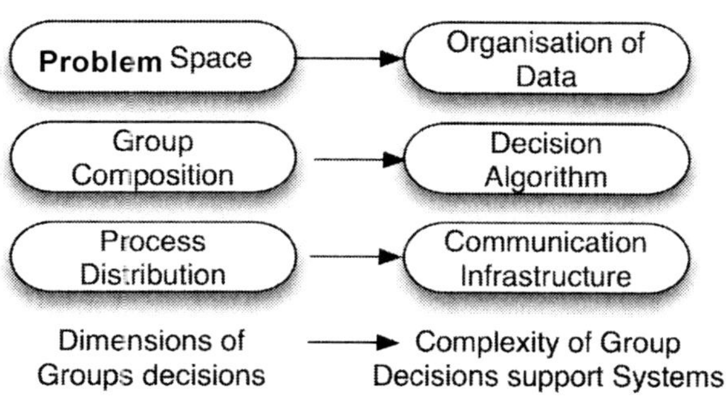

which results in the need for decision support. Counting the votes pro and contra a specific course of action can take a unidimensional decision. Bidimensional decisions require algorithms to rate and compare courses of action. Multidimensional decisions, finally, demand bidirectional communication structures, and algorithms for the ranking of alternatives. Increasing dimensionality of a problem-space increases the data-complexity as well as the complexity of rating mechanisms of the system.

Group Composition

The group composition in-group decision processes is connected to the relation between the decision participants and the dynamism and homogeneity of the group. The first category is a homogeneous group deciding about a specific course of action. In homogeneous groups, all decision participants have the same influence on the decision result and have equal rights (unless formal or informal hierarchical barriers apply). An example for a homogenous group decision process, taking place in a bidimensional problem space, is a political election in a democratic society. The set of elective group members is fixed, and each vote has equal value. Some group decisions take place in heterogeneous groups. In this category, the influence of some group members differs from that of others. An example for a heterogeneous group decision in a bidimensional problem-space is the selection of a new employee in a department. Here, the department staff agrees on a particular ranking of all candidates, yet the ultimate decision is made by the head of the department. The third category in-group composition is a dynamic group. In this case, the set of decision participants varies over time, as decision participants may join or leave the decision scenario. An example for a dynamic decision group can be found in multiphase selection processes, where a set of participants is choosing multiple alternatives and then reduces this set of alternatives in several stages (e.g., Casting shows, where the set of decision participants varies during the decision process). Increasing group heterogeneity and group dynamicity affects the algorithmic complexity of decision process support.

Decision Distribution

The distribution of the decision process introduces the dimension of mobility. In some decision processes, the decision participants are collocated, instead of spatial or temporal distribution. This is commonly the case in meeting-style scenarios. The other categories are called distributed decisions. Decision participants or other decision-relevant resources are spatially distributed. Participants of a decision are not located in the same place, some of them are abroad, or some of the decision relevant resources (e.g., required experts, eternal data sources) are inaccessible from the place where the decision is to be taken. Temporally distributed decisions (or long lasting decisions), on the other hand, take place over a course of time, which requires synchronization between the decision participants. The distribution of decisions influences the communication complexity of support systems in two manners. Networks need to be introduced in order to overcome spatial distances, and synchronization mechanisms are required in order to manage temporal distribution.

As a prerequisite for identifying the potential for mobile computing support, a set of criteria[1] are identified and applied in the analysis of the mobility potential (Gruhn & Koehler, 2004). The chosen criteria fall into three dimensions: The first two focus on the *distribution and uncertainty of the process* comprising distribution in time and space relating to Dix (1994). The third focuses on *interaction requirements* for electronic decision support systems comprising *collaboration, communication, and coordination*, applying the ideas of Teufel, Sauter, Mühlherr, and Bauknecht (1995).

- **Time (T)** is an important aspect, since decision processes may be spread over time or may be conducted parallely, requiring *synchronization* at later points in time. *Spontaneous* user interaction also implies time constraints such as the extension of timelines until a task has to be completed. Spontaneous user interaction implies *temporal uncertainty* and thus flexibility within the decision process. Both time synchronization and temporal flexibility can be supported by mobile computing means since flexible process control and maintenance of task dependencies can be enforced
- **Spatial distribution (S)** refers to both the *physical distribution of artefacts* in the real world as well as to the *virtual distribution of information*, both of which are required for the decision-making process. Physical distribution can be overcome by bringing the computational support to the actual location of the physical artefacts. Virtual distribution is relaxed by telecommunication support enabling electronic access to distributed information sources and to services based on wireless communication technology.
- **Interaction requirements (I)** refer to collaborations indicated by the quantity and complexity of interaction. With the increase of the amount of information at the place where a decision occurs, it may become more and more difficult for humans to process and store this information without use of adequate mobile computing support. Increasing complexity of information requires more and more flexible structuring and derivation mechanisms of information. Coordination efforts increase dramatically as the *number of participating actors* increases. Mobile computing empowers actors to efficiently coordinate their actions with multiple partners, while being enabled to cope more flexibly with issues of scheduling, resource management, or protocols. Interaction with partners in any process implies the need for communication in order to transfer and exchange information. Communication efforts increase as the number and type of communication partners increases and as the means of communication vary. Mobile communication means such as wireless communication and ad-hoc networking empower the user to conduct communication with multiple partners *more efficiently* because they are now able to maintain the

Table 1. Problem space and group composition formings of decisions

Problem space / **Group composition**	**Unidimensional**	**Bidimensional**	**Multidimensional**
Homogenous	Simple poll	Ranking	Idea finding
Heterogeneous	Dominated selection	Weighted ranking	Creative consultation
Dynamic	Dynamic advice	Dynamic ranking	Collective creativity

required *communication flexibility*. In view of frequently occurring media changes (e.g. from paper material to electronic data), additional resources are required in order to cope with redundant duplications. Eliminating media brakes with consequent use of electronic support can significantly increase quality and efficiency of any decision process.

Table 1 shows the varying forms decision scenarios can take, based on their respective problem space and group composition. The following examples clarify each of the forms, including variants for spatial and/or temporal distribution of the entire scenario.

GROUP DECISION SUPPORT SYSTEMS

A group decision support system (GDSS) is an interactive, computer-based system that facilitates the solution of unstructured and semi-structured problems by a set of decision-makers working together in a group. A GDSS aids groups in analyzing problem situations and in performing group decision-making tasks. According to DeSanctis and Galup (DeSanctis & Galup, 1987), a group decision support system can support groups on three levels. It provides process facilitation (technical features), operative process support (group decision techniques), and logical process support (expert knowledge). This research presented in this chapter focuses on process facilitation and operative process support.

According to Power (2003), a *communications driven* GDSS supports more than one person working on a shared task, it includes decision models such as rating or brainstorming, and provides support for communication, cooperation, and coordination. *Data driven* GDSS emphasize access to and manipulation of a time-series of internal company data and external data. *Document driven* GDSS manage, retrieve, summarize, and manipulate unstructured information in electronic formats. *Knowledge driven* GDSS provide expertise in problem solving. *Model driven* GDSS emphasize statistical financial optimization, and provide assistance for analyzing a situation (Power & Kaparthi 2002).

The research described in this chapter concentrates on communication driven aspects of GDSS. Group decision support systems improve the process of decision making by removing common communication barriers, by providing techniques for structuring decision analysis, and by systematically directing the pattern, timing, and content of discussion and deliberation activities (Crabtree, 2003).

Decision support tools basically address the need of decision participants to get in contact with each other (Koch, Monaci, Cabrera, Huis, & Andronico, 2004). They communicate and present each other with information regarding personal preferences and attitudes. Furthermore, they share task-relevant fact knowledge (e.g., surveys and statistics concerning the decision topic. Decision participants need full control over presentation and propagation of the information. In decision scenarios, information is not limited to factual knowledge. It includes actual information about the involved decision items. Information regarding the actual decision state is also assumed as process-relevant knowledge. With increasing complexity of the decision situation, the information becomes less manageable for the participants. There is thus a need for communication between the decision participants via shared media.

Group decision support systems also provide mechanisms to aggregate decision data. Aggregated data manifests the actual decision state and therefore presents a sum of the decision participants' sub goals (the actual group opinion). Continuous visualization of the actual decision state assists effective discussion and therefore facilitates progress in the decision process.

Business intelligence systems as GDSS define tools and platforms that enable the delivery of information to decision makers. The information delivered comes from relational data sources or from other enterprise applications (such as enterprise resource planning, customer relationship management, supply chain.) Technologies typically used for this include online analytical processing and key performance indicators presented through scorecards/dashboards (i.e., OLAP systems as Cognos power play, SAP, Oracle...).

Generally, a decision support system provides actors in decision processes with an objective, with independent tool for using databases, and with models for evaluating alternative actions and outcomes.

MOBILE GROUPS

Following Frehmuth, Tasch, and Fränkle (2003), a group of people equipped with mobile technology linked together in a working process with a common task or goal is defined as a mobile group. Mobile groups do not necessarily emerge from an existing (or fixed) organizational structure. Technologically founded flexibility allows people to generate ad hoc groups, as they are necessary in an actual decision situation. Frehmuth et al. (2003) as well as Bellotti and Bly (1996) mention various terms of mobile and virtual communities.

A group's common goal addressed in this research is defined as an economic goal in a business environment, and does not approach other mobile group scenarios such as everyday mobile communication (Ling & Haddon, 2001) or mobile entertainment.

The technology support allows the group members to fulfill their common task independently of their distribution in space (spatial flexible) and time (temporal flexible). Their common ground (on the basis of which the community is founded) is based on their common access to common and shared resources. Remarks on the social organization of space and place can be found in Crabtree (2003).

Groups working towards a common goal are characterized by their relative degree of coupling. Loosely coupled groups have low interdependencies and require access to shared resources for their collaborations; their need for synchronous communication is limited (e.g., insurance salesmen that support a particular customer group need access to the central register of insurance contracts). Tightly coupled groups organize their workflow with strong interdependencies and a strong need to access shared resources and synchronous communication (e.g., medical staff that care for patients and need access to their data or in emergency cases require immediate synchronous communication with a doctor). By definition, mobile groups are loosely coupled (Pinelle & Gutwin, 2003a). Autonomy of each participant and strict partitioning of work makes a common goal achievement feasible. Strict process analysis leads to optimized usage of mobile technology for task fulfillment. Interdependencies of group members in task fulfillment require asynchronous awareness of group members and their actual states (in the sense of availability and state of task fulfillment).

In existing group decision tools, support for mobile groups is limited, as they mainly address stationary users in fixed working environments. The notion of stationary users, however, does not exclude distributed decision scenarios. Yet the (intrinsic) mobility of (sub-) processes and decision participants is commonly not addressed. Web-based decision support tools allow mobility of decision participants up to a specific limit, in the sense of the support of spatially distributed groups of decision participants (Kirda, Gall, Reif, Fenkam, & Kerer, 2001; Schrott & Gluckner 2004).

Existing tools either focus on communication needs for group decisions, or on the sharing of mainly static artifacts. In traditional working environments with static located decision participants, there is no need for the support of mobile workers and explicitly asynchronous communication with mobile technology.

Informal and subtle aspects of social interaction are critical for accomplishing work, and consequently these issues need to be taken into account in the design of technological support systems for mobile team workers. (Sallnäs & Eval-Lotta 1998).

A tool to support peer and group knowledge discovery collaboration in virtual workspaces is presented by MayBury, with a focus on messaging (chat), member awareness (users in room users online), shared data, private data shared browsing, and a shared whiteboard (MayBury, 2004).

Generally, the goals of mobile groupware are: Improving interpersonal communication and cooperation; Encouraging knowledge sharing; Ubiquitous and transparent access to the organizations information and service network from fixed and mobile nodes; Shared access to different integrated engineering services; Supporting local site dependent activities and mobile working; Constant and timely update of the distributed corporate knowledgebase with many sites acting as potential users of information as well as potential information providers; And lastly efficient information sharing across a widely distributed enterprise environment (Kirda et al., 2001).

GROUP DECISION AS APPLICATION DOMAIN FOR MOBILE TECHNOLOGY

GDSS appear to be suited for mobile technology support because their demands hold characteristics of mobility. The nature of mobility is characterized by flexibility in time and place. Mobile technology as the set of applications, protocols, and devices that enable ubiquitous information access and exchange (Pandaya, 2000) consequently can be seen as facilitators for group decision scenarios (Schmidt, Lauf, & Beigl, 1998; Schrott & Gluckner, 2003).

The use of mobile technology in the application domain of group decisions respects properties of mobility (e.g., spatial distribution) in specific sub-processes of group decisions. Natural limitations of mobile devices, such as small input and output interfaces and limited operation time (and therefore limited availability) might prevent the use of mobile technology during the whole range of a particular process. Applying the criteria for mobility potential will show process parts in which mobile technologies are most suitable.

MOBILE TECHNOLOGY

Mobility is based on the spatial difference of the place of information origin, information processing and information use. For this research, a division into three forms of mobility is essential: user mobility, device mobility and service mobility (Kirda et al., 2001; Pandaya, 2000,). A different notion of mobility, fragmented in micro- and macro mobility, is mentioned at (Luff & Heath, 1998)

Saugstrup and Henten (2003) define parameters of mobility as follows: Geographic parameters (Farnham, Cheley, McGeeh, & Kawal, 2000) (wandering, visiting, traveling, roaming possibilities, place dependencies), time parameters (time dependencies, synchronous asynchronous), contextual parameters (individual or group context, private or business context) and organizational aspects (mobile cooperation, knowledge sharing, reliability).

Mobile multimedia allows the adaptation of information technology to the increasing mobile work practice (BenMoussa, 2003) with location independent access to information resources (Perry, OHara, Sellen, Brown, & Harper, 2001).

The spatial flexibility in decision scenarios requires ubiquitous access to information and communication resources (BenMoussa, 2003). Mobile groups can fulfill tasks, independent of fixed locations and in courses of action that are simultaneous yet spatially disparate, which is demanded by the spatial and temporal flexibility of mobile groups. The arising of information takes place at various places forced by the spatially flexible nature of mobile groups. Mobile groups capture information independently from respective location of the group. Process-relevant information must be available anywhere, including in situations where a group member is moving between various locations (BenMoussa, 2003).

Temporal flexibility brings with it the need for explicit asynchronous communication via shared media. Optimal profit of group (organizational) knowledge as shared resource depends on clear ownership of data and artifacts. Especially in dynamic group composition, ownership of information is decision relevant.

Ubiquitous communication facilities encourage spontaneous interaction and the building of ad-hoc decision groups. They need mobile access to their decision-relevant resources. If the decision participants are spatially distributed, they need additional communication facilities (also provided by mobile technology). With mobile technology, decision participants can collaborate as productive entities. They benefit from each other by enhancing the amount of available resources (mainly knowledge), and by sharing these resources (information use). Not only the mobility of group members needs to be considered, the use of mobile (digital) artifacts relevant for task fulfillment is of equal importance. Micro- and macro-mobility needs to be represented in mobile group support (Luff & Heath, 1998).

IMPLICATIONS ON MOBILE DSS

Mobile technology is suited for group decision scenarios. It offers solutions for continuous collaboration, despite temporal and spatial distribution (Kirda et al., 2001; Schrott & Gluckner, 2003). Wireless connectivity of mobile devices allows ubiquitous information exchange and access. Using mobile devices and services in-group decision scenarios enables ad hoc communication between the decision participants. Traceability of decision processes enhances decision performance and therefore group productivity.

Expected improvements of the described scenarios can be achieved with mobile technology, for example:

- Higher level of consensus in-group decisions (Watson, DeSanctis, & Poole, 1988). A permanent visualization of the actual decision state can be introduced to remind the decision participants of their common goal
- Detailed information about the actual decision state (aggregated data about the decision) and its composition offers functionality for decision retrieval. Looking deeper into an actual decision state (e.g., who decided for which alternative) leads to a more directed type of communication between the decision participants

- More directed communication allows for faster agreement on certain alternatives because others do not need to be discussed any more
- A decision participant can query the actual decision state down to its atomic components and as a result is able to force a higher level of knowledge concerning the actual decision
- Private access to ones own preferences in form of an individual ranking presents the dissimilarity of decision goals (public view)
- The social bias in decision scenarios can be overcome by rendering the decision participants anonymous (Davis, Zaner, Farnham, Marcjan, & McCarthy, 2002)

The technical support of mobile communities needs to focus on their very special needs (Gruhn & Koehler 2004; Kronsteiner & Schwinger 2004). The core needs and therefore the basic criteria for support functionality can easily be found in the mobility itself (portability, low power consumptions, wireless network access, independence) and flexibility. Actual technologies to support mobile decision scenarios include:

- Web services for mobile devices (Schilit, Trevoe, Hilbert, & Khiau Koh, 2002)
- Mobile messaging as benefit for groups (Schrott & Gluckner, 2003)
- Social activity indicators (Farnham et al., 2000)
- Content representation and exchange (Tyevainen, 2003)
- Distributed multimedia (Coulouris, Dollimore, & Kindberg, 2002)
- Distributed collaborative visualization (Brodlie, Duce, Gallop, Walton, & Wood, 2004)

EXEMPLARY SCENARIO

As a prerequisite for identifying the potential for mobile technology support, a set of indicators is identified which is applied in the analysis of the mobility potential. Similar to Gruhn and Koehler (Gruhn & Koehler, 2004), this research also presupposes the prepending analysis of the entire work process. In contrast to the proposed "process landscaping," not only the spatial and temporal distribution of sub processes and the accompanying mobility of services is taken into account, and the dimensionality of the decision space and the group composition also need to be respected in the mobility analysis.

The decision phases are split to sub processes (according to Simon, 1960). For each sub process it needs to be analyzed, how the sub process meets mobility indicators to determine a need for mobility support.

For an exemplary analysis, a prototype was built using a laboratory experiment (see also Van der Heijden, Van Leeuwen, Kronsteiner, & Kotsis, 2004). The experiment setting concentrated on the design- and the choice phase in-group decisions and emphasises the interaction demand on GDSS.

In this experiment, we assumed a group of three people deciding on the division of funding for social projects as a decision of a homogenous group in a bidimensional problem space. (Table 2 shows activities and the affected dimensions in the particular process phases.) The funding budget was assumed to be 500.000• and needed to be divided over six projects (proj A..proj F).

Analyzing the scenario led to a set of implementation requirements:

- **Interaction requirement (I):** Depending on the requirements of the communication style (synchronous/asynchronous

Table 2. Activities and dimensions in ranking scenarios

Phase	Activities	Dimensions
Predecision	An organization decides to spend 500 T• for social projects and elects a group of people (jury) for this decision. (I)	Interaction
Intelligence	Running social projects are analyzed and a set of six social projects is created, including potential arguments for each alternative. (T) Additional information about the social projects needs to be found out directly at the organizations responsible for the particular project (S)	Temporal distribution Spatial distribution
Design	The jury members (decision participants) evaluate each alternative in a free discussion and assign funding for each social project to bring their preference into the decision. The proposed funding is discussed in a face-to-face meeting. (I)	Interaction
Choice	The amounts suggested by the jury members are aggregated to reach a result value for each social project. (I)	Interaction
Postdecision	The funding dedicated to each social project is documented and published. (I)	Interaction

and media type), different technologies are required. In mobile scenarios, synchronous communication demands wireless networking infrastructure, while asynchronous communication demands access to central resources (BBS-, e-mail servers). Depending on the media type, different IO-devices are needed. Collaborative technologies extend communication technologies to shared editable resources (databases, shared document editors, shared artifacts in general). For decision scenarios, the primary requirements of collaborative environments are shared databases collecting and the deployment of decision information (information about alternatives, voting states). In decision scenarios, the coordination concentrates on the decision task as the set of alternatives to manage/evaluate, as well as on the decision participants and their voting. Coordination does not only include the planning of the decision task, but also the execution of the workflow (which decision participant already gave his vote in the actual decision) and alert systems (the decision state has changed, the set of participants has changed,) need to be considered

- **Spatial and temporal distribution (ST):** Decision scenarios that are spatially or temporally distributed require asynchronous access to information resources in an ubiquitous manner. For mobile environments, this leads to wireless wide area networks that allow ubiquitous access to information resources required for the decision-making. Concerning the temporal distribution, it is important to take into account that group members participating in the decision process usually have to divide their attention between several different tasks. Therefore, the information exchange needs to be asynchronous and available on demand. Collaboration and communication in temporally distributed scenarios require the possibility of asynchronous message exchange via shared resources during the process phase in order to allow asynchronous collaboration. In decision scenarios, communication during the design phase cannot be limited to asynchronous message exchange. Access to shared databases is required in order to manage the decision relevant information and to define decision states based on the actual votes of the participants with respect to their heterogeneity.

The focus of this experiment was the application of mobile technology during the design and choice phase. The group forming (288 undergraduate students in groups of three persons respectively) of the predecision phase, and the explanation of the six alternative projects (intelligence phase) is conducted by the experimenter. In the given scenario, the design phase is the discussion of the alternatives and the argumentation pre and against it. The decision participants specify and communicate their actual preferences regarding the decision via mobile devices (architecture and screenshots in Figure 4). The choice is communicated by filling a form with the discussed decision.

Lastly, the personal preferences of each decision participant (after the discussion) were compared to the group decision and the group consensus (Watson et al., 1998) is then calculated to evaluate the decision (post decision phase).

In each decision loop (the recurring task of allocating the money to six projects), the input module accepts the user-preferences (votes). The message assembler serializes the preference values into a tagged dataset. The input module proofs the validity of the data so that the maximum figure of 500 cannot be exceeded during the discussion process.

The transmission of the tagged messages is done via a TCP/IP connection to the Web server. The connection requires an internet connection, but for workload issues a connection to the Web server running the database is only needed during data transmission (each time the input module changed the values and stores them with the save-command). The Web server receives the tagged messages as parameters of an http request calling a server-side script module.

The message-parser module on the Web server is a server-side script that dissects the

Figure 4. Architecture and screenshots of the GDSS prototype

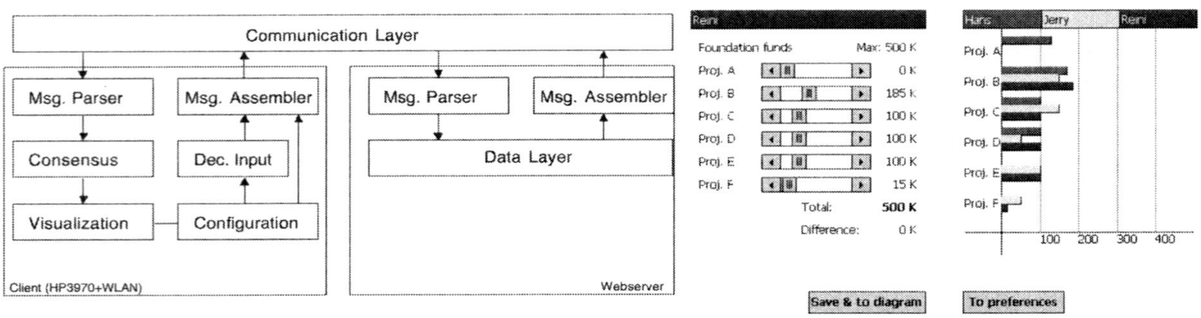

tagged message and uses them for update queries on the datalayer. The datalayer stores the transmitted decision-values for further computation, and provides the participants with actual information. The message assembler on the server side produces tagged messages on request. Such a request is generated upon each refresh loop initiated by the clients. The message parser on the client side dissects the tagged messages and stores them for further computation. Incomplete messages should be discarded. The client side's consensus engine derives the group consensus from the received messages and from stored personal decision-preferences. Ultimately (in further experiments and scenarios), the system is planned to work in an ad-hoc fashion, and the computation load has thus been left to the client.

The visualization-module uses the received values to display bar chart-diagrams of the actual decision situation. These diagrams are automatically refreshed frequently/regularly/ upon request. The derived group-consensus or other decision performance indicators can be displayed. The experiment showed that the participants preferred fixed-scale bar charts for their discussions, and did not accept displayed consensus- measurements.

CONCLUSION

Mobile GDSS tools tend to respect the mobile context in-group decision situations, and can therefore potentially influence the entire decision process. Existing tools support the process by providing multimedia communication facilities. Other improvement is to be found in:

- Clear process steering mechanisms
- Use of mathematical models for alternative rankings
- Avoidance of communication deadlocks
- Structuring of personal and public information

With the use of mobile GDSS group members can overcome spatial distances while accomplishing their task. Process steering mechanisms allow them to structure the communication flow and encourage the members to an equal-footing participation on the discussion process (regardless of group-internal hierarchies and offensive communication behavior on the parts of particular group members). The automatic accompanying process documentation can be analyzed to improve future decision scenarios (e.g., changing the group setup, introducing other creative techniques, other information bases, etc.). Finally, the decision documentation could also improve the development of further decision tools, based on insights gained from the failures and delays of decision processes observed in experiments such as the one described previously.

REFERENCES

Belloti, V., & Bly, S. (2003). Walking away from the desktop computer: Distributed collaboration and mobility in a product design team. *Proceedings CSCW 96*. Cambridge: ACM.

BenMoussa, C. (2003, May). Workers on the move: New opportunities through mobile commerce. *Proceedings of the IADIS Conference*.

Brodlie, K. W., Duce, D. A., Gallop, J. R., Walton, J. P. R. B., & Wood, J. D. (2004). Distributed and collaborative visualization. *Computer Graphics Forum*, *23*(2), 223-251. Oxford: Eurographics Association and Plackwell Publishing.

Coulouris, G., Dollimore, J., & Kindberg, T. (2002). *Verteilte Systeme: Konzepte und Design,* (pp. 703-732). Pearson Studium Munchen.

Crabtree, A. (2003). Remarks on the social organisation of space and place. *Homo Oeconomicus, 19*(4), 591-605.

Davis, J., Zaner, M., Farnham, S., Marcjan, C., & McCarthy, B. P. (2002). Wireless brainstorming: Overcoming status effects in small group decisions. *Proceedings of the 36th HICSS03*. IEEE.

DeSanctis, G., & Gallupe, R. B. (1987, May). A foundation for the study of group decision support systems. *Management Science, 33*(5). INFORMS, Maryland.

Dix, A. (1994). *Cooperation without communication: The problems of highly distributed working* (Tech. Rep. 9404) University of Huddersfield.

Farnham, S., Cheley, H. R., McGeeh, D. E., & Kawal, R. (2000). Structured online interactions: Improving the decision making process of small discussion groups. *ACM Conference on Computer Supported Kooperative Work (CSCW2000)* (pp. 299-308). Philadelphia, December.

Frehmuth, N., Tasch, A., & Fränkle, M. (2003). Mobile communities–New business opportunities for mobile network operators. *Proceedings of the 2nd Interdisciplinary World Congress on Mass Customization and Personalization (MCPC).*

Gruhn, V., & Köhler, A. (2004). Analysis of mobile business processes for the design of mobile information systems. In K. Bauknecht, M. Bichler, & B. Pröll (Ed.), *Lecture notes in computer science 3182. E-commerce and Web technologies* (pp. 238-247). August 30 - September 3. Zaragoza, Spain: Springer.

Janis, I. L., & Mann, L. (1979). *Decision making: A psychological analysis of conflict, choice, and commitment.* New York: Collier MacMillan Publishers.

Kakihara, M., & Sorensen, C. (2002). Mobility: An extended perspective. *Proceedings of HICSS 2002.*

Kirda, E., Gall, H., Reif, G., Fenkam, P., & Kerer, C. (2001, June). Supporting mobile users and distributed teamwork. *Proceedings of ConTEL 2001, 6th International Conference on Telecommunications,* Zagram, Croatia.

Koch, M., Monaci, S., Cabrera, A. B., Huis, M., & Andronico, P. (2004). Communication and matchmaking support for physical places of exchange. *Proceedings of the International Conference of Web Based Communities (WBC2004),* Lisbon (pp. 3-10).

Kronsteiner, R., & Schwinger, W. (2004). Personal decision support through mobile computing. *Proceedings of MOMM 2004* (pp. 321-330).

Ling, R., & Haddon, L. (2001). *Mobile telephony, mobility, and the coordination of everyday life.* Machines that Became Us Conference at Rutgers University. Transaction Publishers.

Luff, P., & Heath, C. (1998). Mobility in collaboration. *Proceedings of CSCW 98*, Seattle.

MayBury, T. M. (2004). *Exploitation of digital artefacts and interactions to enable P2P knowledge management.* 1st International Workshop on P2P Knowledge Management. Boston.

Pandaya, R. (2000). *Mobile and personal communication systems and services.* IEEE Series on digital and mobile communication. IEEE Press.

Perry, M., OHara, K., Sellen, A., Brown, B., & Harper, R. (2001, December). Dealing with mobility. *ACM Transactions on Human Computer Interaction, 8*(4), 323-347.

Pinelle, D., Dyck, J., & Gutwin, C. (2003b). Aligning work practice and mobile technologies: Groupware design for loosely coupled mobile groups. *Proceedings of Mobile HCI 2003* (pp. 177-192).

Pinelle, D., & Gutwin, C. (2003a). Designing for loose coupling in mobile groups. *Proceedings of 2003 International ACM SIGGROUP Conference on Supporting Group Work* (pp. 75-84).

Power, D. J., & Kaparthi, S. (2002). Building Web-based decision support systems. *Studies in Informatics and Control, 11*(4), 291-302.

Sallnäs, E. L., & Eval-Lotta. (1998). *Mobile collaborative work.* Workshop on handheld CSCW 98, Seattle, WA, November.

Saugstrup, D., & Henten, A. (2003). *Mobile service and application development in a mobility perspective.* The 8th International Workshop on Mobile Multimedia Communications, Munich, October 5-8.

Schilit, N. B., Trevoe, J., Hilbert, D. M., & Khiau Koh, T. (2002, October). Web interaction using very small internet devices. *IEEE Computer, 35*(10), 37-45.

Schmidt, A., Lauf, M., & Beigl, M. (1998). *Handheld CSCW.* Workshop on Handheld CSCW at CSCW '98. Seattle, WA, September 14.

Schrott, G., & Gluckner, J. (2003). What makes mobile computer supported cooperative work mobile? Towards a better understanding of cooperative mobile interactions. *International Journal of Human Computer Studies.*

Simon, H. A. (1960). *The new science of management decision.* New York: Harper and Row.

Teufel, S., Sauter, T., Mühlherr, T., & Bauknecht, K. (1995). *Computerunterstützung für die gruppenarbeit.* Bonn: Addison-Wesley.

Tyevainen, P. (2003). Estimating applicability of new mobile content format to organisational use. *Proceedings of HICS 2003.*

Van der Heijden, H., Kotsis, G., & Kronsteiner, R. (2005). Mobile recommendation systems for decision making on the go. *Proceedings of M-Business Conference.*

Van der Heijden, H., Van Leeuwen, J., Kronsteiner, R., & Kotsis, G. (2004). Ubiquitous group decision support for preference allocation decision in three person groups. *Proceedings of ECIS 2004.*

Watson, R. T., DeSanctis, G., & Poole, M. S. (1998, September). Using a GDSS to facilitate group consensus: Some intended and unintended consequences. *MIS Quarterly, 12*(3), 463-478.

KEY TERMS

Group Decision: Communication process in which a set of more than two people try to reach a common result in answering a question or in solving a problem.

Group Decision Support System (GDSS): Interactive, computer-based system that facilitates the solution of unstructured and semi-structured problems by a set of decision-makers working together as a group.

Mobile Multimedia: Set of protocols and stands that enables ubiquitous information access and exchange.

ENDNOTE

1 The letters in parenthesis after the criteria are the references used in the subsequent mobility potential analyses step.

Chapter VIII
Spatial Data on the Move

Wee Hyong Tok
National University of Singapore, Singapore

Stéphane Bressan
National University of Singapore, Singapore

Panagiotis Kalnis
National University of Singapore, Singapore

Baihua Zheng
Singapore Management University, Singapore

ABSTRACT

The pervasiveness of mobile computing devices and wide-availability of wireless networking infrastructure have empowered users with applications that provides location-based services as well as the ability to pose queries to remote servers. This necessitates the need for adaptive, robust, and efficient techniques for processing the queries. In this chapter, we identify the issues and challenges of processing spatial data on the move. Next, we present insights on state-of-art spatial query processing techniques used in these dynamic, mobile environments. We conclude with several potential open research problems in this exciting area.

INTRODUCTION

The pervasiveness of wireless networks (e.g., Wi-Fi and 3G) has empowered users with wireless mobility. Coupled with the wide-availability of mobile devices, such as laptops, personal digital assistants (PDAs), and 3G mobile phones, it enables users to access data anytime and anywhere. Applications that are built to support such data access often need to formulate queries (often spatial in nature) and send the queries to a remote server in order to either retrieve the results or retrieve the data, which is then processed locally by the mobile device. Due to the mobility of the users and limited resources available on the devices used, it compels the need for efficient and scalable query processing techniques that can address the challenges on handling *spatial data on the move*.

Mobile devices (e.g., PDAs, laptops) connect to the servers via wireless networks (e.g., WiFi, 3G, CDMA2000), and have limited resources (power, CPU, memory). Hence, it is necessary to optimize the resources usage. Existing wireless technology suffers from the problem of low-bandwidth (compared with the wired networks) and the range. The maximum bandwidth for WiFiMax, WiFi, and 3G are 75Mbps, 54Mbps and 2Mbps respectively. Also, as the network is susceptible to interference (from other wireless devices, obstructions, etc.), the achievable bandwidth is usually much lower. To reduce unnecessary communication overheads between the server and the clients, it is important to transfer only the required data items. In addition, the query processing techniques would need to adapt to the unpredictable nature of the underlying networks, and yet ensure that data is delivered continuously to the clients.

As the users carrying the mobile devices move, the queries pose might move based on the users' current location. Query processing algorithms need to tackle these mobility challenges. For example, a mobile device might issue the following k-nearest neighbor (kNN) query: *Retrieve the five nearest fast food restaurants*. However, as the user who is carrying the mobile devices move, the results of the kNN query changes. Thus, many existing algorithms designed for static environment, which assumes that the query is static cannot be used directly. In addition, many existing indices are optimized for static datasets, and cannot be directly used for indexing moving data, due to the overheads from updates, and deletions due to expiration of queries or data items. This compels the need for new indices, designed to handle issues introduced due to mobility.

Notably, long-running continuous spatial queries are relatively more common in a mobile environment compared to ad hoc queries and pre-canned queries. For example, users might be interested in monitoring specific regions for activities over an extended period of time, or predict the number of objects at a region in the future. The distinction between queries and data objects is thus relatively blurred. Another observation is that the number of queries is usually relatively smaller than the number of data objects especially over an extended period of time. Thus, to process queries efficiently, it might be more efficient to index the query instead of the data objects.

In this chapter, we present a comprehensive survey on the state-of-art techniques that have been proposed for handling these queries in a wireless mobile environment. We focus on the spatial access method and query processing techniques that have been developed for spatio-temporal and location-aware environment domain.

Chapter Organization

The next few sections are organized as follows: *Background, Querying Spatial Data, Data Dissemination, and Conclusion*. We first present a framework for understanding the various query processing techniques. Next, we present the state-of-art query processing techniques for handling the following type of queries: point and range queries (we look at access methods and data structures), nearest neighbor queries, spatial joins, aggregation, and predictive queries. Then, we look at data dissemination methods used in the mobile environment. We conclude in the last section.

BACKGROUND

In this section, we provide a generic framework for studying the different query processing techniques discussed in the later section. In the framework, we consider the nature of queries and objects, the types of queries and ad hoc vs. continuous queries.

Nature of Queries and Objects

The first aspect of the framework addresses the nature of queries and data objects. The four scenarios characterizing queries and data objects are presented in Figure 1. Most queries posed in a spatial database context would fall into Case A. Case B refers to the scenario where there are moving objects, and the query is static. Case C refers to a moving query window, and the objects are static. In Case D, both objects and queries are moving. In this chapter, we focus on Case B, C and D.

Types of Queries

We consider the types of queries that are commonly used in spatial and spatio-temporal databases, namely: range and nearest neighbor (NN) queries, spatial join, and aggregate queries.

A spatial range query consists of a query window, which specifies the region of interest. Depending on the spatial predicates used, the results that arises from a spatial range query might contains either regions that overlap the query window, regions contained within the query window, or regions that are not in the query window. For example, we could be interested in the locations of all the shopping malls in the Orchard Road area. The results retrieved are all the shopping malls contained within the query window denoting the Orchard Road region. In a spatial-temporal database, the query would also specify the time interval in which the results are valid.

A NN query (Korn, Sidiropoulos, Faloutsos, Siegel, & Protopapas, 1996) retrieves the nearest data object with respect to a query object. An extension of the problem looked at retrieving the *k* nearest neighbor of an object. The reverse nearest neighbor (RNN) of a point *p*, *RNN (p)* are points which have *p* as their 1-nearest neighbor. Many types of NN and kNN queries have been proposed. In this chapter, we focus on NN and kNN queries that are used for processing data on the move.

A spatial join query finds all object pairs from two data sets that satisfy a spatial predicate. The spatial predicate specifies the relationship between the object pairs in the result set. One of the most common spatial predicate used is the *intersect* predicate (i.e., overlap), in which all object pairs in the result set intersect each other. One of the variants is the spatial distance join. In a spatial distance join (Hjaltason & Samet, 1998), all object pairs that are within a specified distance to one another are retrieved. Generalizing the distance join problem, the similarity join was proposed in Bohm and Krebs (2004), where all object pairs from two data sets are returned if they are *similar* to one another. The notion of *similarity* includes: distance range, k-distance and k-nearest neighbor.

In a spatial aggregate query, the count for the total number of objects in a user-specified

Figure 1. Types of queries

Query \ Data	Static	Dynamic
Static	A	B
Dynamic	C	D

region is returned. In a spatial-temporal aggregate query, besides specifying the region of interest, the query also includes a time interval. For example, a spatial aggregate query, the total number of cars in the Orchard Road car park (i.e., user-specified region) at the instance the query is issued would be computed. A spatial-temporal aggregate query might retrieve the total number of cars in the Orchard Road car park between 2pm and 4pm. Note the additional time dimension introduced.

Ad hoc vs. Continuous Queries

The third aspect of the framework considers whether the query processing technique supports ad hoc or continuous queries. In an ad hoc query, the query is issued once, and when the results are returned, the query terminates. In a continuous query, the queries is continuously evaluated when input changes. Due to the limited resources available, most query processing technique that process continuous queries consider the use of either a time-based or count-based window for limiting the amount of data processed.

Ad hoc queries that are used for processing spatial data on the move can be categorized as follows: (1) non-predictive, (2) predictive, and (3) location-aware. Non-predictive queries are queries that are posed against a set of static or moving objects. The results are valid on data that is readily available. In predictive queries, based on past and current data, queries are posed to find out about the future location or count of objects in a future time interval. A location-aware query is interested in the objects that are relevant to the user's location. Thus, the results of the queries are affected both by the mobility of the mobile device, as well as the data objects. To reduce unnecessary communication to the server (due to the need to frequently update the server of a new location) and redundant computations, many recent works (Stanoi, Agrawal, & El Abbadi, 2000; Xiong, Mokbel, Aref, Hambrusch, & Prabhakar, 2004; Zhang, Zhu, Papadias, Tao, & Lee, 2003) considered the identification of an *invariant region*, in which the results do not change even if the data objects or queries moved within this region.

Continuous queries are queries that are constantly evaluated over time. The outputs of continuous queries would also change over time, as new data arrives or old data expires. The continuous query would terminate either the time interval specified by the query has lapsed, or a condition on the result or query window has been met. Most continuous query processing techniques use either a window-based or a count-based approach to bind the inputs, as well as to be able to ensure incremental delivery of results. It was noted by Tao and Papadias (2003) that most continuous spatio-temporal queries can be expressed as a time-parameterized (TP) query which will return *<R, ET, C>*. *R* denotes the results of the spatial query, *ET* is the time in which *R* is valid, and *C* denotes the set of changes that will cause *R* to expire. Many of the conventional queries discussed prior have a TP counterpart (e.g. TP Window query, TP k-nearest neighbors query, TP Spatial Join).

QUERYING SPATIAL DATA

Spatial Access Methods

Spatial access methods (SAMs) are built to facilitate efficient access to the spatial data. Amongst these various spatial access method, the R-tree (Guttman, 1984) is the most popular, and form the basis for many later hierarchical indexing structures, such as R+-tree (Sellis, Roussopoulos, & Faloutsos, 1987) and R*-tree

(Beckmann, Kriegel, Schneider, & Seeger, 1990). Another popular spatial access method is the PMR quad-tree (Nelson & Samet, 1987). Most of the SAMs were designed to handle static spatial data sets, and need to be extended in order to handling queries used on spatial data on the move. In a mobile environment, both the data and queries could be dynamic in nature, and the SAMs would need to handle frequent updates as well as ensuring that the results produced are not out-dated and accurate.

R-tree-Based Indices for Moving Objects/Queries

Many extensions have been made to the R-tree to support query processing in mobile environment. We present several types of novel indices which extend the R-tree for supporting the indexing of mobile data objects and queries. These includes the spatial-temporal r-tree (STR-tree) and trajectory bundle tree (TB-tree), time-parameterized tree (TPR), TPR*-tree and R^{EXP} tree.

Two spatial access methods, the STR-tree and TB-tree were proposed in Pfoser, Jensen, and Theodoridis (2000) to handle a rich set of spatio-temporal trajectory-based such as topological and navigational queries. Topological queries deals with the complete or partial trajectory of an object, and is usually very expensive to compute. Navigational queries deals with the derived information (e.g., speed, direction of objects). In addition, the proposed technique also allowed for the processing of a combination of coordinate-based (point, range and nearest-neighbor queries) and trajectory-based queries. In the proposed methods, sampling is used to obtain the movement of the data objects, and linear interpolation is used to consider the points between the samples. The STR-tree is essentially an R-tree, with new insertion/split strategy introduced to handle the trajectory orientation information, without causing a deterioration of the overall quality of the R-tree. However, in an STR-tree (and also all other R-tree variants), the geometries of the inserted objects (and line segments) are considered to be independent. However, trajectories consist of multiple line segments which are not independent. Thus, due to the inherent structure of the STR-tree, the knowledge of multiple line segments belong to trajectories cannot be fully exploited. The TB-tree considered the notion of *trajectory preservation*, and ensures that the leaf node contains line segments belonging to the same trajectory. Therefore, it can also be seen as bundling the trajectories (i.e., hence the name trajectory-bundle). In essence, the TB-tree sacrifices on its space discrimination property for *trajectory preservation*.

The time-parameterized R-tree (TPR-tree) (Saltenis, Jensen, Leutenegger, & Lopez, 2000) is an extension of the R*-tree, designed for indexing the current and predicted future position of moving points. It supports time-slice, window and moving queries, up to 3-dimensional space. The construction algorithm is similar to the R*-tree. The main difference is that instead of using the original R*-tree criteria (i.e., minimizing area, overlap between MBRs in the same node, distance between the centroid of the MBR to the node containing it) for ensuring the overall quality of the tree, the TPR-tree replaces these with its time-parameterized counterpart. During query processing using a TPR-tree, the extents of the MBRs are computed at runtime, and evaluated against the query window. For example, the MBR of Node *n* might not intersect the query window at current time. However Node *n* must still be visited because its MBR computed at runtime intersect with the window query. Tao and Papadias (2003) provides a comprehensive study of the performance of the TPR-tree and time-parameterized (TP) versions of conventional spatial queries (TP Window queries, TP k-nearest neighbors queries, and TP spatial join). Also, Tao, Papadias, and Sun (2003) provided a cost model for predicting the performance of the TPR-tree. Subsequently, the TPR*-tree was proposed to address the deficiencies of the original TPR-tree.

Noting that the TPR-tree is unable to effectively handle the expiry of moving objects, the R^{EXP} tree was proposed in Saltenis and Jensen (2002). Similar to the TPR-tree, the R^{EXP} also uses time-parameterized bounding rectangles. In a R^{EXP} tree, the expiration time is stored in the leaf index, and a lazy scheme is adopted to remove the expired entries. In the lazy scheme, expired entries in a node are moved only when the node is modified and written to disk. In general, the R^{EXP} outperforms the TPR-tree by a factor of two, for cases where the expirations of duration of objects are not large.

Nearest Neighbor Queries

The k-nearest neighbors (kNN) problem has been well-studied in spatial database. Hjaltason and Samet (1999) and Roussopoulos, Kelley, and Vincent (1995) uses an R-tree for finding the kNN. An incremental nearest neighbor algorithm was proposed in Hjaltason and Samet (1999), and uses the R-tree. Due to the mobility of mobile clients, both data objects and queries could be dynamic, and compels the design of new techniques.

Many techniques for handling continuous kNN (CKNN) queries in a mobile environment were also proposed. Unlike snapshot KNN queries which identifies the nearest-neighbors for a given query point, a continuous KNN query must update its result set regularly in order to ensure that the motion of the data objects and queries are taken into consideration. Most existing works modelled moving points as linear function of time. Whenever an update occurs, the parameters of the function need to be changed.

The problem of finding the k-nearest neighbor for moving query points (k-NNMP) was first studied in Song and Roussopoulos (2001). Subsequently, Tao, Papadias, and Shen (2002) considered the problem of continuous nearest neighbor (CNN) query for points on a given line segment using a single query to retrieve the whole results. For example, the following query retrieves the nearest neighbor of every point on a line segments: *Continuously find all the nearest restaurants as I travel from point A to point B*. It was noted in Tao et al. (2002) that the goals of a CNN query is to locate the set of nearest neighbor of a segment $q=[s,e]$, where s and e denotes the start and end point respectively. In addition, the corresponding list of split points, *SL*, would also need to be retrieved.

Iwerks, Samet, and Smith (2003) considered the problem of processing CKNN queries on moving points with updates. To represent a moving object, the *Point Kinematic Object* (PKO) was introduced, and is modelled by the function $p(t) = \vec{x_0} + (t - t_0)\vec{v}$, where $\vec{x_0}$ denotes the starting location of the object, and t_0 is the start time, and $\vec{v}$ denotes the velocity vector. The continuous windowing kNN algorithm (CW) was proposed for processing window queries on moving points

Another related line of work deals with location-aware queries. In a location-aware environment, the system would need to handle a large number of moving data objects and multiple continuous queries. Without any optimization, the performance of the server would degrade as more data objects and queries are introduced into the system. Motivated by the need for a scalable and efficient algorithm for processing queries in a location-aware environment, Mokbel, Xiong, and Aref (2004) and Xiong, Mokbel, and Aref (2005) proposed novel algorithms for tackling multiple continuous spatial-temporal queries. In Mokbel et al. (2004), a scalable incremental hash-based algorithm (SINA) was proposed to handle concurrent continuous spatio-temporal range queries. In addition, the notion of *positive* and *negative* updates was introduced for conserving network bandwidth by sending only updates, rather than the entire result set. In addition, SINA introduced the notion of a *no-action region*. In a *no-action* region, moving objects can move in a specific region without affecting the results, entity can move in without affecting the results. Xiong et al. (2005) addressed the need to handle a richer combination of moving/stationary queries and moving/stationary data objects. Similar to SINA, a shared execution paradigm was used. The shared-execution algorithm (SEA-CNN) was proposed to answer multiple concurrent CKNN queries. In order to narrow the scope of a re-evaluation in SEA-CNN, search region is associated with each CKNN query. The key features of in these algorithms are: (1) incremental evaluation and (2) shared execution. Incremental evaluation ensures that only queries that are affected by the motion of data objects or queries are re-evaluated, whereas shared execution process the multiple CNKK queries by performing a spatial join between the queries and a set of moving objects

A family of generic and progressive (GPAC) algorithms were proposed in Mokbel and Aref (2005) for evaluating continuous range and k-nearest neighbor queries over mobile queries over spatio-temporal streams. GPAC algorithms are designed to be online, deliver results progressively, and also provide fast response to a rich set of continuous spatio-temporal queries. One of the key features in GPAC is the use of *predicate-based* windows, where only objects that satisfies a query predicate are stored in memory. Whenever objects become invalid (i.e. does not satisfy the query predicate), they are expired. GPAC also introduced the notion of *anticipation*, where the results of a query are anticipated before they are needed, and stored into a cache.

Spatial Joins

Over the past decade, many spatial join algorithms (Brinkhoff, Kriegel, & Seeger, 1996; Brinkhoff, Kriegel, Schneider, & Seeger, 1994; Hoel & Samet, 1992; Huang, Jing, & Rundensteiner, 1997; Lo & Ravishankar, 1994) were proposed. Many of the conventional spatial join algorithms were designed to handle static data sets, and are mostly blocking in nature. In addition, the join algorithms were highly optimized in both Input/Output (I/O) and CPU for the delivery of the entire result sets. None of these conventional spatial join algorithms are able to handle the demands of mobile applications. As noted in Lee and Chen (2002), in a mobile computing environment, there is a disparity between the resources available to the mobile client with respect to the remote servers. The remote servers often have more resources, greater transmission bandwidth and have much smaller transmission cost. This prevents query processing techniques originally developed for distributed databases to be directly applied. In addition, most of the existing works on handling joins between mobile clients focus primarily on relational data. Hence, it compels the need for new query processing techniques to be developed for handling the spatial join. In a later section, we discuss how spatial joins can be performed on a mobile device.

To the best of our knowledge, there is little work done on continuous spatial joins for mobile environment. Related to the work on spatial joins, Bakalov, Hadjieleftheriou, Keogh, and Tsotras (2005) noted that the need to identify similarities amongst several moving object trajectories, which can be modelled as *trajectory joins* Bakalov et al. (2005) examined issues on performing a *trajectory join* between two datasets, and proposed a technique based on symbolic representation using strings.

Aggregation

Another important type of queries in spatio-temporal databases is aggregation queries. A spatial-temporal aggregation returns a value, with respect to an aggregation function, regarding the data objects in a user-specified query window *qr*, and interval *qt*. Typical aggregation function includes *sum* and *count*. In a *sum* query, each data object is associated with a measure, and the query returns the total of the measures for data objects that fall within *qr* during *qt*. In a *count* query, the total number of objects in a given *qr* during *qt* is computed. It is important to note that value returned by typical aggregation queries are with respect either the current time, or a historical interval of which historical data are kept. In contrast, another interesting type of spatial-temporal queries is range aggregate (RA) queries. A RA query returns the aggregated value for a future timestamp.

In a *count* query, the objects that appear within a given *qr* within *qt* are counted, and the total returned. However, existing approaches that deals with spatial-temporal count queries suffer from the *distinct count* problem (i.e., objects that appear within multiple consecutive timestamps are counted multiple times). Compel by the need to efficiently count the number of *distinct* objects in a given region within a time interval Tao, Kollios, Considine, Li, and Papadias (2004) proposed to perform spatial-temporal aggregation using sketches (Flajolet & Martin, 1985). In addition, a *sketch index* was used for efficient retrieval of the sketches.

Tao, Papadias, Zhai, and Li (2005) tackled issues on approximate RA query processing using a technique called *Venn Sampling*, which provides estimation for a set of *pivot* queries, which reflect the distribution of actual queries. In addition, the notion of a *Venn area* was also introduced. Compared with other sampling approaches (which requires $O(2^m)$ samples), *Venn sampling* was able to achieve perfect estimation using only O(m) samples.

Predictive Queries

When processing spatial data and queries on the move, another important type of queries is *predictive queries*, which are used to predict the future location of the data objects that falls within a query window at a future timestamp. Most existing methods for handling predictive queries use linear function to describe object movements. However, in the real-world, object movements are more complex, ane hence cannot be easily expressed as a linear function of time. Noting this problem, Tao, Faloutsos, Papadias, and Liu (2004) introduces a generic framework for monitoring and indexing moving objects. The notion of a *recursive motion function* was proposed which allows more complex motion patterns to be described. The key idea in *recursive motion function* is to relate an object's location to the objects' recent past locations, instead of its initial location. The spatio-temporal prediction (STP) tree was proposed for efficient processing of predictive queries without false misses.

Sun, Papadias, Tao, and Liu (2004) proposed techniques for answering past, present, and future spatial queries A stochastic approach was adopted for the answering of predictive queries. In addition, the *adaptive multi-dimensional histogram* (AHM) and the *historical synopsis* were introduced for handling approximate query processing of present-time queries, and historical queries respectively. In addition, the authors considered the use of several indices, namely: packed B-tree, 3D R-tree. The *historical synopsis* consists of the AHM containing the currently valid buckets and the *past index*, and is used to answer both historical and present-time queries. Predictive queries on the future are answered by using an exponential smoothing technique which uses both present and the recent past data.

DATA DISSEMINATION

We consider two main types of data dissemination techniques: client-server and data broadcast. Most of the proposed techniques assume a client-server model. Even though in the relational domain, data-dissemination techniques have been widely studied (e.g., broadcast disk), data broadcast for spatial data on the move is only starting to emerge as another promising model for query processing.

In a client-server model (also known as the on-demand model), the mobile device first sends the query to the server, and the server then processes the query, and returns the result to the mobile device. The mobile device is usually treated as a dumb device and most of the processing is done by the server. However, there are works that performs computation (e.g., joins) on the mobile device. The connection between the mobile device and the server is usually one-to-one.

In a data broadcast model, data are broadcast on one or several wireless channels. When a mobile device needs to answer a users' query, it will tune to the appropriate wireless channel, and then retrieve the data that meets the query criteria. The data broadcast model can be further categorized into *broadcast push* and *broadcast pull*. The main difference is that in the *broadcast push* method, the server periodically puts data onto the channel without explicit client requests, and clients would just look for the data they need on the channel. In the pull method, the client explicitly requests for data, and the server then decides the best strategy on which data to be put onto the channel, as well as its repeating frequency. Zheng, Lee, and Lee (2004b) provides a comprehensive discussion on spatial query processing in a wireless data broadcast environment.

Client-Server

One of the key considerations of query processing algorithms in a client-server model is to reduce the amount of data sent to the mobile client. Motivated by the need for more optimal usage of network bandwidth, Mamoulis, Kalnis, Bakiras, and Li (2003) noted that some service providers of spatial data have limited capabilities. In addition, a query issued by mobile users might involve multiple service providers. Hence, there is no single provider that can process all the data, and return the results back to the mobile client. Compelled by this need, Mamoulis et al. (2003) proposed a framework, called *MobiHook*, for handling complex distributed spatial operations on mobile devices. The key idea behind *MobiHook* is to make use of a cheap aggregation queries to find out the overall distribution of the datasets. Based on the additional knowledge, the join algorithm, called *MobiJoin* can then avoid downloading data that might not produce any join results. In addition, Lo, Mamoulis, Cheung, Ho, and Kalnis (2004) considered the issues of performing *ad hoc joins* on mobile devices, namely: (1) Independent data providers, (2) Limited memory on the mobile device, and (3) Need for transfer cost-based optimization. The *recursive and mobile join algorithm (RAMJ)* was proposed to address these issues, and performs the join on the mobile device with data coming from two independent data providers. The key idea in RAMJ is to first obtain statistics of the data to be joined from the data providers, and then selectively download the data to be joined.

MobiEyes, a grid-based distributed system, was proposed in Gedik and Liu (2004) to deal with continuous range queries. *MobiEyes* pushes part of the computation to the mobile clients, and the server is primarily used as a mediator for the mobile clients. The notion of *monitoring regions* of queries was introduced to ensure that objects receive information about the query (e.g., position and velocity). When objects enter or leave the monitoring region, it will notify the server. By using monitoring regions, objects only interact with queries that are relevant, and hence conserve precious resources (i.e., storage and computation).

Yu, Pu, and Koudas (2005) considered the problem of monitoring k-nearest neighbor queries over moving objects. Each NN query that is installed in the system needs to be re-evaluated periodically. To support the evaluation, three grid-based methods were proposed to efficiently monitor the kNN of moving points, namely: (1) object-indexing (single-level), (2) object-indexing (hierarchical), and (3) query-indexing. In object-indexing, the index structure consists of cells, denoted by *(i,j)*. Each cell have an object list, denoted by *PL(i,j)* which contains the identifiers (IDs) of all objects that are enclosed by *(i,j)*. When processing a query q at time t, an initial rectangle R_0, centred at the cell containing q, with size l is identified. The value of l is progressively increased until R_0 contains at least k objects. As the algorithm needs to re-compute the kNNs at each time t, it is also known as the *overhaul* algorithm. When the number of queries is small and the number of objects is relatively larger, then the grid can be used to index the queries instead of the objects (i.e., query-indexing). In addition, to tackle the problems introduced by non-uniform distribution of data objects, the *hierarchical object-indexing*, which uses multi-levels of cells and sub-cells to partition the data space, was also introduced.

Hu, Xu, and Lee (2005) noted the deficiencies in the assumption made by existing works on continuous query monitoring Mokbel et al. (2004), Prabhakar, Xia, Kalashnikov, Aref, and Hambrusch (2002), and Yu et al. (2005), which assumes that the moving client would provide updates on its current location. One of the deficiencies noted is that location updates are query-blind (i.e., the location needs to be updated irregardless on the existence of queries). In addition, it was noted that deviations might exist between the servers and the actual results, since the object's location might have changed in between the updates. Also, synchronization of location updates on the server with multiple moving objects would cause an imbalance in the server node, To address these deficiencies, Hu, Xu, and Lee (2005) proposed a framework for monitoring of spatial queries over moving points. The notion of a server-computed *safe region* is introduced. A *safe region* is a rectangular area around an object which ensures that all queries remain valid as long as the object is within its own safe region. A client updates it location to the server whenever it moves out of the safe region. Thus, using the safe regions, the moving clients become query aware and will only report their location changes when they are likely to alter results, thus greatly reducing unnecessary transmitting of location information to the server.

In Papadias, Mouratidis, and Hadjieleftheriou (2005), *conceptual partitioning* (CPM) was proposed for efficient monitoring of continuous NN queries. The space around each query q is divided into several conceptual partitions (each rectangular in shape), and is associated with a *direction* as well as a *level number*. A *direction* (e.g., Up, Down, Left, and Right) indicates the position of the rectangle with respect to q, and the *level number* indicates the number of rectangles between itself and the query. The role of the conceptual partitions is to restrict the

NN retrieval and efficient result maintenance of objects that are in the neighbourhood of q.

Another important type of queries that seek to optimize the bandwidth used is location-based queries. Mobile devices are increasingly equipped with location-aware mechanism (either via cellular triangulation or GPS signals). Location-based queries are queries that continuously output results based on the user (i.e., mobile device) current location. When the user moves, the results will change. The results to a location-based spatial query are constrained to the region in which the query is posed (i.e., position of the mobile device). When the mobile device moves out of the valid region, the results would change. For example, a user could ask the following query: *Give me the names of the restaurants that are within 200m of my current location*. When the user moves, the results (i.e., names of restaurant) could be different since the user is now in a new position.

When a location-based query is evaluated based on the user's current location, there exists a region around the current location in which the results remain valid. By exploiting the characteristics of this region, redundant processing can thus be avoided. Zhang et al. (2003) introduces the notion of *validity regions* for efficient processing of location-based spatial queries. When the mobile client issues a new query at another location, the validity region belonging to the previous query is then check. If the mobile client is still within the validity region, then the results from the previous query can be re-used, hence avoiding redundant re-computation. In addition, the notion of the *influence object* was introduced.

Data Broadcast

Most existing indices focus on access efficiency (i.e., response time, I/Os). In a static environment, this suffices. However, in a mobile environment, where the mobile devices have limited power availability, we need to optimize power consumption. We consider how indices can be used in a data broadcast environment for efficient data access.

In a wireless broadcast environment, an index called an *air index* is commonly used to facilitate power saving of the mobile devices. A mobile device can make use of the air index to predict the arrival time of the desired data, so that it can reduce power consumption by switching to *doze* mode for the time interval in which there are no desired data objects arriving, and when the desired data arrives, it switches back to an *active* mode. The key to an air index is to interleave the index items with the data objects being broadcast. Imielinski, Viswanathan, and Badrinath (1997) provides a comprehensive discussion on accessing data in a broadcast environment and air indices.

Zheng, Lee, and Lee (2004a) proposed two air indexing techniques for the wireless data broadcast model, namely (1) Hilbert curve air index and (2) R-tree air index. Using the two air indices, Zheng, Lee, and Lee (2004a) shows how they can be used to support continuous nearest neighbor (CNN) queries in a wireless data broadcast environment. Two criteria, *access latency* and *tuning time* are also introduced to evaluate the performance of the indices. Access latency refers to the time the mobile client spent on listening on the broadcast channel and is proportional to the power consumption of the mobile device. If the mobile client is in active mode and continuously listen to the wireless channel for the desired data objects, there would incur significant power usage. Tuning time refers to the time interval between data is requested and data is retrieved.

Sequential access is usually used in a data broadcast environment, where the mobile client is able to retrieve data objects in the channels if they become available. When the mobile client misses a data object, it will have to wait for the next cycle before the desired data object can be retrieved. Thus, a linear way of representing spatial data is needed in order to put the spatial data onto the wireless channel to facilitate such sequential access. A common technique used to reduce multi-dimensional space to a one-dimensional (1D) space is to make use of a space-filling curve (e.g., z-order, Hilbert curve). A space filling curve, such as the Hilbert curve would be able to preserve spatial locality. Hence, an air index can be built based on the Hilbert curve. Thus, a linear index structure based on the Hilbert curve air index was proposed in Zheng, Lee, and Lee (2003).

The Hilbert curve air index can be used to process a window query and a kNN query. In a window query, the Hilbert value for the first and last points corresponding to the query window is first computed. Intuitively, the Hilbert values for the start and end points denote a range. A set of candidate objects can be retrieved, in which their Hilbert values are within the range. A filtering step is then applied to find out the objects that are part of the result set.

In a kNN query, the kNN objects which lies along the Hilbert curve with respect to the query point are first identified, and bounded using a minimal circle centered at the query point. The minimum bounding rectangle (MBR) which bounds the circle is then used as the search range. Due to spatial locality property of the Hilbert curve, the results for the kNN query should be near the query point along the Hilbert curve.

The *distributed spatial index* (DSI) was proposed in Lee and Zheng (2005), which distributes the index information over the entire broadcast cycle. DSI is designed to provide sufficient index information to a mobile client, irregardless of when the client tunes into the channel. The key idea behind DSI is to first divide the data objects into frames, and then associate an *index table* with each frame. The *index table* provides information on the Hilbert curve values of the data objects to be broadcast, and when they would be broadcast.

CONCLUSION AND FUTURE WORK

In this chapter, we presented the issues and challenges in processing *spatial data on the move*. In order to understand the rich variety of query processing algorithms proposed, we presented a framework for understanding and studying the algorithms. We discussed various state-of-art query processing techniques that have been proposed. We also presented data dissemination techniques that are commonly used in such mobile environment. With increased usage of mobile devices, and advancement in networking technology, query processing for *spatial data on the move* is an emerging area, which continuously presents new challenges that must be addressed.

REFERENCES

Arge, L. A., Procopiuc, O., Ramaswamy, S., Suel, T., & Vitter, J. S. (1998, 24-27). Scalable sweeping-based spatial joIn in. *Proceedings of International Conference on Very Large Data Bases (VLDB)* (pp. 570-581).

Bakalov, P., Hadjieleftheriou, M., Keogh, E., & Tsotras, V. J. (2005). Efficient trajectory joins using symbolic representations. In P. K. Chrysanthis, & F. Samaras (Eds.), *Mobile data management*. ACM Press.

Beckmann, N., Kriegel, H. P., Schneider, R., & Seeger, B. (1990). The R*-tree: An efficient and robust access method for points and rectangles. In *Proceedings of the ACM SIGMOD International Conference on Management of Data* (pp. 322-331). New York: ACM Press.

Bohm, C., & Krebs, F. (2004). The nearest neighbor join: Turbo charging the kdd process. *Knowledge of Information Systems*, *6*(6), 728-749.

Brinkhoff, T., Kriegel, H. P., Schneider, R., & Seeger, B. (1994). Multi-step processing of spatial joins. In *Proceedings of the ACM 14 SIGMOD International Conference on Management of Data* (pp. 197-208).

Brinkhoff, T., Kriegel, H. P., & Seeger, B. (1993, May). Efficient processing of spatial joins using R-trees. In *Proceedings of the ACM SIGMOD International Conference on Management of Data*. New York: ACM Press.

Brinkhoff, T., Kriegel, H. P., & Seeger, B. (1996). Parallel processing of spatial joins using R-trees. In *Proceedings of International Conference on Data Engineering.*

Flajolet, P., & Martin, G. N. (1985). Probabilistic counting algorithms for database applications. *Journal of Computer Systems Science*, *31*(2), 182-209.

Gedik, B., & Liu, L. (2004). Mobieyes: Distributed processing of continuously moving queries on moving objects in a mobile system. *Proceedings of International Conference on Extending Database Technology* (pp. 67-87).

Guttman, A. (1984, Aug). R-trees: A dynamic index structure for spatial searching. In *Proceedings of the ACM SIGMOD International Conference on Management of Data.* New York: ACM Press.

Hjaltason, G. R., & Samet, H. (1998). Incremental distance join algorithms for spatial databases. In *Proceedings of the ACM SIGMOD International Conference on Management of Data* (pp. 237-248). New York: ACM Press.

Hjaltason, G. R., & Samet, H. (1999). Distance browsing in spatial databases. *ACM Transactions Database Systems*, *24*(2), 265-318.

Hoel, E. G., & Samet, H. (1992). A qualitative comparison study of data structures for large linear segment databases. In *Proceedings of the ACM SIGMOD International Conference on Management of Data* (pp. 205-214). New York: ACM Press.

Hu, H., Xu, J., & Lee, D. L. (2005). A generic framework for monitoring continuous spatial queries over moving objects. In *Proceedings of the ACM SIGMOD International Conference on Management of Data.* New York: ACM Press.

Huang, Y. W., Jing, N., & Rundensteiner, E. (1997). Spatial joins using R-trees: Breadth-first traversal with global optimizations. In *Proceedings of International Conference on Very Large Data Bases (VLDB)* (pp. 396-405).

Imielinski, T., Viswanathan, S., & Badrinath, B. R. (1997, May-June). Data on air—organization and access. *IEEE Transactions on Knowledge and Data Engineering (TKDE)*, *9*(3), 353-372.

Iwerks, G. S., Samet, H., & Smith, K. (2003). Continuous k-nearest neighbor queries for continuously moving points with updates. In *Proceedings of International Conference on Very Large Data Bases (VLDB)* (pp. 512-523).

Iwerks, G. S., Samet, H., & Smith, K. (2004). Maintenance of spatial semijoin queries on moving points. In *Proceedings of International Conference on Very Large Data Bases (VLDB)* (pp. 828-839).

Kifer, D., Ben-David, S., & Gehrke, J. (2004). Detecting change in data streams. In *Proceed-*

ings of International Conference on Very Large Data Bases (VLDB) (pp. 180-191).

Korn, F., & Muthukrishnan, S. (2000). Influence sets based on reverse nearest neighbor queries. In W. Chen, J. F. Naughton, & P. A. Bernstein (Eds.), *Proceedings of the ACM SIGMOD International Conference on Management of Data* (pp. 201-212). New York: ACM Press.

Korn, F., Sidiropoulos, N., Faloutsos, C., Siegel, E., & Protopapas, Z. (1996). F nearest neighbor search in medical image databases. In *Proceedings of International Conference on Very Large Data Bases (VLDB)* (pp. 215-226).

Lee, C. H., & Chen, M.-S. (2002). Processing distributed mobile queries with interleaved remote mobile joins. *IEEE Trans. Computers, 51*(10), 1182-1195.

Lee, W. C., & Zheng, B. (2005). Dsi: A fully distributed spatial index for wireless data broadcast. In *Proceedings of International Conference o n Data Engineering* (pp. 417-418).

Lo, E., Mamoulis, N., Cheung, D. W., Ho, W. S., & Kalnis, P. (2004). Processing ad-hoc joins on mobile devices. In *Proceedings of International Conference on Database and Expert Systems Applications (DEXA), LNCS* (pp. 611-621).

Lo, M. L., & Ravishankar, C. V. (1994). Spatial joins using seeded trees. In *Proceedings of the ACM SIGMOD International Conference on Management of Data*. New York: ACM Press.

Lo, M. L., & Ravishankar, C. V. (1996, May). Spatial hash-joins. In *Proceedings of the ACM SIGMOD International Conference on Management of Data*. New York: ACM Press.

Mamoulis, N., Kalnis, P., Bakiras, S., & Li, X. (2003). Optimization of spatial joins on mobile devices. In *Proceedings of International Symposium on Advances in Spatial and Temporal Databases* (pp. 233-251).

Mamoulis, N., & Papadias, D. (1999). Integration of spatial join algorithms for joining multiple inputs. In *Proceedings of the ACM SIGMOD International Conference on Management of Data* (pp. 1-12). New York: ACM Press.

Mokbel, M. F., & Aref, W. G. (2005). GPAC: Generic and progressive processing of mobile queries over mobile data. In P. K. Chysanthis & F. Samaras (Eds.), *Mobile data management*. ACM Press.

Mokbel, M. F., Xiong, X., & Aref, W. G. (2004). SINA: Scalable incremental processing of continuous queries in spatio-temporal databases. In *Proceedings of the ACM SIGMOD International Conference on Management of Data* (pp. 623-634). New York: ACM Press.

Nelson, R. C., & Samet, H. (1987). A population analysis for hierarchical data structures. In U. Dayal, & I. L. Traiger (Eds.), *Proceedings of the ACM SIGMOD International Conference on Management of Data* (pp. 270-277). New York: ACM Press.

Papadias, D., Mouratidis, K., & Hadjieleftheriou, M. (2005). Conceptual partitioning: An efficient method for continuous nearest neighbor monitoring. In *Proceedings of the ACM SIGMOD International Conference on Management of Data*. New York: ACM Press.

Papadias, D., Tao, Y., Kalnis, P., & Zhang, J. (2002). Indexing spatio-temporal data warehouses. In *Proceedings of International Conference on Data Engineering* (pp. 166-175).

Patel, J. M., & DeWitt, D. J. (1996, May). Partition based spatial-merge join. In *Proceedings of the ACM SIGMOD International Conference on Management of Data*. New York: ACM Press.

Pfoser, D., Jensen, C. S., & Theodoridis, Y. (2000). Novel approaches in query processing for moving object trajectories. In *Proceedings of International Conference on Very Large Data Bases (VLDB)* (pp. 395-406). Morgan Kaufmann.

Prabhakar, S., Xia, Y., Kalashnikov, D., Aref, W., & Hambrusch, S. (2002, October). Query indexing and velocity constrained indexing: Scalable techniques for continuous queries on moving objects. *IEEE Transactions on Computers, 51*(10), 1124-1140.

Roussopoulos, N., Kelley, S., & Vincent, F. (1995). Nearest neighbor queries. In M. J. Carey, & D. A. Schneider (Eds.), *Proceedings of the 15th ACM SIGMOD International Conference on Management of Data* (pp. 71-79). ACM Press.

Saltenis, S., & Jensen, C. S. (2002). Indexing of moving objects for location-based services. In *Proceedings of International Conference on Data Engineering* (pp. 463-472).

Saltenis, S., Jensen, C. S., Leutenegger, S. T., & Lopez, M. A. (2000). Indexing the positions of continuously moving objects. In *Proceedings of the ACM SIGMOD International Conference on Management of Data* (pp. 331-342). New York: ACM Press.

Sellis, T., Roussopoulos, N., & Faloutsos, C. (1987). R+-tree: A dynamic index for multidimensional objects. In *Proceedings of International Conference on Very Large Data Bases (VLDB)*.

Smid, M. (2000). Closest-point problems in computational geometry. In J. R. Sack, & J. Urrutia (Eds.), *Handbook of computational geometry* (pp. 877-935). Amsterdam: Elsevier Science Publishers B. V. North-Holland.

Song, Z., & Roussopoulos, N. (2001). K-nearest neighbor search for moving query point. In *Proceedings of International Symposium on Advances in Spatial and Temporal Databases* (pp. 79-96). London: Springer-Verlag.

Stanoi, I., Agrawal, D., & El Abbadi, A. (2000, May). Reverse nearest neighbor queries for dynamic databases. In D. Gunopulos, & R. Rastogi (Eds.), *Proceedings ACM SIGMOD Workshop on Research Issues in Data Mining and Knowledge Discovery,* Dallas, TX (pp. 44-53).

Sun, J., Papadias, D., Tao, Y., & Liu, B. (2004). Querying about the past, the present, and the future in spatio-temporal. In *Proceedings of International Conference on Data Engineering* (pp. 202-213).

Tao, Y., Faloutsos, C., Papadias, D., & Liu, B. (2004). Prediction and indexing of moving objects with unknown motion patterns. In *Proceedings of the ACM SIGMOD International Conference on Management of Data* (pp. 611–622). New York: ACM Press.

Tao, Y., Kollios, G., Considine, J., Li, F., & Papadias, D. (2004). Spatio-temporal aggregation using sketches. In *Proceedings of International Conference on Data Engineering* (pp. 214-226).

Tao, Y., & Papadias, D. (2003). Spatial queries in dynamic environments. *ACM Transaction Database System, 28*(2), 101-139.

Tao, Y., Papadias, D., & Shen, Q. (2002). Continuous nearest neighbor search. In *Proceedings of International Conference on Very Large Data Bases (VLDB)* (pp. 287-298).

Tao, Y., Papadias, D., & Sun, J. (2003). The TPR* tree: An optimized spatio-temporal access method for predictive queries. In *Proceedings of International Conference on Very Large Data Bases (VLDB)*.

Tao, Y., Papadias, D., Zhai, J., & Li, Q. (2005). Venn sampling: A novel prediction technique for moving objects. In *Proceedings of International Conference on Data Engineering.*

Xiong, X., Mokbel, M. F., & Aref, W. G. (2005). SEA-CNN: Scalable processing of continuous k-nearest neighbor queries in spatiotemporal databases. In *Proceedings of International Conference on Data Engineering* (pp. 643-654).

Xiong, X., Mokbel, M. F., Aref, W. G., Hambrusch, S. E., & Prabhakar, S. (2004). Scalable spatio-temporal continuous query processing for location-aware services. In *Proceedings of the International Conference on Scientific and Statistical Database Management* (pp. 317-326).

Yu, X., Pu, K. Q., & Koudas, N. (2005). Monitoring k-nearest neighbor queries over moving objects. In *Proceedings of International Conference on Data Engineering* (pp. 631-642).

Zhang, J., Zhu, M., Papadias, D., Tao, Y., & Lee, D. L. (2003). Location-based spatial queries. In *Proceedings of the ACM SIGMOD International Conference on Management of Data* (pp. 443-454). New York: ACM Press.

Zheng, B., Lee, W. C., & Lee, D. L. (2003). Spatial index on air. In *Proceedings of the 1st IEEE International Conference on Pervasive Computing and Communications (PERCOM)* (pp. 297). Washington, DC: IEEE Computer Society.

Zheng, B., Lee, W. C., & Lee, D. L. (2004a). Search continuous nearest neighbors on the air. In *MobiQuitous '04: Proceedings of the 1st International Conference on Mobile and Ubiquitous Systems: Networking and Services* (pp. 236-245).

Zheng, B., Lee, W. C., & Lee, D. L. (2004b). Spatial queries in wireless broadcast systems. *Wireless Networks, 10*(6), 723-736.

KEY TERMS

Aggregation: An aggregation is an operation in databases which returns a summarized value, with respect to an aggregation function. Examples of aggregation function includes sum and count.

Continuous Spatial Queries: Continuous spatial queries are queries that are installed once in a system, and executed over an extended period of time against spatial datasets.

Hilbert Curve: A Hilbert curve is part of the family of plane-filling curve. It is commonly used to transform multi-dimensional data to a single dimension.

Histogram: A histogram maintains statistics on the frequency of the data.

Location-Aware Applications: Location-aware applications refer to a class of applications which are unable to recognize and react to the location the user is currently in. The results of the queries changes as the user moves.

Nearest Neighbor (NN) Queries/k-Nearest Neighbor (kNN) Queries: A kNN query retrieves the k nearest data object with respect to a query object. When k = 1, it is called a NN query.

Spatial Join: A spatial join query finds all object pairs from two data sets that satisfy a spatial predicate. A common spatial predicate used in a spatial join is intersection.

Spatio-Temporal Databases: Spatio-temporal databases deal with objects that change their location and/or shape over time.

Chapter IX
Key Attributes and the Use of Advanced Mobile Services:
Lessons Learned from a Field Study

Jennifer Blechar
University of Oslo, Norway

Ioanna D. Constantiou
Copenhagen Business School, Denmark

Jan Damsgaard
Copenhagen Business School, Denmark

ABSTRACT

Advanced mobile service use and adoption remains low in most of the Western world despite impressive technological developments. Much effort has thus been placed on better understanding the behavior of advanced mobile service users. Previous research efforts have identified several key attributes deemed to provide indications of the behavior of consumers in the m-services market. This chapter continues with this line of research by further exploring these key attributes of new mobile services. Through a field study of new mobile service use by 36 Danish mobile phone users, this chapter illustrates the manner in which users' perceptions related to the key attributes of service quality, content-device fit and personalization were adversely affected after approximately three months of trial of the services offered.

INTRODUCTION

Investments in mobile multimedia technologies and services continue to increase. Yet, as has been illustrated in the past, market success does not always follow positive technological gains (Baldi & Thaung, 2002; Funk, 2001). For example, even though the quality and proliferation of mobile phones with photographing capabilities remains on the rise, adoption and use of mobile multimedia messaging services (MMS) continues to dwindle among mobile phone users in Western countries. As investments in mobile applications and services continue, it thus becomes increasingly important to better understand the process whereby users either accept or reject the use of new technology in the mobile arena.

Much research effort has been undertaken on the study of technology acceptance and use over the last two decades. Of primary concern in many existing models and theories related to technology acceptance, such as the diffusion of innovations theory (Rogers, 1983), the technology acceptance model (TAM) (Davis, 1989) and the theory of reasoned action (TRA) (Ajzen & Fishbein, 1980), is the identification of specific elements or factors which are seen to impact individuals' or aggregate group intentions to adopt and use a new technology. As research on the acceptance and use of new multimedia technologies has progressed, emphasis has also been placed on the identification of key attributes deemed to drive consumer behavior related to m-service actions (see Vrechopoulos, Constantiou, Mylonopoulos, & Sideris, 2002)

Through a field study of new mobile service use by 36 Danish mobile phone users, this chapter illustrates the manner in which users' perceptions of some key attributes of new mobile services offered has changed after approximately three months of use. These key attributes have been found to relate to the actual behavior of consumers in the m-service market (Vrechopoulos et al., 2002). In this study we obtain a better understanding of how users' perceptions of these attributes may change during initial technology trial thus providing a more rounded picture of the m-services market. In addition, increased knowledge regarding user perceptions of key m-service attributes offers useful insights related to the manner in which new mobile services should be released and promoted to consumers in the market. The next section of this chapter includes background information on the key attributes and existing related research in the m-service arena. This is followed by an introduction to the field study and a discussion of the results. The conclusions are then presented, summarizing the main findings of this chapter.

LITERATURE INSIGHTS

Many studies have been conducted in various settings in order to investigate the use and uptake of new technology including advanced mobile services. This includes studies rooted in the domains of technology acceptance (Ajzen, 1985, 1991; Davis, 1989; Taylor & Todd, 1995; Venkatesh, Morris, Davis, & Davis, 2003), diffusion of innovations (Rogers, 1995), Domestication (Ling & Haddon, 2001; Pedersen & Ling, 2003; Silverstone & Haddon, 1996), and various studies conducted from the industry perspective (Sharma & Nakamura, 2004). Several perspectives have thus been proposed related to the factors or elements influencing successful adoption of new technologies, ranging from perceptions of technological characteristics such as *ease of use* or *perceived usefulness* (e.g., Davis, 1989), to social factors such as age or gender (e.g., Ling, 2004).

Through the work of Vrechopoulos, Constantiou, Sideris, Doukidis, and Mylonopoulos, (2003) key attributes influencing consumers behavior related to the acceptance and use of new mobile services have been identified. The attributes that were found to be the most significant influences for consumer behavior included:

- Ease of use interface
- Security
- Service quality
- Price

- Personalization
- Content-device fit

These key attributes of m-service acceptance and use have also been explored by other researchers over the last few years. In particular, *ease of use interface* has been underlined by Massoud and Gupta (2003) in the analysis of consumer's perceptions and attitudes to mobile communications and the role of *security* has been highlighted by Andreou et al (2005), Bai, Chou, Yen, & Lin (2005) and Massoud and Gupta (2003). These efforts have also pointed to the design of mobile services whereby the above work has indicated that consumers perceived design to be of low importance. Moreover, *quality of services* has been investigated in the context of mobile multimedia services (Andreou et al., 2005), as well as *pricing* of mobile services which is also underlined by Bai et al (2005). Finally, mobile services *personalization* (Bai et al., 2005) has been explored as well as *content-device fit* both in terms of usability (AlShaali & Varshney, 2005) and in terms of mobile service's design (Chen, Zhang, & Zhou, 2005; Schewe, Kinshuk, & Tiong, 2004).

While many elements have been proposed in the literature related to acceptance and use of new technology including mobile services as mentioned above, the key attributes proposed by Vrechopoulos et al. (2002; Vrechopoulos et al., 2003) encompass both elements of user's cognitive processes (for example, related to pricing decisions) and elements of the technology (such as the security). Thus, we believe these attributes are beneficial in order to investigate the overall process of technology acceptance and use of m-services. While most existing literature has explored these key attributes in a static manner (e.g., via a one time online survey), this chapter investigates how users' perceptions of these attributes may change over time through exposure and trial of new mobile services.

THE FIELD STUDY

In a period of three months from November 2004 to March 2005, 36 Danish consumers were provided with state-of-the-art mobile phones with pre-paid SIM cards granting access to a variety of advanced mobile services. These included services under service categories such as directories, dating, messaging, downloading of content, and news. Participants could use the pre-paid amount of approximately 35 euros per month as they wished (e.g., for voice, SMS, MMS, and use of the advanced data services). During the project period, participants' use of the mobile phones and services was monitored and their feedback was gathered through a variety of means including surveys, focus groups, and interviews. Surveys ranged in focus from the initial survey gathering demographic information to the final survey which gathered participants overall perceptions and attitudes of the project, phones and services offered. Questions on the survey were both of qualitative (e.g., open-ended) and quantitative (e.g., fixed response) nature. The results presented in this chapter are based on the quantitative data gathered through these surveys.

In order to explore participants behaviors related to the acceptance and use of the advanced mobile services offered, participants were queried based on the six key attributes identified in previous research (among other items), both at the onset of the project and once the project was completed. This allowed for a comparison of these attributes and the potential changes in user perspectives prior to trial of the services offered and after users gained first hand experience with those services. Partici-

Table 1. Questions related to the key attributes explored with participants

Indicate to what extent you agree or disagree that mobile services are:
• Complicated to use
• Lack security
• Have poor service quality
• Are too expensive
• Are not adequately personalized
• Are not adequately fitted to mobile use (because of small screen, typing possibilities etc.)

pants responded to questions on a five point scale where 1 = disagree completely and 5 = agree completely (see table 1 for the queries posed to participants).

Indicate to what extent you agree or disagree that mobile services are:

- Complicated to use
- Lack security
- Have poor service quality
- Are too expensive
- Are not adequately personalized
- Are not adequately fitted to mobile use (because of small screen, typing possibilities etc.)

In addition, participants were further queried regarding their feelings related to each of the specific service categories available. As such, series of questions related to the key attributes were explored in further detail. These questions explored the derived value from each of the services, the assessment of content available and participants general intentions to continue to use the services in the future. They were distributed to participant's mid-trial of the services and allowed for responses on the same five-point scale used for the key attributes (see Table 3 for the questions related to the results presented in this chapter).

The Hypothesis

To investigate participants perceptions related to the key attributes of the new mobile services and whether they have changed after actual use of the mobile services offered, we test the following hypothesis:

$H_{0:}$ The participants' perceptions of the key attributes of the new mobile services do not significantly differ before and after trial of the mobile services.

MAIN FINDINGS

Upon exploring the proposed key attributes of the new mobile services by performing pair wise t-tests of data before and after trial of the services, our research indicates that there are significant differences in participants' perceptions related to service quality, personalization and content-device fit (see Table 2). In particular, after trial of the new mobile services, participants perceived the services to be of lower quality as compared to prior to trial. They also indicated that the services lacked personalization and that the content lacked the desired fit with the device.

According to Table 2 the largest differences appear in the case of service quality and con-

Table 2. Pair wise t-tests of key attributes before and after trial

	Mean * Before Trial	Mean After Trial	Means Difference	t-test	p value	Hypothesis Test
Complicate to use	2.62	3.04	-0.42	-1.62	$p>0.05$	Cannot reject H_0
Security	2.42	2.85	-0.42	-1.30	$p>0.05$	Cannot reject H_0
Service quality	2.69	4.00	-1.31	-3.48	$p<0.05$	Reject H_0
Price	3.38	3.96	-0.58	-1.36	$p>0.05$	Cannot Reject H_0
Personalization	2.88	3.81	-0.92	-3.04	$p<0.05$	Reject H_0
Content-device fit	3.04	4.35	-1.31	-3.69	$p<0.05$	Reject H_0

**Means from 1: Strongly Disagree to 5: Strongly Agree*

tent-device fit. For content-device fit, it seems that participants were quite dissatisfied with the use of the m-services provided after trial. This is an underlining issue that has been prominent since the launch of advanced mobile services. The relatively small screen size and the abundance of information displayed through a mobile portal have been highlighted as a potential obstacle for adoption and use (Vrechopoulos et al., 2002). This is also a challenge as the companies responsible for content provisioning can have little to no influence on the devices themselves, and vice versa. It appears that in our study, content-device fit remains an issue for the participants.

In the case of mobile service quality, at the onset of the project participants indicated that they did not agree with the statement that new mobile services have poor quality of service. Yet after trial, participants' perceptions changed. They agreed with this statement indicating that they were disappointed with the quality of the services offered after they had experienced those services first hand. Thus, whereas participants had generally positive expectations related to the quality of the services, these expectations were not met once the service were actually tried and experienced by them. This may relate to the difficulties encountered by mobile operators to serve data traffic on GPRS networks where priority is given to voice services and many times mobile users experience slow network services that directly reflect to their use of content services (e.g., low speed when downloading content). Regardless, this is clearly not positive results for mobile operators offering such services.

Similarly, participants perceptions related to the personalization of the services were generally positive prior to trial, where participants were primarily neutral with the statement that the services would not be adequately personalized. Yet, after trial, they agreed with this statement indicating again that their fairly positive perceptions of the services before trial changed to negative after trial. This is a key issue since mobile services are not yet customized to cover specific needs of mobile users. By combining this attribute with pricing considerations, which remained important and did not change significantly after trial, it seems that mobile users cannot see the value of new mobile services and consequently are not willing to pay in order to use them.

However, there were no significant differences in participants' perceptions on ease of use interface or security. The first observation may relate to the relatively high technical knowledge of participants in our study and the second to the fact that they were not using advanced services that required online transactions or revelation of sensitive private information where the role of security might be more prominent.

In order to obtain a better understanding of participants' perceptions related to the key attributes which differed significantly prior to and post trial (service quality, personalization, and content device fit), we turn to investigate the participants' assessment of certain aspects related to some of the specific categories of services offered.

According to Table 3 we can further explore participants perceptions of the services offered related to the key attributes. Indications related to the service quality attribute are provided by participants perceptions of the quality of each of the services offered. In relation to the overall feelings of the services available through the portal, participants' responses were somewhat neutral. In particular, participants perceptions of the actual quality of the downloadable content and search and find services were relatively neutral, while many participants indicated their general dissatisfaction with the speed of the portal itself. The latter might have further driven their low satisfaction levels of the quality of the services.

In terms of personalization, we obtained indications that participants could not clearly see the value of the services in their everyday life. Here participants provided neutral assessments of the value of the services under the categories of events and messaging but neutral to positive for news and search and find. This indicates that some services did not meet the needs of the participants in our study and that these services need to be better customized in order to be appreciated as valuable in everyday life. Without such added value, these types of services will simply not be accepted and used by consumers. In addition, participants' general evaluation of the mobile portal and the fit of this portal to their service preferences also remained low, while participants generally agreed that the portal would need to be better adapted to their needs.

Table 3. Participants perceptions on the services' attributes

Service Quality	***Mean****
In general, the downloadable contents offer good value	2.54
The downloadable content has good quality	2.58
The search and find services have good quality	2.64
The portal and services download with sufficient speed	2.03
There are many good services available over <the mobile portal>	2.84
Personalization	
<The mobile portal> needs to be better adapted to fit my service preferences	3.83
The news services provide everyday value	3.00
In general, I feel that the message services provide everyday value	2.69
The search and find services provide everyday value	3.21
The events services provide everyday value	2.53
Content-Device fit	
The news services are well adjusted to the mobile phone	3.21
The graphical layout of <the mobile portal> is attractive	2.97

**Means from 1: Strongly Disagree to 5: Strongly Agree*

Finally, in terms of content-device fit, where participants expressed their increased dissatisfaction after trial of the services, we observe that the current mobile portal layout and the services adjustment to mobile devices are relatively neutral for participants. Thus by focusing on the most popular content service category such as news we observe moderate rating indicating that many participants felt that the content of the service was not necessarily very well adjusted to the device itself. In addition, participants indicated that they were neutral on whether the actual layout of the portal (containing access to all the services offered) was well suited for their needs.

CONCLUSION

This chapter has explored several key attributes of mobile services which have been previously identified as providing indications related to consumers behavior in the m-services market. Through a field study of mobile service use in Denmark, indications have been provided that despite fairly neutral initial perceptions related to the key attributes of ease of use, security, service quality, price, personalization and content-device fit, participants were dissatisfied with the quality of the services, personalization and content-device fit after trial of the services offered. This is the opposite of traditional wisdom, where the main challenge is typically to ensure people sample a service or new technology. Here it is the adverse that appears to be true; as long as consumers have not experienced the new services their expectations and attitudes are fairly positive. These are surprising results as despite the impressive technological evolution of mobile communications markets it seems that some of the key attributes affecting consumers' adoption and use identified at the initial stage of service evolution are still prominent issues.

It seems that mobile operators and other stakeholders have not focused their development efforts to consumer needs but have rather kept on investing on new technologies. In particular, the need for better adjustment of mobile content to available devices, the demand for higher service quality and personalization of services have been highlighted in this chapter. Without better adjustment of mobile services to the needs of consumers, mobile operators and service providers run the risk of non-adoption of services offered and a lack of return on their investments. Given the high infrastructural investments made by many operators in European countries for the third generation of mobile telephony, this indicates that mobile operators must adjust their service provisioning strategies if they are to regain their investments. The naturally raising question is why key players do not react to the repeated calls for addressing consumer needs and requirements?

ACKNOWLEDGMENT

This research was conducted as part of the Mobiconomy project at Copenhagen Business School (www.mobiconomy.dk). Mobiconomy is partially supported by the Danish Research Agency, grant number 2054-03-0004.

REFERENCES

Ajzen, I. (1985). From intentions to actions: A theory of planned behavior. In J. Kuhl & J. Beckman (Eds.), *Action control: From cognition to behavior*. New York: Springer.

Ajzen, I. (1991). The theory of planned behavior. *Organizational Behavior and Human Decision Processes, 50*(2), 179-211.

Ajzen, I., & Fishbein, M. (1980). *Understanding attitudes and predicting social behavior*. Englewood Cliffs, NJ: Prentice Hall.

AlShaali, S., & Varshney, U. (2005). On the usability of mobile commerce. *International Journal of Mobile Communications, 3*(1), 29-37.

Andreou, A. S., Leonidou, C., Pitisillides, A., Samaras, G., Schizas, C. N., & Mavromoustakos, S. M. (2005). Key issues for the design and development of mobile commerce and applications. *International Journal of Mobile Communications, 3*(3), 303-323.

Bai, L., Chou, D. C., Yen, D. C., & Lin, B. (2005). Mobile commerce: Its market analyses. *International Journal of Mobile Communications, 3*(1), 66-81.

Baldi, S., & Thaung, H. P. P. (2002). The entertaining way to m-commerce: Japan's approach to the mobile internet—A model for Europe. *Electronic Markets, 12*(1), 6-13.

Chen, M., Zhang, D., & Zhou, L. (2005). Providing Web services to mobile users: The architecture of an m-service portal. *International Journal of Mobile Communications, 3*(1), 1-18.

Davis, F. D. (1989). Perceived usefulness, perceived ease of use, and user acceptance of information technology. *MIS Quarterly, 13*(3), 319-339.

Funk, J. (2001). *The mobile Internet: How Japan dialed up and the west disconnected.* Kent, UK: ISI Publications.

Ling, R. (2004). *The mobile connection. The cell phone's impact on society* (3rd ed.). San Francisco: Morgan Kaufmann Publishers.

Ling, R., & Haddon, L. (2001, April 18-19, 2001). *Mobile telephony, mobility, and the coordination of everyday life*. Paper presented at the "Machines That Become Us", Rutgers University, New Jersey.

Massoud, S., & Gupta, O. K. (2003). Consumer perception and attitude toward mobile communications. *International Journal of Mobile Communications, 1*(4), 390-408.

Pedersen, P. E., & Ling, R. (2003). *Modifying adoption research for mobile internet service adoption: Cross-disciplinary interactions.* Paper presented at the 36th Hawaii International Conference on Systems Science (HICSS), Big Island, Hawaii.

Rogers, E. M. (1983). *Diffusion of innovations* (3rd ed.). New York: The Free Press.

Rogers, E. M. (1995). *Diffusion of innovations* (4th ed.). New York: The Free Press.

Schewe, K. D., Kinshuk, & Tiong, G. (2004). Content adaptivity in wireless Web access. *International Journal of Mobile Communications, 2*(3), 260-270.

Sharma, C., & Nakamura, Y. (2004, January). The DoCoMo Mojo. *J@pan Inc, 51*, 44-49. LINC Media, Inc.

Silverstone, R., & Haddon, L. (1996). Design and the domestication of information and communication technologies: Technical change and everyday life. In R. Mansell, & R. Silverstone (Eds.), *Communication by design* (pp. 44-74). Oxford: Oxford University Press.

Taylor, S., & Todd, P. A. (1995). Understanding information technology usage: A test of competing models. *Information Systems Research, 6*(2), 144-176.

Venkatesh, V., Morris, M., Davis, G. B., & Davis, F. D. (2003). User acceptance of information technology: Towards a unified view. *MIS Quarterly, 27*(3), 425-478.

Vrechopoulos, A. P., Constantiou, I. D., Mylonopoulos, N., & Sideris, I. (2002). Critical success factors for accelerating mobile commerce diffusion in Europe. Paper presented at the *Proceedings of 15th Bled E-Commerce Conference, e-Reality: Constructing the e-Economy*, Bled, Slovenia.

Vrechopoulos, A. P., Constantiou, I. D., Sideris, I., Doukidis, G., & Mylonopoulos, N. (2003). The critical role of consumer behavior research in mobile commerce. *International Journal of Mobile Communications, 1*(3), 329-340.

KEY TERMS

3G: Specification for the third generation of mobile communications technology that promises increased bandwidth, up to 384 Kbps.

Advanced Mobile Services: A general term describing data and media rich mobile services such as the downloading of music or video.

GPRS: Often referred to as the 2.5 generation of mobile telephony, GPRS is a packet-based wireless communication service running on the GSM network with data rates from 56 up to 114 Kbps.

Technology Acceptance: A body of research which investigates how new technology is adopted by consumers, focusing on key constructs said to influence, directly or indirectly, intentions, and attitudes towards technology adoption.

Section II

Standards and Protocols

The key feature of mobile multimedia is to combine the Internet, telephones, and broadcast media into a single device. Section II, which consists of eight chapters, explains the enabling technologies for mobile multimedia with respect to communication networking protocols and standards.

Chapter X
New Internet Protocols for Multimedia Transmission

Michael Welzl
University of Innsbruck, Austria

ABSTRACT

This chapter will introduce three new IETF transport layer protocols in support of multimedia data transmission and discuss their usage. First, the stream control transmission protocol (SCTP) will be described; this protocol was originally designed for telephony signaling across the Internet, but it is in fact broadly applicable. Second, UDP-Lite (an even simpler UDP) will be explained; this is an example of a small protocol change that opened a large can of worms. The chapter concludes with an overview of the datagram congestion control protocol (DCCP), a newly devised IETF protocol for the transmission of unreliable (typically real-time multimedia) data streams.

INTRODUCTION

For decades, two transport layer protocols of the TCP/IP suite were almost exclusively used: TCP and UDP. The services that these protocols provide are entirely different, and easy to grasp: while the latter simply makes the "best effort" service of the Internet accessible to applications, TCP reliably transfers a stream of bytes across the network. UDP only has port numbers that make it possible to distinguish between several communicating entities which share the same IP address and a checksum that ensures data integrity, but TCP encompasses a large number of additional functions:

- **Stream-based in-order delivery:** Packets are ordered according to sequence numbers, and only consecutive bytes are delivered
- **Reliability:** Missing packets are detected and retransmitted

- **Flow control:** The receiver is protected against overload with a sliding window scheme
- **Congestion control:** The network is protected against overload by appropriately limiting the window of the sender
- **Connection handling:** Since TCP is a connection oriented protocol, it must have the ability to explicitly set up and tear down connections
- **Full-duplex communication:** An acknowledgment (ACK) can also carry user data; this is usually referred to as "piggybacking"

The importance of these mechanisms varies. A protocol could, for instance, easily do without the full-duplex communication capability; on the other hand, some form of end-to-end congestion control has been identified as an indispensable element of any protocol that is to be used on the Internet (Floyd & Fall, 1999). This does however not mean that there is only one way to carry out congestion control: TCP uses an "additive increase, multiplicative decrease" strategy which essentially probes for the available bandwidth by linearly increasing the rate until a limit is hit (causing a packet to be dropped or a congestion signaling bit to be set), whereupon the rate is reduced by half. There are proposals for congestion control that is fair towards TCP ("TCP-friendly") yet more suitable for multimedia applications because the rate fluctuations are less severe. One notable example is "TCP-friendly rate control (TFRC)" (Floyd, Handley, Padhye, & Widmer, 2000).

TCP does not provide the flexibility that today's applications need: it is neither possible to disable any of its aforementioned functions (in particular reliability, which adds delay but is typically not needed by real-time multimedia applications), nor can a user change the way they work (e.g., influence how congestion control is carried out). UDP, on the other hand, allows for more flexibility, but its feature set is so small that any additional protocol function must be implemented directly within the application that uses it. Sometimes, this is unacceptable — realizing TCP-friendly congestion control, for instance, is difficult, and may not be worth the effort from the perspective of a single application designer. Indeed, even the popular streaming media applications "RealPlayer" and "Windows Media Player" do not appear to properly adapt their rate in response to congestion (Hessler & Welzl, 2005).

In this chapter, we will take a look at three novel IETF protocols that change this situation somewhat: the "stream control transmission protocol (SCTP)," "UDP-Lite," and the "datagram congestion control protocol (DCCP)." While SCTP could also be regarded as some sort of a "TCP++," these three protocols share one notable property: they can emulate the behavior of TCP (or UDP, in the case of UDP-Lite), but with *less* features. The ability to effectively *disable* TCP features is therefore a feature in itself; this gives new meaning to the saying "*less is more*." Historically, SCTP is by far the oldest of these protocols; its main specification (Stewart et al., 2000) was published in 2000, and it is now going through the difficult post-standardization phase of achieving large-scale Internet deployment. Notably, the IETF recommends this protocol for authentication, authorization, and accounting (AAA) in any future IP service networks, and SCTP has been required by the 3rd Generation Partnership Project (3GPP) (Stewart & Xie, 2002, p. 17). UDP-Lite was recently published as a "Proposed Standard" — the same status as SCTP — by the IETF (Larzon, Degermark, Pink, Jonsson, & Fairhurst, 2004), and DCCP has not even reached this status yet; at the time of writing, its specification (Kohler, Handley, & Floyd, 2005) was still an Internet-draft, which is a preliminary type of IETF document. The protocol can be expected to become a Proposed Standard RFC in the near future, and its impact could then become quite significant.

THE STREAM CONTROL TRANSMISSION PROTOCOL (SCTP)

SCTP is the result of an effort to develop an efficient Internet transport protocol for telephony signaling. As such, its features are not directly related to the transmission of multimedia data; it was however understood that it is a protocol of broad use, and SCTP can certainly be advantageous for mobile multimedia if the data are suitable and the protocol is used in an intelligent manner. This is because delay is always an important issue for real-time multimedia applications, and reduced delay is exactly what SCTP can give you. In what follows, we will take a closer look at its main features.

Reliable Out-Of-Order but Potentially Faster Data Delivery

TCP suffers from a problem that is called "head-of-line blocking delay": when packets 0, 2, 3, 4, and 5 reach a TCP receiver, the data contained in packets 2 to 5 will not be delivered to the application until packet 1 arrives. This effect is caused by the requirement to deliver data in order. By allowing applications to relax this constraint (this is reasonable for telephony signaling), SCTP can deliver data faster while providing the reliability that UDP lacks.

Preservation of Message Boundaries

Faster delivery of out-of-order packets is only possible if the data blocks can be clearly identified by the protocol. In other words, embedding such a function in a TCP receiver would not be possible because of its byte stream-oriented nature. Moreover, giving the application the power to control the elementary data units that are transferred ("application layer framing (ALF)") can yield more efficient programs (Clark & Tennenhouse, 1990). This is shown in Figure 1. Here, four application chunks are transmitted in four packets. Without ALF, it is possible that just a couple of bytes from chunk 2 end up in packet 1; if packet 2 (which contains the rest of chunk 2) is lost, however, these bytes are of no use at the receiver until the retransmitted packet 2 arrives. Similarly, the loss of packet 2 can affect chunk 3, rendering the correctly received packet 3 useless until the retransmitted packet 2 arrives.

Efficiently choosing the size of packets as a function of the application chunk size does of course not mean that packets have to be exactly as large as chunks — the same advantage can be gained if the packet size is an integral multiple of the chunk size or vice versa.

Figure 1. An inefficient choice of packet sizes

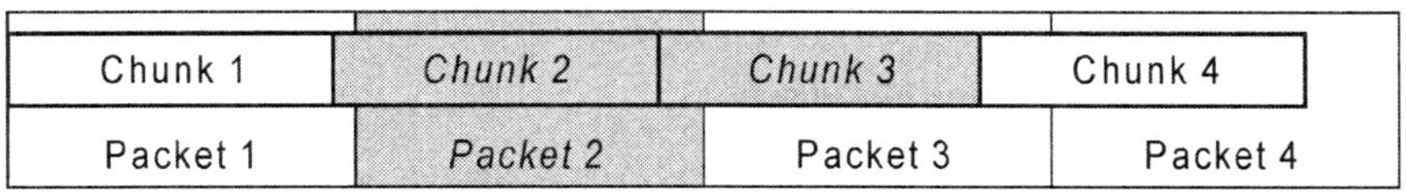

Support for Multiple Separate Data Streams

Sometimes, an application may have to transfer more than one logical data stream. Mapping multiple data streams onto a single TCP connection requires some effort from an application and can be inefficient. Figure 2 shows an example scenario where packets are reordered inside the network (this is indicated via the bold numbers underneath the TCP sender and receiver buffers, which represent TCP sequence numbers). Clearly, even when the streams themselves call for in-order data delivery, this is not necessarily the case for segments that belong to different streams, and head-of-line blocking delay can occur — in the figure, chunk 2 from application stream 1 can only be delivered when chunk 1 from the otherwise unrelated application stream 2 arrives. This problem is eliminated by the multiple stream support feature of SCTP.

Another common solution to this problem is to simply use multiple TCP connections for multiple application streams, but this also means that connection setup and teardown are carried out several times (thereby adding network traffic and increasing delay), and that congestion control is independently executed for each connection, rendering the total behavior of the source more aggressive than it should be (Balakrishnan, Rahul, & Seshan, 1999).

Multihoming

While TCP connections are uniquely identified via two IP addresses and two port numbers, SCTP connections are identified via two sets of IP addresses and two port numbers, and they are actually called "associations" instead of "connections." Multihoming at the transport layer is a powerful concept; it can enable an application to switch from one IP address to another when the communication fails without even noticing it. From the perspective of an application, the transport layer simply becomes more robust when multiple IP addresses are used for an association endpoint. The possible failure is not limited to the machine at the other end—SCTP can also switch when the communication flow is interrupted because of a problem inside the network. This can be used to shorten the time it takes for the network to "repair" an error (e.g., bypassing a failed link—since routing updates are typically sent every 30 seconds, the convergence time of Internet

Figure 2. Transmitting two data streams over one TCP connection

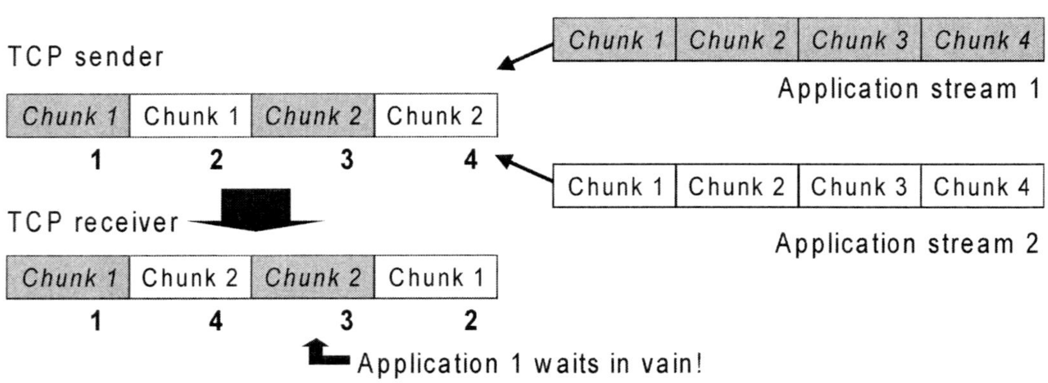

routing protocols can be quite long); SCTP can switch to an alternate address in the meantime and switch back when the problem has been solved. Multihoming may be particularly useful for mobile applications of any kind, where significant handover delays are known to be a common problem.

Partial Reliability

This feature, which was recently added to SCTP in a separate document (Stewart, Ramalho, Xie, Tuexen, & Conrad, 2004), makes it possible for an application to specify how persistent the protocol should be in attempting to deliver a message, including totally unreliable data transfer. This allows for multiplexing of unreliable and reliable data streams across the same connection; the ability to unreliably transfer data with congestion control functionality in place makes the service provided by this usage mode of SCTP quite similar to DCCP with TCP-like behavior (we will get to that later in this chapter), but with the additional benefit of features like multihoming.

UDP-LITE

If we regard SCTP as "TCP++," then UDP-Lite is "UDP++"—or actually "UDP--"—because its only feature is the possibility to restrain or even disable the original UDP checksum (Larzon et al., 2004). The reason to do so is easily explained: there are video and audio codecs that can deal with bit errors (which can, for example, be caused by link noise in a wireless environment). However, even if only a single bit is wrong, the UDP checksum will fail, causing the receiver to drop the whole packet from the stack. The codec then ends up with a large number of bytes missing, as potentially useful data that actually made it to the receiver were discarded by the operating system.

The UDP-Lite header is very similar to the UDP header—just the "Length" field, which is redundant because the length of a datagram is contained in the IP header, was replaced with a field called "Checksum Coverage." It represents the number of bytes, counting from the beginning of the UDP-Lite header, that are covered by the checksum. Such partial coverage can be useful for certain codecs—the "adaptive multi-rate" and "adaptive multi-rate wideband" audio codecs, for example (Sjoberg, Westerlund, Lakaniemi, & Xie, 2002). In any case, it is mandatory to have the header checked because, without knowing that the header is correct, even the port numbers can be wrong and the whole communication flow becomes meaningless (it is actually possible to disable the checksum altogether in standard UDP, but this feature is rather useless).

Despite its simplicity and its seemingly obvious advantages, UDP-Lite caused a lot of discussions in the IETF. The main problem is the fact that UDP-Lite does not yield any benefits whatsoever unless a link layer technology actually hands over corrupt data. Since it is the first IETF development to have that requirement, link layer technologies were so far optimized for protocols that require data integrity. Typically, there is a strong checksum, and often, corrupt frames are retransmitted with a certain persistence and eventually dropped and not forwarded by the link layer (Fairhurst & Wood, 2002); this is, for example, the case with standard 802.11 wireless LAN systems.[2] UDP-Lite can be seen as being at odds with the notion "IP over everything," as it enables application programmers to write an application that works well in one environment (where there is a small loss ratio) and does not work at all in another. These issues are actually quite intricate; more details can be found in Welzl (2005). In any case, from the perspective of a mobile multimedia application programmer, UDP-Lite is probably an attractive protocol, and after a couple of years of discussion, it has been published as a "Proposed Standard" RFC by the IETF. Since it was designed to be downward compatible with UDP, there is not much harm in using it even though the benefits can only be attained if an underlying link layer hands over corrupt data.

THE DATAGRAM CONGESTION CONTROL PROTOCOL (DCCP)

Multimedia applications are supposed to adapt their rate to the allowed transmission rate of the network in order to prevent Internet congestion collapse; ideally, this should be done in a way that is fair towards TCP ("TCP-friendly"). This is easier said than done: simply using TCP is usually not an option, as most real-time multimedia applications put timely delivery before reliability (i.e., users normally accept some noise in a telephone conversation, but having the sound excessively delayed is intolerable). Thus, such applications use UDP instead of TCP—but UDP provides no congestion control, leaving the task up to its user.

Adapting the rate is generally a difficult issue at the application level: data can be layered, compression factors can be tuned, encodings can be changed, but the outcome is not always precisely predictable. The additional requirement of being fair towards TCP and embedding a complete congestion control mechanism within the application may just be too much for most developers. Moreover, there is an incentive problem here—while TCP-friendliness is definitely desirable from the network point of view, it is questionable whether implementing this functionality is worth the effort for a single multimedia application developer. Finally, a user space congestion control implementation is just not ideal because precise timing may be necessary.

The IETF seeks to counter these problems with the datagram congestion control protocol (DCCP), which embodies congestion control functions for applications that do not need reliability. This protocol should be used as a replacement for UDP by networked multimedia applications and could be regarded as a framework for TCP-friendly mechanisms; due to a wealth of additional functions, DCCP is indeed an attractive alternative. According to its main specification (Kohler et al., 2005), one way of looking at the protocol is as TCP minus byte stream semantics and reliability, or as UDP plus congestion control, handshakes, and acknowledgments—but in fact, DCCP, includes much more than these three functions. In what follows, we will take a closer look at the most important elements of the protocol.

Connection Handling

Despite being unreliable, DCCP is connection oriented. The main reason for embedding this function in the protocol is to facilitate traversal of middle boxes such as firewalls which can selectively admit or reject communication flows when packets are associated with connections.

Reliable ACKs

Congestion control requires feedback. As in TCP, this takes the form of acknowledgment packets (ACKs) in DCCP—but with different semantics. In TCP, "ACK 2000" means "I received everything up to byte 1999, and I would like to have byte number 2000". Since DCCP never retransmits a packet, such cumulative ACKs would not make much sense here; thus, DCCP ACKs only acknowledge the reception of individual packets. This means that a sender has to maintain state regarding all the packets that were ACKed, and that it is hard for a sender to decide when to remove the state (at a TCP sender, the reception of a cumulative ACK can be used as an indication to remove any state regarding previous packets). It was therefore decided to make ACKs reliable in DCCP, in other words, retransmit them until the ACKs themselves are successfully ACKed. This has the additional advantage that congestion control can be carried out along the backward path—something that is hard to achieve with TCP, where data packets are reliably transferred, but ACKs are not. This fact makes DCCP superior to TCP in highly asymmetric environments such as satellite access links, where incoming data are streamed across a satellite and ACKs are often sent across a low-bandwidth modem link.

Feature Negotiation

In the DCCP specification, the word "feature" refers to a variable which is used to identify whether a DCCP endpoint uses a certain function. The "congestion control ID (CCID)" is an example of such a feature—endpoints must be able to negotiate which congestion control mechanism is to be used. The specification describes exactly how this is done; this includes the possibility to specify a "preference list," which is like saying "I would like CCID 2, but otherwise, please use CCID 1. If you cannot even give me CCID 1, let us use CCID 3." This procedure is another example of a reliable process that is embedded in this otherwise unreliable protocol. Features are specific to one endpoint, which means that a full-duplex DCCP communication flow can use one congestion control mechanism in one direction and another one in the other.

Checksums

As with UDP-Lite, the DCCP checksum can be restricted or even completely disabled. Additionally, the "data checksum" option can be used to distinguish between corruption-based loss and other loss events even when it is unacceptable to deliver erroneous data to applications. With this mechanism, a DCCP congestion control mechanism can therefore bypass the well-known TCP problem of misinterpreting any kind of packet loss as a sign of congestion (Balan et al., 2001).

Full Duplex Communication

Applications such as VoIP or video conferencing tools may require a bidirectional data stream—here, the full duplex communication capability of DCCP can make things more efficient by piggybacking ACKs onto data packets. Additionally, having only one logical connection for two unidirectional flows facilitates middle box traversal (i.e., firewalls require less state—and making the job easy for firewall developers fosters deployment of the protocol).

Explicit Congestion Notification (ECN) Support

With ECN, routers are given the option to set a bit in packets that they would normally drop (Ramakrishnan et al., 2001); the underlying idea is that a receiver should inform a sender when the "congestion experienced (CE)" bit in the IP header was set by a router, and a sender should react as if the packet had been dropped. With UDP, where the application programmer is responsible of implementing proper congestion control, ECN support could lead to unfairness—after all, who could prevent an application programmer from simply ignoring the bit?

Thus, the fact that DCCP makes use of ECN could be seen as a function that makes it somewhat superior over UDP.

Security

DCCP was designed to be at least as secure as a state of the art TCP implementation; modern TCP functions like ECN Nonces (a mechanism that prevents a receiver from lying about the congestion state) (Spring, Wetherall, Ely, 2003) and "appropriate byte counting (ABC)" (Allman, 2003) as well as "cookies" that can reduce the chance for a TCP-SYN-like DoS attack to succeed are therefore part of the protocol.[3] In TCP, sequence numbers automatically yield some protection against hijacking attacks; due to its unreliable nature, this had to be taken care of by means of a special sequence number synchronization procedure in DCCP.

Mobility

Whether to support mobility or not was discussed at length in the DCCP working group; eventually, neither mobility nor multihoming were included in the main document, and the specification was postponed. A rudimentary mechanism that slightly diverges from the original DCCP design rationale of not using cryptography is currently in the works. It is disabled by default, and an endpoint that wants to use this mechanism must negotiate enabling the corresponding feature. The scheme is simpler than mobility support in SCTP and resembles Mobile IP as specified in RFC 3344 (Perkins, 2002); at this point in time, it is unclear if (or when) it will be published as an RFC.

CONCLUDING REMARKS

The sudden appearance of new transport protocols for the Internet may help to make things more efficient, but it certainly will not make them easier to handle. How is an application programmer supposed to know whether, say, SCTP with parameters chosen for unreliable transmission or DCCP with TCP-like behaviour is more suitable for a certain situation? Also, these new transport protocols may face severe deployment problems—there must be a clear incentive for an application programmer to use a new protocol, which always has the potential risk of not penetrating an outdated firewall. Requiring to update all DCCP-based applications whenever a new CCID becomes defined also does not seem to be very attractive. Many more questions appear on the horizon—for instance, how do RTP and DCCP go together?[4] Finally, experiences with these protocols in mobile environments is quite limited.

While development of these protocols has progressed nicely and already reached a certain level of maturity, their use is still in its infancy. It may take a while until the radical change from TCP and UDP to a total of five transport protocols is welcomed by the majority of application developers; in any case, at this point in time, putting some research efforts into studying their usage in different scenarios seems to be a good idea.

REFERENCES

Allman, M. (2003). *TCP congestion control with appropriate byte counting (ABC)* (Tech. Rep. No. RFC 3465). Internet Engineering Task Force (IETF).

Balakrishnan, H., Rahul, H. S., & Seshan, S. (1999). An integrated congestion management architecture for Internet hosts. In *Proceedings of SIGCOMM,* Cambridge, MA (pp. 175-187).

Balan, R. K., Lee, B. P., Kumar, K. R. R., Jacob, J., Seah, W. K. G., & Ananda, A. L. (2001). TCP HACK: TCP header checksum option to improve performance over lossy links. *20th IEEE Conference on Computer Communications (INFOCOM).*

Clark, D., & Tennenhouse, D. (1990). Architectural considerations for a new generation of protocols. In *Proceedings of SIGCOMM,* Philadelphia (pp. 200-208).

Fairhurst, G., & Wood, L. (2002). *Advice to link designers on link* (Tech. Rep. No. RFC 3366). Automatic Repeat reQuest (ARQ).

Floyd, S., & Fall, K. (1999). Promoting the use of end-to-end congestion control in the Internet. *IEEE/ACM Transactions on Networking,* 7(4), 458-472.

Floyd, S., Handley, M., Padhye, J., & Widmer, J. (2000). Equation-based congestion control for unicast applications. In *Proceedings of ACM SIGCOMM,* Stockholm, Sweden (pp. 43-56).

Hessler, S., & Welzl, M. (2005). An empirical study of the congestion response of RealPlayer, Windows MediaPlayer, and Quicktime. In *Proceedings of the 10th IEEE Symposium on Computers and Communications (ISCC),* La Manga del Mar Menor, Cartagena, Spain.

Kohler, H., Handley, M., & Floyd, S. (2005). *Datagram Congestion Control Protocol (DCCP).* Internet-draft draft-ietf-dccp-spec-11.txt. Retrieved from http://www.icir.org/kohler/dccp/

Larzon, L. A., Degermark, M., Pink, S., Jonsson, L. E. & Fairhurst, G. (2004). *The lightweight user datagram protocol (UDP-Lite)* (Tech. Rep. No. RFC 3828). Internet Engineering Task Force (IETF).

Perkins, C. (2002). *IP mobility support for IPv4* (Tech. Rep. No. RFC 3344). Internet Engineering Task Force (IETF).

Ramakrishnan, K., Floyd, S., & Black, D. (2001). *The addition of explicit congestion notification (ECN) to IP* (Tech. Rep. No. RFC 3168). Internet Engineering Task Force (IETF).

Sjoberg, J., Westerlund, M., Lakaniemi, A., & Xie., Q. (2002). *Real-time transport protocol (RTP) payload format and file storage format for the adaptive multi-rate (AMR) and adaptive multi-rate wideband (AMR-WB) audio codecs* (Tech. Rep. No. RFC 3267). Internet Engineering Task Force (IETF).

Spring, N., Wetherall, D., & Ely, D. (2003). *Robust explicit congestion notification (ECN) signaling with nonces* (Tech. Rep. No. RFC 3540). Internet Engineering Task Force (IETF).

Stewart, R., Ramalho, M., Xie, Q., Tuexen, M., & Conrad, P. (2004). *Stream control transmission protocol (SCTP) partial reliability extension* (Tech. Rep. No. RFC 3758). Internet Engineering Task Force (IETF).

Stewart, R., & Xie, Q. (2002). *Stream control transmission protocol (SCTP). A reference guide.* Boston: Addison-Wesley.

Stewart, R., Xie, Q., Morneault, K., Sharp, C., Schwarzbauer, H., Taylor, T., Rytina, I., Kalla, M., Zhang, L., & Paxson, V. (2000). *Stream control transmission protocol* (Tech. Rep. No. RFC 2960). Internet Engineering Task Force (IETF).

Welzl, M. (2005). Passing corrupt data across network layers: An overview of recent developments and issues. *EURASIP Journal on Applied Signal Processing, 2005*(2), 242-247.

KEY TERMS

Application Layer Framing (ALF): Putting an application in control of block sizes that are transferred across the network.

DCCP: Datagram congestion control protocol.

Head-of-Line Blocking Delay: Delay that is caused by the requirement to deliver data chunks in order.

Multihoming: Associating a single logical connection endpoint with multiple IP addresses.

SCTP: Stream control transmission protocol

ENDNOTES

1. Internet Engineering Task Force–the technical standardization body of the Internet.
2. WiMAX (802.16) is a counter-example: here, it is possible to disable the checksum, albeit for reasons of compatibility with ATM.
3. Cookies can also be found in the SCTP association setup procedure (Stewart & Xie, 2002).
4. The answer to this question is: while DCCP functions could theoretically be implemented on top of RTP, it was decided that having RTP run over DCCP would be the right way to proceed.

Chapter XI
Location-Based Network Resource Management

Ioannis Priggouris
University of Athens, Greece

Evangelos Zervas
TEI-Athens, Greece

Stathes Hadjiefthymiades
University of Athens, Greece

ABSTRACT

The vision that wireless technology in the near future will provide mobile users with at least similar multimedia services as those available to the fixed hosts is quite established today. Towards this direction, extensive research efforts are underway to guarantee Quality-of-service (QoS) in mobile environments. An important factor that affects the provisioning of resources in such environments is the variability of the environment itself. From the user's perspective, this variability is a direct consequence of the user's movement and, at any given time, a function of his position. Exploiting the user's location to optimally manage and provision the resources of the mobile network is likely to enhance both the capacity of the network and the offered quality of service. In this chapter, we aim to provide a general introduction to the emerging research area of mobile communications, which is generally known as location-based network resource management.

INTRODUCTION

This chapter aims at presenting, in a concise form, state of the art material in the field of location-based network resource management. The current section acts as a general introduction to the evolution of mobile wireless networks, services, and the need for network resource management, so that the readers can familiarize themselves with the issues involved and acquire the global picture of the problem.

Mobile Wireless Networks' and Services' Evolution

Two broad categories can be discerned in the realm of mobile wireless networks: Wireless networks that have a well-defined infrastructure (e.g., cellular networks) and ad hoc (infrastructureless) networks. Although there has been a growing interest in the area of ad hoc networks in recent years, in this chapter we concentrate mainly on cellular mobile wireless networks. Since the inception of cellular networks in the early 1980 (the idea of frequency reuse is much older and it can be attributed to D. H. Ring, Bell Laboratories [1947]), the mobile networks have passed several phases. The first generation included the analog systems such as the North American system AMPS (advance mobile phone service), the Nordic system NMT (Nordic mobile telephone), the British system TACS (total access communication system), the Japanese system NAMTS (Nippon advanced mobile telephone system), the German system Netz-C and D, the French system Radiocom 2000, and the Italian system RTMI/RTMS just to name a few. These systems were designed primarily for the transmission of analog voice although there were capable of transmitting digital data in low rates.

The transition from analog to digital (second generation) systems was an imperative need in order to fix problems such as regional incompatibilities, low data rates, high blocking probabilities and low security levels while increasing systems' capacity. In the sphere of second generation systems, we can distinguish the systems GSM (global system for mobile), ADC (American digital cellular or IS-54), PDC (personal digital cellular), DCS-1800 (digital communication system at 1800 MHz) and lower tier cordless systems as DECT (digital European cordless telephone), CT2 (cordless telephone 2), PACS (personal access communication systems) and PHS (personal handy phone system). Second generation systems inherited the circuit-switching feature of analog systems but the users' demand for high-data-rate wireless access applications such as mobile IP, multimedia communications and network providers' demand for high-frequency utilization, pointed to packet-switching technologies. The twiddle of the switching technology towards third generation systems was obtained using intermediate (2.5 generation) systems such as HSCSD (high speed circuit switched data), GPRS (general packet radio service), and EDGE (Enhanced Data rate for GSM Evolution).

Third generation systems "3G" such as the Japanese system ARIB, the European system UMTS and the North American cdma2000 will be based on an all-IP network architecture to deliver the promised broadband services with QoS guarantees. 3G cellular systems will be enhanced by complementary WLAN systems such as IEEE802.11b and HIPERLAN, which offer high-data rate wireless access for low mobility users. Integrated 3G/WLAN network architecture provides a vehicle for the future generation of mobile communications. The next generation of mobile communications, termed 4G, foresees a heterogeneous infrastructure comprising different wireless/wired access technologies, where users will enjoy ubiquitous access to applications in an "always best connected" mode regardless of their mobility. This system will be capable of supporting the provision of higher data rates in localized service areas and seamless inter-system mobility.

The explosion of new radio technologies and network architectures in the past few years was fueled by users' insatiable thirst for advanced data services. Voice is not anymore the key service as in the first and second generation mobile systems and the humble 9.6KBps data rate, offered by GSM, is not sufficient for services like Web browsing or video conferencing. A wider range of broadband wireless services, from mobile business applications to mobile entertainment, has emerged in the last years. For network and service providers, the successful delivery of mobile data services is critical to subscriber growth and thus the increase of average revenue per user. A term frequently used to describe the successful delivery of services is this of QoS (quality of service). QoS provisioning takes different forms depending on the service and the underlying system. For example, from the perspective of cellular systems, QoS settles to measures like call blocking probability, call dropping probability and security whereas for IP networks, QoS means reliable delivery of packets or delay guarantees. The integrated IP-cellular networks, foreseen for 3rd and future generation systems, must melt down these system-wise aspects of QoS requirements. Moreover, the integration of the heterogeneous mobile wireless systems to a packet-switched common architecture must reconcile service specific requirements in terms of bit rate, delay, jitter and packet loss with the available resources of intervening systems.

There are several means that network and service providers can adopt to deliver these QoS guarantees to end users, such as the deployment of new application servers, sophisticated scheduling mechanisms, and signaling protocols and, most of all, efficient network resource management.

Network Resources

Identifying the network resources of the targeted systems is imperative in order to proceed with the discussion of how to efficiently handle them. To a certain point, network resources in mobile environments are similar to these met in fixed infrastructures. However, there are some additional resources, which have to be considered in the case of mobile networks. In the following, we provide a rough enumeration of the manageable resources, available in mobile networks. A wise manipulation of these resources is expected to improve the performance of the network; a goal that is aimed by every network management activity. Resources are classified, in those referring to physical entities inside the network, which are denoted as basic resources, and to others, that are somehow abstract, which are expected to implicitly affect the performance of the network.

Basic Resources

Basic resources correspond to measurable quantities inside the mobile network. Efficient handling of these resources is expected to have significant impact to the behaviour of the network and to the QoS experienced by its users. Resources of this category include, among others:

- **Bandwidth:** In fixed networks bandwidth is a term that refers to the transfer capability between the nodes of the network, as well as between the network and other external networks. It is usually dependable on the hardware equipment of the network. The same concept applies to mobile networks also; however in these networks, bandwidth is translated in parameters like timeslots and frequencies. Usually the fixed part of a mobile network has superior bandwidth capabilities compared to the radio interface. Hence, we assume that the radio segment is the element, which restricts the overall bandwidth capabilities of a wireless system. Using this assumption, bandwidth management and radio resources management are, in most cases, assumed one and the same for mobile networks.

- **Power:** Power is a resource which plays an important role in mobile networks. Terminals and base stations in these systems transmit their data using a certain power level, so as the overall signal to noise ratio (SNR) remains over an acceptable threshold. The power of the transmission is a fundamental resource, since mobile devices have limited power resources. Wasting these resources may reduce further the autonomy of the device and cause disruption or even interruption of the communication.
- **Storage:** Storage resources refer to the capacity of the various buffering elements inside the network. Such elements exist in each network entity (e.g., routers and switches), and their role is to cope with potential bursts of data, which cannot be directly handled by the switching capabilities of these entities. In such cases, the storage elements temporarily buffer incoming packets, thus eliminating the probability of loosing data.
- **Processing:** Normally, it measures the computing power of the various network elements (e.g., routers, servers, etc.). Practically, processing (or processing capacity as usually called) determines the capabilities of the hardware involved in the delivery of the network services. High processing capacity in the intermediate nodes of the network can provide significant improvement of its performance. Data packets are parsed faster, protocols run faster, and this applies to every operation that involves computer processing. Consuming all available processing resources of a network node, may lead to its inability of serving new requests or may slow its operation, thus, degrading the overall performance of the network.

Implicit Resources

Implicit resources are those that, at first glance, do not seem to affect the performance of the network. However, practice shows that their management is important, as well and the benefits from their efficient handling can be huge for the network.

- **Cache:** Caching is a data management technique used by many systems for enhancing the performance of a network. It is the process of replicating part of the information residing to a remote server, in the local system or in systems geographically dispersed inside the network. In this way, the users of the network who perform requests for retrieving data from the remote server, can be redirected to a local mirror and retrieve the same information. Caching concerns data management mainly; though an efficient handling of the caching parameters (e.g., cache sizes, position of caching servers etc.) can provide significant improvements to the overall network performance, reducing data transfers and freeing the core network resources (bandwidth, buffers, etc.).
- **Protocols:** Although not usually considered as resources, protocols play an important role inside the network. An efficient protocol implementation or even configuration can offer significant enhancements to the network's performance, by providing better usage of basic resources like bandwidth, storage, etc. There are different protocols, handling different operations inside the same network. But, even, for the same operation two or more protocols may exist, each suitable for a different situation (e.g., classic IP vs. mobile IP or RSVP vs. MRSVP). In cer-

tain cases, the same protocol may be configured (e.g., adjust TCP's window size) to enable superior performance depending on the type of the communication link.

- **Signaling:** Signaling refers to the specific protocols that are used for handling internal network operations. Connecting to the wireless network, handing over active calls and many others, comprise examples of such operations. Each of these operations requires that certain messages be exchanged between specific nodes of the network. In packet networks inband signaling is typically used, which means that signaling messages consume part of the useful bandwidth. In cellular mobile systems radio signaling is transferred through specific radio channels, either dedicated to a call or common to all terminals that exist in the same geographical area. Excessive or unnecessary signaling can congest the network and degrade severely its performance.

Goals — Objectives

Having, already, defined the various resources available in a network system, in this section we elaborate further on the purpose of network resource management and on how it can affect the network's operation. As already mentioned the main objective behind resource management is the improvement of the performance of the network. However, there are, also, more specific objectives that are hidden behind this primary objective. Specific goals targeted by resource management can be classified in two categories depending on the perspective they are looked from. To make this clearer, one should consider the two basic actors involved in the network's operation: the customer who makes use of the services offered by the network and the operator who owns the network and offers the services. What the former will expect from a high-performance network is good quality of service, small delay, and high availability to mention, only, some of his expectations. On the other hand, the operator will further consider issues like the capacity of the network or the fair distribution of the load experienced throughout the whole network.

User Point of View

The final consumer of the network services is the end-user. In mobile environments the end-user corresponds to the physical person, which owns the mobile device and uses it to connect to the network and access its services. There are three things that such a user expects from a high performance network:

1. The possibility of connecting to the network whenever he likes. This is expressed by the well-known **blocking probability** factor, which should be as low as possible.
2. The possibility of being continuously served by the network, while he moves inside the area covered by the network. In other words, the **dropping probability** for a connected user should be low and, at the same time, the **periods of interruptions** in his connections few and of small duration.
3. The **quality of service** (QoS) offered to the user should be stable enough and should not degrade during his movement inside the network. Usually, QoS includes parameters like: **allocated bandwidth, experienced delay** and **bit error rate**. However, QoS is a broad concept that covers many aspects of the network (i.e., different layers and network components)

and its supported services. Therefore, in many cases it is difficult to assess the offered QoS in a deterministic manner.

Other user-specific goals, which are not so obvious, include:

4. **High autonomy:** Mobile devices do not have a continuous power supply. Mobility restricts their possibility to connect to electric power; therefore their periodic recharging is necessary. To increase the terminal's autonomy efficient power management schemes should be used. Transmitting in high power levels, when there is no need, not only does not offer any benefit to the quality of the communication but also reduces the autonomy of the device.
5. **Health safety**, which means that the user expects to use the network services at no risk to his health. Arguments about the health risks imposed by the use of mobile terminals still continues; however everyone agrees that their power transmission as well as that of the base station should be configured at the minimum acceptable level, in order for such risks to be minimized.

Network Point of View

Looking from the network's perspective, we can claim that all the user-centric goals discussed above are also targeted by the operator of the network. This is rational, as the operator aims to maintain a high level of satisfaction among its users in order to keep them in his network. Moreover, a high level of satisfaction is likely to attract new users in his network, as good word of mouth about the performance of the network spreads. However, there are additional goals that an operator targets for his network. More analytically:

1. The increase of the capacity of his network. Capacity refers to the number of users that can be potentially served by the network simultaneously. It is evident why an operator desires **maximum capacity** for his network. More capacity means more users and more users more money in return. However, it should be considered, that increasing the capacity of the network implicitly assists to the fulfilment of some of the user-centric goals, as well. For example, blocking and dropping probabilities are decreased even further. A problem that may come up, as a result of capacity increase is the degradation of the experienced QoS, as large numbers of users consume more resources and produce higher loads for the network.
2. **High utilization** of the network resources is another thing anticipated by the operator. An operator would not, normally, accept his network to experience high load in certain parts, while it remains idle or underused in others. A balanced operation, where load is efficiently distributed between the various parts of the network is desirable, as it can free storage, processing, and bandwidth resources in congested parts of the network. Complete **load balancing**, of course, is not always possible, as certain parts of the network are certainly more prone to high accumulations of users than others (e.g., areas in town centres compared to suburban areas). However, careful management decisions can, surely, improve the situation and provide for both the increase of the utilization as well as the maintenance of an acceptable QoS level towards the users of the network.

Figure 1. Network resource management: Goals and resources

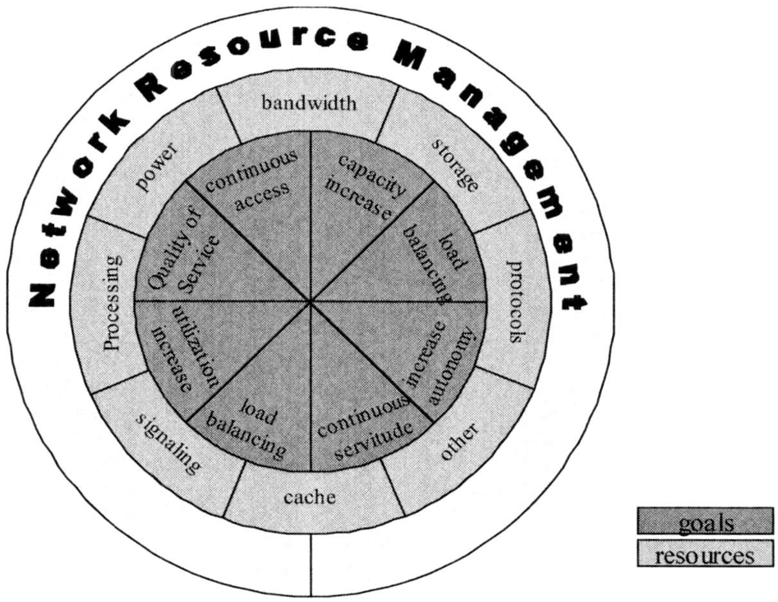

Setting Up the Scene

In this section, we will identify the deeper needs that call for managing network resources. Examples from everyday life show that management of resources is a fundamental process, which applies in many of its aspects. Water supplies, oil, money, are some of the resources that men or communities have to manage in their real life. The need for managing resources, mainly, stems from the fact that they are limited; therefore caution is needed in their usage in order to avoid problems that may come as a result of their overuse.

In mobile networks, resources are limited, as well. If we could have a network with infinite capacity and bandwidth, no need for managing its resources would exist. But this is not the case and therefore this need do exists and is also considered indispensable. Moreover, network resources have two characteristics that have allowed the development of several mechanisms and strategies for their systematic management; they are definite and they can be reused.

The term network resource management refers, exactly, to this process of manipulating the resources of the network. As already stated, the main objective of this process is the improvement of the performance of the network; side goals towards this target do exist and were thoroughly discussed in previous sections. Nevertheless, we saw that what performance enhancement means for a network can vary, depending on the point of view looked from and the criteria used.

In mobile networks, an additional imponderable factor exists, which imposes extra difficulties in the process of managing network resources. This factor is summarized in a single word: mobility. Users in such networks do not have a fixed point of connection but can roam inside the network moving from one connection point to another. The need for dynamically allocating resources within mobile networks is bigger than that in any other type of network. Efficient management of network resources is essential to satisfy both the users' and the operators' needs. And here rises the question: Could we possible exploit mobility in our favor? Could we find a specific characteristic that may assist us in the network resource management process? In order to answer this question we have to answer another one. Which are the parameters that characterize the user during his movement inside the network? The answer is simple: Location. Location not only comprises a fundamental parameter that is always inherent to the mobile user but its change is, which imposes the need for reallocating resources in mobile environments. In other words mobility results from the change of location but location can be used to model mobility as well. There are others parameters also that are important, such as the velocity of the user or its direction of movement, but as it will be shown later in this chapter they can both considered as location-related parameters.

Contemporary mobile networks have built-in capabilities for determining the location of their users. Positioning mechanisms have experienced an unprecedented boom in the recent years and they have matured enough, to provide accurate estimates of the user's location under any circumstances. If knowledge of the user's location is to be used for supporting network resource management in mobile networks the time is now. Location-aware network resource management and the possibilities it offers is the subject of this chapter and in the following sections we provide a thorough study of this specific topic.

LOCATION-BASED RESOURCE MANAGEMENT

Location Estimation

The primary requirement for applying location-aware resource management schemes is the existence of an accurate mechanism for estimating the location of the mobile user. Systems that determine location of a mobile user can be divided into two major categories: tracking and positioning (Schiller, 2004). In tracking a network equipped with suitable sensing devices determines the location of the user. The latter has to wear a specific tag or badge that allows the network to track his position. The location information is not directly available to the user, but only to the network. In order for the user to become aware of his position the network has to transfer him the corresponding location data, through a wireless link.

In positioning systems, it is the mobile system itself that determines the location. No sensor infrastructure is necessary in such systems. The infrastructure, which is used, consists mostly of active components that transmit specific signals carrying location-specific information (e.g., beacons, radio or ultrasound transmitters). Moreover, the location information is directly available at the mobile system and does not have to be transferred wirelessly.

All location systems, irrespective of their category, are based on a small set of basic techniques, some of which are used in combination:

1. **Cell of origin (COO):** A technique used in cellular networks. Its main principle is the differentiation of each cell, through the use of a unique cell identifier, which is transmitted by the base station that covers the cell boundaries.

2. **Time of arrival (TOA):** This technique measures the time window between sending a signal and receiving it to compute the spatial distance between the transmitter and the receiver. A variation of the method uses the time difference between the receptions of two signals to produce more accurate results. This variation is known as Time difference of arrival (TDOA) or enhanced observed time difference (E-OTD).
3. **Angle of arrival (AOA):** This method uses a fixed set of directional antennas and measures the direction (angle) of the signal received. At least two angles have to be determined from two different antennas towards the same mobile object, in order to correctly estimate the location of an object.
4. **Signal strength measurement (SSM):** Given a specific signal strength level, distance from the source can be easily computed by solving the signal attenuation equation. However, in most cases, the space between the transmitter and the receiver is populated with obstacles that affect measurements. Consequently, this method rarely produces accurate results.

Specific positioning methods, whose origins run back in geometry and trigonometry, have been developed in order to estimate the exact position of a mobile object. **Triangulation**, **trilateration** and **traversing** (Schiller, 2004) are well known methods used for this purpose. These methods use the distances and/or angles between the mobile object and two or more fixed points in order to produce accurate location estimations. Distances and angles are determined using the basic techniques listed in the previous paragraph. An alternative method, which borrows principles from stochastic theory and probabilities, is the **location fingerprinting**.

Fingerprinting refers to the matching of one set of measurements with another "reference" set contained in a database. In other words, a mobile device takes a "snapshot" of signals from visible base stations/access points for comparison with reference points stored in the database. A common signal modeling approach is to record samples of wireless signals from points in a large grid, drawn to encompass either, the entire area, covered by the mobile network, or specific segments within it; a process known as training phase for the discussed method. The smaller the grid cell size, the more samples are stored in the database. Location fingerprinting is a common technique used in indoor environments, where the area of coverage is limited and an abundance of signals from different access points exists. In large cellular systems (e.g., GSM), the use of location fingerprinting is very difficult and cumbersome, since the area that needs to be mapped in the database during the training phase can be very large.

In the following couple of sections, we will provide a brief overview of the most popular and commonly used positioning systems that are operational today. Existing systems can be classified to indoor and outdoor systems, depending on their applicability in the respective environment. Another categorization, where the diversification criterion is the positioning technology used, separates them to satellite and terrestrial-infrastructure systems. Finally, many times, in the bibliography, you will see another categorization, based on the type of the location information returned (symbolic location and absolute location systems). Each of the abovementioned categories is further divided in two or more classes, which can be further divided in other classes etc. An object diagram, that shows the different categories of positioning systems, along with their relations can be seen in Figure 2.

Figure 2. Categorisations of positioning systems

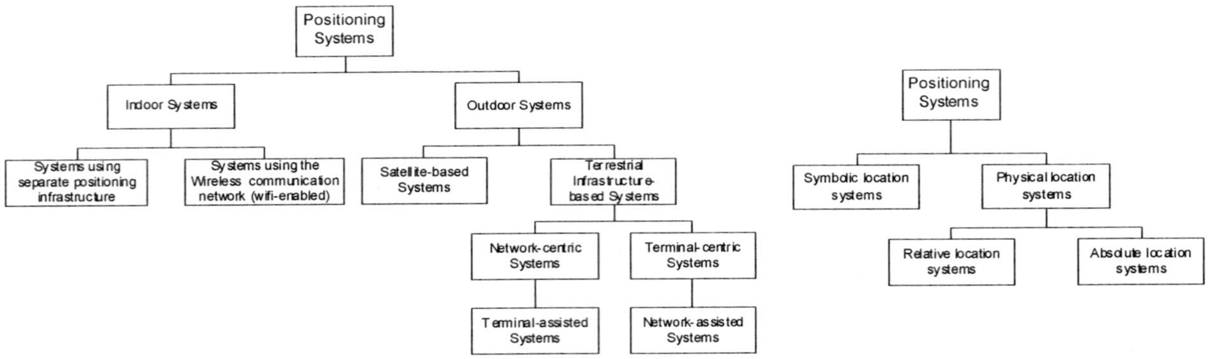

In our overview that follows, we will mainly focus on outdoor positioning systems and present indicative representatives from each sub-category. As seen in Figure 2, outdoor systems are divided in two major categories: satellite systems and terrestrial-infrastructure systems.

Satellite Positioning Systems

The idea of using satellites for positioning goes back to the 1960s. However, more than 30 years, had to pass, for the technology to mature. The first commercial satellite-based system became operational in 1995 and is the well known to everybody global positioning system (GPS) (Schiller, 2004), operated by NASA, the Department of Defense, and the Department of Transportation of the United States. Positioning in GPS relies on the signals transmitted by satellites and the estimated distances and corresponding angles of the received signals. We will not elaborate further on the way GPS operates, as this falls out of the scope of this chapter; however we will see some of its basic characteristics, which apply also, to all satellite-based systems. GPS provides a global positioning service, freely available to the public, with accuracies in the range of 25-43 meters. Greater accuracy is possible, but only for military and governmental purposes. At least three satellite signals are needed for locating a mobile target, while more signals can further enhance the accuracy of the positioning service. Enhancements of traditional GPS, have been proposed in order to increase the achieved positioning accuracy. Differential GPS and the wide area augmentation system (WAAS), use a combination of base stations, GPS satellites, and geostationary satellites in order to improve the precision of the positioning service in the range of 3-meters. However, both systems are limited to a small geographical region.

Other available satellite positioning systems include GLONASS, EGNOS and GALILEO. GLONASS is the Russian counterpart to GPS and provides similar precision with GPS. However, financial problems, led to inability of its maintenance by the Russian government, thus resulting to its early withdrawal. The European geostationary navigation overlay system (EGNOS) is a system similar to WAAS, which enhances GPS and GLONASS precision and provides European coverage. Finally, GALILEO is the European counterpart to GPS.

Its full operability is planned for 2008 and its positioning accuracy is expected to be similar or better than that of GPS.

Summarizing the advantages of satellite positioning systems we can pinpoint the following:

- High precision
- Global availability of the positioning service
- Minimal influence from environmental and weather conditions

Satellite positioning systems have certain disadvantages also, including:

- Considerable cost for creating and supervising the satellites' infrastructure
- Inability of producing location information in indoor environments
- Need for specific equipment (GPS receiver), on the mobile terminal, which can be expensive

Terrestrial Infrastructure Positioning Systems

Terrestrial infrastructure positioning systems exploit the infrastructure of the mobile network in order to estimate the location of a mobile object. These systems are much more inexpensive than their satellite counterpart, as the same infrastructure, which is for data transfer is used for determining the location of the user. The two most known and used systems that fall in this category are: the GSM and the WLAN.

In GSM, position estimation can be achieved in several ways. All techniques mentioned in the beginning of this section can be applied to GSM positioning in order to get the location of a mobile object. The exact way GSM positioning operates, is not within the scope of this chapter and therefore will not be analyzed further. What does matter, however, is the accuracy GSM positioning can achieve. Depending on the underlying positioning method, location information can be really rough, if the COO method (also known as cell global identity — CGI) is used, while more sophisticated mechanism, like Time Advance (TA) TOA, E-OTD or AOA can provide significant improvements over the achieved precision. For the COO method the precision can lie anywhere between less than 1 km to 35 km. The rest can provide accuracy in the range of a few tens or hundreds of meters. Although not so accurate as satellite positioning, GSM positioning bears significant advantages over the latter. First of all it does not need additional equipment on the terminal side; and secondly it can be used for determining the location of a mobile object both in indoor and outdoor environments.

In WLANs measuring the signal strengths from the various access points dispersed within the network's coverage area, and performing the appropriate calculations can provide a good estimate of the location of a moving object. Access Point's identities can be used in order to get a rough estimation of the user's location, as well. However given the fact that WLAN systems have limited coverage, such estimation is of little use. More precise location information can be achieved with systems, such as Nibble and Ekahau, or even better by using tracking location systems (e.g., systems based on a separate sensor infrastructure). In the next couple of paragraphs we provide a brief overview of the Nibble and Ekahau; tracking systems will not be analyzed further, as they comprise, mostly, proprietary solutions and they are too many to be cited here. For those interested in these systems Priggouris, Hadjiefthymiades and Marias (2005) can provide a good information source.

Nibble was developed by UCLA and uses Bayesian filtering in order to distinguish a cer-

Figure 3. Positioning accuracy

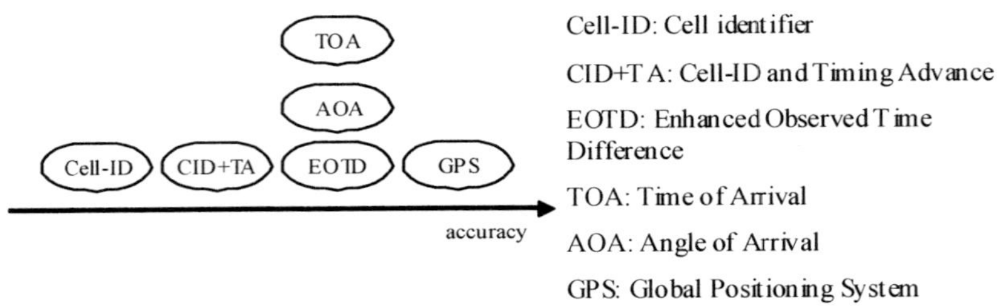

tain location from others with different signal quality characteristics. Nibble exploits the location fingerprinting method for estimating the location of a moving object, which, in turn, necessitates a training phase to be carried out before being able to produce any results. Nibble can generate location information with precision in the range of three meters. However, due to the fact, that signals from access points can significantly fluctuate, depending on the presence of moving objects inside the covered area, estimates can sometimes be much worst (e.g., in the case of a crowded area, with many moving objects). Results are improved, if the numbers of APs covering the area increases. Produced location, usually, comes in symbolic format (e.g., a room identifier), but coordinates can be provided as well, relative to a reference point. In the latter case an exhaustive coverage of the WLAN area, during the training phase, is needed in order for the produced coordinates to be accurate enough.

The Ekahau Positioning Engine™ (EPE), developed by Ekahau, is a commercial product, which combines the Bayesian networks with other complex stochastic methods in order to estimate the location of mobile objects, with accuracies ranging from 1-3 meters. EPE, uses a centralized location server in order to provide its location services and requires that each mobile object can receive signals from at least two access points in order to produce an accurate location estimation. Just like Nibble, EPE can provide either symbolic location information or coordinates relative to a reference point.

Location Prediction–Other Location-Related Parameters

Position information, is a wide concept, which, apart from location, may encapsulate additional information as well. It is evident that in continuously changing environments, such as those covered by mobile networks, locating a user is important; on the other hand location is a temporal characteristic, which after a few minutes may be of little interest to the network. What is important, however, is to know where the user will be in the future; knowledge of the future location enables the network to perform the necessary actions, in order to avert potential undesirable situations (e.g., dropping a call, unavailability of resources, etc.). Predicting the future location of a mobile object, based on its current location, usually requires knowledge of parameters like velocity and direction. Various methods and techniques have been proposed and used for solving the problem of predicting the movement of a mobile object. Some of them use the aforementioned parameters to feed their algorithms; others rely on the history of movement or on principles from the information or probability theory. In the following of this section, we briefly discuss research efforts on movement prediction.

A probabilistic model of the user's movement based on the history of handover behavior is proposed in Choi and Shin (1998). The model considers the aggregate history of all handovers that occurred in a given cell. Two stages are foreseen, namely, the handoff estimation and the predictive-adaptive bandwidth reservation. In the first stage, each BS, involved in handovers, caches quadruplets in the form (T_{event}, prev, next, T_{soj}) for a roaming terminal. Such entries are called "hand-off event quadruplets." T_{event} is the time when the terminal departed from the current cell, prev is the index of the previous visited cell, next is the index of the next cell, T_{soj} is the cell sojourn (residence) time of the terminal. From the cached quadruplets, the BS builds a handoff estimation function (HOE), which describes the estimated distribution of the next cell and sojourn time of a mobile, depending on the cell the mobile came from.

In Bhattacharya and Das (1999), the mobility-tracking problem in a cellular network has been considered from an information theoretic point of view. Comparison of user mobility models has been based upon the concept of entropy. A dictionary of user's path updates is built and maintained by the proposed scheme. Such dictionary supports an adaptive online algorithm that learns the profiles of subscribers. This technique is based on ideas and concepts coming from the area of lossless compression (i.e., the Lempel-Ziv algorithm). The algorithm is called "LeZi-update" and is exploited to reduce the location update related costs while its predictive power is used to reduce paging cost.

The algorithm discussed in Liu and Maguire (1996) is based on mobile motion prediction (MMP) scheme for the prediction of the future location of a roaming user according to his movement history patterns. The scheme consists of regularity-pattern detection (RPD) algorithms and motion prediction algorithm (MPA). Regularity detection is used to detect specific patterns of user movement from a properly structured database (IPB: Itinerary Pattern Base). Three classes of matching schemes are used for the detection of patterns namely the state matching, the velocity or time-matching and the frequency matching. The prediction algorithm (MPA) is invoked for combining regularity information with stochastic information (and constitutional constraints) and

Figure 4. Predictive mobility management algorithm

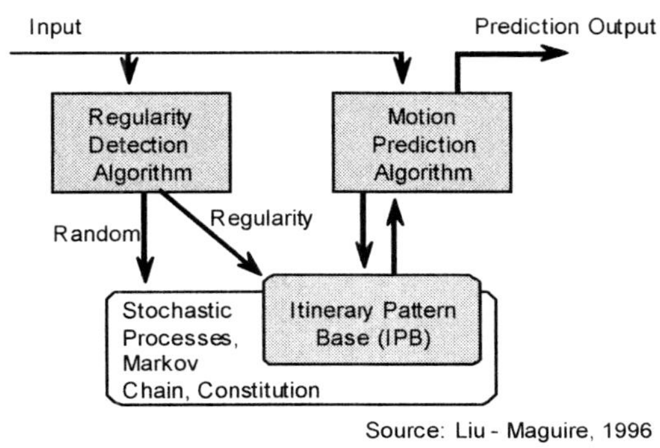

Figure 5. Mobility prediction (Liu, Bahl, & Chlamtac, 1998)

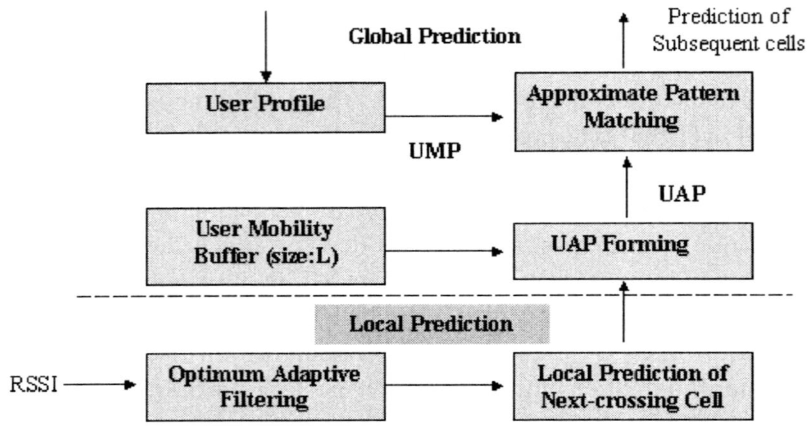

thus, reach a decision—prediction for the future location (or locations) of the terminal. Figure 4 provides an overview of the suggested scheme.

The work presented in Liu, Bahl, and Chlamtac (1998) uses pattern matching techniques and extended, self learning, Kalman filters to estimate the future location of mobile terminals. User mobility patterns (UMB) are stored in a database and fed to an approximate pattern matching algorithm to allow estimation (global prediction, GP) of a terminal's inter-cell movement direction (deterministic model). The Kalman estimator deals with the randomness in user movement by tracking intra-cell trajectory (stochastic model — local prediction, LP). The two models are combined together (Hierarchical Location Prediction) for the derivation of a semi-random movement trajectory (Figure 5). Simulation of the algorithm has shown that it accomplishes a high degree of prediction accuracy as soon as the Kalman filter becomes stable.

A first-order auto-regressive filtering technique is used in Aljadhai and Znati (2001), in order to predict the most likely to be visited cell. The direction-prediction is based on the history of the terminal's movement. The algorithm is little affected by small deviations of the mobile direction and converges rapidly to the new direction of the mobile terminal. Network operators determine the current location of the terminal using radio measurements or satellite positioning (GPS). At any specific time, the directional-probability of any cell being visited next by a mobile terminal can be derived based on (a) angle ratios related to the current cell where the mobile resides, and (b) the estimated direction of the mobile unit at this specific time. The basic property of this probability distribution is that for a given direction, the cell that lies on the estimated direction from the current cell has the highest probability of being visited in the future.

Artificial intelligence techniques were used in Hadjiefthymiades and Merakos (1999) in order to predict the next cell for a terminal and use such information for increasing the quality of mobile service provision. Specifically, a learning automaton (LA) has been used. LA is based on a state transition matrix, which comprises the one-step state transition probabilities and follows a linear reward-penalty (LR-P) scheme. If the LA decision is correct a positive feedback is received from the environment and the probability of the respective state transition is increased ("rewarded"). The rest of the probabilities are evenly reduced ("penalized") in order to balance the increase. If the response is wrong the state transition is "penalized" and the rest of the transitions are "rewarded" accordingly. The path prediction algorithm is executed at the home registry of the terminal. There is an itinerary database for each user with spatio-temporal information. When prediction is requested a set of entries are examined and the one with the highest probability is signaled as the algorithm's prediction output. If that response is correct or not then the procedures mentioned above are invoked. Should no relevant entries be found in the database, a new entry is introduced in the database and a random decision is taken.

Location–Aware Resource Management

In this section, we discuss the exploitation of the terminal's position information toward the management of network resources. The instantaneous recording of the terminal's position facilitates certain types of resource management schemes pertaining to the current status of the terminal/network (synchronous management — Figure 6). A more enhanced scheme, involves the sampled or continuous recording of the terminal's position (or the historical movement patterns) and the inference of information like velocity, acceleration and direction. Such information is very useful for the proactive management of network resources (asynchronous management), which will be used by the terminal or the network in the near future.

Typically, the exact location of the terminal is information that can otherwise be derived from the wireless network. The mobile terminal and the network know the base station (or access point) that currently controls the terminal and can position the terminal in a known, broader geographical area surrounding the base station. Since the knowledge of the area of the base station is of little use to a fine-grained network resource management scheme, the interpretation of time- or power-related information, contained in beacon messages broadcast by the base station, helps in achieving a more accurate positioning of the terminal within the given cell. Similar information from adjacent base stations greatly facilitates the positioning process and increases accuracy. Information derived from cell identifiers and beacons (network-based implicit position determi-

Figure 6. Synchronous/asynchronous network resource management

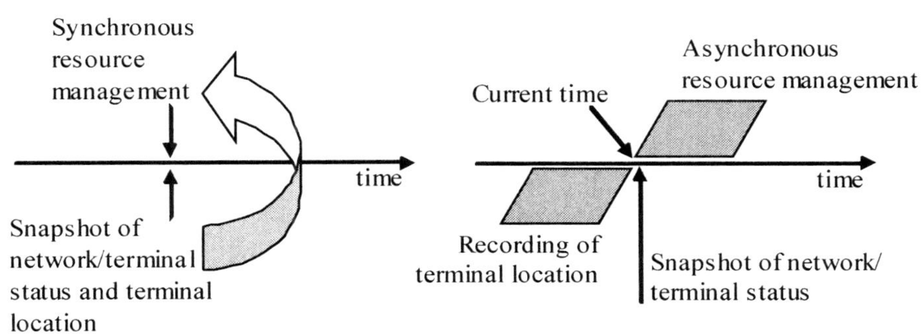

nation, NIPD) can be of low accuracy or reflect a temporary situation (e.g., sudden appearance of obstacles), thus hindering the network resource management mechanisms. Therefore, an important input parameter to the resource management scheme is the absolute position of the terminal as provided by a satellite-positioning scheme or enhanced terrestrial positioning mechanisms. Such information could be exploited supplementary to the NIPD to enhance the quality of network resource management schemes.

Location-dependent network resource management schemes could be classified as follows:

- **Short-term resource management (SRM):** Exploitation of the instantaneous values of terminal position, user sessions, and network status for optimum resource management. We refer to these terms as control input to the resource management problem. This family of management schemes can be considered as re-active in the sense that the management activity is an immediate reaction to the assessment of the current conditions of the user-network dipole.
- **Long-term (pro-active) resource management (LRM):** In this resource management type, the velocity and direction of the user are taken into account (possibly together with historical movement patterns), along with the control input required in SRM. Such information allows a properly structured control mechanism to predict the future position of the terminal and perform, intelligently, advance resource reservation.

Short-Term Resource Management

Examples of resource management schemes that fall in the first category (SRM) include:

- **Admission control:** The network knows the exact position of a number of users that are either idle or have active sessions and are currently roaming in the current cell. The network can decide whether to accept a new call judging from the present location of the user. If the user is on the boundary of two or more cells, the admission control process may refuse the call initiation as this can be handled through an adjacent base station (Figure 7). Otherwise, subject to the availability of network resources the network grants the requested session initiation to the interested user.
- **Network reconfiguration:** The network knows the exact position of a number of users roaming (with or without active sessions) in a cluster of cells. Through such information, the network is capable of calculating an anticipated load in each cell (through session initiation/termination probabilities). If, after this calculation, some cells are found (potentially) congested, the network initiates an internal re-organization / reconfiguration process to properly handle the foreseen load. Such process involves the (silent) reassignment of resources between cells and base stations (e.g., frequencies are temporarily borrowed by adjacent cells to cater for increased load—Figure 8). Even inside the same cell a reconfiguration of resources may take place, depending on the experienced load conditions. For example in cells with low user density common channels (e.g., RACH, PCH in GSM) may be reconfigured to operate in reduced capacity mode (i.e., use less timeslots per time unit). Leftover slots can be used for other signaling needs. Another option that falls in this category of resource management is to treat users as network resources. Instead of shifting resources, like fre-

Figure 7. Connection of user A to BS1 refused by the network

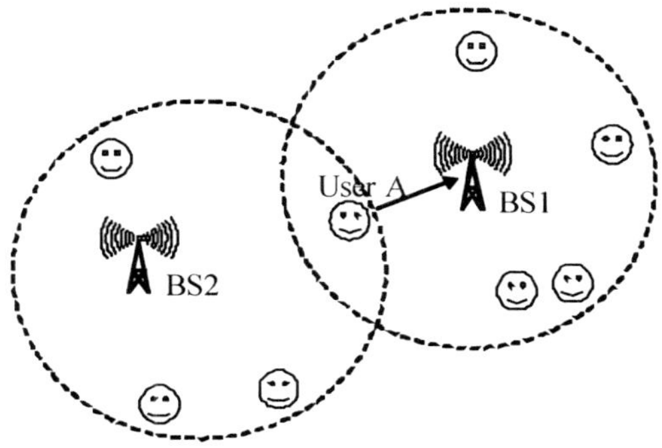

quencies, between cells and base stations the network could rearrange the users' population in order to optimally distribute the load and maximize utilization. In this scheme, the user is provided with specific relocation proposals on how to reach other cells where traffic load is less and better QoS can be attained (Figure 9).

- **Handover:** This scheme is a combination of the mechanisms discussed above. The network knows the exact position of a number of users roaming with active sessions in the current cell. As the user is found close to the boundary of the cell, and the load in the adjacent cell is lighter, the user terminal is instructed to switch communication (i.e., perform a forced handover) to the indicated base station. Alternatively to load balancing objectives, the rationale behind a forced handover could be the support of specific QoS requirements of the user and the avoidance of session termination. In this scenario, no physical relocation of the involved user is required.
- **Routing:** In ad hoc mobile networks, with quasi-stationary nodes, the relative position of nodes, which is known to the nodes through location advertising procedures, may be used to design efficient routing energy aware routing schemes. Such schemes require that a continuous monitoring of the network's structure (e.g., location of the mobile nodes) is performed

Figure 8. BS1 borrows f1 and f3 from BS2

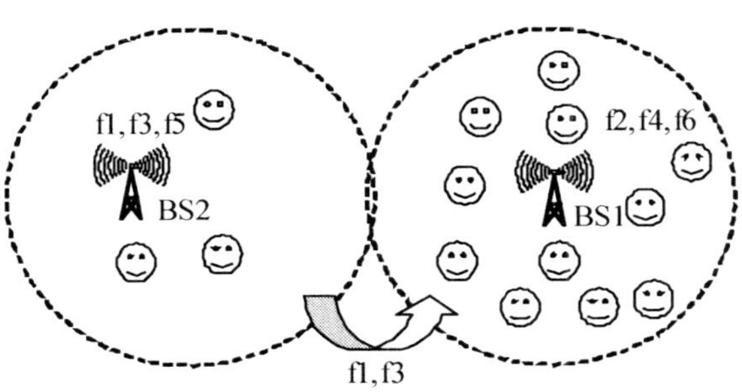

Figure 9. Users directed to BS2

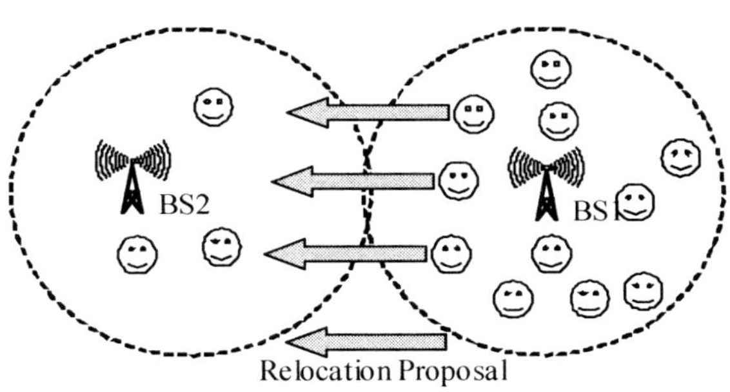

and routing tables are updated accordingly, to reflect the changes imposed by the movement of each node. The objective of this energy-aware management activity is the minimization of the power needed for transmitting data between two end nodes. This, in turn, increases the autonomy of the mobile node.

Long-Term Resource Management

Examples of resource management schemes that fall in the second category (LRM-proactive) include:

- **Fine-grained pre-reservation of resources:** The occurrence of handovers in cellular networks is a very important issue that drives the design of resource management algorithms. In the recent past, pro-active resource management schemes, involving movement prediction, have been adopted for overcoming handover-induced problems like session discontinuation. The network mechanisms, acting before the occurrence of the handover, may reserve resources in the best candidate (i.e., the most likely to be visited) cell of the current cell's neighborhood. After the occurrence of the handover, the terminal does not compete for finite network resources but enjoys a pre-arranged configuration. Hence, due to the pro-active resource management, the user does not experience service discontinuations (increased drop probability), or low service quality. A session (call) may have to be terminated when the mobile terminal is handed off to a new base station, which does not have adequate resources to support the QoS requirements of the particular session. This type of session termination is referred to as handoff blocking, and is very annoying for the user. The handoff blocking probability may be reduced through the use of proactive resource reservation in the neighborhood of the present cell. The more efficient of such reservation schemes use path prediction algorithms to find the most likely neighboring cell the terminal is going to move to. Performance may be further improved by more elaborate reservation schemes that take into account the timing and the criticality of the resource reservation. A taxonomy of such wireless resource management schemes is given in Figure 10.
The proactive resource management, as it involves reservation or reassignment of finite resources which could otherwise be used by e.g. local, stationary users, should be performed in a thorough manner with

Figure 10. (Proactive) Wireless resource management schemes taxonomy

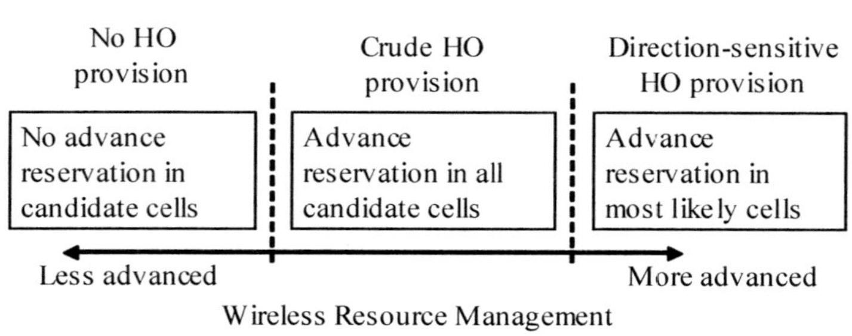

careful time scheduling. Performing a resource pre-reservation too early will lead to undesired waste of resources and low network utilisation. Conversely, a delayed pre-reservation scheme may end-up with fewer resources than required, thus, forcing the termination of sessions and low experienced QoS. This last option may reduce to the "No HO provision" case as illustrated in Figure 10. The terminal location information could be fully exploited in this respect to derive accurate estimates of handover occurrence times.

- **Protocol management:** The determination of exact terminal location and correlation of such information to network spatial availability (radio/network map of the considered area) could facilitate advanced pro-active schemes in heterogeneous infrastructures. Specifically, in 4G infrastructures, the terminal could perform an advance protocol reconfiguration (and/or downloading) to cater for another network, which will, shortly, assume control of the roaming user. A protocol module or a whole protocol stack could be substituted (or differently configured) to efficiently operate in the oncoming network. For example, a terminal could execute a TCP variant (e.g., adopting Explicit Loss Notification, ELN) in a GSM-like network and need to switch to plain-vanilla TCP in anticipation of WLAN connectivity. A dual protocols stack scenario is not feasible in this case due to memory and computing capacity restrictions in the mobile terminal. The discussed protocol reconfiguration needs to be performed pro-actively to facilitate seamless connectivity. The discussed scheme involves operations that are, typically, performed within the mobile terminal. To facilitate operations like protocol downloading (software-based radio), the fixed network could proactively manage resources like protocol bundles/components. To reduce the download time and handover disruption probability, the network (proactively) pushes components that will be requested by the terminal to its forefront (e.g., nodes very close to base stations/access points). Another option for protocol management is the tuning of protocol parameters subject to the current location of the terminal and known, local conditions.

CASE STUDIES AND PROPOSED TECHNIQUES

Several studies can been found in the literature that tackle the problem of location-based network resource management. This section aims at quoting the most representative proposed techniques, which nevertheless encompass a wide spectrum of manageable resources and a variety of goals. As in Section 2, we differentiate among re-active (short-term network management) and pro-active (long-term network management) schemes.

SRM Examples

Employing a user tracking system to reduce the paging signaling load has been proposed in Bhattacharya and Das (1999). In this work, a Markov model is used to capture mobility characteristics of a user, with the transitions between wireless cells as input to a Markov model. As users move between cells, or stay in a cell for a long period of time, the model is updated and the network has to try fewer cells to successfully deliver a call.

The authors in Rodoplu and Meng (1999) describe a distributed position-based network protocol optimized for minimum energy consumption in ad hoc networks. Each node is equipped with a GPS and starts a search by sending out a beacon signal that includes its position. The transmitting node also listens for signals from nearby nodes and finds out their positions. This enables it to determine the relay regions for the neighboring nodes. Simulation results for a stationary network show that as the number of nodes increase the average power expenditure per node reaches a minimum value. The protocol can be applied for mobile nodes as well due to the localized nature of its search algorithm. In the mobile network case, synchronization can be achieved using the absolute time information provided by GPS. Another use of position information provided by GPS is demonstrated in Fleming et al. (1997). In this work, GPS has been considered for reducing the overhead of TCP enhancement protocols like SNOOP, that requires neighboring base stations to cache data information for mobile terminals associated with a particular cell.

Routing and fast handover protocols for ad hoc networks using location information provided by GPS has been proposed in Ergen et al. (2002). In this scenario, sensors form a mesh network and connect to a mobile node. The mobile nodes form an ad hoc network and connect to a fixed base station. Base stations and mobile nodes are GPS equipped. The mobile bases roam in the sensor-scattered area, thus forming smaller sensor networks, gather information from the sensors in its vicinity and send it over multi-hop wireless networks to the fixed base stations. Geographical information for the mobile nodes and the fixed-based stations can be used in two ways. One is to improve handover performance by allowing the current access point, serving a mobile node, to send packets only to these access points that are more likely to be visited by the mobile node instead to all of its neighboring access points. Each access point knows its location and the location of the other access points through a mechanism of exchanging location advertisement messages. The second utilization of geographical information is for efficient routing. Each node has the means to find the position of the destination node and routes packets to nodes known to be close to the destination. Along the routing path, as the data packets get closer to destination, the nodes are more knowledgeable about the destination network topology and route packets more efficiently.

In Naghian (2001) a location-sensitive resource management technique is proposed. The location-sensitive handoff (L-SH) method, as the proposed scheme is called, targets future mobile systems (e.g., UMTS, WCDMA, etc.) and comprises of an improved handover algorithm, which does not rely only on the conventional handoff criteria (i.e., signal quality, traffic load, etc.), but uses specific location information for each user in order to assist the handover process. The new method necessitates the availability of accurate location information either at the network side (network-based or mobile assisted positioning) or at the mobile's side (e.g., using a GPS receiver). Such information is likely to be available in the future UMTS and WCDMA mobile systems, thus making the implementation of the proposed mechanism feasible. L-SH does not concentrate only on the management of bandwidth and power resources but it tackles other resources also, such as signaling. In this sense, it differs from the conventional handoff method, not only regarding its criteria but regarding its objectives as well.

According to LS-H, in order for a handoff to take place two different criteria should be met. The first involves the location of the user, while the second the signal strength. In brief, the algorithm is as follows: the location of the mobile terminal is determined (this can be done periodically or on demand) and a location-specific criterion is checked. For example such criterion may be if the distance of the terminal from its home cell, has surpassed a certain threshold. If the first criterion is met, the decision mechanism proceeds with examining the second criterion, which checks the signal strength level. Only if both criteria are met the handover is executed.

Moreover, LS-H can be applied to hierarchical cellular systems, consisting of overlay pico-cells, micro-cells and macro-cells and provide for significant reduction of the needed signaling, by decreasing the number of required handovers for each mobile terminal. Additional location-sensitive information, such as direction and velocity can be used for efficiently handing off the mobile from a pico-cell to a macro-cell or vice-versa, resulting to both superior quality of service, as well as less signaling overhead for the network. For example, a mobile can be connected to the neighbouring cell, towards which it is moving; fast moving user can be handed over to a macro-cell in order to reduce the possibility of new handover, etc.

The MITOS system (Alyfantis, Hadjiefthymiades, & Merakos, in press) addresses the occurrence of short-term local congestion in WLAN environments where user population is dense. Congestion adversely impacts the network and the user. Users found in a congested access point experience degraded QoS. At the same time, there may be other APs in the vicinity that are significantly less loaded, as fewer users are present in their coverage areas. The MITOS system balances the traffic load across the WLAN, so that users take advantage of the overall wireless bandwidth. With such a system, the operator could optimally exploit the infrastructure and maximise its return, while the users receive better QoS. If a MITOS-like system is not adopted, network operators, in order to support the user requirements during short-term congestion, need to over-provision the network resources. In the problem discussed above, the co-operation between users and the network may prove beneficial to both parties. Specifically, if users agreed to move to appropriately indicated locations, they could enjoy improved QoS; at the same time, the provider would not need to resort to network over-provision. MITOS is a Smart Spaces system that influences the user locations to balance the traffic load across a WLAN installation, and improve user QoS. The MITOS platform is capable of discovering whether congestion takes place in a certain segment of the network, and is aware of user locations. If congestion occurs, MITOS urges affected users to move to another location (relocation proposal, RP), where bandwidth reserves are higher. MITOS also issues navigation instructions for this transition. Under certain circumstances, owing to user behavior, the efficiency of MITOS may be compromised. To alleviate such a risk, the system is enhanced with game theoretic mechanisms.

An approach for congestion relief in WLAN hot-spots is discussed in (Balachandran, Bahl, & Voelker, 2002) to maximize user bandwidth allocation and overall network utilization. In case of local congestion, the terminal finds a less congested AP in vicinity to associate with, making a trade-off between available bandwidth and signal strength. If no neighboring AP can guaranty connection improvement, a network-monitoring server provides feedback to the user, indicating a less loaded, yet distant, AP. Such explicit network feedback does not cater for those situations where congestion affects numerous users. Users in this system are assumed cooperative (i.e., their actions are assumed coordinated to avoid side effects of the system feedback). The scheme of channel switching relies on the assumption of overlapping, non-congested, cells and on specialized client network equipment (i.e., network infrastructure dependent). Work in Balachandran, Bahl, & Voelker (2002) also assumes a QoS-sensitive MAC layer in terminals in order to meet user Service Level Agreements (SLAs).

LRM Examples

The LRM type techniques are more involved, since prediction of the user's future location is required. Several interesting proposals can be found in this area. The authors in Sparacino (2002) propose the use of infrared beacons to create individualized models of museum visitors allowing each exhibit to present custom audiovisual narrations to each user. Thus, the provided service is personalized and resources of the network are used accordingly. The authors in Liu and Maguire (1995), describe a generalized network architecture that incorporates prediction with the goal of supporting mobile computing. Mobile units wirelessly communicating with the network provide updates of their locations and a predictive model is created, allowing services and data to pre-cached at the most likely future locations. The prediction algorithm is based on a pattern matching technique that exploits the regularity of the users' movement patterns.

Another predictive scheme based on GPS can be found in Chiu and Bassiouni (1999). The use of GPS is considered in predictive radio channel resource allocation algorithms. Simulation results show that the handoff blocking probability is reduced while not affecting drastically the new call blocking probability if the mobile's location information is employed to reserve resources for it during handover.

In Liu and Maguire (1995), the authors propose a mobile motion prediction algorithm, which is based on a two-tier hierarchical location algorithm. The algorithm is used to provide the necessary information for advance resource reservation in wireless ATM networks. The higher-tier prediction scheme uses an approximate pattern matching technique to track intercell movements, whereas the lower-tier intracell tracking component is used to predict the trajectory within a cell and estimate the next cell to be crossed. Although in Liu and Maguire (1995), the latter scheme involves RSSI (received signal strength indication) measurements, which are filtered through an extended self-learning Kalman filter to obtain estimates of distances, velocities and accelerations, the whole process can be simplified if direct location measurements are performed by the mobile unit. Not only location estimates will be more accurate, since the extended Kalman filter is not optimal and may diverge due to the nonlinearity of the system, but the computational load of the Kalman filtering will be diminished at the mobile unit.

In Aljadhai and Znati (2001), a framework is proposed that integrates mobility prediction and CAC, to provide support for predictive timed-QoS guarantees, where each call is guaranteed its QoS requirements for the time interval that the mobile unit is expected to spend within each cell it is likely to visit during the lifetime of the call. The support for predictive timed-QoS is achieved based on an accurate estimate of mobile's trajectory as well as the arrival and departure times for each cell along the path. Using these estimates, the network can determine if enough resources are available in each cell along the mobile's path to support QoS requirements of the call. The basic components of the work proposed in Aljadhai and Znati (2001) are: (1) a predictive service model to support timed-QoS guarantees (2) a mobility model to determine the mobile's most likely cluster; and (3) a CAC model to verify the feasibility of supporting a call within the most likely cluster.

The authors in Liang and Haas (2003) employ a mobile's location and velocity information, inferred by measurements reported by the mobile itself, to predict the future location of the mobile. Location predictions are used to reduce the mobility management cost associated with paging, location updates, and location inspection. There is a tradeoff between location updating and mobile paging with both procedures consuming network or mobile resources. Frequent location updates result in a more precise network's knowledge about the mobile location and therefore the number of paging messages can be reduced considerably. However, frequent location updates consume mobile's limited energy supply, channel's bandwidth and induce a burden at the location databases. In Liang and Haas (2003), the mobile checks its position periodically and performs a location update only if the distance between the predicted and the measured location exceeds a threshold. Location prediction is based on a Gauss-Markov model, which can represent different degrees of mobility ranging from a constant velocity model (fluid-flow) to a random-walk model. The parameters of the Gauss-Markov process are estimated and updated using samples of the mobile's velocity taken by the mobile unit. Defining a total cost of mobility management per call arrival as the sum of three terms, the location inspection cost, the location update cost and the page cost, the authors in Liang and Haas (2003) demonstrate a mobility management cost reduction of about 50% compared to other non-predictive distance-based schemes.

The CELLO project[1] (CELLO project, 2005) uses location information for assisting the network resource management process. CELLO, proposes the introduction of a new subsystem inside the mobile network, which handles location-related information. The main component of this system is the mobile network geographic information server (MGIS), which stores and analyses location-related information for all users attached to the mobile network. Such information includes information originating at the terminal (e.g., for terminals equipped with a GPS receiver), information produced by the network infrastructure (e.g., from location servers) and information deduced through estimations (e.g., based on a variety of models and methods). Additional information stored in the MGIS includes performance data about the cellular network as well as static geographical information regarding the area covered by the network. The information from the server is then used for assisting resource management processes such as handover, network planning and mobility management.

The location-aided handover (LAH), proposed by CELLO, consists of a set of algorithms, which aim to efficiently tackle the handover problem. Based on the information available in the MGIS the used algorithms, have to decide the most appropriate base station for handing over a mobile terminal. Consulting the MGIS, the LAH escapes from conventional handover algorithms, where decisions are based exclusively on the RSSI value. LAH algorithms can identify critical areas, monitor user movement, and take intelligent handover decisions, thus eliminating many of the shortcomings, imposed by conventional handover methods. For example if it detects that the mobile terminal is moving across the borders of a cell, it may delay the handover in order to avoid a possible "ping-pong" effect; If more than one candidate target cells exist, location information will help to choose the optimal target cell; Even typical handovers from an overlay macro-cell to the underlying micro/pico-cells can benefit from the accurate location information maintained in the MGIS, and assist the system to choose the most appropriate target cell. Furthermore, by analyzing information available in the MGIS the network can, possibly, estimate the direction of the user's movement and reserve resources in the potential target cells, which the mobile user may inhabit in the near future. Resources influenced by LAH include mainly bandwidth and power. Moreover, signaling is also affected in the sense that efficient handover means that unnecessary handovers are decreased and therefore the corresponding signaling traffic is reduced, as well.

Location-aided planning (LAP) aims to improve planning for the covered network so that radio resources are distributed between different areas in an optimum manner. Location-related information from the MGIS together with the retrieved performance data is analyzed to determine problematic areas inside the network. The accumulated knowledge can be used for creating alternative network plans depending on the traffic conditions (e.g., allocating more radio channels in specific areas experiencing congestion), thus increasing the capacity of the network and the offered QoS.

Finally, location-aided mobility management (LAM) is proposed by CELLO, as a mean to support vertical handover and interworking of different networks and systems. The LAM algorithms, takes into account location-specific information for the mobile, specific service requirements of the user and static location information (e.g., existing access points, antennas, etc.) in the neighboring area, all stored in the MGIS, and may inform the user of nearby infrastructures, that can support his needs. For example, suppose that a user wants to access a wideband service and the network that is currently attached to cannot support his needs. This may be either because the network does not have such capabilities or due to lack of resources. Suppose also that a nearby WLAN access point, which can support the requested service, exists. The LAM algorithm will notify the user of the presence of such a capable infrastructure and prompt him to use the WLAN system instead.

CONCLUSION

A survey study on state of the art techniques employing user location information for efficient network resource management was presented in this chapter. The study begun with a short description of the basic principles of network resource management and the identification of the specific problems that mobility imposes sets in resource management for wireless environments. Mobilized, by the evolution of position estimation systems, in the recent years, we analyzed the possibility of exploiting the location of the user — either directly or as an input to movement prediction — for making the network resource management process more efficient. A variety of mechanisms and approaches that facilitate location-aware network resource management were discussed, along with several implementation examples from bibliography.

ACKNOWLEDGMENT

This work is supported by the PYTHAGORAS programme of the Greek Ministry of National Education and Religious Affairs (University of Athens Research Project No. 70/3/7411).

REFERENCES

Aljadhai, A., & Znati, T. F. (2001). Predictive mobility support for QoS provisioning in mobile wireless environments. *IEEE Journal on Selected Areas in Communications*, *19*(10), 1915-1930.

Alyfantis, G., Hadjiefthymiades, S., & Merakos, L. (in press). *An overlay smart spaces system for load balancing in wireless lans.* To be published in ACM/Kluwer MONET, Special Issue on Internet Wireless Access: 802.11 and Beyond.

Balachandran, A., Bahl, P., & Voelker, G. (2002). Hot-spot congestion relief and user service guarantees in public-area wireless networks. In *Proceedings of the 4th IEEE Workshop on Mobile Computing Systems and Applications* (p. 70).

Bhattacharya, A., & Das, S. K. (1999). LeZi update: An information theoretic approach to track mobile users in PCS networks. In *Proceedings of ACM/IEEE Mobicom '99*, Seattle, WA.

CELLO Project. (2005). *CELLO Project Web site*. Retrieved June 2005, from http://www.telecom.ntua.gr/cello/

Chiu, M. H., & Bassiouni, M. (1999). Predictive channel reservation for mobile cellular networks based on GPS measurements. In *Proceedings of the IEEE International Conference on Personal Wireless Communications (ICPWC'99).*

Choi, S., & Shin, K. G. (1998) predictive and adaptive bandwidth reservation for hand-offs in QoS-sensitive cellular networks. In *Proceedings of ACM SIGCOMM '98*, Vancouver.

Ergen, M., Coleri, S., Dundar, B., Jain, R., Puri, A., & Varaiya, P. (2002). Application of GPS to mobile IP and routing in wireless networks. In *Proceedings of IEEE Vehicular Technology Conference (VTC)* (Vol. 2, pp. 1115-1119).

Fleming, K. et al. (1997). *Handoffs using GPS in mobile environment.* Pittsburgh: Information Networking Institute, Carnegie Mellon University.

Hadjiefthymiades, S., & Merakos, L. (1999). ESW4: Enhanced scheme for WWW computing in wireless communication environments. *ACM SIGCOMM Computer Communication Review, 29*(5), 24-35.

Liang, B., & Haas, Z. J. (2003). Predictive distance-based mobility management for multidimensional PCS networks. *IEEE/ACM Transactions on Networking, 11*(5), 718-732.

Liu, G. Y., & Maguire, G. Q. (1995). Efficient mobility management support for wireless data services. In *Proceedings of 45th IEEE Vehicular Technology Conference*, Chicago.

Liu, G. Y., & Maguire, G. Q. (1996). A class of mobile motion prediction algorithms for wireless mobile computing and communications. *MONET, 1*(2), 113-121.

Liu, T., Bahl, P., & Chlamtac, I. (1998). Mobility modeling, location tracking, and trajectory prediction in wireless ATM networks. *IEEE JSAC, 16*(6), 922-936.

Naghian, S. (2001). *Location-sensitive radio resource management in future mobile systems. The book of visions* (Vol. 1). Wireless World Research Forum (WWRF).

Priggouris, I., Hadjiefthymiades, S., & Marias, G. (2005). Location-based services. In N. Passas, A. Salkintzis, & Wiley (Eds.), *Emerging wireless multimedia services and technologies* (Chap. 14). West Sussex, UK: John Wiley & Sons, Inc.

Rodoplu, V., & Meng, T. H. (1999). Minimum energy mobile wireless networks. *IEEE JSAC, 17*(8), 1333-1344.

Schiller, J., & Voisard, A. (2004). *Location-based services.* San Francisco: Morgan Kaufman Publishers, Elsevier.

Sparacino, F. (2002). The museum wearable: Real time sensor driven understanding of visitors' interests for personalized visually-augmented museum experiences. In *Proceedings of Museums and the Web*, Boston.

KEY TERMS

Admission Control: The process of restricting access to a system (e.g., network or application), based on certain criteria.

GPS: Global positioning system. A satellite-based system for estimating the location of a moving object.

Handover or Handoff: The process by which a mobile's terminal conversation is transferred from on Base Station to another, when the user is in motion.

Location-Aware: Consideration of the user's location for performing various operations.

Network Resource Management: The process of manipulating resources of a network (e.g., bandwidth, storage, etc.), in order to improve the performance of the network.

Positioning: The process of estimating the location of a moving object.

Pre-Reservation: The process of reserving network resources for a specific user proactively (e.g., before the user actually needs them).

Quality of Service (QoS): A term that refers to the quality of network services provided by a specific network.

ENDNOTE

1. Implemented in the context of EU IST framework.

Chapter XII
Discovering Multimedia Services and Contents in Mobile Environments

Zhou Wang
Fraunhofer Integrated Publication and Information Systems Institute (IPSI), Germany

Hend Koubaa
Norwegian University of Science and Technology (NTNU), Norway

ABSTRACT

Accessing multimedia services from portable devices in nomadic environments is of increasing interest for mobile users. Service discovery mechanisms help mobile users freely and efficiently locating multimedia services they want. The chapter first provides an introduction to the topic service discovery and content location in mobile environments, including background and problems to be solved. Then, the chapter presents typical architectures and technologies of service discovery in infrastructure-based mobile environments, covering both emerging industry standards and advances in the research world. Their advantages and limitations, as well as open issues are discussed, too. Finally, the approaches for content location in mobile ad hoc networks are described in detail. The strengths and limitations of these approaches with regard to mobile multimedia services are analyzed.

INTRODUCTION

Recently, the advances in mobile networks and increased use of portable devices deeply influenced the development of multimedia services. Mobile multimedia services enable users to access multimedia services and contents from portable devices, such as laptops, PDAs, and even mobile phones, at anytime from anywhere. Various new applications, that would use multimedia services on portable devices from both the fixed network backbone and peer mobile devices in its proximity, are being developed, ranging from entertainment and information services to business applications for M-Commerce, fleet management, and disaster management.

However, to make mobile multimedia services become an everyday reality, some kinds of service infrastructures have to be provided or enhanced, in order to let multimedia services and contents on the network be discovered and utilized, and simultaneously allow mobile users to search and request services according to their own needs, independently of the physical places they are visiting and the underlying host platforms they are using. Particularly, with the explosive growth of multimedia services available in the Internet, automatic service discovery is gaining more and more significance for mobile users. In this chapter we focus on the issue of discovering and locating multimedia services and contents in mobile environments. After outlining necessary background knowledge, we will take an insight into mobile multimedia service discovery. Major service discovery architectures and approaches in infrastructure-based networks and in mobile ad hoc networks will be investigated. We present also a detailed analysis of their strengths and limitations with regard to mobile multimedia services.

DISCOVERING MOBILE MULTIMEDIA SERVICES AND CONTENTS IN INFRASTRUCTURE-BASED ENVIRONMENTS

Overview

In order to use various multimedia services on the network, the first necessary step is to find the exact address of service providers that implement the service. In most cases, end users might only know what kind of service (service type) and some service characteristics (e.g., data format, cost) they want, but without having the server address. Currently, browsing is one often-used method to locate relevant information. As the num-ber and diversities of services on the network grow, mobile users may be overwhelmed by the sheer volume of available information, particularly in an unacquainted environment. On the other side, user mobility presents new challenges for service access. Mobility means that users probably change their geographic locations frequently. Consequently, services available to users will appear or disappear dynamically while users move here and there. Moreover, mobile users are often interested in the services, (e.g., malls, restaurants) in the close proximity of his or her current place. Therefore, unlike classical distributed environments where location is often kept transparent, applications often need to dynamically obtain information that is relevant to their current location. The service search procedure should be customized according to user's context, (e.g., in terms of when (i.e., time) and where (i.e., location) a user is visiting).

Since most current multimedia services are designed for stationary environments, they do not address these issues. Recently, a number of service discovery solutions are developed. These solutions range from hardware-based technologies such as Bluetooth SDP, to single protocols, (e.g., SLP and SDS) to frameworks such as UPnP and Jini. From architectural point of view, we observed three models are used to discover services in different network environments (Wang, 2003): the broadcast model, the centralized service directory model, and the distributed service directories model. Next, we will investigate these paradigms in detail.

Broadcast Model

The simplest architecture for service discovery is using broadcast to locate services and contents. The conceptual scheme of the broadcast model is depicted in Figure 1. In this model, clients and servers talk directly with each other through broadcast or multicast.

According to who initiates the announcement and who lis-tens, two strategies are differentiated. The first strategy is the *pull strategy* where a client announces his requests, while all servers keep listening to requests. The ser-vers that match the search criteria will send responses (using either unicast or multicast) to the client. The other strategy is the *push strategy*. The servers adver-tise themselves periodically. Clients who are interested in certain types of services listen to the service advertisements, and extract the appropriate information from service advertise-ments. Of course, hybrid strategies are applied by some approaches.

The **simple service discovery protocol (SSDP)** is one typical approach based on the broadcast model (Goland, Cai, Leach, Gu, & Albright, 1999). The SSDP builds upon HTTP and UDP-multicasting protocols, and employs a hybrid structure combin-ing client announcement and service announcement. When a device is newly added to the network, it multicasts an "ssdp:alive" message to advertise its presence. Simi-larly, when a client wants to discover services, it multicasts a discovery message and awaits responses.

Figure 1. Broadcast model

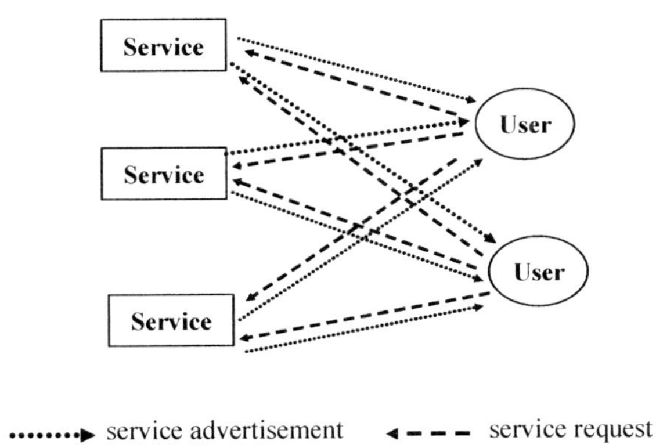

Figure 2. Centralized directory model

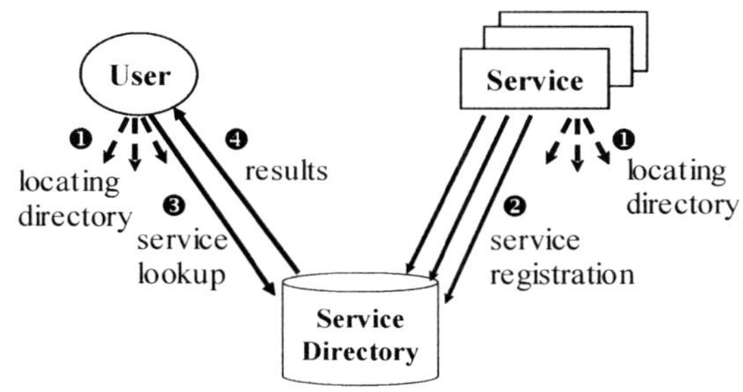

The broadcast model works well in small simple net-works, such as home and small office. The primary advantage of such systems is that they need "zero" or little configuration and administration. Besides, they accommo-date well to frequent service join/leave actions in a dynamic environment. However, they usually generate heavy network traffic due to broadcast, and thus have only minimal scalability.

In order to improve scalability and performance, an additional entity, service directory, is introduced. Two different models use the service directory: the centralized service discovery model and the distributed service directories model. Both models will be presented in the following sections.

Centralized Service Directory Model

The conceptual scheme of the centralized directory model is shown in Figure 2. The service directory becomes the key component in the search discovery architecture, because it stores information about all available services.

The service discovery procedure consists usually of the following steps:

1. **Locating directory:** Either clients or servers should determine the address of the service directory before they utilize or advertise services. The directory could be located by manual configuration, by querying a well-known server, or through broadcast/multicast requests/replies.
2. **Service registration:** Before a service can be found by clients, it must be registered in the appropriate directory. A service provider explicitly initiates a registration request to the directory, and the directory stores the service data in its database. The service description data include service type, service attributes, server address, etc.
3. **Service lookup:** As a client searches for a particular service, he describes his requirements, e.g. service type and desired characteristics, in a query request, and sends it to the directory.
4. **Searching:** The directory searches services in its database according to the criteria provided by the client. When services are found, the server addresses and other information of qualified services are sent back to the client.

The centralized directory model has been used by several service discovery approaches. In this section we will examine some of them.

Service Location Protocol (SLP)

The service location protocol (SLP) is an example of centralized directory-based solution, and is now an IETF standard (Guttman, Perkins, Veizades, & Day, 1999). The current version is SLP Version 2 (SLPv2). The SLP uses DHCP options, or UDP-based multicasting to locate the service directory (known as directory agent (DA)), without manual configuration on individual clients and services (known as user agents (UAs), service agents (SAs) respectively). A multicast convergence algorithm is adopted in SLP to enhance multicast reliability.

Service registration and lookup are performed through UDP-based unicast communication between UAs/SAs and DAs. In addition, SLP can operate without DAs. In this mode, SLP works in the same way as the broadcast model. A service in SLP is described with service type in the form of a character string, the version, the URL as server address, and a set of attribute definitions in the form of key-value pairs.

To improve performance and scalability, more DAs can be deployed in network. How-ever, SLPv2 does not provide any synchronization mechanisms to keep DAs consistent, but leaves this responsibility to SAs which should register with each DA they detect. Recently, (Zhao & Guttman, 2000) proposed a mesh enhancement for DAs to share known services between one another. Each SA needs to register only with a single DA, and its registration is automatically propagated among DAs.

Generally, SLP is a flexible IP-based service discovery protocol which can operate in networks ranging from a single LAN to an enterprise network. However, it is intended to function within networks under cooperative administrative control, and thus does not scale for the Internet.

JINI

Sun's Jini provides a similar architecture as SLP for delivering services in a net-work (Sun Microsystems Inc., 2003), but it is tightly bound to the Java environ-ment and needs Java Virtual Machine (JVM) support. The protocols in

JINI are implemented as Java APIs. For this reason, the JINI client is not as lightweight as the SLP client. However, JINI is more than a discovery protocol. It provides further facilities for service invocation, for transaction, and for distributed events.

INS

Adjie-Winoto, Schwartz, Balakrishnan, and Lilley, (1999) proposes a resource discovery system named *intentional naming system* (INS). The main idea is that resources or services are named using an ordered list of attribute-value pairs. Since ser-vice characteristics can be described by the service name itself, the service discovery procedure is equal to name resolving which is accompanied by the *intentional name resolver* (INR). The INR is actually a service directory that holds the global knowledge about names in the whole network. INS is different from other naming services (e.g., DNS), in that the name describes service attributes and values, rather than simple network locations of objects.

In conclusion, most centralized directory-based architectures have been designed for local net-works or enterprise-wide networks which are under a common administration. The primary issue for these systems is scalability. As the number of services and clients increases, a centralized directory, even replicated, will not be feasible to accommodate a large number of registrations and lookups. In this context, the distributed repositories model has been suggested.

Distributed Service Directories Model

In the distributed directories model, the whole service domain is split into partitions, possibly according to organizational boundary, network topology, geographic locations etc. In each partition, there are one or more directories. The conceptual scheme of the distributed directories model is shown in Figure 3. The distributed directories model is different from the centralized directory model in that no directory has a complete global view of services available in the entire domain. Each directory holds only a collection of ser-vices in its partition, and is responsible for interaction with clients and services in the partition.

The service registration and query submis-sion in the distributed model remain similar to that in the centralized directory model. But the service search operation becomes more complicated. If required services can be found by local directories, the discovery procedure is akin to that in the centralized directory model. But if not, the directories in other partitions

Figure 3. Distributed directories model

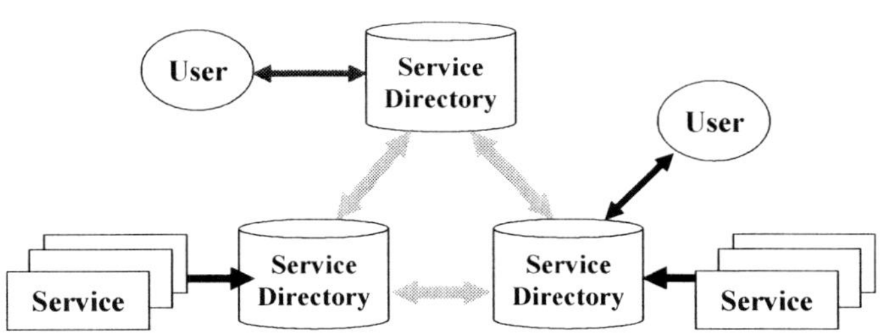

should be asked, to ensure that a client can discover any service offers in the entire domain.

The directories in this model are organized in some way to achieve cooperation. As stated in (Wang, 2003), the directories can be organized in a hierar-chy structure or in a mesh structure. While in the hierarchy structure there is a "belong to" relationship between directly connected directories, directories in the mesh architecture are organized in a flat interconnected form without hierarchy. The interconnection structure might have strong implications on query routing. In the hierarchy structure queries are passed along the hierarchy, either upward or downward, thus the routing path is inherently loop free. But the rigid hierarchy obstructs to shortcut the routing path in some cases. On the other hand, the mesh structure is advantageous for optimizing the routing path, but might rely on some mechanisms to avoid loop circles or repeated queries.

A typical example of distributed directories-based architecture is **service discovery service (SDS)**, developed in Berkeley (Hodes, Czerwinski, Zhao, Joseph, & Katz, 2002). The SDS is based on the hierarchy model which is maintained by periodic "heartbeat" messages between parent and child nodes. Each SDS server pushes service announcements to its parent. By this means, each SDS server gathers a complete view of all services present in its underlying tree. The significant feature of SDS is the hierarchical structure with lossy aggregation to achieve better scalability and reachability. The SDS server applies multiple hash functions (e.g., MD5), to various subsets of tags in the ser-vice description and uses the results to set bits in a fixed-size bit vector. The parent node ORs all bit vectors from its children to summarize available services in the underlying tree.

The hierarchical structure with lossy aggregation helps SDS to reach better scalability, while ensuring users to be able to discover all services on all servers. However, the SDS is more favorable for applying in stationary net-work environments since it requires additional overheads to maintain the hierarchi-cal structure and to propagate index updates. If services change attributes rapidly or join/leave frequently, it will generate too much communication burden. Moreover, the OR-operation during aggregation may cause "false positive" answers in query routing. Although it does not sacrifice correctness, it will lead to unneeded additional query forwarding.

The media gateway discovery protocol (MeGaDiP) is developed especially for discover-ing media gateways that act as proxy for transforming or caching data between media source and end users (Xu, Nahrstedt, & Wichadakul, 2000). In MeGaDiP the discovery procedure starts from the local directory, and forwards the query to directories along the routing path of the network layer between media source and destination. This idea is driven by the heuristics that a media gateway on or close to the end-to-end path is likely to find more bandwidth and/or to incur smaller end-to-end delay.

Other Issues in Service Discovery

The architectural models and various approaches presented above solved the service discovery problem to some extent. However, in order to let users comfortably and effectively locate mobile multimedia services and contents, there are still some issues to be addressed. From our point of view, interoperability, asynchronous service discovery, and semantic service discovery are the most important.

Interoperability

As previously stated, a number of service discovery approaches have been proposed. Despite that most of them provide similar functionality, namely automati-cally discovering services based on service characteristics, they have differ-ent features and are not compatible with each other. This incompatibility is one of the biggest obstacles for mobile users to really benefit from service discovery. From our point of view, it is more useful to make different approaches interoperable, than to design a new protocol to cover functionalities of existing protocols. So far, some solutions have been proposed to bridge service discovery mechanisms, but they are limited to pair-wise bridges, such as JINI to SLP (Guttman & Kempf, 1999). Authors in Friday, Davies, and Catterall (2001) proposed a general solution on a modified form of the Structured Query Language (SQL). However, no implementation details are presented in the paper. More generally, Wang and Seitz (2002) addressed this issue by providing an intermediary layer between mobile users and underlying service discovery protocols. The intermediary layer on the one hand provides clients with a general consistent view of service configuration and a universal means to formulate search requests, on the other hand is capable of talking with various types of service discovery protocols and handling service requests from users.

Asynchronous Service Discovery

Apart from the heterogeneous environments, most of the existing approaches rarely take the issues of thin client and poor wireless link into consideration. For example, synchro-nous operation is one of the intrinsic natures of most exist-ing service discov-ery ap-proaches, such as SLP, Jini, and SDS. Although synchronous operation simplifies protocol and application design, it is fastidious for mobile environments. The unexpected but fre-quent disconnec-tions and possible long delay of wireless link greatly influence the useful-ness and efficiency of synchronous calls. To relax the communication restrains in wireless environments, (Wang & Seitz, 2002) proposed in their CHAPLET system an approach to achieve asynchronous service discovery by adopting mobile agents. The asynchronous service discovery allows mobile users to submit a service request, without having to wait for results, nor continuously keeping the permanently active connection in the process of service discovery.

Semantic Service Discovery

Most existing service discovery approaches support only syntactic-level searching (i.e., based on attribute comparison and exact value matching). However, it is often insufficient to represent a broad range of multimedia services in real world, and lacks of capability to apply inexact matching rules. Therefore, there is need to discover services in a semantic manner. Chakraborty, Perich, Avancha, and Joshi (2001) proposes in the DReggie project to use the features of DAML to reason about the capabilities and functionality of different services. They designed a DAML-based language to describe service functionality and capability, enhanced the Jini Lookup Service to enable semantic matching process, and provided a reasoning engine based on Prolog. Yang (2001) presents a centralized directory-based framework for semantic service discovery. However, the semantic-based service discovery is still in its infancy. To promote wide development of semantic service discovery, more research efforts should be devoted.

DISCOVERING MULTIMEDIA SERVICES AND CONTENTS IN AD HOC ENVIRONMENTS

Overview

There are two well-known basic variants of mobile communication networks: infrastructure-based networks and ad hoc networks. Mobility support described in the previous sections relies on the existence of some infrastructure. A mobile node in the infrastructure-based networks communicates with other nodes through the access points which act as bridge to other mobile nodes or wired networks. Normally, there is no direct communication between mobile nodes. Compared to infrastructure-based networks, ad hoc networks do not need any infrastructure to work. Nodes in ad hoc networks can communicate if they can reach each other directly or if intermediate nodes can forward the message. In recent years, mobile ad hoc networks are gaining more and more interest both in research and industry. In this section we will present some typical approaches that enable discover and locate mobile multimedia services and contents in ad hoc environments. First we present broadcast-based approaches, and then the geographic service location approach is discussed. Next, a cluster-based approach is introduced. Finally, we present a new service or content location solution that addresses the scalability problem in multi-hop ad hoc networks.

Broadcast-Based Approaches

Considering the fact that no infrastructure is available in ad hoc environments, service directory-based solutions are unusable for service discovery in ad hoc networks. Instead, assuming that network supports broadcasting, service discovery through broadcast is one of most widely adopted solutions. Two broadcast-based approaches are possible: (1) broadcasting client requests and (2) broadcasting service announcements. In the first approach, clients broadcast their requests to all the nodes in the ad hoc network. Servers hosting requested services reply back to the clients. In the second approach, servers broadcast their services to all the nodes in the network. Each client is thus informed about the location of every service in the ad hoc network. Since these both approaches are mainly based on broadcasting, their efficiency strongly depends on the broadcast efficiency. The service location problem in that context can be reduced to the broadcast problem in ad hoc networks. For this reason, in the following, we present a summary of proposed approaches for broadcasting in ad hoc networks. These broadcast approaches are not designed specifically for service location but we believe that a broadcast-based service location protocol has to be informed about how broadcast is carried out. This will help in deploying a cross layer-based service location protocol.

The broadcast techniques can be categorized into four families: Williams and Camp (2002), simple flooding, Jetcheva, Hu, Maltz, and Johnson (2001), probabilistic broadcast, Tseng, Ni, Chen, and Sheu (1999), location-based broadcast, and neighbor information broadcast, Lim and Kim (2000) and Peng and Lu (2000). Flooding represents a simple mechanism that can be deployed in mobile ad hoc networks. Using flooding, a node having a packet to be broadcasted sends this packet to his neighbors who have to retransmit it to their own neighbors. Every node receiving the packet for the first time has to retransmit it. To reduce the number of transmissions used in broadcasting, other broadcast approaches are proposed. The probabilistic broadcast is similar to flooding except that nodes have to retransmit the broadcast packet with a predetermined probability. Randomly choosing the nodes that have to retransmit can improve the bandwidth use without influencing the reachability. In the case of location-based broadcast techniques, a node x retransmits the broadcast packet received from a node y only if the distance between x and y exceeds a specific threshold.

The information on the neighborhood can also be used to minimize the number of nodes participating in the broadcast packet retransmission. Lim and Kim (2000) uses the information about the one hop neighborhoods. Node A, receiving a broadcast packet from node B, compares its neighbors to those of B. It retransmits the broadcast packet only if there are new neighbors that will be covered and that will receive the broadcast packet. Other broadcast protocols are based on the 2 hop neighborhood information. The protocol used in Peng and Lu (2000) is similar to the one proposed in Lim and Kim (2000). The difference is that in Lim and Kim (2000) the neighborhood information is sent within HELLO packets, whereas in Peng and Lu (2000), the neighborhood information is enclosed within the broadcast packet.

The study carried out in Williams and Camp (2002) showed that the probabilistic and location broadcast protocols are not scalable in terms of the number of broadcast packet retransmissions. The neighborhood-based broadcast techniques perform better by minimizing the number of nodes participating to the broadcast packet retransmission. The most significant disadvantage of these protocols is that they are sensitive to mobility.

Geographic Service Location Approaches

A more interesting service location approach than broadcasting the whole network is to restrict broadcasting to certain regions. These regions can be delimited on the basis of predefined trajectories. In fact, recently, geometric trajectories are proposed to be used for routing (Nath & Niculeson, 2003) and content location in location-aware ad hoc networks (Aydin & Shen, 2002; Tchakarov & Vaidya, 2004). Aydin and Shen (2002) and Tchakarov and Vaidya (2004) are closely related where content advertisements and queries are propagated along four geographical directions based on the physical location information of the nodes. At the intersection point of the advertising and query trajectories the queries will be resolved. Moreover, Tchakarov and Vaidya (2004) improves the performance by suppressing update messages from duplicate resources. However, basically they still rely on propagating advertisements and queries through the network.

Cluster-Based Solutions

Besides enhancements in broadcast, clustering can also be used to improve the performance of service discovery in mobile ad hoc networks. An interesting cluster-based service location approach designed for ad hoc networks is proposed in Koubaa and Fleury (2001) and Koubaa (2003). The proposed approach involves four phases: (1) the servers providing services are organized within clusters by using a clustering protocol. The cluster-heads, elected on the basis of an election protocol, have the role of registering the addresses of the servers in their neighborhoods (clusters). (2) A reactive multicast structure gathering the cluster-heads to which participate the cluster-heads of the created clusters is formed at the application layer. Each client or a server in the network is either a part of this structure or one hop away from at least one of the multicast structure members. (3) Clients send their request inside this multicast structure. (4) An aggregation protocol is used to send the replies of the cluster-heads within the multicast structure. The aim of the aggregation protocol is to avoid using different unicast paths for reply transmission by using the shared paths of the multicast structure.

A study comparing broadcast approaches to the cluster-based approach is carried out in Koubaa and Fleury (2002). This comparison study showed that clustering reduces the overhead needed for clients to send their requests and for servers to send back their replies. This reduction is noticeable when we increase the number of clients, the number of servers, and the number of nodes in the ad hoc network. The multicast structure used in Koubaa (2003) consists of a mesh structure which is more robust than a tree structure. The density of the mesh structure is dynamically adapted to the number of clients using it. The key idea of this dynamic density mesh structure is that the maintaining of the mesh is restricted to some clients called effective clients. Indeed, when the network is dense or the number of clients is high there is no need that all clients participate the multicast structure maintaining. This new mesh structuring approach is compared to ODMRP (Koubaa, 2003) where all the multicast users participate in the mesh maintaining. The comparison study showed that the proposed dynamic density mesh is more efficient than ODMRP. Compared to the tree-based multicast structure, the mesh-based multicast structure shows better server reply reachability performance but using more bandwidth.

Scalability Issue in Service Location

Currently it is well known that ad hoc networks are not scalable due to their limited capacity. The scalability problem is mainly related to the specific characteristics of the radio medium limiting the effective ad hoc network capacity. Even though, we think that designing specific solutions for scalable networks can help us at defining how much scalable is an ad hoc network. In the context of service location, authors in Koubaa and Wang (2004) state the problem of scalable service location in ad hoc networks and propose a new solution inspired by peer-to-peer networks called HCLP (hybrid content location protocol). The main technical highlights in approaching this goal include: (1) the hash function for relating content to zone, (2) recursive network decomposition and recomposition, and (3) content dissemination and location-based on geographical properties.

The hashing technique is used in HCLP both for disseminating and locating contents. But unlike the approaches in peer-to-peer systems where the content is mapped to a unique node, the hash function in HCLP maps the content to a certain zone of the network. A zone means in HCLP a certain geographical area in the network. The first reason for mapping content into zone, i.e. a subset of nodes, instead of an individual node, is mainly due to the fact that it could be expensive in radio mobile environments to maintain a predefined rigid structure between nodes for routing advertisements and queries. For example, in Stoica, Morris, Karger, Kaashoek, and Balakrishnan (2001), each joining and leaving of nodes has to lead to an adjustment of the Chord ring. Moreover, the fact that the routing in ad hoc networks is far less efficient and less robust than in fixed networks makes the adjustments more costly if there is node movement. The second reason for relating content to zone is that it is more robust to host a content within many nodes inside a zone than to host it within an individual node.

The underlying idea of network decomposition in HCLP is to achieve load distribution by maintaining the zone structure. It is well known that if the number of the nodes and contents in an unstructured and decentralized zone is beyond a certain limit, the network overhead related to content advertisement/location would become unsatisfactory. Therefore, to ensure a favorable performance and to achieve a better load distribution in HCLP, a zone could be divided into sub-zones recursively if the cost related to content advertisement/location using unstructured approaches in the zone exceeds a certain threshold.

To enable network decomposition in different zones a protocol is deployed to make it possible to nodes on the perimeter of the network exchanging their geographical locations. This will help estimating the position of the centre of the network. Knowing the locations of the nodes on the perimeter and the location of the network centre, a simple decomposition of the network into four zones is used. Each of these zones can also be decomposed again into four zones, etc.

In HCLP, for disseminating or locating a content in the network, a user first sends out its announcement or query request along one of four geographical directions (north, south, east, and west) based on geographic routing. In a dense network, the announcement or the request will then be caught on the routing path by a node that knows the central region of the network, in the worst case by a perimeter node on the network boundary. This node will then redirect the request into the direction of the central region, again by geographic routing. The node that belongs to the central region and receives this query message has the responsibility to decide whether to resolve the request directly within the zone or whether to redirect the request to the next level of the zone hierarchy, until the content is discovered.

Such a content dissemination and location scheme works completely decentralized. Moreover, only a small portion of nodes is involved in routing and resolving advertisement or query messages. Because not all nodes are necessary for maintaining routing information nor a global knowledge of the whole network is required, HCLP can be expected to be well scalable to large ad hoc networks.

CONCLUSION

The prevalence of portable devices and wide deployment of easily accessible mobile networks promote the usage of mobile multimedia services. In order to facilitate effectively and efficiently discovering desirable mobile multimedia services and contents, many research efforts have been done. In this chapter, we discussed existing and ongoing research work in the service discovery field both for infrastructure-based mobile networks and mobile ad hoc networks. We introduced three main architectural models and related approaches for service discovery in infrastructure networks, and pointed out some emerging trends. For discovering services and contents in ad hoc networks, we presented and compared proposed approaches based on either broadcast or cluster, and discussed the scalability issue in detail. We believe that service discovery will play an important role for successful development and deployment of mobile multimedia services.

REFERENCES

Adjie-Winoto, W., Schwartz, E., Balakrishnan, H., & Lilley, J. (1999). The design and implementation of an intentional naming system. In *Proceedings of the 17th ACM Symposium on Operating Systems Principles (SOSP ´99).*

Aydin, I., & Shen, C. (2002, October). *Facilitating match-making service in ad hoc and sensor networks using pseudo quorum*. In the 11th IEEE International Conference on Computer Communications and Networks (ICCCN).

Chakraborty, D., Perich, F., Avancha, S., & Joshi, A. (2001, October). *DReggie: Semantic service discovery for m-commerce applications*. In the Workshop on Reliable and Secure Applications in Mobile Environment, in Conjunction with 20th Symposium on Reliable Distributed Systems (SRDS).

Friday, A., Davies, N., & Catterall, E. (2001, May). Supporting service discovery, querying, and interaction in ubiquitous computing environments. In *Proceedings of the 2nd ACM International Workshop on Data Engineering for Wireless and Mobile Access,* Santa Barbara, CA (pp. 7-13).

Goland, Y., Cai, T., Leach, P., Gu, Y. & Albright, S. (1999). *Simple service discovery protocol*. IETF Draft, draft-cai-ssdp-v1-03.txt.

Guttman, E., & Kempf, J. (1999). Automatic discovery of thin servers: SLP, Jini, and the SLP-Jini Bridge. In *Proceedings of the 25th Annual Conference of IEEE Industrial Electronics Society (IECON'99),* Piscataway, USA.

Guttman, E., Perkins, C., Veizades, J., & Day, M.(1999). *Service location protocol, version 2. IETF* (RFC 2608). Retrieved from http://www.ietf.org/rfc/rfc2608.txt

Hodes, T. D., Czerwinski, S. E., Zhao, B. Y., Joseph, A. D., & Katz, R. H. (2002, March/May). An architecture for secure wide-area service discovery. *ACM Wireless Networks Journal, 8*(2-3), 213-230.

Jetcheva, J., Hu, Y., Maltz, D., & Johnson, D. (2001, July). *A simple protocol for multicast and broadcast in mobile ad hoc networks*. Internet Draft draft-ietfmanet-simple-mbcast-01.txt, Internet Engineering Task Force.

Koubaa, H. (2003). *Localisation de services dans les réseaux ad hoc*. PhD thesis, Université Henri Poincaré Nancy,1, Mars 2003.

Koubaa, H., & Fleury, E. (2001, November). *A fully distributed mediator based service location protocol in ad hoc networks*. In IEEE Symposium on Ad hoc Wireless Networks, Globecom, San Antonio, TX.

Koubaa, H., & Fleury, E. (2002, July). *Service location protocol overhead in the random graph model for ad hoc networks*. In the IEEE Symposium on Computers and Communications, Taormina/Giardini Naxos, Italy.

Koubaa, H., & Wang, Z. (2004, June). *A hybrid content location approach between structured and unstructured topology*. In the 3rd Annual Mediterranean Ad hoc Networking Workshop, Bodrum, Turkey.

Lim, H., & Kim, C. (2000, August). *Multicast tree construction and flooding in wireless ad hoc networks*. In ACM MSWiM, Boston.

Nath, B., & Niculescu, D. (2003). Routing on a curve. *SIGCOMM Computer Communication Review, 33*(1), 155-160.

Peng, W., & Lu, X. (2000, August). *On the reduction of broadcast redundancy in mobile ad hoc networks*. In the 1st ACM International Symposium on Mobile Ad hoc Networking and Computing (MobiHoc), Boston.

Stoica, I., Morris, R., Karger, D., Kaashoek, M. F., & Balakrishnan H. (2001). Chord: A scalable peer-to-peer lookup service for internet applications. In *Proceedings of the 2001 Conference on Applications, Technologies, Architectures, and Protocols for Computer Communications* (pp. 149-160). ACM Press.

Sun Microsystems Inc. (2003). *Jini technology core platform specification, version 2.0*. Retrieved June, 2003, from http://www.jini.org/

nonav/standards/davis/doc/specs/html/core-title.html

Tchakarov, T., & Vaidya, N. (2004, January). *Efficient content location in wireless ad hoc networks*. In the IEEE International Conference on Mobile Data Management (MDM).

Tseng, Y., Ni, S., Chen, Y., & Sheu, J. (1999, August). The broadcast storm problem in a mobile ad hoc network. *5th Annual International Conference on Mobile Computing (MOBICOM)*, Washington, DC, *31*(5), 78-91.

Wang, Z. (2003). *An agent-based integrated service platform for wireless and mobile environments*. Aachen, Germany: Shaker Verlag.

Wang, Z., & Seitz, J. (2002). An agent based service discovery architecture for mobile environments. In *Proceedings of the 1st Eurasian Conference on Advances in Information and Communication Technology,* Shiraz, Iran, October (LNCS 2510, pp. 350-357). Springer-Verlag.

Wang, Z., & Seitz, J. (2002, October). Mobile agents for discovering and accessing services in nomadic environments. In *Proceedings of the 4th International Workshop on Mobile Agents for Telecommunication Applications,* Barcelona, Spain (LNCS 2521, pp. 269-280). Springer-Verlag.

Williams, B., & Camp. (2002, June). *Comparison of broadcasting techniques for mobile ad hoc networks*. In the 3rd ACM International Symposium on Mobile Ad hoc Networking and Computing (MobiHoc), Lausanne, Switzerland.

Xu, D., Nahrstedt, D., & Wichadakul, D. (2000). *MeGaDiP: A wide-area media gateway discovery protocol*. In the 19th IEEE International Performance, Computing, and Communications Conference (IPCCC 2000).

Yang, X. W. (2001). A framework for semantic service discovery. In *Proceedings of the Student Oxygen Workshop, MIT Oxygen Alliance, MIT Computer Science and Artificial Intelligence Laboratory, 2001*. Retrieved from http://sow.csail.mit.edu/2001/proceedings/yxw.pdf

Zhao, W., & Guttman, E. (2000). *mSLP–Mesh enhanced service location protocol*. Internet Draft draft-zhao-slp-da-interaction-07.txt.

KEY TERMS

Aggregation: A process of grouping distinct data. Two different packets containing different data can be aggregated into a single packet holding the aggregated data.

Broadcast: A communication method that sends a packet to all other connected nodes on the network. With broadcast, data comes from one source and goes to all other connected sources at the same time.

Clustering: Identifying a subset of nodes within the network and vest them with the responsibility of being a cluster-head of certain nodes in their proximity.

Hash: Computing an address to look for an item by applying a mathematical function to a key for that item.

Mobile Ad Hoc Network: A kind of self-configuring mobile network connected by wireless links where stations or devices communicate directly and not via an access point. The nodes are free to move randomly and organize themselves arbitrarily, thus, the network's topology may change rapidly and unpredictably.

Multicast: A communication method that sends a packet to a specific group of hosts.

With multicast, a message is sent to multiple destinations simultaneously using the most efficient strategy that delivers the messages over each link of the network only once and only creates copies when the links to the destinations split.

Scalability: The ability to expand a computing solution to support large numbers of components without impacting performance.

Service: An abstraction function unit with clearly defined interfaces that performs a specific functionality. Users, applications, or other services can use the service functionality through well-known service interfaces without having to know how it is implemented.

Service Directory: An entity in service discovery architecture that collects and stores information about a set of services within a certain scope, which is used for searching and/or comparing services during the service discovery procedure. Service directory is also known as service repository or directory agent. Service directory can be organized in central or distributed manner.

Service Discovery: The activity to automatically find out servers in the network based on the given service type and service attributes. The service discovery is, therefore, a mapping from service type and attributes to the set of servers.

Chapter XIII
A Fast Handover Method for Real-Time Multimedia Services

Jani Puttonen
University of Jyväskylä, Finland

Ari Viinikainen
University of Jyväskylä, Finland

Miska Sulander
University of Jyväskylä, Finland

Timo Hämäläinen
University of Jyväskylä, Finland

ABSTRACT

Mobile IPv6 (MIPv6) has been standardized for mobility management in the IPv6 network. When a mobile node changes its point of attachment in the IPv6 network, it experiences a time due MIPv6 procedures when it cannot receive or send any packets. This time called the handover delay might also cause packet loss resulting undesired quality-of-service degradation for various types of applications. The minimization of this delay is especially important for real-time applications. In this chapter we present a fast handover method called the flow-based fast handover for Mobile IPv6 (FFHMIPv6) to speed up the MIPv6 handover processes. FFHMIPv6 employs flow information and IPv6-in-IPv6 tunneling for the fast redirection of the flows during the MIPv6 handover. Also, FFHMIPv6 employs a temporary hand-off-address to minimize the upstream connectivity. We present the performance results comparing the FFHMIPv6 method to other fundamental handover methods with Network Simulator 2 (ns-2) and Mobile IPv6 for Linux (MIPL) network.

INTRODUCTION

In the last few years, the number of mobile devices as well as a variety of possible access technologies have increased. More importantly, the mobile device will have several integrated access technologies. Already, new mobile phones have integrated IEEE 802.11b Wireless

LAN and Bluetooth interfaces, in addition to traditional cellular systems such as GSM (Global System for Mobile Communications) and GPRS (General Packet Radio Service). These different access technologies have different characteristics related to Quality-of-Service (e.g., bandwidth), coverage area, cost, power consumption, etc. (Frodigh, Parkvall, Roobol, Johansson, & Larsson, 2001). The access technologies might also provide their specific link layer handover mechanisms. But, for the mobile terminal to be always globally accessible, some upper layer **mobility management** technique is necessary, such as **Mobile IPv6** (MIPv6) (Johnson, Perkins, & Arkko, 2004). This is emphasized as the IP protocol seems to be the enabling technology to both applications and access networks (Berezdivin, Brenig, & Topp, 2002). IP technology integrates all access technologies to one heterogeneous **All-IP** network (i.e., integration of traditional cellular networks and IP data networks is inevitable).

Although MIPv6 enables the mobility at the IP layer, the processes related to MIPv6 mobility management result in a short period of time when the mobile node (MN) cannot receive or send any packets. This time, called the handover delay, degrades the performance of especially real-time applications such as multimedia or voice over IP (VoIP). In current days, technologies such as **IP-TV** and **VoIP** phone calls are becoming more and more popular because of low prices and integration features of the IP protocol. For example the Skype VoIP call program has 120 million downloads thus far. Also cellular manufacturers constantly introduce more efficient phones with new multimedia software. Mobile TV is shortly becoming reality.

Even though today these VoIP calls and multimedia streaming services are usually used between static desktop machines, the inevitable direction is towards mobile terminals with wireless access. This requires efficient IP-based mobility management when the user is traveling between IP subnets. Thus the mobility management problem has been under heavy research for many years and as a result, several Mobile IPv6 enhancements have been proposed in the academic literature.

In this chapter we present a **flow-based fast handover for Mobile IPv6** (FFHMIPv6) method (Sulander, Hämäläinen, Viinikainen, & Puttonen, 2004) as a solution to the handover delay problem. In FFHMIPv6, the flows of the mobile node (MN) can be redirected to the current location simultaneously with the address registration process by using flow state information and IPv6-in-IPv6 tunneling. The Mobile IPv6 specifies that the MN can send upstream data only after receiving a binding acknowledgment (BACK) to the binding update (BU) from the home agent (HA). In FFHMIPv6, this is solved by assigning a temporary hand-of-address (HofA) to the MN during the handover process (Viinikainen, Kašák, Puttonen, & Sulander, in press).

In this chapter we will present the Mobile IPv6 protocol and its most important enhancements for handover delay minimization, **hierarchical MIPv6** (Soliman, Castellucia, El-Malki, & Bellier, 2004) and fast handovers for Mobile IPv6 (Koodli, 2004), but the main emphasis is on the FFHMIPv6 method. The FFHMIPv6 method is presented and compared mainly against the Mobile IPv6. The idea is to give the reader a glimpse of the Mobile IPv6 protocol and the constant research that is being performed around it.

BACKGROUND

Applications with network usage need to bind themselves a network socket with a specified destination address. When a MN moves and changes its point of attachment to the network, it usually changes its IP subnet. This affects the applications in use, because the IP address in use is not valid in the new IP subnet anymore. Mobility support in IPv6 (Johnson et al., 2004) enables transparent routing of packets to the MN. This is enabled by the usage of two separate addresses: one for the applications and communicating parties and one related to the current IP subnet. In the home network, MN is assigned a permanent home address (HoA), which is used in every application layer connection. When MN changes its IP layer (OSI Layer 3) attachment in the network, it acquires a new local address from the foreign network to be again accessible. This address, called the care-of-address (CoA), is registered through a binding update (BU) process to a special router in the home network called the home agent (HA). The HA maintains a binding cache, in which the HoA-CoA bindings are stored, and employs tunneling to redirect the flows to the current CoA of the MN. Now, all of the MN's flows (both incoming and outgoing) are directed via the HA. The MN is able to receive and send packets from and to the corresponding nodes (CNs) after the binding process to HA is finished.

The Mobile IPv6 enables the applications' connections to be intact when the MN is moving between different subnets. But, the two-way tunneling between MN and HA has some problems. For example, the end-to-end communication delay between MN and its CNs is not optimal and this architecture causes one weak point to the system, the HA.

The most significant enhancement in MIPv6, compared to the mobility support in IPv4, is route optimization. Now, MN also registers its new CoA to the CNs, which also maintain a binding cache. Thus, CNs can send packets directly to the MN without two-way tunneling to the HA, therefore improving the end-to-end communication delay. This happens at the expense of the handover delay, because of an extra round-trip time caused by the BUs to CNs.

The efforts to reduce the MIPv6 handover delay can be divided into three parts: IP layer movement detection, CoA acquiring, and CoA registering delay to the HA and CN(s). When MN changes its point of attachment to another network, it receives a router advertisement from the new access router (nAR). Using the stateless or the stateful address auto-configuration, MN forms or receives a new reachable CoA. If the address is acquired with stateless address auto-configuration, it needs to be verified for its uniqueness with duplicate address detection (DAD) process. Next, the new CoA needs to be registered to MN's HA and CNs. This requires a two-way BU process to all parties.

The hierarchical Mobile IPv6 (HMIPv6) (Soliman et al., 2004) reduces the registration time of the new CoA and MIPv6 signaling load by introducing a new node called the mobility anchor point (MAP), which separates local and global mobility (also called micro and macro mobility). Mobility management inside the local domain is handled by the MAP, and between separate MAP domains by the HA. MAP acts basically as a local HA in the foreign network tunneling flows to the current location of the MN.

Local mobility is handled with the usage of two CoAs: regional CoA (RCoA) and on-link CoA (LCoA). When MN moves to an entirely new MAP domain, it receives or forms a new RCoA, which is registered to the MAP, HA, and CNs. When MN changes its point of attachment inside the MAP domain, it only needs to inform the MAP about the new on-Link CoA, which matches to the current subnet prefix. MAP intercepts the packets heading for the old LCoA and tunnels them to the new LCoA acting just like HA.

Fast handovers for Mobile IPv6 (FMIPv6) shortens the delay caused by the CoA acquiring phase and employs tunneling to reduce the packet loss during the handover. When MN receives information about the next point of attachment, it sends a router solicitation for proxy (RtSolPr) message to the old AR (oAR) to start the fast handover procedure. With the information provided in the proxy router advertisement (PrRtAdv) message, the MN formulates a prospective new CoA (nCoA) and sends a fast binding update (FBU) message. The purpose of FBU is to authorize oAR to bind previous CoA to nCoA, so that arriving packets can be tunneled to the new location of the MN. FMIPv6 describes two modes of operation, predictive and reactive. In predictive mode, MN sends FBU and receives FBACK via the oAR link, while in the reactive mode via the nAR link.

In addition to HMIPv6 and FMIPv6, numerous other methods have been proposed for reducing the handover delay in Mobile IP networks. Instead of describing all or even some of those, we will just present the key concepts behind them. The methods are based on ideas such as:

- **Local HA and tunneling:** There is a router, which is closer than the HA, that handles the mobility management together with HA. Usually this means extra tunneling, but also reduction of signaling (Thing, Lee, & Xu, 2003)
- **Buffering:** Some element in the communication chain CN, HA, and MN buffers the packets, thus reducing the packet loss during handovers (Omae, Ikeda, Inoue, Okajima, & Umeda, 2002)
- **Prediction:** The MN guesses the next point of attachment, for example on the basis of link layer information or Router Advertisements, thus it can start the MIPv6 handover processes earlier (Yegin, Njedjou, Veerepalli, Montavont, & Noel, 2004).

THE FLOW-BASED FAST HANDOVER FOR MOBILE IPV6

Even though there has been a lot of research and proposals to decrease the handover time of MIPv6, the existing methods have also unwanted features. The Link Layer dependency, as in FMIPv6, requires that all of the link layer technologies support the link information (e.g., Link Up and Link Down) that can be used to anticipate the shortly occurring handover. In hierarchical MIPv6, the MAP can be seen as a weak point in the architecture. The IP layer mobility management technique needs, in addition to being effective, robustness, and simplicity.

We propose flow-based fast handover for Mobile IPv6 (Sulander et al., 2004) for reducing handover delay in the Mobile IPv6 networks. FFHMIPv6 is an interoperable and fully backward compatible enhancement for MIPv6. It uses flow state information and IPv6-in-IPv6 tunneling to enable reception of flows during the BU process. For upstream traffic, the access routers provide temporary addresses for the MNs to be used during address registration processes (Viinikainen et al., in press).

The functionality of the FFHMIPv6 method is described step by step with the help of Figure 1 and Figure 2.

1. The MN is communication two-way with a corresponding node using route optimization (e.g., two-way VoIP call)
2. After link layer handover is performed (e.g., IEEE 802.11), MN detects that it has moved to different IPv6 subnet from router advertisement (RA) messages that the access routers send periodically

Figure 1. The functionality of the FFHMIPv6 method

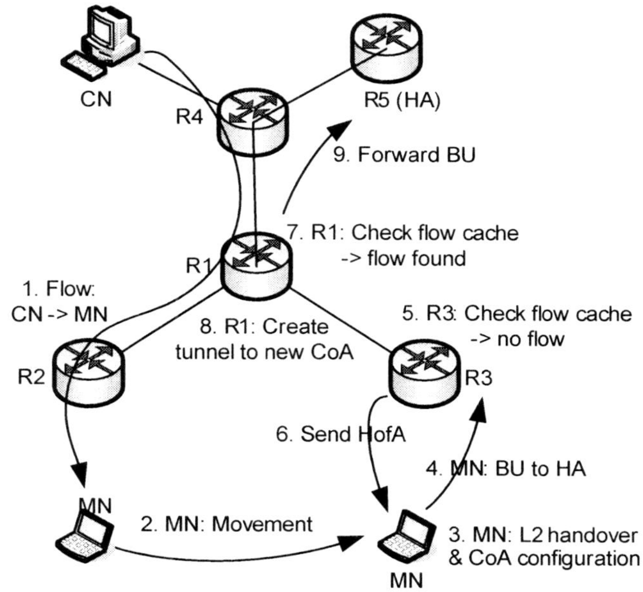

3. The MN configures a new valid CoA with stateless or stateful address auto configuration and possibly performs DAD
4. The MN registers the new CoA to the HA via BU process. In the FFHMIPv6 method the hop-by-hop header, including the old CoA and the addresses of the CNs, is added to the BU register message heading for the HA. The goal of this BU message is to redirect all of the MN's flows to the new location
5. When router R3 receives the BU, it checks its flow cache, if it has routed the mobile node's flows (i.e., CN->oCoA). In this case the flow is not found and the BU is forwarded to the next hop
6. The router R3 responds with a temporary handover address (HofA) in a special type of binding acknowledgment (BACK) message. This address can be used in upstream communication without having to wait for the BACK message from the HA. Now the upstream VoIP traffic is enabled to the CN
7. Router R1 checks its flow cache after receiving the BU and now the correct flow (i.e., CN->oCoA) is found
8. Router R1 creates a tunnel to the new CoA, thus all the packets from CN to old CoA are encapsulated to the new CoA. The CN address is removed from the hop-by-hop header, so that the FFHMIPv6 procedures are not performed twice for the same flow. Now the downstream VoIP traffic is enabled from the CN to the MN
9. Finally the BU message is forwarded towards the HA. With the FFHMIPv6 method the flow is received even before the BU has reached the HA. With the MIPv6 the MN would have to wait for the BACK from HA, return routability procedure to CN and BU process to CN

Figure 2. The functionality of the FFHMIPv6 method in flow chart form

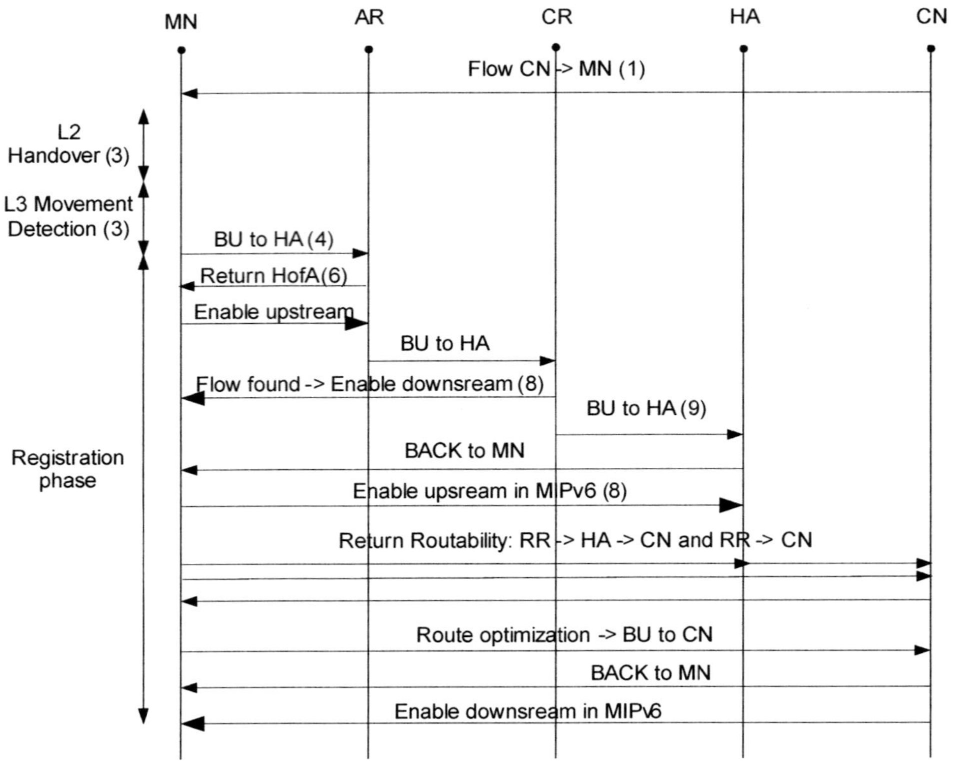

The FFHMIPv6 method is designed to be used as a micro mobility solution. Network topologies are often built hierarchically, so that all of the domains ingress and egress traffic pass through the same router (border router). Given this assumption, the crossover router would very likely be found in most networks. If the flows are not found from the routers' flow cache or the routers do not support FFHMIPv6, normal MIPv6 BU process is applied. In Figure 3 and Figure 4, we have compared the FFHMIPv6 downstream tunneling to Mobile IPv6 in such hierarchical network. Figure 3 corresponds to theoretical **analysis** results (Sulander et al., 2004) and Figure 4 to Network Simulator 2 (ns-2) **simulation** results (Puttonen, Sulander, Viinikainen, Hämäläinen, Ylönen, & Suutarinen, in press). In the optimal case the crossover router is found near the MN, thus the flow is redirected to the new CoA quite fast. In the worst case the crossover router is not found at all, thus the FFHMIPv6 is functioning as effectively as the Mobile IPv6. In the simulative results the MIPv6 is functioning slightly better than was assumed. This is due the fact that the return routability is not implemented in ns-2, so the results related to the MIPv6 are about one third better than in reality.

One benefit of FFHMIPv6 is that the handover delay does not depend on the distance of the CNs. With Mobile IPv6, the handover delay is directly related to the distance of the CNs, because the handover process consists of two-way BU process to the HA, return

Figure 3. Theoretical analysis in the optimal and the worst case

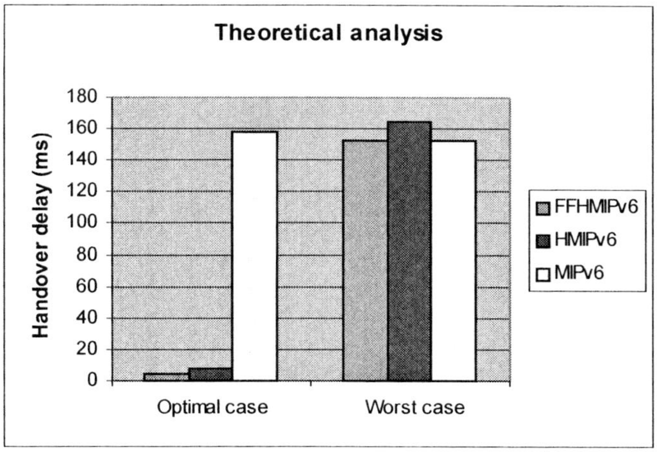

Figure 4. Simulative analysis in the optimal and the worst case

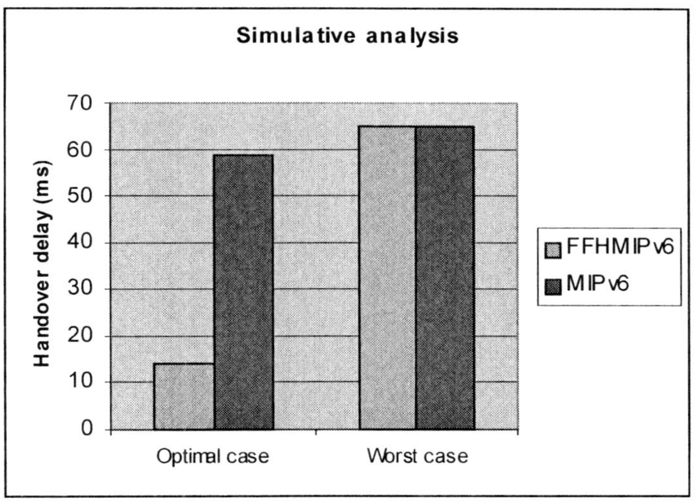

routability procedure, and two-way BU process to CNs. With FFHMIPv6 in the hierarchical scenarios the crossover router is found always quite near, so the MN's flows can be directed with one BU message.

Figure 5 and Figure 6 (Puttonen et al., in press) show the results when the distance of the CN is increased by causing extra delay between the MN and CN. The simulative results have been achieved with Network Simulator 2 and real environment results from Mobile IPv6 for Linux (MIPL) environment. The results clearly state that the downstream redirection is very useful in the typical hierarchical

Figure 5. Simulative analysis comparing CN distance and the handover delay

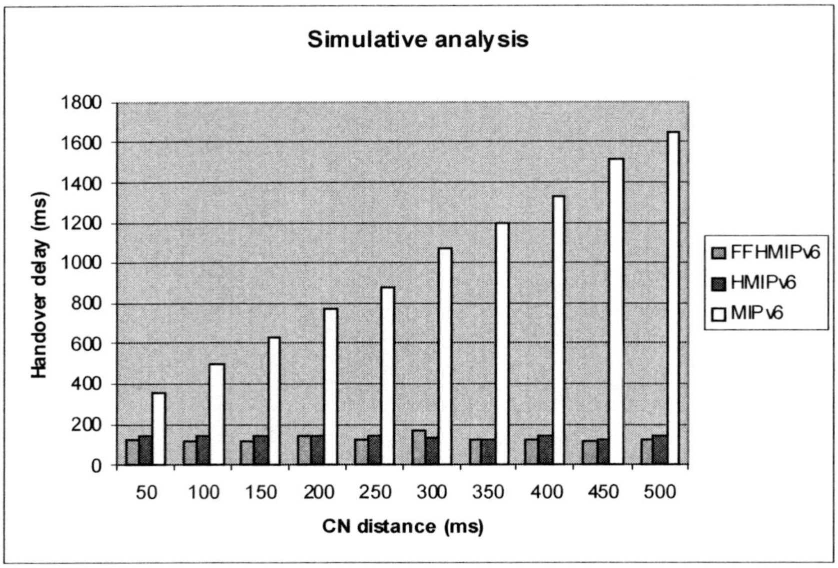

Figure 6. Real environment analysis comparing the CN distance and the handover delay

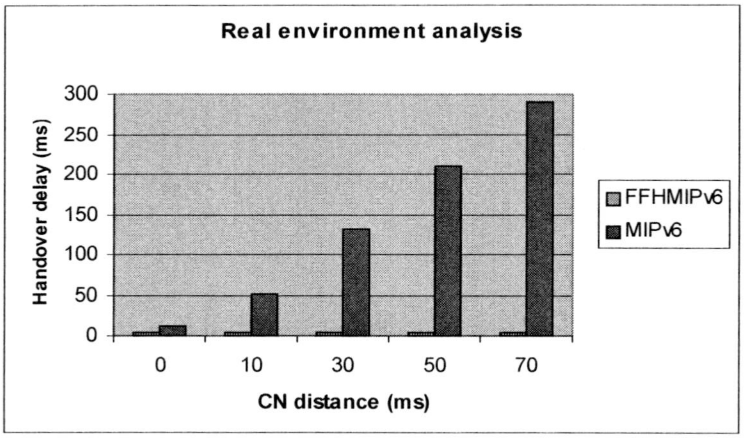

network scenarios. The delay remains almost constant and more importantly independent of the corresponding node distance.

In MIPv6, the upstream traffic of the MN is enabled after a successful binding acknowledgment from the HA. When the distance between the MN and HA is large, the delay might have an negative effect to the two-way communicating applications in use. For example, TCP protocol would not benefit from the pure fast downstream redirection, because the MN cannot acknowledge the packets before the BACK from the HA. Also, voice over IP (VoIP) connections are two-way UDP connections, where fast upstream handover will benefit the communication.

In the fast upstream for FFHMIPv6, the upward communication during address registration process is made by using a temporary hand-off-address (HofA) allocated by the access router. AR handles that there is no possible duplicate addresses in the IP subnet. The HofA and the new AR address is used to encapsulate upstream traffic until the MN receives a BACK from the HA, after which the normal MIPv6 operation is in use.

In Figure 7, we have simulated with ns-2 the effect of fast upstream with UDP-based CBR traffic (Viinikainen et al., in press). The total number of MNs per BS is varied and the L3 packet loss (upstream packet loss) is measured due to L3 handover. In can be seen that even if the overall load in the network increases, the FFHMIPv6 with fast upstream outperforms the MIPv6. This is of course due the fact that the upstream traffic is enabled much faster with the temporary HofA address of FFHMIPv6.

With the advent and increased popularity of mobile and wireless networks has brought some new challenges to the data security area. IP version 6 brings itself new possibilities with integrated IP security (IPSec) support. Thus IPSec can verify the packets integrity and origin. In Mobile IPv6 the location registration procedures (BU processes) are protected with IPSec. For route optimization security the Mobile IPv6 introduces return routability procedure.

In FFHMIPv6 the biggest security threat is to verify the origin of the FFHBU. Without checking this, an unauthorized user would be able to redirect the flows of some user just by sending false FFHBUs to the networks from its IP address. One way to avoid this security threat is for the MN to send its encrypted identification code along with the FFHBU to its HA which is decrypted only by the HA and that the HA can authenticate easily. The false

Figure 7. Packet loss caused by upstream traffic during handover

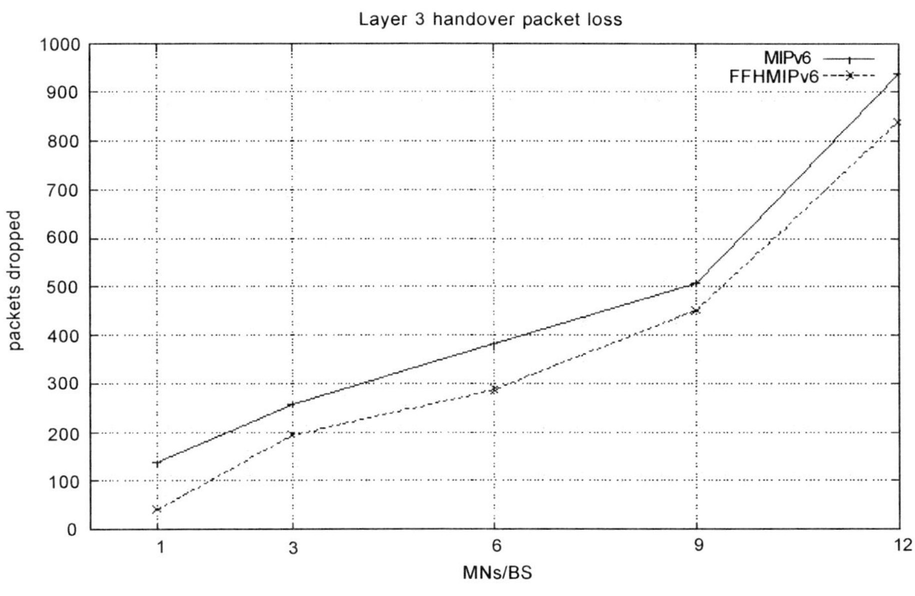

FFHBU is not authenticated hence dropped by the HA. Since all MNs are authorized users of the home network, they are either identified with their MAC/Physical address or user login accounts to their respective networks. An identification code from this information could be generated, by a devoted server, for each device or user at the home network.

FUTURE TRENDS

The trends for mobile multimedia lies in user attractive applications such as IP-based mobile TV and VoIP calls. This chapter has concentrated on Mobile IP, the enabling technology for these streaming applications. Now, we present some future research trends in the field of mobility management to serve the applications and users in better ways.

Even though we have criticized the use of link layer notifications in the handover decision, it seems to be under heavy research and standardization. The link layer triggers can be used to speed up the movement detection procedures and give hints to improve the handover decision. There are several problems to be solved before this can be put into use. Different access technologies function a little differently, so how can we obtain the same information (e.g., LINKUP and LINKDOWN triggers) from them? In the usage of link layer hints, such as signal strength, we must be careful, because due to (e.g., multipath propagation) the signal strength may decrease and increase tens of decibels during short times or distances. These can provide us just hints, not accurate handover information. In both IETF and IEEE there exists working groups that aim to solve these L2 problems.

Even though the Mobile IPv6 provides good integration technology to perform also vertical (i.e., inter-technology) handovers, a lot of research are focusing on how it can be improved to support more intellectual handover decisions. For example, Mäkelä, Hämäläinen, Fekete, and Narikka (2004) aims to find out different ways of using and extending Mobile IPv6 to suite these kinds of issues. The authors address this by introducing a kind of middleware that controls the MIPv6 according to several input parameters (e.g., the link layer notifications and user input).

After successful vertical and horizontal handovers, the next step of mobile communications is multihoming support. This means that the user can use several interfaces (of the same or different technology) simultaneously and different applications can be divided among those intellectually. This requires real-time knowledge of the state and quality of the links and QoS requires of different applications. The Mobile IPv6 protocol needs also some modifications to support multihoming and simultaneous access. Basically it needs multiple CoA-HoA bindings separated by port numbers or some other application tags (Montavont, Noel, & Kassi-Lahlou, 2004).

CONCLUSION

Mobile IPv6 seems to be the mobility management technology in the heterogeneous access environment of the future. It provides unbreakable application level connections independently of the subnet change. Anyway, several procedures of MIPv6 affect the application layer performance. When technologies such as IP-TV and VoIP phone calls increase their popularity, the mobility management needs to be transparent and as seamless as possible. In this chapter we have introduced the flow-based fast handover for Mobile IPv6 networks to reduce the handover delays of the Mobile IPv6 protocol. Both simulative and real network results show that the FFHMIPv6 method decreases the downstream handover delay in hierarchical networks. The fast upstream of FFHMIPv6 works efficiently independently of the network topology.

REFERENCES

Berezdivin, R., Brenig, R., & Topp, R. (2002). Next generation wireless communications concepts and technologies. *IEEE Communications Magazine*, *3*(40), 49-55.

Frodigh, M., Parkvall, S., Roobol, C., Johansson, P., & Larsson, P. (2001). Future generation wireless networks. *IEEE Personal Communications*, *5*(8), 10-17.

Johnson, D., Perkins, C., & Arkko, J. (2004). *Mobility Support in IPv6* (Tech. Rep. No. RFC 3775). IETF. Retrieved June 2005, from http://www.ietf.org/rfc/rfc3775.txt

Koodli, R. (2004). *Fast handovers for Mobile IPv6* (Tech. Rep. No. RFC 4068). IETF. Retrieved July 2005, from http://www.ietf.org/rfc/rfc4068.txt

Montavont, N., Noel, T., & Kassi-Lahlou, M. (2004). Description and evaluation of mobile IPv6 for multiple interfaces. In *Proceedings of Wireless Communications and Networking Conference* (Vol. 1, pp. 144-148).

Mäkelä, J., Hämäläinen, T., Fekete, G., & Narikka, J. (2004). Intelligent vertical handover system for mobile clients. In *Proceedings of the 3rd International Conference on Emerging Telecommunications, Technologies, and Applications* (pp. 151-155).

Omae, K., Ikeda, T., Inoue, M., Okajima, I., & Umeda, N. (2002). Mobile node extension employing buffering function to improve handoff performance. In *Proceedings of the 5th International Symposium on Wireless Personal Multimedia Communications* (Vol. 1, pp. 62-66).

Puttonen, J., Sulander, M., Viinikainen, A., Hämäläinen, T., Ylönen, T., & Suutarinen, H. (in press). Flow-based fast handover for mobile IPv6 environment—implementation and analysis. *Elsevier Computer Communications Special Issue on IPv6*.

Soliman, H., Castellucia, C., El-Malki, K., & Bellier, L. (2004). *Hierarchical mobile IPv6 mobility management (HMIPv6)* (Tech. Rep. No. RFC 4140). IETF. Retrieved August, 2005, from http://www.ietf.org/rfc/rfc4140.txt

Sulander, M., Hämäläinen, T., Viinikainen, A., & Puttonen, J. (2004). Flow-based fast handover method for mobile IPv6 network. In *Proceedings of the IEEE 59th Semi-Annual Vehicular Technology Conference* (Vol. 5, pp. 2447-2451).

Thing, V., Lee, H., & Xu, Y. (2003). Performance evaluation of hop-by-hop local mobility agents probing for mobile IPv6. In *Proceedings of the 8th IEEE International Symposium on Computers and Communication* (pp. 576-581).

Viinikainen, A., Kašák, S., Puttonen, J., & Sulander, M. (in press). Fast handover for upstream traffic in mobile IPv6. In *Proceedings of the 62nd Semi-Annual Vehicular Technology Conference*.

Yegin, A., Njedjou, E., Veerepalli, S., Montavont, N., & Noel, T. (2004). *Link-layer event notifications for detecting network attachments* (Internet draft, expires April 27, 2006). IETF. Retrieved from http://www.ietf.cnri.reston.va.us/internet-drafts/draft-ietf-dna-link-information-03.txt

KEY TERMS

CoA (Care-of-Address): An address of the MN that is valid in the current subnet of the MN.

FFHMIPv6 (Flow-Based Fast Handover for Mobile IPv6): A MIPv6 enhancement, which uses flow state information and tunneling to redirect the flows during the location update process of MIPv6.

HA (Home Agent): A router, which handles the mobility of the MN.

IP-TV (Television over IP): Broadcasting or multicasting television over IP protocol.

MIPv6 (Mobile IPv6): Mobility management protocol for IPv6 networks, which handles mobility at the IP layer.

MIPL (Mobile IPv6 for Linux): An implementation of MIPv6 for Linux operating system.

MN (Mobile Node): A mobile device that has Mobile IPv6 functionality.

ns-2 (Network Simulator 2): A discrete event simulator targeted at networking research.

VoIP (Voiceover IP): Transferring speech over IP protocol.

Chapter XIV
Real-Time Multimedia Delivery for All-IP Mobile Networks

Li-Pin Chang
National Chiao-Tung University, Taiwan

Ai-Chun Pang
National Taiwan University, Taiwan

ABSTRACT

Recently, the Internet has become the most important vehicle for global information delivery. As consumers have become increasingly mobile in the recent years, introduction of mobile/wireless systems such as 3G and WLAN has driven the Internet into new markets to support mobile users. This chapter is focused not only on QoS support for multimedia streaming but also dynamic session management for VoIP applications: As the types of user devices become diverse, mobile networks are prone to be "heterogeneous." Thus, how to effectively deliver different quality levels of content to a group of users who request different QoS streams is quite challenging. On the other hand, mobile users utilizing VoIP services in radio networks are prone to transient loss of network connectivity. Disconnected VoIP sessions should be effectively detected without introducing heavy signaling traffic. To deal with the above two issues, an efficient multimedia broadcasting/multicasting approach is introduced to provide different levels of QoS, and a dynamic session refreshing approach is proposed for the management of disconnected VoIP sessions.

INTRODUCTION

By providing ubiquitous connectivity for data communications, the Internet has become the most important vehicle for global information delivery. The flat-rate tariff structures and low entry cost characteristics of the Internet environment encourage global usage. Furthermore, introduction of mobile/wireless systems such as 3G and WLAN has driven the Internet into new markets to support mobile users. As consumers become increasingly mobile, wireless access to services available from the Internet are strongly demanded. Specifically, mobility, privacy, and immediacy offered by wireless access introduce new opportunities for Internet business. Therefore, mobile/wireless networks are becoming a platform that provides leading edge Internet services.

The existing point-to-multipoint (i.e., multicasting and broadcasting) services for the Internet allow data from a single source entity to be transmitted to multiple recipients. With rapid growth of wireless/mobile subscribers, these services are expected to be used extensively over wireless/mobile networks. Furthermore, as multimedia applications (e.g., video streaming and voice conferencing) are ubiquitous around the Internet world, multimedia broadcasting, and multicasting is considered as one of the most important services in future wireless/mobile communication systems.

As the number of mobile devices and the kinds of mobile applications explosively increases in the recent years, the device types become diverse, and mobile networks are prone to be "Heterogeneous." Multicast/broadcast users with different kinds of mobile devices may request different quality levels of multimedia streams due to (1) users' preferences, (2) service charges, (3) network resources, and (4) device capabilities. Thus, how to effectively deliver different quality levels of content to a group of users who request different QoS streams is quite challenging in the existing/future wireless/mobile communications. In this chapter, an efficient QoS-based multimedia broadcasting/multicasting approach to transmit multimedia streams to the users requesting different levels of service quality would be discussed.

Based on satisfactory and reliable streams delivered over radio network, services provided to fulfill user's strong demand for mobile technologies should then be considered. With the explosive growth of Internet subscriber population, supporting Internet telephony services, also known as voice over IP (VoIP), is considered as a promising trend in telecommunication business. Thus, how to efficiently provide VoIP services over mobile/wireless networks becomes an important research issue. Two major standards are currently used for VoIP products. One is proposed by the ITU-T/H.323, and the other is developed by the IETF/SIP (Internet engineering task force/session initiation protocol). SIP brings simplicity, familiarity, and clarity to Internet telephony that H.323 does not have.

Mobile users roaming in radio networks are prone to transient loss of network connectivity. For example, when a wireless VoIP user in conversation fails to connect the network (e.g., due to abnormal radio disconnection), the failure of this session might not be detected. As resources are still reserved for the failed session, new sessions could not be granted due to the lack of resources. To resolve this problem, one of SIP extensions, *SIP session timer* (Rosenberg, et al., 2002), specifies a keep-alive mechanism for SIP sessions. In this mechanism, the duration of a communicating session is extended by using an UPDATE request sent from one SIP user to the proxy server (then to the other SIP user). A session timer (maintained in the proxy server and the user) records the duration of the session that the user requests to extend. When the session timer nearly expires, the user re-sends an UPDATE request to refresh the session interval. Existing approaches to implement the SIP session timer mechanism are based on static (periodic) session refreshing. The selection of the length for the session timer significantly affects the system performance in the static session refreshing approach due to a tradeoff between resource utilizations and housekeeping traffic. In this chapter, a dynamic session refreshing approach to adjust the session interval according to the network state is discussed. The objective is to efficiently detect session failures without introducing heavy signaling traffic.

BACKGROUND AND RELATED WORK

This section provides a brief summary of specifications and related work regarding to QoS-based multicasting over mobile network and VoIP session management.

3GPP 22.146 has defined a multimedia broadcast/multicast service (MBMS) for universal mobile telecommunications system (UMTS) networks. Both the broadcast and multicast modes are intended to efficiently use radio/network resources, which can be achieved by the multicast tables of the networks nodes such as GGSN (gateway GPRS support node), SGSN (serving GPRS support node) and RNC (radio network controller) (3GPP, 2004; Pang, Lin, Tsai, & Agrawal, 2004). Figure 1 shows an example of MBMS architecture for UMTS networks (Lin, Huang, Pang, & Chlamtac, 2002). The UMTS network connects to the *packet data network* (PDN; see Figure 1a) through the SGSN (see Figure 1b) and the GGSN (see Figure 1c). The SGSN connects to the radio access network. The GGSN provides interworking with the external PDN, and is connected with SGSNs via an IP-based GTP (GPRS Tunneling Protocol) network. To support MBMS, a new network node, *broadcast and multicast service node* (BM-SC; see Figure 1g), is introduced to provide MBMS access control for mobile users. BM-SC communicates with MBMS source located in the external PDN for receiving multimedia data, and connects to the GGSN via IP-based Gmb interface. The *UMTS terrestrial radio access network* (UTRAN) consists of *node Bs* (the UMTS term for base stations; see Figure 1d) and RNCs (see Figure 1e). A *user equipment* (UE) or mobile device (see Figure 1f) communicates with one or more node Bs through the radio interface based on the wideband CDMA radio technology (Holma & Toskala, 2002).

As the number of mobile devices and the kinds of mobile applications explosively increases in the recent years, the device types become diverse, and mobile networks are prone to be "heterogeneous." Applying the scalable-

Figure 1. The 3GPP MBMS architecture

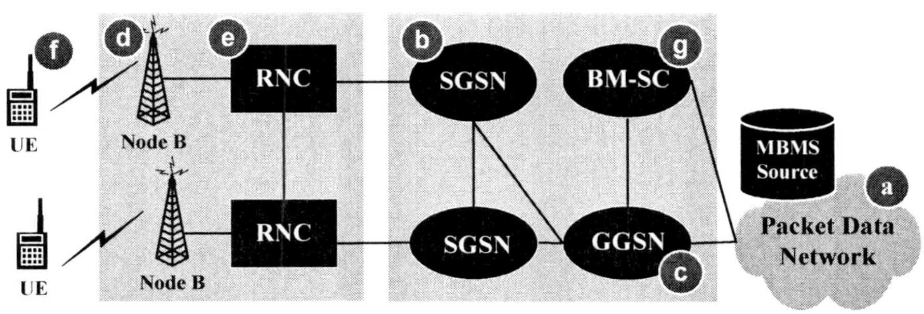

BM-SC: Broadcast and Multimcast Service Center
MBMS: Multimedia Broadcast/Multicast Service
RNC: Radio Network Controller
UTRAN: UMTS Terrestrial Radio Access Network
GGSN: Gateway GPRS Support Node
Node B: Base Station
SGSN: Serving GPRS Support Node
UE: User Equipment

coding technique to wireless transmission has been intensively studied in the literature. In particular, Yang et. al have proposed a TCP-friendly streaming protocol, WMSTFP, to reduce packet loss to improve the system throughput over wireless Internet. Also, the issues for power consumption and resource allocation over wireless channels have been investigated (Lee, Chan, Zhang, Zhu, & Zhang, 2002; Zhang, Zhu, & Zhang, 2002; Zhang, Zhu, & Zhang, 2004). However, little work has been done in multimedia broadcasting/multicasting with scalable-coding support.

Based on satisfactory and reliable multimedia streams delivered over radio network, services provided to fulfill user's strong demand for mobile technologies should then be considered. Supporting Internet telephony services, also known as *voice over IP* (VoIP), is considered as a promising trend in telecommunication business. Recent introduction of mobile/wireless systems (e.g., 3G/GPRS, IEEE 802.11 WLAN, Bluetooth) has driven the Internet into new markets to support mobile/wireless users. Thus, how to efficiently provide VoIP services over mobile/wireless networks becomes an important research issue, which has been intensively studied (Chang, Lin, & Pang, 2003; Garg & Kappes, 2003; Rao, Herman, Lin, & Chou, 2000).

SIP (Rosenberg, et al., 2002) is an application-layer signaling protocol for creating, modifying and terminating multimedia sessions or calls. Two major network elements are defined in SIP: the user agent and the network server. The user agent (UA) that contains both a *user agent client* (UAC) and a *user agent server* (UAS) resides in SIP terminals such as hardphones and soft-phones. The UAC (or calling user agent) is responsible for issuing SIP requests, and the UAS (or called user agent) receives the SIP request and responds to the request. There are three types of SIP network servers: the proxy server, the redirect server, and the registrar. The proxy server forwards the SIP requests from a UAC to the destination UAS. Also, the proxy server is responsible for performing user authentication, service logic execution and billing/charging for a SIP-based VoIP network. The redirect server plays a similar role as the proxy server, except that the redirect server responds to a request issuer with the destination address instead of forwarding the request. To support user mobility, a UA informs the network of its current location by explicitly registering with a registrar. The registrar is typically co-located with a proxy or redirect server.

REAL-TIME MULTIMEDIA DELIVERY FOR ALL-IP MOBILE NETWORKS

QoS Multicasting over Mobile Network

This section is focused on the issue of multicasting multimedia streams with QoS guarantees over mobile wireless networks.

QoS-Based Multimedia Multicasting for UMTS Networks

To support MBMS (i.e., multimedia broadcast/multicast service) for mobile devices with diverse capabilities, 3GPP 23.246 (3GPP, 2004) has proposed a multimedia multicasting1 approach for UMTS networks. In this approach (Approach I; see Figure 2), multimedia (e.g., video and audio) streams are duplicated and encoded as different QoS levels at MBMS source. Then based on users' QoS profiles in the multicast tables maintained by GGSN, SGSNs, and RNCs, the encoded video/audio streams of each QoS level are respectively

Figure 2. The 3GPP 23.246 approach

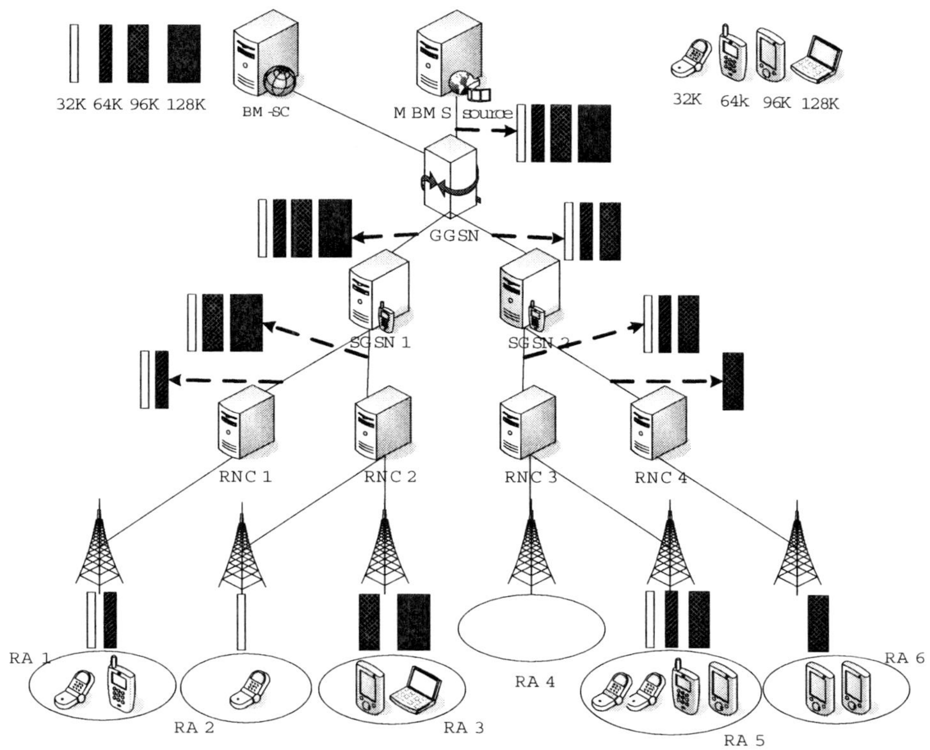

transmitted to the multicast users requesting that quality. As shown in Figure 2, there are two SGSNs in the UMTS network: SGSN1 and SGSN2. SGSN1 covers routing areas RA1, RA2, and RA32. SGSN2 covers routing areas RA4, RA5, and RA6. We assume that four QoS levels (i.e., 32Kbps, 64Kbps, 96Kbps, and 128Kbps) for multimedia streaming are provided to mobile multicast users. To perform QoS-based multicasting in this approach, MBMS source duplicates multimedia streams, and encoded the duplicated streams with four data rates. The four encoded streams are transmitted to the GGSN, and based on the multicast table, the GGSN forwards each stream to the SGSNs covering the multicast users with that quality request. In Figure 2, the streams of 32Kbps, 64Kbps, and 96Kbps (through three GTP tunnels) are delivered to SGSN1, and SGSN2 receives the streams of four QoS levels (via four GTP tunnels). Similarly, the SGSNs relays the proper streams to the accordingly RNCs, and then to the RAs through radio channels.

By using the 3GPP 23.246 approach, the transmitted streams fulfill the QoS level each multicast user requests. However, as the number of supported QoS levels increases (i.e., the number of types of mobile devices increases and the networks become more "heterogeneous"), data duplication becomes more serious, which results in more resource consumption of core and radio networks. Thus based on standard 3GPP MBMS architecture (3GPP, 2004), we propose an efficient multimedia multicasting approach (Approach II) to deliver

Figure 3. The scalable coding technique

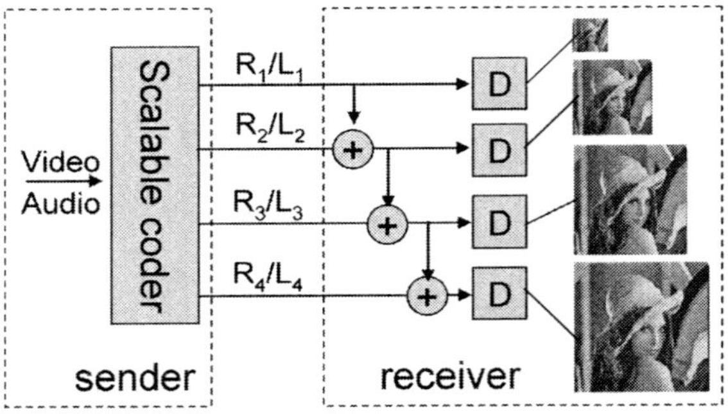

scalable-coded multimedia to a group of users (i.e., multicast users) requesting a specific level of multimedia quality.

The goals of our multimedia multicasting approach are (1) to have a single multimedia stream source (i.e., no duplication at MBMS source), (2) to transmit multimedia streams to all members in the multicast group with satisfied quality, and (3) to effectively utilize the resources of core and radio networks. To achieve such a goal, the existing scalable-coding technique is adopted in our approach to deliver multimedia streams. Figure 3 elaborates on the basic concept for scalable coding. The scalable coding technique utilizes a layered coder to produce a cumulative set of layers where multimedia streams can be combined across layers to produce progressive refinement. For example, if only the first layer (or base layer) is received, the decoder will produce the lowest quality version of the signal. If, on the other hand, the decoder receives two layers, it will combine the second layer (or the enhancement layer) information with the first layer to produce improved quality. Overall, the quality progressively improves with the number of layers that are received and decoded.

With scalable coding, the requirement of single-source multimedia streams is fulfilled, and all multicast users can decode their preferred multimedia packets depending on the devices' capabilities. However, how to effectively utilize the resources of core and radio networks to transmit scalable-coded multimedia streams is still a challenging issue. Thus, we develop two kinds of transmission modes for our scalable-coding enabled multimedia multicasting: "Packed" mode (Mode *A* or Approach II_A) and "Separate" mode (Mode *B* or Approach II_B).

In the packed mode (see Figure 4), all layered multimedia data for one frame are packed into a packet at MBMS source. Then these packed packets are sequentially delivered in one shared tunnel (between GGSN and SGSN, and between SGSN and RNC) and one shared radio channel to all multicast users. As shown in Figure 4, each packed packet (which consists of 4-layered multimedia data of one frame) is sent from GGSN to the SGSNs (i.e., SGSN1 and SGSN2), the RNCs (i.e., RNC1, RNC2, RNC3, and RNC4) and then the RAs (i.e., RA1, RA2, RA3, RA5, and RA6) where the multicast users reside. Upon receipt of 4-layered multimedia data, the multicast users can select certain layers to perform decoding based on their preferences.

For Mode *A*, our QoS-based multimedia multicasting can be easily implemented in UMTS networks without any modification of the existing GGSN, SGSNs, and RNCs. Since the GGSN, SGSNs, and RNCs are not aware of scalable coding and can not differentiate the layers of multimedia streams, 4-layered multimedia streams have to be sent to all multicast users regardless of the QoS levels the users request, which may result in extra resource (i.e., link bandwidth and channelization code) usage of core and radio networks. Also, this kind of transmission leads to the increase of power consumption of mobile devices (e.g., the mobile phone in RA2) requesting low-quality multime-

Figure 4. Transmission mode I for our QoS-based multimedia multicasting

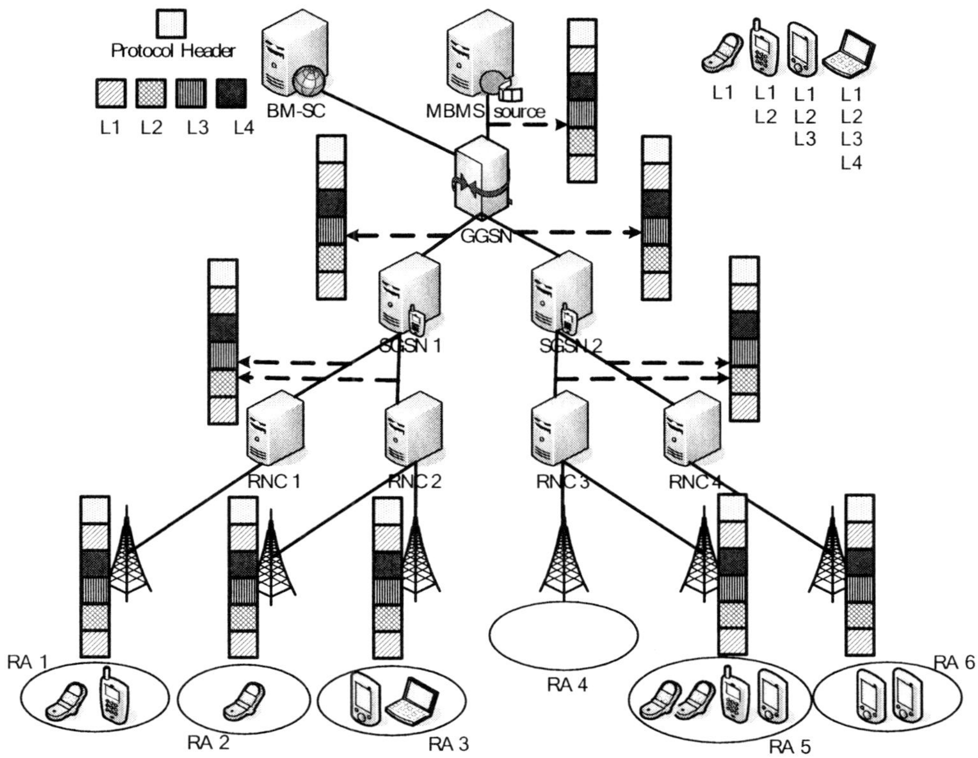

dia streams. Therefore "separate" mode (Mode *II*) is further developed to improve transmission efficiency of scalable coded multimedia streams. Figure 5 shows the scenario of "Separate" mode for scalable-coded multimedia multicasting. In Mode *B*, each layered multimedia data is encapsulated in one GTP packet, and all GTP packets are transmitted through a single tunnel. To effectively deliver the scalable coded multimedia streams, the GGSN, SGSNs, and RNCs would be modified to become aware of scalable coding. Note that these network nodes do not have to understand how scalable coding works. They only need to differentiate the layers of received multimedia streams, which can be accomplished through the tag of GTP packet headers. Since the layer-differentiation can be done by the RNCs, each layer stream would be transmitted by one radio channel, and mobile devices can freely select and receive the preferred layers of multimedia streams, which results in significant reduction of power consumption of mobile devices and channel usage of radio networks.

Based on the above discussion, Table 1 compares our proposed QoS-based multimedia multicasting approach (Approach II_A and Approach II_B) with 3GPP 23.246 approach. The following issues are addressed.

1. Both 3GPP 23.246 approach and Approach II_B select the multicasting path for

Figure 5. Transmission mode II for our QoS-based multimedia multicasting

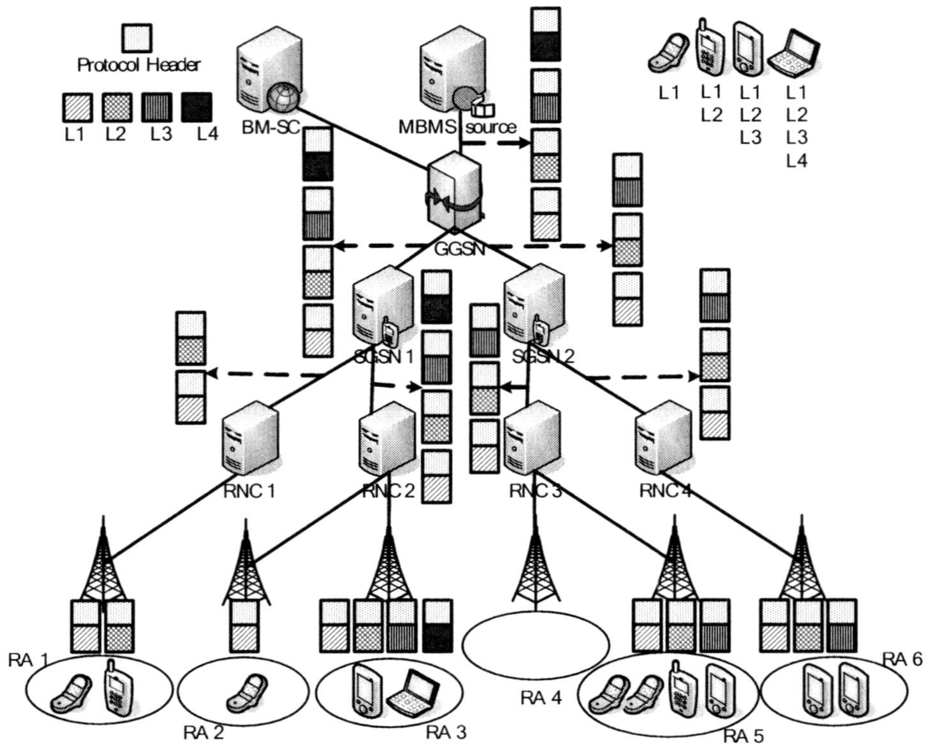

a specific quality of multimedia streams based on the users' QoS profile. Thus, the network nodes such as GGSN, SGSNs, and RNCs in these two approaches have to maintain the QoS requests of mobile users. However, since all scalable-coded layers of multimedia streams are delivered to the multicast users, QoS maintenance for multicast users is not needed in Approach II_A.

2. For Approach *I*, the multimedia streams have to be duplicated and encoded as different qualities at MBMS source. On the other hand, since the scalable coding technique is used in Approach *II*, duplication can be avoided.
3. For Approach II_B, UEs may receive multiple layers of multimedia streams through several channels, which results in the synchronization problem between the received layered streams.
4. Approach II_B is capable of adapting to bandwidth variation especially for the bandwidth reduction of wireless links. When the bandwidth suddenly reduces, the transmission of multimedia streams for high quality can be temporarily suspended. At this time, the mobile devices with ongoing high-quality multimedia transmission can still receive low-quality streams without causing service interruption, which can not be achieved through Approach *I* and Approach II_A.

Table 1. Comparing our proposed QoS-based multimedia multicasting with 3GPP 23.246 Approach

Issues \ Approaches	Approach I (3GPP 23.246)	Approach II_A	Approach II_B
Issue 1: QoS Maintenance for Multicast Users	Yes	No	Yes
Issue 2: Single Source for Heterogeneous Devices	No	Yes	Yes
Issue 3: Synchronization Problems (for UE)	No	No	Yes
Issue 4: Adoption to Bandwidth Variation	No	No	Yes

Performance Evaluation

In this section, we use some numerical examples to evaluate the performance of 3GPP 23.246 approach (Approach *I*) and our QoS-based multimedia multicasting approach (Approach II_A and Approach II_B). In our experiments, two classes of RAs are considered. Class 1 RAs cover urban areas with dense population, and thus with diverse mobile devices. On the other hand, the rural RAs (Class 2 RAs) have a uni-type of mobile devices. Let α be the portion of class 1 RAs, and assume that class 1 and class 2 RAs are uniformly distributed in the UMTS system. Note that our model can be easily extended to analyze other distributions of class 1 and class 2 RAs. Experiments are evaluated in terms of the transmission costs (C_t), which are measured by the following weighted function of bandwidth requirement of multimedia transmission for core and radio networks:

$$C_t = B_g C_g + B_s C_s + B_r C_r + B_b C_b$$

where B_g, B_s, B_r and B_b respectively represent the total bandwidth requirements for multimedia multicasting between the GGSN and the SGSNs, between the SGSNs and RNCs, between the RNCs and node Bs and between the node Bs and UEs. Similarly, C_g, C_s, C_r, and C_b respectively denote the unit transmission costs between the GGSN and the SGSNs, between the SGSNs and RNCs, between the RNCs and Node Bs and between the node Bs and UEs. From Rummler, Chung, and Aghvami (2005), the values of C_g, C_s, C_r, and C_b are set to 0.2, 0.2, 0.5, and 5.

Foreman is used for test sequences, and the number of frames (with the size of 176x144 QCIF) is 400. MPEG-4 FGS and MPEG-4 are respectively used for scalable coding and non-scalable coding, and Codec adopts Microsoft MPEG-4 Reference Software (Wang, Tung, Wang, Chiang, & Sun, 2003). Furthermore, the uni-truncation (with equivalent bit-rate) are used for all enhancement layers of I-Frame and P-Frame. Six levels of service quality are provided in the experiments. For non-scalable coding, the six quality levels are accomplished by 120Kbps, 150Kbps, 180Kbps, 210Kbps, 240Kbps, and 270Kbps bit rates. The experimental results indicated that the bit rate of based layer (L_1) for scalable coding would be

120Kbps, and the bit rates for accordingly enhancement layers (i.e., L_2, L_3, L_4, L_5, and L_6) are 150Kbps, 120Kbps, 105Kbps, 90Kbps, and 75Kbps. Furthermore for Approach II_B, we have t-playing-time multimedia data as a unit, and have each layered data separately encapsulated in one packet. Table 2 shows input parameters and their values used in our experiments.

Figure 6 indicates the effect of α (i.e., portion of class 1 RAs covering diverse mobile devices) on the transmission costs C_T for Approach I, Approach II_A and Approach II_B. In this figure, the C_T value for Approach II_A

Table 2. Input parameters

Variable	Description	Value
N_S	The number of SGSNs	10
K	The number of RNCs covered by each SGSN	10
M	The number of Node Bs covered by each RNC	50
n	The number of QoS levels	6
T	Playing time for test sequences	13.3sec
L_u	Header lengths of UDP	8 bytes
L_i	Header lengths of IP	20 bytes
L_g	Header lengths of GTP	12 bytes
L_p	Header lengths of PDCP	3 bytes

Figure 6. Effect of α on C_r

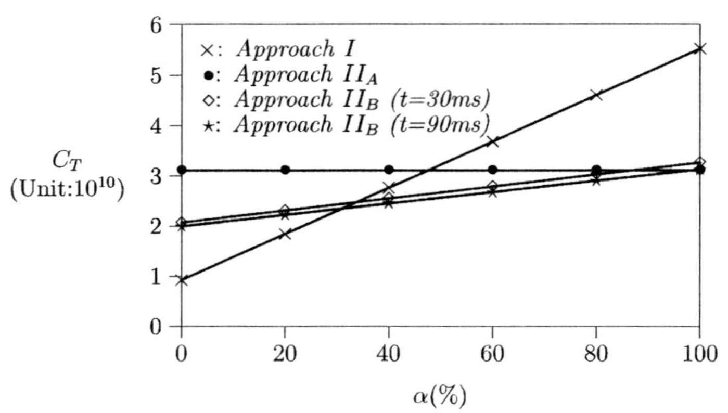

remains the same as α increases (i.e., the number of dense areas increases). On the other hand, the increase of α results in the increase of C_T for Approach *I* and Approach II_B. Specifically, the increasing rate for Approach *I* is much larger than that for Approach *II*. Furthermore, when $\alpha > 40\%$, Approach *II* (for both Mode *A* and Mode *B*) has a small C_T than Approach *I*. From this figure, we observe that when all RAs are class 1, Approach II_A has the lowest C_T. However, when α nearly equals to 0, the performance of Approach *I* is better than that of Approach *II*. Also, this figure indicates that as *t* increases from 30ms to 90ms, the overhead for Approach II_B decreases, and thus C_T slightly decreases.

Session Timer for Wireless VoIP

This section is aimed at the discussions of a resource-efficient session management method for wireless VoIP applications based on session timers.

The Dynamic Session Refreshing Approach

Mobile users roaming in radio networks are prone to transient loss of network connectivity. As resources are still reserved for the failed session, new sessions could not be granted due to the lack of resources. Under the basic SIP specification, a basic SIP proxy server is not able to keep track of the states of sessions and determine whether an established session is still alive or not. To resolve this problem, one of SIP Extensions, *SIP session timer*, specifies a keep-alive mechanism for SIP sessions. In SIP Session Timer, the UA in conversation sends an UPDATE request to extend the duration of the communicating session. The interval for two consecutive UPDATE requests (i.e., the length of the session timer) is determined through a negotiation between the UAC and the UAS. If an UPDATE request is not received before the session timer expires, the session is considered as abnormal disconnection, and will be

Figure 7. The SIP-based VoIP network architecture

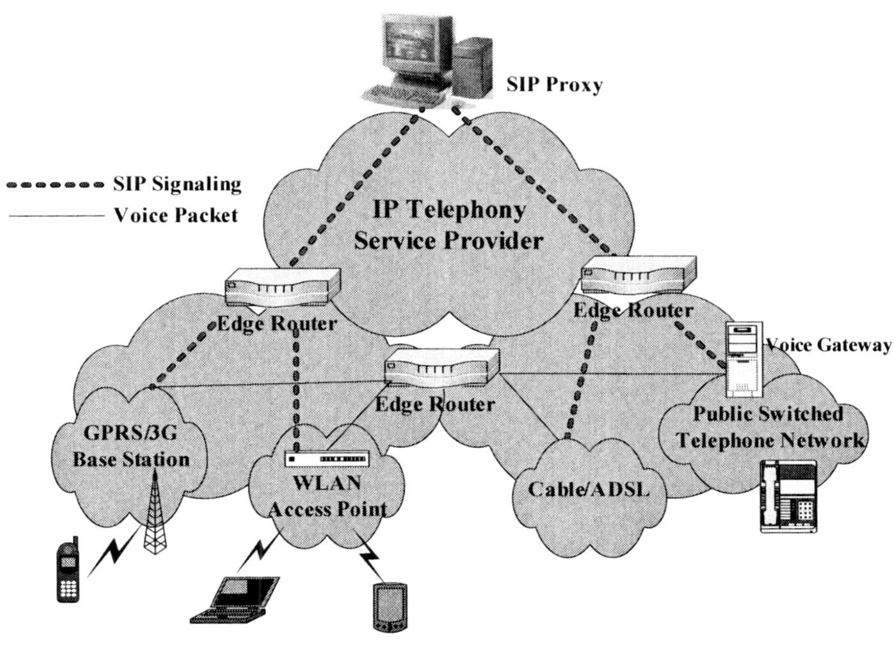

force-terminated. Then the proxy server will release the allocated resources for the failed session.

Based on the network architecture shown in Figure 7, our dynamic session refreshing approach is described below. In this figure, SIP UAs can access IP telephony services via heterogeneous networks including the wireless/mobile networks (e.g., IEEE 802.11 WLAN and GPRS/3G) and the wireline networks (e.g., cable, ADSL, and PSTN). In Figure 7, the dashed and solid lines respectively represent the SIP signaling and RTP(real-time transport protocol)/RTCP(RTP control protocol) voice paths, where the SIP signaling is carried through the proxy servers, and the voice packets are directly transmitted between two communicating UAs. For an established session, abnormal detaching from the network due to the crash of the UA and/or the radio disconnection for one of the participant UAs will result in the session force-termination. By using the SIP Session Timer mechanism, the occurrence of the session force-termination can be detected by the proxy server, and then the proxy server can quickly release the resources allocated for the failed session.

To estimate the state of the radio link for a wireless UA, the data from lower layers (e.g., MAC) should be periodically collected. If the collected data indicate that the *frame error rate* (*FER*) (or packet loss statistics) has been low for a period of time, the network condition is considered as a good state. The period of time denoted by the **Adjusting Window** (*AW*) is used as a history reference to determine the point of the next UPDATE request. All FER values collected within an *AW* are weight-averaged, and its value is denoted by *aFER*.

A low *aFER* value represents a "GOOD" network state with low probabilities of packet loss, and with low probabilities of the radio disconnection. Whether the network state is identified as good or not depends on the **Good Threshold** (*GT*). If *aFER* is equal to or less than *GT*, the network condition is considered as a good state. In this case, to save the network bandwidth, the session timer is increased based on the **Increase Ratio** (*IR*) to avoid sending the UPDATE request frequently. If the network state has been good for a long time, the session interval will become extremely large. Suppose that the session disconnection suddenly occurs. With such a large session timer, the session failure will be detected by the proxy server too slowly. Thus, to prevent the session timer from being over-enlarged, an **Upper Bound** (*UB*) for the session timer is set.

On the contrary, when *aFER* is high (i.e., equal to or larger than the **Bad Threshold** (*BT*), the network condition is considered as a bad state. In this state, the probability of packet loss is high, and the established session will fail due to the radio disconnection very probably. Thus in order to detect the session failure earlier, the UPDATE requests should be sent to the proxy server more frequently by decreasing the session timer based on the **Decrease Ratio** (*DR*). Similar to *UB* in the good state, a **Lower Bound** (*LB*) in the bad state is used to prevent the session timer from being over-reduced, which results in overwhelming signaling traffic and the decrease of the available network bandwidth.

Based on the above descriptions, the session interval can be smoothly increased/decreased with *IR*/*DR* according to the estimated state of the radio link. However, when the network condition rapidly switches between "GOOD" and "BAD" states, the session timer may not be immediately changed to a proper value by using *IR*/*DR*. To further improve the performance of our dynamic session refreshing approach, the situation for the significant change between the network states should be considered. Whether a significant network change

Table 3. The variables used in our dynamic session refreshing approach

Parameter	Description	Value
Adjusting Window (AW)	The window size for collecting radio link information from lower layers	2
Average FER (aFER)	The average FER value within AW	-
Bad Threshold (BT)	Used to check whether the state of the network is bad or not	28%
Decreasing Ratio (DR)	The ratio used to decrease the session timer	1.15
Good Threshold (GT)	Used to check whether the state of the network is good or not	10%
Increasing Ratio (IR)	The ratio used to increase the session timer	1.30
Lower Bound (LB)	A lower limit of the length of the session timer	$1/20\ \mu$
Network Change (NC)	Used to check whether the network state changes or not	18%
Query Number (QN)	The number of queries for retrieving the lower-layer radio link information	-
Session Timer (ST)	The session interval	-
Upper Bound (UB)	An upper limit of the length of the session timer	$1/5\ \mu$

occurs depends on the difference between the previous collected *FER* value (*pFER*) and the current collected *FER* value (*cFER*) from the lower layer. If $|pFER - cFER| > NC$ (**Network Change**), the session interval is adjusted to the initial value instead of slightly increasing/decreasing the current value by *IR*/*DR*.

The steps of our dynamic session refreshing algorithm are described as follows. The variables used in our dynamic session refreshing algorithm are summarized in Table 3. Table 3 also presents the values set for these variables for our experiments in the later section.

- **S0:** When the SIP session is successfully established, the following parameters are initialized.

  ST = default ST,
  cFER = 0,
  pFER = -1,
  FER[i] = 0 for $1 \le i \le AW$.

  Also, the number of query times (*QN*) for radio link information within *AW* is set to zero.
- **S1:** The value of *QN* is increased by 1, and the value of *pFER* is set to that of *cFER*. Then the value of *cFER* is obtained by querying the lower layers, and the value of *FER*[*QN*] is set to that of *cFER*.
- **S2:** If the value of *pFER* is not equal to -1 (i.e., the *pFER* value is not obtained from the first query within *AW*), and the difference between *pFER* and *cFER* is

larger than NC, then go to Step 0. At Step 0, the session timer is set to the default value, and the value of QN is reset. Otherwise, Step 3 is executed.

- **S3:** When the number of query times achieves AW, Step 4 is executed. Otherwise, go to Step 1 for collecting more radio link information.
- **S4:** This step is used to adjust the length of the session timer based on the data collected from above steps.
- **S4.1:** The value of QN is set to zero, and $aFER$ is calculated as below.

$$aFER = \sum_{i=1}^{AW} \left[\frac{2i}{AW(AW+1)} \right] FER[i]$$

- **S4.2:** Check if $aFER$ is less than GT. If yes, go to Step 4.3; otherwise go to Step 4.4.
- **S4.3:** This step is used to adjust the session timer for the good network state. Thus:

  $ST = ST^* \mid IR - aFER \mid$

  If ST is larger than UB after the adjustment, the value of ST is set to UB. Then Step 1 is executed.
- **S4.4:** If $aFER < BT$, no adjustment for the session timer is performed, and the algorithm returns to execute Step 1. On the other hand, if $aFER \geq BT$, the session timer is adjusted as follows:

  $ST = ST^* \mid DR - aFER \mid$

  Similarly, If ST is less than LB after the adjustment, the value of ST is set to LB. Then the algorithm returns to Step 1.

Performance Evaluation

Based on the scenario shown in Figure 8, this section proposes an analytic model, and a simulation models to evaluate the performance of *SIP session timer* for wireless VoIP. Our analytic model has been validated against the simulation experiments. The simulation model follows the approach we developed in (Pang et al., 2004), and the details are omitted.

In Figure 8, the proxy server monitors the state (i.e., dead or alive) of the communicating session between UA1 and UA2 through the *SIP Session Timer* mechanism. We assume that UA1 accesses the IP telephony service via the wireless link such as IEEE 802.11 WLAN and 3G/GPRS, and UA2 is connected to the Internet through the wireline access (e.g., ADSL and cable). After the session is established, UA1 is responsible for issuing the UPDATE request to the proxy server to refresh the session interval. By using UPDATE from UA1, the proxy server is informed about whether the session is dead or alive. Note that by using the quality feedback information carried in RTCP packets, our model can be easily extended to the case where both UA1 and UA2 are the wireless VoIP users. In the remainder of this paper, the term "call" represents the real-time multimedia/voice session.

To model the condition of the wireless link for UA1, three kinds of network states, "GOOD," "BAD" and "DEAD," are considered. Different kinds of network states represent different *frame error rate* (FER) for wireless links. A large FER leads to a high probability of packet loss. When UA1 (i.e., the call that UA1 involves) resides in "GOOD" state, the FER and packet-loss probability are small, and most voice and signaling packets can be successfully transmitted from UA1 to the proxy server and then to UA2. In "BAD" state,

Figure 8. The scenario for the analytic model

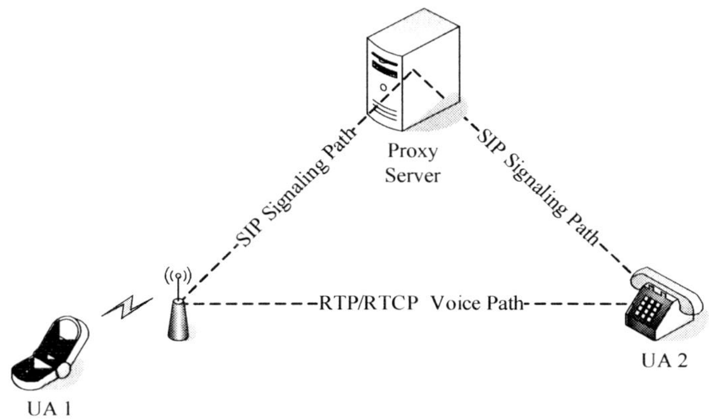

Figure 9. The transition probabilities between the network conditions for the wireless link (G: GOOD, B: BAD and D: DEAD)

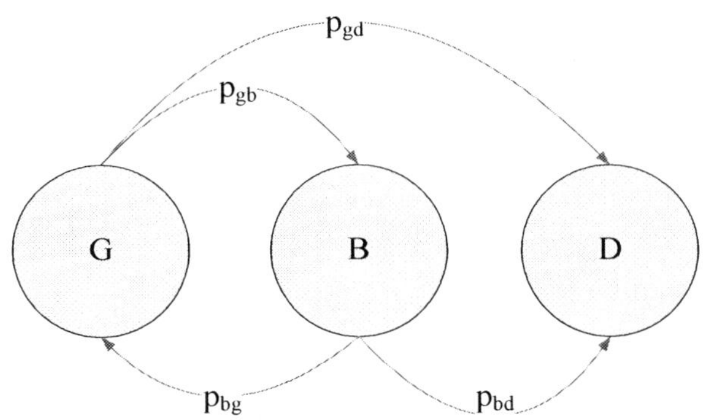

with a large FER, the network condition is unstable, and this results in a large number of lost packets. When the wireless network enters in "DEAD" state, the signaling path (between UA1 and the proxy server) and the voice path (between UA1 and UA2) are force-disconnected, and all packet deliveries from UA1 will fail. Figure 9 shows the transition probabilities between "GOOD", "BAD" and "DEAD" states for the wireless link of UA1, where $P_{bd} + P_{bg} = 1$ and $P_{gd} + P_{gb} = 1$. The time intervals (i.e., t_g and t_b) that UA1 stays in "GOOD" and "BAD" states are assumed to have Exponential distributions with rates λ_g and λ_b, respectively. This assumption will be relaxed to accommodate Gamma distributions for t_g and t_b in our developed simulation model (Chlamtac, Fang, & Zeng, 2003; Fang & Chlamtac, 1999; Kelly, 1979). Also, we assume that the packet loss probabilities for "GOOD" and "BAD" states are respectively P_{lg} and P_{lb}.

Several output measures are defined in this study, and listed as follows.

- P_{d_f}: The probability that the detection event (i.e., UPDATE loss) occurs before the call actually fails or completes. This probability is also called the mis-detection probability.
- N_u: The average number of UPDATE requests transmitted between UA1 and UA2 (via the proxy server) for an established call
- $E[T_B]$: The expected number of Bad Debt. The Bad Debt is defined as the time interval between the time that the failure (i.e., UA1 enters in "DEAD" state) occurs and the time that the proxy server releases the resources for the call.

In our experiments, the default values for the input parameters are set, i.e., $\lambda_g = 3\mu$, $\lambda_b = 5\mu$, $P_{lg} = 10^{-6}$, $P_{lb} = 10^{-3}$, $P_{gd} = 10^{-6}$ and $P_{bd} = 0.05$. Furthermore, the initial value for the session timer (ST) is set to $\frac{1}{10\mu}$, and the query frequency for radio-link information is 30μ.

Effect of λ_g:

Figures 11a and 11b plot the the expected number of UPDATE requests per call (N_u), the expected number of **Bad Debt** ($E[T_B]$), and the mis-detection probability (P_{d_f}) as a function of λ_g, where the input parameters except λ_g are set to the default values. In Figure 10a, as λ_g increases (i.e., the reduction of the average time of the good state where a wireless UA resides), the curves for the static and dynamic session refreshing approaches respectively decrease and increase. For $\lambda_g \leq 4\mu$, the static session refreshing approach has more UPDATE requests than the dynamic one. On the other hand, when λ_g is larger than 4μ, the opposite result is observed. This phenomenon is explained as follows. As λ_g increases, the average time of the bad state for a call relatively increases. Thus, the call suffers from the radio disconnection more probably, and the call holding time decreases due to the increasing force-termination probability. For the static session refreshing approach, the UPDATE request is periodically sent regardless of the network state. As the call holding time decreases, N_u for the static approach decreases.

Figure 10. The effect of λ_g on N_u, $E[T_B]$ and P_{d_f}

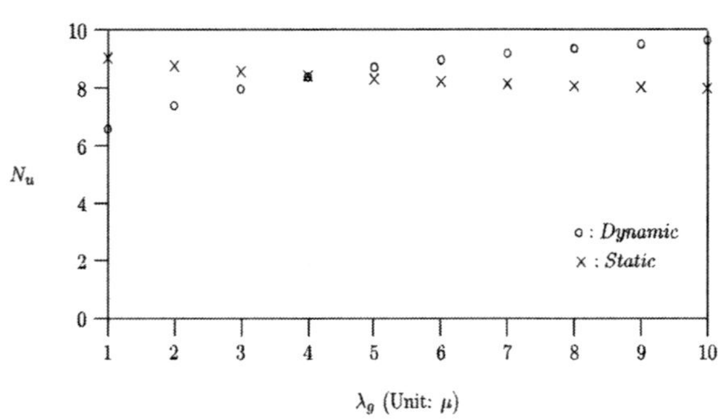

(a) Effect of λ_g on N_u

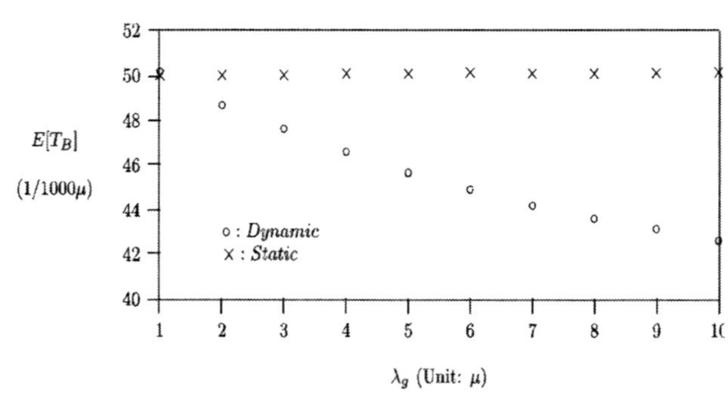

(b) Effect of λ_g on $E[T_B]$

Figure 11. The effect of P_{bd} on N_u and P_{d_f}

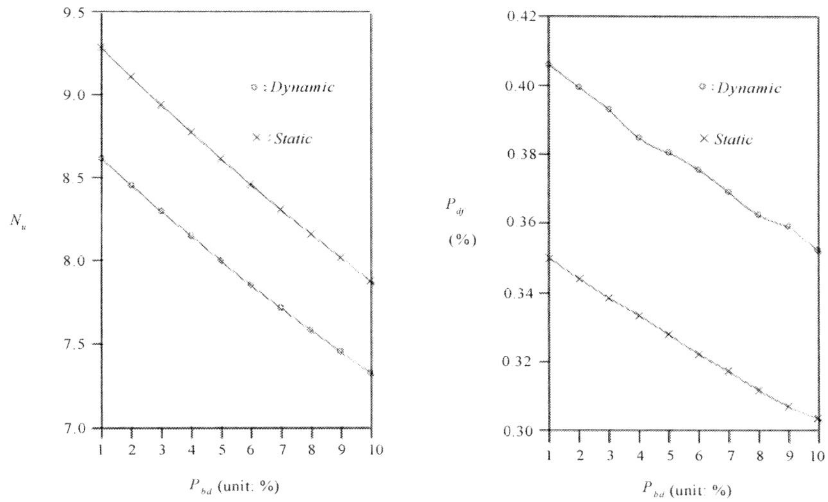

(a) Effect of P_{bd} on N_u

(b) Effect of P_{bd} on P_{df}

On the contrary, the frequency of UPDATE deliveries for our dynamic approach increases when the network state remains bad, and this results in the increase of the session refreshing number N_u. Figure 10b shows that $E[T_B]$ for the static session refreshing approach is not influenced by λ_g. However, for the dynamic session refreshing approach, $E[T_B]$ significantly decreases as λ_g increases, which indicates that our dynamic approach effectively adjusts the session timer especially when the network condition is unstable.

Effect of P_{bd}:

Figure 11 plots N_u and P_{d_f} as a function of P_{bd}. The curve for the effect of P_{bd} on $E[T_B]$ is not presented since **Bad Debt** is irrelevant to the transition probability from the bad state to the dead state for an established call. Figure 11a shows that for both the static and dynamic session refreshing approaches, N_u decreases as P_{bd} increases. The increase of P_{bd} results in more call force-terminations due to the session failure, and thus the decrease of the number of UPDATE deliveries. Furthermore, the curve for static session refreshing is steeper than that for dynamic session refreshing. The decreasing rate of N_u for these two approaches depends on the ratio of t_g to t_b where an established call resides. If $\frac{\lambda_b}{\lambda_g} > 1$, the decreasing rate of N_u for the static approach is faster than that for the dynamic one. On the contrary, an opposite result is observed. Similar to what we observe in Figure 11a, Figure 11b shows that P_{d_f} decreases as P_{bd} increases for both the static and dynamic session refreshing approaches.

CONCLUSION

As IP infrastructure had been successfully driven into wireless and ubiquitous networks as a low cost scheme for global connectivity, the ability of multimedia streaming over wireless network is quickly emerging as a key to the success of the next-generation Internet business. In this chapter, we addressed two challenges in delivering real-time multimedia streams over all-IP mobile networks: QoS guarantees and session management. A scalable-coding-based multicasting technique was introduced to deliver real-time streams so as to meet user preferences and/or capabilities of user equipments. The proposed method could be adopted in existing UMTS with minor modifications and it outperformed existing 3GPP 23.246 approach in terms of transmission costs of core/radio networks. Regarding session management, a dynamic session refreshing approach was presented to adjust the session timer depending on the conditions of radio links for wireless VoIP subscribers. With our dynamic session refreshing approach, the session failure can be efficiently detected without a considerable increase of signaling traffic.

ACKNOWLEDGMENTS

We would like to thank Prof. Tei-Wei Kuo for his helpful comments and suggestions.

REFERENCES

3GPP. (2004). *3rd generation partnership project; technical specification group services and systems aspects; Multimedia Broadcast/Multicast Service (MBMS); Architecture and functional description (Release 6)* (Technical Report 3GPP).

Chang, M. F., Lin, Y. B., & Pang, A. C. (2003) vGPRS: A mechanism for voice over GPRS. *ACM Wireless Networks, 9,* 157-164.

Chlamtac, I., Fang, Y., & Zeng, H. (1999). Call blocking analysis for PCS networks under general cell residence time. *Proceedings of IEEE WCNC.*

Fang, Y., & Chlamtac, I. (1999). Teletraffic analysis and mobility modeling for PCS network. *IEEE Transactions on Communications, 47*(7), 1062-1072.

Garg, S., & Kappes, M. (2003). An experimental study of throughput for UDP and VoIP traffic in IEEE 802.11b networks. *Proceedings of IEEE WCNC.*

Holma, H., & Toskala, A. (Eds.) (2002). *WCDMA for UMTS.* John Wiley & Sons.

Kelly, F. P. (1979). *Reversibility and stochastic networks.* John Wiley & Sons Ltd.

Lee, T. W., Chan, S. H., Zhang, Q., Zhu, W., & Zhang, Y. Q. (2002). Allocation of layer bandwidths and FECs for video multicast over wired and wireless networks. *IEEE Transactions on Circuits and Systems for Video Technology, 12*(12), 1059-1070.

Lin, Y. B., Huang, Y. R., Pang, A. C., & Chlamtac, I. (2002). All- IP approach for third generation mobile networks. *IEEE Network, 16*(5).

Pang, A. C., Lin, Y. B., Tsai, H. M., & Agrawal, P. (2004). Serving radio network controller relocation for UMTS all-IP network. *IEEE Journal on Selected Area in Communications, 22*(4).

Rummler, R., Chung, Y. W., & Aghvami, A. H. (2005). Modeling and analysis of efficient multicast mechanism for UMTS. *IEEE Transactions on Vehicular Technology, 54*(1).

Rao, H., C. H., Lin, Y. B., & Chou, S. L. (2000). iGSM: VoIP service for mobile network. *IEEE Communications Magazine.*

Rosenberg, J. et al. (2002). *SIP: Session Initiation Protocol.* IETF RFC 3261.

Schulzrinne, H. (2004). *The SIP Session Timer.* Technical Report draft-ietf-sipsession-timer-14. Internet Engineering Task Force.

Wang, S. H,, Tung, Y. S., Wang, C. N., Chiang, T., & Sun, H. (2003). *AHG Report on Editorial Convergence of MPEG-4 Reference Software* (Technical Report JTC1/SC29/WG11 MPEG2003/M9632). ISO/IEC.

Zhang, Q., Zhu, W., & Zhang, Y. Q. (2002). Power-minimized bit allocation for video communication over wireless channel. *IEEE Transactions on Circuits and Systems for Video Technology, 12*(6), 398-410.

Zhang, Q., Zhu, W., & Zhang, Y. Q. (2004). Channel-adaptive resource allocation for scalable video transmission over 3G wireless network. *IEEE Transactions on Circuits and Systems for Video Technology, 14*(8), 1049-1063.

KEY TERMS

3G: Third generation wireless format.

GGSN: Gateway GPRS support node.

MBMS: Multimedia broadcast/multicast service.

QoS: Quality of service.

RNC: Radio network controller.

SGSN: Serving GPRS support node.

SIP: Session initiation protocol.

UMTS: Universal mobile telecommunications system.

WLAN: Wireless local area network.

ENDNOTES

1. Broadcasting is a special case of multicasting.
2. We assume that each RA is covered by one node B.

Chapter XV
Perceptual Voice Quality Measurement – Can You Hear Me Loud and Clear?

Abdulhussain E. Mahdi
University of Limerick, Ireland

Dorel Picovici
University of Limerick, Ireland

ABSTRACT

In the context of multimedia communication systems, quality of service (QoS) is defined as the collective effect of service performance, which determines the degree of a user's satisfaction with the service. For telecommunication systems, voice communication quality is the most visible and important aspects to QoS, and the ability to monitor and design for this quality should be a top priority. Voice quality refers to the clearness of a speaker's voice as perceived by a listener. Its measurement offers a means of adding the human end user's perspective to traditional ways of performing network management evaluation of voice telephony services. Traditionally, measurement of users' perception of voice quality has been performed by expensive and time-consuming subjective listening tests. Over the last decade, numerous attempts have been made to supplement subjective tests with objective measurements based on algorithms that can be computerised and automated. This chapter examines some of the technicalities associated with voice quality measurement, presents a review of current subjective and objective speech quality measurement techniques, as mainly applied to telecommunication systems and devices, and describes their various classes.

INTRODUCTION

There is mounting evidence that the quality of the bread-and-butter product of cellular and mobile communication industry, voice that is, isn't really very good. Or, at least not as good as their customers would expect by comparing what they get to what they have traditional been offered. Mobile phone operators today might be trying to convince us that there is much more than just talking which we can do with our handsets. Intimately, though, this is true particularly in view of the present dynamic business environment, where voice services are no longer sufficient to satisfy customers' requirements. However, they also know that their crown jewel has always been, and continue to be, the provision of voice. The problem is, this valuable commodity existed long before the time mobile networks began to spread all over the world, and enjoyed a relatively good reputation in the hands of their previously dominant providers, the local telephone companies.

In a highly competitive telecommunications market where price differences have been minimised, quality of service (QoS) has become a critical differentiating factor. In the context of multimedia communication systems, QoS is defined as the collective effect of service performance, which determines the degree of a user's satisfaction with the service. However, when it comes to telecommunication networks, voice/speech communication quality is the most visible and important aspects to QoS. Thus, the ability to continuously monitor and design for this quality should always be a top priority to maintain customers' satisfaction of quality. Voice quality, also known as voice clarity, refers to the clearness of a speaker's voice as perceived by a listener. Voice quality measurement, also known by the acronym VQM, is a relatively new discipline which offers a means of adding the human, end-user's perspective to traditional ways of performing network management evaluation of voice telephony services. The most reliable method for obtaining true measurement of users' perception of speech quality is to perform properly designed subjective listening tests. In a typical listening test, subjects hear speech recordings processed through about 50 different network conditions, and rate them using a simple opinion scale such as the ITU-T (The International Telecommunication Union — Telecommunication Standardization Sector) 5-point listening quality scale. The average score of all the ratings registered by the subjects for a condition is termed the *mean opinion score* (MOS).

Subjective tests are, however, slow and expensive to conduct, making them accessible only to a small number of laboratories and unsuitable for real-time monitoring of mobile networks for example. As an alternative, numerous objective voice quality measures, which provide automatic assessment of voice communication systems without the need for human listeners, have been made available over the last decade. These objective measures, which are based on mathematical models and can be easily computerised, are becoming widely used particularly to supplement subjective test results. This chapter examines some of the technicalities associated with VQM and presents a review of current voice quality measurement techniques, as mainly applied to telecommunication networks. Following this Introduction, the Background section provides a broad discussion of what voice quality is, how to measure it and the needs for such measurement. Sections *Subjective Voice Quality Testing* and *Objective Voice Quality Measures* define the two main categories of measures used for evaluating voice quality, that is subjective and objective testing, describing, and reviewing the various methods and procedures of both, as well as indicating and comparing these methods' target applications and their advantages/disadvantages. The *Non-Intrusive Objective Voice Quality Measures* section discusses the various approaches employed for non-intrusive measurement of voice quality as required for monitoring live networks, and provides an up-to-date review of developments in the field. The section *Voice Quality of Mobile Networks* focuses on issues related to voice quality of current mobile phone networks, and discusses the findings of a recently reported study on how voice quality offered by cellular networks in the UK compare to traditional fixed line networks. The *Conclusion* section concludes the work by summarising the overall coverage of voice quality measurement in this chapter.

BACKGROUND

In the context of telecommunications, quality of service (QoS) is defined as the collective effect of service performance, which determines the degree of a user's satisfaction with the service. The QoS is thought to be divided into three components (Moller, 2000). The major component is the speech or voice communication quality, and relates to a bi/multi-directional conversation over the telecommunications network. The second component is the service-related influences, which is commonly referred to as the "service performance," and includes service support, a part of service operability and service security. The third component of the QoS is the necessary terminal equipment performance. The voice communication (or transmission) quality is more user-directed and, therefore, provides close insight in the question of which quality feature results in an acceptability of the service from the user's viewpoint.

What Is Voice Quality and How to Measure It?

Quality can be defined as the result of the judgement of a perceived constitution of an entity with regard to its desired constitution. The perceived constitution contains the totality of the features of an entity. For the perceiving person it is a characteristic of the identity of the entity (Moller, 2000). Applying this definition to speech, voice quality can be regarded as the result of a perception and assessment process, during which the assessing subject establishes a relationship between the perceived and the desired or expected speech signal. In other words, voice quality can be defined as the result of the subject's judgement on spoken language, which he/she perceives in a specific situation and judges instantaneously according to his/her experience, motivation, and expectation. Regarding voice communication systems, quality is the customer's perception of a service or product, and voice quality measurement (VQM) is a means of measuring customer experience of voice telephony services. The most accurate method of measuring voice quality therefore would be to actually ask the callers. Ideally, during the course of a call, customers would be interrupted and asked for their opinion on the quality. However, this is obviously not practical. In practice, there are two broad classes of voice quality metrics: *subjective* and *objective*. Subjective measures, known as subjective tests, are conducted by using a panel of people to assess the voice quality of live or recorded speech signals from the voice communication system/device under test for various adverse distortion conditions. Here, the speech quality is expressed in terms of various forms of a mean opinion score (MOS), which is the average quality perceived by the members of the panel. Objective measures, on the other hand, replace the human panel by an algorithm that compute a MOS value using a small portion of the speech in question. Detailed descriptions of both types of methods will be described in the proceeding sections.

Subjective tests can be used to gather first-hand evidence about perceived voice quality, but are often very expensive, time-consuming, and labour-intensive. The costs involved are often well justified, particularly in the case of standardisations or specification tests, and there is no doubt that the most important and accurate measurements of perceived speech quality will always rely on formal subjective tests (Anderson, 2001). However, there are many situations where the costs associated with formal subjective tests do not seem to be justified. Examples of these situations are the various design and development stages of algorithms and devices, and the continuous monitoring of telecommunications networks. Hence, an instrumental (non-auditive) method for evaluation of perceived quality of speech is in high demand. Such methods, which have been of great interest to researchers and engineers for a long time, are referred to as *Objective Speech/Voice Quality Measures* (Moller, 2000). The underlying principle of objective voice quality measurement is to predict voice communication/transmission quality based on objective metrics of physical parameters and properties of the speech signal. Once automated, objective methods enable standards to be efficiently maintained together with effective assessment of systems and networks during design, commissioning, and operation.

A voice communication system can be regarded as a distortion module. The source of the distortion can be background noise, speech codecs, and channel impairments such as bit errors and frame loss. In this context, most current objective voice quality evaluation methods are based on comparative measurement of the distortion between the original and distorted speech. Several objective voice quality measures have been proposed and used for the assessment of speech coding devices as well as voice communication systems. Over the last three decades, numerous different measures based on various perceptual speech analysis models have been developed. Most of these measures are based on an input-to-output or intrusive approach, whereby the voice quality is estimated by measuring the distortion between an "input" or a reference speech signal and an "output" or distorted speech signal. Current examples of intrusive voice quality measures include the Bark spectral distortion (BSD), perceptual speech quality (PSQM), modified BSD, measuring normalizing blocks (MNB), PSQM+, perceptual analysis measurement systems (PAMS) and most recently the perceptual evaluation of speech quality (PESQ) (Anderson, 2001). In 1996, a version of the PSQM was selected as ITU-T Rec. P.861 for testing codec but not networks (ITU-T, 1996b). The MNB was added to P.861 in 1998, also for testing codecs only. However, since P.861 was found unsuitable for testing networks it was withdrawn and replaced in 2001 by P.862 that specifies the PESQ (ITU-T, 2001).

Needs for VQM

There are several reasons for both mobile and fixed speech network providers to monitor the voice quality. The most important one is represented by customers' perception. Their decision in accepting a service is no longer restricted by limited technology or fixed by monopolies, therefore customers are able to select their telecommunications service provider according to price and quality. Another reason is *end-to-end measurement* of any impairment, where end-to-end measurements of voice quality yield a compact rating for whole transmission connection. In this context, voice quality can be imagined as a "black-box" approach that works irrespective of the kind of impairment and the network devices causing it. It is very important that a service provider has state-of-the-art VQM algorithms that allow the automation of speech quality evaluation, thereby reducing costs, enabling a faster response to customer needs, optimising and maintaining the networks. In a competitive mobile communication market, there is an increased interest in VQM by the following parties:

- **Network operators:** Continuous monitoring of voice quality enables problem detection and allows finding solutions for enhancement
- **Service providers:** VQM enable the comparison of different network providers based on their price/performance ratio
- **Regulators:** VQM provide a measurement basis in order to specify the requirements that network operators have to fulfil

SUBJECTIVE VOICE QUALITY TESTING

Voice quality measures that are based on ratings by human listeners are called subjective tests. These tests seek to quantify the range of opinions that listeners express when they hear speech transmission of systems that are under test. There are several methods to assess the subjective quality of speech signals. In general, they are divided in two main classes: (a) conversational tests and (b) listening-only tests. Conversational tests, whereby two subjects have to listen and talk interactively via the transmission system under test, provide a more realistic test environment. However, they are rather involved, much more time consuming, and often suffer from low reproducibility, thus listening-only tests are often recommended. Although listening-only tests are not expected to reach the same standard of realism as conversational tests and their restrictions are less severe in some respect, the artificiality associated with them brings with it a strict control of many factors, which in conversational tests are allowed to their own equilibrium.

In subjective testing, speech materials are played to a panel of listeners, who are asked to rate the passage just heard, normally using a 5-point quality scale. All subjective methods involve the use of large numbers of human listeners to produce statistically valid subjective quality indicator. The indicator is usually expressed as a *mean opinion score (MOS)*, which is the average value of all the rating scores registered by the subjects. For telecommunications purposes, the most commonly used assessment methods are those standardised and recommended by the ITU-T (ITU-T, 1996a):

- Conversational opinion
- Absolute category rating
- Quantal-response detectability
- Degradation category rating
- Comparison category rating

The first method in the above list represents a conversational type test, while the rest are effectively listening-only tests. Among the above-listed methods, the most popular ones are the *absolute category rating (ACR)* and the *degradation category rating (DCR)*. In the ACR, listeners are required to make a single rating for each speech passage using a *listening–quality scale* using the 5-point category-judgement scale shown in Table 1. The rating are then gathered and averaged to yield a final score known as the mean opinion score, or MOS. The test introduced by this method is well established and has been applied to analogue and digital telephone connections and telecommunications devices, such as digital codecs. If the voice quality were to drop during a telephone call by one MOS, an average user would clearly hear the difference. A drop of

Table 1. Listening-quality scale

Quality of speech	Score
Excellent	5
Good	4
Fair	3
Poor	2
Bad	1

Table 2. Recommended MOS terminology

Measurement	Listening-only	Conversational
Subjective	MOS-LQS	MOS-CQS
Objective	MOS-LQO	MOS-CQO
Estimated	MOS-LQE	MOS-CQE

half a MOS is audible, whereas a drop of a quarter of a point is just noticeable (Psytechnics, 2003). A typical public switched telephony network (PSTN) would have a MOS of 4.3. DCR involves listeners presented with the original speech signal as a reference, before they listen to the processed (degraded/distorted) signal, and are asked to compare the two and give a rating according to the amount of degradation perceived.

In May 2003, ITU-T approved Rec. P800.1 (ITU-T, 2003a) that provides a terminology to be used in conjunction with voice quality expressions in terms of MOS. As shown in Table 2, this new terminology is motivated by the intention to avoid misinterpretation as to whether specific values of MOS are related to listening quality or conversational quality, and whether they originate from subjective tests, from objective models or from network planning models.

According to Table 2, the following identifiers are recommended to be used together with the abbreviation MOS in order to distinguish the area of application: LQ to refer to *listening quality*, CQ to refer to *conversational quality*, S to refer to Subjective testing, O to refer to Objective testing using an objective model, and E to refer to Estimated using a network planning model.

OBJECTIVE VOICE QUALITY MEASURES

Objective voice quality metrics replace the human panel by a computational model or an algorithm that compute a MOS value by observing a small portion of the speech in question (Quackenbush, Barnawell, & Clements, 1988). The aim of objective measures is to predict MOS values that are as close as possible to the ratings obtained from subjective tests for various adverse speech distortion conditions. The accuracy and effectiveness of an objective metric is, therefore, determined by its correlation, usually the Pearson correlation, with the subjective MOS scores. If an objective measure has a high correlation, typically >0.8 (Yang, 1999), it is deemed to be effective measure of perceived voice quality, at least for the speech data and transmission systems with the same characteristics as those in the test experiment. Starting from late 1970, researchers and engineers in the field of objective measures of speech quality have developed different objective measures based on various speech analysis models. Based on the measurement approach, objective measures are classified into two classes: *intrusive* and *non-intrusive*, as illustrated in Figure 1. Intrusive measures, often referred to as input-to-output measures, base their measurement on computation of the distortion between the original (clean or input) speech signal and the degraded (distorted or output) speech signal. Non-intrusive measures (also known as output-based or single-ended measure), on the other hand, use only the degraded signal and have no access to the original signal.

Figure 1. Intrusive and non-intrusive voice quality measures

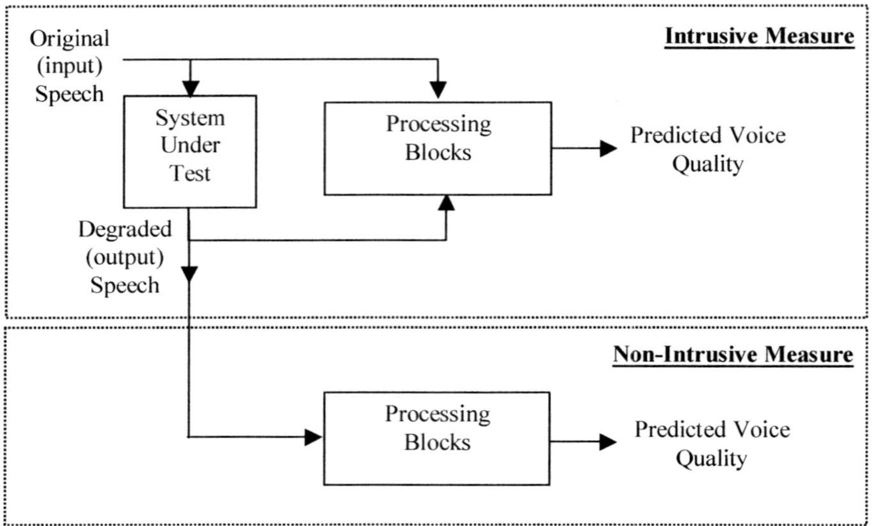

Figure 2. Basic structure of an intrusive (input-to output) objective voice quality measure

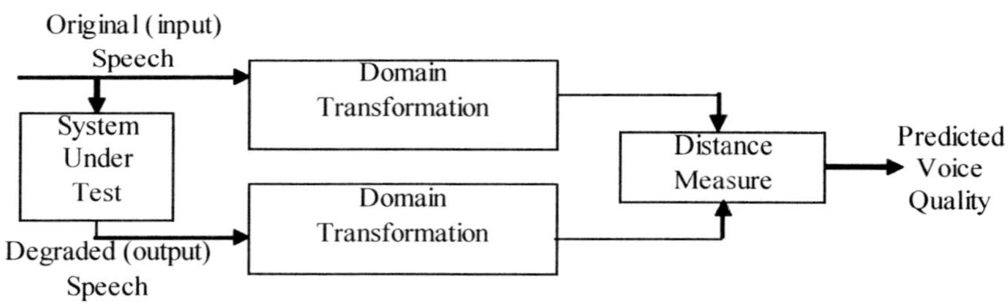

Intrusive Objective Voice Quality Measures

Although there are different types of intrusive (or input-to output) objective speech quality measures, they all share a similar measurement structure that involves two main processes, as shown in Figure 2.

The first process is the domain transformation. In this process, the original (input) speech signal and the signal degraded by the system under test (i.e., the output signal) are transformed into a relevant domain such as temporal, spectral or perceptual domain. The second process involves a distance measure, whereby the distortion between the transformed input and output speech signals is computed using an appropriate quantitative measure.

Depending on the domain transformation used, objective measures are often classified into three categories as shown in Figure 3.

Figure 3. Classification of objective voice quality measures based on the transformation domain

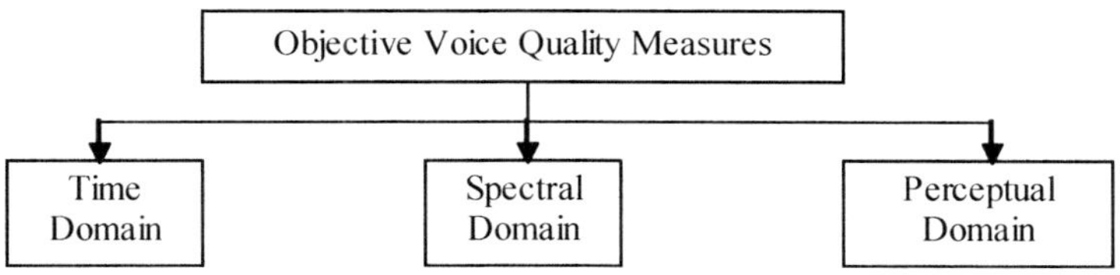

Time Domain Measures

Time domain measures are generally applicable to analogue or waveform coding systems in which the target is to reproduce the waveform. Signal-to-noise ratio (SNR) and *segmental SNR (SNRseg)* are typical time domain measures (Quackenbush et al., 1988). In time domain measures, speech waveforms are compared directly, therefore synchronisation of the original and distorted speech is crucial. If the waveforms are not synchronised accurately the results obtained by these measures do not reflect the distortions introduced by the system under test. Time domain measures are of little use nowadays, since the actual codecs are using complex speech production models which reproduce the same sound of the original speech signal, rather than simply reproduce the original speech waveform.

In Signal-to-Noise Ratio (SNR) measures, "Signal" refers to useful information conveyed by some communications medium, and "noise" to anything else on that medium. Classical SNR, segmental SNR, frequency weighted segmental SNR, and granular segmental SNR are variations of SNR (Goodman, Scagliola, Crochiere, Rabiner, & Goodman, 1979). Signal-to-noise measures are used only for distorting systems that reproduce a facsimile of the input waveform such that the original and distorted signals can be time aligned and noise can be accurately calculated. To achieve the correct time alignment it may be necessary to correct phase errors in the distorted signal or to interpolate between samples in a sampled data system.

It has often been shown that SNR is a poor estimator of subjective voice quality for a large range of speech distortions (Quackenbush et al, 1988), and therefore is of little interest as a general objective measure of voice quality. Segmental signal-to-noise ratio (SNRseg), on the other hand, represents one of the most popular classes of the time-domain measures. The measure is defined as an average of the SNR values of short segments, and can commonly be computed as follows:

$$\mathrm{SNRseg} = \frac{10}{M}\sum_{m=0}^{M-1}\log_{10}\sum_{n=Nm}^{Nm+N-1}\left(\frac{x^2(n)}{(d(n)-x(n))^2}\right) \quad (1)$$

where $x(n)$ represents the original speech signal, $d(n)$ represents the distorted speech reproduced by a speech processing system, N is the segment length, and M represents the number of segments in the speech signal. Classical windowing techniques are used to segment the speech signal into appropriate speech segments.

Performance measure in terms of SNRseg is a good estimator of voice quality of waveform codecs (Noll, 1974), although its performance is poor for vocoders where the aim is to generate the same speech sound rather than to produce the speech waveform itself. In addition, SNRseg may provide inaccurate indication of the quality when applied to a large interval of silence in speech utterances. In the case of a mainly silence segment, any amount of noise will cause negative SNR ratio for that segment which could significantly bias the overall measures of segmental SNR. A solution for this drawback involves identifying and excluding the silent segments. This can be done by computing the energy of each speech segment and setting an energy level threshold. Only the segments with energy level above the threshold are included in the computation of segmental SNR.

Spectral Domain Measures

Spectral domain measures are more credible than time-domain measures as they are less susceptible to the occurrence of time misalignments and phase shift between the original and the distorted signals. Most spectral domain measures are related to speech codecs design and use the parameters of speech production models. Their capability to effectively describe the listeners' auditory response is limited by the constraints of the speech production models. Over the last three decades, several spectral domain measures have been proposed in the literature, including the *log likelihood ratio*, *Itakura-Saito distortion measure* (Itakura & Saito, 1978), and the *cepstral distance* (Kitawaki, Nagabuchi, & Itoh, 1988).

The log likelihood ratio (LLR) measure, or Itakura distance measure, is founded on the difference between the speech production models such as all-pole linear predictive coding models of the original and distorted speech signals. The measure assumes that a speech segment can be represented by a pth order all-pole linear predictive coding model defined by the following equation:

$$x(n) = \sum_{i=1}^{p} a_i x(n-m) + G_x u(n) \qquad (2)$$

where $x(n)$ is the n-th speech sample, a_i (i=1, 2, ..., p) represents the coefficients of the all-pole filter, G_x is the gain of the filter and $u(n)$ is an appropriate excitation source for the filter. LLR measure is frequently presented in terms of the autocorrelation method of linear prediction analysis, in which the speech signal is windowed to form frames with the length of 15 to 30 ms. The LLR measure can be written as:

$$\text{LLR} = \log\left(\frac{a_x R_x a_x^T}{a_d R_d a_d^T}\right) \qquad (3)$$

where a_x represents the linear predictive coding (LPC) coefficient vector (1, $-a_x(1)$, $a_x(2)$, ..., $a_x(p)$) for the original speech , R_x represents the autocorrelation matrix for the original speech and a_d represents the LPC coefficient vector (1, $-a_d(1)$, $a_d(2)$, ..., $a_d(p)$) for the distorted speech and T denotes the transpose operation.

The Itakura-Saito measure (IS) is a variation of the LLR that includes in its computation the gain of the all-pole linear predictive coding model. Linear prediction coefficients (LPC) can also be used to compute a distance measure based on cepstral coefficients known as the cepstral distance measure. Unlike the cepstrum computed directly from speech waveform, one computed from the predictor coefficients provides an estimate of the smoothed speech spectrum.

Perceptual Domain Measures

As most of the spectral domain measures use the parameters of speech production models used in codecs, their performance is usually limited by the constraints of those models. In contrast to the spectral domain measures, perceptual domain measures are based on models of human auditory perception and, hence, have the best potential of predicting subjective quality of speech. In these measures, speech signals are transformed into a perception-based domain using concepts of the psychophysics of hearing, such as the critical-band spectral resolution, frequency selectivity, the equal-loudness curve, and the intensity-loudness power law to derive an estimate of the auditory spectrum (Quatieri, 2002). In principle, perceptually relevant information is both sufficient and necessary for a precise assessment of perceived speech quality. The perceived quality of the coded speech will, therefore, be independent of the type of coding and transmission, when estimated by a distance measure between perceptually transformed speech signals. The following sections give descriptions of currently used perceptual voice quality measures.

Bark Spectral Distortion Measure (BSD)

The Bark spectral distortion (BSD) measure was developed by Wang and co-workers (Yang, 1999) as a method for calculating an objective measure for signal distortion based on the quantifiable properties of auditory perception. The overall BSD measurement represents the average squared Euclidian distance between spectral vectors of the original and coded utterances. The main aim of the measure is to emulate several known features of perceptual processing of speech sounds by the human ear, especially frequency scale warping, as modelled by the Bark transformation, and critical band integration in the cochlea; changing sensitivity of the ear as the frequency varies; and difference between the loudness level and the subjective loudness scale.

The approach in which the measure is performed is shown in Figure 4. Both the original speech record, $x(n)$, and the distorted speech (coded version of the original speech), $d(n)$, are pre-processed separately by identical operations to obtain their Bark spectra, $L_x(i)$ and $L_d(i)$, respectively. The starting point of the pre-processing operations is the computation of the magnitude squared FFT spectrum to generate the power spectrum, $|X(f)|^2$. This is followed by critical-band filtering to model the non-linearity of the human auditory system, which leads to a poorer discrimination at high frequencies than at low frequencies, and the masking of tones by noise.

The spectrum available after critical band filtering is loudness equalised so that the relative intensities at different frequencies correspond to relative loudness in phones rather than acoustical levels. Finally, the processing operation ends with another perceptual non-linearity: conversion from phone scale into perceptual scale of sones. By definition a sone represents the increase in power which doubles the subjective loudness. The ear's non-linear transformations of frequency and amplitude, together with important aspects of its frequency analysis and spectral integration properties in response to complex sounds, is represented by the Bark spectrum $L(i)$. By using the average squared Euclidian distance between two spectral vectors, the BSD is computed as:

$$\mathrm{BSD} = \frac{\frac{1}{M}\sum_{m=1}^{M}\sum_{i=1}^{O}\left[L_x^{(m)}(i) - L_d^{(m)}(i)\right]^2}{\frac{1}{M}\sum_{m=1}^{M}\sum_{i=1}^{O}\left[L_x^{(m)}(i)\right]^2} \tag{4}$$

where M is the number of frames (speech segments) processed, O is the number of critical bands, $L_x^{(m)}(i)$ is the Bark spectrum of the m-th critical frame of original speech, and $L_d^{(m)}(i)$ is the Bark spectrum of the m-th critical frame

Figure 4. Block diagram representation of the BSD measure

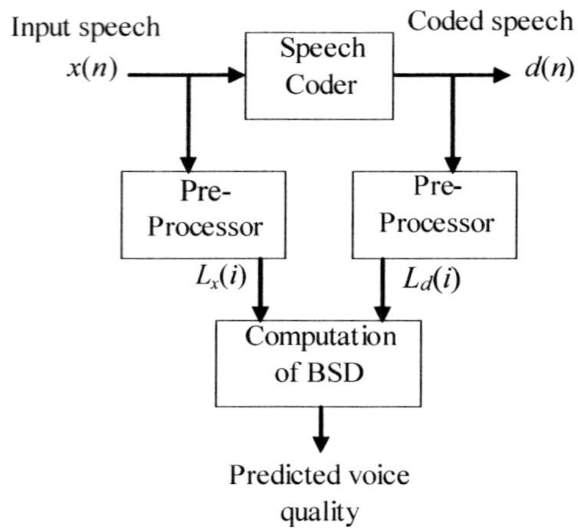

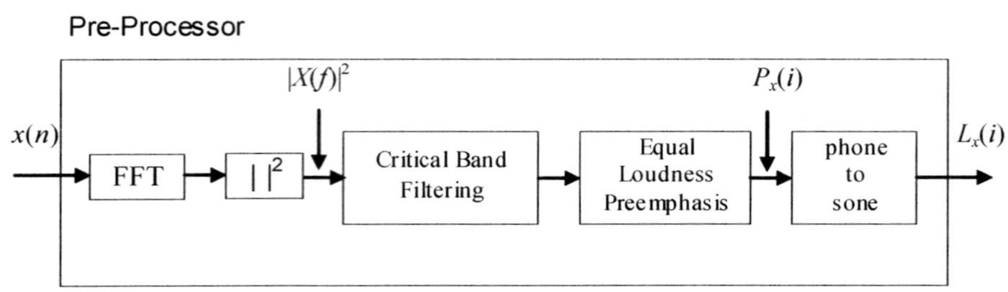

of coded speech. BSD works well in cases where the distortions in voice regions represent the overall distortion because it processes voiced regions only. Hence, voiced regions have to be detected.

Modified and Enhanced Modified Bark Spectral Distortion measures (MBSD & EMBSD)

The modified Bark spectral distortion (MBSD) measure (Yang, 1999) is a modification of the BSD in which the concept of a noise-masking threshold that differentiates between audible and inaudible distortions is incorporated. It uses the same noise-masking threshold as that used in transform coding of audio signals (Johnson, 1988). There are two differences between the conventional BSD and MBSD. First, noise-masking threshold for determination of the audible distortion is used by MBSD, while the conventional BSD uses an empirically determined power threshold. Secondly, the way in which the distortion is computed. While the BSD defines the distortion as the average squared Euclidian distance of estimated loudness, the MBSD defines the distortion as the difference in estimated loudness. Figure 5 describes the MBSD measure. The loudness of the noise-masking threshold is compared to the loudness difference of the original and the distorted (coded) speech to establish any perceptible distortions. When the loudness difference is below the loudness of the noise masking threshold, it is imperceptible and, hence, not included in the calculation of the MBSD.

Figure 5. Block diagram of MBSD measure

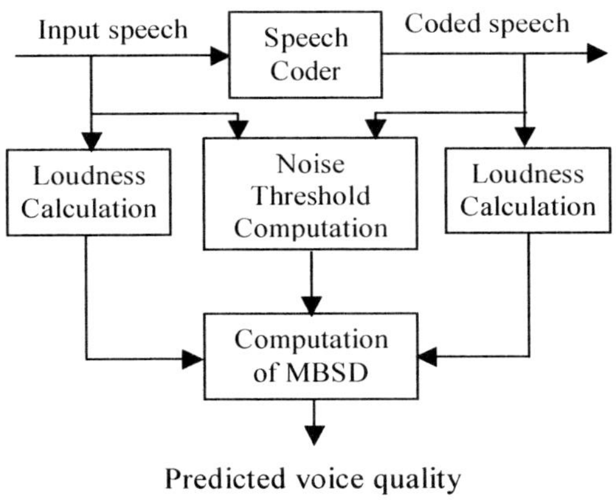

The enhanced modified Bark spectral distortion (EMBSD), on the other hand, is a development of the MBSD measure where some procedures of the MBSD have been modified and a new cognitive model has been used. These modifications involve the followings: the amount of loudness components used to calculate the loudness difference, the normalisation of loudness vectors before calculating loudness difference, the inclusion of a new cognition model based on post masking effects, and the deletion of the spreading function in the calculation of the noise-masking threshold (Yang, 1999).

Perceptual Speech Quality Measurement (PSQM)

To address the continuous need for an accurate objective measure, Beerends and Stemerdink from KPN Research — Netherlands, developed a voice quality measure which takes into account the clarity's subjective nature and human perception. The measure is called the perceptual speech quality measurement or PSQM (Beerends & Stemerdink, 1994). In 1996 PSQM was approved by ITU-T and published by the ITU as Rec. P.861 (ITU-T, 1996b). The PSQM, as shown in Figure 6, is a mathematical process that provides an accurate objective measurement of the subjective voice quality. The main objective of PSQM is to produce scores that reliably predict the results of the recommended ITU-T subjective tests (ITU-T, 1996a). PSQM is designed to be applied to telephone band signals (300-3400 Hz) processed by low bit-rate voice compression codecs and vocoders.

To perform a PSQM measurement, a sample of recorded speech is fed into a speech encoding/decoding system and processed by whatever communication system is used. Recorded as it is received, the output signal (test) is then time-synchronised with the input signal (reference). Following the time-synchronisation the PSQM algorithm will compare the test and

Figure 6. PSQM testing process

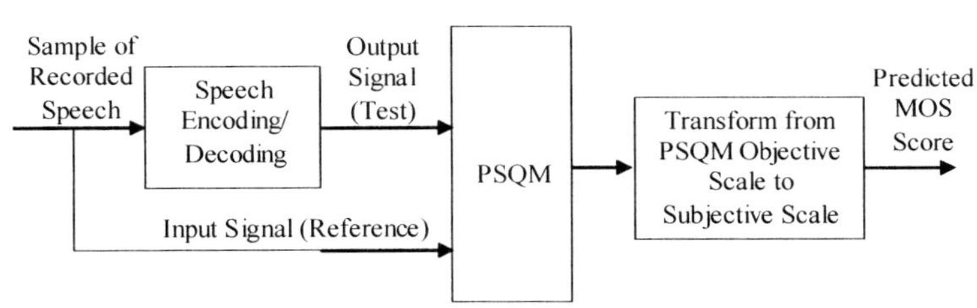

reference signals. This comparison is performed on individual time segments (or frames) acting on parameters derived from spectral power densities of the input and output time-frequency components. The comparison is based on factors of human perception, such as frequency and loudness sensitivities, rather than on simple spectral power densities. The resulting PSQM score representing a perceptual distance between the test and reference signals can vary from 0 to infinity. As an example, 0 score suggests a perfect correlation between the input and output signals, which most of the time is classified as perfect clarity. Higher scores indicate increasing levels of distortion, often interpreted as lower clarity. In practice upper limits of PSQM scores range from 15 to 20. At the final stage, the PSQM scale is mapped from its objective scale to the 1-5 subjective MOS scale. One of the main drawbacks of this measure is that it does not accurately report the impact of distortion caused by packet loss or other types of time clipping. In other words, human listeners reported higher speech quality score than PSQM measurements for such errors.

Perceptual Speech Quality Measurement Plus (PSQM+)

Taking into account the drawbacks of the PSQM, Beerends, Meijer, and Hekstra developed an improved version of the conventional PSQM measure. The new model, which became known as PSQM+, was reviewed by ITU-T Study Group 12 and published in 1997 under COM 12-20-E (Beerends et al., 1997). PSQM+, which is based directly on the PSQM model, represents an improved method for measuring voice quality in network environments. For systems comprising speech encoding only both methods give identical scores. PSQM+ technique, however, is designed for systems which experience severe distortions due to time clipping and packet loss. When a large distortion, such as time clipping or packet loss is introduced (causing the original PSQM algorithm to scale down its score), the PSQM+ algorithm applies a different scaling factor that has an opposite effect, and hence produces higher scores that correlate better with subjective MOS than the PSQM.

Measuring Normalising Blocks (MNB)

In 1997, the ITU-T published a proposed annex to Rec. P.861 (PSQM), which was approved in 1998 as appendix II to the above-mentioned Recommendation. The annex describes an alternative technique to PSQM for measuring the perceptual distance between the perceptually transformed input and output signals. This technique is known as measuring normalising blocks (MNB) (Voran, 1999). Based on the fact that listeners adapt and react differently to spectral deviations that span different time and frequency scale, the MNB defines a new perceptual distance across multiple time and frequency scales. The model as shown in Figure 7 is recommended for measuring the impact of transmission channel errors, CELP and hybrid codecs with bit rates less than 4 kb/s and vocoders. In this technique, perceptual transformations are applied to both output and input signals before measuring the distance between them using MNB measurement. There are two types of MNBs: time measuring normalising blocks (TMNB) and frequency measuring normalising blocks (FMNB) (Voran, 1999). TMNB and FMNB are combined with weighting factors to generate a nonnegative value called auditory distance (AD). Finally, a logistic function maps AD values into a finite scale to provide correlation with subjective MOS scores.

Figure 7. The MNB model

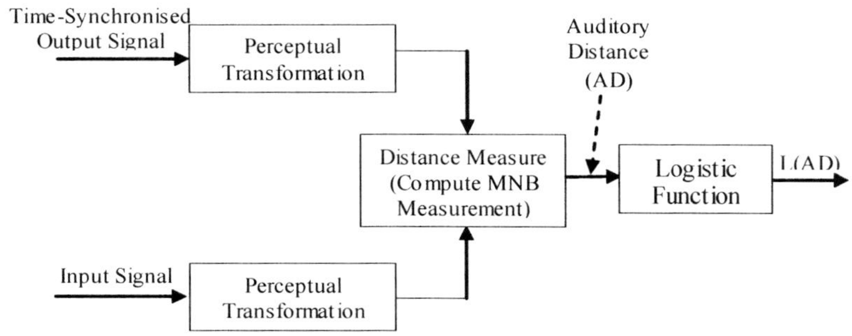

Perceptual Analysis Measurement System (PAMS)

Psytechnics, a UK-based company associated with British Telecommunications (BT), developed an objective speech quality measure called perceptual analysis measurement system (PAMS) (Rix & Hollier, 2000). PAMS uses a model based on factors of human perception to measure the perceived speech clarity of an output signal as compared with the input signal. Although similar to PSQM in many aspects, PAMS uses different signal processing techniques and a different perceptual model (Anderson, 2001). The PAMS testing process is shown in Figure 8.

As shown in Figure 8, to perform a PAMS measurement a sample of recorded human speech is inputted in into a system or network. The characteristics of the input signal follow those that are used for MOS testing and are specified by ITU-T (1996a). The output signal is recorded as it is received. PAMS removes the effects of delay, overall systems gain/attenuation, and analog phone filtering by performing time alignment, level alignment, and equalisation. Time alignment is performed in time segments so that the negative effects of large delay variations are removed. However, the perceivable effects of delay variation are preserved and reflected in PAMS scores. After time alignment PAMS compares the input

Figure 8. PAMS testing process

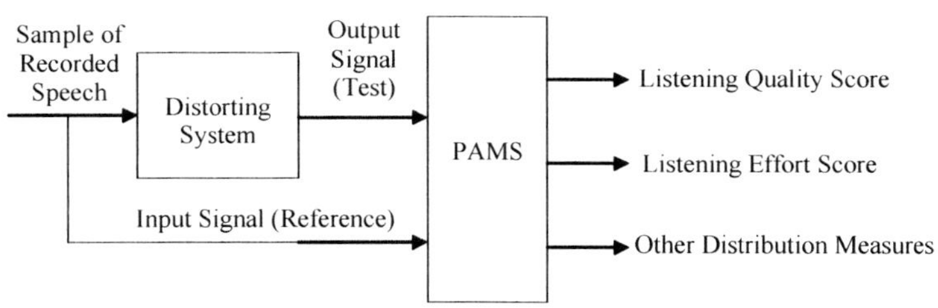

and output signals in the time-frequency domain. This comparison is based on human perception factors. The results of the PAMS comparison are scores that range from 1-5 and that correlate with the same scales as MOS testing. In particular, PAMS produces a listening quality score and a listening effort score that correspond with both the ACR opinion scale in ITU-T Rec. P.800 (ITU-T, 1996b) and P.830 (ITU-T, 1996a), respectively. The PAMS system is flexible in adopting other parameters if they are perceptually important. The accuracy of PAMS is dependent upon the designer intuition in extracting candidate parameters as well as selecting parameters with a training set. It is not simple to optimise both the parameter set and the associated mapped function since the parameters are usually not independent of each other. Therefore, during training extensive computation is performed.

Perceptual Evaluation of Speech Quality (PESQ)

In 1999, KPN Research-Netherlands improved the classical PSQM to correlate better with subjective tests under network conditions. This resulted in a new measure known as PSQM99. The main difference between the PSQM99 and PSQM concerns the perceptual modelling where they are differentiated by the asymmetry processing and scaling. PSQM 99 provides more accurate correlations with subjective test results than PSQM and PSQM+. Later on, ITU-T recognised that both PSQM99 and PAMS had significant merits and that it would be beneficial to the industry to combine the merits of each one into a new measurement technique. A collaborative draft from KPN Research and British Telecommunications was submitted to ITU in May 2000 describing a new measurement technique called Perceptual Evaluation of Speech Quality (PESQ). In February 2001, ITU-T approved the PESQ under Rec. P.862 (ITU-T, 2001). PESQ is directed at narrowband telephone signals and is effective for measuring the impact of the following conditions: waveform and non waveform codecs, transcodings, speech input levels to codecs, transmission channel errors, noise added by system (not present in input signal), and short and long term warping.

The PESQ combines the robust time-alignment techniques of PAMS with the accurate perceptual modelling of PSQM99. It is designed for use with intrusive tests: a signal is injected into the system under test, and the distorted output is compared with the input

Figure 9. The PESQ model

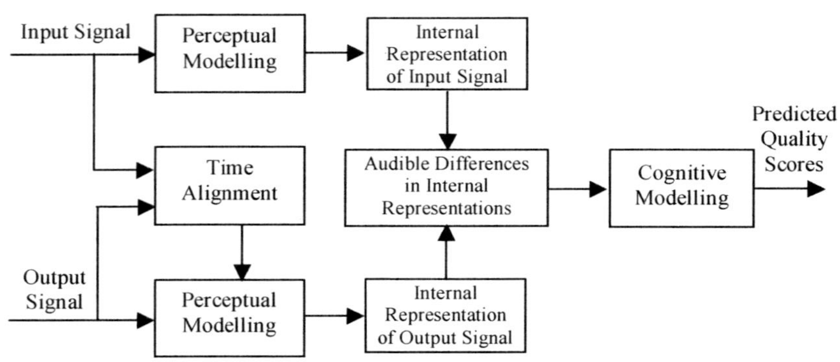

(reference) signal. The difference is then analysed and converted into a quality score. As a result of this process, the predicted MOS as given by PESQ varies between 0.5, which corresponds to a bad distortion condition, and 4.5 which corresponds to no measurable distortion. The PESQ model is shown in Figure 9. PESQ can be used in a wide range of measurement applications, such as codecs development, equipment optimisation and regular network monitoring. Being fast and repeatable, PESQ makes it possible to perform extensive testing over a period of only few days, and also enables the quality of time-varying conditions to be monitored.

In order to align with the new MOS terminology, a new ITU-T Recommendation, Rec. P.862.1 (ITU-T, 2003b) was published. This Recommendation defines a mapping function and its performance for a single mapping from raw P.862 scores to the MOS-LQO (Rec. P.800.1).

NON-INTRUSIVE OBJECTIVE VOICE QUALITY MEASURES

All objective measures presented in the preceding Sections are based on an input-to-output approach, whereby speech quality is estimated by objectively measuring the distortion between the original or input speech and the distorted or output speech. Besides being intrusive, input-to-output speech quality measures have few other problems. Firstly, in all these measures the time-alignment between the input and output speech vectors, which is achieved by automatic synchronization, is a crucial factor in deciding the accuracy of the measure. In practice, perfect synchronization is difficult to achieve due to fading or error burst that are common in wireless systems, and hence degradation in the performance of the measure is inevitable. Secondly, there are many applications where the original speech is not available, as in cases of wireless and satellite communications. Furthermore, in some situations the input speech may be distorted by background noise, and hence, measuring the distortion between the input and the output speech does not provide a true indication of the speech quality of the communication system. In most situations, it is not always possible to have access to both ends of a network connection to perform speech quality measurement using an input-to-output method. There are two main reasons for this: (a) too many connections must be monitored and (b) the far end locations could be unknown. Specific distortions may only appear at the times of peak traffic when it is not possible to disconnect the clients and perform networks tests.

An objective measure which can predict the quality of the transmitted speech using only the output (or degraded) speech signal (i.e., one end of the network, would therefore cure all the above problems and provide a convenient non-intrusive measure for monitoring of live networks. Ideally what is required for a non-intrusive objective voice quality measure is to be able to assess the quality of the distorted speech by simply observing a small portion of the speech in question with no access to the original speech. However, due to non-availability of the original (or input) speech signal such a measure is very difficult to realise. In general, there are two different approaches to realise a non-intrusive objective voice quality measure: *priori-based* and *source-based*.

Priori-Based Approach

This approach is based on identifying a set of well-characterised distortions and learning a statistical relationship between this finite set and subjective opinions. An example of this kind of approach has been reported in (Au & Lam, 1998). Their approach is based on visual features of the spectrogram of the distorted speech. According to early work done on speech spectrograms, it was established that most of the underlying phonetic information could be recovered by visually inspecting the speech spectrogram. The measurement is realised by computing the dynamic range of the spectrogram using digital image processing.

Another example of such non-intrusive approach is the speech quality measure known as ITU Rec. P.562, which uses in-service, non-intrusive measurement devices (INMD) (ITU-T, 2000). An INMD is a device that has access to the voice channels and performs measurements of objective parameters on live call traffic, without interfering with the call in any way. Data produced by an INMD about the network connection, together with knowledge about the network and the human auditory system, are used to make predictions of call clarity in accordance with ITU-T Rec. P.800 (ITU-T, 1996a). Recently ITU-T recommended a new computational model known as the E-model (ITU-T, 2003c), that in connection with INMD can be used for instance by transmission planners to help ensure that users will be satisfied with end to end transmission performance. The primary output from the model can be transformed to give estimates of customer opinion. However, such estimates are only made for transmission planning purposes and not for actual customer opinion prediction.

All the above-described methods can be used with confidence for the types of well-known distortions. However, none of them have been verified with very large number of possible distortions. Most recently, the ITU-T approved a new model as Rec. P.563: "Single ended method for objective speech quality assessment in narrow-band telephony applications" (ITU-T, 2004). The P.563 approach is the first recommended method for single-ended non-intrusive voice quality measurement applications that takes into account the full range of distortions occurring in public switched telephony networks (PSTN) and that is able to predict the voice quality on a perception-based scale MOS–LQO according to ITU-T Rec. P.800.1. The validation of this method included all available experiments from the former P.862 (PESQ) validation process, as well as a number of experiments that specifically tested its performance by using an acoustical interface in a real terminal at the sending end. Furthermore, the P.563 algorithm was tested independently with unknown speech material by third party laboratories under strictly defined requirements. The reported experimental results indicate that this non-intrusive measure compares favourably with the first generation of intrusive perceptual models such as PSQM. However, correlation of its quality predicted scores and the MOS-LQS is lower than the second generation of intrusive perceptual models such as PESQ. ITU-T recommended that P.563 be used for voice quality measurement in 3.1 kHz (narrow-band) telephony applications only.

Source-Based Approach

This approach represents a more universal method that is based on a prior assumption of the expected clean signal rather than on the distortions that may occur. The approach permits to deal with ample range of distortion types, where the distortions are characterised by comparing some properties of the degraded signal with a priori model of these properties for clean signal.

Initial attempt to implement such an approach was reported by (Jin & Kubichek, 1995). The proposed measure was based on an algorithm which uses perceptual-linear prediction (PLP) model to compare the perceptual vectors extracted from the distorted speech with a set of perceptual vectors derived from a variety of undegraded clean source speech material. However, the measure was computationally involved since it was based on the use of a basic vector quantization (VQ) technique. In addition, it has a number of drawbacks: (a) the size and structure of the codebook as created by the VQ technique was not optimised, (b) the search engine used was based on a basic full-search technique which represents one of the slowest and most inefficient search techniques, and (c) the method was tested with a relatively small number of distortion conditions only, most of which are synthesised, and therefore its effective ness was not verified for a wide range of applications. In 2000, Gray, Hollier, and Massara (2000) reported a novel use of the vocal-tract modelling technique, which enables the prediction of the quality of a network degraded speech stream to be made in a non-intrusive way. However, athough good results were reported, the technique suffers from the followings drawbacks: (a) its performance seems to be affected by the gender of the speaker gender, (b) its application is limited to speech signals with a relatively short duration in time, (c) its performance is influenced by distorted signals with a constant level of distortions, and (d) the vocal-tract parameters are only meaningful when they are extracted from a speech stream that is the result of glottal excitation illuminating an open tract.

Recently, the authors proposed a new perception-based measure for voice quality evaluation using the source-based approach. Since the original speech signal is not available for this measure, an alternative reference is needed in order to objectively measure the level of distortion of the distorted speech. As shown in Figure 10, this is achieved by using an internal reference database of clean speech records. The method is based on computing objective distances between perceptually-based parametric vectors representing degraded speech signal to appropriately matching reference vectors extracted from a pre-formulated reference codebook, which is constructed from the database of clean speech records. The computed distances provide a reflection of the distortion of the received speech signal. In order to simulate the functionality of a subjective listening test, the system maps the measured distance into an equivalent subjective opinion scale such as the mean opinion score (MOS). The method has been described in detail in Picovici and Mahdi (2004). Its performance has been compared to that of the ITU-T Rec. P.862 (PESQ). Presented evaluation results show that the proposed method offers a high level of accuracy in predicting the subjective MOS (MOS-LQS) and compares favourably with the

Figure 10. Non-intrusive perception-based measure proposed by the authors for voice quality evaluation

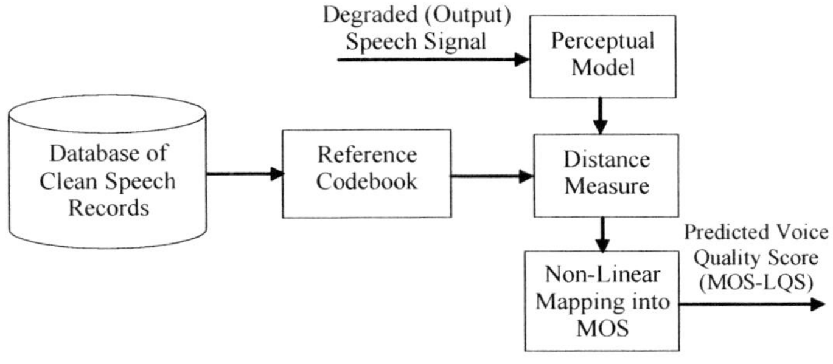

second generation of intrusive perceptual models such as PESQ.

VOICE QUALITY OF MOBILE NETWORKS

Mobile Quality — Speak Up, I Can't Hear You

Over the last few years, the mobile phone market has experienced sharp growth throughout the world, with many recent market analyses indicating virtual market saturation. In this situation, possible market growth for a mobile phone operator will either come from acquiring competitors' customers, by attracting more PSTN network users, or by increasing the average revenue per existing user. In the UK in 2001/02, for example, more than 310 billion minutes worth of voice calls were made from fixed line phones, compared to just over 46 billion minutes of calls made from mobile phones (Psytechnics, 2003). The average mobile user has a bill of approximately £20 per month, a figure which has changed little over the last three years. However, with 73% of UK adults still considering a fixed line at home to be their main calling making/receiving method compared to only 21% who use their mobiles as primary method, operators are facing a tough challenge. There are a number of commonly held perceptions that influence attracting new customers and/or persuading existing ones to use their mobiles more. Firstly, using a mobile is still perceived to be more expensive compared to using a fixed line phone. Secondly, there is a perception that mobile networks provide poorer quality service than PSTNs, an issue acknowledged by the industry experts. Even when there is full signal strength showing on the handset, mobile voice quality can still be affected by (Psytechnics, 2003):

- Voice compression commonly used in GSM to reduce data rate
- Radio link coverage: proximity to base station and effect of buildings and surrounding landscape
- Interference from other traffic on the same network
- Handsets: for example some handsets have built-in noise reduction, type and location of aerial
- Noise in the user's environment

Mobile Voice Quality Survey

A study to find out how exactly the voice quality offered by cellular networks in the UK compared to each other and to traditional PSTNs was carried by the UK-based company Psytechnics in September 2003 (Psytechnics, 2003). Psytechnics measured the performance of the five main UK mobile operators to assess their overall voice quality when receiving a full strength signal. The measurement was based on the PESQ, which is currently the interna-

Table 3. Typical MOS-LQO measured using PESQ (Psytechnics, 2003)

MOS-LQO (PESQ)	Conditions
4.3	High-quality fixed network (PSTN)
4.0-4.1	GSM/3G network in ideal conditions (GSM-EFR codec with no noise or interference)
3.5	GSM-FR codec (older handsets prior to 2000)
2.9-4.1	Typical GSM network operating range (GSM-EFR codec)

tional standard for measuring the voice quality recommended by the ITU-T. The networks were tested in 20 urban locations using an average of 150 calls in total for each operator covering typical conditions experienced by customers using the chosen handset. Eight different mobile handsets, most of which are currently available in the UK, were tested. Table 3 shows typical overall MOS-LQO as measured by Psytechnics using PESQ.

Regarding how much worse exactly voice quality is between the cellular networks and the PSTNs, the study provided a resounding answer: a 0.8 of a MOS point when the average overall performance of the five operators is considered. The testing showed the following facts:

- Voice quality scores for all operators fell below the PSTN accepted level of 4.3 MOS
- Voice quality varies considerably between different operators, and voice quality can vary during the course of a call, despite the indicative signal strength showing 'full bars' on the handset
- Handsets have an important influence on voice quality, and the voice quality varies considerably between different handsets, with a difference of almost 1 MOS between the best and the worst performing handsets. Also, higher cost does not necessarily equate to better voice quality
- The uplink voice quality tends to be poorer than the downlink, with the worst case being the uplink from mobile to PSTN

CONCLUSION

In this chapter, we have presented a detailed review of currently used metrics and methods for measuring the user's perception of the voice quality of telephony networks. Descriptions of various internationally standardised subjective tests that are based on ratings by humans were presented, with particular emphasis on those approved by the ITU-T. Limitations of subjective testing were then discussed, paving the ground for a comprehensive review of various objective voice quality measures, highlighting in a comparative manner their historical evolution, target applications and performance limitations. In particular, two main categories of objective voice quality measures were described: intrusive or input-to-output measures and non-intrusive or single-ended measures, providing an insight into advantages/disadvantages of each. Finally, issues related to the voice quality of mobile phone networks were discussed in view of current status of the mobile market and the findings of a recent industrial study on how voice quality offered by cellular networks in compare to traditional PSTNs. As in any fast-paced industry, it seems that innovation has led the mobile market, and up until few years ago the focus of cellular operators was on making services available and then looking at customer retention and revenue generation. However, times move on, industries in their infancy suddenly mature and customers' expectations grow with every new development, particularly regarding quality of service and there is nothing more important in this regard than voice quality.

REFERENCES

Anderson, J. (2001). *Methods for measuring perceptual speech quality*. White paper, Agilent technologies, USA. Retrieved from http://www.agilent.com

Au, O. C., & Lam, K. H. (1998). *A novel output-based objective speech quality measure for wireless communication*. New York: Prentice Hall.

Beerends, J. G., Meijer, E. J., & Hekstra, A. P. (1997). *Improvement of the P. 861 perceptual speech quality measure*. Contribution to COM 12-20, ITU-T Study Group 12, International Telecommunication Union, CH-Geneva.

Beerends, J. G., & Stemerdink, J. A. (1994). A perceptual speech quality measure based on a psychoacoustic sound representation. *Journal of Audio Engineering Society, 42*(3), 115-123.

Goodman, D. J., Scagliola, C., Crochiere, R. E., Rabiner, L. R., & Goodman, J. (1979). Objective and subjective performance of tandem connections of waveform coders with an LPC vocoder. *Bell Systems Technical Journal, 58*(3), 601-629.

Gray, P., Hollier, M. P., & Massara, R. E. (2000). Non-intrusive speech quality assessment using vocal-tract models. *IEE Proceedings—Vision Image Signal Processing, 147*(6), 493-501.

Itakura, F., & Saito, S. (1978). Analysis synthesis telephony based on the maximum likelihood method. In *Proceedings of the 6th International Congress on Acoustics*, Tokyo, Japan (C17-C-20).

ITU-T. Recommendation P.800. (1996a). *Methods for subjective determination of transmission quality*. International Telecommunication Union, CH-Geneva.

ITU-T. Recommendation P.861. (1996b). *Objective quality measurement of telephone-band (300-3400 Hz) speech codecs*. International Telecommunication Union, CH-Geneva.

ITU-T. Recommendation P.562. (2000). *Analysis and interpretation of INMD voice-service measurements*. International Telecommunication Union, CH-Geneva.

ITU-T. Recommendation P.862. (2001). *Perceptual evaluation of speech quality (PESQ), an objective method for end-to-end speech quality assessment of narrowband telephone networks and speech codecs*. International Telecommunication Union, CH-Geneva.

ITU-T. Recommendation P.800.1. (2003a). *Mean opinion score (MOS) terminology*. International Telecommunication Union, CH-Geneva.

ITU-T. Recommendation P.862.1. (2003b). *Mapping function for transforming P.862 raw result scores to MOS-LQO*. International Telecommunication Union, CH-Geneva.

ITU-T. Recommendation G. 107. (2003c). *The e-model, a computational model for use in transmission planning*. International Telecommunication Union, CH-Geneva.

ITU-T. Rec. P.563. (2004). *Single ended method for objective speech quality assessment in narrow-band telephony applications*. International Telecommunication Union, CH-Geneva.

Jin, C., & Kubichek, R. (1995). Output-based objective speech quality using vector quantization techniques. In *Proceedings of ASILOMAR. Conference on Signals, Systems, and Computers* (pp. 1291-1294).

Johnson, J. D. (1988). Transform coding of audio signals using perceptual noise criteria. *IEEE Journal on Selected Areas in Communications, 6*(2), 314-323.

Kitawaki, N., Nagabuchi, H., & Itoh, K. (1988). Objective quality evaluation for low-bit-rate speech coding systems. *IEEE Journal on Selected Areas in Communications, 6*(2), 242-248.

Moller, S. (2000). *Assessment and prediction of speech quality in telecommunications*. Boston: Kluwer Academic Publishers Group.

Noll, A. M. (1974). Cepstrum pitch determination. *Journal of the Acoustical Society of America, 41*(2), 293-309.

Picovici, D., & Mahdi, A. E. (2004). New output-based perceptual measure for predicting subjective quality of speech. In *Proceedings of International Conference on Acoustics, Speech, and Signal Processing (ICASSP-2004),* Toronto, Canada (pp. 633-636).

Psytechnics. (2003). *Mobile quality survey.* Case study report prepared by Psytechnics, UK. Retrieved from http://www.psytechnics.com/psy_frm01.html

Quackenbush, S. R., Barnawell, T. P., & Clements, M. A. (1988). *Objective measures of speech quality.* New York: Prentice Hall.

Quatieri, T. E. (2002). *Discrete-time speech signal processing: Principles and practice.* New Jersey: Prentice Hall PTR.

Rix, A. W., & Hollier, M. P. (2000). The perceptual analysis measurement system for robust end-to-end speech quality assessment. In *Proceedings of International Conference on Acoustics, Speech, and Signal Processing (ICASSP-2000),* Istanbul, Turkey (pp. 1515-1518).

Voran S. (1999). Objective estimation of perceived speech quality—Part I: Development of the measuring normalizing block technique. *IEEE Transactions on Speech and Audio Processing, 7*(4), 371-382.

Yang, W. (1999). *Enhanced Modified Bark Spectral Distortion (EMBSD).* PhD Thesis. Philadelphia: Temple University.

KEY TERMS

Intrusive Objective Voice Quality Measure: Objective voice quality measure that bases its measurement on computation of the distortion between the original speech signal and the degraded speech signal. Such measure is often referred to as *input-to-output* or *two-ended measure*.

Mean Opinion Score (MOS): Average value of all the rating scores registered by the human listeners (conducting a subjective voice quality test) for a given test condition.

Non-Intrusive Objective Voice Quality Measure: Objective voice quality measure that uses only the degraded speech signal and have no access to the original speech signal. Such measure is often referred to as *output-based* or *single-ended* measure.

Objective Voice Quality Measure: Metric based on a computational model or an algorithm that computes MOS voice quality values that are as close as possible to the ratings obtained from subjective tests, by observing a small portion of the speech in question.

Quality-of-Service (QoS): The set of those quantitative and qualitative characteristics of a distributed multimedia system, which are necessary in order to achieve the required functionality of an application.

Subjective Voice Quality Test/Measure: Voice quality test/measure that is based on ratings by human listeners.

Voice Quality: Result of a person's judgement on spoken language, which he/she perceives in a specific situation and judges instantaneously according to his/her experience, motivation, and expectation. Regarding voice communication systems, voice quality is the customer's perception of a service or product.

Voice Quality Measurement (VQM): Means of measuring customer experience of voice communication services (systems/devices).

Chapter XVI
Modular Implementation of an Ontology-Driven Multimedia Content Delivery Application for Mobile Networks

Robert Zehetmayer
University of Vienna, Austria

Wolfgang Klas
University of Vienna, Austria

Ross King
Research Studio Digital Memory Engineering, Austria

ABSTRACT

Today, mobile multimedia applications provide customers with only limited means to define what information they wish to receive. However, customers would prefer to receive content that reflects specific personal interests. In this chapter we present a prototype multimedia application that demonstrates personalised content delivery using the multimedia messaging service (MMS) protocol. The development of the application was based on the multimedia middleware framework METIS, which can be easily tailored to specific application needs. The principle application logic was constructed through three indepdent modules, or "plug-ins" that make use of METIS and its underlying event system: the harverster module, which automatically collects multimedia content from configured RSS feeds, the news module, which builds custom content based on user preferences, and the MMS module, which is reponsible for broadcasting the resulting multimedia messages. Our experience with the implementation demonstrated the rapid and modular development made possible by such a flexible middleware framework.

INTRODUCTION

Multimedia messaging service (MMS) has not achieved a similar market acceptance and customer adoption rate as *short message service* (SMS), but is nevertheless one of the primary drivers of new income streams for telecommunication companies and is, in the long run, on the way to becoming a true mass market (Rao & Minakakis, 2003). It provides new opportunities for customised content services and represents a significant advance for innovative mobile applications (Malladi & Agrawal, 2002).

Until now, however, mobile operators have failed to deliver meaningful focused mobile services to their users and customers. Telecommunication companies have made considerable investments (license, implementation costs) into third generation (3G) mobile networks but have not yet generated compensating revenue streams (Vlachos & Vrechopoulos, 2004). Customers are often tired of receiving information from which they get no added value, because the information does not reflect their personal interests and circumstances (Sarker & Wells, 2003). The goal is instead to establish a one-to-one relationship with the user and provide costumers with relevant information only. Through personalisation, the number of messages the customer receives will decrease significantly, thus reducing the number of irrelevant and unwanted messages (Ho & Kwok, 2003).

Currently available MMS subscription services (e.g., Vodafone, 2005) allow customers to define what kind of information they want to receive in a very limited way. Broad categories like *Sports*, *Business*, or *Headline News* can be defined, but there is no generic mechanism for the selection of more specific concepts within a given domain of interest. The personalised and context-aware services demanded by savvy customers require a mediation layer between the users and content that is capable of modelling complex semantic annotations and relationships, as well as traditional and strongly-typed metadata. These will be defining characteristics of next-generation multimedia middleware.

This paper describes the modular development of a mobile news application, based on a custom multimedia middleware framework. The application supports ontology-driven semantic classification of multimedia content gathered using a widespread news markup language. It allows users to subscribe to content within a particular domain of interest and filters information according to the user's preferences. Moreover it delivers the content via MMS. The example domain of interest is the Soccer World Cup 2006 for which a prototypical ontology for personal news feeds has been developed. However, the middleware framework enables mobile multimedia delivery that is completely independent from the underlying domain-specific ontology.

BACKGROUND AND RELATED WORK

Related Work

At this time, there are no readily available systems that combine the power of ontology-based classification, published syndicated content, and a personalised MMS delivery mechanism. There are however a number of proposals and applications that make use of principles and procedures that are similar to those presented in this chapter.

Closely related to the classification aspect of the presented MMS news application are hierarchical textual classification procedures such as D'Alessio, Murray, Schiaffino, and Kreshenbaum (2000). These approaches mostly consider the categorisation of Web pages and e-mails (see also Sakurai & Suyama, 2005) and classify content according to a fixed classification scheme.

Ontologies that can provide classification in the form of concepts and relationships within a particular domain are used by Patel, Supekar, and Lee (2003) for similar purposes. The idea behind their work is to use a hierarchical ontology structure in order to suggest the topic of a text. Terms that are extracted from a specific textual representation are mapped on to their corresponding concepts in an ontology. The use of ontologies is one step ahead of the use of general classification schemes as they introduce meaningful semantics between classified items. Similar in this respect is the work of Alani et al. (2003), which attempts to automatically extract ontology-based metadata and build an associated knowledge base from Web pages. The reverse method is also possible as demonstrated by Stuckenschmidt and van Harmelen (2001), who built an ontology from textual analysis instead of classifying the text according to an ontology. Schober, Hermes, and Herzog (2004) go one step further by extending the ontological classification scheme from textual information to images and their associated metadata.

Even more closely related to the topics presented in this paper are the techniques employed in the *news engine Web services* application (News, 2005), which is currently under development. It is based on the news syndication format PRISM and ontological classification, and its goal is to develop news intelligence technology for the semantic Web. This application should enable users to access, select, and personalise the reception of multimedia news content using semantic-based classification and associated inference mechanisms (Fernandez-Garcia & Sanchez-Fernandez, 2004).

News Markup Languages and Standards

News syndication is the process of making content available to a range of news subscribers free of charge or by licensing. This section briefly sketches three current technologies and standards in the field of news syndication: RSS, PRISM, and NewsML.

Our MMS application employs RSS feeds in order to harvest news data, due to the volume and free availability of these types of feeds. Of course this would raise serious copyright issues in a commercial application; however, our approach provides an initial proof of concept, allows the harvesting of significant volumes of data for testing classification algorithms, and is easily upgradeable to a commercially appropriate standard, thanks to the modular nature of the system architecture. For this reason, we describe the RSS standard in more detail than the other more commercially significant standards.

Rich Site Summary (RSS)

First introduced by Netscape in 1999, RSS (which can stand for *RDF site summary*, *rich site summary*, or *really simple syndication* depending on the RSS version) is a group of free lightweight XML-based (quasi) standards that allow users to syndicate, describe and distribute Web site and news content, respectively. Using these formats, content providers distribute headlines and up-to-date content in a brief and structured way. Essentially, RSS describes recent additions to a Web site by making periodical updates. At the same time, consumers use RSS readers and news aggregators to manage and monitor their favourite feeds in one centralised program or location (Hammersley, 2003).

RSS comes in three different flavours: relatively outdated RSS 0.9x, RSS 1.0 and RSS 2.0. RSS 2.0 is currently maintained by the *Berkman Center for Internet and Society* at Harvard. On the other hand RSS 1.0 is a *World Wide Web Consortium* (W3C) standard and was developed independently. Thus RSS 2.0 is not an advancement of RSS 1.0, despite what the version numbers might suggest. The line of RSS development was split into two rival branches that are only marginally compatible. The main difference is that RSS 1.0 is based on the W3C *resource description framework* (RDF) standard, whereas the other types are not (Wustemann, 2004). In our MMS news application scenario the focus is on RSS 2.0 channels, because of their special characteristics relating to multimedia content and the general availability of feeds of this type in contrast to RSS 1.0.

The top level of an RSS 2.0 document is always a single RSS element, which is followed by a single channel element containing the entire feed's content and its associated metadata. Each channel element incorporates a number of elements providing information on the feed as a whole and furthermore item elements that constitute the actual news and their corresponding message bodies. Items consist of a title element (the headline), a description element (the news text), a link (for further reading), some metadata tags and one or more optional enclosure elements. Enclosures are particularly important in the context of multimedia applications, as they provide external links to additional media files associated with a message item. Enclosures can be images, audio or video files, but also executables or additional text files, and they are used for building up the multimedia base of our MMS news application.

Publishing Requirements for Industry Standard Metadata (PRISM)

Publishing Requirements for Industry Standard Metadata (PRISM, 2004) is a project to build standard XML metadata vocabularies for the publishing industry to facilitate syndicating, aggregating and processing of news, book, magazine and journal content of any type. It provides a framework for the preservation and exchange of content and of its associated metadata through the definition of metadata elements that describe the content in detail.

The impetus behind PRISM is the need for publishers to make effective use of metadata to cut costs from production operations and to increase revenue streams as well as availability for their already produced content through new electronic distribution methods. Metadata in this context makes it possible to automate processes such as content searching, determining rights ownership and personalisation.

News Markup Language (NewsML)

News Markup Language (NewsML) is an open XML-based electronic news standard developed and ratified by the *International Press Telec Council* (IPTC) and lead-managed by the world's largest electronic news provider Reuters (IPTC, 2005). According to Reuters (2005), NewsML could revolutionise publishing, because it allows publishers and journalists to deliver their news and stories to a range of different devices including cell phones, PDAs, and desktop computers. At the same time, it allows content providers to attach rich metadata so that customers only receive the most relevant information according to their preferences.

NewsML is extensible and flexible to suit individual user's needs. The goal is to facilitate the exchange of any kind of news, be it text, photos or other media, accurately and quickly, but it may also be used for news storage and publication of news feeds. This is achieved by bundling the content in a way that allows highly automated processing (NewsML, 2003).

Multimedia Messaging Service and Mobile Network Architecture

Multimedia messaging service (MMS) is an extension to the *short message service* (SMS) protocol, using the *wireless application protocol* (WAP) as enabling technology that allows users to send and receive messages containing any mixture of text, graphics, photographic images, speech and music clips or video sequences. High-speed communication and transmission technologies, such as *general packet radio services* (GPRS) and *universal mobile telecommunications system* (UMTS), provide support for powerful and fast messaging applications (Sony Ericsson Developers Guidelines, 2004).

MMS Network Architecture

An MMS-enabled mobile device communicates with a WAP gateway using WAP transport protocols over GPRS or UMTS networks.

Data is transported between the WAP Gateway and the MMS Centre (MMSC) using the HTTP protocol as indicated in Figure 1. The MMSC is the central and most vital part of the

Figure 1. MMS network architecture (Nokia Technical Report, 2003)

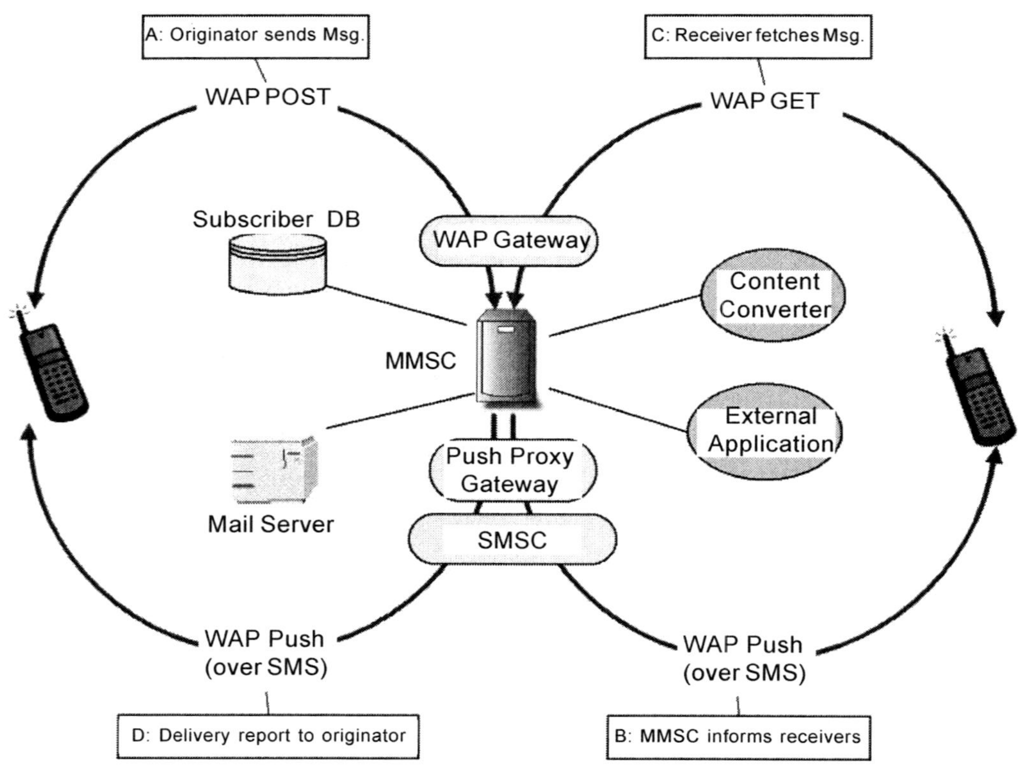

architecture and consists of an MMS Server and an MMS Proxy-Relay. Amongst other functions it stores MMS messages, forwards and routes messages to external networks (external MMSCs), delivers MMS via e-mail (using the SMTP protocol), and performs content adaptation according to the information known about the receiver's mobile phone. This is managed via so-called *user agent profiles* that identify the capabilities of cell phones registered in a provider's network (Sony Ericsson Developers Guidelines, 2004). Leveraging the content-adaption capability of the MMSC is a key feature of our MMS application.

MMS and SMIL

The *Synchronized Multimedia Integration Language* (SMIL) is a simple but powerful XML-based language specified by the W3C that provides mechanisms for the presentation of multimedia objects (Bulterman & Rutledge, 2004). The concept of SMIL as well as MMS presentations in general includes the ordering, layout, sequencing, and timing of multimedia objects as the four important functions of multimedia presentations. Thus a sender of a multimedia message can use SMIL to organise the multimedia content and to define how the multimedia objects are displayed on the receiving device (OMA, 2005).

A subset of SMIL elements must be used (and are used by our application) to determine the presentation format of an MMS message. Listing 1 shows an example SMIL document defining 2 slides (<par> elements), each containing a text, an image, and an audio element, as it would be the case in typical MMS.

Listing 1. SMIL XML example

```
<?xml version="1.0" encoding="UTF-8"?>
<smil xmlns="http://www.w3.org/2001/SMIL20/Language">
 <head>
      <meta name="title" content="vacation photos" />
      <meta name="author" content="Danny Wyatt" />
      <layout>
       <root-layout width="160" height="120"/>
       <region id="Image" width="100%"
            height="80" left="0" top="0" />
       <region id="Text" width="100%"
            height="40" left="0" top="80" />
      </layout>
 </head>
 <body>
      <par dur="8s">
       <img src="FirstImage.jpg" region="Image"/>
       <text src="FirstText.txt" region="Text"/>
       <audio src="FirstSound.amr"/>
      </par>
      <par dur="7s">
       <img src="SecondImage.jpg" region="Image"/>
       <text src="SecondText.txt" region="Text"/>
      </par>
 </body>
</smil>
```

MMS Message Structure and WSP Packaging

MMS is implemented on top of WAP 1.2.1 (as of October 2004) and supports messages of up to 100 Kbytes, including header information and payload. In order to transmit an MMS message, all of its parts must be assembled into a multi-part message associated with a corresponding MIME (*multipurpose Internet mail extensions*) type, similar to the manner in which these types are used in other standards such as HTML or SMTP. What is actually sent are so-called MMS *protocol data units* (PDUs). An example of which is shown in Figure 2. In the next step, PDUs are passed into the content section of a *wireless session protocol* (WSP) message, in the case of most mobile networks, or a HTTP message otherwise (Nokia Technical Report, 2003).

One of three possible content type parameters is associated with these content sections, specifying the type of the MMS (Sony Ericsson Developers Guidelines, 2004):

- **Application/vnd.wap.multipart.related:** This type is used if there is a SMIL part present in the MMS. The header must then also include a type parameter application/smil on the first possible position
- **Application/vnd.wap.multipart.mixed:** Used if no SMIL part is included in the MMS
- **Application/vnd.wap.multipart.alternative:** Indicates that the MMS contains alternative content types. The receiving device will choose one supported MIME type from the available ones

The Multimedia Middleware Framework METIS

The following sections give an overview of the METIS multimedia framework, its generic data model, and methods for the extension of its basic functionality by developing semantic modules and kernel plug-ins. An introduction to the template mechanism that is extensively used in our application is also provided.

System Overview

The METIS framework (King, Popitsch, & Westermann, 2004) provides an infrastructure for the rapid development of multimedia applications. It is essentially a classical middleware application located between highly customisable persistence and visualisation layers. Flexibility was one of the primary design criteria for METIS. As can be seen in Figure 3, this crite-

Figure 2. Example MMS PDUs

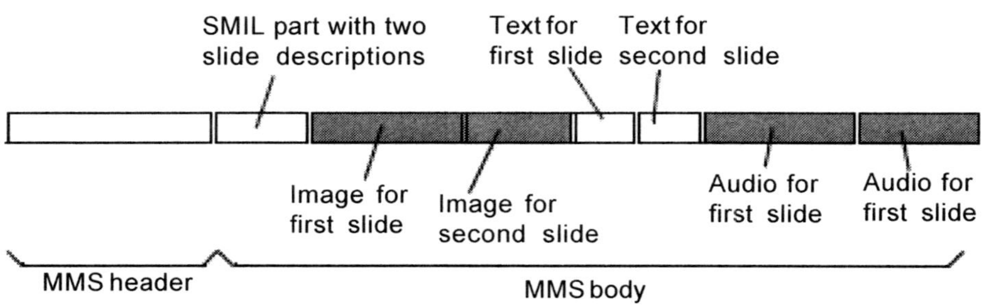

Figure 3. METIS system architecture

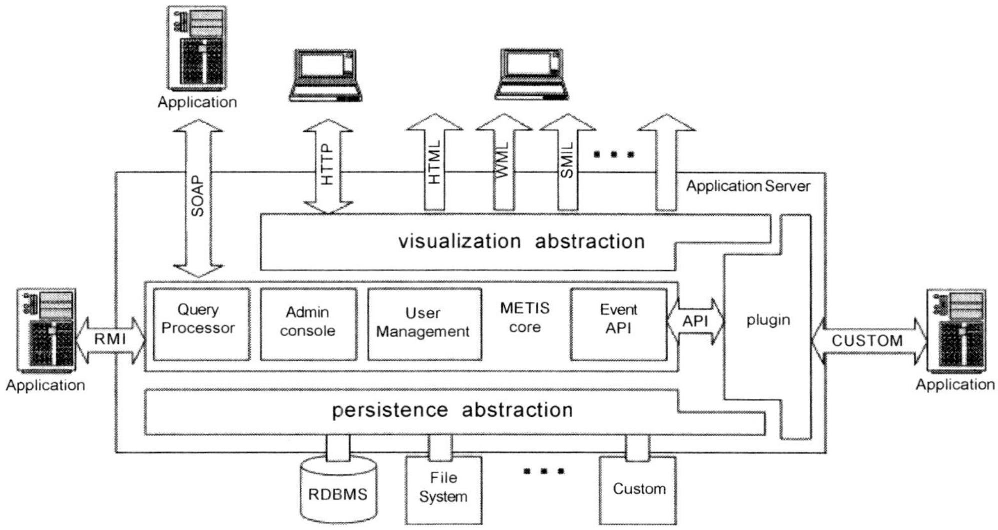

rion especially applies to the back-end and front-end components of the architecture as well as to the general extensibility through kernel plug-ins and semantic modules. The design as a whole offers a variety of options for the adaptation to specific application needs.

METIS Data Model

The METIS data model provides the basis for complex, typed metadata attributes, hierarchical classification, and content virtualisation. Application developers need only consider their specific data models at the level of ontologies (specified, for example, by RDFS or OWL) which can then be easily mapped to the METIS data model using existing tools. Object relational modelling is handled by the framework and the developer need never concern himself with relational tables or SQL statements.

Figure 4 illustrates the basic building blocks of the model and their relationships. Media in METIS are represented as a so-called *single media objects* (SMOs), which are abstract, logical representations of one or more physical media items. Media items are attached to a SMO as *media instances* and connected to the actual media data via *media locaters*, which are in turn a kind of pointer to the data, allowing METIS to address transparently media items in a variety of distributed storage locations such as file systems, databases or Web servers.

As a foundation for semantic classification, media objects can be organised in logical hierarchical categories, known as *media types*. Media types can take part in multiple inheritance as well as multiple instantiation associations. *Metadata attributes* are connected to media types, can be as simple or complex as desired, and can be shared among multiple media types with different cardinalities, default values, and ranges.

Finally, media objects can be connected to each other by binary directed relationships (so-called *associations*). The semantics of these associations are defined by *association types* that are freely configurable within an application domain.

As mentioned previously, there exist simple tools with which domain semantics can be

Figure 4. METIS Core Data Model (King et al., 2004)

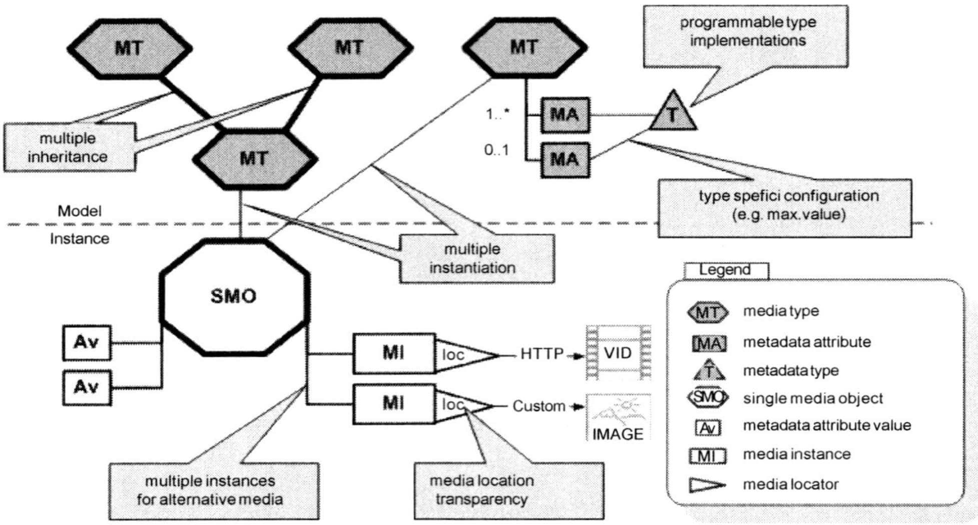

packaged as semantic modules (also called *semantic packs*) that can be dynamically loaded in a given METIS instance and thereby provide the required domain-specific customisations.

Complex Media Objects and Templates

For modelling specific media documents that are made up of several media items, the METIS data model provides *complex media objects* (CMOs). CMOs are quite similar to SMOs when it comes to instantiating media types, taking part in associations and being described by metadata attributes. The crucial difference is that they serve as containers for other media objects, either SMOs or other CMOs. Complex media objects can be rendered in specific visualisation formats by applying the METIS template mechanism (King et al., 2004). A template is an XML representation of a specific multimedia document format (such as SMIL, HTML or SVG), enriched by *placeholders*. When a visualisation of the CMO is requested, these placeholders are dynamically substituted by specific data extracted from the CMO employing that template, using a format-specific XSLT style sheet. Our MMS application makes use of this template mechanism in order to define the format of MMS messages, by employing the SMIL-based mechanism described in a previous section.

Semantic Modules and Kernel Plugins and the Event Framework

Kernel plug-ins constitute the functional components of an application that extend the basic functionalities provided by the METIS core. These plug-ins not only have access to all customisation frameworks within METIS, but also to the event system, which provides a basic publish/subscribe mechanism. Through the METIS framework, plug-ins can subscribe to certain predefined METIS events and can easily implement their own new application-specific events. This loose coupling between functional extensions provided by the event framework allows large modular applications to be implemented with METIS.

THE MMS NEWS APPLICATION

The METIS framework is used extensively to implement our modular application for content delivery in mobile networks. This *MMS news application* illustrates two strengths of METIS: extensibility and fast implementation time.

In order to demonstrate these advantages and the core functionalities, the prototype news application implements a showcase in the wider area of the Soccer World Cup 2006 in Germany. We present an ontology for this domain, which allows a relatively confined set of topics and their relationships to be modelled. However, the system is designed to be as open and extensible as possible and allows mobile multimedia content delivery that is completely independent from the underlying domain-specific ontology.

System Architecture

An overview of principal components of the *MMS news application*'s modular architecture is given in Figure 5. The implementation is split into three functional parts: the RSS import module, the news application module (containing the main application logic), and the MMS output and transmission module. Each module is implemented as a kernel plug-in, and each module is loosely connected with other plug-ins through the METIS event mechanism.

This approach makes it possible to cleanly separate functionalities into logical modules. It is therefore simple to integrate various functional units into the application's context and substitute existing plug-ins with newly implemented ones whenever changes in the application's environment are required. The interface to which all these plug-ins must adhere is defined by the various events that are issued by components that adopt a given role in the application.

Figure 5. MMS news application architecture (simplified)

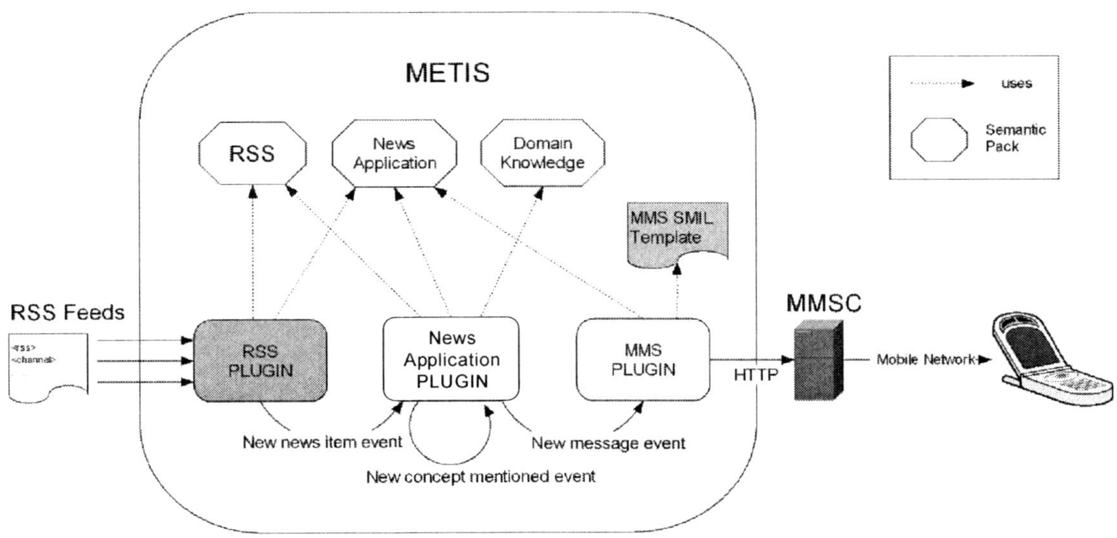

From a high-level perspective, the RSS plug-in takes the role of the multimedia content source that loads multimedia news items into the system. Obviously, RSS would not be the choice for a commercial application; the previously mentioned NewsML and PRISM standards, whose feeds are not normally free of charge, would be more powerful alternatives.

RSS was chosen for the prototype as it allows the demonstration of the essential strengths and advantages of the presented approach with no associated costs. Furthermore it is easy to implement and allows the testing of the whole application on large datasets. In the future, additional multimedia content source plug-ins based on NewsML or PRISM could be quickly developed to replace the RSS plug-in.

The *NewsApplication* plug-in is the core module of the whole application. It integrates the surrounding plug-ins and uses their provided functionalities to create personalised news content. This plug-in itself offers flexibility in the mechanisms used to find topics mentioned in news items as well as in the creation of messages for specific users.

The MMS plug-in fulfils the role of the content delivery mechanism within the *MMS news application* by linking the application to mobile network environments. In the current prototype it is used to send MMS messages via an associated MMSC to a user's mobile handset. Once again, the MMS plug-in offers a variety of extension possibilities and is very flexible when it comes to the system used for the actual MMS transmission. It could be easily substituted by other content delivery plug-ins that target different receiving environments and devices. For example, one might consider a SMS delivery mechanism or a mechanism that delivers aggregated news feeds about certain topics to Web-based news reader applications.

Data Model

The data model and domain-specific semantics of a METIS-based application must be specified through the *semantic pack* mechanism. Semantic packs are to a large extent quite similar to ontologies that define the semantics of specific domains of knowledge by modelling classes, attributes, and relationships. The *MMS news application* is based on three independent semantic packs:

- **RSS semantic pack:** This module maps the RSS 2.0 element and attribute sets to the METIS environment, and supports import from previous RSS versions including RSS 1.0 (without additional modules). Media types included are *news feed*, *aggregated news item*, *news content* with corresponding attributes (e.g., *title*, *description* or *publication date*) as well as general purpose media types such as *image*, *text*, *audio,* and *video* that are all child elements of *news content*. Associations between these elements are defined as well. Generally, this semantic pack is intended to be as independent as possible from the underlying publishing standard that is used, and as extensible as possible in order to facilitate the implementation of other types of import plug-ins
- **News application semantic pack:** This module provides the application-specific management ontology. It defines media types and metadata attributes that are required by the internal logic of the application in order to store and differentiate between application-specific media objects. Media types in this category are *user*, *created message,* and *searchable*. A *User* normally subscribes to multiple

searchable media object instances (SMOs) that are supplied by the domain-specific semantic pack, and associations of type *subscribed news topic* are created between these. Furthermore, associations of type *received message* are instantiated between a *user* and all the *created messages* he has received as a result of his subscriptions

- **Domain-specific semantic pack:** This module constitutes the domain-specific component of the application used for the subscription services and the applied ontology-based classification method. The application's internal logic is completely independent of the domain of interest that is defined by this semantic pack. As a demonstrator, an ontology for soccer was implemented, but additional domains can be implemented and plugged into the existing application with minimal effort

The general dependencies between the three semantic packs and the specific media and association types are presented in Figure 6.

Domain-Specific Semantics and Knowledge Base

The domain-specific semantic pack contains key concepts and their relationships within a specific domain of interest, and defines the structure of a knowledge base containing specific instances of defined classes that must be instantiated. The *MMS news application* is independent of the domain of interest supplied by this semantic pack; any ontology satisfying the basic requirement of having a single parent class from which all other classes are directly or indirectly derived can be loaded into the system and used as a basis for the subscription mechanism.

Domain concepts or classes are stored in the METIS environment as media types. A concept instance is modelled as a SMO of the corresponding concept's media type. In our prototype, all classes are direct or indirect subclasses of the abstract base class *Football Ontology*. Example classes (media types) are *Field Player*, *Trainer*, *National Team*, *Club,* and *Referee*. Furthermore, an application-spe-

Figure 6. MMS news application semantic pack dependencies

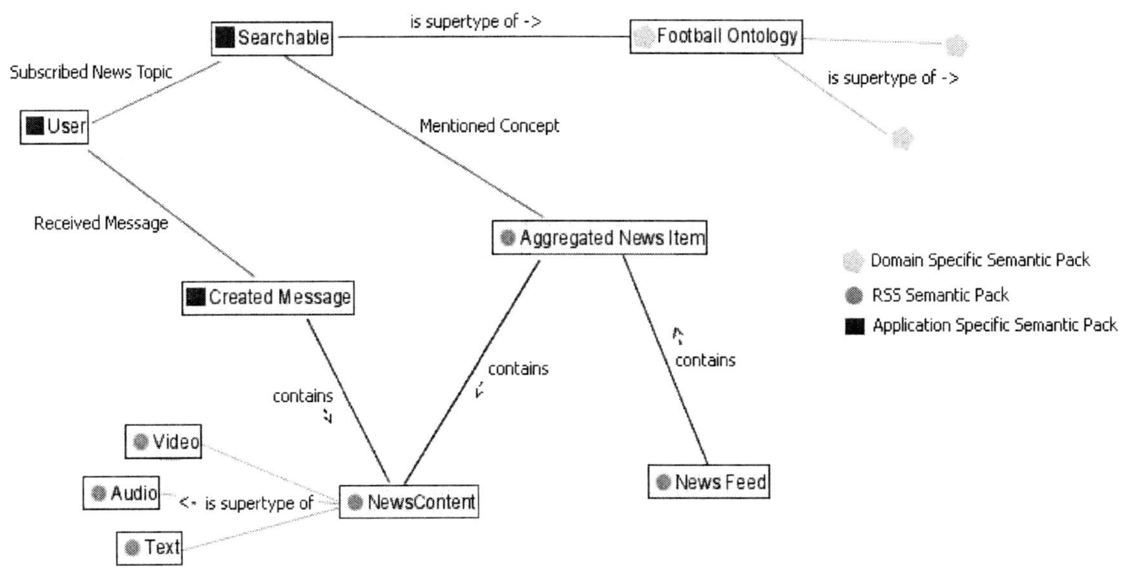

cific media type *searchable* is included, which provides the required *search term* metadata attribute. This search term enables the textual identification of the instance through the presently simple algorithm based on matching regular-expressions. Domain associations form the basis for the semantic classification algorithm as they relate concepts (i.e. classes) and establish meaningful relationships between them.

Instances (e.g., *David Beckham*) of concepts (e.g., *Field Player*) can be included within the semantic pack itself or defined via the *news application's* user interface. The only constraint that instances within an imported ontology must satisfy is that they must supply at least one identifying search term string attribute for the ontology-based classification mechanism.

Every instance added to the system becomes visible to end-users, who can then subscribe to specific concept instances and receive MMS messages associated with them.

In the case of our prototype, a knowledge base of about 250 instances and their associations was developed in approximately 4 hours. This suggests that it is possible to implement other domains of interest and to adapt the whole application to other application scenarios in a reasonably short time.

Module Integration and Event Mechanism

The RSS plug-in provides all RSS-related mechanisms. RSS news feeds typically contain news items that contain the actual messages. Whenever an item is added and stored, the RSS plug-in informs all interested system components of this fact via a *new news item event*. The only subscriber to this event in the current architecture is the news application plug-in, which is subsequently activated. It searches the new news item for occurrences of domain-specific concept instances (e.g., the instance *David Beckham*) contained in the domain-specific knowledge base. Whenever such an occurrence is found, a *new concept mentioned event* is issued. The news application then attempts to find subscribers to the discovered concept instance (i.e., users who want to receive messages about it) as well as subscribers of associated instances. Associated concept instances in this respect mean instances that are directly connected to the discovered concept through a relationship in the domain-specific ontology. If a user has chosen to receive messages from related instances (by default, a user would receive messages only directly related to the subscribed concept), he will also be added to the set of found users. As an example, consider a subscriber of the instance *English national team* who also chooses to receive messages from related concept instances; he would, for example, also receive messages about *David Beckham*, because *Beckham* is a member of that team. In this case, the user would be an *indirect subscriber* to the Beckham concept instance.

Whenever direct or indirect subscribers are found, the plug-in creates a new CMO (of type *created message*) containing various SMOs such as a news text, suitable images, video or audio items. It is important to note that this newly created message is not a one-to-one translation of the news item contained in the RSS feed. The news application searches the multimedia document base and tries to find media instances that are associated with the discovered concept instance and may be suitable for the newly created message.

The architecture is designed to be as open and extensible as possible. Implementations of new algorithms for ontology-based classification and the associated message-creation mechanism can be easily upgraded within in the application.

Having assembled this message, a *new message event* is issued and the MMS plug-in, as a subscriber to this event, sends the message as a MMS to the users' mobile phones. Outgoing messages are formatted using the METIS template mechanism in conjunction with a predefined MMS SMIL template. The application could be easily extended to allow users to choose from a variety of templates and define the final format of their received messages.

RSS Import

The RSS import plug-in fulfils the role of an RSS input parser and news aggregator that manages multiple RSS feeds simultaneously and makes their content available to the other components of the application.

Using media types and attributes specified in the RSS semantic pack, the RSS plug-in maps feeds to corresponding METIS media objects by parsing these and extracting media and metadata. In general, a feed is represented as a METIS CMO as depicted in Figure 7. The *FEED CMO* (type: *news feed*) can incorporate several *News ITEM CMOs* (type *aggregated news item*), which in turn include multiple *media SMOs* (subtypes of *news content*) that map RSS media enclosures included in the feed.

Figure 7. News FEED complex media object containing CMOs and SMOs

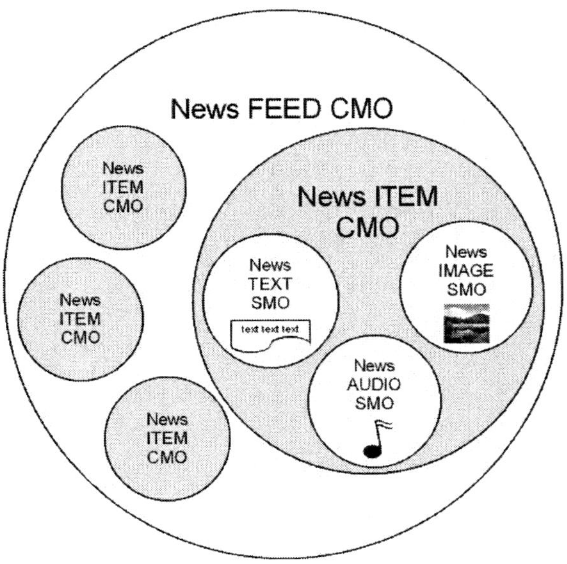

By regularly searching and updating the stored feeds, a multimedia document base is gradually constructed over time.

The RSS plug-in also functions as a common RSS newsreader and aggregator by providing an HTML visualisation of the created *News FEED* CMO. This again demonstrates the power and adaptability of the METIS approach, as the RSS plug-in can already serve as a standalone application without including it the context of the *MMS news application*.

Ontology-Driven Message Creation

The *news application* plug-in provides core functionalities in the areas of ontology-based classification and discovery of specific media objects, as well as message creation from these search results.

The search terms provided by the knowledge base are used to identify textual occurrences of concept instances in *news ITEM CMOs*. We make the simplifying assumption that all other *media SMOs* included in this *news ITEM CMO* are also related to the discovered instance.

The news application plug-in uses this classification mechanism to relate concept instances to news items and their included media objects. A simple strategy based on regular expressions that searches all *news TEXT SMOs* for the occurrences of concept instance search terms defined in the knowledge base is currently implemented. This approach allows us to easily test the whole modular application on large datasets. Different search strategies can be

Figure 8. Concept mentioned association example

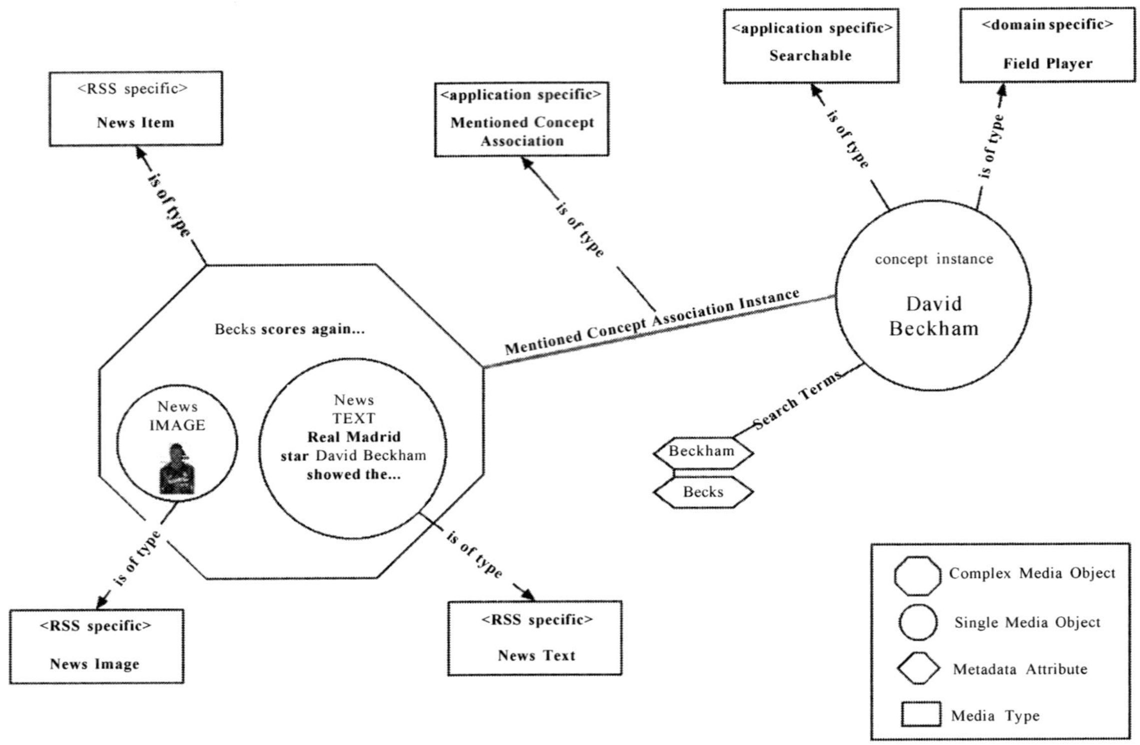

utilised in this context and new ones can be added easily. For example, advanced full text analysis approaches could be employed in the application; this is a subject for our future research.

When a search term is found, a METIS association (of type *Mentioned Concept*) between the news text's *news ITEM CMO* and the concept instance SMO in the knowledge base is created, as depicted in Figure 8. This in turn fires a *new mentioned concept event* that triggers the message creation mechanism.

Created messages are stored in a new container CMO of type *created message*. In most cases, news items contain only textual headlines; information and suitable media objects must be added in order to create a multimedia message for MMS delivery.

Once again, the domain-specific ontology provides valuable information about the relationships between a specific concept instance and other instances. As instances are bound to news items, the relationships can be derived for these news items as well. Media SMOs can thus be harvested from concept instances not bound to them, but bound to a closely related instance. Consider an example in which there are no images of the instance *David Beckham* available — in this case an image could be taken from an instance of *English National Team* as the latter is related to the former via an association of type *team member*. Only directly related concepts are taken into account, because we assume that the further apart two instances are, the more likely it is that unsuitable *media SMOs* will be chosen.

MMS Creation and Content Delivery

The purpose of the MMS plug-in is to assemble a MMS message from a message CMO (of type *created message*) and transmit it to subscribed users. This plug-in employs the METIS template mechanism to create suitable SMIL-based MMS slideshow presentations, including media objects supplied by the *created message*. The template includes placeholders that are dynamically replaced by the actual multimedia object instance data.

During the next step, the MMS message is packaged as a binary stream (because the MMS format does not allow any links to external media) consisting of the actual media data referenced by the included SMOs and the generated SMIL file. General message attributes such as the receiver's phone number, the MMS title and subject, as supplied by the *created message* CMO, are also included in the header. That package is then sent to a MMSC, which continues by sending the MMS to the corresponding mobile device over a carrier's network.

This architecture has some specific advantages over other methods of sending MMS messages. First of all, the MMSC usually offers a mechanism for content adaptation and conversion according to a mobile phone's capabilities. This frees the METIS MMS plug-in from any consideration of the supplied media items in terms of conversion and adaptation to specific mobile devices. The second reason is that this design makes it possible to switch between MMSC implementations quite easily. Thus it is possible to adapt the *MMS news application* to any provider's or carrier's network architecture with a minimum amount of effort. Live environments that can send thousands of messages per second, compared to 2-4 messages in the testing environment, are therefore a future possibility.

CONCLUSION AND FUTURE WORK

Today, mobile multimedia applications provide customers with only limited means to define what kind of information they want to receive. Customers would prefer to receive information that reflects their specific personal interests, and this requires a mediation layer between the users and content that is capable of modelling complex semantic annotations and relationships. This will be a crucial characteristic of next-generation multimedia platforms.

In this chapter we have presented a prototype multimedia application that demonstrates this type of personalised content delivery. The development of the application was based on a custom multimedia middleware framework, METIS, which can be easily tailored to specific application needs. Our experience with the implementation demonstrated the rapid and modular development made possible by such a flexible middleware framework.

The example domain chosen to illustrate our approach is the Soccer World Cup. An ontology for personal news feeds from this domain was developed, and our experience indicates that similar ontologies and the corresponding knowledge bases for other domains can be created with very little effort. In any case, the application architecture is independent of the specific application domain.

The first module of our prototype application harvests media information from RSS feeds. As a result of the modular application architecture, one could easily integrate additional content sources (for example, encoded in NewsML) that are commercially available from many news agencies, in order to create a commercial application.

In the second module, harvested news items are classified according to the concepts given by the ontology. In our demonstrator application we employed simple text classification techniques, but again thanks to flexible system architecture, more advanced classification techniques can be developed without altering other system components. Future work will focus on more advanced methods of content classfication and on measuring the quality of aggregated media content.

In the final application module, multimedia news messages are composed and delivered to users, according to preferences specified during the subscription process. In the demonstrator we composed and delivered SMIL-based MMS messages to the mobile phones of registered users using a local MMSC. However, the integration with commercial MMSCs, enabling mass transmission of MMS messages, would require no additional implementation and minimal configuration effort.

In conclusion, we believe that the guiding principles for future mobile multimedia applications must be derived from personalised services (i.e., "personalised content is king.") Through personalisation, such applications can provide the possibility for mobile service providers to improve customer retention and usage patterns through the created added value for the customer.

ACKNOWLEDGMENTS

This work was supported by the Austrian Federal Ministry of Economics and Labour.

REFERENCES

Alani, H., Kim, S., Millard, D. E., Weal, M. J., Hall, W., Lewis, P. H., & Shadbolt, N. (2003). Automatic ontology-based knowledge extraction and tailored biography generation from the Web. *IEEE Intelligent Systems*, *18*(1), 14–21.

Bulterman, D. C. A., & Rutledge, L. (2004). *SMIL 2.0. Interactive multimedia for Web and mobile devices series*. Heidelberg, Germany: X.media Publishing.

D'Alessio, D., Murray K., Schiaffino, R., & Kreshenbaum, A. (2000). Hierarchical text categorization. *Proceedings of the RIAO2000.*

Fernandez-Garcia, N., & Sanchez-Fernandez, L. (2004). Building an ontology for news applications. Poster Presentation. *Proceedings of the International Semantic Web Conference ISWC-2004*, Hiroshima, Japan.

Hammersley, B. (2003). *Content syndication with RSS*. Sebastopol, CA: O'Reilly.

Ho, S. Y., & Kwok, S. H. (2003). The attraction of personalized service for users in mobile commerce: An empirical study. *SIGecom Exchanges*, *3*(4), 10-18.

IPTC. (2005). *International Press Telec Council (IPTC) Web site*. Retrieved May 15, 2005, from http://www.iptc.org

King, R., Popitsch, N., & Westermann, U. (2004). METIS—A flexible database solution for the management of multimedia assets. *Proceedings of the 10th International Workshop on Multimedia Information Systems (MIS 2004).*

Malladi, R., & Agrawal, D. P. (2002). Current and future applications of mobile and wireless networks. *Communications of the ACM*, *45*(10), 144-146.

News. (2005). *NEWS (News Engine Web Services) Project Web Site*. Retrieved May 15, 2005, from http://www.news-project.com

NewsML. (2003). *NewsML Specification 1.2*. Retrieved May 15, 2005, from http://www.newsml.org/pages/spec_main.php

Nokia Technical Report. (2003). *How to create MMS services*. Retrieved May 15, 2005, from http://www.forum.nokia.com/main/1,,040,00.html?fsrParam=2-3-/main.html&fileID=3340

OMA. (2005). *Multimedia Messaging Service—Architecture overview*. Version 1.2. Open Mobile Alliance. Retrieved May 15, 2005, from http://www.openmobilealliance.org/release_program/docs/MMS/V1_2-20050301-A/OMA-MMS-ARCH-V1_2-20050301-A.pdf

Patel, C., Supekar, K., & Lee, Y. (2003). Ontogenie: Extracting ontology instances from WWW. *Proceedings of the ISWC2003*.

Prism. (2004). *Publishing Requirements for Industry Standard Metadata (PRISM) Specification 1.2*. IDEAlliance. Retrieved May 15, 2005, from http://www.prismstandard.org/specifications

Rao, B., & Minakakis, L. (2003). Evolution of mobile location-based services. *Communications of the ACM, 46*(12), 61-65.

Reuters. (2005). *Reuters NewsML Showcase Website*. Retrieved May 15, 2005, from http://about.reuters.com/newsml

Sakurai, S., & Suyama, A. (2005). An e-mail analysis method based on text mining techniques. *Applied Soft Computing*. In Press.

Sarker, S., & Wells, J. D. (2003). Understanding mobile handheld device use and adoption. *Communications of the ACM, 46*(12), 35-40.

Schober, J. P., Hermes, T., & Herzog, O. (2004). Content-based image retrieval by ontology-based object recognition. *Proceedings of the KI-2004 Workshop on Applications of Description Logics (ADL-2004)*. Ulm, Germany.

Sony Ericsson Developers Guidelines. (2004). *Multimedia Messaging Service (MMS)*. Retrieved May 15, 2005, from http://developer.sonyericsson.com/getDocument.do?docId=65036

Stuckenschmidt, H., & van Harmelen, F. (2001). Ontology-based metadata generation from semi-structured information. *K-CAP 2001: Proceedings of the International Conference on Knowledge Capture* (pp. 163-170). New York.

Vlachos, P., & Vrechopoulos, A. (2004). Emerging customer trends towards mobile music services. *ICEC '04: Proceedings of the 6th International Conference on Electronic Commerce* (pp. 566-574). New York.

Vodafone. (2005). *Vodafone live! UK—MMS Sports Subscription Services*. Retrieved May 15, 2005, from http://www.vizzavi.co.uk/uk/sportsfootball.html

Wustemann, J. (2004). RSS: The latest feed. *Library Hi Tech, 22*(4), 404-413.

KEY TERMS

3G Mobile: Third generation mobile network, such as UMTS in Europe or CDMA2000 in the U.S. and Japan.

METIS: METIS is an intermedia middleware solution facilitating the exchange of data between diverse applications as well as the integration of diverse data sources, demantic searching and content adaptation for display on various publishing platforms.

MMS: Multimedia Messaging Service is a system used to transmit various kinds of multimedia messages and presentations over mobile networks.

NewsML: News Markup Language is an open XML-based electronic news standard used by major news providers to exchange news and stories and to facilitate the delivery of these to diverse receiving devices.

News Syndication: Is the process of making content available to a range of news subscribers free of charge or by licensing.

Ontology: A conceptual schema representing the knowledge of a certain domain of interest.

PRISM: Publishing Requirements for Industry Standard Metadata is a standard XML metadata volabulary for the publishing industry to facilitate syndicating, aggregating, and processing of content of any type.

Semantic Classification: Is the classification of multimedia objects and concepts and their interrelationships using semantic information provided by a domain schema (i.e., ontology).

SMIL: Synchronized Multimedia Integration Language is a XML-based language for integrating sets of multimedia objects into a multimedia presentation.

RSS: Really Simple Syndication (also Rich Site Summary and RDF Site Summary) is a XML-based syndication language that allows users to subscribe to news services provided by Web sites and Weblogs.

Chapter XVII
Software Engineering for Mobile Multimedia: A Roadmap

Ghita Kouadri Mostéfaoui
University of Fribourg, Switzerland

ABSTRACT

The abstract should be changed to this new abstract: Research on mobile multimedia mainly focuses on improving wireless protocols in order to improve the quality of services. In this chapter, we argue that another perspective should be investigated in more depth in order to boost the mobile multimedia industry. This perspective is software engineering which we believe it will speed up the development of mobile multimedia applications by enforcing reusability, maintenance, and testability. Without any pretense of being comprehensive in its coverage, this chapter identifies important software engineering implications of this technological wave and puts forth the main challenges and opportunities for the software engineering community.

INTRODUCTION

A recent study by Nokia (2005) states that about 2.2 billion of us are already telephone subscribers, with mobile subscribers now accounting for 1.2 billion of these. Additionally, it has taken little more than a decade for mobile subscriptions to outstrip fixed lines, but this still leaves more than half the world's population without any kind of telecommunication service. The study states that this market represents a big opportunity for the mobile multimedia industry.

Research on mobile multimedia mainly focuses on improving wireless protocols in order to improve the quality of service. In this chapter, we argue that another perspective should be investigated in more depth in order to boost the mobile multimedia industry. This perspective is software engineering which we believe it will speed up the development of mobile multimedia applications by enforcing reusability, maintenance, and testability of mobile multimedia applications. Without any pretense of being comprehensive in its coverage, this chapter identifies important software engineering implications of this technological wave and puts forth the main challenges and opportunities for the software engineering community.

ORGANIZATION OF THIS CHAPTER

The next Section presents the state of the art of research in mobile multimedia. The section "What Software Engineering Offers to Mobile Multimedia?" argues on the need for software engineering for mobile multimedia. The section "Contributions to 'Mobile' Multimedia Software Engineering" surveys initiatives in using software engineering techniques for the development of mobile multimedia applications. The section "Challenges of Mobile Multimedia Software Engineeering " highlights the main challenges of mobile multimedia software engineering. Some of our recommendations for successfully bridging the gap between software engineering and mobile multimedia development are presented. The last section concludes this chapter.

STATE OF THE ART OF CURRENT RESEARCH IN MOBILE MULTIMEDIA

I remember when our teacher of "technical terms" in my Engineering School introduced the term "Multimedia" in the middle of the 1990s. He was explaining the benefits of Multimedia applications and how future PCs will integrate such capabilities as a core part of their design. At this time, it took me a bit before I could understand what he meant by integrating image and sound for improving user's interactivity with computer systems. In fact, it was only clear for me when I bought my first "Multimedia PC."

Multimedia is recognized as one of the most important keywords in the computer field in the 1990s. Initially, communication engineers have been very active in developing multimedia systems since image and sound constitute the langua franca for communicating ideas and information using computer systems through networks. The broad adoption of the World Wide Web encouraged the development of such applications which spreads to other domains such as remote teaching, e-healthcare, and advertisement. People other than communication engineers have also been interested in multimedia like medical doctors, artists, and people in computer fields such as databases and operating systems (Hirakawa, 1999).

Mobile multimedia followed as a logical step towards the convergence of mobile technologies and multimedia applications. It has been encouraged by the great progress in wireless technologies, compression techniques, and the wide adoption of mobile devices. Mobile multimedia services promote the realization of the ubiquitous computing paradigm for providing anytime, anywhere multimedia content to mobile users. The need for such content is justified by the huge demand for a quick and concise form of communication–compared to text—formatted as an image or an audio/video file. A recent study driven by MORI, a UK-based market researcher (LeClaire, 2005), states that the demand for mobile multimedia services is on the rise, and that the adoption of mobile multimedia services is set to take off in the coming years and will drive new form factors. The same study states that 90 million mobile phones users in Great Britain, Germany, Singapore, and the United States, are likely to use interactive mobile multimedia services in the next two years.

We are looking at the cell phone as the next big thing that enables mobile computing, mainly because phones are getting smarter" Burton Group senior analyst Mike Disabato told the E-Commerce Times. "We'll see bigger form factors coming in some way, shape or form over the next few years. Those form factors will be driven by the applications that people want to run.

In order to satisfy such a huge demand, research has been very active in improving current multimedia applications and in developing new ones driven by consumers' needs, such as mobile IM (Instant Messaging), group communication, and gaming, along with speed and ease of use. When reviewing efforts in research on mobile multimedia, one can observe that most of the contributions fall into the improvement of wireless protocols and development of new mobile applications.

Mobile Networks

Research on wireless protocols aims at boosting mobile networks and Internet to converge towards a series of steps:

- **WAP:** In order to allow the transmission of multimedia content to mobile devices with a good quality/speed ratio, a set of protocols have been developed and some of them have been already adopted. The wireless application protocols (WAP), aim is the easy delivery of Internet content to mobile devices over GSM (global system for mobile communications), is published by the WAP Forum, founded in 1997 by Ericsson, Motorola, Nokia, and Unwired Planet. The WAP protocol is the leading standard for information services on wireless terminals like digital mobile phones and is based on Internet standards (HTML, XML, and TCP/IP). In order to be accessible to WAP-enabled browsers, Web pages should be developed using WML (Wireless Markup Language), a mark-up language based on XML and inherited from HTML.
- **GPRS:** The General Packet Radio Service is a new non-voice value added service that allows information to be sent and received across a mobile telephone network (GSM World, 2005). GPRS has been designed to facilitate several new applications that require high speed such as collaborative working, Web browsing, and remote LAN access. GPRS boosts data rates over GSM to 30-40 Kbits/s in the packet mode.
- **EDGE:** The Enhanced Data rates for GSM Evolution technology is an add-on to GPRS and therefore cannot work alone. The EDGE technology is a method to increase the data rates on the radio link for GSM. It introduces a new modulation technique and new channel coding that can be used to transmit both packet-switched and circuit-switched voice and data services (Ericsson, 2005). It enjoys a data rate of up 120-150 Kbits/s in packet mode.
- **UMTS:** Universal Mobile Telecommunications Service is a third-generation (3G) broadband, packet-based transmission of text, digitized voice, video, and multimedia at data rates up to 2 megabits per second (Mbps) that offers a consistent set of services to mobile computer and phone users no matter where they are located in the world (UMTS, 2005).

Research on wireless protocols is still an active field supported by both academia and leading industry markets.

Mobile Multimedia Applications

With the advantages brought by third-generation (3G) networks like the large bandwidth, there are many chances that PDAs and mobile phones will become more popular than PCs since they will offer the same services with mobility as an added-value. Jain (2001) points out that important area where we can contribute important ideas is in improving the user's experience by identifying the relevant applications and technology for mobile multimedia.

Currently, the development of multimedia applications for mobile users is becoming an active field of research. This trend is encouraged by the high demand of such applications by mobile users from different fields of applications ranging from gaming, rich-information delivery, and emergencies management.

WHAT SOFTWARE ENGINEERING OFFERS TO MOBILE MULTIMEDIA?

Many courses on software engineering multimedia are taught all over the world. Depicting the content of these courses shows a great focus on the use of multimedia APIs for human visual system, signal digitization, signal compression, and decompression. Our contribution, rather, falls into software engineering in its broader sense including software models and methodologies.

Multimedia for Software Engineering vs. Software Engineering for Multimedia

Multimedia software engineering can be seen in two different, yet complementary roles:

1. The use of multimedia tools to leverage software engineering
2. The use of software engineering methodologies to improve multimedia applications development

Examples of the first research trail are visual languages and software visualization. Software Visualization aims at using graphics, pretty-printing, and animation techniques to show program code, data, and dependencies between classes and packages. Eclipse (Figure 1), TogetherSoft, and Netbeans are example tools that use multimedia to enhance code exploration and comprehension.

The second research trail is a more recent trend and aims at improving multimedia software development by relying on the software engineering discipline. An interesting paper by Masahito Hirakawa (1999) states that software engineers do not seem interested in multimedia. His guess is that "they assume multimedia applications are rather smaller than the applications that software engineers have traditionally treated, and consider multimedia applications to be a research target worth little." He argues that the difference between multimedia and traditional applications is not just in size but also the domain of application. While there is no disagreement on this guess, it would be more appropriate to expand. We claim that there is a lack of a systematic study that highlights the benefits of software engineering for multimedia. Additionally, such study should lay down the main software approaches that may be extended and/or customized to fit within the requirements of "mobile" multimedia development.

Due to the huge demand of software applications by the industry, the U.S. President's Information Technology Advisory Committee (PITAC) report puts "Software" as the first four priority areas for long-term R&D. Indeed, driven by market pressure and budget constraints, software development is characterized by the preponderance of ad-hoc development approaches. Developers don't take time to investigate methodologies that may accelerate software development because learning

Figure 1. A typical case tool

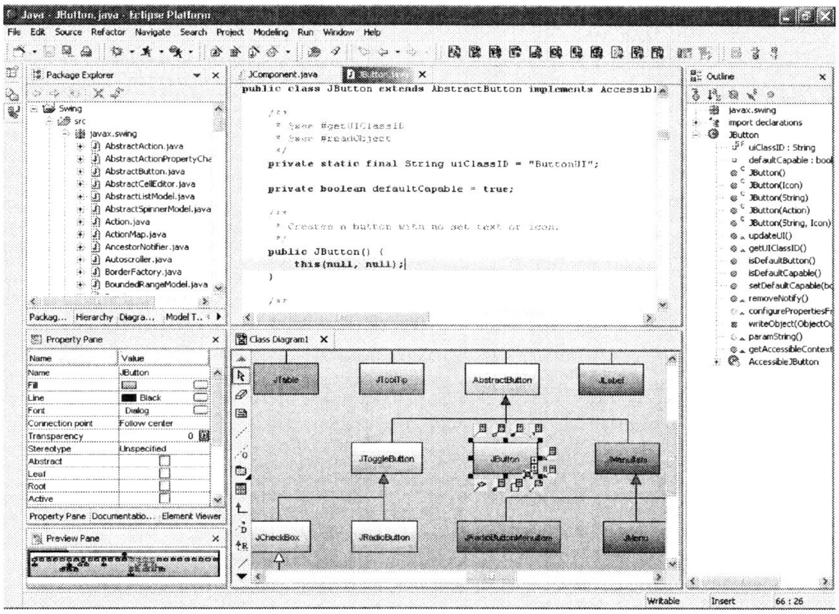

these tools and methodologies itself requires time. As a result, software applications are very difficult to maintain and reuse, and most of the time related applications-domains are developed from scratch across groups, and in the worst case in the same group.

The demand for complex, distributed multimedia software is rising; moreover, multimedia software development suffers from similar pitfalls discussed earlier. In the next section, we explore the benefits of using software engineering tools and methodologies for mobile multimedia development.

Software Engineering for Leveraging Mobile Multimedia Development

Even if mobile multimedia applications are diverse in content and form, their development requires handling common libraries for image and voice digitization, compression/decompression, identification of user's location, etc. Standards APIs and code for performing such operations needs to be frequently duplicated across many systems. A systematic reuse of such APIs and code highly reduces development time and coding errors. In addition to the need of reuse techniques, mobile multimedia applications are becoming more and more complex and require formal specification of their requirements. In bridging the gap between software engineering and mobile multimedia, the latter domain will benefit from a set of advantages summarized in the following:

- **Rapid development of mobile multimedia applications:** This issue is of primordial importance for the software multimedia industry. It is supported by reusability techniques in order to save time and cost of development.

- **Separation of concerns:** A mobile multimedia application is a set of functional and non-functional aspects. Examples are security, availability, acceleration, and rendering. In order to enforce the rapid development of applications, these aspects need to be developed and maintained separately.
- **Maintenance:** This aspect is generally seen as an error correction process. In fact, it is broader than that and includes software enhancement, adaptation, and code understanding. That's why, costs related to software maintenance is considerable and mounting. For example, in USA, annual software maintenance has been estimated to be more than $70 billion. At company-level, for example, Nokia Inc. used about $90 million for preventive Y2K-bug corrections (Koskinen, 2003).

In order to enforce the requirements previously discussed, many techniques are available. The most popular ones are detailed in the next Section including their concrete application for mobile multimedia development.

CONTRIBUTIONS TO "MOBILE" MULTIMEDIA SOFTWARE ENGINEERING

This section explores contributions that rely on software design methodologies to develop mobile multimedia applications. These contributions have been classified following three popular techniques for improving software quality including the ones outlined above. These techniques are: middleware, software frameworks, and design patterns.

Middleware

An accustomed to conferences in computer science has with no doubt attended a debate on the use of the word "middleware." Indeed, it's very common for developers to use this word to describe any software system between two distinct software layers, where in practice; their system does not necessarily obey to middleware requirements.

According to (Schmidt & Buschmann, 2003) middleware is software that can significantly increase reuse by providing readily usable, standard solutions to common programming tasks, such as persistent storage, (de)marshalling, message buffering and queuing, request demultiplexing, and concurrency control. The use of middleware helps developers to avoid the increasing complexity of the applications and lets them concentrate on the application-specific tasks. In other terms, middleware is a software layer that hides the complexity of OS specific libraries by providing easy tools to handle low-level functionalities.

CORBA (common object request broker architecture), J2EE, and .Net are examples middleware standards that emerge from industry and market leaders. However, they are not suitable for mobile computing and have no support for multimedia.

Davidyuk, Riekki, Ville-Mikko, and Sun (2004) describe CAPNET, a context-aware middleware which facilitates development of multimedia applications by handling such functions as capture and rendering, storing, retrieving and adapting of media content to various mobile devices (see Figure 2). It offers functionality for service discovery, asynchronous messaging, publish/subscribe event management, storing and management of context information, building the user interface, and handling the local and network resources.

Figure 2. The architecture of CAPNET middleware (Davidyuk et al., 2004)

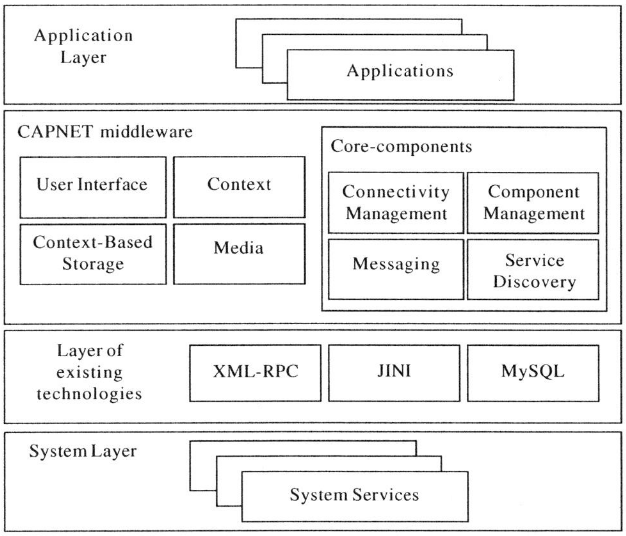

Mohapatra et al. (2003) propose an integrated power management approach that unifies low level architectural optimizations (CPU, memory, register), OS power-saving mechanisms (dynamic voltage scaling) and adaptive middleware techniques (admission control, optimal transcoding, network traffic regulation) for optimizing user experience for streaming video applications on handheld devices. They used a higher level middleware approach to intercept and doctor the video stream to compliment the architectural optimizations.

Betting on code portability, Tatsuo Nakajima (2002) describes a java-based middleware for networked audio and visual home appliances executed on commodity software. The high-level abstraction provided by the middleware approach makes it easy to implement a variety of applications that require composing a variety of functionalities.

Middleware for multimedia networking is currently a very active area of research and standardization.

Software Frameworks

Suffering from the same confusion in defining the word middleware, the word "framework" is used to mean different things. However, in this chapter, we refer to frameworks to software layers with specific characteristics we detail in the following. Software frameworks are used to support design reuse in software architectures. A framework is the skeleton of an application that can be customized by an application developer. This skeleton is generally represented by a set of abstract classes. The abstract classes define the core functionality of the framework, which also contains a set of concrete classes that provide a prototype application introduced for completeness. The main characteristics of frameworks are their provision of high level abstraction; in contrast to an application that provides a concrete solution to a concrete problem, a framework is intended to provide a generic solution for a set of related problems. Plus, a framework captures the programming expertise: necessary to solve a particular class of problems. Programmers purchase or reuse frameworks to obtain such problem-solving expertise without having to develop it independently.

Such advantages are exploited in Scherp and Boll (2004) where a generic java-based software framework is developed to support personalized (mobile) multimedia applications for travel and tourism. This contribution provides an efficient, simpler, and cheaper development platform of personalized (mobile) multimedia applications.

The Sesame environment (Coffland & Pimentel, 2003) is another software framework built for the purpose of modeling and simulating heterogeneous embedded multimedia systems.

Even if software frameworks are considered as an independent software technique, they are very often used to leverage middleware development and to realize the layered approach.

Design Patterns

Design patterns are proven design solutions to recurring problems in software engineering. Patterns are the result of developers' experience in solving a specific problem like request to events, GUIs, and on-demand objects creation. In object-oriented technologies, a design pattern is represented by a specific organization of classes and relationships that may be implemented using any object-oriented language. The book by Gamma, Helm, Johnson, and Vlissides (1995) is an anchor reference for design patterns. It establishes (a) the four essential elements of a pattern, namely, the pattern name, the problem, the solution and the consequences and (b) a preliminary catalog gathering a set of general purposes patterns. Later, many application-specific software patterns have been proposed such as in multimedia, distributed environments and security.

Compared to software frameworks discussed earlier, patterns can be considered as

Figure 3. Architecture of MediaBuilder patterns (Van den Broecke & Coplien, 2001)

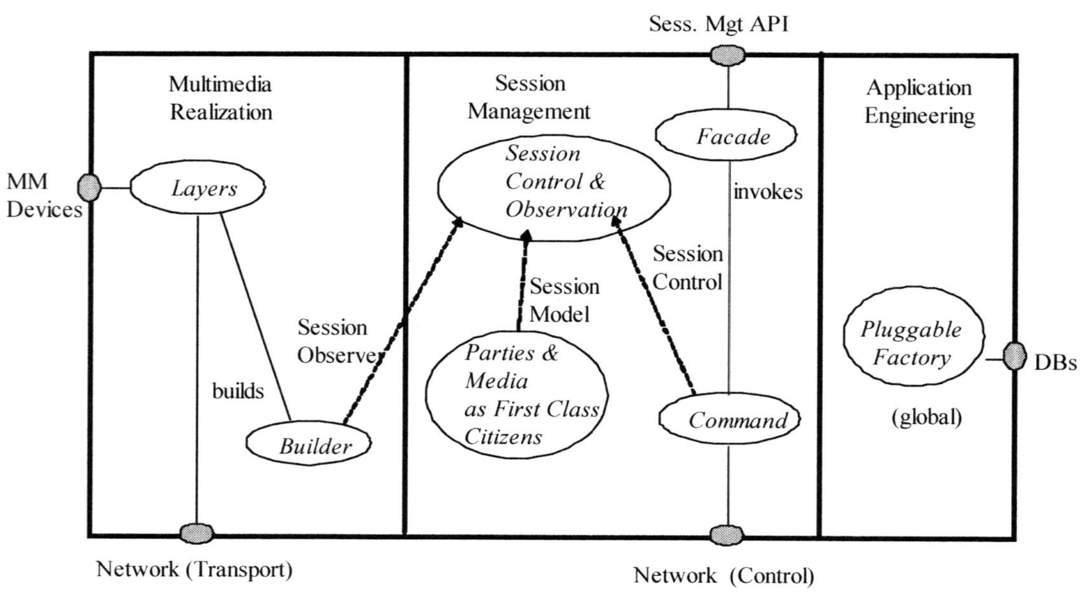

micro software frameworks; a partial program for a problem domain. They are generally used as building blocks for larger software frameworks.

MediaBuilder (Van den Broecke & Coplien, 2001) is one of most successful initiatives to pattern-oriented architectures for mobile multimedia applications. MediaBuilder is a services platform that enables real-time multimedia communication (i.e., audio, video, and data) between end-user PC's. It supports value-added services such as multimedia conferencing, telelearning, and tele-consultation, which allows end-users at different locations to efficiently work together over long distances. The software architecture is a set of patterns combined together to support session management, application protocols, and multimedia devices. Figure 3 summarizes the main patterns brought into play in order to determine the basic behavior of MediaBuilder. Each pattern belongs to one of the functional areas, namely; multimedia realization, session management, and application engineering.

The use of design patterns for mobile multimedia is driven by the desire to provide a powerful tool for structuring, documenting, and communicating the complex software architecture. They also allow the use of a standard language making the overall architecture of the multimedia application easier to understand, extend, and maintain.

The synergy of the three techniques previously discussed is depicted in Schmidt and Buschmann (2003). This synergy contributes to mobile multimedia development by providing high quality software architectures.

CHALLENGES OF MOBILE MULTIMEDIA SOFTWARE ENGINEERING

While system support for multimedia applications has been seriously investigated for several years now, the software engineering community has not yet reached a deep understanding of the impacts of "mobility" for multimedia systems. The latter has additional requirements compared to traditional multimedia applications. These requirements are linked to the versatility of the location of consumers and the diversity of their preferences. In the following, we address the main research areas that must be investigated by the software engineering community in supporting the development of mobile multimedia applications. These areas are not orthogonal. It means that same or similar research items and issues appear in more than one research area. We have divided the research space into four key research areas: (1) mobility, (2) context-awareness, and (3) real-time embedded multimedia systems.

Mobility

For the purpose previously discussed, the first trail to investigate is obviously "mobility." It is viewed by Roman, Picco, and Murphy (2000) to be the study of systems in which computational components may change location. In their roadmap paper on software engineering for mobility, they approach this issue from multiple views including models, algorithms, applications, and middleware. The middleware approach is generally adopted for the purpose of hiding hardware heterogeneity of mobile platforms and to provide an abstraction layer on top of specific APIs for handling multimedia content.

However, current investigations of software engineering for mobility argue that there is a lack of well-confirmed tools and techniques.

Context-Awareness

Context has been considered in different fields of computer science, including natural language processing, machine learning, computer vision, decision support, information retrieval, pervasive computing, and more recently computer security. By analogy to human reasoning, the goal behind considering context is to add adaptability and effective decision-making.

In general mobile applications, context becomes a predominant element. It is identified as any information that can be used to characterize the situation of an entity. Where an entity is a person, or object that is considered relevant to the interaction between a user and an application, including the user and application themselves (Dey, 2001). Context is heavily used for e-services personalization according to consumers' preferences and needs and for providing fine-grained access control to these e-services. In the domain of mobile multimedia, this rule is still valid. Indeed, multimedia content whether this content is static (e.g., jpeg, txt), pre-stored (e.g., 3gp, mp4) or live, must be tuned according to the context of use. Mobile cinema (Pan, Kastner, Crowe, & Davenport, 2002) is an example, it is of great interest to health, tourism, and entertainment. Mobile cinema relies on broadband wireless networks and on spatial sensing such as GPS or infrared in order to provide mobile stories to handled devices (e.g., PDAs). Mobile stories are composed of media sequences collected from media spots placed in the physical location. These sequences are continually rearranged in order to form a whole narrative. Context used to assemble mobile stories are mainly time and location but can be extended to include information collected using bio-sensors and history data.

Multimedia mobile service (MMS) is a brand new technology in the market but rapidly becomes a very popular technique used to exchange pictorial information with audio and text between mobile phones and different services. Häkkilä and Mäntyjärvi (2004) propose a model for the combination of location—as context—with MMS for the provision of adaptive types of MM messages. In their study, the authors explore user experiences on combining location sensitive mobile phone applications and multimedia messaging to novel type of MMS functionality. As they state in Häkkilä & Mäntyjärvi (2004), the selected message categories under investigation were presence, reminder, and notification (public and private), which were selected as they were seen to provide a representing sample of potentially useful and realistic location related messaging applications.

Coming back to the software perspective and based on a review of current context-aware applications, Ghita Kouadri Mostéfaoui (2004) points up to the lack of reusable architectures/mechanisms for managing contextual information (i.e., discovery, gathering, and modeling). She states that most of the existing architectures are built in an ad hoc manner with the sole desire to obtain a working system. As a consequence, context acquisition is highly tied up with the remaining infrastructure leading to systems that are difficult to adapt and to reuse.

It is clear that context-awareness constitute a primordial element for providing adaptive multimedia content to mobile devices. Even if currently, location is the most used source of contextual information, many other types can be included such users' preferences. Thus, we argue that leveraging mobile multimedia software is tied up with the improvement of software engineering for context-awareness. The latter constitutes one of the trails that should be considered for the development of adaptive mobile multimedia applications.

Real-Time Embedded Multimedia Systems

Real-time synchronization is an intrinsic element in multimedia systems. This ability requires handling events quickly and in some cases to respond within specified times. Real-time software design relies on specific programming languages in order to ensure that deadlines of system response are met. Ada is an example language; however, for ensuring a better performance, most real-time systems are implemented using the assembler language. The mobility of multimedia applications introduces additional issues in handling time constraints. Such issues are management of large amount of data needed for audio and video streams. In Oh and Ha (2002), the authors present a solution to this problem by relying on code synthesis techniques. Their approach relies on buffer sharing. Another issue in real-time mobile multimedia development is software reusability. Succi, Benedicenti, Uhrik, Vernazza, and Valerio (2000) point to the high importance of reusability for the rapid development of multimedia applications by reducing development time and cost. The authors argue that reuse techniques are not accepted as a systematic part of the development process, and propose a reusable library for multimedia, network-distributed software entities.

Software engineering real-time systems still present many issues to tackle. The main ones are surveyed by Kopetz (2000) who states that the most dramatic changes will be in the fields of composable architectures and systematic validation of distributed fault-tolerant real-time systems.

Software engineering mobile multimedia embraces all these domains and therefore claims for accurate merging of their respective techniques and methodologies since the early phases of the software development process.

Bridging the Gap Between Software Engineering and Mobile Multimedia

Different software engineering techniques have been adopted to cope with the complexity of designing mobile multimedia software. Selecting the "best" technique is a typical choice to be made at the early stage of the software design phase. Based on the study we presented earlier, we argue that even if the research community has been aware of the advantages of software engineering for multimedia, mobility of such applications is not yet considered at its own right. As a result, the field is still lacking a systematic approach for specifying, modeling and designing, mobile multimedia software. In the following, we stress a preliminary set of guidelines for the aim to bridging the gap between software engineering and mobile multimedia.

- The mobile multimedia software engineering challenges lie in devising notations, modeling techniques, and software artifact that realize the requirements of mobile multimedia applications including mobility, context-awareness, and real-time processing
- The software engineering research can contribute to the further development of mobile multimedia by proposing development tools that leverage the rapid design and implementation of multimedia components including voice, image, and video
- Training multimedia developers to the new software engineering techniques and methodologies allows for the rapid detection of specific tools that leverage the advance of mobile multimedia
- Finally, a community specializing in software engineering mobile multimedia should be established in order to (1) gather such efforts (e.g., design patterns for mobile multimedia) and (2) provide a concise guide for multimedia developers (3) to agree on standards for multimedia middleware, frameworks and reusable multimedia components

CONCLUSION

In this chapter, we highlighted the evolving role of software engineering for mobile multimedia development and discussed some of the opportunities open to the software engineering community in helping shape the success of the mobile multimedia industry. We argue that a systematic reliance on software engineering methodologies since the early stages of the development cycle is one of the most boosting factors of the mobile multimedia domain. Developers should be directed to use reuse techniques in order to reduce maintenance costs and produce high-quality software even if the development phase takes longer.

REFERENCES

Coffland, J. E., & Pimentel, A. D. (2003). A software framework for efficient system-level performance evaluation of embedded systems. *Proceedings of the 18th ACM Symposium on Applied Computing, Embedded Systems Track,* Melbourne, FL (pp. 666-671).

Davidyuk, O., Riekki, J., Ville-Mikko, R., & Sun, J. (2004). Context-aware middleware for mobile multimedia applications. *Proceedings of the 3rd International Conference on Mobile and Ubiquitous Multimedia* (pp. 213-220).

Dey, A. (2001). Supporting the construction of context-aware applications. In *Dagstuhl Seminar on Ubiquitous Computing.*

Ericsson. (2005). *EDGE introduction of high-speed data in GSM/GPRS networks,* White paper. Retrieved from http://www.ericsson.com/products/white_papers_pdf/edge_wp_technical.pdf

Gamma, E., Helm, R., Johnson, R., & Vlissides, J. (1995). *Design patterns: Elements of reusable object-oriented software.* Reading, MA: Addison-Wesley.

GSM World. (2005). *GPRS Platform.* Retrieved from http://www.gsmworld.com/technology/gprs/intro.shtml#1

Häkkilä, J., & Mäntyjärvi, J. (2004) User experiences on combining location sensitive mobile phone applications and multimedia messaging. *International Conference on Mobile and Ubiquitous Multimedia,* Maryland (pp. 179-186).

Hirakawa, M. (1999). Do software engineers like multimedia? *Proceedings of the International Conference on Multimedia Computing and Systems,* Florence, Italy (pp. 85-90).

Jain, R. (2001). Mobile Multimedia. *IEEE MultiMedia, 8*(3), 1.

Kopetz, H. (2000). Software engineering for real-time: A roadmap. *Proceedings of the Conference on the Future of Software Engineering.*

Koskinen, J. (2003). *Software maintenance costs.* Information Technology Research Institute, ELTIS-project, University of Jyväskylä.

Kouadri Mostéfaoui, G. (2004). *Towards a conceptual and software framework for integrating context-based security in pervasive environments.* PhD thesis. University of Fribourg and University of Pierre et Marie Curie (Paris 6), October 2004.

LeClaire, J. (2005). Demand for mobile multimedia services on rise. *E-Commerce Times.* Retrieved from http://www.ecommercetimes.com/story/Demand-for-Mobile-Multimedia-services-on-Rise-40168.html

Mohapatra, S., Cornea, R., Nikil, D., Dutt, N., Nicolau, A., & Venkatasubramanian, N. (2003). Integrated power management for video streaming to mobile handheld devices. *ACM Multimedia 2003* (pp. 582-591).

Nakajima, T. (2002). Experiences with building middleware for audio and visual networked home appliances on commodity software. *ACM Multimedia 2002* (pp. 611-620).

Nokia Inc. (2005). *Mobile entry.* Retrieved from http://www.nokia.com/nokia/0,6771,56483,00.html

Oh, H., & Ha, S. (2002). *Efficient code synthesis from extended dataflow graphs for multimedia applications.* Design Automation Conference.

Pan, P., Kastner, C., Crowe, D., & Davenport, G. (2002). M-studio: An authoring application for context-aware multimedia. *ACM Multimedia 2002* (pp. 351-354).

Roman, G. C., Picco, G. P., & Murphy, A. L. (2000). Software engineering for mobility: A roadmap. In A. Finkelstein (Ed.), *Future of software engineering. ICSE'00,* June (pp. 5-22).

Scherp, A., & Boll, S. (2004). *Generic support for personalized mobile multimedia tourist applications.* Technical Demonstration for the ACM Multimedia 2004, New York, October 10-16.

Schmidt, D. C., & Buschmann, F. (2003). Patterns, frameworks, and middleware: Their synergistic relationships. *Proceedings of the 25th International Conference on Software Engineering (ICSE 2003)* (pp. 694-704).

Succi, G., Benedicenti, L., Uhrik, C., Vernazza, T., & Valerio, A. (2000). Reuse libraries for real-time multimedia over the network. *ACM SIGAPP Applied Computing Review, 8*(1), 12-19.

UMTS. (2005). *UMTS.* Retrieved from http://searchnetworking.techtarget.com/sDefinition/0,,sid7_gci213688,00.html

Van den Broecke, J. A., & Coplien, J. O. (2001). Using design patterns to build a framework for multimedia networking. *Design patterns in communications software* (pp. 259-292). Cambridge University Press.

KEY TERMS

Context-Awareness: Context awareness is a term from computer science that is used for devices that have information about the circumstances under which they operate and can react accordingly.

Design Patterns: Design patterns are standard solutions to common problems in software design.

Embedded Systems: An embedded system is a special-purpose computer system, which is completely encapsulated by the device it controls.

Middleware: Middleware is software that can significantly increase reuse by providing readily usable, standard solutions to common programming tasks, such as persistent storage, (de)marshalling, message buffering and queuing, request de-multiplexing, and concurrency control.

Real-Time Systems: Hardware and software systems that are subject to constraints in time. In particular, they are systems that are subject to deadlines from event to system response.

Software Engineering: Software engineering is a well-established discipline that groups together a set of techniques and methodologies for improving software quality and structuring the development process.

Software Frameworks: Software frameworks are reusable foundations that can be used in the construction of customized applications.

Section III

Multimedia Information

Multimedia information as combined information presented by various media types (text, pictures, graphics, sounds, animations, videos) enriches the quality of the information and represents the reality as adequately as possible. Section III contains ten chapters and is dedicated to how information can be exchanged over wireless networks whether it is voice, text, or multimedia information.

Chapter XVIII
Adaptation and Personalization of User Interface and Content

Christos K. Georgiadis
University of Macedonia, Thessaloniki, Greece

ABSTRACT

Adaptive services based on context-awareness are considered to be a precious benefit of mobile applications. Effective adaptations however, have to be based on critical context criteria. For example, presence and availability mechanisms enable the system to decide when the user is in a certain locale and whether the user is available to engage in certain actions. What is even more challenging is a personalization of the user interface to the interests and preferences of the individual user and the characteristics of the used end device. Multimedia personalization is concerned with the building of an adaptive multimedia system that can customize the representation of multimedia content to the needs of a user. Mobile multimedia personalization especially, is related with the particular features of mobile devices' usage. In order to fully support customization processes, a personalization perspective is essential to classify the multimedia interface elements and to analyze their influence on the effectiveness of mobile applications.

INTRODUCTION

Limited resources of mobile computing infrastructure (cellular networks and end user devices) set strict requirements to the transmission and presentation of multimedia. These constraints elevate the importance of additional mechanisms, capable of handling economically and efficiently the multimedia content. Flexible techniques are needed to model multimedia data adaptively for multiple heterogeneous networks and devices with varying capabilities. "Context" conditions (the implicit information about the environment, situation and surrounding of a particular communication) are of great importance.

Adaptive services based on context-awareness are indeed a precious benefit of mobile applications: in order to improve their provided service, mobile applications can actually take advantage of the context to adjust their behaviors. An effective adaptation has to be based on certain context criteria: presence and availability mechanisms enable the system to decide when the user is in a certain locale and whether the user is available to engage in certain actions. Hence, mobile applications aim to adapt the multimedia content to the different end user devices.

However, typically each and every person receives the same information under the same context conditions. What is even more challenging is a personalization of the user interface (UI) to the interests and preferences of the individual user and the characteristics of the user end device. The goal of mobile applications is to increasingly make their service offerings more personalized toward their users. Personalization has the ability to adapt (customize) resources (products, information, or services) to better fit the needs of each user. Personalization in mobile applications enables advanced customized services such as alerts, targeted advertising, games, and improved, push-based mobile messaging. In particular, multimedia personalization is concerned with the building of an adaptive multimedia system that can customize the representation of multimedia content to the needs of a user.

Multimedia personalization enlarges the application's complexity since every individual's options have to be considered and implemented. It results in a massive amount of variant possibilities: target groups, output formats, mobile end devices, languages, locations, etc. Thus, manual selection and composition of multimedia content is not practical. A "personalization engine" is needed to dynamically create the context-dependent personalized multimedia content. General solution approaches concerning the personalization engine, include personalization by transformation (using XML-based transformations to produce personalized multimedia documents), adaptive multimedia documents (using SMIL-like presentation defined alternatives), personalization by constraints (optimization problem—constraint solving), personalization by algebraic operators (algebra to select media elements and merge them into a coherent multimedia presentation), or broader software engineering approaches.

Mobile multimedia (M3) personalization especially, is related with the particular features of mobile devices' usage. Because of their mobility and omnipresence, mobile devices have two characteristics worth noticing. First, users have limited attention as they operate their mobile devices (this is because they usually are concerned at the same time in other tasks, (e.g., car driving)). Second, users tend to treat their mobile devices in a quite personal way, seeking for personal services and personalized content. The preferences of users are therefore noticeably affected. In many cases, they favor content and services which do not require transmitting large quantities of information. Thus, low-intensity content (e.g., ring tones, weather reports, and screen icons) proved to be very popular. This is not only because of the low availability of mobile devices' resources which complicates the processing of large volumes of information. Users demand further individually customized content on the mobile Internet because its personalization level is higher than that of the fixed Internet.

Detailed issues concerning M3 personalization can be described, analyzing UI design issues. Existing mobile applications offer a reasonably easy, browser-based interface to help user access available information or services. In order to support adaptation and personalization mechanisms they should be also as far as possible concentrated on the individual prerequisites of the human in contact with it. In this chapter, after the presentation of background topics we discuss critical issues of the mobile setting (characteristics of mobile applications and mobility dimensions in user interactions) that influence adaptation and personalization technologies. Then, as an application case, we focus on m-commerce applications and customer interfaces. All current research studies tend to acknowledge that the design rules of wired Internet applications are only partially useful. They should not be directly adopted in mobile computing area, because of the considerably different user requirements and device constraints. On the other hand, experience gained from the fixed Internet formulated as the well-accepted 7C framework, is always welcomed. Hence, we classify the multimedia interface elements and we analyze their influence on m-commerce site's effectiveness from a personalization perspective.

BACKGROUND

Adaptation Objectives

The diversity of end device and network capabilities in mobile applications along with the known multimedia challenges (namely, the efficient management of size, time, and semantics parameters of multimedia), demand media content and service to be flexible modeled for providing easy-to-use and fast multimedia information. Multimedia adaptation is being researched to merge the creation of the services so that only one service is needed to cover the heterogeneous environments (Forstadius, Ala-Kurikka, Koivisto, & Sauvola, 2001). Even though adaptation effects could be realized in a variety of ways, the major multimedia adaptation technologies are adaptive content selection, and adaptive presentation. Examples of adaptation include "down-scaling" the multimedia objects and changing the style of multimedia presentation according to user's context conditions.

In general, adaptive hypermedia and adaptive Web systems belong to the class of user-adaptive systems. A user model—the explicit representation of all relevant aspects of a user's preferences, intentions, etc.—forms the foundation of all adaptive systems (Bauer, 2004). The user model is used to provide an adaptation effect, which is tailoring interaction to different users. The first two generations (pre-Web and Web) of adaptive systems explored mainly adaptive content selection and adaptive recommendation based on modeling user interests. Nowadays, the third (mobile) generation extends the basis of the adaptation by adding models of context (location, time, bandwidth, computing platform, etc.) to the classic user models and explores the use of known adaptation technologies to adapt to both an individual user and the context of the user's work (Brusilovsky & Maybury, 2002).

Personalization Objectives and Mechanisms

Personalization is a special kind of adaptation of the UI which focuses on making a Web application more receptive to the unique and individual needs of each user (Cingil, Dogac, & Azgin, 2000). Personalization mechanisms presuppose two phases. First, the accumulation of user information, in order to build up a profile that illustrates a set of descriptors essential to administrators (e.g., visitor's interest, navigation paths, entitlements and roles in an organization, purchases, etc.). The second phase is the analysis of user information to recommend actions specific to the user.

Figure 1. Analyzing mobile setting

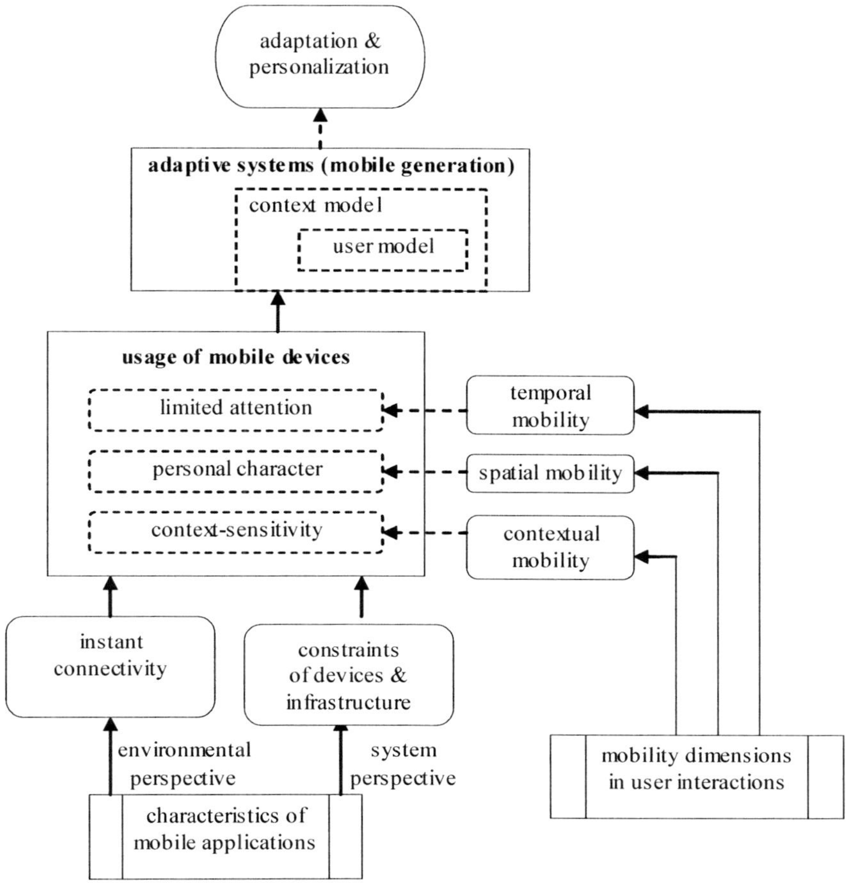

To develop the best recommendation, rule-based practices (allowing administrators to specify principles for their applications to thrust personalization) are usually combined with filtering algorithms which analyze user profiles (Pierrakos, Paliouras, Papatheodorou, & Spyropoulos, 2003). Simple filtering techniques are based on predefined groups of users, classifying their accounts by age groups, asset value etc. Content-based filtering can be seen as locating objects comparable to those a user was fond of in the past. Finally, collaborative filtering builds up recommendations by discovering users with similar preferences.

CONTENT ADAPTATION AND PERSONALIZED USER INTERFACE

Analyzing Mobile Setting

The characteristics of the mobile Internet applications can be appreciated from three different viewpoints: system, environment, and user (Chae & Kim, 2003). From the system's viewpoint, mobile applications present disadvantages, because they provide a lower level of available system resources. Mobile devices, especially cellular phones, have lower multimedia processing capabilities, inconvenient input/output facilities (smaller screens/keyboards), and lower network connection speeds than desktop computers. However, from the environmental viewpoint there is an uncontested benefit: they enable users to access mobile Internet content anywhere and anytime. The term "instant connectivity" is used for mobile browsing to describe actually the fact that it is possible to do it at the moment of need.

User's perspective characteristics must be regarded rather differently, because they are to a certain degree consequences of the system and of the environment. In addition, the multi-dimensioned concept of "mobility" influences on them in many ways. Mobile users perform their tasks in terms of place, time and context. Different terms are used by the research community to describe user's mobile setting, and their interactions within it, but these converge at the ones described below (Kakihara & Sorensen, 2001; Lee & Benbasat, 2004):

- Spatial mobility denotes mainly the most immediate dimension of mobility, the extensive geographical movement of users. As users carry their mobile devices anywhere they go, spatiality includes the mobility of both the user and the device
- Temporal mobility refers to the ability of users for mobile browsing while engaged in a peripheral task
- Contextual mobility signifies the character of the dynamic conditions in which users employ mobile devices. Users' actions are intrinsically situated in a particular context that frames and it is framed by the performance of their actions recursively

Because of their mobility (and in correspondence with its dimensions), we distinguish three attributes regarding mobile device usage:

1. Users have a tendency to treat their mobile device in a quite personal and emotional way (Chae & Kim, 2003). They prefer to access more personalized services when are involved in mobile browsing. Spatial mobility must be considered as the major reason behind this behaviour, which is quite normal considering user's perspective: the mobile phone is a portable, ubiquitous and exposed on everybody's view gadget, able to signify user's aesthetic preferences and personality
2. Users have limited attention as they manage their mobile devices (Lee & Benbasat, 2004). This is because they usually are involved at the same time in other tasks (e.g., walking). Temporal mobility is the reason of this phenomenon
3. Users manage their mobile devices in broadly mixed environments that are relatively unsteady from one moment to the next. Contextual mobility requires context-sensitivity on mobile device operations. So, mobile device is able to detect the user's setting (such as location and resources nearby) and subsequently to propose this information to the mobile application. In this way, mobile device practically may offer task-relevant services and information.

Application Case: User Interfaces in M-Commerce Applications

Mobile Commerce Applications

The mobile sector is creating exciting new opportunities for content and applications developers. The use of wireless technologies extends the nature and scope of traditional e-commerce by providing the additional aspects of mobility (of participation) and portability (of technology) (Elliot & Phillips, 2004). One of the most rapidly spreading applications within the m-commerce world is the mobile Internet: the wireless access to the contents of the Internet using portable devices, such as mobile phones. Undoubtedly, delivering personalized information is a critical factor concerning the effectiveness of an m-commerce application: the organization knows how to treat each visitor on an individual basis and emulate a traditional face-to-face transaction. Thus, has the ability to treat visitors based on their personal qualities and on prior history with its site. M-commerce applications support mechanisms to learn more about visitor (customer) desires, to recognize future trends or expectations and hopefully to amplify customer "loyalty" to the provided services.

Personalized Multimedia in Interfaces of M-Commerce Applications

The goal of adaptive personalization is to increase the usage and acceptance of mobile access through content that is easily accessible and personally relevant (Billsus, Brunk, Evans, Gladish, & Pazzani, 2002). The importance of interface design has been commonly acknowledged, especially regarding mobile devices adoption: interfaces characteristics had been identified as one of the two broad factors (along with network capabilities), affecting the implementation and acceptance of mobile phones emerged (Sarker & Wells, 2003). Devices adoption is a critical aspect for the future of m-commerce, because without widespread proliferation of mobile devices, m-commerce can not fulfill its potential.

Lee and Benbasat (2004) describe in detail the influence of mobile Internet environment to the 7C framework for customer interfaces. This framework studies interface and content issues based on the following design elements: customization (site's ability to be personalized), content (what a site delivers), context/presentation (how is presented), connection (the degree of formal linkage from one site to others), communication (the type of dialogues between sites and their users), community (the interaction between users), and commerce (interface elements that support the various business transactions) (Rayport & Jaworski, 2001). A generic personalized perspective is presented in (Pierrakos et al., 2003) with a comprehensive classification scheme for Web personalization systems. Based on all these works, we focus on multimedia design issues concerning personalized UIs for m-commerce applications. We present a reconsideration of the 7C framework from an M3 customization aspect, in which we distinguish the following mobile multimedia adaptation/personalization categories:

M3 content is the main category. It contains the parts of the 7C's "content" and "commerce" design elements, which deal with the choice of media. "Multimedia mix" is the term that is used in 7C framework regarding exclusively the 'content' element. However, in our approach multimedia elements regarding shopping carts, delivery options etc. are also belong here because they share a lot of commons concerning adaptation and personalization. It is commonly accepted that large and high visual fidelity images, audio effects, and motion on interfaces are multimedia effects which might lead to a higher probability of affecting users' decisions in e-commerce environments (Lee & Benbasat, 2003). However, in m-commerce setting things are different because we can not assume that the underlying communication system is capable of delivering an optimum quality of service (QoS). The bandwidth on offer and the capabilities of devices are setting limitations. Therefore, a central issue to the acceptance of multimedia in m-commerce interfaces is the one of quality. The longer the response delay, the less inclined will the user be to visit that specific m-commerce site, resulting in lost revenue. Obviously, an end-to-end QoS over a variety of heterogeneous network domains and devices is not easily assured, but this is where adaptation principle steps in.

Dynamic content adaptation of the media quality to the level admitted by the network is a promising approach (Kosch, 2004). Content adaptation can be accomplished by modifying the quality of a media object (resolution and its play rate); so, it can be delivered over the network with the available bandwidth and then it can be presented at the end device (satisfying its access and user constraints).

An essential issue for effective content adaptation is the perceptual quality of multimedia. Quality of perception (QoP) is a measure which includes not only a user's satisfaction with multimedia clips, but also his ability to perceive, analyze, and synthesize their informational content. When a "personalization engine" is called out to adapt multimedia content, the perceptual impact of QoS can be extremely valuable, and it can be summarized by the following points (Ghinea & Angelides, 2004):

- Missing a small number of media units will not be negatively perceived, given that too many such units are not missed consecutively and that this incident is infrequent
- Media streams could flow in and out of synchronization without substantial human displeasure
- Video rate variations are tolerated much better than rate variations in audio
- Audio loss of human speech is tolerated quite well
- Reducing the frame rate does not proportionally reduce the user's understanding (user has more time to view a frame before changes)
- Users have difficulty absorbing audio, textual and visual information concurrently, as they tend to focus on one of these media at any one moment (although they may switch between the different media)
- Highly dynamic scenes have a negative impact on user understanding and information assimilation

Another important issue regarding M3 content adaptation (both for quality and for the selection of media items), is the usage patterns for the mobile Internet. Users purchase more low-risk products (e.g., books) than high-risk ones, because they can not pay full attention to their interactions with mobile devices. Also, users tend to subscribe to content with low information intensity more than to content with high information intensity (e.g., education), because mobile devices have inferior visual displays.

Device's constraints and personalization requirements emphasize the need for additional effective content adaptation methods. Personalization mechanisms allow customers to feel sufficiently informed about products and services they are interested in, despite the limited multimedia information delivered by a restricted display device. They can be considered as filters which reject the delivery of multimedia content that users don't appreciate. More and more, mobile applications exploit positioning information like GPS to guide the user on certain circumstances providing orientation and navigation multimedia information, such as location-sensitive maps. To facilitate personalized adaptation, multimedia content is desirable to include personalization and user profile management information (in the form of media descriptors) (Kosch, 2004). In this way, adaptive systems can utilize information from the context (or user model) in use. Especially personalized UIs are able to exercise all kinds of personalization mechanisms (rule-based practices and simple, content-based or collaborative filtering), to locate or predict a particular user's opinion on multimedia items.

M3 presentation is also an important adaptation/personalization category. It contains all parts from 7C's "context," "commerce," and "connection" design elements related to multimedia presentation. M3 presentation refers to the following aspects:

- The aesthetic nature of multimedia in interfaces (i.e., the visual and audio characteristics such as color schemes, screen icons, ring melodies, etc.). These multimedia UI elements are certainly used by mobile users in order to make their phones more personal
- The operational nature of multimedia in interfaces, including internal/external link issues and navigation tools (in what ways the moving throughout the application is supported). An important issue here deals with the limited attention of users when interacting with their mobile devices. So, minimal attention interface elements, able to minimize the amount of user attention required to operate a device are welcomed. For example, utilizing audio feedback in order to supplement users' limited visual attention is considered in general a desirable approach in mobile setting (Kristoffersen & Ljungberg, 1999). There is also an additional point to take into consideration regarding M3 presentation adaptation and personalization: how to

Figure 2. Mobile multimedia adaptation/personalization categories

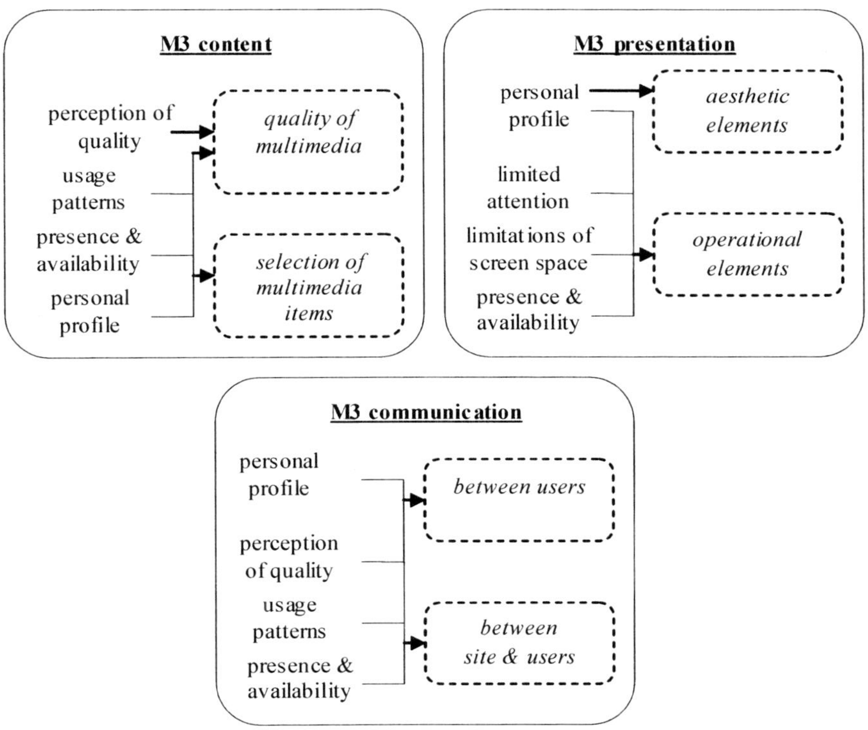

overcome the limitations due to the lack of screen space. Certainly, visual representations of objects, mostly through graphic icons, are easier to manipulate and retain than textual representations. But, small screens need not set aside a large portion of their space for infrequently used widgets. In this context, potential adaptations can be made by substituting visual elements with non-speech audio cues (Walker & Brewster, 1999), or by using semi-transparent screen-buttons, that overlap with the main body of content in order to make the most of a small screen (Kamba, Elson, Harpold, Stamper, & Sukaviriya, 1996).

All users are not having the same context conditions and preferences. Personalization mechanisms are used for both the aesthetic and the operational nature of multimedia in interfaces. Obviously, multimedia personalization engine must be able to provide context-sensitive personalized multimedia presentation. Hence, when a distracting user setting is acknowledged, the adapted multimedia presentations on the interface should call for only minimal attention in order to complete successfully critical transaction steps. Moreover, context-awareness of mobile devices may influence M3 presentation adaptation/personalization regarding connection issues. Indeed, the recommendation of a particular external link among a set of similar ones may depend not only from its content, but also from its availability and efficiency under the specific conditions of user's setting.

M3 communication contains all parts from 7C's "communication" and "community" design elements related to multimedia. In our approach, they belong to the same adaptation/personalization category because they deal with multimedia enriched communication and interaction services. Mobile devices are inherently communication devices. Location and positioning mechanisms provide precise location information, enabling them to better interact with applications to deliver greatly targeted multimedia communication services. The perceptual quality of multimedia and relative previously discussed issues, are also important factors for effective multimedia communication adaptation.

With M3 communication personalization, m-commerce administrators are able to make use of information about users' mobile setting to catch the right type of multimedia communication for the right moment (taken into account also the preferences of each user about the most wanted type of communication between him or her and the site). In addition, supporting adaptive (interactive or non-interactive) multimedia communication between users enables opinion exchange about current transactions and network accesses. Undoubtedly, such functionality may provide useful information for collaborative filtering techniques, resulting in more successful personalized sites.

FUTURE TRENDS

Providing adaptation and personalization affects system performance, and this is an open research issue. A basic approach to improve performance is to cache embedded multimedia files. However, when personalized multimedia elements are used extensively, multimedia caching can not maximize performance. The trend is therefore to provide personalization capabilities when server-usage is light and disallow such capabilities at periods of high request. Alternatively, users can have a personalized experience, even at times of high system load, if they pay for the privilege (Ghinea & Angelides, 2004). In any case, the design of a flexible context (or user) model, capable of understanding the characteristics of mobile setting in order to facilitate multimedia adaptation and personalization processes, it appears as an interesting research opportunity.

In a multi-layered wireless Web site, more sophisticated adaptation and personalization mechanisms are introduced as we get closer to the database layer. From that point of view, multimedia database management system (MMDBMS) emerging technology may support significantly the (mobile) multimedia content adaptation process. Existing multimedia data models in MMDBMSs are able to partially satisfy the requirements of multimedia content adaptation because contain only the basic information about the delivery of data (e.g., frame rate, compression method, etc.). More sophisticated characteristics such as the quality adaptation capabilities of the streams are not included. This information would be of interest to the end user. Consequently, a lot of research deals with extending the functionalities of current MMDBMSs by constructing a common framework for both the quality adaptation capabilities of multimedia and for the modeling/querying of multimedia in a multimedia database (Dunkley, 2003; Kosch, 2004).

CONCLUSION

The advances in network technology, together with novel communication protocols and the considerably enhanced throughput bandwidths of networks, attracted more and more consumers to load or stream multimedia data to their mobile devices. In addition, given the limited display space, the use of multimedia is recommended so that display space can be conserved. However, mobile setting's limitations regarding multimedia are serious. In fact, enhancing the mobile browsing user experience with multimedia is feasible only if perceptual and contextual considerations are employed.

The major conclusion of previously presented issues is that efficient delivery, presentation and transmission of multimedia has to rely on context-sensitive mechanisms, in order to be able to adapt multimedia to the limitations and needs of the environment at hand, and even more to personalize multimedia to individual user's preferences.

REFERENCES

Bauer, M. (2004). Transparent user modeling for a mobile personal assistant. *Working Notes of the Annual Workshop of the SIG on Adaptivity and User Modeling in Interactive Software Systems of the GI* (pp. 3-8).

Billsus, D., Brunk, C. A., Evans, C., Gladish, B., & Pazzani, M. (2002). Adaptive interfaces for ubiquitous Web access. *Communications of the ACM, 45*(5), 34-38.

Brusilovsky, P., & Maybury, M. T. (2002). From adaptive hypermedia to adaptive Web. *Communications of the ACM, 45*(5), 31-33.

Chae, M., & Kim, J. (2003). What's so different about the mobile Internet? *Communications of the ACM, 46*(12), 240-247.

Cingil, I., Dogac, A., & Azgin, A. (2000). A broader approach to personalization. *Communications of the ACM, 43*(8), 136-141.

Dunkley, L. (2003). *Multimedia databases.* Harlow, UK: Addison-Wesley–Pearson Education.

Elliot, G., & Phillips, N. (2004). *Mobile commerce and wireless computing systems.* Harlow, UK: Addison Wesley–Pearson Education.

Forstadius, J., Ala-Kurikka, J., Koivisto, A., & Sauvola, J. (2001). Model for adaptive multimedia services. *Proceedings SPIE, Multimedia Systems, and Applications IV* (Vol. 4518).

Ghinea, G., & Angelides, M. C. (2004). A user perspective of quality of service in m-commerce. *Multimedia Tools and Applications, 22*(2), 187-206. Kluwer Academic.

Kakihara, M., & Sorensen, C. (2001). Expanding the "mobility" concept. *ACM SIGGROUP Bulletin, 22*(3), 33-37.

Kamba, T., Elson, S., Harpold, T., Stamper, T., & Sukaviriya, P. (1996). Using small screen space more efficiently. In R. Bilger, S. Guest, & M. J. Tauber (Eds.), *Proceedings of CHI 1996 ACM SIGCHI Annual Conference on Human Factors in Computing Systems* (pp. 383-390). Vancouver, WA: ACM Press.

Kosch, H. (2004). *Distributed multimedia database technologies*. Boca Raton, FL: CRC Press.

Kristoffersen, S., & Ljungberg, F. (1999). Designing interaction styles for a mobile use context. In H. W. Gellersen (Ed.), *Proceedings of the 1st International Symposium on Handheld and Ubiquitous Computing (HUC — 99)* (pp. 281-288).

Lee, W., & Benbasat, I. (2003). Designing an electronic commerce interface: Attention and product memory as elicited by Web design. *Electronic Commerce Research and Applications, 2*(3), 240-253. Elsevier Science.

Lee, Y. E., & Benbasat, I. (2004). A framework for the study of customer interface design for mobile commerce. *International Journal of Electronic Commerce (1086-4415/2004), 8*(3), 79-102.

Pierrakos, D., Paliouras, G., Papatheodorou, C., & Spyropoulos, C. (2003). Web usage mining as a tool for personalization: A survey. *User Modeling and User-Adapted Interaction, 13*(4), 311-372. Kluwer Academic.

Rayport, J., & Jaworski, B. (2001). *Introduction to e-commerce*. New York: McGraw-Hill.

Sarker, S., & Wells, J. D. (2003). Understanding mobile handheld device use and adoption. *Communications of the ACM, 46*(12), 35-40.

Walker, A., & Brewster, S. (1999). Extending the auditory display space in handheld computing devices. *Proceedings of the 2nd Workshop on Human Computer Interaction with Mobile Devices*.

KEY TERMS

Content Adaptation: The alteration of the multimedia content to an alternative form to meet current usage and resource constraints.

MMDBMS: Multimedia database management system is a DBMS able to handle diverse kinds of multimedia and to provide sophisticated mechanisms for querying, processing, retrieving, inserting, deleting, and updating multimedia. Multimedia database storage and content-based search is supported in a standardized way.

Personalization: The automatic adjustment of information content, structure and presentation tailored to an individual user.

QoS: Quality of service notes the idea that transmission quality and service availability can be measured, improved, and, to some extent, guaranteed in advance. QoS is of particular concern for the continuous transmission of multimedia information and declares the ability of a network to deliver traffic with minimum delay and maximum availability.

Streaming: Breaking multimedia data into packets with sizes suitable for transmission between the servers and clients, in order to allow user to start enjoying the multimedia without waiting to the end of transmission.

UI: (Graphical) user interface is the part of the computer system which is exposed to users. They interact with it using menus, icons, mouse clicks, keystrokes and similar capabilities.

Chapter XIX
Adapting Web Sites for Mobile Devices – A Comparison of Different Approaches

Henrik Stormer
University of Fribourg, Switzerland

ABSTRACT

With the rise of mobile devices like cell phones and personal digital assistants (PDAs) in the last years, the demand for specialized mobile solutions grows. One key application for mobile devices is the Web service. Currently, almost all Web sites are designed for stationary computers and cannot be shown directly on mobile devices because of their limitations. These include a smaller display size, delicate data input facilities and smaller bandwidth compared to stationary devices. To overcome the problems and enable Web sites also for mobile devices, a number of different approaches exist which can be divided into client and server based solutions. Client based solutions include all attempts to improve the mobile device, for example by supporting zoom facilities or enhance the data input. Server based solutions try to adapt the pages for mobile devices. This chapter concentrates on server-based solutions by comparing different ways to adapt Web sites for mobile devices. It is assumed that Web sites designed for stationary devices already exist. Additionally, it concentrates on the generation of HTML pages. Other languages, designed especially for mobile devices like WML or cHTML, are not taken into account simply because of the improvement of mobile devices to show standard HTML pages. The following three methods are generally used today: Rewrite the page, use an automatic generator to create the page, or try to use the same page for stationary and mobile devices. This chapter illustrates each method by adapting one page of the electronic shop software eSarine. Afterwards, the methods are compared using different parameters like the complexity of the approach or the ease of integration in existing systems.

INTRODUCTION

Mobile devices have become more and more popular in the last years. The most popular device is the cell phone. The Forrester (2003) statistic shows that 71% of all Europeans owned a cell phone in 2003. Other mobile devices are personal digital assistants (PDAs), mostly used to organize address books and calendars, or to write down short notes. Interesting developments are smart phones. A smart phone is a mobile device with PDA, as well as cell phone functionalities. On the one hand, there exists cell phones with PDA functionalities, on the other hand, there are PDAs, which can be used as a cell phone. With the starting of faster network solutions like UMTS new applications will become possible. One application is the use of the Internet Web service to access Web sites.

However, mobile devices have some disadvantages compared to stationary computers. These are:

- **Small display size:** The display size of mobile devices vary from small cell phones 96×65 pixel or less to 320×480 pixel on foldable smart phones. Even these displays are small compared to typical stand alone computer sizes with up to 1280×1024 pixel
- **Delicate data input:** On mobile devices, data input is done mainly with a small keyboard or by using a touch screen. Both ways are not as convenient as input on standalone systems using a keyboard and mouse
- **Small bandwidth:** Today's mobile networks offer a small bandwidth. Users find often no more than 9600 bits per second where a 50 Kbytes Web site needs more than 40 seconds to load
- **Lower memory size:** Mobile devices have a RAM size of 16 to 64 MB whereas stationary computers come with 512 MB equipped

These disadvantages have a large impact on mobile Internet usage. Therefore, it is problematic to use the same solutions, in this case Web sites, for stationary and mobile devices. The Web sites should be adapted in order to be usable on a mobile device.

Web site adaptation can be done on the client or on the server. In the first case, the (non-adapted) page is sent to the client and adapted there. This can be done by extending the navigation facilities of the client. Typical solutions usually work with zoom capabilities (Bederson & Hollan, 2003) or reordering to show one part of a site. These solutions can also be found in most Web browsers designed for mobile devices today. However, the problem of scrolling through the site remains. Additionally, the bandwidth problem cannot be solved using this approach because the non-adapted page is sent completely to the client. Therefore, this chapter concentrates on the server site adaptation which is usually done by the Web administrator of the pages.

The remainder has the following structure: The next section gives some background information for adapting Web pages. Afterwards, the adaptation scenario is presented which shows the Web shop eSarine and the test environment. The following section shows the three adaptation solutions that were used for this test. In the comparison part, all three solutions are compared and some guidance is given. The Conclusion finishes the chapter and takes a look at future work.

BACKGROUND

When adapting pages both for mobile and stationary devices, the solution must fulfil the following two steps:

1. Identify if the client is a mobile or stationary device
2. Eventually generate the adapted page, afterwards send the page to the device

For both problems, different approaches (or combinations) already exist. In step one, the Web server has to determine if the client is a mobile device and needs the adapted page or not. For this problem, a number of approaches exist:

- **Use a different domain name/URL:** This is a simple solution that returns the problem to the user of the page. The non adapted pages are returned when a default URL is requested (i.e., http://www.sport1.de), the adapted pages are sent when a different URL is requested (i.e., http: //pda.sport1.de). The major problem of this approach is that the user has to know that there are specialized pages. This can be achieved by adding a special entry page where the user can choose the URL.
- **Use a client cookie:** The solution of cookie setting is usually implemented together with the customization approach (see following description of adaptation solutions). The user can choose which Web elements he wants to retrieve on the client. Afterwards, his choice is stored on the client by putting this information in a cookie and sending it to a client device. Using this approach, the user can have a different look on a stationary and mobile device. This solution works only if the client accept cookies.
- **Parse the HTTP string:** Whenever a Web browser is requesting a Web site from a Web server, it sends some client information to the Web server. This typically includes the operating system and the Web browser. Using this information, the Web server can try to determine the client. This approach has two disadvantages: The user can edit this information and some browsers do not send enough information for a correct determination.
- **Use CSS media types** This approach will be presented in more detail in 4.2. In fact, the automatic determination is one of the advantages of solution 2.
- **Retrieve client profiles:** The Mobile Web Initiative from the W3C aims to define a standard to support the Web service for mobile devices. For the detection of the client, they proposed the Composite Capability/Preference Profiles (CC/PP) (Klyne et al., 2005). These profiles are sent from the mobile device to the Web server and can be used to identify the client device and to specify the user preferences. CC/PP defines some common attributes, for example the number of pixels of the display or the ability to show colors. It can be extended by further attributes (i.e., location information). However, right now only very few browsers support this profile.

Besides the presented three solutions in this chapter, the adaptation of Web pages can be done using the following approaches:

- **Try to create a page that works well on all devices:** The W3C has released the Authoring Challenges for Device Independence (ACDI) document that deals with Web site adaptation for different devices. It provides information on how authors of Web pages should define adaptable Web pages.
- **Use a proxy:** Some researchers propose to use a special Web server, a so called proxy, that acts as an intermediary for

mobile devices. The proxy retrieves a complete Web site but delivers only a predefined part of it to the mobile client. Note that this approach does not solve the question on how the predefined part should be extracted.

- **Let the user configure the page:** Customization (Lei, Motta, & Domingue, 2003) is another approach that can be used to solve the small display problem and to a further extend also the bandwidth problem when Web pages are adapted for mobile devices (Steinberg & Pasquale, 2002). This approach lets the user define a personalized page by providing an online editor comparable to a graphical user interface (GUI). Some Web sites already offer a way for a user to configure a Web site and to apply a special design to it. When the user enters the site, it is presented using the predefined style. One example is the Excite search engine that offers a "My Excite Start Page."
- **Try to reorder the page:** Another approach deals with the reordering of a large Web page by defining elements on the page and letting them display in a special look. An element could be a search bar containing a search input object and a button or navigation bar. An element could be displayed in another way, for example by using special features if the client supports these features, or by displaying the objects in a tab row (Magnusson & Stenmark, 2003).
- **Use personalization:** Personalization (Vassiliou, Stamoulis, Spiliotopoulos, & Martakos, 2003) usually goes in combination with other presented approaches (Anderson, Domingos, & Weld, 2001). Personalization helps to find out which Web elements on a page are needed by the user and which not. This information is used by all approaches that try to generate the pages dynamically. Two examples are customisation (Coener, 2003) or the CSS approach (solution 2) (Stormer, 2004).

ADAPTING PAGES FOR ESARINE

eSarine

The eSarine online shop is designed to offer goods of any kind on the Internet (Werro, Stormer, Frauchiger, & Meier, 2004). It is developed in Java using the Model 2-based Struts framework (Husted, Dumoulin, Franciscus, & Winterfeldt, 2003). Like most Web shops, eSarine is divided into a storefront and a storeback. In the storeback, the whole Web shop can be managed, including products, users, and payment. In the storefront, the products and services are offered to the customers.

The main advantage of eSarine is the modular architecture and the use of the Struts framework to part the business logic from the view part. Therefore, it is a good platform to test different approaches as only the view part has to be adapted.

Adaptation Scenario

The aim of this chapter is to describe how two eSarine pages were adapted to mobile devices using three different approaches. Figure 1 shows both pages on a stationary device. As you can see, the product list site (top) itemizes different products among and beneath each other. This can be the result of a product search or a navigation using categories. For each product, a small picture as well as a short description is presented. The "more" link at the end of the description can be used to navigate to the detailed product view site (bottom). This page presents much more information of one prod-

Figure 1. The (non-adapted) pages on a stationary device

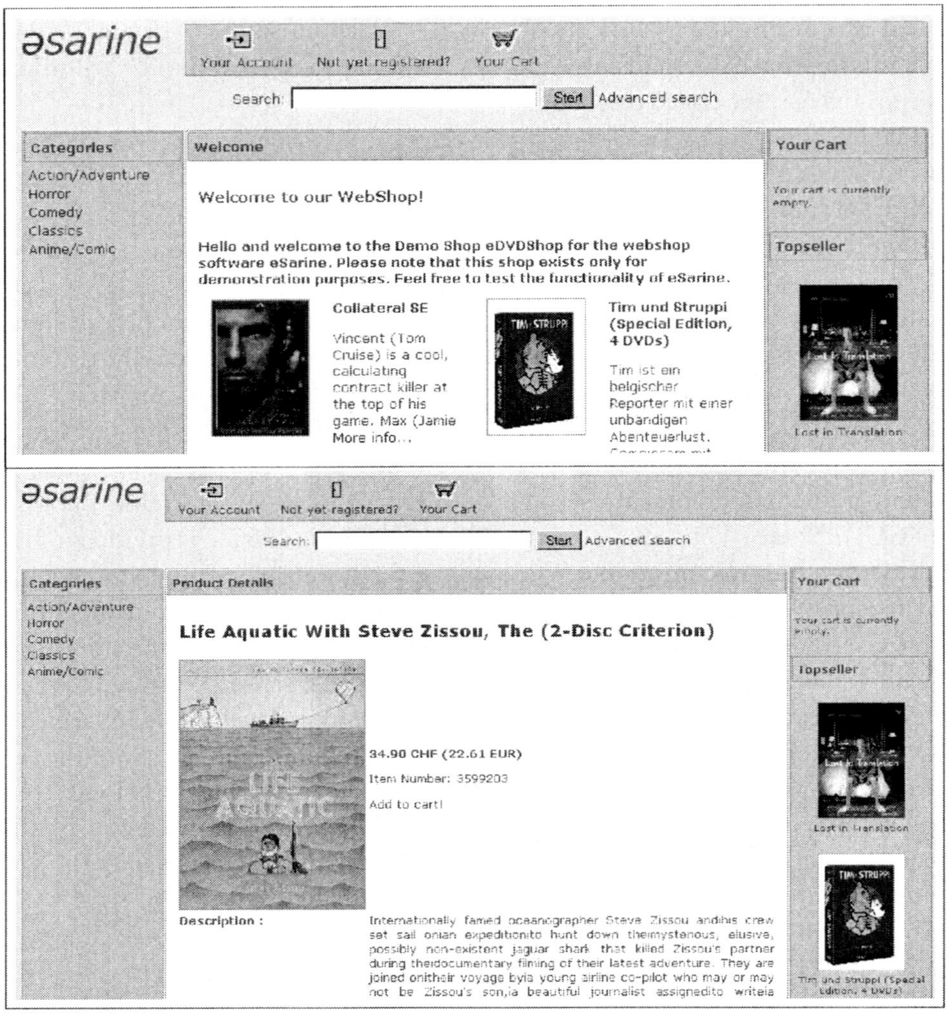

uct. Both pages are typical and can be found in almost all online shops.

The test was done using two different mobile devices. The first device is a Siemens S65 cell phone running a siemens self-developed operating system, the second one a QTec 8080 running Windows smart phone 2003. The Siemens S65 is equipped with an Openwave (www.openwave.com) Mobile Browser Version 7.001, the QTec runs the popular Opera (www.opera.com) Mobile Browser Version 7.60 beta 3.

Figure 2 shows the non-adapted pages on the Siemens (left) and QTec (right). The red rectangle shows the display size of the mobile devices. Both devices try to format the page by ordering the elements among each other to avoid horizontal scrolling. This leads to the strange menu presentation of the Siemens. Additionally, not all style sheet commands are

Figure 2. The non-adapted pages on the Siemens (left) and QTec (right) mobile devices. Both mobile browser have problems displaying the pages correctly.

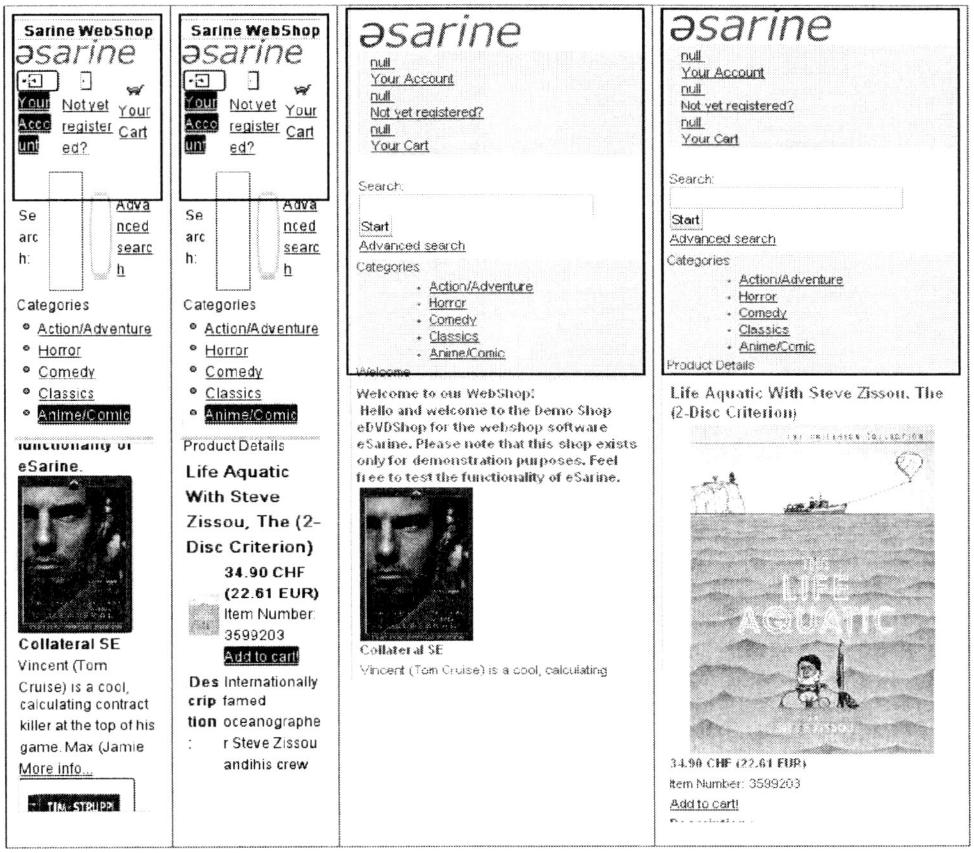

interpreted by the browsers, for example the list bullets are not hidden (*list-style:none*).

For the adaptation of the pages, three different solutions are presented that will be shown in detail in the next section.

Three Different Solutions for Adapting Web Pages

Solution 1: Rewrite the Page

Rewriting the page is the simplest form of adapting the page to a mobile device. The first pages available where rewritten using special languages like the Wireless Markup Language (WML) or compact HTML (cHTML). However, these languages did not have a large success. Because of the growing ability of mobile devices to display HTML pages, this chapter concentrates on HTML. However, HTML pages must have a special design to be displayed well on a mobile device. The previous section showed the non-adapted pages, using HTML tables for layouting. This is quite common today, but not very elegant. In a first step, the original pages where copied and the table

Figure 3. The adapted pages look very nicely on both devices. The category is presented vertical, the search bar is correctly formatted and also the picture has a better size to improve the bandwidth.

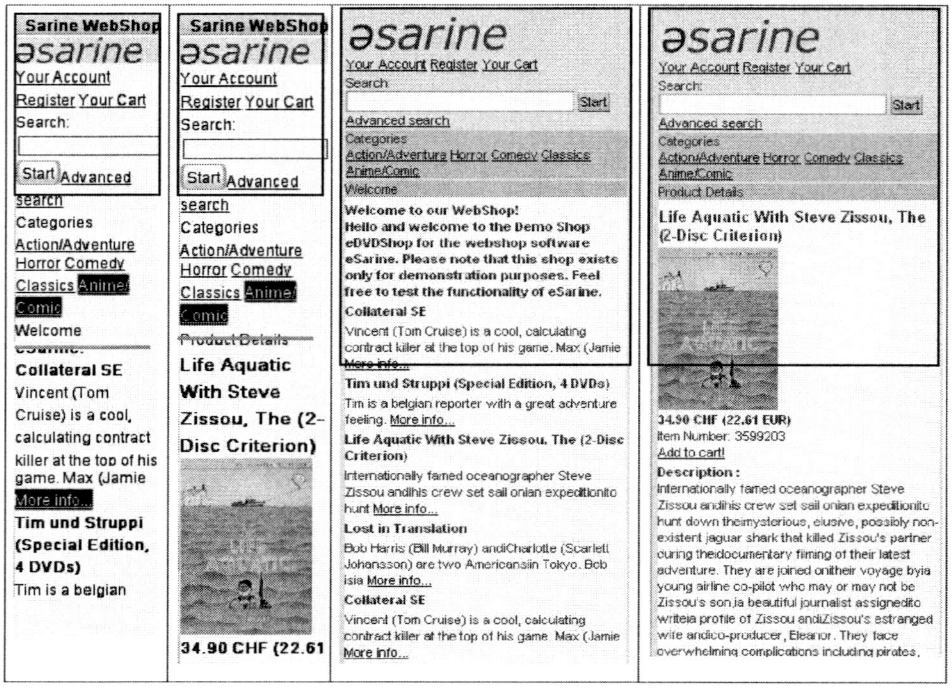

layout was replaced by using block elements. This works well on stationary and mobile devices. Additionally, the Top-Seller and Cart menu was deleted to save space and bandwidth. Further, all product images on the product list where removed and on the detailed product view, the picture was resized. The resulting pages are shown in Figure 3.

Solution 2: Adapt the Page

The World Wide Web Consortium (W3C) has developed a standard called cascading style sheets (CSS) (Bos, Celik, Hickson, & Lie, 2004; Lie& Bos, 1999). This technology can be used to adapt a Web site for mobile devices.

The first version (CSS level 1) (Lie & Bos, 1999) was developed in 1996 and is supported by a large number of current Web browsers. The main idea behind CSS is to part the content from the representation of a Web site. Older Web sites included the content and the representation information in one file. CSS can be used to move the representation to a new file, the CSS file. Typically, CSS files are included in the header's HTML file using the command:

```
<link rel="stylesheet" type="text/css"
href="layout.css" >
```

With this directive, the HTML file stores the representation information in the layout.css file.

The Web browser typically loads the HTML file first. Afterwards, the style information is received by loading the CSS file. In February 2004, the W3C has introduced a new version of CSS (CSS level 2.1) (Bos et al., 2004). This version supports so called media types to present a solution for adapting HTML pages to mobile devices. The idea is to create multiple CSS files, one for each device class. Then, the browser chooses the correct CSS file depending on the current device where it is executed. In the HTML file, all different CSS files are included. If this command:

```
<link rel="stylesheet" type="text/css"
href="stationary.css" media="screen">

<link rel="stylesheet" type="text/css"
href="mobile.css" media="handheld">
```

is inserted in the header of a HTML file, two different CSS files are included. If the browser is running on a stationary device, the stationary.css file is loaded. If it is executed on a mobile device, the mobile.css file is loaded. This solution was included in CSS to support the adaptation of Web pages to different devices. Right now, not only mobile devices are supported, there exist a list of more then 12 different media types for different devices. To use eSarine with CSS adaptation, the table layout of the two pages has first to be removed (the same step as in solution 1). Afterwards, a new mobile.css file was created and added to the header of the HTML files. Then, the pages were adapted by hiding the left and right panel. This can be done by adding a display:none entry in the CSS file. Additionally, the width of the search bar is reduced and the topmenu is refor-

Figure 4. With CSS, not all things can be done. Therefore, the resulting page looks more similar to the stationary one (in fact, it is the complete stationary page). However, be disabling some menues and reformatting other menus, the result is not bad, at least much better then the non-adapted pages.

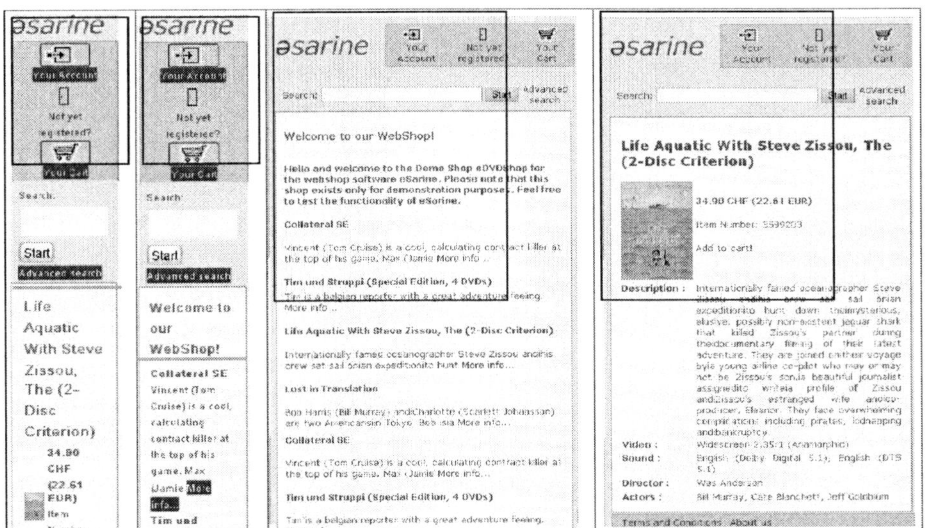

matted (all by adapting the mobile.css). The result is shown in Figure 4.

There is still a problem with the large image that is loaded and displayed. To overcome this problem, the following approaches can be used today:

- The client browser resizes the image. This can be achieved by adding a width: 10%; height: 10%; entry in the CSS file. This approach has some weaknesses: The image is still loaded completely by the client. Furthermore, the calculation for resizing the image takes time on a mobile device. Also, in our tests, not all browsers on mobile devices used today were able to resize the image properly.
- The image is not displayed by using a display:none entry. This approach has also some weaknesses: Typically, an online shop should display a product image. Furthermore, the image is still loaded completely by the client machine which is annoying because the CSS standard level 2.1 defines in paragraph 9.2.4 clearly that there is no need to load the image when it is not displayed: "This value causes an element to generate no boxes in the formatting structure."
- CSS can be used to display background images. If the width and height of the image is known, this approach can be used to move the inclusion of the image to the CSS file. Then, a smaller image can be included for the mobile device. In the example, the HTML <img src=...> entry is replaced by a <div class="product-img"/>. Then, in the CSS file, the class is defined for the stationary device:

```
.product-img{
background: url(12-1.jpg);
width:92px; height:130px;}
```

For the stationary device, the large product picture is included:

```
.product-img{
background: url(12-2.jpg);
width:200px; height:282px;}
```

The solution works fine on stationary and mobile devices. However, the inclusion of images in the HTML has to be changed and images with a different size require that new entries with the correct width and height values have to be inserted.

For eSarine, approach (3) was used, which can be seen by comparing the picture sizes of Figures 2 and 4.

This approach can be extended by choosing the elements to show or hide using personalization techniques (Stormer, 2004).

Solution 3: Use XML to Transform the Page

If the originating HTML page is written using the XHTML standard or directly created from XML sources, a further conversion is possible by using the Extensible Stylesheet Language (XSL) (Lie & Bos, 2004). This language provides mechanisms to parse an XML file. The following small XML document

```
<?xml version="1.0" encoding="ISO-8859-1"?>
<products>
<product>
<id>12</id>
<name>Collateral SE</name>
<short-description>
Vincent (Tom Cruise) is a cool, calculating
contract killer at the top of his game.
</short-description>
</product>
```

```
</products>
</xml>
```

can be transformed by an XSL transformator. As an input, the transformator needs an XSL document that provide the rules on how to do the formatting. An example XSL document would be:

```
<?xml version="1.0"?>
<xsl:stylesheet
xmlns:xsl=http://www.w3.org/1999/XSL/Transform
version="1.0">
 <xsl:template match="/">
  <html>
   <head>
    <style>body {font-size: 0.8em;}</style>
   </head>
   <body>
    <xsl:apply-templates select="products" />
   </body>
  </html>
 </xsl:template>
 <xsl:template match="products">
  <p>
   <xsl:apply-templates select="product" />
  </p>
 </xsl:template>
 <xsl:template match="product">
  <h1><xsl:value-of select="name" /></h1>
  <xsl:value-of select="short-description" />
 </xsl:template>
 </xsl:stylesheet>
```

The resulting file is nearly HTML compliant (to save space, some obligatory HTML elements like the DOCTYPE are not presented) and has the following structure:

```
<html>
<head>
<style>body {font-size: 0.8em;}</style>
</head>
<body>
<p>
<h1>Collateral SE</h1>
Vincent (Tom Cruise) is a cool, calculating
contract killer at the top of his game.
</p>
</body>
</html>
```

eSarine does not create XML documents by default. Instead, it uses Java Server Pages (JSP) to generate the resulting HTML pages. Therefore, the first step was to replace the JSP part with an XSL transformator. It is also possible to rewrite the JSP pages for the creation of another view. However, the usage of XML was decided to show the power of XSL which is used in more and more Web applications. To combine Struts with XML, stxx was used. Stxx (stxx.sourceforge.net/) is a Struts extension to support XML transformation using XSL without changing the functionality of Struts. Additionally, the Struts action files had to be changed, because with stxx, XML files are generated instead of JavaBeans. Stxx can be used on top of the already existing classes. For the tests, only the two sites where changed.

Fortunately, XML documents are already created to export product information in product catalogues. The generation of XML documents was realized using these methods. Then, for both pages, XSL documents where written that transformed the XML documents to HTML (like the example above).

Basically, XSL is powerful and in combination with the XML generating action part, it is possible to do everything. It was possible to generate exactly the stationary and mobile pages of solution 1 (c.f., Figure 3), but this time only one base and two XSL processors have to be managed. However, if the XML document and the resulting HTML file differ strongly, the XSL document becomes quite large.

SOLUTIONS COMPARISON

All three solutions where implemented using eSarine. Figure 5 shows the differences between the solutions.

The less complex method is to use solution 1 and rewrite the page. However, this is only preferred if either the pages are static or the application does not have the preconditions for solutions 2 or 3.

Solution 2 fits well if the application is not Model 2-based or does not have an XML output. Typical examples are (older) PHP scripts or simple Content Management Systems. The integration of Solution 2 is quite easy, however the result is somewhat limited.

Solution 3 promises the most flexible way to adapt pages, but on the other hand requires some preconditions. If an XML output is already provided by the Web application, this solution is best. Because of these advantages, new application developers should think about a Model-2 architecture that uses XML output. The tested Struts environment using the stxx module is a good choice.

CONCLUSION

Adapting Web sites for mobile devices will become more and more important in the future. This chapter should help to decide which solution to use when an adaptation is to be done.

The customization approach, which was described in the related work part was not included in the comparison. This is because of

Figure 5. Differences between the presented solutions

	Solution 1 Rewrite	Solution 2 CSS	Solution 3 XML
automatically determine if client is stationary or mobile	not directly possible	integrated in solution	not directly possible
complexity of solution	no complexity	little complexity	high complexity
preconditions for the stationary Web server	none	special layout, eventually custom-build picture inclusion	must use XML generation, best with Model-2 architecture
maintenance costs	high	low	middle
integration of other languages like cHTML or WML	possible by adding new pages	not possible	easy
possibilities for adaption	boundless	limited	boundless
bandwidth reduction	full	limited	full

the large effort needed to enable customization in a Web application today. However, it is planned to add this feature to eSarine in the future.

The tests have also shown that mobile devices and their ability to show pages differ. Therefore, another interesting addition, especially when solution 3 is used, is the creation of more than one mobile device page. This could be done by collecting parameters from the user or by parsing profiles like CC/PP (cf Background) when available. To gain parameters of the user's device, a small Web site could be presented where the user can enter more information.

REFERENCES

Anderson, C., Domingos, P., & Weld, D. (2001). Personalizing Web sites for mobile users. *Proceedings of the 10th International WWW Conference*.

Bederson, B. B., & Hollan, J. D. (1994). Pad++: A zooming graphical interface for exploring alternate interface physics. *Proceedings of ACM User Interface Software and Technology Conference (UIST)* (pp. 17-26).

Bos, B., Celik, T., Hickson, I., & Lie, H. W. (2004). *Cascading style sheets, level 2, revision 1* (Technical Report). World Wide Web Consortium (W3C).

Coener, A. (2003). Personalization and customization in financial portals. *The Journal of American Academy of Business*, *2*(2), 498-504.

Forrester. (2003). *Consumer-Technographics — Study Europe Benchmark*. Retrieved from http://www.forrester.com

Husted, T., Dumoulin, C., Franciscus, G., & Winterfeldt, D. (2003). *Struts in action*. Greenwich: Manning Publications Co.

Klyne, G., Reynolds, F., Woodrow, C., Ohto, H., Hjelm, J., Butler, M. H., & Tran, L. (2005). *Composite capability/preference profiles (CC/PP): Structure and vocabularies* (Technical Report). World Wide Web Consortium (W3C), Working Draft

Lie, H. W., & Bos, B. (1999). *Cascading style sheets, level 1* (Technical Report). World Wide Web Consortium (W3C).

Lie, H. W., & Bos, B. (2004). *Extensible stylesheet language (xsl) version 1.1*. (Technical report). World Wide Web Consortium (W3C)

Lei, Y., Motta, E., & Domingue, J. (2003). Design of customized Web applications with OntoWeaver. *Proceedings of the K-CAP'03*.

Magnusson, M., & Stenmark, D. (2003). Mobile access to the Internet: Web content management for PDAs. *Proceedings of the 9th Americas Conference on Information Systems (AMCIS)*.

Steinberg, J., & Pasquale, J. (2002). A Web middleware architecture for dynamic customization of content for wireless clients. *Proceedings of the International World Wide Web Conference (WWW 2002)*, Honolulu, Hawaii, USA.

Stormer, H. (2004). Personalized Web sites for mobile devices using dynamic cascading style sheets. *Proceedings of the 2nd International Conference on Advances in Mobile Multimedia (MoMM)*.

Vassiliou, C., Stamoulis, D., Spiliotopoulos, A., & Martakos, D. (2003). Creating adaptive Web sites using personalization techniques: A unified, integrated approach and the role of evaluation. In N. V. Patel (Ed.), *Adaptive evolutionary information systems* (pp. 261-286). Hershey, PA: Idea Group Publishing.

Werro, N., Stormer, H., Frauchiger, D., & Meier, A. (2004). eSarine—A Struts-based Web shop for small and medium-sized enterprises. *Proceedings of the Workshop Information Systems in E-Business and E-Government (EMISA).*

KEY TERMS

CSS: Cascading stylesheet (CSS) can be used to separate the content and design of a Web page. The content is defined in the HTML file, the design is done using a CSS file.

Dynamic Web Site Generation: With dynamic Web site generation, a Web application generates the resulting Web site dynamic. Different parameters can be used for this generation step.

eSarine Web Shop: The eSarine Web shop was developed by the IS search group of the University of Fribourg. It is a Java-based Web application that can be used to offer products and services on the internet.

Mobile Device: A mobile device is a small and lightweight computer that can be carried by the owner easily. The most popular mobile device is the cell phone.

Server Side Solutions: server side solutions are used by a Web application to generate different Web sites for different devices by the server. In contrast, a client side solution always sends the same Web page. The page is then transformed by the client to fit the device needs.

Web Site Adaptation: Web site adaptation defines the process of generating a Web site for different client devices.

XSL: The Extensible Stylesheet Language (XSL) contains a number of tools for handling XML documents. It can be used to parse and transform XML documents.

Chapter XX
Ensuring Task Conformance and Adaptability of Polymorph Multimedia Systems

Chris Stary
University of Linz, Austria

ABSTRACT

This chapter shows how specifications of mobile multimedia applications can be checked against usability principles very early in software development through an analytic approach. A model-based representation scheme keeps transparent both, the multiple components of design knowledge as well as their conceptual integration for implementation. The characteristics of mobile multimedia interaction are captured through accommodating multiple styles and devices at a generic layer of abstraction in an interaction model. This model is related to context representations in terms of work tasks, user roles and preferences, and problem-domain data at an implementation-independent layer. Tuning the notations of the context representation and the interaction model enables, prior to implementation, to check any design against fundamental usability-engineering principles, such as task conformance and adaptability. In this way, also alternative design proposals can be compared conceptually. Consequently, not only the usability of products becomes measurable at design time, but also less effort has to be spent on user-based ex-post evaluation requiring re-design.

INTRODUCTION

Mobile and wireless applications provide essential benefits for their users (Siau & Shen, 2003): They allow them to do business anytime and anywhere; Data can be captured at the source or point of origin, and processed for simultaneous delivery in multiple codalities and ways, including multimedia applications. The emergent utility of those applications is based on the flexible coupling the business the logic of applications with multiple interaction devices and styles (Nah, Siau, Sheng, 2005). Consequently, developers have to:

1. Construct multiple user interfaces to applications
2. Ensure user-perceivable quality (usability) for each user interface

The first objective means for multimedia applications not only the development of stationary user interfaces, but also the provision of user interfaces for a diverse set of devices and related styles of interaction. A typical example for this setting is the access to public information systems via mobile devices (WAP (wireless application protocol)-based cell phones, palmtops, PDAs (personal digital assistants), tablet PCs, handhelds, etc.), as well as stationary user-interface software, such as kiosk browsers at public places (airports, cine-plexes, malls, railway stations, etc.). Hence, the same data (content) and major functions for navigation, search, and content manipulation facilities should become available through various presentation styles and interaction features.

This kind of openness while preserving functional consistency does not only require the provision of different codalities of information and corresponding forms of presentation, such as text and audio streams for multimedia applications, but also different ways of navigation and manipulation. For instance, WAP is designed to scale down well on simple devices, such as the mobile phone with less memory and display capacity. Due to the nature of the devices and their context of use, it is not recommended to transfer Web application directly to WAP applications without further analyses. For WAP applications, it is highly recommended to use menus as much as possible to save the user from using the limited keypad facilities. In addition, any text should be short and consistent in its structure. Wording should be easy to understand, and fit within the display. Finally, obligatory user-typed data input, such as providing telephone numbers through touch-typing, should be avoided, since entering text or numbers on a phone keypad is a difficult and error prone operation (Arehart et al., 2000). Besides these usability issues, accessibility matters (e.g., to support aging users) (Ziefle & Bay, 2005)

Here the second objective comes into play: It is the set of users that decide primarily whether a product is successful or not. In more detail, it is the user-perceived quality (in terms of usability principles, such as adaptability) that has to be ensured through development techniques and tools. They have to comprise or include some measurement of usability principles at design time to avoid time- and cost-consuming rework of coded user interfaces based on the results of a posteriori evaluation.

When setting the stage to meet both objectives, we have to define design representations (i.e., models) that include application context. The design knowledge represented in this way can then be processed to check the implementation of generic properties of usability principles. Such an approach recognises, both, the need for:

- Techniques and tools ensuring the openness of the application logic towards various interaction devices and styles while preserving functional consistency
- Early usability testing—focusing on algorithms checking design representations against generic characteristics of quality parameters. For instance, the shortest paths (from the organizational perspective) in a menu-tree should be specified in

a design representation, in order to achieve task conformance

Although the subjective measurements of user interfaces cannot be replaced in this way, an indicative answer can be provided to the designer's question "Will the product be usable from the perspective of work task accomplishment for a target user population?" In that sense, designers can receive feedback without involving users from their first proposal on. In case this analytic evaluation is not successful, additional design variants can be developed. Designers might also compare various designs analytically for selection before users are involved. Analytical a-priori testing addresses design time and means both, usability testing before implementation and before user involvement.

The chapter is structured as follows: Subsequently, the concept of a priori-usability testing with respect to mobile multimedia computing is detailed. Several requirements for the representation and the processing of design knowledge are set up. Some related work from usability engineering and model-based user interface development to meet the identified requirements is reviewed in the follow-up section. Then a dedicated representation scheme and its capability to support testing task conformance and adaptability is reviewed. These two usability principles have been selected due to their close relationship: (1) to the context of applications (i.e., the tasks users want to accomplish), and (2) to the characteristics of polymorph applications requiring the openness of an application logic to various user interfaces. In the final section of this chapter a summary and research outlook are given.

A-PRIORI-USABILITY TESTING AND MOBILE MULTIMEDIA APPLICATIONS

According to ISO DIS 9241-11 "*usability* is defined as the extent to which a product can be used by specified users to achieve specified goals with effectiveness, efficiency and satisfaction in a specified context of use" (ISO DIS 9241-11, 1997), thus, being in line with the emergent utility of mobile applications (Siau et al., 2003). The *context of use* is defined in ISO (1997) as "the nature of the users, tasks and physical and social environments in which a product is used." The notion of context has been recognised as a crucial design factor for any type of computational system (Abowd, & Mynatt, 2000; Norman, & Draper, 1986, Stephanidis et al., 1998). However, in the majority of software developments context is not represented explicitly, and bound to what is normally encountered by "typical" users (Nardi, 1996). This normative and functional perspective does not allow for testing usability in the course of development prior to some kind of implementation. Traditional development techniques rather enforce ex-post evaluations through checking non-functional or usability principles.

The concept of software quality helps to bridge the gap between function-oriented design and ex-post evaluations based on non-functional parameters stemming from usability principles. Although the term has been introduced with various meanings (Bevan, 1995; Garvin, 1984), there have been attempts to embody quality management into the construction process of software (e.g., ISO, 1987, 1991, 1994). As such, software quality entails the consideration of functional and non-functional attributes. The latter characterise the use of technology artefacts by humans. Hence, the notion of quality goes beyond functional performance measures, such as interoperability. It focuses on principles such as task conformance, which are not measurable by traditional techniques from software engineering when looking at performance parameters (Dumke, Rautenstrauch, Schmietendorf, & Scholz, 2001).

Since the usability of a product is not an attribute of the product alone, but rather an attribute of the interaction with a product in a context of use (see previous and also Karat, 1997), for design and testing several content elements have to be represented and processed (Adler & Winograd, 1992):

- User characteristics
- The organisation of the environment (tasks and problem domain data) (e.g., specified as a set of goals in a specified context of use) such as work or edutainment
- Technical features (the user interface)
- Their intertwining

In case of polymorph applications different users utilise different devices and interaction styles to access common information and functions operating on and processing that information. Hence, the development of polymorph multmedia applications implies a construction process that allows for dynamic adaptability to various devices and interaction styles. It emphasises polymorphism with respect to user interaction as a design concern, in contrast to construct user interfaces as an afterthought.

To this end, it is important that not only the needs of the possible user population (establishing part of the context of use), but also the technological capabilities to present and manipulate information are taken into account in the early design phases of (new) products and services, and thus, become explicit part of design representations. Such representations might be processed to indicate whether the tasks and/or users of an application has been designed for can be supported or not. Hence, a-priori-usability testing aims for processing design representations to avoid lengthy rework of already programmed applications.

Such kind of processing might also help to overcome the common misconception of the development process of highly distributed and interoperable software: Mobile or Web applications are assumed to be accessible for "all" due to the common user-interface features provided by browsers. Empirical results show that Web designers are still typically developing Web applications for marketing appeal rather than for usability provision (Selvidge, 1999). However, if usability deficiencies cannot be removed, neither the acceptability by the majority of users nor the sustainable diffusion of multimedia applications in e-markets seems very likely (Silberer & Fischer, 2000).

Taking care about usability makes it an inherent part of any mobile multimedia product development process. According to Rubin (1994) testing usability has to occur at several stages of development (see Figure 1 — dotted lines) to lead to a user-oriented system. Developers learn through empirical evidence and shape the product to finally meet the users' abilities, expectations, and attitudes.

Subsequently, it is suggested to complement the exploratory and assessment tests with automated analytic tests of high-level and detailed design specifications against generic properties of usability principles *prior* to the construction of a product. As a result, the inputs to empirical tests are already the result of previously performed analytical measurements (stemming from processing design representations). This processing, however, requires the provision of analytical measures derived from quality-in-use attributes, such as task conformance. Accurate measures have to be provided for each usability principle (i.e., quality-in-use attribute)

Figure 1. Product development stages including usability-testing activities

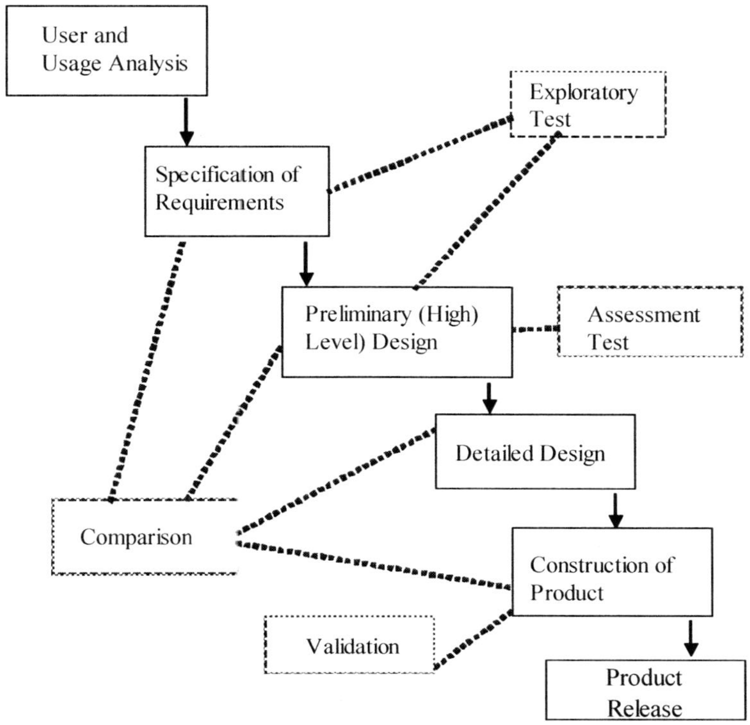

for analytical testing. They have to be set in relation to domain- and context-specific design elements, such as work tasks.

RELATED WORK

Decoupling the application logic from the user interface has not only tradition in software engineering (however, in terms of adding user interface software after functional development), but also in Web engineering and distributed computing (Coulouris, Dollimore, & Kindberg, 1995). However, these approaches are related closer to performance than to usability engineering. Hence, in the following some fundamental approaches to represent non-functional design knowledge, and to process this type of knowledge, are reviewed.

As a general principle it has been stated that good user interface design can only be achieved through taking fully into account the users' previous experiences and the world they live in (e.g., Beyer & Holtzblatt, 1998; Hackos & Redish, 1998; Nielsen, 1993; Norman, 1998). However, there have been different approaches to implement these ideas: at the level of design support, and at the level of software architecture.

With respect to design support, one approach has been to address design work and to set up and check dedicated information spaces, so-called design spaces. The questions, options, and criteria (QOC)-notation (MacLean,

Young, Bellotti, & Moran, 1991) and a corresponding procedure helping to formalise and record design decisions documents the rationale of design decision making. It is based on structuring relationships between options and their context of use, namely through making explicit the criteria for evaluating the options. As such, QOC allows open product development (as required for polymorph multimedia computing) and intertwining functional and non-functional properties of software. Although the principles become transparent, there is no direct relationship to high-level and detailed design representations. Moreover, this kind of approaches does not refer to operational definitions of usability principles so far.

ISO DIS 13407 (see also Bevan, 1997, p. 6) provides guidance on achieving quality in use by incorporating user-centred design activities throughout the life cycle of interactive computer-based systems. There are 4 user-centred design activities that need to take place at all stages during a project, namely to:

- Understand and specify the context of use
- Specify the user and organisational requirements
- Produce design solutions
- Evaluate designs against requirements

The latter refers to the usability-testing scenario shown in Figure 1 and the objective of this work. Experiencing the need for linking task requirements and software development (Rosson & Alpert, 1990; Rosson & Carroll, 1995) several projects have been started to tackle this issue from the methodology and tool perspective. For instance, in the Adept project (Johnson, Wilson, Markopoulos, Pycock, 1993; Wilson & Johnson, 1996) different models have been identified, namely for task definition, the specification of user properties and the user interface. The task model does not only comprise a representation of existing tasks, but also envisioned ones. An (abstract) interface model contains both, guidelines for feature design, and the features required to implement the envisioned tasks through a GUI.

Other approaches incorporated evaluation activities, such as Humanoid (Luo, Szekely, & Neches, 1993). This model-based approach starts with a declarative model of how the interface should look like and behave (in the sense of the above mentioned envisioned task model), and should be refined to a single application model that can be executed. Each user interface is handled according to five semi-independent perspectives, namely:

(1) the application semantics which is captured through domain objects and operations
(2) the presentation part emphasising the visual appearance of interface elements
(3) the behaviour part capturing the input operations(e.g., mouse clicks) that can be applied to presented objects, and their effects on the state of the application and the interface
(4) constraints for executing operations that are specified through dialogue sequencing
(5) triggers that can be defined through specifying operational side-effects

The lifecycle in Humanoid corresponds to iterations of design-evaluation-redesign activities based on interpretations of the executable model of the interface. Although such kind of model-based approaches took up the idea of viewpoint-driven specification, as detailed in Kotonya (1999), they still lack operational means for analytic usability testing.

Software-engineering projects in the field of polymorph multimedia computing do not focus on different perspectives on design knowledge at all. They rather focus on modular or layered software architectures for multi-modal interaction handling. With respect to multi-modality the technical integration of signals of different sources (e.g., Cohen et al., 1997), seems to be at the centre of interest rather than conceptual or methodological issues for development. The Embassi-project (Herfet, Kirste, & Schnaider, 2001) provides a generic architecture to aggregate modalities dynamically and switch between them. Although Embassi supports a variety of devices, and consequently, polymorph application development, it does not support designers in evaluating their proposals or comparing design alternatives.

ANALYTIC USABILITY TESTING

As already mentioned analytic a-priori-usability testing requires (1) design representations capturing the context of use, and (2) algorithms which implement operational definitions of non-functional usability principles. In order to meet both objectives we will use the experiences from model-based interactive software development as already indicated in the previous section. In the first sub section the model-based representation scheme for the design of polymorph applications is given. The subsequent section introduces the algorithms for the operational definitions of task conformance and adaptability.

Representing Design Knowledge

Following the tradition of model-based design representations (Puerta, 1996; Stary, 2000; Szekely, 1995) several models are required to define (executable) specifications of interactive software:

- **A *task* model:** That part of the situational setting (organisation, public domain, etc.) the interactive computer system is developed for
- **The *user* model:** The set of user roles to be supported, both, with respect to functional roles (e.g., search for information), and with respect to individual interaction preferences and skills (e.g., left-handed pointing)
- **A *problem domain (data)* model:** The data and procedures required for task accomplishment
- **The *interaction (domain)* model:** Providing those interaction styles that might be used for interactive task accomplishment

Of particular interest for polymorph multimedia system development is the separation of the interaction model from the problem domain (data) model, as an application should be separated from the set of possible interaction devices and interaction styles to allow the dynamic assignment of several devices and styles to application functions. The separation of the task and user model from the data and interaction model is also of particular importance, since it enables to implement an operational definition of task conformance and adaptability, and thus, analytic evaluation.

The task model provides the input to specify software designs and enables to derive the user and problem-domain data model. For user-interface development, devices and modalities for interaction have to be provided (e.g., in terms of GUI-platform-specific specifications). An integrated structure and behaviour specification (sometimes termed *application* model since it captures the integrated perspective on the different models) then provides a sound basis to check usability metrics.

With respect to the context of use, an explicit and high-level representation of (work) tasks, such as looking for film before going to a cinema, facilitates designing the relationship of functional elements of an application (such as a search function on a film database) to generic interaction features, such as browser buttons for information retrieval. The categorisation of tasks has been initiated by Carter (1985). He has grouped user-oriented system functions according to their generic (task) capabilities in interaction distinguishing functions for control in/output from those for data in/output. The group of functions for control contains execution functions and information functions. It comprises activities that are session-specific (execution functions) and situation-specific, such as user guidance, help, and tutoring (information functions). The group of functions for data manipulation contains so-called data functions and information functions. Data functions comprise all functions for the manipulation of data structures and data values, such as insert. Information functions as part of data functions address user task-related activities, such as searching for information.

This framework is helpful for designers and evaluators, since it provides a level of abstraction that allows to refine design knowledge as well as to evaluate usability principles for a particular application domain.

The information functions as part of data functions can further be refined to particular sub task types. According to Ye (1991) information search can be classified into two major categories: known-item and unknown-item search. In a known-item search, users know precisely the objects for which they are looking for, whereas in an unknown-item search they do not know exactly the items they are looking for. Both types might be relevant for applications accessible for a diverse set of users. In a typical unknown-item search applications suggest a set of terms that is assumed to be relevant for (a variety of) users or in a particular application domain. Unknown-item search poses a greater challenge to the user-interface designer than known-item search, in particular in cases where a search task requires users to traverse multiple search spaces with a limited knowledge about the application domain. That challenge requires to rethink the representation and navigation through the search and information space of an application.

While performing a task analysis for Web applications Byrne et al. (1999) came up with a Web-specific taxonomy of user tasks. Based on research in the field of information finding and termed "taskonomy," it comprises six generic tasks: Use information, locate on page, go to page, provide information, configure browser, and react to environment. Each task contains a set of sub tasks related to the task. For instance, "provide information" captures provision activities such as search string, shipping address and survey response. Such a taskonomy can be utilised in the course of model-based design and the construction of task-based and adaptable user interfaces. It also serves as a reference model for evaluation, since the tasks to be performed are understood from the perspective of users and usage rather than from the perspective of technology and implementation.

How can such taxonomies be embedded in design representations? In the following a solution is proposed that results from our experiences with the model-based TADEUS-environment and design methodology (Stary, 2000) and its successor ProcessLens. Following the tradition of context-aware and multi-perspective user-interface development four different models can be identified: Task, (problem domain) data, user, and interaction (domain) model. In the course of design these models have to be refined and finally, migrated to an application model. Each of the models is composed of a structure and behaviour specification using the same object-oriented notation facilitating the migration process.

Figure 2. Generic structure specification of GUIs and browsers

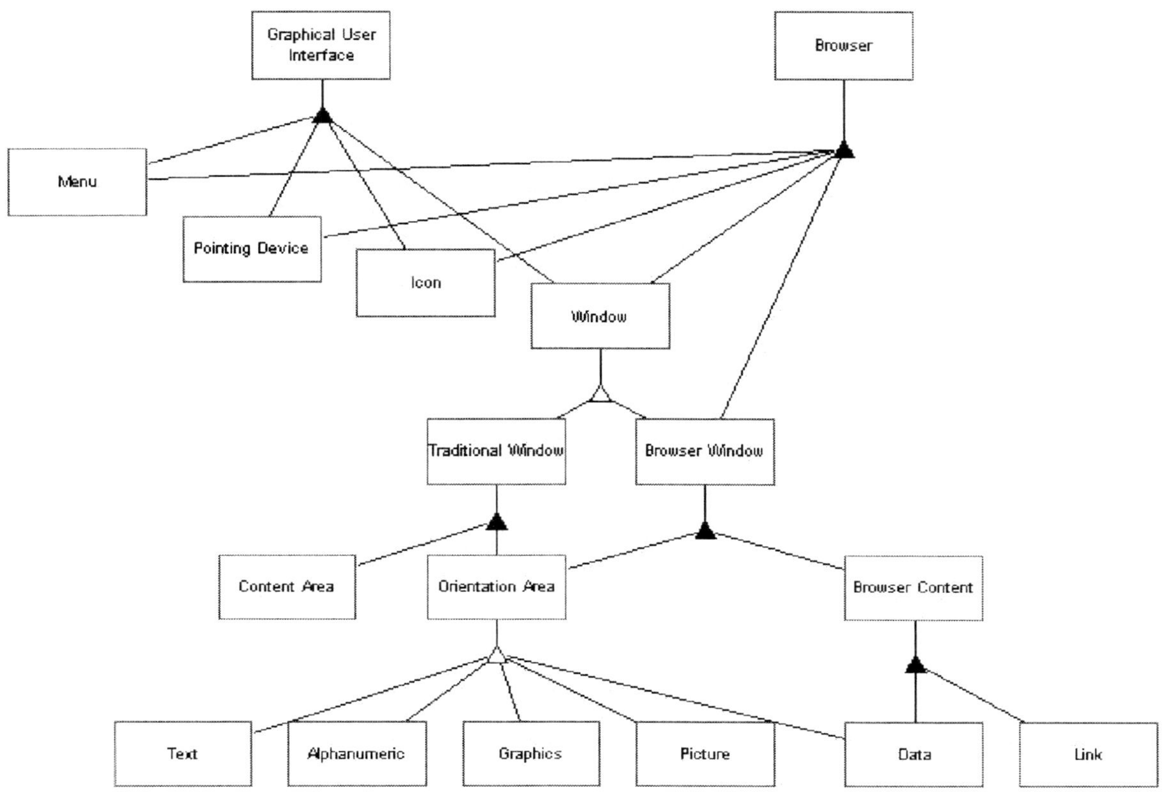

Taskonomies for applications can be embedded at different levels of specification, namely at the level of task specification, and the level of interaction specification. At the level of task specifications generic task descriptions can be denoted as context of use. Assume the search for a film. At a generic level for interactive search tasks two ways to accomplish the search can be captured (e.g., either search via newsgroups or search via the Web). In the first case, a request is posted to a newsgroup, in the latter case an URL and options are provided (see also Figure 6). This example demonstrates one of the core activities in designing user interfaces at a generic level and in evaluating their design specification analytically. The design specification contains both behaviours, the newsgroup posting and the Web search. The evaluation procedures checks whether there is more than a single path in the behaviour specification that enables the search and leads to identical task results.

Since users might have different understandings of how to accomplish tasks, not only a variety of different procedures might lead to identical (work) results, but also the (visual or acoustic) arrangement of information and the navigation at the user interface might differ significantly from user to user and from device to device. This brings us to the second level of utilising taskonomies for mobile multimedia computing: the level of interaction. Figure 2 shows a sample of generic structuring of interaction facilities. In this case, GUI elements and browsing facilities have been conceptually integrated. The boxes denote classes, the white triangles aggregations, the black ones specialisations.

The dynamic perspective (i.e., behaviour specification can also contain generic information) namely navigation and interaction tasks. Some of the entries might become part of an application specification, as shown in the state-transition diagram in Figure 3 for the specification of the search task-based on selected items from the browser taxonomy. The direct-surf branch specifies a-kind-of known-item search.

Switching between modalities of interaction is facilitated through generic task specification at the interaction level. Assume the user interface should be able to provide access through browsing for visually impaired and visually enabled persons. In that case, taskonomies of browsing systems could capture mapping of function keys to the basic interaction features for browser-based interaction, according to empirical data (e.g., Zajicek, Powell, & Reeves, 1998): F1—Load a URL etc. Using that kind of mechanism, modality—or style switching can be lifted to a more conceptual, since implementation—independent level of abstraction, and does not remain implicitly (since hard-coded). In this way, the adaptability of user interface technology can be achieved at design time (if not run time, in case the user-interface platform executes design specifications—see next paragraph), enriching the interaction space. Analytic a-priori-evaluation procedures can immediately process that knowledge.

Processing Design Knowledge

The crucial part in order to achieve automated analytic evaluation of design representations is to define an operational metrics (i.e., to define algorithms which check whether usability crite-

Figure 3. Behaviour specification derived from a taskonomy of Web tasks

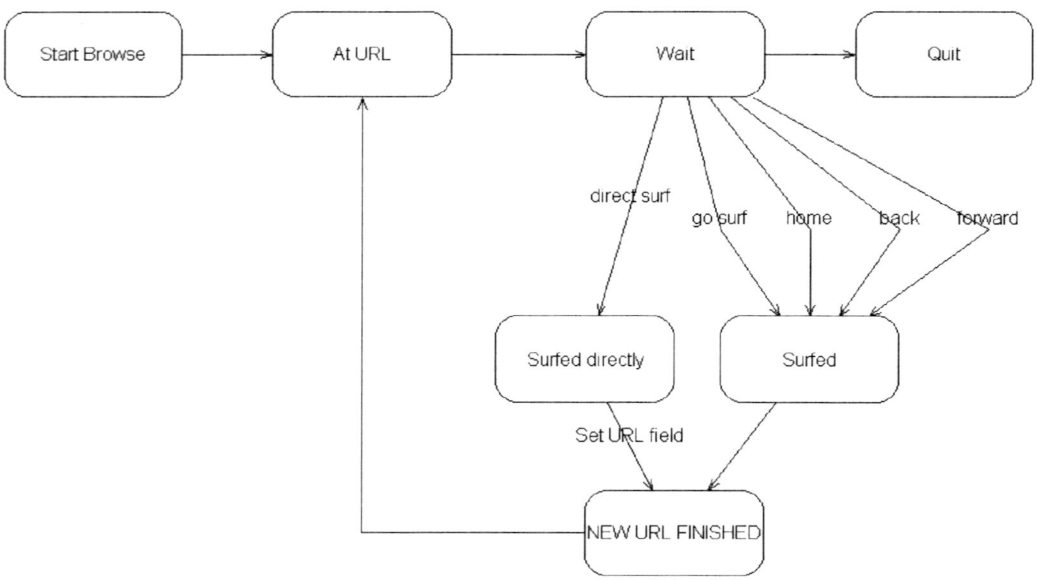

ria are met). In the following examples for operationalising, task conformance and adaptability are given.

Task conformance is defined as follows: "A dialogue supports task conformance, if it supports the user in the effective and efficient completion of the task. The dialogue presents the user only those concepts which are related to the task" (ISO 9241 Part 10). It can be measured in an analytical way through two metrics: completeness with respect to task accomplishment (effectiveness), and efficiency. The first metrics means that at the system level there have to be both, a path to be followed for user control (i.e., navigation procedures), and mechanisms for data presentation and manipulation, in order to accomplish a given task. The second metrics is required to meet the principle in an optimised way. High efficiency in task accomplishment can only be achieved, in case the users are not burdened with navigation tasks and are able to concentrate on the content to be handled in the course of task accomplishment.

In order to minimise the mental load the navigation paths have to be minimal. At the specification level an algorithm has to check whether the shortest path in the application model for a given task, device and style of interaction has been selected for a particular solution, in order to accomplish the control and data-manipulation tasks. TADEUS is based on the OSA-notation (Embley, Kurtz, & Woodfield, 1992), whereas ProcessLens uses UML (Rumbaugh, Jacobson, & Booch, 1999). Consider the case of arranging a cinema visit and its model-based design representation, as shown in Figure 4 (object relationship diagram ORD). The enriched notation allows tracing how the task is related to data which is a prerequisite for task accomplishment. It also allows the assignment to interaction styles, such as form-based interaction for acquiring data.

The model-based approach supports a variety of relationships, in order to capture relationships between models and model entities. For each of the relationships a corresponding environment provides algorithms and allows the creation of such algorithms to process further semantic relationships (Eybl, 2003; Vidakis & Stary, 1996). The relationships between objects and/or classes are used (i) for specifying the different models, and (ii) for the integration of these models. For instance, "before" is a typical relationship being used within a model (in this case, the task model), whereas "handles" is a typical relationship connecting elements of different models (in this case the user and the task model). In both cases algorithms specific for each relationship are used checking the semantically correct use of the relationships.

For processing, each model is described in an object-oriented way, namely through a structure diagram (i.e., an ORD (object relationship diagram)), and a behaviour (state/transition) diagram (i.e., an OBD (object behaviour diagram)). Structural relationships are expressed through relationships, linking elements either within a model or stemming from different models. Behaviour relationships are expressed through linking OBDs at the state/transition level, either within a model or between models. Both cases result in OIDs (object interaction diagrams). In Figure 5, the different types of specification elements at the model and notation level are visualised.

In order to ensure the semantically and syntactically correct use of specification elements, activities are performed both at the structure and behaviour level:

1. **Checking ORDs:** It is checked whether each relationship
   - Is used correctly (i.e., whether the entities selected for applying the relationship correspond to the defined

Figure 4. A sample task, role, and data model structure specification

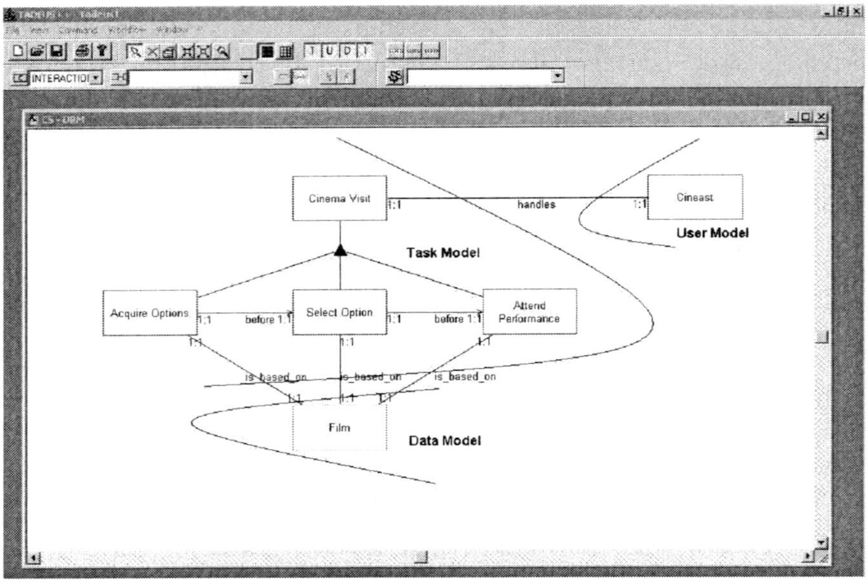

semantics) (e.g., task elements have to be given in the task model to apply the "before" relationship)

- Is complete (i.e., all the necessary information is provided to implement the relationship) (e.g., checking whether there exist task-OBDs for each task being part of a "before"-relationship.

2. **Checking OBDs:** It is checked whether the objects and/or classes connected by the relationship are behaving according to the semantics of this relationship (e.g., the behaviour specification of data objects corresponds to the sequence implied by "before").

The corresponding algorithms are designed in such way that they scan the entire set of design representations. Basically, the meaning of constructs is mapped to constraints that concern ORDs and OBDs of particular models, as well as OIDs when OBDs are synchronised. The checker indicates an exception, as long as the specifications do not completely comply to the defined semantics. For instance, the correct use of the "before" relationship in the task model requires to meet the following constraints: (1) at the structure layer: "Before" can only be put between task specifications, (2) at the behaviour layer: The corresponding OBDs are traced whether all activities of task 1 (assuming task 1 "before" task 2) have been completed before the first activity of task 2 is started. Hence, there should not occur synchronisation links of OIDs interfering the completion of task 1 before task 2 is evoked.

The same mechanism is used to ensure task conformance and adaptability. According to the ISO-definition above task conformance requires: (a) the provision of an accurate functionality by the software system to accomplish work tasks, and (b) minimal mental load for performing interaction tasks at the user inter-

Figure 5. Interplay between models at the specification level

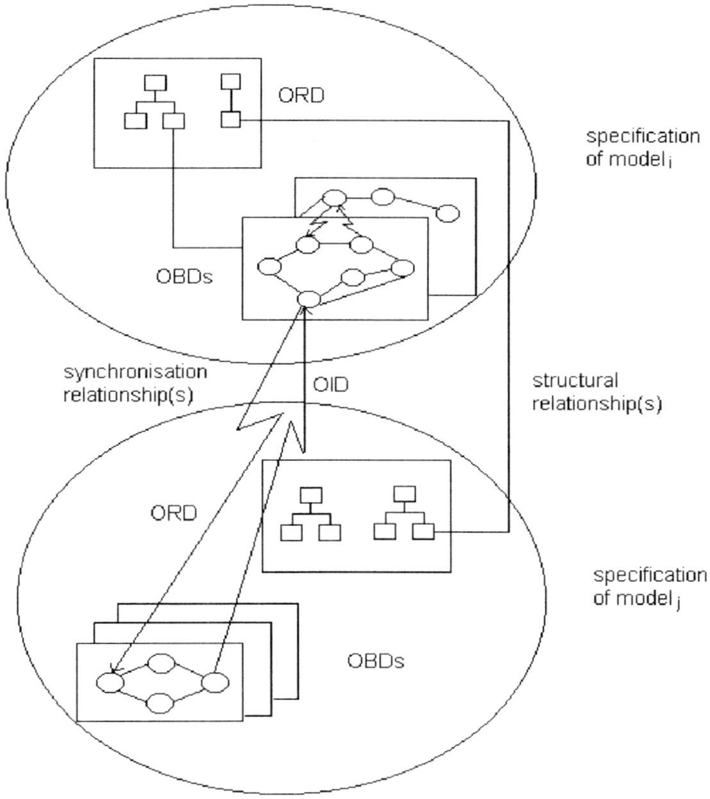

face to accomplish these tasks. In terms of specifications this understanding of task conformance means to provide task specifications (task models) and refinements to procedures both, at the problem domain level (data and user models) and at the user interface level (interaction and application models). In order to achieve (b), based on an adequate modality the number of user inputs for navigation and manipulation of data has to be minimised, whereas the software output has to be adjusted to the task and user characteristics. The latter is performed at the user, interaction, and application-model level. As a result, the task-specific path of an OBD or OID should not contain any procedure or step that is not required for task accomplishment. It has to be minimal in the sense, that only those navigation and manipulation tasks are activated that directly contribute to the accomplishment of the addressed work task. The sample task-model OBD in Figure 6 provides two different task-conform paths for search tasks.

Consequently, in order to ensure task conformance two different types of checks have to be performed: (1) completeness of specification, and (2) minimal length of paths provided for interactive task accomplishment (only in case check (1) has been successful). Check (1) ensures that the task is actually an interactive task (meaning that there exists a presentation

Figure 6. OBD for the task acquire options

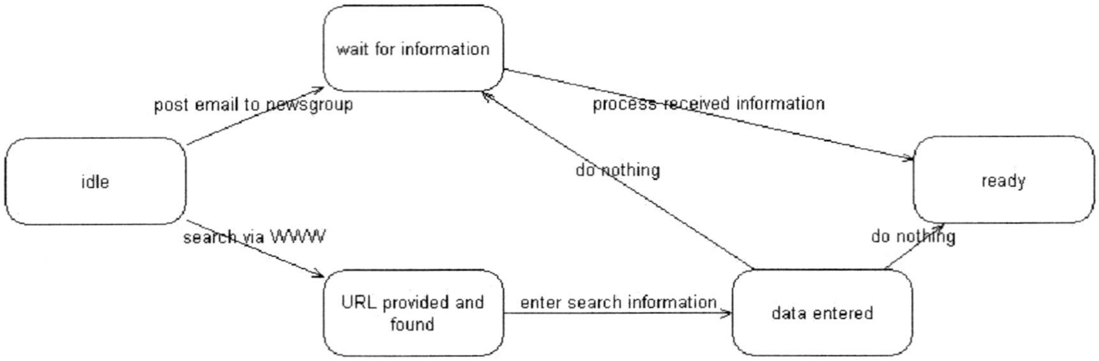

of the task at the user interface) as well as there exists an interaction modality to perform the required interactive data manipulation when accomplishing the addressed task.

For (2), initially, the checker determines whether the task in the task-model OBD is linked to an element of an ORD of the interaction model to present the task at the user interface (through the relationship "is-presented"). Then it checks links throughout the entire specification whether the task is linked to data that have to be manipulated interactively in the course of task accomplishment. In order to represent that relationship between tasks and problem-domain data structures, the "is-based-on"-relationship has to be used. Each interactive task also requires a link to interaction-model elements which present the work-task data for interactive manipulation to the user. Finally, each task requires a link to a user role (expressed through "handles"). Once the completeness of a task specification has been checked, each path related to that task in an OBD is checked, whether there occurs a path that is not related to the data manipulation of that task. This way, the specification is minimal in the sense, that the user is provided with only those dialogues that are necessary to complete the addressed task through the assigned role. Note, that there might be different paths for a task when handled through different roles. This property relates to adaptability which is discussed after the implementation sample for an algorithm.

The algorithms have been implemented as inherent part of the ProcessLens environment, namely in the set of model editors for designers and the prototyping engine. In the following the algorithm for the check of consistent task model hierarchies (task model check — static view) is exemplified in Java pseudo code (Eybl, 2003; Heftberger et al., 2004; p. 137f).

```
checkTaskModelConsistency () {
    taskModel = get TaskModel();
    allBeforeRelations = new Vector();
    enum = taskModel.getElements();
    while (enum.hasMoreElements()) {
    element = enum.nextElement();
    enum2 = bmod.getRelations();
    error = false;
    while (enum2.hasMoreElements()) {
    relation = enum2.nextElement();
    if (relation instanceof BeforeRelation) {
        if (!(allBevorRelations.contains(relation)))
    allBevorRelations.addElement(relation);
```

```
            }
        }
    }
    enum3 = allBevorRelations.elements();
    while (enum3.hasMoreElements()) {
        rel = enum3.nextElement();
        if (rel instanceof BevorRelation) {
            checkConsistency2(rel.getStartElement(),
                              rel.getEndElement());
            checkConsistency3();
        }
    }
    if (error) println(„inconsistent 'before'-relationship in
task model");
}
```

checkConsistency2 (Element start, Element end) {

```
    startVector = new Vector();
    endVector = new Vector();
    startVector.addElement(start);
    endVector.addElement(end);
    fill startVector with all elements that have direct and
indirect is_part_of
    relation with element ‚start'
    fill endVector with all elements that have direct and
indirect is_part_of
    relation with element ‚end'
}
```

checkConsistency3 () {

```
    enum = startVector.elements();
    while (enum.hasMoreElements()) {
        source = enum.nextElement();
        enum2 = endVector.elements();
        while (enum2.hasMoreElements()) {
            destination = enum2.nextElement();
            checkBeforeCycles(source, destination);
        }
    }
}
```

checkBeforeCycles(Element precessor, Element successor) {

```
    enum = successor.getAssociationRelations();
    while (enum.hasMoreElements()) {
        association = enum.nextElement();
        if (association instanceof BeforeRelation) {
            beforeRelation = association;
            next = beforeRelation.getEndElement();
            if (!(next.equals(successor))) {
        if (next.equals(precessor)){
                error = true;
                return;
            }
            checkBeforeCycles (precessor, next);
        }
    }
  }
}
```

In CheckConsistency( ) those elements of the task model that are not leave nodes in the task hierarchy are processed. When being part of a before-relation the method CheckConsistency2(start, end) is executed. "Start" denotes the start element of the "before"-relation, and "end" the end element of the "before"-relation. In CheckConsistency2(start, end) all elements are stored in a vector which have a direct or indirect is part of-relation with the "start" element (startVector). Another vector has to be set up for the "end" element (i.e., the endVector). In CheckConistency3( ) it is checked whether an element of endVector is part of a before-relation involving an element of startVector. If such an element can be found, the task model has inconsistent before-relationships and an error is reported.

The ProcessLens environment does not only support the task-conform specification of user interfaces, but also the execution of specifications. The designer might create a user interface artifact, as soon as task conformance has been checked successfully. Due to the capability of handling multiple task interaction procedures (adaptability), task conformance can be ensured even for individual users (rather than user groups defined through functional role concepts).

In ISO 9241, Part 10 "Dialogue systems are said to support suitability for individualization if the system is constructed to allow for *adaptation* to the user's individual needs and skills for a given task." Hence, in general, interactive applications have to be adaptable in a variety of ways: (1) to the organisation of tasks; (2) to different user roles; (3) to various interaction styles, and (4) to assignments. Adaptability means to provide more than a single option, and to be able to switch between the various options for each of the issues (1)–(4).

Adaptability with respect to the organisation of tasks means that a particular task might be accomplished in several ways. Hence, a specification enabling flexibility with respect to tasks contains more than a single path in an OBD or ORD within a task (implemented through "before"-relationships) to deliver a certain output given a certain input. Flexibility with respect to user roles means that a user might have several roles and even switch between them, eventually leading to different perspectives on the same task and data. Adaptability with respect to interaction devices and styles (i.e., the major concern for polymorph multimedia application development) does not only require the provision of several devices or styles based on a common set of dialog elements, as shown in Figure 2 for Graphical User Interfaces and browser-based interaction, but also the availability of more than a single way to accomplish interaction tasks at the user interface, for instance direct manipulation (drag & drop) and menu-based window management. The latter can be checked again at the behaviour level, namely through checking whether a particular operation, such as closing a window, can be performed along different state transitions. For instance, closing a window can be enabled through a menu entry or a button located in the window bar.

Adaptability of assignments involves (1)–(3) as follows: In case a user is assigned to different tasks or changes roles (in order to accomplish a certain task) assignments of and between tasks and roles have to flexible. In case a user wants to switch between modalities or to change interaction styles (e.g., when leaving the office and switching to mobile interaction, the assignment of interaction styles to data and/or tasks has to be modified accordingly. Changing assignments requires links between the different entities that are activated or de-activated at a certain point of time. It requires the existence of assignment links as well as their dynamic manipulation. Both has been provided in the environment for modeling user interaction, either through semantic relationships (e.g., "handles" between roles and tasks (linking different models), or the runtime environment of the prototyping engine.

Actually, to ensure adaptability (1) the designer has to set up the proper design space, and (2) modifications have to occur in line with the semantics of the design space. The first objective can be met through providing relationships between design items, both, within a model (e.g., "before"), and between models (e.g., "handles"). The first relationship enables flexibility within each of the perspectives mentioned above (e.g., the organisation of tasks) whereas the second relationship allows for flexible tuning of components, in particular the assignment of different styles of interaction to tasks, roles, and problem domain data.

The second objective can be met allowing to manipulate relationships according to the syntax and semantics of the specification language, and providing algorithms to check the correct use of the language as well as the changes related to the manipulation of relationships. For instance, the "before" relationship can only be set between sub tasks in the task model, since the algorithm for "before" described previously processes the design representation: In case the relationship is modified (e.g., "before" set between other sub tasks) the restrictions that applied to the OBDs are lifted, and novel ones, according to the semantics of the relationship, have to be enforced, however, using the same processing scheme.

CONCLUSION

Design support for mobile and stationary multimedia applications requires preserving the consistency of specifications while allowing polymorph applications. Consistent specifications can be processed for analytical measurement of usability principles. In this chapter a scheme for design representations (i.e., models) has been introduced that allows task-based and polymorph multimedia application development. Its algorithms process design knowledge, in order to check the implementation of generic properties of usability principles. Using the algorithms the designer should receive early feedback whether the specified proposal is usable *in principle* for the target user population, even without involving users. In case this analytic evaluation does not lead to satisfying results, alternative designs might be developed.

Analytical a-priori-testing reduces design time and avoid development results not acceptable by users. Consequently, further research will not only focus on the optimised implementation of the scheme and its algorithms, but also on extending the set of usability principles. On the long run, the approach should help to meet the requirements of the quality-of-use standards through automated analytical procedures at design time.

REFERENCES

Abowd, G. D., & Mynatt, E. D. (2000). Charting past, present, and future research in ubiquitous computing. *ACM TO-CHI, 7*(1), 29-58.

Adler, P. S., & Winograd, T. (1992). *Usability: Turning technologies into tools.* Oxford, New York: Oxford University Press.

Arehart, C. et al. (2000). *Professional WAP.* WroxPress.

Bevan, N. (1995). Measuring usability as quality of use. *Software Quality Journal, 4,* 115-130.

Bevan, N., & Azuma, M. (1997). Quality in use: Incorporating human factors into the software engineering lifecycle. *Proceedings ISESS'97 International Software Engineering Standards Symposium, IEEE* (pp. 169-179).

Beyer, H., & Holtzblatt, K. (1998). *Contextual design. Defining customer-centered systems.* San Francisco: Morgan Kaufmann.

Byrne, M. D., John, B. E., Wehrle, N. S., & Crow, D. C. (1999). The tangled Web we wove: A taskonomy of WWW use. *Proceedings CHI'99* (pp. 544-551). ACM.

Carter, J. A., Jr. (1985). A taxonomy of user-oriented data processing functions. In R. E. Eberts, & C. G. Eberts (Eds.), *Trends in ergonomics/human factors II* (pp. 335-342). Elsevier.

Cohen, P., Johnston, M., McGee, D., Oviatt, D., Pittman, J. A., Smith, I., Chen, L., & Clow, J. (1997). Quickset: Multimodal interaction for distributed applications. *Proceedings of the 5th International Multimedia Conference* (pp. 31-40). ACM.

Coulouris, G., Dollimore, J., & Kindberg, T. (1995). *Distributed systems. Concepts and design.* Wokingham: Addison Wesley.

Dumke, R., Rautenstrauch, C., Schmietendorf, A., & Scholz, A. (Eds.). (2001). *Performance engineering within the software development. Practices, techniques and technologies* (LNCS, Vol. 2047). Berlin: Springer.

Embley, D. W., Kurtz, B. D., & Woodfield, S. N. (1992). *Object-oriented systems analysis. A model-driven approach*. Englewood Cliffs, NJ: Yourdon Press.

Eybl, A. (2003). *Algorithms ensuring consistency in BILA* (in German). Master Thesis, University of Linz.

Garvin, M. (1984). What does "product quality" really mean? *Sloan Management Review*, 25-48.

Hackos, J. T., & Redish, J. C. (1998). *User and task analysis for interface design*. New York: Wiley.

Heftberger, S., & Stary, C. H. (2004). *Participative organizational learning. A process-based approach* (in German). Wiesbaden: Deutscher Universitätsverlag.

Herfet, T., Kirste, T., & Schnaider, M. (2001). Embassi–Multimodal assistance for infotainment and service infrastructure. *Computer & Graphics*, *25*(5).

ISO DIS 13407. (1997). A user-centred design process for interactive systems. *International Standards Organisation*, Geneva.

ISO 9001. (1987). International Standard 9001, Quality systems — model for quality assurance in design, development, production, installation and servicing. *International Standards Organisation*, Geneva.

ISO 8402. (1994). Draft International Standard (DIS) 8402, quality vocabulary. *International Standards Organisation*, Geneva.

ISO IEC 9126. (1994). Software product evaluation—quality characteristics and guidelines for their use. *International Standards Organisation*, Geneva.

ISO 9241. (1993). Ergonomic Requirements for Office Work with Visual Display Terminals (VDTs), Part 10-17, Dialogue Principles, Guidance Specifying and Measuring Usability, Framework for Describing Usability in Terms of User-Based Measures, Presentation of Information, User Guidance, Menu Dialogs, Command Dialogs, Direct Manipulation Dialogs, Form Filling Dialogs. *International Standards Organisation,* Geneva.

ISO DIS 9241-11. (1997). Draft International Standard (DIS) 9241-11, ergonomic requirements for office work with visual display terminals, Part 11, Guidance on Usability. *International Standards Organisation*, Geneva.

Karat, J. (1997). User-centered software evaluation methodologies. In M. Helander et al. (Eds.), *Handbook of human-computer interaction* (pp. 689-704). North Holland, Amsterdam: Elsevier Science.

Kotonya, G. (1999). Practical experience with viewpoint-related requirements specification. *Requirements Engineering*, *4*, 115-133.

Luo, P., Szekely, P., & Neches, R. (1993). Management of user interface design in humanoid. *Proceedings INTERCHI'93* (pp. 107-114). ACM/IFIP.

MacLean, A., Young, R., Bellotti, V., & Moran, T. (1991). Questions, options, and criteria: Elements of design space analysis. *Human-Computer Interaction, 6*, 201-250.

Nah, F. F. H., Siau, K., & Sheng, H. (2005). The value of mobile applications: A utility company study. *Communications of the ACM, 48*(2), 85-90.

Nardi, B. (1996). *Context and consciousness: Activity theory and human-computer interaction*. Cambridge, MA: MIT Press.

Nielsen, J. (1993). *Usability engineering*. Boston: Academic Press.

Norman, A. D., & Draper, W. S. (1986). *User-centred system design: New perspectives in human-computer interaction*. Hillsdale, NJ: Lawrence Erlbaum.

Norman, D. (1993). *Things that make us smart*. Reading, MA: Addison-Wesley.

Norman, D. (1998). *The design of every-day things*. London: MIT.

Puerta, A. R. (1996). The Mecano Project: Comprehensive and integrated support for model-based interface development. *Proceedings CADUI'96: Second International Workshop on Computer-Aided Design of User Interfaces,* Namur, Belgium (pp. 10-20).

Rosson, M. B., & Alpert, S. H. R. (1990). The cognitive consequences of object-oriented design. *Human-Computer Interaction*, *5*(4), 345-380.

Rosson, M. B., & Carroll, J. M. (1995). Integrating task and software development for object-oriented applications. *Proceedings CHI'95* (pp. 377-384). ACM.

Johnson, P., Wilson, S. T., Markopoulos, P., & Pycock, J. (1993). ADEPT—advanced design environments for prototyping with task models. *Proceedings INTERCHI'93* (pp. 56). ACM/IFIP.

Rubin, J. (1994). *Handbook of usability testing: How to plan design and contact effective tests*. New York: Wiley and Sons.

Rumbaugh, J., Jacobson, I., & Booch, G. (1999). *The unified modeling language reference manual*. Reading, MA: Addison-Wesley.

Selvidge, P. (1999). Reservations about the usability of airline Web sites. *Proceedings CHI'99 Extended Abstracts* (pp. 306-307). ACM.

Siau, K., & Shen, Z. (2003). Building customer trust in mobile commerce. *Communications of the ACM, 46*(4), 91-94.

Silberer, G., & Fischer, L. (2000). Acceptance, implications, and success of Kiosk systems (in German), In G. Silberer, L. Fischer, Gabler, & Wiesbaden (Eds.), *Multimediale Kiosktermínals* (pp. 221-245).

Stary, C. H. (2000). TADEUS: Seamlesss development of task-based and user-oriented interfaces. *IEEE Transactions on Man, Systems, and Cybernetics*, *30*, 509-525.

Stephanidis, C., Salvendy, G., Akoumianakis, D., Bevan, N., Brewer, J., Emiliani, P. L., Galetsas, A., Haataja, S., Iakovidis, I., Jacko, J., Jenkins, P., Karshmer, A., Korn, P., Marcus, A., Murphy, H., Stary, C., Vanderheiden, G., Weber, G., & Ziegler, J. (1998). Toward an information society for all: An international R&D agenda. *International Journal of Human-Computer Interaction*, *10*(2), 107-134.

Szekely, P., Sukaviriya, P., Castells, P., Muthukumarasamy, J., & Salcher, E. (1995). Declarative interface models for user interface construction tools, the mastermind approach. *Proceedings EHCI'95*, North Holland, Amsterdam.

Vidakis, N., & Stary, C. H. (1996). Algorithmic support in TADEUS. *Proceedings ICCI'96,* IEEE.

Weiser, M. (1991). The computer for the 21st century. *Scientific American*, *265*(3), 94-104.

Wilson, S. T., & Johnson, P. (1996). Bridging the generation gap, from work tasks to user interface design. *Proceedings CADUI'96* (pp. 77-94).

Ye, M. M. (1991). System design and cataloging meet the users: User interface design to online public access catalogs. *Journal of American Society for Information Science, 42*(2), 78-98.

Zajicek, M., Powell, C., & Reeves, C. (1998). A Web navigation tool for the blind. *Proceedings of the 3rd International Conference on Assistive Technologies*, ACM.

Ziefle, M., & Bay, S. (2005). How older adults meet complexity: Aging effects on the usability of different mobile phones. *Behaviour & Technology*.

Chapter XXI
Personalized Redirection of Communication and Data

Yuping Yang
Heriot-Watt University, UK

M. Howard Williams
Heriot-Watt University, UK

ABSTRACT

One current vision of future communication systems lies in a universal system that can deliver information and communications at any time and place and in any form. However, in addition to this, the user needs to be able to control what communication is delivered and where, depending on his or her context and the nature of the communication. Personalized redirection is concerned with providing the user with appropriate control over this. Depending on the user's preferences, current context and attributes of the communication the user can control its delivery. This chapter provides an understanding of what is meant by personalized redirection through a set of scenarios. From these, it identifies the common features and requirements for any system for personalized communications, and hence the essential functionality required to support this. It goes on to describe in detail two systems that aim to provide a personalized redirection service for communication and information.

INTRODUCTION

The computing landscape of the future will be an environment in which computers and applications are autonomous and provide largely invisible support for users in their everyday lives. One aspect of this vision is universal access to information and communication. The rapid development of the Internet and the proliferation of networks and devices, such as mobile phones and pager networks, is improving prospects for universal access by providing increasing coverage for access to information and data. Such communication-intense environments will enable users to access content ubiquitously through a variety of networks and stationary or mobile devices despite a growing range of different data formats.

Thus the vision of future communication lies in a user-oriented universal communication system which can accommodate versatile communication needs (Abowd & Mynatt, 2000, Abu-Hakima, Liscano, & Impey, 1998; Satyanarayanan, 2001). It should be able to deliver information at any time, in any place and in any form. But ubiquitous access is not enough. With such access comes an increasing need for users to have more control over when, where and how communications are delivered. This will depend on the context of the user at the time. Thus any future system will need to cater for user requirements relating to user control and maintain information on the current user context.

The design and implementation of such a system is challenging due to the variety of networks, devices and data, the preservation of user privacy, administration and management overheads, system scalability, and so on, and is the subject of this chapter.

The rest of the chapter is structured as follows. The next section provides an understanding of what is meant by personalized redirection. This is followed by a brief discussion of related work, based on commercial systems and research projects, and a section on the essential functionality for personalized redirection. The following two sections describe two prototype systems — the PRCD system and the Daidalos system, and explain how these map onto this essential functionality. It discusses the integration technologies and an example is used to demonstrate how personalized redirection works in Daidalos. The final section sums up the chapter.

WHAT IS PERSONALIZED REDIRECTION?

In order to understand what is meant by personalized redirection, this section describes several scenarios where redirection might be useful. From these it extracts the common features and arrives at a set of requirements for systems to support personalized redirection.

Doctor Scenario

A maternity patient who is deemed to be at risk of having a premature delivery, is at home wearing a foetal heart rate (FHR) monitor. This is a sensor that monitors the heart rate of the unborn baby. Suppose that it can be connected to the telecommunications network via the patient's home PC.

Now suppose that the patient is concerned for some reason. She calls the doctor. He needs to see the data currently being produced by the FHR monitor. Because the doctor prefers the FHR data to be displayed graphically, the data needs to be converted to a graph and then to an image using an appropriate software package. The doctor currently has access to a desktop, laptop, TV set, telephone, and mobile phone, but prefers his desktop. In this case the data need to be redirected via the conversion software to the desktop. (See Figure 1) On the other hand, the doctor may have been visiting another patient or may be in the hospital, and may not have access to a computer but only his mobile phone. In this case the data from the FHR monitor needs to be routed to the software package to convert to a graph and then an image and then sent to the mobile phone.

The doctor may want to compare the FHR graph against another stored in a database accessible via the Web. This previous FHR must be traced from the relevant database and the FHR data fetched. Again following the doctor's preferences, an appropriate conver-

Figure 1. A scenario of personalized redirection

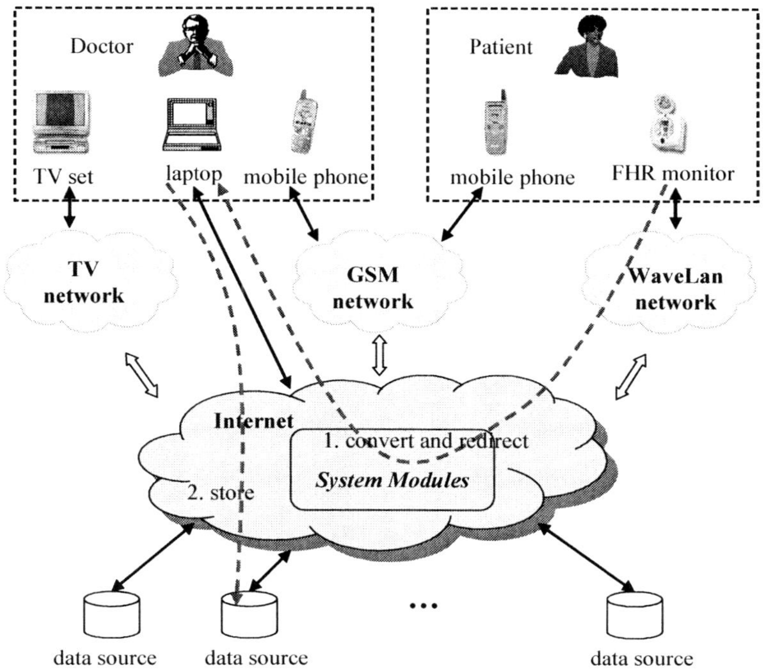

sion is determined which may be different from the previous one, and an appropriate graphics package is selected whose output, if necessary, should be converted to a suitable form for the doctor's current device. The graph is then displayed, overlapped with the current trace.

After the doctor has finished work he goes to play golf. From this time on, he does not want to be interrupted by calls from his patients — instead these should be rerouted to the doctor who is on-call. On the other hand, if there is a message from his wife, he would really like to be notified of this on his mobile phone so that he can respond at once if required. All other messages should be sent to his e-mail box.

Burglar Alarm Scenario

A young lady has installed a Web camera in her home to monitor the security of the house in case an intruder should break in. She may want to be informed of any problem by sending her an instant message (IM) to her current device. Suppose that one day when she is at work, the window of her house is broken. This is detected by the camera and a message is sent to her mobile phone to warn her. If for some reason she has switched off her phone then depending on her location, an e-mail may be sent to her e-mail box, a warning message to her desktop, or a voice mail to the office phone. If she is not accessible at all, a message may be sent instead to her husband's mobile phone or to her neighbour's phone. If she receives the message, she may use her mobile phone, PDA, or laptop to retrieve related video clips from the Web camera in the house and hence decide whether or not to call the police.

Music Scenario

A youngster has been a customer at a music shop and signed up for promotions. When he or she passes the shop next, an SMS is sent concerning the latest song released by his/her favourite artist. The youngster goes in and buys it, and loads it onto his or her mobile device (e.g., PDA). The youngster may decide to play it back on the device wherever he or she is. However, on returning home, he or she may wish to redirect the sound to a hi-fi system or a video clip to the TV set. The data stream may even be split so that the video part is redirected to the TV set while the audio component is sent to the hi-fi system.

Common Features and Requirements

These three scenarios illustrate the close connection between data and communication and the need to direct either to the appropriate device(s) for the appropriate person at the appropriate place/time, performing whatever conversions are necessary to achieve this. From these scenarios, one can identify several key features that any system for personal communication must provide:

1. Users need an intuitive and convenient way to specify their preferences. A user-friendly mechanism is required for users to enter new preferences or update existing ones, which is easy for non-experts to understand and use
2. The system should be able to locate where user preferences are stored and know how to represent and process them
3. Communications should be routed to the recipient regardless of where he or she is and whether the sender has direct access to the same kind of network, device, or application as the recipient
4. The system should be able to determine the user's location and device state to decide whether the communication can be delivered to him or her
5. In the light of user preferences and the characteristics of data and communication, incoming communication should be redirected to the preferred devices of the user or other users specified by the user
6. Data should be displayed according to user preferences (e.g., user's preferred format) or using a format suitable for the user's preferred device. Thus appropriate conversions are required to convert data from the original form to the final form
7. To achieve conversions, the system must be capable of selecting an appropriate conversion routine. It must be aware of which conversion routines can perform the task, where they are, and how they are related to each other to construct a feasible conversion path
8. Different devices deal with different data types and a single device can support multiple data formats. The system needs to determine which formats are appropriate for devices under current circumstances
9. Where different routines or conversion paths can be used to effect the same conversion or a device can support different formats, the system may have several options to choose from. Hence a decision process is needed for deciding between different options
10. Generally, the user may have access to several devices, each of which has a corresponding name. To provide device name independence, it is necessary for integration to have name mapping between user and his or her devices
11. The need for privacy and security for the users should be implicit in the design of such a system

EXISTING WORK

A growing number of commercial communities have put effort into providing integration of communication services. These include services such as e-mail/SMS message integration (SMSMate), e-mail/voice integration (SonicMail), text/SMS integration (SMS Messenger), etc. But each of these has tended to be fairly limited in the range of different data sources that can be integrated. Some systems (CallXpress3, OcteLink) address the integration of incoming communication from different sources and the accessibility of them across heterogeneous networks. They combine very simple message filtering into their systems.

This section describes briefly several research projects related to personal communication.

Seamless Personal Information Networking (SPIN)

The SPIN (Liscano et al., 1997) project has designed a seamless messaging system to intercept, filter, convert and deliver messages. Its objective is to manage personal messages of multiple mode formats including voice, fax and e-mail messages. However, it assumes that various data formats can be transformed into a standard text format, which leads to two problems. One is that it is difficult to convert some data formats, such as images, to the standard text format. The other is that a new converter needs to be written to convert an added new data format to the standard SPIN format. In addition, the SPIN project makes user's location information available throughout the system, and thus it does not protect user's privacy.

Telephony Over Packet Networks (TOPS)

TOPS (Anerousis et al., 1998) is an architecture used for redirecting incoming communication by a terminal-tracking-server. With telephony-like applications being its target, TOPS aims at providing both host and user mobility for telephony over packet networks, where real-time voice and/or video are the predominant content types. In TOPS, all filtering functionality is pushed into the directory service. In addition, TOPS exposes a person's point of attachment to others and requires all end-user applications to be rewritten. It emphasizes user preference management and name translation, but lacks functions for data conversion.

Universal Mobile Telecommunications System (UMTS)

UMTS (Samukic, 1998) is a third generation mobile system developed by ETSI. It seeks to extend the capability of current mobile technologies, and personal mobility across device end-points is one of its features. Intelligent Network components are used for realizing its functionality (Faggion & Hua, 1998). However, there are no explicit components in UMTS for redirection or data conversion based on preferences. In addition, due to its SS7-based architecture, there are implications on the high cost of entry to adding novel functionality (Isenberg, 1998).

Mobile People Architecture (MPA)

The MPA architecture (Appenzeller et al., 1999; Roussopoulos et al., 1999) addresses the challenge of personal mobility. Its main goal is to put the person, not the device that the person uses, at the endpoints of a communication session. To tackle this problem, a Personal Proxy is introduced, which maintains the list of devices or applications through which a person is currently accessible and transforms messages into a format preferred by the recipient. Each person is identified by a globally unique personal online ID. The use of the personal proxy protects a person's privacy by blocking unwanted messages and hiding his/her location. One problem in MPA is that all data must go through the user's home network, which performs the necessary functions on the data, and routes it to the user. This can cause additional delay if the user is far from his/her home network. There are also restrictions in extending and scaling the MPA architecture due to its tightly coupled components, which are not implemented as reusable network service.

Internet-Core nEtwork BEyond the thiRd Generation (ICEBERG)

The ICEBERG (Raman et al., 1999; Wang et al., 2000,) project has provided a composable service architecture founded on Internet-based standards for flow routing. Its functionality has a heavy dependency on a pre-existing networking infrastructure which involves a large number of nodes called iceberg access points (IAP). Correspondents are required to locate an IAP or have a local IAP. In each type of network supported, IAPs need to be installed. This requires modifying switches or base stations for PSTN (public switched telephone network) and cellular networks, which is practically difficult and makes it hard to have a broad deployment of ICEBERG.

2K and Gaia

2K is a research project carried out at the University of Illinois. It is an adaptable, distributed, component-based, network-centric operating system for the next Millennium (2K, 2001). It manages and allocates distributed resources to support a user in a distributed environment. The basis of the 2K architecture is an application- and user-oriented service model in which the distributed system customizes itself in order to better fulfil the user and application requirements. Research results from adaptable, distributed software systems, mobile agents, and agile networks are integrated to produce an open systems software architecture for accommodating change. The architecture encompasses a framework for architectural awareness—the architectural features and behaviour of a technology are refined and encapsulated within the software. Adaptive system software, because it is aware of these features and behaviour, is able to support applications which form the basis for adaptable and dynamic QoS (quality of service), security, optimization, and self-configuration (Roman & Campbell, 2000, 2002).

ESSENTIAL FUNCTIONALITY FOR PERSONALISED REDIRECTION

Any general architecture for personalised redirection should include functionality which encompasses the following functions.

- **Preference registry:** Since each user can specify his/her own preferences, a mechanism for storing and processing user preference profiles is needed. Some form of Preference Registry is needed to manage the uploaded preference profiles and authenticate users to update them. In addition, it should process queries to access the user's current preferences, such as a request for the current preferred format in which to display an image.

- **User context:** The context of a user changes with time, and the user's requirements may depend on the current context. Obvious examples of context are the user's location, his or her current activity and the state of a device to which the user has access—for example, is his or her mobile phone switched off, busy or idle? Thus, another aspect of the profile of the user is needed to keep track of the user's context and state of devices, and the functionality to manage this will be referred to here as *user context*. This tracks a user's context, and cooperates with the *preference registry* to provide his/her current accessibility information.
- **Converter selection:** One of the main problems with data communication is that data often comes in a form that is not useful to the recipient or not suitable for the recipient's device. A common solution to this problem is to convert the data to an acceptable format. Thus, a mechanism is needed to determine what converters are needed to implement specific transformations on the incoming communication.
- **Converter:** One obviously needs a number of converters that convert from one format to another. A simple example is the conversion between different image formats (e.g., gif, bmp, etc.) while a more complex example is the conversion from audio format into text. Ideally, one might have a single converter to convert between any pair of formats although in practice this may not be feasible.
- **Directory server:** A directory service associates names with objects and also allows such objects to have *attributes*. This enables one to look up an object by its name or to search for the object based on its attributes. Network directory services conveniently insulate users from dealing with network addresses. To allow directory servers to be fast and efficient, a directory service is required to locate a user's service agents and map the user's device id to his or her person id.
- **Protocol parser and device manager:** To receive incoming communication from or send out the resulting communication to an application-specific end-point, components are needed to provide this functionality. The *protocol parser* parses incoming communication and the *device manager* sends out the resulting communication.

A SYSTEM FOR PERSONALIZED REDIRECTION OF COMMUNICATION AND DATA (PRCD)

The systems described in this section and the next share some goals with those mentioned earlier. However, they aim at building a general architecture for personalized redirection of communication and data. In the first model, more attention is given to user preferences, and hence much of the work has been focused on intelligent data conversion. Format transformation, information filtering, and data splitting are all important aspects of the architecture. This enables users to interact flexibly in ways that suit them.

The first system is known as the PRCD system. The design of the architecture and technology of the implementation are presented here. This system provides a basis to investigate the mechanisms required to support personalized redirection of communication and data from a variety of devices, documents and so on, and explore how to mediate among diverse data formats. The overall goal is to create a general architecture/system in which any type of communication and data can be accessed from anywhere in whatever form the user wants it.

In terms of the functionality outlined in the previous section, the PRCD architecture includes a **preference registry**, a user context module (**user context tracking**), and a **directory server**, as well as **protocol parser** and **device manager**. A set of **converters** is maintained although the approach to handling conversions is a general one. Instead of assuming a single converter to convert between any pair of formats, the system attempts to find an appropriate sequence of converters to convert the input to the required output format.

The **conversion plan generator** is responsible for constructing a conversion plan which strings together a sequence of converters to achieve an appropriate data-flow and conversion between any two end-points. It must plan and invoke a sequence of converters that implement specific transformations on the incoming communication.

Well-defined converters and corresponding data-flow can be used to compose plans easily. However, this process needs to take account of different possible end-formats, different user preferences for end-formats depending on the circumstances, and different ways of achieving those end-formats. Conversion plan generation is the process of doing this composition automatically by choosing the right subset of converters to connect any two end-points.

When a user asks for particular information which is stored in some subset of data sources, the system should be able to find this informa-

Figure 2. System architecture

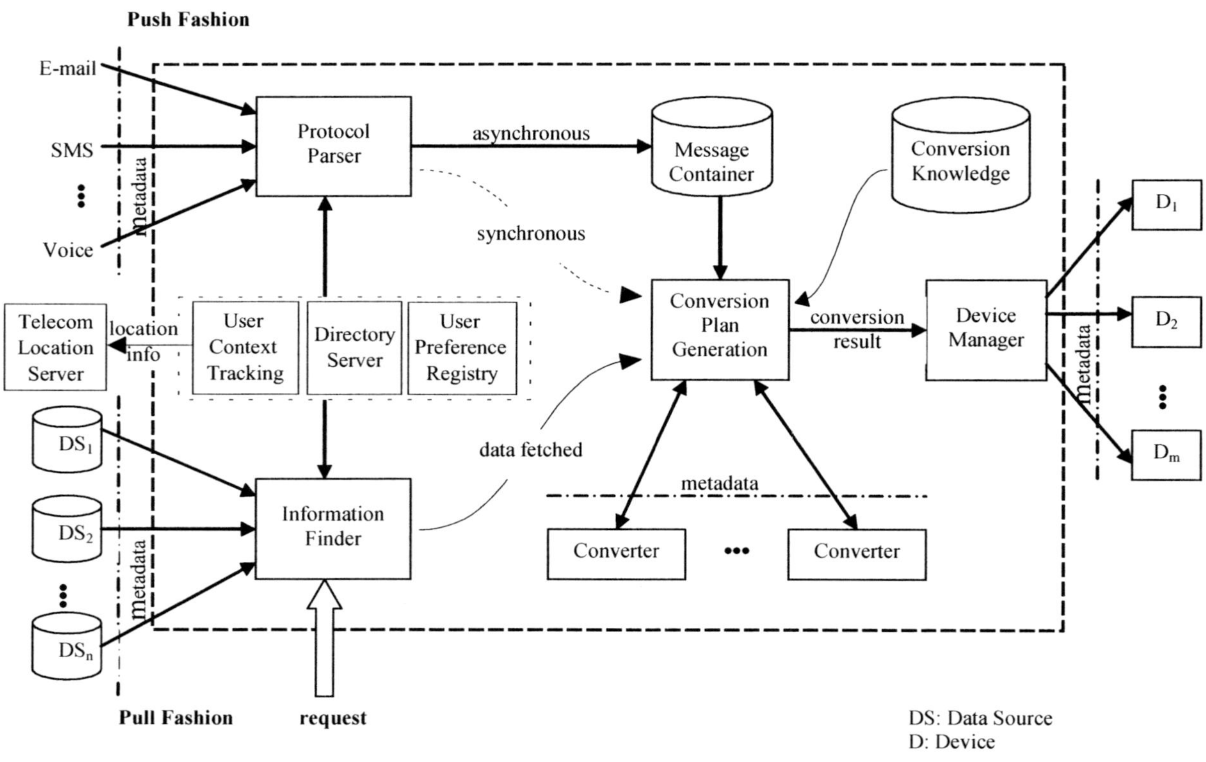

tion. A component referred to as the **information finder** is used to handle this request. It is responsible for the integration of distributed, heterogeneous, and autonomous data sources that involve structured, semi-structured, and unstructured data.

Figure 2 illustrates the various components of the system architecture and their relationships.

Original Scenarios

The scenarios described earlier are revisited here. For the youngster scenario, devices used to carry out the experiments include a laptop, a mobile phone simulator, and a speaker. In order to test the redirection of a song to the youngster's preferred device, the following rule was set through the GUI for specifying user preference rules:

```
ON occurrence of an audio
IF (Message-Component.type = audio) AND (location IS home)
THEN SEND_TO Laptop
```

Splitting of a video clip and the redirection of the generated two parts to appropriate devices is illustrated by the following rule:

```
ON occurrence of a video
    IF (Message-Component.type = video) AND
(location IS home)
    THEN SPLIT(VideoPlayer, HifiSystem)
```

For the doctor scenario, devices used consist of a desktop, a PDA simulator, a mobile phone simulator, and a microphone. The following two rules were set for communications to be redirected to a device preferred by the doctor in a certain situation:

```
ON occurrence of an e-mail
    IF (Message-Component.type = e-mail) AND
    (schedule IS PlayGolf)
        AND (sender IS family)
    THEN CONVERT_TO voice and SEND_TO
    MobilePhone

ON occurrence of an audio
    IF (Message-Component.type = audio) AND
    (schedule IS VisitPatient)
        AND (MobilePhone IS (busy or
        SwitchedOff))
    THEN SEND_TO EmailBox
```

Rules specifying the doctor's favourite image format are given below:

```
ON occurrence of an image
    IF (Message-Component.type = image) AND
    (location IS home)
    THEN DOWNLOAD_CONVERTER ToGIF and
    SEND_TO PDA

ON occurrence of an image
    IF (Message-Component.type = image) AND
    (sender IS patient)
    THEN CONVERT_TO JPG and
    CONVERSION_QUALITY>0.8 and SEND_TO
    Desktop

ON occurrence of bit stream
    IF (Message-Component.type = bitstream)
        AND (schedule IS (WorkingDay AND
        LunchTime))
    THEN SEND_TO OfficePhoneOfSecretary
```

The experimental results showed that appropriate conversion plans were constructed and the images were displayed in the user's favourite formats satisfying his or her requirements for conversion quality. The incoming communications were directed to appropriate devices and the user can later use any of the mobile phone, PDA, and computer to retrieve the data.

For the security scenario, a Web camera, as well as some other devices, were used. The following rules were given in order to show that when the Web camera detects something unexpected happening in the house, an instant message is sent to the user or the person specified by the user:

```
ON occurrence of SMS
    IF (Message-Component.type = SMS) AND
    (SendingDevice IS HouseWebCam)
    THEN SEND_TO MobilePhone and DIS
    PLAY 'There may be a burglar in your
    house!!!'

ON occurrence of SMS
    IF (Message-Component.type = SMS) AND
    (SendingDevice IS HouseWebCam)
        AND (MobilePhone IS (busy OR
        SwitchedOff))
    THEN SEND_TO MobilePhoneOfNeighbour
                AND DISPLAY 'There may be a
                burglar in your neighbour Jane's
                house!!!'
```

When executed, the message was sent to the appropriate device with corresponding content displayed. After receiving the message, the user was able to retrieve the live stream from the Web camera in the house and could see clearly what was happening in the house.

PERSONAL COMMUNICATION IN A PERVASIVE ENVIRONMENT

This section introduces how personal communication is taken into account in a pervasive computing environment such as that being developed in the Daidalos project. The main aim of Daidalos[1] (which stands for Designing Advanced Interfaces for the Delivery and Administration of Location independent Optimised personal Services) is to develop and demonstrate an open architecture based on a common network protocol (IPv6), which will combine diverse complementary network technologies in a seamless way and provide pervasive and user-centred access to these services.

In the overall Daidalos architecture there are two types of platform. The pervasive service platform (PSP) lies at the top level. It cooperates with the underlying service provisioning platforms (SPPs) to achieve its main task: the provision of pervasive services to the user.

The SPPs support the end-to-end service delivery across many different networks. In particular, the SPP subsystems are focused on E2E network protocol management. The purpose of an SPP is to provide full telecommunication support for real-time and non-real-time session management, including establishing, managing, and terminating sessions in a multi-provider federated network. It also interacts with other parts of the Daidalos architecture in brokering the QoS (quality of service), A4C (authentication, authorisation, accounting, auditing, and charging) and other enabling services on behalf of the PSP and the user (including personalization of the enabling services based on service context and user profile).

The architecture of the PSP (Farshchian et al., 2004) comprises six main software components, namely:

- **The context manager:** This manages information relating to the user's current situation. This includes location, personal preferences, available services and networks, etc
- **Personalization module:** This is responsible for handling personalisation at various points in the process of providing user services. These include the selection and composition of services, redirection of

messages and learning of new user preferences

- **Pervasive service management:** Central to the provision of a pervasive environment is a module to discover, select, and compose services in a dynamic and pervasive way that protects the user from the complexity of the underlying networks, devices, and software
- **Event manager:** The dynamically changing context is tracked by firing an event whenever a change occurs. This triggers the Event Manager, which notifies the appropriate component (generally the Rule Manager)
- **Rule manager:** This module is responsible for maintaining the set of rules that drive the overall control of the system, based on individual user's personal preferences
- **Security and privacy manager:** This is responsible for ensuring privacy in relation to application and network providers

In mapping the essential functionality of personalised redirection into Daidalos the roles of converter selection and converters reside in the infrastructure provided by the SPPs where a single converter is assumed for each conversion. Part of the user preference registry and user context components are subsumed in the context manager. The remainder of the user preference registry and the protocol parser and device manager now form part of the personalization module. The function of the rule engine is currently handled by the rule manager.

One aspect to handling privacy is to allow each user to have a set of virtual identities, each with its own user preferences. The redirection function has been enhanced by combining with different services in different situations (e.g., redirect communications via networks with the best quality when incoming calls are of high priority). It also takes account of the virtual identity of the user and redirects communications to appropriate devices according to the user preferences associated with the appropriate virtual identity.

SIP Protocol

One major difference between the PRCD and Daidalos systems lies in the protocol used for session handling. The lack of a standard session initiation protocol has long been hindering the achievement of real unified messaging. In response to the problem of various proprietary standards, the IETF (Internet Engineering Task Force) community has developed SIP, which stands for session initiation protocol (IETF). SIP is a text-based application-layer control protocol, similar to HTTP and SMTP, for creating, modifying, and terminating interactive communication sessions between users. Such sessions include voice, video, chat, interactive games, and virtual reality. It is a signalling protocol for Internet conferencing, telephony, presence, events notification, and instant messaging. SIP is not tied to any particular conference control protocol and is designed to be independent of the lower-layer transport protocol. SIP was first developed within the IETF MMUSIC (multiparty multimedia session control) working group, whose work has been continued by IETF SIP working group since September 1999. It is currently specified as proposed RFC 2543. As the latest emerging standard to address how to combine data, voice and mobility into one neat package, SIP may make unified messaging finally come true with its simple and integrated approach to session creation.

In Daidalos, SIP is used for all multimedia applications. Non-SIP applications are also considered and they are called legacy applications. The MMSPP (multimedia service provisioning platform) is part of the SPP which supports all functions related to SIP-based services (including establishing multimedia sessions, handling requests from clients, etc.). The core of the personalized redirection function resides on the PSP in the form of a service, and it is called PRS (personalized redirection service).

An Example of Redirection of a SIP Call

Figure 3 gives an example in which a SIP call is redirected to the user's preferred device taking into account his current context including preferences. It is elaborated below.

Bart is at home. He has two terminals on each of which there is a SIP-based VoIP (voice over IP) application running. Some time during the day a call comes in from his boss. The boss calls Bart at his general SIP address sip:bart@daidalos.org. The device Bart's boss is ringing from (sip:bossPDA@daidalos.org) forwards an INVITE sip:bart@daidalos.org to the MMSPP on the network. The MMSPP checks with the PRS Bart's preferred device in the current situation. The PRS knows that Bart, when staying at home during weekends, wants all calls from his Boss to be redirected to his PDA and all other calls to be diverted to the voicemail server. So the PRS determines the device to which the call should be redirected, i.e., sip:bartPDA@daidalos.org, and informs the MMSPP of it. MMSPP updates itself with the information and instructs the VoIP application on Boss's PDA to send an INVITE to that device.

Figure 3. An example of redirection of SIP call

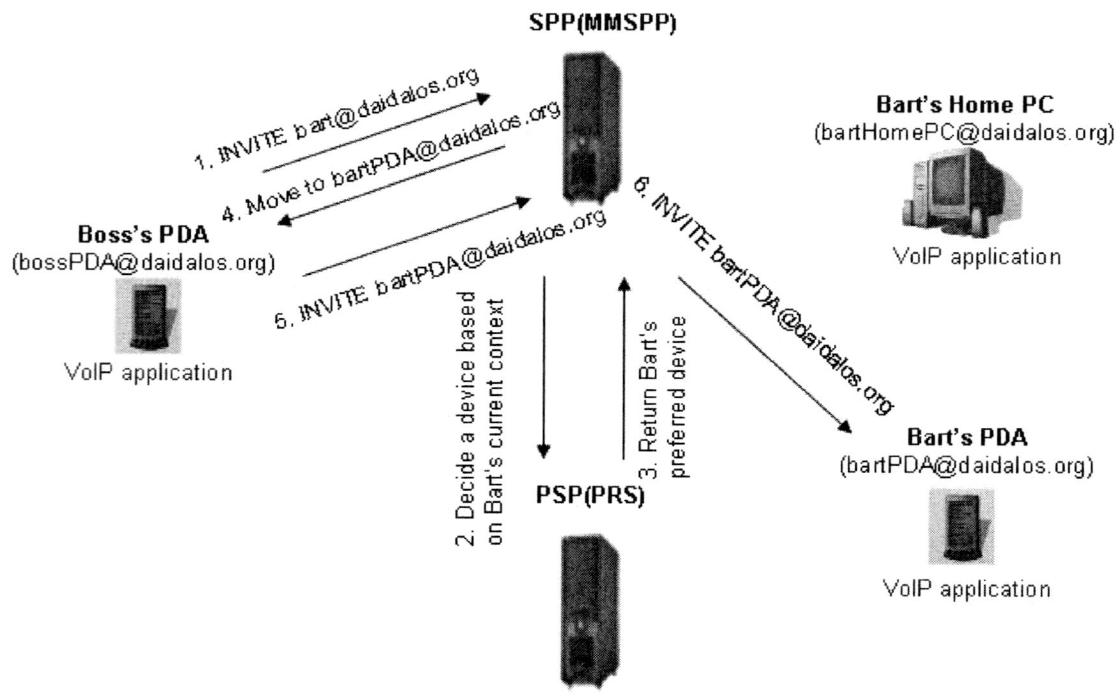

SUMMARY

This chapter has demonstrated the main ideas of how to build a personalized redirection system that could route communications and data to the user's preferred devices in his or her desired form at any time wherever he or she may be. It shows that using appropriate service components, a personalized communication system can be built that gives users control over the delivery and presentation of information. Two systems, PRCD and Daidalos, have been introduced in this chapter.

ACKNOWLEDGMENT

This work has been partially supported by the Integrated Project Daidalos, which is financed by the European Commission under the Sixth Framework Programme. The authors thank all our colleagues in the Daidalos project developing the pervasive system. However, this chapter expresses the authors' personal views, which are not necessarily those of the Daidalos consortium. Apart from funding the Daidalos project, the European Commission has no responsibility for the content of this chapter.

REFERENCES

2K. (2001). An operating system for the next millennium. Retrieved from http://choices.cs.uiuc.edu/2k

Abowd, G., & Mynatt, E. (2000). Charting past, present, and future research in ubiquitous computing. *ACM Transactions on Computer-Human Interaction, Special Issue on HCI in the New Millennium*, *7*(1), 29-58.

Abu-Hakima, S., Liscano, R., & Impey, R.(1998). A common multi-agent testbed for diverse seamless personal information networking applications. *IEEE Communications Magazine*, *36*(7), 68-74.

Anerousis, N., Gopalakrishnan, R., et al. (1998). The TOPS architecture for signaling, directory services, and transport for packet telephony. *Proceedings of the 8th International Workshop on Network and Operating Systems Support for Digital Audio and Video (NOSSDAV),* Cambridge, UK (pp. 41-53).

Appenzeller, G., Lai, K., et al. (1999). *The mobile people architecture* (Tech. Rep. No. CSL-TR-00000). Stanford University.

CallXpress3 Product Information sheet, Applied Voice Technology. Kirkland, Washington, 1996.

Faggion, N., & Hua, C. T. (1998). Personal communications services through the evolution of fixed and mobile communications and the intelligent network concept. *IEEE Network*, *12*(4), 11-18.

Farshchian, B., Zoric, J., et al. (2004). Developing pervasive services for future telecommunication networks. *WWW/Internet 2004* (pp. 977-982). Madrid, Spain.

IETF. Session Initiation Protocol. Retrieved from http://www.ietf.org/html.charters/sip-charter.html

Isenberg, D. S. (1998). The dawn of the stupid network. *ACM Networker*, *2*(1), 24-31.

Liscano, R., Impey, R., et al. (1997). Integrating multi-modal messages across heterogeneous networks. *Proceedings of the IEEE International Conference on Communications*, Montreal, Canada. Retrieved from http://www.dgp.toronto.edu/~qyu/papers/NRC-40182.pdf

OcteLink (1996). The OcteLink Network Service Product Information Sheet, Octel Communications, Milpitas, California.

Raman, B., Katz, R. H., et al. (1999). *Personal mobility in the ICEBERG integrated communication network*. Technical Report CSD-99-1048, University of California Berkeley.

Roman, M., & Campbell, R. H. (2000). Gaia: Enabling active spaces. *Proceedings of ACM SIGOPS European Workshop,* Kolding, Denmark (pp. 229-234).

Roman, M., & Campbell, R. H. (2002). *A user-centric, resource-aware, context-sensitive, multi-device application framework for ubiquitous computing environments* (Tech. Rep. No. UIUCDCS-R-2002-2284 UILU-ENG-2002-1728). University of Illinois at Urbana-Champaign.

Roussopoulos, M., Maniatis, P., et al. (1999). Person-level routing in the mobile people architecture. *Proceedings USENIX Symposium on Internet Technologies and Systems,* Boulder, CO (pp. 165-176).

Samukic, A. (1998). UMTS universal mobile telecommunications system: Development of standards for the third generation. *IEEE Transactions on Vehicular Technology*, *47*(4), 1099-1104.

Satyanarayanan, M. (2001). Pervasive computing: Vision and challenges. *IEEE PCM*, *8*(4), 10-17.

SonicMail. (n.d.). *SonicMail: E-mail and Voice Messages*. Retrieved from http://www.sonicmail.com/

SMS Messenger. (n.d.). *SMS Messenger: text to SMS Messages*. Retrieved from http://rasel.hypermart.net/

SMSMate. (n.d.). *SMSMate: E-mail to SMS Messages*. Retrieved from http://www.ozzieproductions.co.uk/

Wang, H. J., Raman, B., et al. (2000). ICEBERG: An Internet-core network architecture for integrated communications. *IEEE Personal Communications Magazine*, *7*(4), 10-19.

KEY TERMS

Communication Control: This allows users to access communications flexibly under a range of different circumstance according to their preferences.

Personal Communication: This is the ability to access many types of communications (e.g., e-mail, voice call, fax and instant messaging) with different types of devices (e.g., mobile phones, PC, fax machine).

Personalized Redirection: This is the mechanism to control the delivery of incoming communication and data to a user's preferred devices (or persons specified by the user) at any time in his/her preferred form taking into account user context. It intercepts, filters, converts and directs communications, thereby giving the user control over the delivery and presentation of information.

Pervasive Computing: As a major evolutionary step, following on from two distinct earlier steps–distributed systems and mobile computing, it is concerned with universal access to communication and information services in an environment saturated with computing and communication capabilities, yet having those devices integrated into the environment such that they "disappear."

Universal Access: This is the mechanism for providing access to information wherever the user may be, adapting content to the constraints of the client devices that are available.

User Context: User context is any relevant information that can be used to characterize the situation of a user. There are three important aspects of user context: where the user is, whom the user is with, and what resources are nearby. Typically, user context consists of user's location, profile, people nearby, the current social situation, humidity, light, etc.

User Preferences: This consists of a set of personal data indicating what to do with incoming communications and which device to use under which circumstances (e.g., data format, location, etc.). The user can modify these preferences as often as desired. User preferences could be in the form of rules. A rule is composed of a set of conditions (on caller identity, location and time) and an action (accept, delete or forward): when the conditions are met, the action is executed.

ENDNOTE

[1] Daidalos is a project funded under the European Sixth Framework Programme. Further details on Daidalos can be found on the Web site www.ist-daidalos.org.

Chapter XXII
Situated Multimedia for Mobile Communications

Jonna Häkkilä
Nokia Multimedia, Finland

Jani Mäntyjärvi
VTT Technical Centre of Finland, Finland

ABSTRACT

This chapter examines the integration of multimedia, mobile communication technology, and context-awareness for situated mobile multimedia. Situated mobile multimedia has been enabled by technological developments in recent years, including mobile phone integrated cameras, audio-video players, and multimedia editing tools, as well as improved sensing technologies and data transfer formats. It has potential for enhanced efficiency of the device usage, new applications, and mobile services related to creation, sharing, and storing of information. We introduce the background and the current status of the technology for the key elements constructing the situated mobile multimedia, and identify the existing development trends. Then, the future directions are examined by looking at the roadmaps and visions framed in the field.

INTRODUCTION

The rapid expansion of mobile phone usage during last decade has introduced **mobile communication** as an everyday concept in our lives. Conventionally, **mobile terminals** have been used primarily for calling and employing the short message service (SMS), the so-called text messaging. During recent years, the multimedia messaging service (MMS) has been introduced to a wide audience, and more and more mobile terminals have an integrated camera capable of still, and often also video recording. In addition to imaging functions, audio features have been added and many mobile terminals now employ (e.g., an audio recorder and an MP3 player). Thus, the capabilities of creating, sharing, and consuming multimedia items are growing, both in the sense of integrating more advanced technology and reaching ever-increasing user groups. The introduction of third generation networks, starting from Japan in October 2001 (Tachikawa, 2003), has put more emphasis on developing services requiring faster data transfer, such as streaming audio-video content, and it can be anticipated that the role of multimedia will grow stronger in mobile communications.

The mobile communications technology integrating the **multimedia** capabilities is thus expected to expand, and with this trend both the demand and supply of more specific features and characteristics will follow. In this chapter we concentrate on describing a specific phenomenon under the topic of mobile multimedia — namely, integrating **context awareness** into mobile multimedia.

Context-awareness implies that the device is to some extent aware of the characteristics of the concurrent usage situation. Contextual information sources can be, for instance, characteristics of the physical environment, such as temperature or noise level, user's goals and tasks, or the surrounding infrastructure. This information can be bound to the use of mobile multimedia to expand its possibilities and to enhance the **human computer interaction**. Features enhancing context awareness include such things as the use of context-triggered device actions, delivery of multimedia-based services, exploiting recorded metadata, and so on.

In this chapter we will look into three key aspects—mobile communications, multimedia, and context-awareness—and consider how they can be integrated. We will first look at each key element to understand the background and its current status, including identifying the current development trends. Then the future directions will be examined by looking at the roadmaps and visions framed in the field. The challenges and possibilities will then be summarized.

BACKGROUND

The development of digital multimedia has emerged in all segments of our everyday life. The home domain is typically equipped with affiliated gadgets, including digital TV, DVD, home theaters, and other popular infotainment systems. The content of the digital multimedia varies from entertainment to documentation and educative material, and to users' self-created documents. Learning tools exploiting digital multimedia are evident from kindergarten to universities, including all fields of education (e.g., language learning, mathematics, and history). Digital multimedia tools are used in health care or security monitoring systems. So far, the platforms and environments for the use of digital multimedia have been non-mobile, even "furniture type" systems (i.e., PC-centered or built around a home entertainment unit). The PC, together with the Internet, has been the key element for storing, editing, and sharing multimedia content. User-created documents have involved gadgets such as video cameras or digital cameras, from where the data needs to be transferred to other equipment to enable monitoring or editing of the produced material.

However, the role of mobile multimedia is becoming increasingly important both in the sense of creating, and sharing and monitoring the content. The increased flexibility of use following from the characteristics of a mobile communication device—it is a mobile, personal, and small-size gadget always with the user—has expanded the usage situations and created possibilities for new applications; the connection to the communication infrastructure enables effective data delivery and sharing. Adding the situational aspect to mobile multimedia can be utilized using context awareness, which brings information of the current usage situation or preferred functions, and can

Figure 1. Integrating multimedia, mobile communication technology, and context awareness

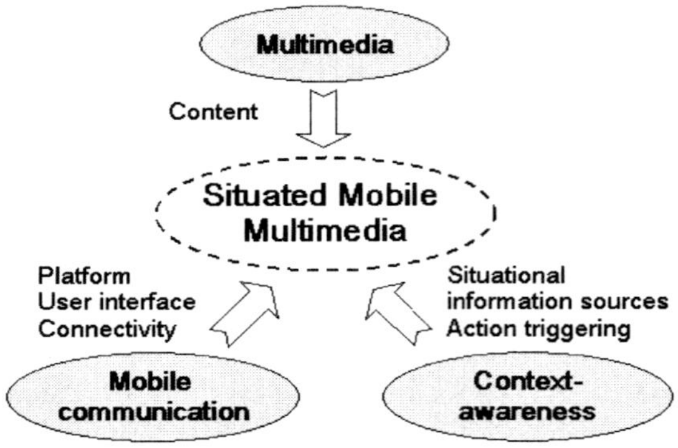

be used for action triggering for instance (Figure 1). Here, the multimedia content is dealt with in a mobile communication device, most often a mobile phone, which offers the physical platform and user interface (UI) for storing, editing, and observing the content.

In the following sections, we first look at the use of mobile communication technology and then the concept of context awareness more closely.

Mobile Phone Culture

During recent decades, mobile communication has grown rapidly to cover every consumer sector so that the penetration rates approach and even exceed 100% in many countries. Although the early mobile phone models were designed for the most basic function, calling, they soon started to host more features, first related primarily to interpersonal communication, such as phonebook and messaging applications accompanied by features familiar to many users from the PC world (e.g., electronic calendar applications and text document creation and editing). These were followed by multimedia-related features, such as integrated cameras, FM radios, and MP3 players, and applications for creating, storing, and sharing multimedia items.

Defining different audio alert profiles and ringing tones, and defining distinct settings for different people or caller groups, expanded the user's control over the device. Personalization of the mobile phone became a strong trend supported by changeable covers, operator logos, and display wallpapers. All this emphasized the mobile phone as a personal device.

The text messaging culture was quickly adopted by mobile phone users, especially by teenagers. The asynchronous communication enabled new styles of interaction and changed the way of communicating. Text messaging has been investigated both from the viewpoint of how it is associated with everyday life situations, the content and character of messaging, and the expression and style employed in messaging (Grinter & Eldridge, 2003). When looking at teenagers' messaging behavior, Grinter and Eldridge (2003) report on three types of messaging categories: chatting, planning activities, and coordinating communications in which the messaging leads to the use of some other communication media, such as face-to-face meetings or phone calls. Text messaging has also created its own forms of expression (e.g., the use of shortened words, mixing letters, and number characters, and the use of acronyms that are understood by other heavy SMS users or a certain group).

Due to the novelty of the topic, mobile communication exploiting multimedia content has so far only been slightly researched. Cole and Stanton (2003) report that mobile technology capable of pictorial information exchange has been found to hold potential for a youngster's collaboration during activities, for instance, in story telling and adventure gaming. Kurvinen (2003) reports a case study on group communication, where users interact with MMS and picture exchange for sending congratulations and humorous teasing. Multimedia messaging has been used, for example, as a learning tool within a university student mentoring program, where the mentors and mentees were provided with camera phones and could share pictorial information on each other's activities during the mobile mentoring period (Häkkilä, Beekhuyzen, & von Hellens, 2004). In a study on camera phone use, Kindberg, Spasojevic, Fleck, and Sellen (2005) report on user behavior in capturing and sharing the images, describing the affective and functional reasons when capturing the photos, and that by far the majority of the photos stored in mobile phones were taken by the user him or herself and kept for a sentimental reason.

Context Awareness for Mobile Devices

In short, context awareness aims at using the information of the usage context for better adapting the behavior of the device to the situation. Mobile handsets have limited input and output functionalities, and, due mobility, they are used in altering and dynamically varying environments. Mobile phones are treated as personal devices, and thus have the potential to learn and adapt to the user's habits and preferences. Taking these special characteristics of mobile handheld devices into account, they form a very suitable platform for context-aware application development. Context awareness has been proposed as a potential step in future technology development, as it offers the possibilities of smart environments, adaptive UI's, and more flexible use of devices.

When compared with the early mobile phone models of the 1990s, the complexity of the device has increased dramatically. The current models have a multiple number of applications, which, typically, must still be operated with the same number of input keys and almost the same size display. The navigation paths and the number of input events have grown, and completing many actions takes a relatively long time, as it typically requires numerous key presses. One motivation for integrating context awareness into mobile terminals is to offer shortcuts to the applications needed in a certain situation, or automating the execution of appropriate actions.

The research so far has proposed several classifications for contextual information sources. For example, the TEA (Technology for Enabling Awareness) project used two general categories for structuring the concept of context: human factors and physical environment. Each of these has three subcategories: human factors divides into information on the user, his or her social environment, and tasks, and physical environment distinguishes location, infrastructure, and physical conditions. In addition, orthogonal to these categories, history provides the information on the changes of context attributes over time Schmidt, Beigl, and Gellersen (1999). Schilit, Adams, and Want (1994) propose three general categories: user context, physical context, and computing context. Dey and Abowd (2000) define the context as "any information that can be used to characterize the situation of an entity."

In Figure 2, we present the contextual information sources as they appear from the mobile communication viewpoint, separating the five different main categories—physical environment, device connectivity, user's actions, preferences, and social context—which emphasize the special characteristics of the field. These categories were selected as we saw them to represent the different aspects especially important to the mobile communication domain, and they are briefly explained in the following. The proposed categories overlap somewhat

Figure 2. Context-aware mobile device information source categories and examples

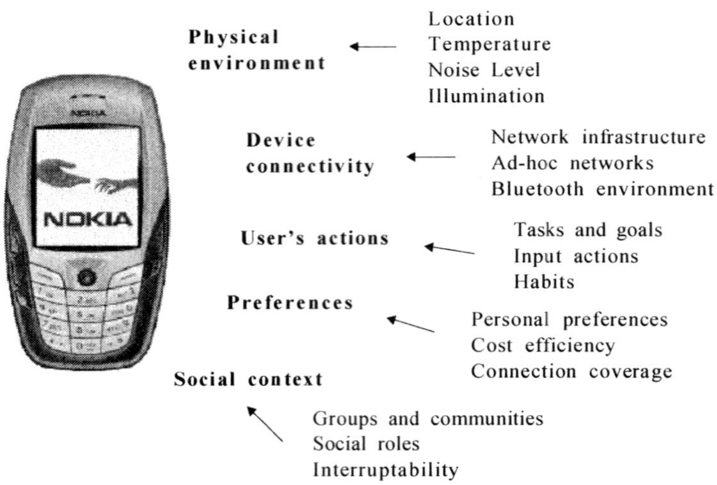

and individual issues often have an effect on several others. Thus, they are not meant as strictly separate matters but aim to construct an entity of overall contextual information sources.

Physical environment is probably the most used contextual information source, where the data can be derived from sensor-based measurements. Typical sensor data used in context-aware research includes temperature, noise, and light intensity sensors and accelerometers. Location, a single attribute that has resulted in the most research and applications in the field of mobile context-aware research, can be determined with GPS or by using the cell-ID information of the mobile phone network. *Device connectivity* refers to the information that can be retrieved via data transfer channels that connect the device to the outside world, other devices, or the network infrastructure. This means not only the mobile phone network, such as GSM, TDMA, or GPRS connections, but also ad hoc type networks and local connectivity systems, such as Bluetooth environment or data transfer over infrared. A certain connectivity channel may enable different types of context-aware applications: for example, Bluetooth can be used as a presence information source due its short range. The category *user's actions* implies the user's behavior, which here covers a wide range of different aspects from single input events, such as key presses and navigation in the menus, to prevailing tasks and goals, and general habits typical of the individual user.

Contrary to the previous categories, which are more or less typical of the research in the field of context awareness, we propose that the last two categories have an important role when using a mobile communication device. By *preferences* and *social context,* we refer to the factors relating to the use situations especially important from the end-user perspective. The preferences cover such issues as cost-efficiency, data connection speed, and reliability, which are important to the end-user and which relate closely to the connectivity issues dealing with handovers and alternative data transfer mediums. But, not only technical issues affect the usage. The user's personal preferences, which can offer useful information for profiling or personalizing mobile services, are also important. *Social context* forms an important information category as mobile terminals are still primarily used for personal communication and are often used in situations where the presence of other people cannot be avoided. This category forms a somewhat special case among the five classes (Figure 2) as it has a strong role both as an input and an output element of a context-aware application, so it can—and should—be taken into account both as an information source and in the consequent behavior of a context-aware system. By inferring the prevailing social context, one can not only gain information on the preferred functions but also vice versa–the social context also has an effect on how we wish the device to react in terms of interruptability or privacy.

Contextual information can be used for automating certain device actions: when specified conditions are fulfilled, the detected context information triggers a pre-defined action to take place. As an example, a mobile phone ringing tone volume could be automatically set to maximum if the surrounding noise exceeded 90 dB. In addition to automated device behavior, semi-automated and manual execution of actions has been suggested to ensure an appropriate level of user control (Mäntyjärvi, Tuomela, Känsälä, & Häkkilä, 2003). Previous work in the field of mobile context-aware devices has implemented location-aware tour guides and reminders (Davies, Cheverst, Mitchell, & Efrat, 2001), where accessing a certain location triggers the related information or reminder alert to appear on the device screen, or the automated ringing tone profile changes (Mantyjärvi & Seppänen, 2003), as does screen layout adaptation (Mäntyjärvi & Seppänen, 2003) in certain environmental circumstances.

CURRENT STATUS

Technology Enablers

With the term "technology enabler," we mean the state-of-the-art technology that is mature enough and commonly available for building systems and applications based on it. When looking at the current status of the technology enablers for situated multimedia (Figure 3) it can be seen that there are several quite different factors related to the three domains of multimedia, mobile technology and context awareness that are still quite separated from each other. Recent developments have brought mobile technology and multimedia closer to each other, integrating them into mobile phone personalization, multimedia messaging, and imaging applications. Context awareness and mobile communication technology have moved closer to each other mainly on the hardware frontier, as GPS modules and integrated light-intensity sensors and accelerometers have been introduced for a number of mobile phones. Altogether, one can say that the areas presented in Figure 3 are under intensive development, and the features are overlapping more and more across the different domains. The development in hardware miniaturization, high-speed data transfer, data packaging, artificial intelligence, description languages, and standardization are all trends that lead toward seamless integration of the technology with versatile and robust applications, devices, and infrastructures.

A closer look at the current status of context awareness shows that the applications are typically built on specific gadgets, which include such device-specific features as modified hardware. However, the development is toward implementing the features on commonly used gadgets, such as off-the-self mobile phones or PDA's, so that platform-independent use of

Figure 3. Technology enablers and their current status for situated mobile multimedia

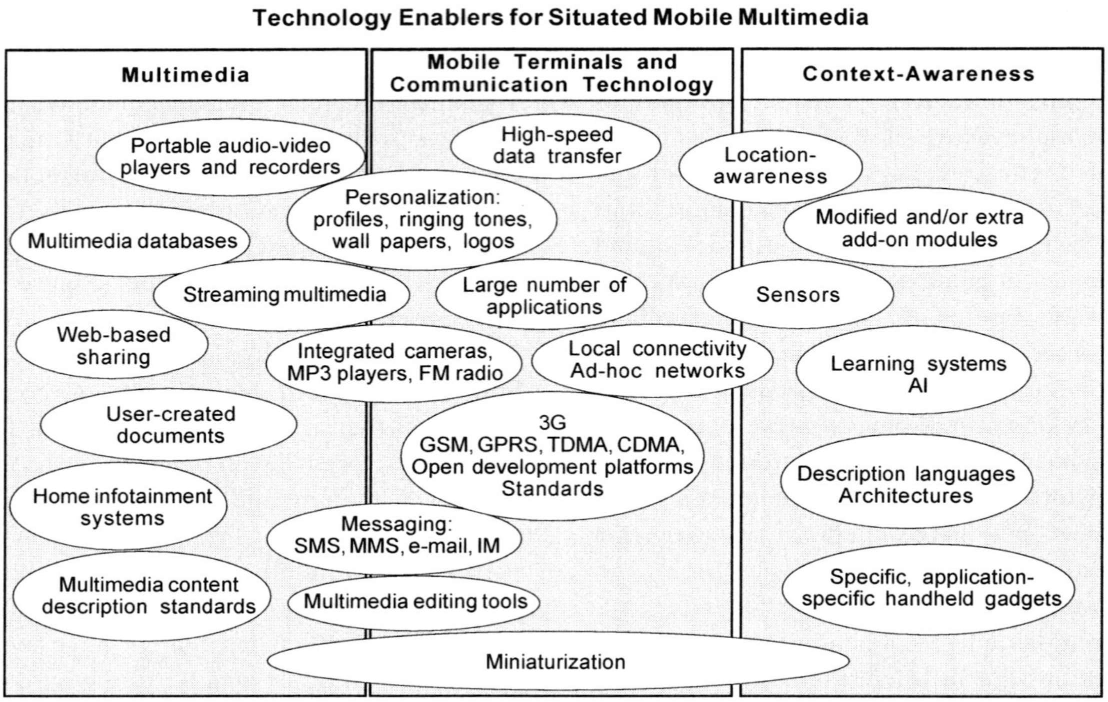

applications and services is possible. Also, applications utilizing location-awareness have typically concentrated on a defined, preset environment, where the beacons for location-detection have been placed across a limited, predefined environment, such as a university campus, distinct building, or certain area of the city. Thus, the infrastructure has not yet been generally utilized, and expanding it would require an extra effort. So far, the accuracy of GPS or cellular-ID-based recognition has often been too poor for experimenting with location-sensitive device features or applications requiring high location detection resolution. In mobile computing, it is challenging to capture a context, as a description of a current (e.g., physical situation) with a full confidence (==1). Various machine intelligence and data analysis-based methods such as self organizing neural networks (Flanagan, Mäntyjärvi, & Himberg, 2002), Bayesian approach (Korpipää, Koskinen, Peltola, Mäkelä, & Seppänen, 2003), fuzzy reasoning (Mäntyjärvi & Seppänen, 2003), and hidden Markov models (Brand, Oliver, & Pentland, 1997), to mention a few, have been studied to. In most approaches, quite good context recognition accuracies (~70-100%) are presented. However, it must be noted that all results are obtained with different and relatively limited data sets and results are quite subjective. The mobile context aware computing research, particularly context recognition, still lacks the systematic approach (e.g., benchmark datasets).

When examining the current technological status in mobile multimedia, the strongest trend during recent years has been the introduction of the multimedia messaging service (MMS), which has now established its status as an "everyday technology" with widespread use of so-called smart phones and camera phones. In Europe, it is estimated that people sent approximately 1.34 billion multimedia messages in 2005. This shows that MMS is a considerable potential technology that end users have become to adopt, although this is only a small fraction of the number of SMSs sent, which has been estimated to be 134,39 billion in 2005 (Cremers & de Lussanet, 2005).

Personalization of mobile phones, which so far has been executed manually by the user, has taken the form of changing ringing tones, operator logos, and wallpapers. Multimedia offers further possibilities for enhancing the personalization of the mobile device — both from the user's self-expression point of view when creating his or her own items and as the receiving party when the user may access multimedia content via peer-to-peer sharing, information delivery, or mobile services.

Toward Situated Mobile Multimedia

The research in the area of situated mobile multimedia is still in its early development stage and many of the current projects are very limited, still concentrating mainly on textual information exchange. The most common contextual information source used in mobile communication is the location. In addition to the information bound to the physical location, information on the current physical location and distance may provide useful data for (e.g., time management and social navigation). E-graffiti introduces an on-campus location-aware messaging application where users can create and access location-associated notes, and where the system employs laptop computers and wireless network-based location detection (Burrell & Gay, 2002). InfoRadar supports public and group messaging as a PDA application, where the user interface displays location-based messages in a radar-type view showing their orientation and distance from the user (Rantanen, Rantanen, Oulasvirta, Blom, Tiitta, & Mäntylä, 2004). The applications exploiting multimedia elements are typically location-based museum or city tour guides for tourists (see e.g., Davies & al., 2001).

Multimedia messaging has become a popular technique for experimentation within the field since there is no need to set up any specific infrastructure and standard mobile phones can be used as the platform. The widespread use of mobile phones also enables extending the experiments to large audiences, as no specific gadgets need to be distributed. These aspects are used in the work of Koch and Sonenberg (2004) for developing an MMS-based location-sensitive museum information application utilizing Bluetooth as the sensing technology. In the Rotuaari project carried out in Oulu, Finland, location-aware information and advertisements were delivered to mobile phones in the city center area by using different messaging applications, including MMS ("Rotuaari," n.d.).

Use of context can be divided into two main categories: push and pull. In the push type of use, the context information is used for automatically sending a message to the user when he or she arrives at a certain area, whereas with the pull type, the user takes the initiative by requesting context-based information, such as recommended restaurants in the prevailing area.

Currently, most of the experiments concentrate on the push type of service behavior. This is partially due to the lack of general services and databases, which, in practice, prevents the use of the request-based approach. A general problem is the shortage of infrastructure supporting sensing, and the absence of commonly agreed principles for service development. Attempts to develop a common framework to enable cross-platform application development and seamless interoperability exist, but so far there is no commonly agreed ontology or standard.

In Häkkilä and Mäntyjärvi (2005) we have presented a model for situated multimedia and how context-sensitive mobile multimedia services could be set, stored, and received, and examined users' experiences on situated multimedia messaging. The model combines multimedia messaging with the context awareness of a mobile phone and phone applications categorized into three main groups, notification, reminder, and presence, which were seen to form a representative group of applications relevant to a mobile handset user. The composing entity (i.e., a person or a service) combines the device application information, the multimedia document, and the context information used for determining the message delivery conditions to a situated multimedia message. After sending, the message goes through a server containing the storage and context inferring logic. The server delivers the message to the receiving device, where it has to pass a filter before the user is notified of the received message. The filter component prevents the user from so-called spam messages and enables personalized interest profiles.

FUTURE TRENDS

In order to successfully bring new technology to the use of a wide audience, several conditions must be fulfilled. As discussed before, the technology must be mature enough so that durable and robust solutions can be provided at a reasonable price, and an infrastructure must be ready to support the features introduced. The proposed technological solutions must meet the end users' needs, and the application design has to come up with usable and intuitive user interfaces in order to deliver the benefits to the users. Importantly, usable development environments for developers must exist. Usability is a key element for positive user experience. In the ISO 13407 (3.3), standard on human-centred design processes for interactive systems, usability has been defined to be the "extent to which a product can be used by specified users to achieve specified goals with effectiveness, efficiency, and satisfaction in a specified context of use" (Standard ISO 13407, 1999).

In this section, we discuss the near and medium-term future concepts focusing on situated mobile multimedia enabled by the development trends in context awareness, multimedia technologies, and mobile terminals.

Context Awareness

Context awareness has been recognized as one of the important technology strategies on the EC level (ITEA, 2004). The key factors for the human-system interaction of mobile devices recognized by ITEA are: simple, self-explanatory, easy to use, intelligent, context-aware, adaptive, seamless, and interoperable behaviour of user interfaces. The main driving forces for the development of context-awareness for mobile devices are that the user interface of mobile terminals is limited (by the small physical size), and there is an enormous expected growth in mobile applications and services to be accessed by terminals in near future. So, it can be said that the need for context-awareness is evident and the need is strongly market and industry-driven. The main enablers for the context-awareness of mobile devices are: ease of use and an open development environment for smart phones, architectures for mobile context-awareness enabling advanced context reasoning, miniaturized and low-power sensing technologies and data processing, and suitable languages for flexibly describing the context information. These main enablers are recognized in the research field and the research is rapidly advancing globally toward true context-awareness.

However, promising technological solutions often have some drawbacks. In context awareness, they are related to user experience. While the nice scenarios of context awareness provide a better tomorrow via intelligently behaving user interfaces—the right information in the right situation—the reality might be a bit different. For example, entering a new context may cause the adapted UI to differ radically from what it was a moment ago, and a user may find him or herself lost in the UI. On the other hand, the device may incorrectly recognize the situation and behave in an unsuitable manner. These are just examples of horror usability scenarios, but the idea is that the responsibility for the functionality of the device is taken away from the user, and when a user's experience of a product is negative, the consequences for the terminal business may be fatal.

There are ways to overcome this problem. One—and a careful—step toward successful context-awareness for mobile terminals is to equip the devices with all the capabilities for full context-awareness and provide an application, a tool by which a user him or herself may configure a device to operate in a context-aware manner (Mäntyjärvi et al., 2003). Obviously, this is an opportunity to accomplish context-aware behavior, but on the other hand, the approach sets more stress on the user when he or she must act as an engineer (so-called end user programming), and continuous configuration may become a nuisance. However, this approach is more attractive for the terminal business when the user him or herself is responsible for any possible unwanted behavior of a device instead of the device manufacturers.

Future Development Trends

In the very near future, we are to witness a growth in basic wireless multimedia applications and services (Figure 4). By basic services and applications we refer to local services (push multimedia kiosks, mobile mm content, etc) and mobile multimedia services, including downloadable content: ringtones, videos, skins, games, and mobile terminal TV. Streaming multimedia is strongly emerging, and mobile TV in particular has raised expectations as the next expanding application area. At the moment, several broadcasting standards are dominating the development in different parts of the world: ISDB-T in Japan; DVB-H, particularly in Europe and the US; and DBM, especially in Korea, where several mobile terminals on the market already support it. Although the status of mobile TV is not reaching a wide audience of end users yet, a number of trials are in existence, such as a half-year trial in Oxford (UK) with NTL Broadcast and O2 starting in the summer of 2005, when 350 users will access several TV channels with the Nokia 7710 handset (BBC News, 2005).

The increased growth in MM services is supported by the maturation of various technologies, including 3G-networks, extensive content communication standards such as MPEG 7, multimedia players and multimedia editing tools for mobile terminals. In addition, an increase in the amount of mobile multimedia is expected to support the stimulation of the mobile service business.

The increase in the amount of personal multimedia is expected to be considerable, mainly because of digital cameras and camera phones. The explosion of mobile personal multimedia has already created an evident need for physically storing data—by which we mean forms of memory: memory sticks and cards, hard-drive discs, etc.

However, another evident need is end user tools and software for managing multimedia content—digital multimedia albums, which are already appearing in the market in the form of home multimedia servers enabling communica-

Figure 4. Future development trends enabled by context-awareness, mobile terminals, and communications technology and multimedia

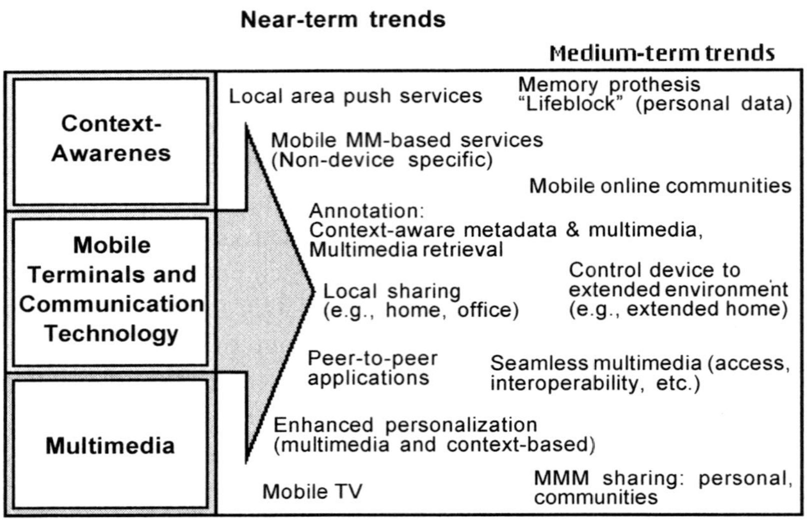

tion of multimedia wirelessly locally at home, and personal mobile terminal MM album applications.

The next overlapping step in the chain of development is the mobile online sharing of the multimedia content of personal albums. People have always had a need to get together (i.e., the need to form communities) (e.g., around hobbies, families, work, etc). Today, these communities are in the Web and the interaction is globally online. Tomorrow, the communities will be mobile online communities that collaborate and share multimedia content online.

In the near future, as the personal multimedia content is generated with context-aware mobile phones, the content will be enhanced semantic context-based annotations, with time, place, and social and physical situation. The created metadata will be sketched with effective description languages enabling more effective information retrieval in the mobile semantic web and more easily accessible and easy to use multimedia databases.

Even though we have only identified a few near-term concepts enabled by the combination of context awareness and mobile multimedia technologies, we can also see the effects in the long term. The role of the mobile terminal in creating, editing, controlling, and accessing personal and shared multimedia will emphasize. The emerging standards in communication and in describing content and metadata enable seamless access and interoperability between multimedia albums in various types of systems. The personal "My Life" albums describing the entire life of a person in a semantically annotated form are become commonplace.

CONCLUSION

In this chapter we have examined the concept of situated mobile multimedia for mobile communications. We have introduced the characteristics of the key elements — multimedia, context awareness, and mobile communications — and discussed their current status and future trends in relation to the topic. Linked to this, we have presented a new categorization for contextual information sources, taking account of the special characteristics of mobile communication devices and their usage (Figure 2).

Integrating context awareness into mobile terminals has been introduced as a potential future technology in several forums. The motivation for this arises from the mobility characteristics of the device, its limited input and output functionalities, and the fact that the complexity of the device and its user interface is constantly growing. Context awareness as such has potential to functions such as automated or semi-automated action executions, creating shortcuts to applications and device functions, and situation-dependent information and service delivery. In this chapter we have limited our interest to examining the possibilities of combining the multimedia content to the phenomenon.

Currently, most of the mobile devices employing context awareness are specifically designed for the purpose, being somehow modified from the standard products by adding sensor modules. The lack of commonly agreed ontologies, standards, and description languages, as well as the shortage of suitable, commonly used gadgets as application platforms, has hindered the development of generally available, wide-audience services, and applications.

Multimedia, especially in the form of camera phones and Multimedia Messaging Service, has become a solid part of mobile communication technology during recent years. MMS enables easy delivery of multimedia content to a broad audience in a personalized manner. With situated multimedia, this means information delivery with multimedia content, such as information on local services or current events. Context awareness can also be exploited for executing underlying actions hidden from the user, such as selecting the data transfer medium for lower-cost streaming or better connection coverage.

REFERENCES

BBC News. (2005, May 11). *Mobile TV tests cartoons and news*. Retrieved June 15, 2005, from http://news.bbc.co.uk/1/hi/technology/4533205.stm

Brand, M., Oliver, N., & Pentland, A. (1997). Coupled hidden Markov models for complex action recognition. In *Proceedings of the 1997 Conference on Computer Vision and Pattern Recognition*.

Burrell, J., & Gay, G. K. (2002). E-graffiti: Evaluating real-world use of a context-aware system. *Interacting with Computers, 14*(4), 301-312.

Cole, H., & Stanton, D. (2003). Designing mobile technologies to support co-present collaboration. *Personal and Ubiquitous Computing, 7*(6), 365-371.

Cremers, I., & de Lussanet, M. (2005, March). *Mobile messaging forecast Europe: 2005 to 2010*. Cambridge, MA: Forrester.

Davies, N., Cheverst, K., Mitchell, K., & Efrat, A. (2001). Using and determining location in a context-sensitive tour guide. *IEEE Computer 34*(8), 35-41.

Dey, A. K., & Abowd, G. D. (2000). Toward a better understanding of context and context-awareness. In the *CHI 2000 Workshop on the What, Who, Where, When, Why, and How of Context-Awareness*.

Flanagan, J., Mäntyjärvi, J., & Himberg, J. (2002). Unsupervised clustering of symbol strings and context recognition. In *Proceedings of the IEEE International Conference of Data Mining 2002* (pp. 171-178).

Gellersen, H.W., Schmidt, A., & Beigl, M. (2002). Multi-sensor context-awareness in mobile devices and smart artefacts. *Mobile Networks and Applications, 7*(5), 341-351.

Grinter, R.E., & Eldridge, M. (2003). Wan2tlk?: Everyday text messaging. *CHI Letters, 5*(1), 441-448.

Häkkilä, J., Beekhuyzen, J., & von Hellens, L. (2004). Integrating mobile communication technologies in a student mentoring program. In *Proceedings of the IADIS International Conference of Applied Computing 2004* (pp. 229-233).

Häkkilä, J., & Mäntyjärvi, J. (2005). Combining location-aware mobile phone applications and multimedia messaging. *Journal of Mobile Multimedia, 1*(1), 18-32.

ITEA Technology Roadmap for Software-Intensive Systems (2nd ed.). (2004). Retrieved June 15, 2005, from http://www.itea-office.org/newsroom/publications/rm2_download1.htm

Kindberg, T., Spasojevic, M., Fleck, R., & Sellen, A. (2005, April-June). The ubiquitous camera: An in-depth study on camera phone use. *Pervasive Computing, 4*(2), 42-50.

Koch, F., & Sonenberg, L. (2004). Using multimedia content in intelligent mobile services. In *Proceedings of the WebMedia & LA-Web 2004* (pp. 41-43).

Korpipää, P., Koskinen, M., Peltola, J., Mäkelä, S. M., & Seppänen, T. (2003). Bayesian approach to sensor-based context-awareness. *Personal and Ubiquitous Computing, 7*(2), 113-124.

Kurvinen, E. (2003). Only when Miss Universe snatches me: Teasing in MMS messaging. In *Proceedings of DPPI'03* (pp. 98-102).

Mäntyjärvi, J., & Seppänen, T. (2003). Adapting applications in mobile terminals using fuzzy context information. *Interacting with Computers, 15*(4), 521-538.

Mäntyjärvi, J., Tuomela U., Känsälä, I., & Häkkilä, J. (2003). Context studio–tool for personalizing context-aware applications in mobile terminals. In *Proceedings of OZCHI 2003* (pp. 64-73).

Rantanen, M., Oulasvirta, A., Blom, J., Tiitta, S., & Mäntylä, M. (2004). InfoRadar: Group and public messaging in the mobile context. In *Proceedings of NordiCHI 2004* (pp. 131-140).

Rotuaari. (n.d.) *Rotuaari.* Retrieved June 15, 2005, from http://www.rotuaari.net/?lang=en

Schilit, B., Adams, N., & Want, R. (1994) Context-aware computing applications. In *Proceedings of IEEE Workshop on Mobile Computing Systems and Applications* (pp. 85-90).

Schmidt, A., Beigl, M., & Gellersen, H. (1999). There is more context than location. *Computers and Graphics Journal, 23*(6), 893-902.

Standard ISO 13407. (1999). *Human-centred design processes for interactive systems.*

Tachikawa, K. (2003, October). A perspective on the evolution of mobile communications. *IEEE Communication Magazine, 41*(10), 66-73.

KEY TERMS

Camera Phone: Mobile phone employing an integrated digital camera.

Context Awareness: Characteristic of a device that is, to some extend, aware of its surroundings and the usage situations

Location Awareness: Characteristic of a device that is aware of its current location.

Multimedia Messaging Service (MMS): Mobile communication standard for exchanging text, graphical, and audio-video material. The feature is commonly included in so-called camera phones.

Situated Mobile Multimedia: Technology feature integrating mobile technologies, multimedia, and context awareness.

Chapter XXIII
Context-Aware Mobile Capture and Sharing of Video Clips

Janne Lahti
VTT Technical Research Centre of Finland, Finland

Utz Westermann[1]
VTT Technical Research Centre of Finland, Finland

Marko Palola
VTT Technical Research Centre of Finland, Finland

Johannes Peltola
VTT Technical Research Centre of Finland, Finland

Elena Vildjiounaite
VTT Technical Research Centre of Finland, Finland

ABSTRACT

Video management research has been neglecting the increased attractiveness of using camera-equipped mobile phones for the production of short home video clips. But specific capabilities of modern phones — especially the availability of rich context data — open up new approaches to traditional video management problems, such as the notorious lack of annotated metadata for home video content. In this chapter, we present MobiCon, a mobile, context-aware home video production tool. MobiCon allows users to capture video clips with their camera phones, to semi-automatically create MPEG-7-conformant annotations by exploiting available context data at capture time, to upload both clips and annotations to the users' video collections, and to share these clips with friends using OMA DRM. Thereby, MobiCon enables mobile users to effortlessly create richly annotated home video clips with their camera phones, paving the way to a more effective organization of their home video collections.

INTRODUCTION

With recent advances in integrated camera quality, display quality, memory capacity, and video compression techniques, people are increasingly becoming aware that there mobile phones can be used as handy tools for the spontaneous capture of interesting events in form of small video clips. The characteristics of mobile phones open up new ways of combining traditionally separated home video production and management tasks at the point of video capture: The ability of mobile phones to run applications allows video production tools that combine video capture and video annotation. The classic approach of using video annotation tools to provide metadata for the organization and retrieval of video long after capture lacks user acceptance leading to the characteristic lack of metadata in the home video domain (Kender & Yeo, 2000). Context data about video capture available on mobile phones can be exploited to ease annotation efforts, which users try to avoid even at the point of capture (Wilhelm, Takhteyev, Sarvas, van House, & Davis, 2004). Time, network cell, GPS position, address book, and calendar can all be used to infer events, locations, and persons possibly recorded.

Furthermore, mobile phone-based video production tools can combine video capture with video upload and video sharing. With the ability to access the Internet via 2G and 3G networks from almost anywhere, phone users can directly load their clips to their home video collections stored on their PCs or by service providers disencumbering the limited memory resources of their phones. They also can share clips instantly with their friends via multimedia-messaging services. Digital rights management platforms like OMA DRM give users rigid control over the content they share preventing unwanted viewing or copying of shared clips.

However, video management research so far has mainly regarded mobile devices as additional video consumption channels. There has been considerable work concerning mobile retrieval interfaces (e.g., Kamvar, Chiu, Wilcox, Casi, & Lertsithichai, 2004), the generation of video digests for mobile users (e.g., Tseng, Lin, & Smith, 2004), and adaptive video delivery over mobile networks (e.g., Böszörményi et al., 2002), but a comprehensive view that considers the use of mobile phones as video production tools is still missing.

In this chapter, we present MobiCon: a context-aware mobile video production tool. Forming a cornerstone of the Candela platform, which addresses mobile home video management from production to delivery (Pietarila et al., 2005), MobiCon allows Candela users to record video clips with their camera phones and to semi-automatically annotate them at the point of capture in a personalized fashion. After recording, MobiCon extracts context data from the phone and passes it to an annotation Web service that derives reasonable annotation suggestions. These do not only include time- or position-based suggestions such as the season, city, or nearby points of interest possibly documented by the video; they also include personal calendar- and address book-based suggestions such as likely documented events and known locations like a friend's house. Besides these suggestions, the user can select concepts from a personal ontology with little manual effort or enter keywords for additional annotation.

MobiCon is further capable of uploading clips and their annotations to the users' private video collections in Candela's central video database directly after capture and permits users to immediately share these clips with friends, granting controlled access via OMA DRM.

Thus, MobiCon enables mobile phone users to create and share richly annotated home video clips with little effort, paving the way towards the more effective organization of their home video collections. The extensible architecture of the annotation Web service allows us to embrace and incrementally integrate almost any method for the generation of annotation suggestions based on context without having to change the MobiCon application.

In the following, we first illustrate the use of MobiCon in an application scenario. We then relate MobiCon to state-of-the-art mobile home video production tools. After a brief coverage of the Candela platform, we provide a technical description of the MobiCon tool. We provide a discussion and outline future developments, before we come to a conclusion.

MOBICON APPLICATION SCENARIO

In this section, we want to provide an intuitive understanding of MobiCon by illustrating its usage for home video clip production and sharing in a typical application scenario.

In the scenario, MobiCon is used to produce two video clips of a birthday barbecue and sauna party. Figure 1 depicts a sequence of screenshots of the basic steps involved when using MobiCon to capture, annotate, and share a video clip showing some guests having a beer outdoors; Figure 2 shows a similar sequence for an indoor clip showing guests leaving the sauna that is created by a different user, who also wants to restrict the playback of the shared clip via DRM protection. After the capture of both video clips (Figure 1(a) and Figure 2(a)), the users can immediately annotate them. MobiCon gathers context data from each phone and passes it to an annotation Web service operated by the Candela platform. Based on this data, the Web service infers possible annotations that are suggested to the users (Figure 1(b) and Figure 2(b)). Suggestions do not only include rather simple ones inferred from the capture time like "April" and "evening" (Figure 2(b)); when a mobile phone is connected to a GPS receiver that MobiCon can access, they also include location annotations like "Oulu" (town) and "Peltokatu" (the street name) the

Figure 1. Basic video capture, annotation, and sharing with MobiCon

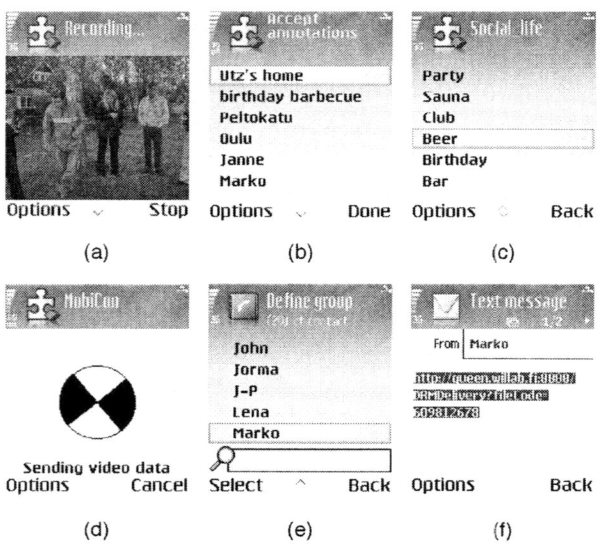

Web service derived from the GPS position of the capture using a reverse-geocoder (Figure 1(b)). The availability of a current GPS position also suggests that a clip covers an outdoor event (not shown in Figure 1(b)).

There are further highly personalized suggestions derived from phone address books and calendars, which can be synchronized with the Web service. Matching derived location information from the entries in a user's address book, the Web service can suggest known locations like "Utz's home" as annotations (Figure 1(b)); matching the capture time with the entries in a user's calendar, the Web service can suggest documented events like "birthday barbecue" (Figure 1(b)) along with event locations like "Utz's garden" (Figure 2(b)) and participants like "Janne" and "Marko" (Figure 1(b)) provided with the calendar entries. Users can correct the suggestions of the annotation Web service. In Figure 1(b), for instance, the user can remove the name "Marko" because he does not appear in the video.

In addition to the automatically generated annotation suggestions, MobiCon allows users to provide personalized manual clip annotations. Users can select concepts from personal, hierarchically organized home video ontologies that cover the aspects of their daily lives that they frequently document with video clips. The creator of the first video clip likes to have beers with friends, so his personal ontology contains the concept "beer" as a sub concept of "social life" (Figure 1(c)) that he can simply select for the annotation of his clip. The ontology of the creator of the second clip can contain different concepts due to different interests, such as the concept "camp fire" depicted in Figure 2(c). For the annotation of situations not covered by a user's personal ontology, MobiCon permits the entry of arbitrary keywords with the phone's keyboard as a last resort (Figure 2(d)).

After annotation, MobiCon uploads video clips and annotations to the users' personal video collections on the Candela platform (Figure 1(d)). Furthermore, MobiCon allows users to share freshly shot clips with contacts from their phone address books (Figure 1(e)). MobiCon then sends a text message with a link pointing to the shared clip in the user's collec-

Figure 2. DRM-protected sharing of clips

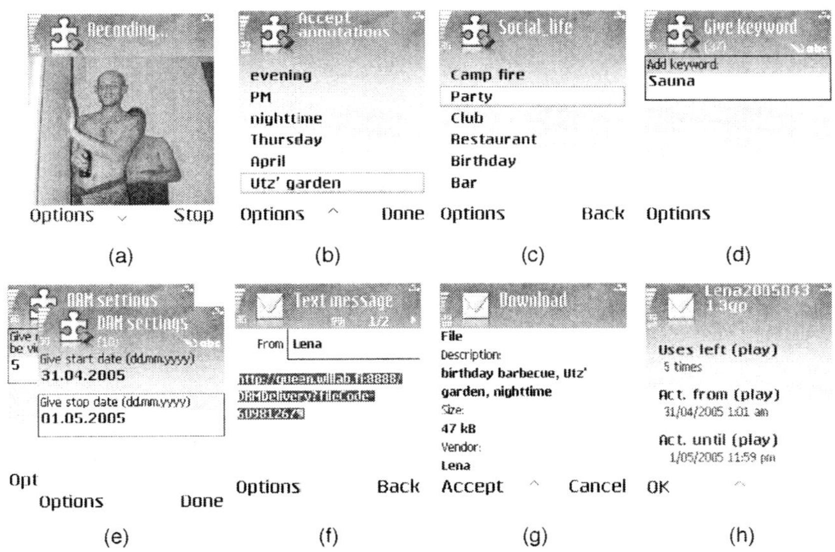

tion to each selected contact, as depicted by Figure 1(f). When the recipient selects the link, the phone will download and play the clip. The second video clip shows the somewhat delicate situation of two party guests coming out of the sauna. While the creator of this clip still wants to share it with a friend, she wants to impose usage restrictions. Utilizing MobiCon's DRM support, she restricts playback of the shared clip to five times within the next 24 hours on the phone of her friend (Figure 2(e)). MobiCon makes the Candela platform prepare a copy of the clip that encodes these limitations using OMA DRM. The link to the video contained in the text message that is then sent to the friend points to the DRM-protected copy (Figure 2(f)). After selecting the link, the recipient sees a description of the clip and is asked for permission to download Figure 2(g)). If download is accepted, the OMA-DRM-compliant phone recognizes and enforces the restrictions imposed upon the clip and displays the corresponding DRM information before starting playback (Figure 2(h)).

RELATED WORK

The previous section illustrated MobiCon's different functionalities from a user's perspective in a typical application scenario. We now compare MobiCon to existing approaches in the field of mobile video production tools, thereby showing how it exceeds the state-of-the-art. In particular, we relate MobiCon to mobile video capture tools, mobile video editing applications, mobile video annotation tools, and tools for mobile content sharing.

Mobile Video Capture Tools

Probably every modern mobile phone with an integrated camera features a simple video capture tool. MobiCon goes beyond these tools by not only allowing the capture of a video clip but also allowing for immediate annotation for later retrieval, its immediate upload to the user's home video clip collection, as well as its immediate sharing controlled via OMA DRM.

Mobile Video Editing Tools

Mobile video editing tools like Movie Director (n.d.) or mProducer (Teng, Chu, & Wu, 2004) facilitate simple and spontaneous authoring of video clips at the point of capture on the mobile phone. Unlike MobiCon, the focus of these tools lies on content creation and not on content annotation, uploading, and sharing.

Mobile Video Annotation Tools

While there are many PC-based tools for video annotation as a post-capturing processing step (e.g., Abowd, Gauger, & Lachenmann, 2003; Naphade, Lin, Smith, Tseng, & Basu, 2002), mobile tools like MobiCon permitting the annotation of video clips at the very point of capture, when users are still involved in the action, are rare. M4Note (Goularte, Camancho-Guerrero, Inácio Jr., Cattelan, & Pimentel, 2004) is a tool that allows the parallel annotation of videos on a tablet PC while they are being recorded with a camera. Unlike MobiCon, M4Note does not integrate video capture and annotation on a single device. Annotation is fully manual and not personalized; context data is not taken advantage of for suggesting annotations. M4Note does not deal with video upload and sharing.

Furthermore, mobile phone vendors usually provide rudimentary media management applications for their phones that — compared to MobiCon and its support for annotation suggestions automatically derived out of context data and personalized manual annotation using concepts from user-tailored ontologies and keywords — offer only limited video annotation capabilities. As an example, Nokia Album (n.d.) allows the annotation of freshly shot clips with descriptive titles. As a form of context-awareness, Nokia Album records the time stamps of video captures but does not infer any higher-level annotations out of them.

The lack of sophisticated mobile video annotation tools constitutes a contrast to the domain of digital photography. Here, research has recently been investigating the use of context data such as time and location to automatically cluster photographs likely documenting the same event (Cooper, Foote, Girgensohn, & Wilcox, 2003; Pigeau & Gelgon, 2004) and to automatically infer and suggest higher-level annotations, such as weather data, light conditions, etc. (Naaman, Harada, Wang, Garcia-Molina, & Paepcke, 2004). Compared to MobiCon, these approaches do not present the inferred annotation suggestions to users at the point of capture for immediate acceptance or correction; inference takes place long afterwards when the photographs are imported to the users' collections.

For the annotation of photographs at the point of capture, Davis, King, Good, and Sarvas (2004) have proposed an integrated photo capture and annotation application for mobile phones that consults a central annotation database to automatically suggest common annotations of pictures taken within the same network cell. Apart from its focus on video, MobiCon mainly differs from this approach by offering a different and broader variety of derivation methods for context-based annotation suggestions and by addressing content upload and sharing.

Mobile Content Sharing Tools

Mobile content sharing applications like PhotoBlog (n.d.), Kodak Mobile (n.d.), and MobShare (Sarvas, Viikari, Pesonen, & Nevanlinna, 2004) allow users to immediately share content produced with their mobile phones, in particular photographs. Compared to MobiCon, there are two major differences. Firstly, these applications realize content sharing by uploading content into central Web albums, in which users actively browse for shared content with a Web browser. In contrast, MobiCon users view shared content by following links in notification messages they receive. Also, MobiCon gives users more control over shared content by applying DRM techniques.

Secondly, current content sharing systems offer rather restricted means for content annotation, mainly allowing content to be manually assigned to (usually flat) folder structures and attaching time stamps for folder- and timeline-based browsing. Nokia Lifeblog (n.d.) goes a bit beyond that by automatically annotating content with the country where it has been created, which is obtained from the mobile network that the phone is currently logged in to. But compared to MobiCon, these still constitute very limited forms of context-based annotations.

THE CANDELA PLATFORM

Facing the increasingly popular use of mobile devices for home video production, we have developed the Candela mobile video management platform. Incorporating MobiCon, it provides support for all major process steps in the mobile home video management chain, ranging from mobile video creation, annotation and sharing to video storage, retrieval, and delivery using various mobile and stationary terminals connected to the Internet via various types of networks like GPRS/EDGE, 3G/UMTS, WLAN, and fixed networks. In the following, we briefly describe the platform's key elements and their relationship to MobiCon.

Figure 3 illustrates the interplay of the different components of the Candela platform. As explained before, the MobiCon mobile phone-based video production application permits the integrated capture, personalized, context-aware annotation, upload, and DRM-controlled sharing of video clips. To this end, MobiCon interacts closely with the central Candela server, namely with its ontology manager, annotation Web service, and upload gateway components.

The RDF-based ontology manager stores the personal home video ontologies of Candela's users. When MobiCon starts for the first time, it loads the ontology of the current user from the manager so that its concepts can be used for the personalized annotation of videos.

The annotation Web service is called by MobiCon during clip annotation, passing context data such as capture time, GPS position, and user information. The Web service derives annotation suggestions based on this data, which MobiCon then presents to the user.

The upload gateway is used to transfer clips and their annotation after capture from MobiCon to the users' video collections. The gateway receives the clips in 3GP format and clip metadata including user annotations and context data in MPEG-7 format. The clips are passed on to the video manager for storage and transcoding into suitable formats for the video players of different devices and for different network speeds. The video manager also prepares OMA DRM-enhanced clip variants when MobiCon users define usage restrictions for the video clips that they are about to share. The clip metadata is stored in a database implemented on top of the Solid Boost Engine distributed relational database management system for

Figure 3. Candela platform architecture

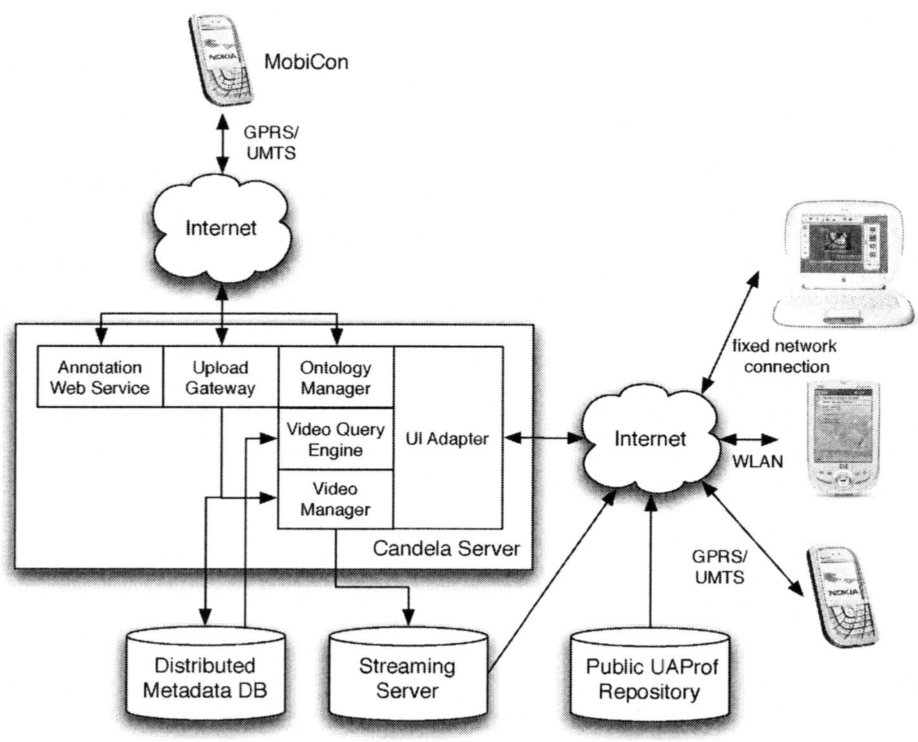

scalability to large numbers of users and videos.

Via its UI adapter, video query engine, and video manager components, the Candela server also provides rich video retrieval facilities. While MobiCon is a standalone mobile phone application, the video retrieval interfaces of the Candela platform are Web browser-based. Thus, we can apply Web user interface adaptation techniques to give users access to their video collections from a variety of user terminals and networks.

The UI adapter is implemented on top of an Apache Cocoon Web-development framework. Using XSLT stylesheets, it generates an adaptive video browsing and retrieval interface from an abstract XML-MPEG7 content, considering the capabilities of the user devices obtained from public UAProf repositories. For example, when using a PC Web browser, the adapter creates a complex HTML interface combining keyword queries, ontology-based video browsing, as well as the display and selection of query results into a multi-frame page. When using a mobile phone browser, the adapter splits the same interface into several HTML pages.

For performing video browsing and content-based retrieval, the UI adapter interacts with the video query engine, which supports the use of time, location, video creators, and keywords as query parameters. The video query engine translates these parameters into corresponding SQL statements run on the metadata database and returns a personalized ranked result list in MPEG-7 format, which the UI adapter then integrates into the user interface. The engine interacts with the ontology manager for personalized keyword expansion. For example, the search term "animal" will be expanded to all subconcepts of "animal," (e.g., "cat" and "dog") in querying user's personal ontology.

When a video clip is selected for viewing, the video manager takes care of its delivery. It selects the format and compression variant most appropriate to the client device and network, again exploiting the device capability profiles in the public UAProf repositories–especially the information about screen size, and the video manager supports HTTP-based download of a clip as well as streaming delivery via the Helix DNA streaming server.

MOBICON

MobiCon is a Java 2 Micro Edition/MIDP 2.0 application that runs on Symbian OS v8.0 camera phones with support of the Mobile Media, Wireless Messaging, and Bluetooth APIs. We now provide details on the video production and management tasks—video capture, annotation, upload, and sharing—combined by MobiCon.

Video Capture

When MobiCon is started for the first time, the user is authenticated by the Candela platform. Upon successful authentication, MobiCon receives the user's personal ontology from the ontology manager and stores it along with the user's credentials in the phone memory for future use making use of MIDP record management, as it is assumed that the user stays the same. MobiCon still permits re-authentication for a different user.

After successful login, users can start capturing clips. For this purpose, MobiCon accesses the video capture tool of the mobile phone via the Mobile Media API. The captured content is delivered in 3GP format, using AMR for audio encoding and H.263/QCIF at 15 frames per second and 174x144 pixels resolution for video encoding. MobiCon stores the captured video clip in the phone's memory. Users can view the captured or another stored clip, capture another clip, or start annotating a stored clip as explained in the following.

Video Annotation

For the annotation of video clips, MobiCon provides automatic, context-based annotation suggestions as well as the option to manually annotate clips with concepts of personal home video ontologies or keywords. We now provide more details on the generation of context-based annotation suggestions and the use of personal ontologies for annotation.

Context-Based Annotation Suggestions

For the generation of appropriate annotation suggestions, MobiCon gathers context data that is available about the capture of a video clip on the mobile phone. In particular, MobiCon collects the username, capture time, and duration of the clip. Additionally, MobiCon is able to connect via the Bluetooth API to GPS receivers that support the NMEA protocol. If such a receiver is connected to the phone, MobiCon polls for the current GPS position and stores it along with a timestamp as a measure for its age. Given these context data, MobiCon invokes the annotation Web service running on the Candela server as a Java servlet via an HTTP request, opening a connection to the Internet via UMTS or GPRS if not yet established.

The reasons for outsourcing the derivation of annotation suggestions to a Web service are mainly ease of prototyping and deployment. We can incrementally add new methods for annotation suggestions to the Web service while keeping the MobiCon client unchanged, thus saving on update (re)distribution costs. Also, a Web service allows the reuse of the context-based annotation suggestion functionality on devices other than mobile phones.

A drawback of this design is the costs incurred by remotely invoking a Web service from a mobile phone. But given the costs accrued anyway by uploading and sharing comparably high-volume video clips, these are negligible. A further problem is how to provide the Web service with access to personal user data for the generation of annotation suggestions, such as phone calendars or address books; passing the whole address book and calendar of a user as parameters to the Web service with each invocation is certainly not feasible. Leaving privacy issues aside, we circumvent this problem by allowing users to upload their calendars and address books to a central directory on the Candela server in iCalendar and vCard formats via a MobiCon menu option. From this directory, this data can be accessed from the Web service with user names as keys.

Figure 4 presents an overview of the design of the annotation Web service. When the Web service receives an annotation request, it publishes the context data carried by the request on the annotation bus. The annotation bus forms a publish/subscribe infrastructure for annotation modules that are in charge of actually deriving annotation suggestions. The annotation modules run concurrently in their own threads, minimizing response times and maximizing the utilization of the Web service's resources when processing multiple annotation requests.

The annotation modules listen to the bus for the data they need for their inferences, generate annotation suggestions once they have received all required data for a given annotation request, and publish their suggestions back to the bus, possibly triggering other annotation modules. The annotation Web service collects all suggestions published to the bus for a request, and, once no more suggestions will be generated, returns the results to MobiCon.

This results in a modular and extensible design: the annotation modules used for the

Figure 4. Annotation Web service design

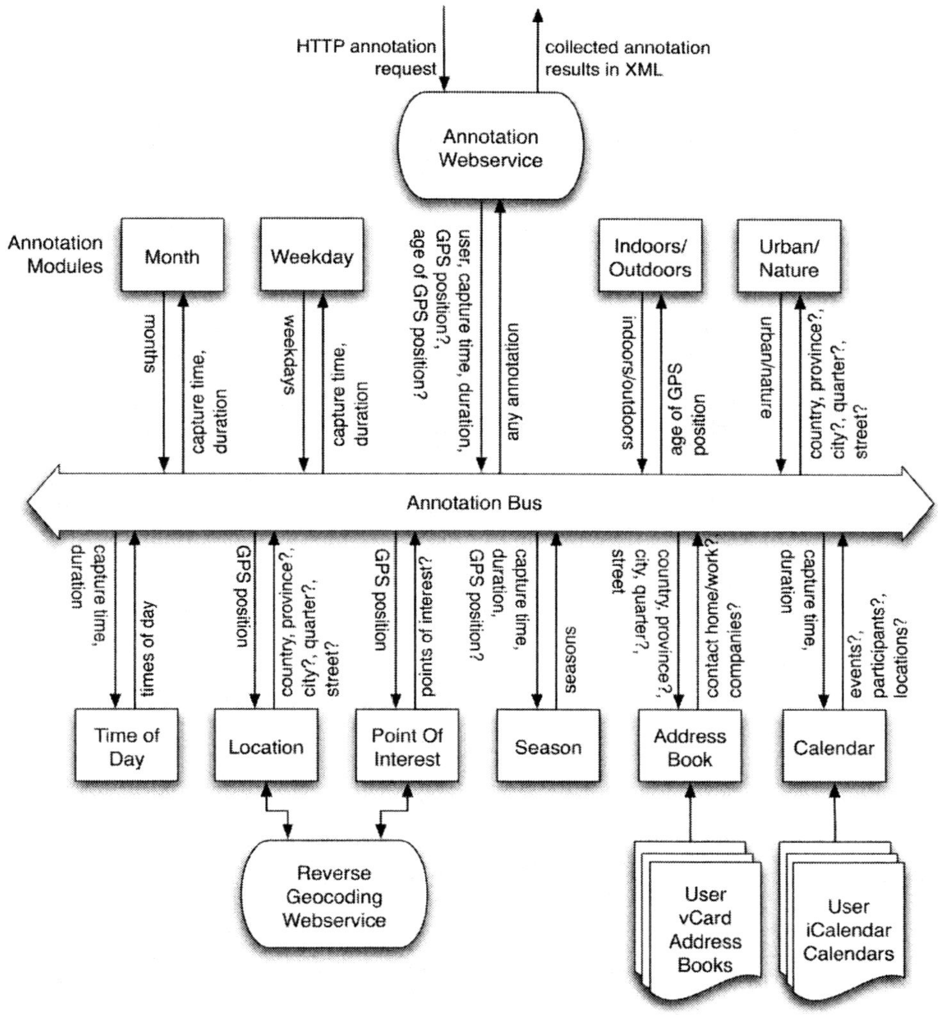

generation of annotation suggestions can be selected to suit the needs of an individual application and new modules can be dynamically added to the system as they become available without having to reprogram or recompile the Web service.

Figure 4 also provides information about the annotation modules currently implemented, along with the types of data on which they base their inferences and the types of suggestions they publish. In the following, we highlight some of the more interesting ones:

The location and point of interest annotation modules suggest address and points of interests probably captured by the clip being annotated based on GPS position utilizing the commercial

ViaMichelin reverse-geocoding Web service. The calendar annotation module searches the user calendar for events that overlap with the capture time, suggesting event names, locations, and participants as annotations. The address book annotation module searches the user address book for the home or work addresses of contacts or company addresses matching the address data derived by any other annotation module, suggesting them as location annotations. The indoors/outdoors annotation module suggests whether a clip has been shot outdoors or indoors, utilizing the fact that GPS signals cannot be received indoors and thus the age of the GPS position will exceed a threshold in this case. Depending on the level of detail of address data derived by other modules, the urban/nature annotation module suggests whether a clip shows an urban environment or nature. If information about a city or street is missing, it suggests nature, otherwise an urban environment is assumed.

Ontology-Based Annotations

MobiCon permits an inexpensive manual annotation of content using hierarchically structured ontologies with concepts from the daily lives of users. Instead of having to awkwardly type such terms with the phone keyboard over and over again, users can simply select them by navigating through MobiCon's ontology annotation menu as illustrated in Figure 5 (a-c). Without imposing a single common ontology onto every user, MobiCon permits each user to have a personal ontology for home video annotation, merely predefining two upper levels of generic concepts that establish basic dimensions of video annotation (Screenshots (a) and (b) of Figure 5). Below these levels, users are free to define their own concepts, such as those depicted in Screenshot (c). MobiCon's user interface permits the entry of new concepts at any level at any time during the annotation process in Screenshot (d).

The rationale behind this approach is as follows: firstly, it allows users to optimize their ontologies for their individual annotation needs, so that they can reach the concepts important to them in few navigation steps and without having to scroll through many irrelevant concepts on a small phone display on the way. Our experiences from initial user trials indicate that precisely because users want to keep annotation efforts low, they are willing to invest some efforts into such optimization.

The concepts that are important for clip annotation differ very much between people: a person often enjoying and documenting sauna events might introduce "sauna" as a subconcept of "social life" to his or her ontology, whereas an outdoor person might need a subconcept "camp fire", and so on. Differences also occur in the hierarchical organization of concepts: users frequently visiting bars might consider the concept "bar" as a subconcept of "social life" (like in Screenshot (c)), while a bar's owner might see it as a subconcept of "work activity."

Secondly, by imposing a common set of top-level concepts (used for representation of profiles of users' interests) onto the personal ontologies of the users, we establish a common foundation for the querying and browsing of video collections, making it easier to find interesting clips also in the collections of other users.

MobiCon receives the personal ontology of a user from the ontology manager in RDF format after successful authentication and caches it for successive use in the phone's memory.

Figure 5. MobiCon ontology user interface

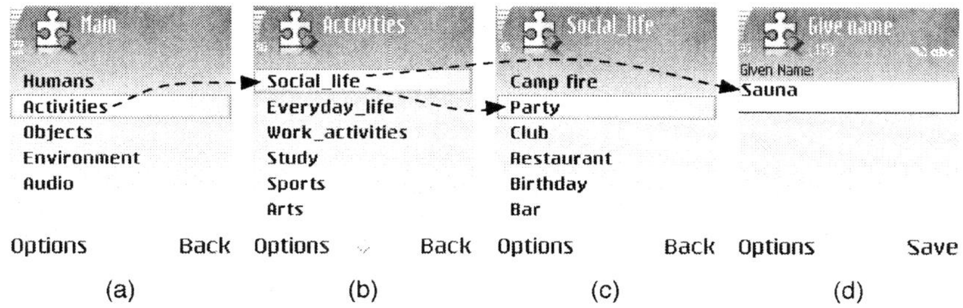

Video Upload and Storage

After annotation, MobiCon gives the user an opportunity to upload the video clip and its annotations to his or her video collection on the Candela server via the upload gateway. As already explained, the video clip is handed over to the video manager which transcodes it to different formats at different bit rates in order to provide a scaleable service quality for different devices and network connections: Real Video, H.264, and H.263 encodings are used for delivering video content to mobile devices, as well as MPEG4 for desktop computers. In the future, scalable video codecs will remove the need of transcoding.

The clip metadata is represented in MPEG-7 format that mainly constitutes a profile of the video and video segment description schemes defined by the standard. Figure 6 gives a sample of this format. It incorporates context data about the clip's capture including the creator's name, GPS position, region and country, date and time of day, and length of the video clip, as well as the clip annotations embedded in free text annotation elements. This includes the suggestions generated by the annotation Web service, the concepts selected from the user's personal home video ontology, and the keywords manually provided by the user.

Video Sharing

Users can share uploaded clips with the contacts in their address book, defining usage restrictions according to the OMA DRM standard if desired. The standard offers three approaches to content protection: forward-lock, combined delivery, and separate delivery. Forward-lock thwarts the forwarding of content to a different device, while combined delivery allows one to impose further restrictions, such as a limited number of playbacks or a permissible time interval for playback. In both approaches, the protected content is embedded by the content provider in a DRM packet along with the specification of the usage restrictions. Under separate delivery, the restrictions and the content are delivered separately and integrated on the playback device.

MobiCon supports the protection of video clips via forward-lock and combined delivery. For reasons of implementation, usage complexity, and the requirements imposed onto client devices, we have chosen not to support separate delivery at this stage.

When the user has specified the desired usage restrictions for a clip being shared, MobiCon uses a secure connection to contact the video manager, which employs the Nokia Content Publishing Toolkit to put a copy of the video clip into a DRM packet with the specified

Figure 6. The MobiCon metadata format

```
<?xml version="1.0" encoding="iso-8859-1"?>
<Mpeg7 xmlns="urn:mpeg:mpeg7:schema:2001"
       xmlns:xsi="http://.../XMLSchema-instance"
<Description xsi:type="ContentEntityType">
  <DescriptionMetadata>
    <Creator>
      <Role href="creatorCS">
        <Name>Utz</Name>
      </Role>
    </Creator>
    <CreationLocation>
      <GeographicPosition>
        <Point latitude="65.0057"
               longitude="25.4654"/>
      </GeographicPosition>
      <Region>FI</Region>
      <Country>Finland</Country>
    </CreationLocation>
    <CreationTime>
      2005-04-07T20:36:00+03:00
    </CreationTime>
    <Instrument>
      <Tool>
        <Name>MobiCon v1.0</Name>
      </Tool>
    </Instrument>
  </DescriptionMetadata>
  <MultimediaContent xsi:type="VideoType">
    <Video>
      <MediaLocator>
        <MediaUri>
          mobileUpload/Utz200543118.3gp
        </MediaUri>
      </MediaLocator>
      <MediaTime>
        <MediaTimePoint>
          T00:00:00:0F30000
        </MediaTimePoint>
        <MediaDuration>
          PT0H0M24S0N30000F
        </MediaDuration>
      </MediaTime>
      <TemporalDecomposition>
      <VideoSegment>
        <TextAnnotation confidence="1"
                        relevance="1"
                        type="scene">
          <FreeTextAnnotation>
            Peltokatu
          </FreeTextAnnotation>
          <FreeTextAnnotation>
            Janne
          </FreeTextAnnotation>
            ...
          <FreeTextAnnotation>
            beer
          </FreeTextAnnotation>
        </TextAnnotation>
        <MediaTime>
          <MediaTimePoint>
            T00:00:00:0F30000
          </MediaTimePoint>
          <MediaIncrDuration>
            ...
          </MediaIncrDuration>
        </MediaTime>
      </VideoSegment>
      </TemporalDecomposition>
    </Video>
  </MultimediaContent>
</Description>
</Mpeg7>
```

restrictions. The video manager also creates a key pair for each recipient of the clip. One key of every pair remains with the DRM packet, while the other is returned to MobiCon.

Using the Wireless Messaging API, MobiCon then sends a text-message to each recipient containing URL-link with a key pointing to the DRM protected clip. When the recipient of the message selects the link, the phone establishes an HTTP connection to the video manager. Using the recipient's key, the video manager checks whether access to the DRM protected clip can be granted by pairing the key with the right clip. If a matching clip is found, a download descriptor with basic information about the clip like creator, length, and description is returned to the recipient's mobile phone and the used key pair is removed, in order to prevent re-usage. After deciding to really download the packet, the user can finally watch the protected video clip, but only on the paired device and within the limits of the usage restrictions.

DISCUSSION

Having given a technical description of the MobiCon application for the combined production, context-aware annotation, and sharing of home video clips with mobile phones at the point of capture, we now provide a critical discussion and outline future developments.

The ways in which the annotation Web service can utilize temporal and spatial context data for the generation of annotation suggestions are not limited to those described in the previous section: weather or light conditions probably documented by a video can be obtained from meteorological databases given capture time and location (Naaman et al., 2004), annotations from other videos shot at the same time and place can be suggested using clustering methods (Davis et al., 2004; Pigeau & Gelgon, 2004), and much more. We want to support these uses for time and location context data with MobiCon as well. For that purpose, we benefit from the extensible design of the annotation Web service, as it enables us to incrementally develop and integrate modules for these kinds of annotation suggestions without having to modify the MobiCon application itself.

Reasonable annotation suggestions cannot only be derived from context data, from content analysis, or a combination of both. We plan to integrate an audio classifier that is capable of identifying segments of speech, music, and different kinds of environmental noises within videos with high degree of reliability. The results of such an audio classification can be used to enhance our simplistic indoors/outdoors and urban/nature annotation modules, which so far are solely based on the age of the last available GPS position and the level of detail of the address returned by the reverse-geocoder for that position.

Integrating content analysis with the current centralized annotation Web service design is problematic. As an annotation module using content analysis methods needs access to the full video clip being annotated, the clip has to be uploaded to the Web service before any suggestions can be created. The incurring delay will hamper the capture and annotation process. Therefore, we want to distribute the annotation Web service, permitting annotation modules to run on the server and on the mobile phone. This will not only allow us to perform content analysis on the mobile phone avoiding upload delays; we will also be able to perform annotations based on sensitive personal data like address books and calendars directly on the phone, avoiding the privacy issues raised by moving such data to a central server as done currently.

Beyond improving the generation of annotation suggestions, MobiCon's user interface for annotating video clips on the basis of personal ontologies will also require some improvement. So far, users only have very limited means of modifying their ontologies in the middle of the video capture and annotation process, merely being able to add new subconcepts. Larger modifications must be performed outside of MobiCon using Candela's Web front-end. Moreover, MobiCon's DRM-based video sharing functionality is limited, allowing the sharing of clips only right after capture. We are currently investigating the integration of a user interface into MobiCon that allows users to share any clip existing in their collections.

Finally, we want to improve the video capturing and editing functionalities of MobiCon by integrating it with a mobile video editing application.

CONCLUSION

This chapter has introduced MobiCon, a video production tool for mobile camera phones that exploits specific characteristics of mobile phones— in particular the ability to run applications, the availability of context data, and access to the Internet from almost anywhere — to integrate traditionally separated home video production and management tasks at the point of video capture. MobiCon assists mobile phone users in capturing home video clips, uses context data after capture to suggest reasonable annotations via an extensible annotation Web service, supports personalized manual annotations with user-specific home video ontologies and keywords, uploads video clips to the users' video collections in Candela's central video database, and facilitates the controlled sharing of clips using OMA.

Initial experiences we have been able to gain so far from our personal use of MobiCon are encouraging. With MobiCon, the provision of useful annotations for home video clips is largely automatic and not overly intrusive to the general video capturing process, effectively resulting in the better organization of home video clips without much additional overhead. We are in the process of subjecting this personal experience towards a user study.

This work was done in the European ITEA project "Candela", funded by VTT Technical Research Centre of Finland and TEKES (National Technology Agency of Finland). Support of Finnish partners Solid Information Technology and Hantro Products is greatly acknowledged.

REFERENCES

Abowd, G. D., Gauger, M., & Lachenmann, A. (2003). The family video archive: An annotation and browsing environment for home movies. *Proceedings of the 11th ACM International Conference on Multimedia*, Berkeley, CA.

Böszörményi, L., Döller, M., Hellwanger, H., Kosch, H., Libsie, M., & Schojer, P. (2002). Comprehensive treatment of adaptation in distributed multimedia systems in the ADMITS project. *Proceedings of the 10th ACM International Conference on Multimedia*, Juan-les-Pins, France.

Cooper, M., Foote, J., Girgensohn, A., & Wilcox, L. (2003). Temporal event clustering for digital photo collections. *Proceedings of the 11th ACM International Conference on Multimedia*, Berkeley, CA.

Davis, M., King, S., Good, N., & Sarvas, R. (2004). From context to content: Leveraging context to infer multimedia metadata. *Proceedings of the 12th ACM International Conference on Multimedia*, New York.

Goularte, R., Camancho-Guerrero, J. A., Inácio Jr., V. R., Cattelan, R. G., & Pimentel, M. D. G. C. (2004). M4Note: A multimodal tool for multimedia annotations. *Proceedings of the WebMedia & LA-Web 2004 Joint Conference*, Ribeirão Preto, Brazil.

Kamvar M., Chiu P., Wilcox L., Casi, S., & Lertsithichai, S. (2004). MiniMedia Surfer: Browsing video segments on small displays. *Proceedings of the 2004 Conference on Human Factors and Computing Systems (CHI 2004)*, Vienna, Austria.

Kender, J. R., & Yeo, B. L. (2000). On the structure and analysis of home videos. *Proceedings of the 4th Asian Conference on Computer Vision (ACCV 2000)*, Taipei, Taiwan.

Kodak Mobile (n.d.). Retrieved May 3, 2005, from http://www.kodakmobile.com

Movie Director (n.d.). Retrieved May 3, 2005 from http://www.nokia.com/nokia/-0,6771, 54835,00.html

Naaman, M., Harada, S., Wang, Q. Y., Garcia-Molina, H., & Paepcke, A. (2004). Context data in geo-referenced digital photo collections. *Proceedings of the 12th ACM International Conference on Multimedia*, New York.

Naphade, M., Lin, C. Y., Smith, J. R., Tseng, B., & Basu, S. (2002). Learning to annotate video databases. *Proceedings of the SPIE Electronic Imaging 2002 Symposia (SPIE Volume 4676)*, San Jose, California.

Nokia Album (n.d.). Retrieved May 3, 2005, from http://www.nokia.com/nokia/-0,6771, 54835,00.html

Nokia Lifeblog (n.d.). Retrieved May 3, 2005, from http://www.nokia.com/lifeblog

PhotoBlog (n.d.). Retrieved May 3, 2005, from http://www.futurice.fi

Pietarila, P., Westermann U., Järvinen, S., Korva J., Lahti, J., & Löthman, H. (2005). Candela — storage, analysis, and retrieval of video content in distributed systems — personal mobile multimedia management. *Proceedings of the IEEE International Conference on Multimedia & Expo (ICME 2005)*, Amsterdam, The Netherlands.

Pigeau, A., & Gelgon, M. (2004). Organizing a personal image collection with statistical model-based icl clustering on spatio-temporal camera phone meta-data. *Journal of Visual Communication & Image Retrieval*, *15*(3), 425-445.

Sarvas, R., Viikari, M., Pesonen, J., & Nevanlinna, H. (2004). MobShare: Controlled and immediate sharing of mobile images. *Proceedings of the 12th ACM International Conference on Multimedia*, New York.

Teng, C. M., Chu, H. H., & Wu, C. I. (2004). mProducer: Authoring multimedia personal experiences on mobile phones. *Proceedings of the IEEE International Conference on Multimedia & Expo (ICME 2004)*, Taipei, Taiwan.

Tseng, B. L., Lin, C. Y., & Smith, J. R. (2004). Using MPEG-7 and MPEG-21 for personalizing video. *IEEE MultiMedia*, *11*(1), 42-52.

Wilhelm, A., Takhteyev, Y., Sarvas, R., van House, N., & Davis, M. (2004). Photo annotation on a camera phone. *Proceedings of the 2004 Conference on Human Factors and Computing Systems (CHI 2004)*, Vienna, Austria.

KEY TERMS

3GP Format: Mobile phone video file format produced by mobile phone video recording applications.

Annotation: Extra information or note associated with a particular object.

Candela: A two-year EUREKA/ITEA project researching content analysis, delivery, and architectures.

DRM: Digital rights management is a method for licensing and protecting digital media.

GPS (Global Positioning System): A global satellite-based navigation system.

Metadata: Metadata is the value-added information of data, for example, describing a content of picture, video, or document.

MIDP 2.0 (Mobile Information Device Profile Version 2.0): A Java runtime environment for mobile devices.

MPEG-7 (Multimedia Content Description Interface): MPEG-7 is an ISO/IEC standard developed by MPEG (Moving Picture Experts Group) to describe multimedia content.

OMA DRM (Open Mobile Alliance's Digital Rights Management): A standard developed by the OMA organization for the management of digital rights in mobile phones.

Ontology: A description of the concepts and relationships of objects in a formal way using a controlled vocabulary.

ENDNOTE

1 This work was carried out under the tenure of an ERCIM fellowship.

Chapter XXIV
Content-Based Video Streaming Approaches and Challenges

Ashraf M. A. Ahmad
National Chiao Tung University, Taiwan

ABSTRACT

Video streaming poses significant technical challenges in quality of service guarantee and efficient resource management. Generally, it is recognized that end-to-end quality requirements of video streaming application can be reasonably achieved only by integrative study of advanced networking and content processing techniques. However, most existing integration techniques stop at the bit stream level, ignoring a deeper understanding of the media content. Yet, the underlying visual content of the video stream contains a vast amount of information that can be used to predict the bit-rate or quality more accurately. In the content-aware video streaming framework, video content is extracted automatically and used to control video quality under various manipulations and network resource requirements.

INTRODUCTION

Video has been an essential element for communications and entertainment for many years. Initially video was captured and transmitted in analog shape. The emergence of digital integrated circuits and computers led to the digitization of video, and digital video enabled a revolution in the compression and communication of video. Video compression (Mitchell, Pennebaker, Fogg, & LeGall 1996) and transmission became an important area of research in the last two decades and enabled a variety of applications including video storage on DVD and Video-CD, video broadcasting over digital cable, satellite and terrestrial digital television (DTV), high definition TV (HDTV), video conferencing and videophone over circuit-switched networks. The drastic growth and popularity of the Internet motivated video communication over best-effort packet networks. Video over best-effort packet networks is complicated by a number of factors including unknown and time varying bandwidth, delay, and packet losses, as well as many additional issues such as how to fairly share the network resources amongst many flows and how to efficiently perform one-to-many communication for popular content "congestion control," and so forth. The Internet disseminates enormous amounts of information for a wide variety of applications all over the world. As the number of active users on the Internet has increased, so has the tremendous volume of data that is being exchanged between them, resulting in periods of transient congestion on the network. On the transmitted data over the internet regards, some researchers estimates (Chandra, & Ellis 1999; Ortega, Carignano, Ayer, & Vetterli, 1997) about 77% of the data bytes accessed on the Web are in the form of multimedia objects.

This chapter examines the challenges that face simultaneous delivery and playback, or streaming of video on a content awareness basis. We explore *approaches and systems* that enable streaming of pre-encoded or live video over packet networks such as the Internet in content aware manner.

First, we try to describe and discus some of the basic approaches and key challenges in video streaming. Generally the most straightforward approach for video delivery in the Internet is by an approach similar to a file download, but we refer to it as video download to keep in mind that it is a video and not a general file type. Specifically, video download is similar to a file download, but it is a very large file. This scheme allows the use of established delivery mechanisms, for example TCP as the transport layer, FTP, HTTP, or HTTPS at the application layers. However, this scheme has a number of drawbacks. Since videos generally correspond to very large files, the download approach usually requires long download times and large storage spaces. These are all crucial practical limitation. In addition, the entire video file must be downloaded before viewing can start. This requires patience on the client part and also reduces flexibility in certain scenarios. In one scenario, if the client is unsure of whether he wants to view the video, he must still download the entire video before viewing it and making a decision. In another scenario, the user may not be aware about the exact disk space on his machine, therefore he might start to download a large video file which takes few hours, then an error message would pop up stating disk insufficiency. The user wasted hours for nothing. These scenarios and other scenarios cause great obstacles in the video file download scheme. Video delivery by video streaming attempts to overcome the problems associated with the video file download scheme, and also provides a significant amount of additional capabilities "viewing flexibility." The basic idea behind video streaming is to make "make simultaneous delivery and playback" which splits the video into portions, transmits these portions in succession, and enabled the receiver to decode and playback the video as these parts are received, without having to wait for the entire video to be delivered. Video streaming enables simultaneous delivery and playback of the video. This is in contrast to file download where the entire video must be delivered before playback can begin. In video streaming there usually is a short latency (usually on the order of 10-15 seconds) between the start of delivery and the beginning of playback at the client. Video streaming provides a number of advantages including low delays before viewing starts, and low storage requirements since only a small portion of the video is stored at the client at any point in time. The storage issues can be enhanced by deploying some caching strategies as well. For video streaming data, any data that is lost in transmission cannot be used at the receiver. Furthermore, any data that arrives

Figure 1. General video streaming architecture

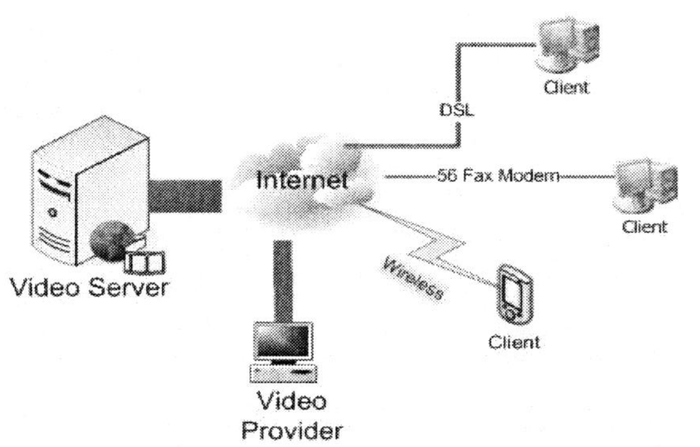

late is also useless. Especially, any data that arrives after its decoding and display deadline is too late to be displayed, which CODEC technology referred to it as time_stamp constraint (Mitchell, et al. 1996). Reader may note that certain data may still be useful even if it arrives after its display time. For example, if subsequent data depends on this "late" data (e.g., the relation between I-frame and P-frame in MPEG GOP) (Mitchell, et al. 1996). Therefore, an important goal of video streaming is to perform the streaming in a manner so that time constraints are met. A general architecture for video streaming is presented in Figure 1. You may notice the video streaming could be transmitted among different users with different paths and environments, such as a mobile user in wireless network, a video client in DSL network and modem, and so forth.

VIDEO STREAMING OBSTACLES AND CONTENT AWARE PRINCIPLE

There are a number of basic problems that afflict video streaming, as video streaming over the Internet is difficult because the Internet only offers best effort service. That is, it provides no guarantees on bandwidth, delay jitter, or loss rate. Moreover, these characteristics are unknown and dynamic. Therefore, a key goal of video streaming is to design a system to reliably deliver high-quality video over the Internet when dealing with unknown and dynamic bandwidth, delay jitter and loss rate. The bandwidth available between two points in the Internet is generally unknown and time-varying. If the server transmits faster than the available bandwidth then congestion occurs, packets are lost, and there is a severe drop in video quality. If the server transmits slower than the available bandwidth then the receiver produces suboptimal video quality. The goal to overcome the bandwidth dilemma is to estimate the available bandwidth and than match the transmitted video bit rate to the available bandwidth. Additional considerations that make the bandwidth problem very challenging include accurately estimating the available bandwidth, matching the pre-encoded video to the estimated channel bandwidth, transmitting at a rate that is fair to other concurrent flows in the Internet, and solving this problem in a multicast situation where a single sender streams data to multiple receivers where each may have a different available bandwidth. The end-to-end delay that a packet experiences may propagate from packet to packet. This variation in end-to-end delay is referred to as the delay jitter. Delay jitter is a problem because the receiver must receive, decode and display frames at a constant rate, and any late frames resulting from the delay jitter can produce problems in the reconstructed video. This problem is typically addressed by including a playout buffer at the receiver. While the playout buffer can compensate for the delay jitter, it also introduces additional delay. The third fundamental obstacle is packet losses. A number of different types of losses may occur, depending on the particular network under consideration. For example, wired packet networks such as the Internet are afflicted by packet loss, where an entire packet is lost. On the other hand, wireless channels are typically afflicted by bit errors or burst errors. Losses can have a very destructive effect on the reconstructed video quality. To overcome the effect of losses, a video streaming system is designed with error control.

Many of traditional video streaming systems which are trying to overcome the aforementioned limitation and constraints, consider videos as low-level bit streams, ignoring the underlying visual content. Unfortunately, current video applications adapt to fit the available network resources without regard to the video content. Content-aware video streaming is a new framework that explores the strong correlation between video content, video data (bit rate), and quality of service. Such a framework facilitates new ways of quality modeling and resource allocation in video streaming.

We refer the video content to the high-level multimedia features that can analyzed by the computer. Examples include visual features (such as color, texture, and shape) and motion information (such as motion vectors). In addition, video scenes or object features (e.g., motion, complexity, size, spatio-temporal relationships, and texture). These features can be systematically analyzed. The video content can be used for controlling the video generation to facilitate the network-wise scalability. It can be used in selecting the optimal transcoding architecture and content filtering.

The content-aware video streaming framework is based on the recognition of strong correlation among video content, required network resources (bandwidth), and the resulting video quality. Such correlation between the video content and the traffic has been reported in our prior work (Ahmad & Lee, 2004; Ahmad, Ahmad, Samer, & Lee, 2004; Ahmad, Talat, Ahmad, & Lee, 2005) in which a conceptual model for content aware-based video streaming has been proposed. Figure 2 states the content aware video streaming concept. You can note the content scaler would perform certain scaling mechanism based upon the content analyzer result and network conditions.

Among successful content-aware video streaming frameworks are joint source-channel coding (Ortega, & Khansari, 1995), adaptive media scaling, and resilience coding (Reyes, Reibman, Chuang, & Chang, 1998), object and texture aware video streaming (Ahmad & Lee, 2004; Ahmad et al., 2004; Ahmad et al., 2005.

Figure 2. Content aware scaling

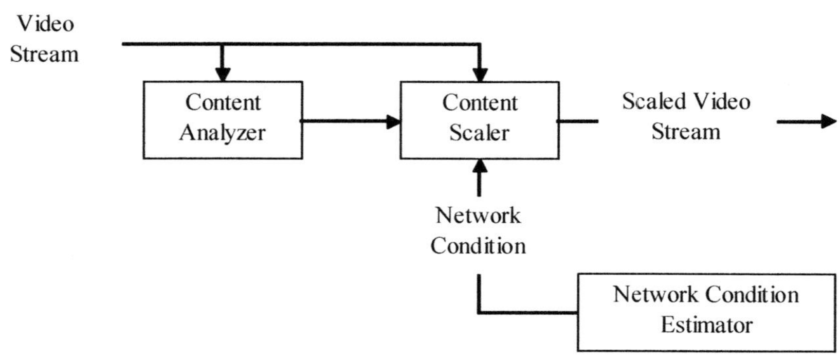

CONTENT AWARE VIDEO STREAMING APPROACHES

First, we need to discuss the point of view of video communication protocols. To overcome short-term network condition changes and avoid long term congestion collapse, various network condition changes control strategies have been built into the Transmission Control Protocol (TCP). For video traffic, TCP is not video protocol of choice. Unlike traditional data flows, video flows do not necessarily require a completely reliable transport protocol because they can absorb a limited amount of loss without significant reduction in perceptual quality (Claypool & Tanner, 1999). On the other hand, SVFTP (Shigeyuki, Yasuhiro, Yoshinori, Yasuyuki, Masahiro, & Kazuo, 2005) some researchers try to exploit TCP protocol in order to deliver video content by setting up many TCP connections at the same time. Thus, they can accomplish video content delivering, but have inefficient network performance. According to our earlier discussion video flows have fairly strict delay and delay jitter requirements. Video flows generally use the user datagram protocol (UDP). This is significant since UDP does not have a network condition changes control mechanism built in, therefore most video flows are unable to respond to network congestion and adversely affect the performance of the network as a whole.

While proposed multimedia protocols like (Floyd, Handley, Padhye, & Widmer, 2000) and (Floyd & Jacobson, 1993) respond to congestion by scaling the bit rate, they still require a mechanism at the application layer to semantically map the scaling technique to the bit rate. In times of network condition changes, the random dropping of frames by the router (Floyd & Jacobson, 1993) (Lin & Morris 1997) may seriously degrade multimedia quality since the encoding mechanisms for multimedia generally bring in numerous dependencies between frames (Mitchell et al., 1996).

For instance, in MPEG encoding (Mitchell et al., 1996) dropping an independently encoded frame will result in the following dependent frames being presented as useless since they cannot be displayed and would be better off being dropped, rather than occupying unnecessary bandwidth. A multimedia application that is aware of these data dependencies can drop the least important frames much more efficiently than can the router (Hemy, Hangartner, Steenkiste, & Gross, 1999; Ahmad & Lee,

Figure 3. General content aware video streaming architecture

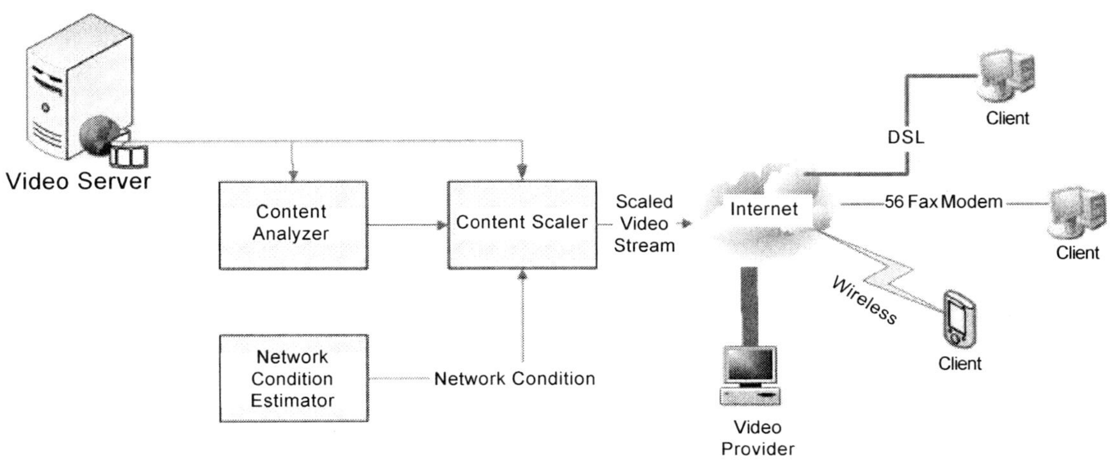

2004; Ahmad et al., 2004). Such application specific data rate reduction is classified as content aware video streaming. Figure 3 states a general architecture for content aware video streaming system.

Clearly, Content aware video streaming is a combination of video content analyzer, network condition estimator and a scaling mechanism to respond for the network conditions after studying the video content. The estimator part is clearly stated in the literatures of computer networks (Miyabayashi, Wakamiya, Murata, & Miyahara, 2000; Rejaie & Estrin 1999). The video content analyzer is proposed in many papers (Ahmad & Lee, 2004; Ahmad et al., 2004). It has been shown that the content of the stream can be an important factor in influencing the video streaming mechanism. Regarding video scaling or transcoding techniques for video to be used in the content aware streaming systems can be broadly categorized as follows (Bocheck, Campbell, Chang, & Lio, 1999; Mitchell et al., 1996; Tripathi & Claypool, 2002):

1. **Spatial scaling:** In spatial scaling, the size of the frames is reduced by transmitting fewer pixels and increasing the pixel size, thereby reducing the level of detail in the frame
2. **Temporal scaling:** In temporal scaling, the application drops frames. The order in which the frames are dropped depends upon the relative importance of the different frame types. In the case of MPEG, the encoding of the I-frames is done independently and they are therefore the most important and are dropped last. The encoding of the P-frames is dependent on the I-frames and the encoding of the B-frames is dependent on both the I-frames and the P-frames, and the B-frames are least important since no frames are encoded based upon the B-frames. Therefore, B-frames are most likely to be the first ones to be dropped.
3. **Quality scaling:** In quality scaling, the quantization levels are changed, chrominance is dropped or DCT and DWT coefficients are dropped. The resulting frames are of a lower quality and may have fewer colors and details.

In sum, it has been shown that the content of the stream can be an important factor in influencing the choice of the scaling scheme for under processing video (Ahmad & Lee, 2004; Ahmad et al., 2004; Ahmad et al., 2005; Mitchell et al., 1996; Tripathi & Claypool, 2002).

We are going to explore many different approaches in the area of content aware video streaming. A fine grained, content-based, packet forwarding mechanism (Shin, Kim, & Kuo, 2000) has been developed for differentiated service networks. This mechanism assigns relative priorities to packets based on the characteristics of the macroblocks contained within it. These characteristics include the macroblock encoding type, the associated motion vectors, the total size in bytes and the existence of any picture level headers. Their proposed scheme requires some mechanisms for queue management and weighted fair queuing to provide the differentiated forwarding of packets with high priorities and therefore will not work in today's Internet.

A basic mechanism that uses temporal scaling for MPEG streams is suggested in Chung and Claypool (2000). In case of network condition changed, the frame rate is reduced by dropping frames in a predefined precedence (first B-frames and then P-frames) until the lowest frame rate, where only the I-frames are played out, is reached or the minimum bandwidth requirement matches the availability.

An adaptive MPEG Streaming player based on similar schemes techniques was developed

(Walpole, Koster, Cen, & Yu, 1997). These systems have the capabilities for dynamic rate adaptation but do not support real-time, automatic content detection and analyzing. Automatic adaptive content-based scaling may significantly improve the perceptual quality of their played out streams. The above mechanisms, while considering the specific characteristics of streaming flows behavior, do not take into account the content of the video flows when scaling based on network condition changes.

Based on the following phenomena, if a video clip shot has fast motion and had to be scaled then it would look better if all the frames were played out albeit with lower quality, some expert design their streaming systems. That would imply the use of either quality or spatial scaling mechanisms. On the other hand, if a video clip scene has low motion and needed to be scaled it would look better if a few frames were dropped but the frames that were shown were of high quality. Such a system has been suggested in (Tripathi & Claypool, 2002). Relevant approach (Yeadon, Garcia, & Hutchinson, 1996) has developed a filtering mechanism for video applications capable of scaling video streams. Using these filters it is possible to change the characteristics of video streams by dropping frames, dropping colors, changing the quantization levels etc. (Mitchell et al., 1996; Tripathi & Claypool, 2002) utilize these filtering mechanisms in conjunction with a real-time content analyzer that measures the motion in an MPEG stream in order to implement a content-aware scaling system. Tripathi and Claypool (2002) conduct a user study where the subjects rate the quality of video clips that are first scaled temporally and then by quality in order to establish the optimal mechanism for scaling a particular stream. They find the content aware system can improve perceptual quality of video by as much as 50%.

Protocol related solutions have various limitation and capabilities. Various mechanisms have been proposed for video protocols to respond to network condition changes on the Internet (Tripathi & Claypool, 2002). Floyd et al. (2000) is a mechanism for equation-based network condition changes control for unicast traffic. Unlike TCP, Floyd et al. (2000) refrains from reducing the sending rate in half in response to a single packet loss. Therefore, traffic such as best-effort unicast streaming multimedia could find use for this TCP-friendly congestion control mechanism. A TCP-friendly protocol (Miyabayashi et al., 2000) was implemented and evaluated for fairness in bandwidth distribution among the TCP and the Miyabayashi et al. (2000) flows. Rejaie and Estrin (1999) is a TCP-friendly rate adaptation protocol, which employs an additive increase, multiplicative decrease scheme. Its main goal is to be fair and TCP-friendly while separating network congestion control from application level reliability. Content aware video-scaling can make the most effective use of bandwidth from these protocols.

Another approach to media scaling uses a layered source coding algorithm (McCanne, Vetterli, & Jacobsen, 1997) with a layered transmission system (McCanne, Jacobsen, & Vetterli, 1996). By selectively forwarding subsets of layers at constrained network links, each user may receive the best quality signal that the network can deliver. In the receiver-driven layered multicast scheme suggested, multicast receivers can adapt to the static heterogeneity of link bandwidths and dynamic variations in network capacity. However, this approach may have problems with excessive use of bandwidth for the signaling that is needed for hosts to subscribe or unsubscribe from multicast groups and fairness issues in that a host might not receive the best quality possible on account of being in a multicast group with low-end users.

A protocol that uses a TCP congestion window to pace the delivery of data into the network has also been suggested to handle video network condition changes (Jacobs & Eleftheriadis, 1998). However other TCP algorithms, like retransmissions of dropped packets, etc. that are detrimental to real time multimedia applications have not been incorporated deeply. This solution is closed to SVFTP (Shigeyuki et al., 2005) solution and has the same limitations.

Some approaches tend to use the video object as a unit for measure and scaling video.

Ahmad and Lee (2004) and Ahmad et al. (2004) have proposed an efficient object-based video streaming system. A motion vector-based object detection is used to dynamically detect the objects. To utilize the bandwidth efficiently, the important object can be real time detected, encoded, and transmitted with higher quality and higher frame rate than those of background. The experimental results show that the proposed object-based streaming is indeed effective and efficient. Therefore, it can fit for the real time streaming application. Figure 4 states their approach clearly.

CONCLUSION

Content aware video streaming overcomes the significant technical challenges in quality of service guarantee and efficient resource management for video streaming. We conclude that end-to-end quality requirements of video streaming application can be reasonably achieved only by integrative study of advanced networking and content processing techniques.

However, most existing integration techniques stop at the bit stream level, ignoring a deeper understanding of the media content. Yet, the underlying visual content of the video stream contains a vast amount of information that can be used to stream video in a semantic manner. We explore different approaches that are thought to be content aware video streaming. Some approaches were more like network-centric approaches to solving the problems of unresponsiveness in video flows. Others were classified as either protocol-related solution or based on its features (i.e., solution used object to do the scaling was classified as object-based video streaming system). The rest are classified upon the scaling mechanism itself (e.g., frame dropping-based) and so forth. We believe content aware video streaming is a very

Figure 4. Object-based video streaming approach

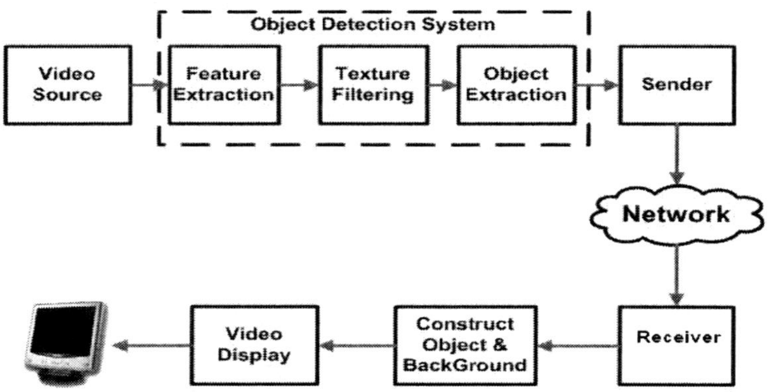

promising field for both video and communication societies. And still it has a lot of room for investigation and developing.

REFERENCES

Ahmad, A. M. A., Ahmad, B. M. A., Talat, S. T., & Lee, S. (2004). Fast and robust object detection framework for object-based streaming system. In G. Kotsis, D. Taniar, & I. K. Ibrahim (Eds.), *The 2nd International Conference on Advances in Mobile Multimedia* (pp. 77-86). Bali: Austrian Computer Society.

Ahmad, A. M. A., & Lee, S. (2004). Novel object-based video streaming technique. In M. A. Ahmad (Ed.), *2nd International Conference on Computing, Communications, and Control Technologies*. (pp. 255-300). Austin: The University of Texas at Austin and The International Institute of Informatics and Systemics (IIIS).

Ahmad, A. M. A., Samer, T., & Ahmad, B. M. A. (2005). A novel approach for improving the quality of service for mobile video transcoding. In G. Kotsis, D. Taniar, S. Bressan, I. K. Ibrahim, & S. Mokhtar (Eds.), *The 3rd International Conference on Advances in Mobile Multimedia* (pp. 119-126). Kuala Lumpur: Austrian Computer Society.

Bocheck, P., Campbell, A., Chang, S. F., & Lio, R. (1999). Utility-based network adaptation for MPEG-4 systems. In C. Kalmanek (Ed.), 9th *International Workshop on Network and Operating System Support for Digital Audio and Video* (pp. 55-67). AT&T Learning Center: AT&T Press.

Chandra, S., & Ellis, C. (1999). JPEG compression metric as a quality aware image transcoding. In D. Klein (Ed.), *Second Usenix Symposium on Internet Technologies and Systems* (pp. 81-92). Boulder: USENIX Assoc.

Chung, J., & Claypool, M. (2000). Better-behaved, better-performing multimedia networking. In F. Broeckx, & L. Pauwels (Eds.), *Euromedia Conference* (pp. 388-393). Antwerp: European Publishing House.

Claypool, M., & Tanner, J. (1999). The effects of jitter on the perceptual quality of video. In M. Steenstrup (Ed.), *ACM Multimedia Conference* (pp. 115-118). New York: ACM Press.

Floyd, S., Handley, M., Padhye, J., & Widmer, J. (2000). Equation-based congestion control for unicast applications. In C. Partridge (Ed.), *ACM the Special Interest Group on Data Communication Conference* (pp. 45-58). New York: ACM Press.

Floyd, S., & Jacobson, V. (1993). Random early detection gateways for congestion avoidance. *IEEE/ACM Transactions on Networking, 1*(4), 397-413.

Hemy, M., Hangartner, U., Steenkiste P., & Gross T. (1999). MPEG system streams in best-effort networks. In A. Basso (Ed.), *International Packet Video Workshop* (pp. 33-39). New York: IEEE Press.

Jacobs, S., & Eleftheriadis, A. (1998). Streaming video using dynamic rate shaping and TCP congestion control. *Journal of Visual Communication and Image Representation, 9*(3), 211-222.

Lin, D., & Morris, R. (1997). Dynamics of random early detection. In M. Steenstrup (Ed.), *ACM the Special Interest Group on Data Communication Conference* (pp. 127-137). Cannes: ACM Press.

McCanne, S., Jacobsen, V., & Vetterli, M. (1996). Receiver-driven layered multicast. In

M. Steenstrup (Ed.), *ACM the Special Interest Group on Data Communication Conference* (pp. 117-130). New York: ACM Press.

McCanne, S., Vetterli, M., & Jacobson, V. (1997). Low-complexity video coding for receiver-driven layered multicast. *IEEE Journal on Selected Areas in Communications 16*(6), 983-1001.

Mitchell, J., Pennebaker, W., Fogg, E., Chad, LeGall, J., & Didier. (1996). *MPEG video: Compression standard* (1st ed.). New York: Chapman and Hall.

Miyabayashi, M., Wakamiya, N., Murata, M., & Miyahara, H. (2000). Implementation of video transfer with TCP-friendly rate control protocol. In N. Myung (Ed.), *International Technical Conference on Circuits/Systems, Computers and Communications* (pp. 117-120). Pusan: The Institute of Electronics Engineers of Korea (IEEK).

Ortega, A., Carignano, F., Ayer, S., & Vetterli, M. (1997). Soft caching: Web cache management techniques for images. In Y. Wang, A. R. Reibman, B. H. Juang, T. Chen, & S. Kung (Eds.), *IEEE Signal Processing Society, First Workshop on Multimedia Signal Processing* (pp. 475-480). Princeton, NJ: IEEE Press.

Ortega, A., & Khansari, M. (1995). Rate control for video coding over variable bit rate channels with applications to wireless transmission. In B. Werner (Ed.), *IEEE International Conference on Image Processing* (Vol. 3, pp. 3388-3393). Washington DC: IEEE Press.

Rejaie, R. M., & Estrin, D. (1999). RAP: An end-to-end rate-based congestion control mechanism for real-time streams in the Internet. In B. Werner (Ed.), *IEEE Infocom* (pp. 1337-1345). San Francisco: IEEE press.

Reyes, G. de los, Reibman, A. R., Chuang, J. C. I., & Chang, F. (1998). Video transcoding for resilience in wireless channels. In B. Werner (Ed.), *IEEE International Conference on Image Processing* (pp. 338-342). Chicago: IEEE Press.

Shigeyuki, S., Yasuhiro, T., Yoshinori, K., Yasuyuki, N., Masahiro, W., & Kazuo, H. (2005). Video data transmission protocol "SVFTP" using multiple TCP connections and its application. *IEICE Transaction on Information and Systems, 88*(5), 976-983.

Shin, J., Kim, J., & Kuo, C. J. (2000). Content-based video forwarding mechanism in differentiated service networks. In A. Basso (Ed.), *IEEE International Packet Video Workshop* (pp. 133-139). Sardinia: IEEE Press.

Tripathi, A., & Claypool, M. (2002). improving multimedia streaming with content-aware video scaling. In S. Li (Ed.), *The 2nd International Workshop on Intelligent Multimedia Computing and Networking* (pp. 110-117). Durham: Association for Intelligent Machinery, Inc.

Walpole, J., Koster, R., Cen, S., & Yu, L. (1997). A player for adaptive MPEG video streaming over the Internet. In J. M. Selander (Ed.), 26th *Applied Imagery Pattern Recognition Workshop* (pp. 270-281). Washington, DC: SPIE.

Yeadon, N., Garcia, F., & Hutchinson, D. (1996). Filters: QoS support mechanisms for multipeer communications. *IEEE Journal on Selected Areas in Communications, 4*(7), 1245-1262.

KEY TERMS

Best Effort Network: Describes a network service in which the network does not provide any special features that recover lost or corrupted packets.

CODEC: Coder/decoder equipment used to convert and compress video and audio signals into a digital format for transmission, then convert them back to their original signals upon reaching their destination.

Congestion Control: A technique for monitoring network utilization and manipulating transmission or forwarding rates for data frames to keep traffic levels from overwhelming the network medium.

GOP (Group of Pictures): In MPEG video, one or more I pictures followed by P and B pictures.

I-Frame, P-Frames: Matrix of basic element in MPEG video to represent the temporal domain.

MPEG: Motion Pictures Experts Group, digital audio and video compression standards.

TCP, SVFTP: Main protocols in networks fore reliable transmission.

Video Scaling: Changing the video content in respond to some conditions.

Video Streaming: The transmissions of full-motion video over the Internet without download it.

Chapter XXV Portable MP3 Players for Oral Comprehension of a Foreign Language

Mahieddine Djoudi
Université de Poitiers, France

Saad Harous
University of Sharjah, UAE

ABSTRACT

In this chapter, we present an approach for mobile learning that aims at equipping learners by portable MP3 players. As known, the primary use of this device is to listen to music in MP3 format, but it can be adopted to be a useful tool for the service of teaching/learning of languages. This method is based on an easy to use technology that makes it possible for learners to work, at their own pace/rhythm, the oral comprehension of a foreign language. It is a question of supporting the personalization (but not only) of what audio files (short, long) each user should listen to. These files are created by the teacher and uploaded on a Web based distance-learning platform. So, these audio resources are available permanently on the server and can be downloaded by learners at any time. The proposed method is designed for a diversified population and allows the development and the maintenance of knowledge throughout the life.

INTRODUCTION

In this chapter, we present an approach for mobile learners which aims at equipping learners with portable MP3 player. The primary use of this device is to listen to music in MP3 format, but it can be adopted to be a useful tool for the service of teaching/learning of languages. This method is based on an easy to use technology which makes it possible for learners to work, at their own pace/rhythm, the oral comprehension of a foreign language.

It is a question of supporting the personalization (but not only) of what audio files (short, long) each user should listen to. These files are created by the teacher and uploaded on a Web-based **distance-learning** platform. This will facilitate to the learner the access to these audio resources available permanently and can be down loaded at any time. The proposed method is designed for a diversified population and allows the development and continuous learning throughout the life.

The term mobile learning (mlearning) refers to the use of mobile and handheld information technology devices in teaching and learning. These mobile tools often travel with the learners (Kadyte & Akademi, 2003). Among these tools we can quote the telephone (Attewell & Savill-Smith, 2003), the PDA (Kneebone, 2003), Pocket PC (Holme & Sharples, 2002), the portable computer (Willis & Miertschin, 2004), the Portable MP3 Players (Bayon-Lopez, 2004), etc. Mobile technologies are transforming the educational world. The question is to know how these technologies affect the training environment, pedagogy, and continuing education (Mifsud, 2002).

According to Bryan (2004), mobile technologies and their adoption by the younger generations are going to transform the education itself. It is a question "of modeling learners as creative and communicating participants, rather than passive consumers," and "to describe the world like a service on which one can read and write." The article adopts a broad definition of mobility. It is interested in continuous connectivity, the dynamic combinations of wired and wireless devices, and learners and their environment (Bryan, 2004).

From recent but abundant work in the field of the mobile learning (Cohen & Wakeford, 2005; Keefe, 2003; Kossen, 2001; Lindroth, 2002; Pearson, 2002; Sharples, 2000; Vavoula, 2004), we can raise the following remarks:

- The reconfiguration of the classrooms and the campuses in reconfigurable open spaces, mixing physical presence and distant **collaboration**, seems to be one of the attractive prospects. There is no need any more to equip these spaces in a fixed way. Also, we do not need to limit the learners to a specific area because they are equipped with their own communication devices, the borders are pushed to the infinity
- The continuous co-operation, independent of the place, could transform the way in which research is undertaken on the ground or the training experiments are done. One can imagine dispersed teams which exchange and publish their results and analyses in real time
- Finally, the mlearning could become the way to follow in order to have a lifelong learning. In this approach, any person could, at any given place and time, choose a particular subject, find a learning community that is learning this topic. He/she can join this group for while and leave when his/her objectives are achieved

ANALYTICAL SCHEME OF LANGUAGE CAPACITIES

In order to understand the problem being considered in this chapter, it is of primary importance to know what are the capacities concerned during a learning process of a foreign language. We point out that the capacities in learning a language represent the various mental operations that have to be done by a listener, a reader, or a writer in an unconscious way, for example: to locate, discriminate, or process the data. One distinguishes in the analytical diagram, basic capacities which correspond to linguistic activities and competence in communication that involve more complex capacities.

Language Basic Capacities

The use of a language is based on four competences (skills). Two of these skills are from comprehension domain or what Shannon attributes to the receiver in his communication diagram. These are oral and written comprehension. The last two concern the oral and written expression (or production) or the source according to Shannon's scheme (Shannon, 1948). A methodology can give the priority to one or two of these competences or it can aim at the teaching/learning of these four competences together or according to given planned program.

On one hand, oral comprehension corresponds to the most frequent used competence and can be summarized in the formula "to hear and deduce a meaning." Chronologically, it is always the one that is confronted first, except in exceptional situations (people only or initially confronted with the writing, defective in hearing, study of a dead language (a language that is not in use any more), study of a language on the basis of the writing for an autodidact). On the other hand, the written expression paradoxically is the component in which the learner is evaluated more often. It is concerned with the most demanding phase of the training by requiring an in depth knowledge of different capacities (spelling, grammatical, graphic, etc.)

Communication Competence

The evolutions of linguistics and didactic favor the introduction of new communication methodologies which emphasize the concept of communication competence. In these methodologies the way the information is conveyed is more important than form, based on the practice of the authentic documents and open to social variations. Indeed, to communicate, it is not enough to know the language, the linguistic system; in addition one needs to know how to make use of the language according to the social context and to know rules of social use of the language.

Capacities of different nature than the ones mention earlier intervene in the language activity, but which constitute as many competence components as communication. We mention only the ones which play the most important part in **language learning** and practice. These capacities are the sociolinguistic capacities, the discursive capacities, the cultural and sociocultural capacities, and the different strategic capacities.

Table 1. The four basic concepts

	Oral	**Written**
Comprehension	Listening	Reading
Expression	Speaking	Writing

Work Context

The use of the mobile tools in language learning has been developing at very high speed these last years. Thus, we are witnessing many research and development projects, methodologies, and scientific publications (Norbrook & Scott, 2003; Sharples, 2003). However, the interest in research related to oral comprehension competence remains relatively low. Our idea is based on a study of the current situation of foreign languages oral training. We propose a simple and original approach which utilizes the MP3 player to enhance learners' oral comprehension.

Oral Training Status

Learner's lack of oral practice of the language affects the learning process in a negative way. This phenomenon is caused by several factors which are due to the fact that the classes are overloaded (number of learners per class is high). Also many learners are skeptic about the need of communicating in a foreign language.

These combined elements make that the oral practice restricted to few learners who are at ease with the different situations they face while trying to practice the proposed exercises by the teacher by listening and studying some sound documents.

The oral training is in a disadvantage compared to the read/written training. This imbalance leads us to think that the oral training must be given more attention. It is a help that the learner can not circumvent in language learning. The social and civic dimensions must be taken into account because they play a major part in the training of the individual. Indeed the oral training has the following advantages:

- It reveals phenomena that are hidden if we have only written document: intonations, accents, realization or not of certain vowels, their stamps, etc.
- It gives the instructor the possibility to intervene to explain or raise questions, in order to guide the listener to what is important and to contribute to the perception structuring and the audio recognition
- It asks the listeners to present assumptions for interpretation and discussion

MP3 PLAYER AND ITS USE

Today, the sales of MP3 player is growing much faster than the sale of CD readers. Several design features of MP3 player seem to be particularly interesting to explore within the framework of designing training for foreign languages learning (Sabiron, 2003):

- Device weight and size are extremely reduced
- Absence of mechanical wear which decreases the dysfunction
- Digital sound quality is excellent
- Sound documents' handling is very easy (reading, pause, rewind, etc.)
- Audio files remote loading is fast even through modem connection
- Multi-distribution of documents from the same site is easy to setup
- Device's storage capacities are very sufficient
- Purchase cost is low
- No particular computer skills are required from the learner. The actual MP3 players are connected very easily through the USB port and are very easy to operate
- It is not necessary to have a microcomputer to listen to the audio files: only the regular "reloading" of audio files requires to have access to a fixed station
- Existing devices are directly usable

- Even if these devices evolve regularly, the MP3 player will remain always portable, autonomous and loadable (with or without wire)

MOTIVATION AND APPROACH DESCRIPTION

Motivation

Languages' instructors note that it is very difficult to make the learners practice their oral expressions while studying a foreign language because the class size is often very large, reduced schedules, exams are often written exams, learners have reservation to learn to speak a foreign language at an old age, and they do not see that it is necessary to learn a foreign language because they are not confronted to the language in their daily life, etc.

In general, very few institutions have a language laboratory that can be used by the learners to practice their oral using a computer (one or two learners per computer) but computer access remains always a problem when learners return home. How do learners study for the oral examination when they have very limited access to the tools?

Not all learners have the opportunity to access a computer and even the ones who have, usually do not have access to fast ADSL network connections to quickly download audio files and how to access the support material chosen by the instructor?

Faced with these difficulties, the idea of the approach to equip all learners with MP3 player. Learners can have access in a "guided autonomy" to recordings chosen by the instructor so they can practice using them (Little, 2000). Practice on the audio support can be done in a traditional classroom, using a PC, in language laboratory and especially continued at home (Farmer & Taylor, 2002).

Learners can truly get their ears very well acquainted to the target foreign language. They can listen to the recording as many times as they want. For the instructor the advantage is undeniable as regards to the choice of documents in the target languages. Indeed the direct and free access to sound resources of foreign languages on the Web sites (without copyright problem), makes it possible to expose beginner and continuing learners to the authenticate language which is the first step to being competent in comprehension. This approach seems to be very promising because it offers learners more possibilities to work on their oral expression and to be exposed to the target language.

Approach Description

Current uses of these devices was never intended to correspond entirely to the uses envisaged by the originators or inventors, which results (Pearson, 2002) in speaking about diversion of use, to characterize the share of social and cultural creativity which is and will always remain in the users' hands. Our method draws its originality from the diverted use of a simple device (MP3 player) which was not initially created for teaching, to help increase the learner's oral comprehension of a foreign language.

The diffusion of the sound files on the Web server of the teaching platform designed for this purpose, allows the authorized public a fast remote loading of the sound documents. The platform also provides instructions on the work to be done and corrects exercises for isolated learners. The listening to the sound files is then done starting from the computer itself, and especially by remote loading of these files on an MP3 player for listening independent of the computer.

The innovation is thus done by two successive levels of distribution of the diffusion: initially it sent from the server to the computers

Figure 1. Sound files diffusion

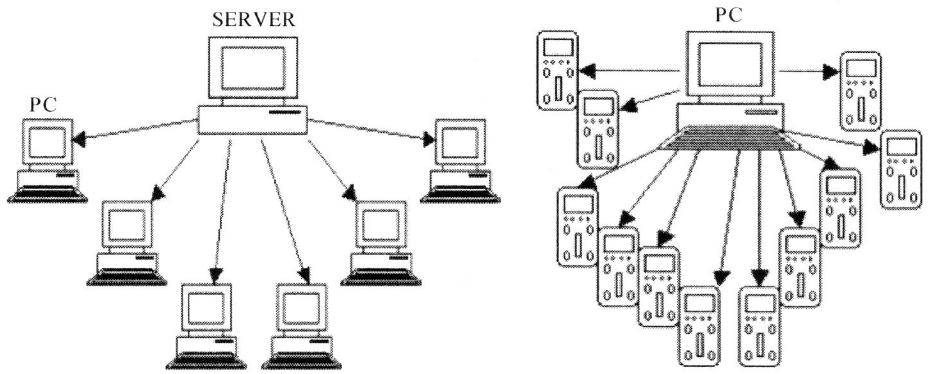

connected to the network, then it sent from each receiving computer to an unlimited number of MP3 players supplied locally (Sabiron, 2003).

Our approach is also based on an evolutionary methodology using a pretest and post-tests where several groups of learners will be monitored in order to quantify the possible impacts and collect statistics about the use and its evolutions.

Learners are going to answer questionnaires, that have been prepared, at different stages of the training to make it possible to measure the impacts of the device use on the learners' behavior, in particular, the performances in the language oral comprehension and the degree of motivation (Norbrook & Scott, 2003). Moreover, regular discussions make it possible to obtain users' profiles and a typology of the uses of the MP3 players.

Our approach which is based on the combination of technologies (Information Technology, Internet and MP3 players), requires little competences in information technology and its financial cost is relatively moderate. It aims on one hand, to quantify and qualify the impacts related to the use of an innovating device dedicated to foreign languages training and on the other hand, to study the processes of adapting a specific technical device. The generalization of the device use in other trainings and/or for other types of learning seems in the short and medium term a prospect (Bayon-Lopez, 2004; Sabiron, 2003).

EXPECTED OBJECTIVES OF THE APPROACH

The approach proposes to the learners new way of learning a language as one of principal objectives. Also it help learners to become speakers who are able to make their idea comprehensible and to progress quickly in learning a foreign language.

It also claims to give a coherence to language training through an exposure geared toward the target language. Thus, the MP3 player is presented as a tool adapted to the achievement of these objectives since it allows:

- Learners to familiarize themselves with a new technological environment, a new workspace, and a different working method integrating communication and information technologies

- To diversify teaching and learning forms of the languages in connection with the committed reforms and within the national programs guidelines
- To propose to the learners training situations which give them confidence and motivation. In this direction, the use of the MP3 player in language training contributes to a positive modification of learner's attitude where a stronger participation of all concerned people is necessary (Norbrook & Scott, 2003)
- To develop learners' autonomy (they have permanent access to their working group information and resources via the platform's means) and to support regular and constant personal work (Little, 2000)
- To modify the work habits of individual learners (Lundin & Magnusson, 2002). Specific tasks are assigned to the learners to carry out every week to support a regular practice of their oral expression

In addition to these objectives of a general nature, the following priorities are added:

- To improve the learner's oral competences which are found in very heterogeneous classes and of rather low level
- To favor the listening and comprehension work on authentic sound documents
- To allow a work in a guided autonomy outside the classroom based on supporting materials prepared by the instructor
- To facilitate access to the sound resources via the means offered by the training platform on the Web
- To support class participation by working on regularly exercises and activities geared toward oral expression and comprehension practices

The approach is an integral part of the general pedagogy framework which aims at making the learner as autonomous as possible and especially, very active: active in his or her trainings, active in the knowledge construction (Mitchell, 2002; Zurita & Nussbaum, 2004). We are using the MP3 player as a tool because it serves our teaching objectives and not because it is technically a powerful object (Little, 2000).

PLATE-FORM PEDAGOGIC

Consideration for Training Nature

Foreign languages represent a special field of research for the design and development of pedagogic platforms. The linguistic and cultural contents are clearly multimode and hypermedia. The design of such platforms is necessarily multi-field, the different needed modeling (targeted domain linguistic, cognitive progression, interaction pedagogic) are relatively complex. The language teaching models tend moreover towards a personalized training. The pedagogic platforms have then as a task to make available for the learners a necessary digital work environment adequate for the users learning and the language practice.

Software Architecture

The teaching platform is a "full Web" application which provides the three principal users (instructor, learner, administrator) a device which has for primary functionality the availability and the remote access to pedagogical contents for language teaching, personalized learning and distance tutoring (Djoudi & Harous, 2002). The platform allows not only the downloading of the resources made available on line (using a standard navigator) but also the diffusion in **streaming** of these same resources.

The sound files are accompanied by textual documents introducing the subject, its use con-

Figure 2. Software architecture of the plate-form pedagogic

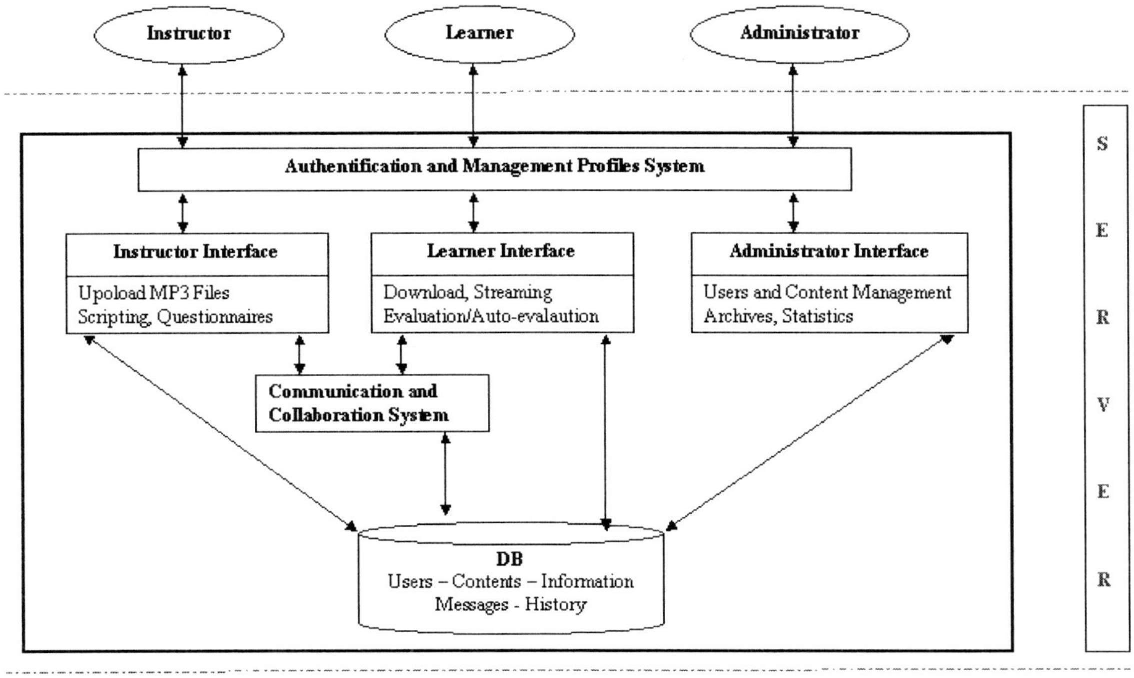

textual, foreign speakers presentation, their phonological variation, in order to make it possible to the individual listener to locate the spoken language characteristics.

Instructor's Interface

The teaching platform allows the instructor, via a dedicated **interface**, to make at the learners' disposal a considerably large amount of compressed digital audio documents, of excellent quality to listen to. These documents are created by the instructors or recovered from Internet. The interface also makes it possible for the teacher to describe in the most complete possible way the sound files. Information relative to each file are: the name, the language, duration, public concerned, expected pedagogic objectives, the period of accessibility, the source, copyright, etc. The documents thus prepared by the instructor are loaded in the database located on the platform server.

If the learner can put his or her own techniques and strategies to understand the oral expression, then the instructor role consists in helping him or her to develop and enrich the learning strategies. It is thus necessary to add to the sound files a work plan to guide the learners on how to practice their listening within the framework of the designed learning methodology.

Learner's Interface

This type of work environment puts, in an unquestionable way, the learner in the center of the task. He has an objective: an oral document to understand, which may contain some obstacles. He has also tools, of standard type, to help in solving the encountered problems. The learner is thus faced with a situation with problems. This is the moment where the learner uses his/her preferred strategies, according to the psycholinguistics "block box" model, since one does not prejudge activities which he or she will have to deploy to understand. We have to consequently recognize the importance of having different accessible tools with respect to the difficulties that each learner may face.

Streaming

Streaming is a technique for transferring data such that it can be processed as a steady and continuous stream. Streaming technologies are widely used in transmitting large multimedia (voice, video, and data) files in a fast way. With streaming, the client browser or plug-in can start displaying the multimedia data before the entire file has been transmitted.

Streaming is a technique of multi-media data transfer in a regular and continuous flow (stream) via Internet based on request and in real time. The streaming is a principle making it possible to the user to progressively read audio and video contents while it is being remotely loaded live or preloaded.

This directly contrasts a static model of data delivery, where all the data is delivered to the client machine prior to actual use. For this whole process to work properly, the client browser must receive the data from the server and pass it to the streaming application for processing. The streaming application converts the data into sounds (or pictures). An important factor in the success of this process is the ability of the client to receive data faster than the application can display the information. Excess data is stored in a buffer — an area of memory reserved for data storage within the application. If the data is delayed in transfer between the two systems, the buffer empties and the presentation of the material will not be smooth.

The streaming server (installed at the same time as the Web server) must know how to manage the adaptation and the optimization of flow and the contents, the quality of service. The adaptation to the network and the terminal must be done in real time. Distribution networks of mobile contents are developed based on the content delivery networks model of the Internet. At the networks borders, close to the user, the multimedia servers manage part of the distribution and adaptation to the user context.

COLLABORATION AND COMMUNICATION TOOLS

The teaching platform, while making it possible to accompany the users (instructors and learners), also offers to the pedagogic team the possibility to set up a true collaborative work based on the multi-use of competences and the resources.

The collaboration and **communication** tools used within the platform are thought to be supporting tools for the learning and not to be an end in themselves. The principal challenge is to proportion well the technologies used and to reach the perfect adequacy between the teaching platform, the studied subjects, and the learning population.

The collaboration and communication system is installed on the platform server and offers the necessary means for the user to communicate with other users, to complete team work, and to take part in discussions. In order to support the co-operative learning, the interfaces are designed in a way to make the presence of the other participants known by providing indices of their availability and their activity on the server (Djoudi & Harous, 2002).

The implementation of the collaboration module within the platform must take into account the problems of cognitive nature relative to each user, in particular:

- Humans are limited in their capacities, in order not to get exhausted in the medium term, they simplify, avoid bothering themselves with things that are not necessary and take pleasure in repetitive procedures
- The user prefers spontaneously the old but well controlled means of communication over the new means and objectively more effective
- The inter-human natural model is badly adapted. The communication man-man is based on a considerable implicit knowledge and on an external redundancy in the dialogue means (gesture, word, attitudes)
- Faced to the machine, there still persists some blockings of the user: feeling of dependence on computer tools, constraints not understood, lack of basic know-how. This explains mainly the under exploitation of the system

In addition to the heterogeneous initial competences in information technology of each one taking part in the training, it was necessary to define some basic strict criteria in the choice of the tools, taking into account, by default, the students with the least developed competences:

- **A simple tool:** The use and the handling must be easy and relatively intuitive. Learning how to use the tool is not the essential goal of the training, it must be fast and optimal
- **A stable tool:** The learner must be able to count on a reliable tool. It is not a question of using an application requiring the modification of many parameters setting
- **A common tool:** As much as possible, the application used must be able to be reused thereafter as a working tool. To this end, the selected tools must be up to date (even implemented by default on the machines) and not be exclusively reserved for the training
- **An adaptable tool:** The communication conveyed via the selected tools must be adaptable to any changes required by the training

Log Book

The goal of the log book is to set up an automatic book keeping of information related to the learner's activity while he or she carries out a scenario on a teaching object (date and duration of each connection, MP3 files downloaded or listened to in streaming, exercises for self **evaluation**, etc). This requires an effort of information structuring and an implementation within the platform. An exploitation of this information by learners can guide them through their training plan.

By analogy with the normal paper which one uses during a traditional training, we will use the metaphor of log book to keep track of the training path the learner is following. The access to the log book via the user interface to explore it according to relevant criteria would be an invaluable help for the learner and the instructor. Last, a statistical analysis of log books of a group of learners that have done the same activity would give a synthetic vision of the group's training, and would be useful to all people involved in the training.

Exercises for Oral Comprehension Improvement

The oral comprehension competence is more difficult to acquire. It develops only with the learners who follow a progressive and regular practices. The first requirement of the practice is to expose the learner very often to the target language.

The approach offers learners the opportunity of multiplying their listening times, by the access to a large volume of audio files in the authentic language (language's tutors collaboration in the institution).

However even if the MP3 player's use is regular and constant, it is not enough to expose learners to a language and expect them to learn it. The instructor's role remains very important and one can not do without. Thus, the teacher must guide the learners in their trainings by means of follow up records and exercises which go hand in hand with the audio files. Loaded by the instructor on the Web server, these follow up records contain work instructions, descriptions of the tasks to be achieved, and correction (to go with the learner), as well as scripts (transcriptions) of the audio files.

Types of Exercises

Here is a non-exhaustive list of the various types of exercises for analyzing the concrete situation of the learners (Bayon-Lopez, 2004).

- Exercises that help the users to locate themselves using only one audio file: this exercise helps the learners to be in a state to recognize language's facts, lexicon simple or specialized with respect to the document or the topic studied in class. The learner is thus put in a state "to observe the language" based on the content, as well as on the syntax. The MP3 player facilitates the learners' task because the possibilities of repeating and rewinding
- Exercises that help the users to locate themselves using two audio files: In this activity one of the two files is a file with missing words. While going from one file to the other learner must be able based on an attentive listening discriminate and locate the missing words

LEARNER EVALUATION

Design of the Evaluation

The protocol used for the evaluation of the progress achieved by learners using MP3 players is based on:

- **Formative evaluation** is done with a small group of people to "test" various aspects of teaching materials. Formative evaluation is typically conducted during the development or improvement of learning. It is conducted, often more than once. Its purpose is to validate or ensure that the teaching's goals are being achieved and to improve the teaching, if necessary, by identifying the problems and subsequent finding a remedy. In other words this evaluation is used to monitor the learners' progress
- **Summative evaluation** provides information on the product's effectiveness (it's ability to do what it was designed to do). For example, did the learners learn what they were supposed to learn after using the instructional module. In a sense, it lets the learner know "how they did," but more importantly, by looking at how the learner's did, it helps you know whether the product teaches what it is supposed to teach. Overall evaluation is typically quantitative, using numeric scores or letter grades to assess learner achievement
- **Learner's self-evaluation** which leads the user to have a self-checking, sometimes the user will decide based on this

evaluation to do more practices to remedy some short comings. The objective is also to enable the user to be self critical and to push him/herself to always better do

We are thinking of applying an evaluation by the languages' instructors who are teaching the classes but we intend to also use the self-evaluation. The self-evaluation has as an objective the involvement of the learner in his or her work, so that he/she becomes an active learner who acts on his own evaluation based on the report which summarizes the various criteria raised.

The purpose of the diagnostic and formative evaluations carried out by the instructor and the tutor on a regular bases will be to check that the grammatical, linguistic and cultural objectives operational indeed were achieved or are in progress of being achieved. We must also measure at the same time the impact of this new method (Harous, Douidi, Djoudi, & Khentout, 2004).

Evaluation Criteria

The principal criteria which we considered for the learners' evaluation are as follows (Bayon-Lopez, 2004; Brindley, 1998):

- To locate and understand the essential points of an audio document
- To seize and identify the general topic of the document
- To isolate and distinguish the elements which will enable the learner to relate different parts of the information to each others
- To understand statements reported at a normal speed or fast
- To locate the various forms of speech, the varieties of languages, and accents
- To identify the attitudes and emotions
- To derive the direction of expressions discovered "based on the state"
- To perceive the implicit (humor, irony, point of view, etc)
- To orally (or written) explain what he or she understood (Buck, 1998)

EXPERIMENT PROTOCOL

Targeted Public

The approach presented here is designed to be used and tested in real teaching situation in collaboration with the instructors and the tutors. The experimentation is planned for a set classes teaching material necessary at the university. It is initially concerned with English oral comprehension for French-speaking public. The objective is to see whether the approach is likely to answer the learners' needs and to increase their interest in the English language.

Before even starting the evaluation, a meeting with the concerned members will be organized. The rules and the conditions of use will be explained and commented on. The pedagogical objectives of the approach will be clearly exposed.

MP3 Player Suitability

Taking into account the design features of the MP3 player and the functionalities which it offers, it is necessary to plan a phase of adaptation of this tool. Approximately, two hours will be devoted to the training of all learners in the class during the first week: to discover the tool's principal functionalities, to be in a concrete situation, to clarify the approach and objectives of the project.

Resources Acquisition

The diffusion of the audio file via the platform, makes it possible for the learners to remotely access the server database, any time of the day and from any station connected to the Internet. There will be a sufficient number of computers available for the learners in rooms with free access.

CONCLUSION

We presented in this chapter an original approach for oral comprehension of a foreign language by using a device whose initial function was for a different purpose. The MP3 player as a nomad object with its characteristics of portability, accessibility, and autonomy is similar to a book.

The approach proposes an innovation which is done on two successive levels. On one hand, the diffusion or provision of sound resources prepared by the instructors on the distance teaching platform and on the other hand, the use of the MP3 player to expose sufficiently to a quality authentic language.

In prospect, the approach aims at developing in the learners other oral competence; namely the oral expression so that they can express themselves in foreign language. The mastering of the language goes through the mastering of elocution. The approach thus envisages to give the learners the opportunity to produce audio files, as a result of their work, by using the "recording" function of the MP3 player. Concrete situations of uninterrupted speech, to give a summary of a lecture, oral comments about documents studied in class, exercises to argument or justify a point of view, facilitate the use and adaptation of the target language (Bayon-Lopez, 2004).

REFERENCES

Attewell, J., & Savill-Smith, C. (2003). *Young people, mobile phones, and learning*. London: Learning and Skills Development Agency.

Bayon-Lopez, D. (2004). Audio NOMADE: un laboratoire de langues virtuel. 3e Journée des langues vivantes: l'oral : stratégies d'apprentissage et enjeux. 24 novembre. CDDP Gironde, Bordeaux, France.

Brindley, G. (1998). Assessing listening abilities. *Annual Review of Applied Linguistics, 18*(1), 11-191.

Bryan, A. (2004). Going nomadic: Mobile learning in higher education. *Educause, 39*(5), 28-34.

Buck, G. (1998). Testing of listening in a second language. In C. M. Clapham, & D. Corson (Eds.), *Language testing and assessment. Encyclopedia of language and education* (Vol. 7, pp. 65-74). Dordrecht: Kluwer Academic Publishers.

Cohen, K., & Wakeford, N. (2005). *The making of mobility, The making of self.* INCITE, University of Surrey in collaboration with Sapient. Retrieved April 22, 2005, from http://www.soc.surrey.ac.uk/incite/AESOP%20Phase3.htm

Djoudi, M., & Harous, S. (2002). An environment for cooperative learning over the Internet. *International Conference on Artificial Intelligence (IC-AI'2002)* (pp. 1060-1066). Las Vegas, NV, USA.

Farmer, M., & Taylor, B. (2002). A creative learning environment (CLE) for anywhere anytime learning. *Proceedings of the European Workshop on Mobile and Contextual Learning,* The University of Birmingham, England.

Harous, S., Douidi, L., Djoudi, M., & Khentout, C. (2004). Learner evaluation system for distance education. *The 6th International Conference on Integration and Web-Based Application & Services (iiWAS2004)* (pp. 579-586). Jakarta, Indonesia.

Holme, O., & Sharples, M. (2002). Implementing a student learning organiser on the pocket PC platform. *Proceedings of MLEARN 2002: European Workshop on Mobile and Contextual Learning,* Birmingham, UK (pp. 41-44).

Kadyte, V., & Akademi, A. (2003). Learning can happen anywhere: A mobile system for language learning. *Proceedings of Mlearn 2003 Conference on Learning with Mobile Devices*, Central London, UK.

Keefe, T. (2003). Mobile learning as a tool for inclusive lifelong learning. *Proceedings of Mlearn 2003 Conference on Learning with Mobile Devices*, London.

Kneebone, R. (2003, May 19-20). PDAs as part of learning portfolio. *Proceedings of Mlearn Conference on Learning with Mobile Devices*, Central London, UK.

Kossen, J. (2001). *When e-learning becomes m-learning*. Palmpower Mag. Retrieved April 22, 2005, from http://www.palmpowerenterprise.com/issues/issue200106/elearning001.html

Lindroth, T. (2002, August). Action, place, and nomadic behavior—A study towards enhanced situated computing. *Proceedings of IRIS25*, Copenhagen, Denmark. Retrieved April 22, 2005, from http://www.laboratorium.htu.se/publikationer/qiziz.pdf

Little, D. (2000). Learner autonomy: Why foreign languages should occupy a central role in the curriculum. In S. Green (Ed), *New perspectives on teaching and learning modern languages* (pp. 24-45). Clevedon: Multilingual Matters.

Lundin, J., & Magnusson, M. (2002, August 29-30). Walking & talking—Sharing best practice. *Proceedings IEEE International Workshop on Wireless and Mobile Technologies in Education,* Växjö, Sweden (pp. 71-79).

Mifsud, L. (2002). Alternative learning arenas—Pedagogical challenges to mobile learning technology in education. *Proceedings IEEE International Workshop on Wireless and Mobile Technologies in Education,* Växjö, Sweden (pp. 112-116).

Mitchell, A. (2002). Developing a prototype microportal for m-learning: A social-constructivist approach. *Proceedings of the European Workshop on Mobile and Contextual Learning*. The University of Birmingham, England.

Norbrook, H., & Scott, P. (2003). Motivation in mobile modern foreign language learning. *Proceedings of Mlearn 2003 Conference on Learning with Mobile Devices*, Central London, UK.

Pearson, E. (2002). Anytime anywhere: Empowering learners with severe disabilities. *Proceedings of the European Workshop on Mobile and Contextual Learning*, The University of Birmingham, England.

Sabiron, J. (2003). Outils techniques et méthodologiques de l'apprenant nomade en langues étrangères. Computer a primavera 2003, Biblioteca regionale—Aosta, Italia.

Shannon, C. E. (1948). A mathematical theory of communication. *Bell System Technical Journal, 27,* 379-423 and 623-656.

Sharples, M. (2000). The design of personal mobile technologies for lifelong learning. *Computers and Education*, *34*, 177-193.

Sharples, M. (2003). Disruptive devices: Mobile technology for conversational learning. *In-*

ternational Journal of Continuing Engineering Education and Lifelong Learning, 12(5/6), 504-520.

Vavoula, G. (2004). *KLeOS: A knowledge and learning organisation system in support of lifelong learning.* PhD Thesis, The University of Birmingham.

Willis, C. L., & Miertschin, L. (2004). Technology to enable learning II: Tablet PC's as instructional tools or the pen is mightier than the board! *Proceedings of the 5th Conference on Information Technology Education,* Salt Lake City, UT (pp. 153-159).

Zurita, G., & Nussbaum, M. (2004). A constructivist mobile learning environment supported by a wireless handheld network. *Journal of Computer Assisted Learning, 20*(4), 235-243.

KEY TERMS

Basic Language Skills: The ability to listen, read, write, and speak in language.

Logbook: Personal learning environment running on a personal computer or mobile device. It integrates and aggregates the learner's activities. A significant element of the tool is its support of activity logging. A combination of automatic and manual log entries enable the learner to simply reflect on their personal learning journey.

MP3: Stands for "MPEG-1 Audio Layer-3." (MPEG is short for "moving picture experts group"). It is the most popular compressed audio file format. An MP3 file is about one tenth the size of the original audio file, but the sound is nearly CD-quality. Because of their small size and good fidelity, MP3 files have become a popular way to store music files on both computers and portable devices.

Portable MP3 Player: A device for storing and playing MP3s. The idea is for it to be small and, thus, portable. It is like a digital music library that you can take anywhere you go.

Streaming: A technique for transferring data such that it can be processed as a steady and continuous stream. Streaming technologies are widely used in transmit large multimedia (voice, video, and data) files quickly. With streaming, the client browser or plug-in can start displaying the multimedia data before the entire file has been transmitted.

Chapter XXVI
Towards a Taxonomy of Display Styles for Ubiquitous Multimedia

Florian Ledermann
Vienna University of Technology, Austria

Christian Breiteneder
Vienna University of Technology, Austria

ABSTRACT

In this chapter, a domain independent taxonomy of sign functions rooted in an analysis of physical signs found in public space is presented. This knowledge is necessary for the construction of future multimedia systems that are capable of automatically generating complex yet legible graphical responses from an underlying abstract information space such as a semantic network. The authors take the presence of a sign in the real world as indication for a demand for the information encoded in that sign, and identify the fundamental types of information that are needed to fulfill various tasks. For the information types listed in the taxonomy, strategies for rendering the information to the user in digital mobile multimedia systems are discussed.

INTRODUCTION

Future mobile and ubiquitous multimedia systems will be even more an integrated part of our everyday reality than it is the case today. A digital layer of information will be available in everyday situations and tasks, displayed on mobile devices, blended with existing contents of the real, physical world. Such an "augmented reality" (Azuma et al., 2001) will put into practice recent developments in the area of mobile devices, wireless networking, and ubiquitous information spaces, to be able to provide the right information to the right person at the right time.

The envisioned applications for these kinds of systems are manifold; the scenarios we are thinking of are based on a dense, spatially distributed information space which can be browsed by the user either explicitly (by using navigation interfaces provided by hardware or software) or implicitly (by moving through space or changing one's intentions, triggering changes in the application's model of the user's *context*). Examples for the information stored in such an information space would be historical anecdotes, routes, and wayfinding information for a tourist guide or road and building information for wayfinding applications. The question of how to encode this information in a suitable and universal way is the subject of ongoing research in the area of semantic modeling (Chen, Perich, Finin, & Joshi, 2004; Reitmayr & Schmalstieg, 2005). For the applications we envision, we will require the information space not only to carry suitable abstract metainformation, but also multimedia content in various forms (images, videos, 3D-models, text, sound) that can be rendered to the user on demand.

Besides solving the remaining technical problems of storage, querying, distribution, and display of that information, which are the subject of some of the other chapters in this book, we have to investigate the consequences of such an omnipresent, ubiquitous computing scenario for the user interfaces of future multimedia applications. Up to now, most research applications have been mainly prototypes targeted towards a specific technical problem or use case; commercial applications mostly focus on and present an interface optimized for a single task (for example, wayfinding). In the mobile and ubiquitous multimedia applications we envision, the user's task and therefore the information that should be displayed cannot be determined in advance, but will be inferred at runtime from various aspects of the user's spatio-temporal context, selecting information and media content from the underlying information space dynamically. To communicate relevant data to the user, determined by her profile, task, and spatio-temporal context, we have to create legible representations of the abstract data retrieved from the information space. A fundamental problem here is that little applicable systematic knowledge exists about the automatic generation of graphical representations of abstract information.

If we want to take the opportunity and clarify rather than obscure by adding another layer of information, the following questions arise: Can we find ways to render the vast amounts of abstract data potentially available in an understandable, meaningful way, without the possibility of designing each possible response or state of such a system individually? Can we replace a part of existing signs in the real world, already leading to "semiotic pollution" (Posner & Schmauks, 1998) in today's cities, with adaptive displays that deliver the information the user needs or might want to have? Can we create systems that will work across a broad range of users, diverse in age, gender, cultural and socio-economical background?

A first step towards versatile systems that can display a broad range of context-sensitive information is to get an overview of which types of information could possibly be communicated. Up to now, researchers focused on single aspects of applications and user interfaces, as for example navigation, but to our knowledge there is no comprehensive overview of what kinds of information can generally occur in mobile information systems. In this article, we present a study that yields such an overview. This overview results in a *taxonomy* that can be used in various ways:

- It can be formalized as a schema for implementing underlying databases or semantic networks
- It can be used by designers to create representative use case scenarios for mobile and ubiquitous multimedia applications
- It can be used by programmers implementing these systems as a list of possible requirements.
- It can be used to systematically search the literature and conduct further research to compile a catalog of display techniques that satisfy the information needs identified. Such a catalog of techniques, taken from available literature and extended with our own ideas, is presented in the second part of the article

BACKGROUND

Augmented reality blends sensations of the real world with computer-generated output. Already in the early days of this research discipline, its potential to not only add to reality, but also subtract from ("diminished reality") (Mann & Fung, 2002) or change it ("mediated reality") has been recognized. Over the past years, we have created prototypes of mobile augmented reality systems that can be used to roam extensive indoor or outdoor environments. The form factor of these devices has evolved from early back-pack systems (Reitmayr & Schmalstieg, 2004), which prohibited usage over longer time periods or by inexperienced users, to recent PDA-based solutions (Wagner & Schmalstieg, 2003), providing us with a system that can be deployed on a larger scale to untrained and unsupervised users and carried around over an extended time span in an extended environment. Furthermore, on the PDA-class devices,

Figure 1. Our outdoor augmented reality wayfinding system

Directional arrows, landmarks, a compass and location information are superimposed on the view of the real world

classical and emerging multimedia content formats can be easily integrated, leading to hybrid applications that can make use of different media, matching the needs of the user.

One of our research applications is concerned with outdoor wayfinding in a city (Reitmayr & Schmalstieg, 2004). As can be seen in Figure 1, the augmented reality display provides additional information like directional arrows, a compass, and an indication of the desired target object. After experiments with early ad-hoc prototypes, it became clear that a structured approach to the design of the user interface would be necessary to make our system usable across a wide range of users and tasks. A kind of "toolbox" with different visualization styles is needed to visualize the information in the most suitable way. To design and implement such a toolbox, we need to have an overview of the information needs that might occur in our applications, and look for techniques that can successfully fulfill these needs in a flexible, context-dependent way.

Plenty of studies exist that evaluate different display techniques for augmented reality systems. However, we found that the majority of these studies present a novel technique and test the usability of the technique, and do not compare different alternatives for satisfying the same information need. Therefore, these studies were of little direct value for us because they didn't allow us to compare techniques against each other or to find the best technique for a given task. We had to focus on identifying and isolating the proposed techniques, and leave the comparison of techniques against each other for future work. In the future, we will implement some of the proposed techniques and conduct user studies and experiments to be able to compare the techniques to each other.

For conventional 2D diagrams, Chappel and Wilson (1993) present a comparison of different diagram types for various informational purposes. They present a table, listing different tasks (such as, for example, "judging accurate values" or "showing relationships") and for each task, they list the best diagram type according to available cognitive psychology literature. The diagrams discussed include only classical diagram types like pie chart, bar chart or graphs, while we need results in a similar form for recently developed display techniques that can be applied to mobile augmented reality systems.

Some research has been done on the generation of automatic layout for augmented reality displays. Lok and Feiner (2001) present a survey of different automated layout techniques, knowledge that is used by Bell, Feiner, and Höllerer (2001) to present a system for view management for augmented reality. The only information type they are using are labels attached to objects in the view of the user. Nevertheless, their techniques can be applied for controlling the overall layout of an application, once the individual rendering styles for different parts of the display have been chosen.

As the found literature in the field of human-computer-interaction and virtual reality does not answer our questions stated in the introduction, we have to look into other, more theoretical disciplines to find guidelines for the generation of appropriate graphical responses for our systems.

Semiotics and Design

The process that transforms the intention of some agent (software or human) into a legible sign that can be read and understood by users and possibly leads to some action on the user side involves a series of steps: creating a suitable graphical representation for the given intention, placing the created media artifact at a suitable location in the world, identification and perception of the sign by the user, interpreting the sign to extract some meaning and acting according to that meaning. Ideally, the original intention is preserved in this process, and the user acts exactly like the creator intended. However, in the real world these processes are complex, and understanding them is the subject of various scientific disciplines (Figure 2):

- Design theory (Norman, 1990) can teach us how to create aesthetically pleasing and legible signs
- Cognitive psychology (Goldstein, 2004) deals with the perceptual issues involved in sensing and reading
- Semiotics (Eco, 1976) is concerned with the transformation of observed facts into meaning

Generally, the research areas previously mentioned are usually concerned with far less dynamic information than present in the ubiquitous digital applications we are looking for. It is therefore not possible to directly implement the information systems we are envisioning based only on existing knowledge — we first have to examine how these aspects could play together in the context-sensitive applications we want to create. As a first step, we need an overview of what kinds of information can possibly be communicated through signs.

Figure 2. Sign creation and interpretation

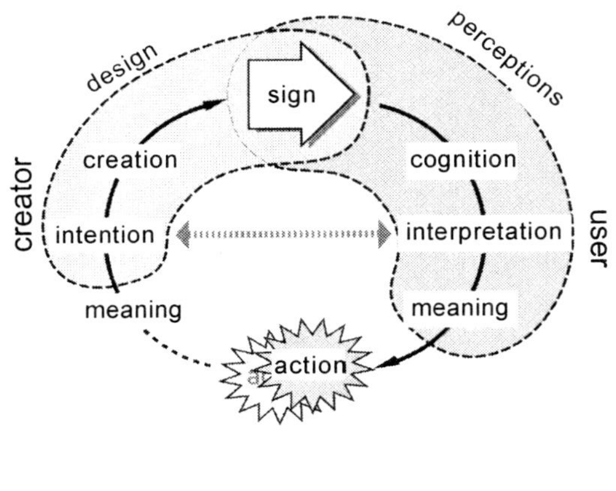

STUDYING REAL-WORLD SIGNS

How can we construct an overview of possible usages of a system we have not yet built? Our hypothesis is that fundamental information needs of our potential users are covered already in the world today, in the form of conventional media and signs. We undertook an exhaustive survey of signs and media artifacts in public space, and from that experience we extracted the core concepts or *atomic functions* of signs in the real world.

Our environments are full of signs—either explicitly and consciously created or left behind without intention. Examples for the first category would be road signs, signposts, labels, and door signs, but also stickers and graffitis, which use public surfaces as ground for articulation and discourse. The signs that are unconsciously created include traces of all kinds, like a path through the grass in a park or the garbage left behind after a barbecue, picnic, or rock concert. Also the design of an object or building can indicate some meaning or suggest some usage that is not explicitly encoded there, but presented as an affordance (Norman, 1990), a feature that is suggesting some way of usage in a more implicit way.

The starting point for our research are signs present in public space. We take existing signs and significant visual features of the environment as indicators for a demand for the information encoded in the sign and/or the individual or political will to create the sign. Therefore, the sign becomes the documentation of the action of its creation, and an indicator of possible actions that can be carried out by using the information that is encoded.

By collecting a large number of examples, we obtained an overview of sign usage in public space and were able to structure intentions and actions into categories, which we could analyze further and relate to each other. In the envisioned ubiquitous augmented reality applications, space and time will be fundamental aspects for structuring the presented information. We therefore focused on signs that are related to spatial or temporal aspects of the world—media created purely for information or the attraction of attention, without any reference to their location or temporal context (like, for example, advertisements) do not fall in this category.

The collection of examples has been gathered in the city of Vienna, Austria, in public space, public transport facilities, and some public buildings. The research was constrained to include only visual information, and most of the examples were originally photographed with a built-in mobile phone camera. This allowed the spontaneous gathering of new example images in everyday situations, and avoided the necessity to embark for specific "signspotting" trips, which would probably have biased the collection in some direction. Some of the images have been replaced by high-resolution images taken with a consumer digital camera on separate occasions; care has been taken to reproduce the original photo as closely as possible. An unstructured collection of example images is shown in Figure 3.

Obviously, the collection of examples is heavily biased by the photographer's view of the city, his routes, tasks, and knowledge. An improved approach would include several persons with different demographical backgrounds, especially age, cultural and professional background and of varying familiarity with the city. However, our study covers a good part of the explicit signs present in urban space, and allows us to draw conclusions that will be valuable for future research by us and others.

FUNDAMENTAL FUNCTIONS OF SIGNS

In this section, we give an overview of all atomic functions identified in our study. While it is impossible to prove that a given set of categories covers all possible examples without examining every single instance, these categories could already be successfully applied to a number of newly found examples. Therefore, there is some indication that the proposed set of

Figure 3. Some examples of images taken in our study: (a) annotated safety button; (b) number plate; (c) signposts; (d) roadsign; (e) graffiti; (f) map

functions covers at least a good part of the use cases that can be found in an urban, public space scenario.

We choose to arrange the functions in five fields, resembling what in our opinion are fundamental aspects of future context sensitive ubiquitous applications: Object metainformation, object relationship information, spatial information, temporal information, and communication. Inside the respective sections, the identified concepts are listed and discussed, together with possible display styles that can be used to render the information in multimedia information systems.

Object Metainformation

Adding metainformation to existing objects in the real world is a fundamental function of both real and digital information systems.

Naming

Naming establishes a linguistic reference for an object in a specific context. The user has to be part of that context to be able to correctly understand the name and identify the referenced object. The context also determines whether the name is unique or not—for example, the name of an institute is unique in the context of a university, but not in a global context. Depending on the user, displayed names have to be chosen appropriately to allow identification.

Identification

Identification is a more technical concept than naming, which allows identifying a specific entity, usually in a global context. Examples would be number plates for cars or street addresses for houses. Note that also in these examples, the identification might need additional parts in a larger context—in a city, the street name is usually unique, but not in a global context, where it has to be prefixed with country and municipality information.

Explanation

Explanation is important if it is not clear from an object's design how to use it, or if the user just wants it for informational purposes. Sometimes it is sufficient to name the object, if the name already implies the mode of operation. A special class of explanation that we identified is *type information*—information about what an object *is*. In contrast to naming, type information denotes the class of an object, and does not provide a reference to a specific instance. (Note that when only a single instance of an object is present in the current context, the type information might also be sufficient to identify the object. Example: "the door" in a room with only a single exit.)

As these three kinds of object-related information mentioned above are mostly textual, the primary problem for displaying it in a digital system is that of automatic layout. The placement, color, and size of labels have to be chosen to be legible, unobtrusive, and not conflicting with other elements of the display. Lok and Feiner (2001) examine different strategies of automatically generating appropriate layouts, knowledge which was used by Bell et al. (2001) to automatically place labels for objects in an augmented reality application.

Accentuation

Accentuation means to emphasize a specific object by increasing its visibility. In the real world, accentuation is mostly performed to permanently improve the visibility of objects or regions for safety reasons by using bright, high contrast colors. In digital systems, image-based

Figure 4. Examples for accentuated objects: (a) fire extinguisher; (b) first step of descending stairs; (c) important announcement in public transport system

methods like partially increasing the contrast or saturation could be used, as well as two- or three-dimensional rendering of overlay graphics. An approach found in some systems (Feiner, Macintyre, & Seligmann, 1993), however never formally evaluated against other techniques, is to superimpose a wireframe model of the object to be highlighted on the object — if the object in question is occluded by other things, dashed lines are used to indicate this. This approach is inspired by technical drawings, where dashed lines are often used to indicate invisible features.

Ownership

While ownership is actually relational information (to be discussed in the next section), linking an owner entity to a specific object, it can often be read as information about the purpose of an object. Examples are the logos of public transport companies on buses. In most cases, the user is not interested in a link to the location of the company, but reads the ownership information as an indication of the object's function.

General Metainformation

Metainformation is often found on device labels to indicate some key properties of the device. Obviously, in digital systems this information can be subject to sophisticated filtering, rendering only the relevant information according to the user's task context. For textual metainformation, the layout considerations discussed above apply.

Status

Display of an object's status is the most dynamic metainformation found in conventional signs—the current state of an object or a subsystem is displayed to the user by using LEDs or alphanumeric displays. In today's cities, this is used for example in public transport systems to display the time until arrival of the next bus.

Status information is, due to its dynamic nature, an example where conventional, physical signs are reaching their limitations. In digital information systems, the possibilities to include dynamic information are much greater. Appropriate filtering has to be applied to prevent information overload and provide only the necessary information to the user. For a discussion of information filtering in an augmented reality context, see Julier, Livingston, Brown, Baillot, and Swan (2000).

Object-Relationship Information

The second type of information we find in various contexts is relating objects to each other. Entities frequently related to each other are people, rooms, buildings, or locations on a map. In most cases, the location of both objects (and the user) determines how the relationship is displayed and what actions can be carried out by the user.

Linking

Linking an object in the real world with another entity is another often-found purpose of signs. In augmented reality applications, one of the two objects (or both) might be virtual objects placed at real world locations. For example, an object in the real world might be linked to a location on a map presented on the user's display.

Rendering a link to the user depends on how the user is supposed to use that information. If the user should be guided from the one object to the other one, arrows can be used to give directional information (see the section on wayfinding below). If the objects are related in some other way, it might be sensible to display the name, an image, or a symbolic representation of the second object, if available, and denote the type of relationship as suitable. If the two objects are close together and both are visible from the users point of view, a straight line can be rendered to connect the objects directly—an approach also used by Bell et al. (2001) to connect labels with the objects they are related to.

Browsing

Browsing means to give the user an overview of all entities that are available for a specific interaction. Real-world examples for browsing opportunities would be signs in the entrance areas of buildings that list all available rooms or persons. The user can choose from that list or look for the name of the entity she is trying to locate.

Computers are frequently used for browsing information. In contrast to the physical world, browsing can be combined with powerful information filtering that passes only relevant information to the user. In most cases, the system will be able to choose the relevant information from the user's context, making browsing only necessary when an explicit choice is to be made by the user.

Spatial Information

The term "navigation" is often used casually for some of the concepts in this section. In our research we found out, however, that we have to break this term down into subconcepts to get an insight into the real motivations and demands of users.

Wayfinding

Wayfinding is what is most often referred to as navigation—finding the way from the current location to a specific target object. Note that for wayfinding only, other aspects of the user's spatial context like overview or orientation can be ignored—the user could be guided by arrows, without having any mental representation of the space she is moving through. In real spaces, wayfinding is supported by arrows and signposts, labeled with the name of the destination object or area. In digital applications, a single, constantly displayed arrow can be used that changes direction as needed.

Overview

Overview supports the ability to build a mental model of the area and is useful for generic wayfinding — finding targets for which no explicit wayfinding information is available, or finding fuzzy targets like areas in a city or district. Also, overview is related to browsing, as it allows looking for new targets and previously unknown locations. Traditionally, overview has been supported by maps (Däßler, 2002). Digital maps offer several new possibilities, like the possibility to mark areas that have been visited by the user before (see the section on trails below).

Orientation

To be useful for wayfinding, overview has to be complemented by orientation, the ability of the user to locate herself on a map or in her mental model of the environment. Maps installed at fixed locations in the world can be augmented with static "You are here" markers, a feature that can be implemented in a dynamic way on a digital map (Vembar, 2004). Overview is also supported by *landmarks*, distinctive visual features of the environment that can be seen from many different locations in the world. Ruddle (2001) points out the important role of landmarks in virtual environments, which often offer too few distinctive features with the consequence of users feeling lost or disoriented.

Marking Territories

Marking of districts or territories is another example for spatially related information. Real-world examples include road signs or marks on the ground marking the beginning and ending of certain zones (see Figure 5 for example images). One of the problems that conventional signs have is that a human needs to keep track of the current state of the zones she is in as she moves through space.

Spatial Awareness

Ideally, the beginning and ending markings are accompanied by information that provides continuous, ambient feedback of which zone the user is in. This can be found in some buildings, where different areas are marked by using differently colored marks on the walls. Obviously, in digital information systems there are

Figure 5. Marking of zones: (a) beginning of a speed-limit zone; (b) dashed border surrounding a bus stop; (c) location awareness by colored marking on the wall

more advanced ways to keep track of and visualize the zones a user is currently in. Continuous feedback, for example in the form of appropriate icons, can be provided to the user on her display, visualizing the currently active zones.

Remote Sensing

A new possibility that emerges with digital multimedia systems is that of remote sensing. By remote sensing, we mean the accessibility of a live video image or audio stream that can be accessed by the user from remote locations. Information provided by remote sensing is less abstract than the other discussed concepts, and opens up the possibility for the user's own interpretation. CCTV cameras installed in public space are an example of remote sensing, although the user group and technical accessibility are limited.

Traces

Traces are often created by crowd behavior and are indicators for usage or demands. Classical examples are paths through the grass in a park, indicating that the provided paths are not sufficient to fulfill the needs of the visitors. In the digital domain, traces can be much more dynamic, collected at each use of the system and annotated with metainformation like date or current task. Some research exists on how traces can be used to aid wayfinding and overview in large virtual environments (Grammenos, Filou, Papadakos, & Stephanidis, 2002; Ruddle, 2005).

Temporal Information

An area where the limitations of conventional signs become clearly visible is information that changes over time. Temporal change has to be marked in advance if the validity of a sign changes over time (for example, parking limitations constrained only to specific times). This additional information can lead to cluttered and overloaded signs (see Figure 3(d)).

Temporal Marking

Temporal marking can be accomplished much easier in digital systems — if the sign is not valid, it can simply be hidden from the users view. Care has to be taken, however, that information that might be relevant for the user in the future (for example, the beginning of a parking limitation) is communicated in advance to allow the user to plan her actions. Which information is relevant to the user in these cases depends highly on the task and activity.

Temporary Change

Similarly, temporary change means the temporary change of a situation (for example, due to construction work) with an undefined ending date. In real world examples, it is usually clearly visible that the change is only temporary and the original state will be restored eventually. If we want to communicate a temporal change in a digital system, this aspect has to be taken into account.

Synchronization

Good examples for synchronization of different parties are traffic lights. Despite their simplicity, traffic lights are among the most complex dynamic information source that can be found in public space. Obviously, the capabilities of future multimedia systems to communicate dynamic information are much greater; therefore, synchronization tasks can probably be adapted dynamically to the current situation.

Sequencing

Synchronization is related to sequencing, where the user is guided through a series of steps to fulfill a task. In real world examples this is usually solved by providing a list of steps that the user is required to take. In digital systems, these steps can be displayed sequentially, advancing to the next step either by explicit user interaction or automatically, if the system can sense the completion of the previous step (for example, by sensing the user's location).

Communication

While signs are always artifacts of communication, signs in the real world are usually only created by legitimate authorities. There are few examples of direct user to user communication—a possibility that can be extended with digital information systems.

Articulation

The surfaces of a city enable articulation in the form of graffiti and posters. While mostly illegal, it is an important property of physical surfaces that they can be altered, extended, or even destroyed. Digital environments are usually much more constrained in what their users are able to do—the rules of the system are often directly mapped to the interaction possibilities that are offered to the user (not taking into account the possibility of hacking the system and bypassing the provided interaction mechanisms). If we replace physical signs by digital content, we should keep in mind that users may want to interact with the information provided, leaving marks, and comments for other users.

Discourse

Discourse through signs and writings involving two or more parties is much rarer observed in public space. The capabilities of networked information systems could improve the ability to support processes of negotiation and communication between multiple parties in public space.

Mapping the Taxonomy

As mentioned above, a linear representation of a taxonomy cannot reproduce a multi-dimensional arrangement of concepts and the relationships between them. To create a more intuitive overview, we have created a 2-dimensional map of the concepts of the taxonomy (Figure 6). Four of the main fields identified above (metainformation, spatial aspects, temporal aspects, communication) are represented in the corners of the map, and the individual concepts are arranged to represent their relation to these fields. In addition, related concepts are linked in the diagram.

RENDERING AND DISPLAY STYLES FOR MOBILE MULTIMEDIA

The following table summarizes the techniques that we identified for the various types of information from the taxonomy. The third column references appropriate literature, where the listed techniques have been discussed or evaluated. The table lists also tasks, for which no appropriate display technique has been presented or evaluated so far. These situations are opportunities for future work, to develop and evaluate techniques that are able to address the communication of the desired information.

Figure 6. An arrangement of the found concepts on a conceptual map

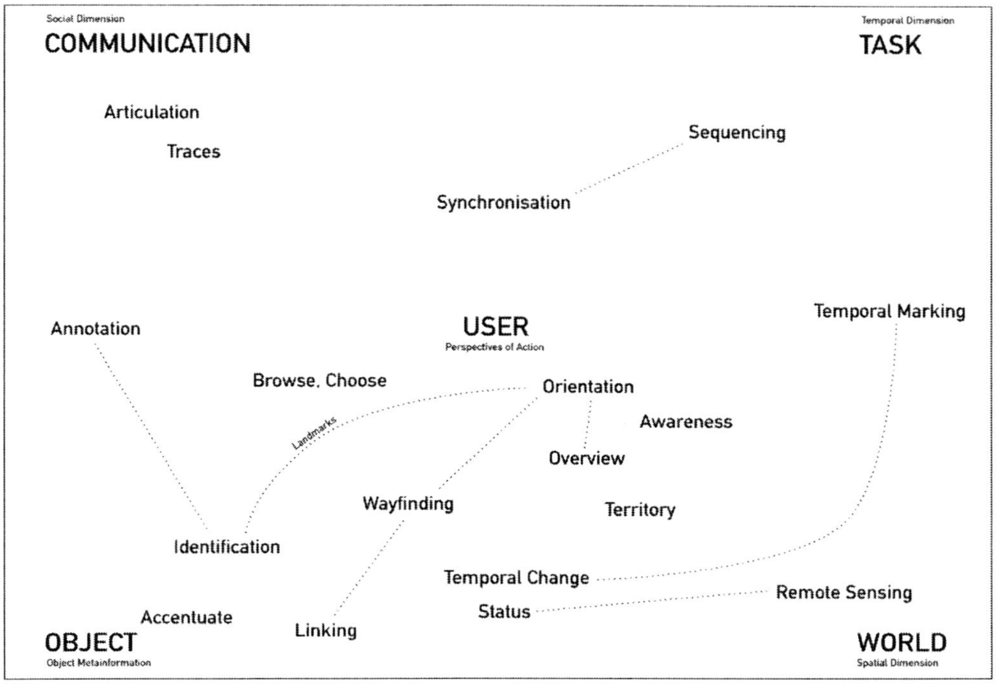

CONCLUSION

To support the systematic design of future ubiquitous multimedia applications, we have provided an overview of the types of information that users may demand or content providers may want to communicate. We rooted that overview in a study of sign usage in the real world, taking existing signs as indications for the demand for the information encoded in the sign. From that analysis, we can extrapolate the consequences of bringing that information into the digital domain, which will result in improved possibilities for the display of dynamic information, changing over time and with the context of the user.

While we could identify techniques for rendering some of the information types in digital systems, for some of the identified types of information further research is needed to identify appropriate ways of displaying them to the user. By identifying these "white spots" on our map of display techniques, we provide the basis for future research in the area, targeting exactly those areas where no optimal techniques have been identified so far.

The overview given by the taxonomy may be used by designers of future information systems as a basis for constructing more complex use cases, choosing from the presented scenarios the elements needed for the specific application context. In a (yet to be developed) more formalized way, the presented taxonomy can lay the ground for formal ontologies of tasks and information needs, which could result in more advanced, "semantic" information systems that are able to automatically choose filtering and presentation methods from the user's task and spatio-temporal context.

Figure 7.

Task	Technique	References
Labeling: Positioning Labels		Bell et al. (2001)
Metainformation	Information Filtering	Julier et al. (2000)
Highlighting: Visible Objects	Wireframe overlay	Feiner et al. (1993)
Highlighting: Occluded Objects	Cutaway View	Furmanski, Azuma, & Daily et al. (2001)
	Dashed wireframe overlay	Feiner et al. (1993)
Highlighting: Out-of-view Objects		
Linking: Objects to Objects		
Linking: Labels to Objects	Connect with line	Bell et al. (2001)
Linking: Objects to Map		
Navigation: Wayfinding	User–aligned directional arrow	Reitmayr and Schmalstieg (2004)
	Landmarks connected by arrows	Reitmayr and Schmalstieg (2004)
Navigation: Overview	World-in-miniature	Stoakley, Conway, & Pausch et al. (1995)
	Viewer-aligned Map	Diaz and Sims (2003)
Navigation: Orientation	Spatial Audio	Darken and Sibert (1993)
	Landmarks	Darken and Sibert (1993)
	Navigation Grid	Darken and Sibert (1993)
	Breadcrumb Markers	Darken and Sibert (1993)
	Coordinate Feedback	Darken and Sibert (1993)
	Viewer-aligned arrow on map	Vembar (2004)
Territory: Marking		
Traces	Dynamic Trails	Ruddle (2005)
	Breadcrumb Markers	Darken and Sibert (1993)
	Virtual Prints	Grammenos et al. (2002)
Temporal marking		

REFERENCES

Azuma, R., Baillot, Y., Behringer, R., Feiner, S., Julier, S., & MacIntyre, B. (2001). Recent advances in augmented reality. *IEEE Computer Graphics and Applications, 21*(6), 34-47.

Bell, B., Feiner, S., & Höllerer, T. (2001). View management for virtual and augmented reality. *Proceedings of the Eurographics Symposion on User Interface Software and Technology 2001 (UIST'01)* (pp. 101-110). New York: ACM Press.

Chappel, H., & Wilson, M. D. (1993). Knowledge-based design of graphical responses. *Proceedings of the ACM International Workshop on Intelligent User Interfaces* (pp. 29-36). New York: ACM Press.

Chen, H., Perich, F., Finin, T., & Joshi, A. (2004). SOUPA: Standard ontology for ubiquitous and pervasive applications. *Proceedings of the International Conference on Mobile and Ubiquitous Systems: Networking and Services*, Boston.

Darken, R. P., & Sibert, J. L. (1993). A toolset for navigation in virtual environments. *Proceedings of the Eurographics Symposion on User Interface Software and Technology 1993 (UIST'93)* (pp. 157-165). New York: ACM Press.

Däßler, R. (2002). Visuelle Kommunikation mit Karten. In A. Engelbert, & M. Herlt (Eds.), *Updates–Visuelle Medienkompetenz*. Würzburg, Germany: Königshauser & Neumann.

Diaz, D. D., & Sims, V. K. (2003). Augmenting virtual environments: The influence of spatial ability on learning from integrated displays. *High Ability Studies, 14*(2), 191-212.

Eco, U. (1976). *Theory of semiotics*. Bloomington: Indiana University Press.

Feiner, S., Macintyre, B., & Seligmann, D. (1993). Knowledge-based augmented reality. *Communications of the ACM, 36*(7), 53-62.

Furmanski, C., Azuma, R., & Daily, M. (2002). Augmented-reality visualizations guided by cognition: Perceptual heuristics for combining visible and obscured information. *Proceedings of the International Symposion on Mixed and Augmented Reality 2002 (ISMAR'02)* (pp. 215-224). Washington, DC: IEEE Computer Society.

Goldstein, B. E. (2004). *Cognitive psychology* (2nd German ed.). Heidelberg, Germany: Spektrum Akademischer Verlag.

Grammenos, D., Filou, M., Papadakos, P., & Stephanidis, C. (2002). Virtual prints: Leaving trails in virtual environments. *Proceedings of the 8th Eurographics Workshop on Virtual Reality (EGVE'02)* (pp. 131-138). Aire-la-Ville, Switzerland: Eurographics Association.

Julier, S., Livingston, M., Brown, D., Baillot, Y., & Swan, E. (2000). Information filtering for mobile augmented reality. *Proceedings of the International Symposion on Augmented Reality (ISAR) 2000*. Los Alamitos, CA: IEEE Computer Society Press.

Lok, S., & Feiner, S. (2001). A survey of automated layout techniques for information presentations. *Proceedings of SmartGraphics 2001* (pp. 61-68).

Mann, S., & Fung, J. (2002). EyeTap devices for augmented, deliberately diminished, or otherwise altered visual perception of rigid planar patches of real-world scenes. *Presence: Teleoperators and Virtual Environments, 11*(2), 158-175.

Norman, D. (1990). *The design of everyday things*. New York: Doubleday.

Posner, R., & Schmauks, D. (1998). Die Reflektiertheit der Dinge und ihre Darstellung in Bildern. In K. Sachs-Hombach, und K. Rehkämper (Eds.), *Bild–Bildwahrnehmung–Bildverarbeitung. Interdisziplinäre Beiträge zur Bildwissenschaft* (pp. 15-31). Wiesbaden: Deutscher Universitäts-Verlag.

Reitmayr, G., & Schmalstieg, D. (2004). Collaborative augmented reality for outdoor navigation and information browsing. *Proceedings of the Symposium on Location Based Services and TeleCartography*.

Reitmayr, G., & Schmalstieg, D. (2005). Semantic world models for ubiquitous augmented reality. *Proceedings of the Workshop towards Semantic Virtual Environments (SVE'05)*, Villars, CH.

Ruddle, R. A. (2001). Navigation: Am I really lost or virtually there? *Engineering Psychology and Cognitive Ergonomics*, *6*, 135-142. Burlington, VT: Ashgate.

Ruddle, R. A. (2005). The effect of trails on first-time and subsequent navigation in a virtual environment. *Proceedings of IEEE Virtual Reality 2005 (VR'05)* (pp. 115-122). Bonn, Germany.

Stoakley, R., Conway, M. J., & Pausch, R. (1995). Virtual reality on a WIM: Interactive worlds in miniature. *Conference Proceedings on Human Factors in Computing Systems* (pp. 265-272). Denver, CO: Addison-Wesley.

Vembar, D. (2004). Effect of visual cues on human performance in navigating through a virtual maze. *Proceedings of the Eurographics Symposium on Virtual Environments 2004 (EGVE04)*. Aire-la-Ville, Switzerland: Eurographics Association.

Wagner, D., & Schmalstieg, D. (2003). First steps towards handheld augmented reality. *Proceedings of the 7th International Conference on Wearable Computers (ISWC2003)*, White Plains, NY.

KEY TERMS

Augmented Reality: Augmented reality (AR) is a field of research in computer science which tries to blend sensations of the real world with computer-generated content. While most AR applications use computer graphics as their primary output, they are not constrained by definition to visual output—audible or tangible representations could also be used. A widely accepted set of requirements of AR applications is given by Azuma (2001):

- AR applications combine sensations of the real world with virtual content.
- AR applications are interactive in real-time
- AR applications are registered in the 3-dimensional space of the real world

Recently, several *mobile* AR systems have been realized as research prototypes, using laptop computers or handheld devices as mobile processing units.

Taxonomy: A taxonomy is a classification of things or concepts, often in a hierarchical manner.

Ubiquitous Computing: The term ubiquitous computing (UbiComp) captures the idea of integrating computers into the environment rather than treating them as distinct objects, which should result in more "natural" forms of interaction with a "smart" environment than current, screen-based user interfaces.

Chapter XXVII
Mobile Fractal Generation

Daniel C. Doolan
University College Cork, Ireland

Sabin Tabirca
University College Cork, Ireland

Laurence T. Yang
St. Francis Xavier University, Canada

ABSTRACT

Ever since the discovery of the Mandelbrot set, the use of computers to visualise fractal images have been an essential component. We are looking at the dawn of a new age, the age of ubiquitous computing. With many countries having near 100% mobile phone usage, there is clearly a potentially huge computation resource becoming available. In the past years there have been a few applications developed to generate fractal images on mobile phones. This chapter discusses three possible methodologies whereby such images can be visualised on mobile devices. These methods include: the generation of an image on a phone, the use of a server to generate the image and finally the use of a network of phones to distribute the processing task.

INTRODUCTION

The subject of Fractals has fascinated scientists for well over a hundred years ever since what is believed to be the discovery of the first fractal by Gregor Cantor in 1872 (The Cantor set). Benoit Mandelbrot (Mandelbrot 1983) first coined the term "Fractal" in the mid 1970's. It is derived from the latin "fractus" or "broken." Before this period, they were often referred to as "mathematical monsters." Fractal concepts can be applied to a wide-ranging variety of application areas such as: art (Musgrave & Mandelbrot, 1991), music generation (Itoh, Seki, Inuzuka, Nakamura, & Uenosono, 1998), fractal image compression

(Lu, 1997), or fractal encryption. The number of uses of fractals is almost as limitless as their very nature (fractals are said to display infinite detail). They also display a self-similar structure, for example, small sections of the image are similar to the whole. Fractals can be found throughout nature from clouds and mountains to the bark of a tree. To appreciate the infinite detail that fractal images posses it is necessary to be able to zoom in on such images. This "fractal zoom" allows the viewer to experience the true and infinite nature of the fractal. One typically cannot fully appreciate the fractals that exist in nature. To fully explore the true intricacies of such images one must visualise them within the computing domain.

The generation of a fractal image is a computationally expensive task, even with a modern day desktop computer, the time to generate such images can be measured from seconds to minutes for a moderately sized image. The chief purpose of this chapter is to explore the generation of such images on mobile devices such as mobile phones. The processing power of mobile devices is continually advancing, this allows for the faster computation of fractal images. The memory capacity so too increasing rapidly allowing for larger sized images to be generated on the mobile device. The current generation of smartphones have processing speeds with a range of 100 to 200Mhz and are usually powered by the ARM9 family of processors. The next generation of smartphones will be powered with the ARM11 processor family and should have speeds of up to 500Mhz. This is clearly a dramatic increase in speed and as such we will see that the next generation of smartphones will be able to run a myriad of applications that current phones are too slow to run effectively.

Certainly, this study can be applied to various visualisation problems that involve large amount of computation on mobile devices. Although the computation power of mobile devices has increased, this may still be insufficient to rapidly generate the image of the object to visualise. Therefore, alternative solutions must be investigated.

FRACTAL GENERATION

In this section, we will outline the algorithms to generate the Mandelbrot and Julia sets. One can use the exact same algorithms to generate a fractal image on a mobile phone with slight differences in the implementation but the overall structure stays the same.

Mandelbrot and Julia sets

The Mandelbrot and Julia sets are perhaps the most popular class of non-linear self-similar fractals. The actual equation and algorithm for generating both the Julia and Mandelbrot like sets are quite similar and generally quite simple. They use the polynomial function $f: C \rightarrow C, f(z) = z^u + c^v$ to generate a sequence of points $\{x_n : n \geq 0\}$ in the complex plane by $x_{n+1} = f(x_n), \forall n \geq 0$. There have been several mathematical studies to prove that the sequence has only two attractors 0 and infinity. The Julia and Mandelbrot sets retain only those initial points that generate sequences attracted by 0 as Equations (1-2) show:

$$J_c = \{x_0 \in C : x_{n+1} = f(x_n), \; n \geq 0 \textit{ are attracted by } 0\} \quad (1)$$

$$M = \{c \in C : x_0 = 0, x_{n+1} = f(x_n), \; n \geq 0 \textit{ are attracted by } 0\} \quad (2)$$

The most important result (Mandelbrot, 1983) on this type of fractal shows that the set M is an index for the sets J (see Figure 1). In this case,

Figure 1. Relation between the Julia and Mandelbrot sets

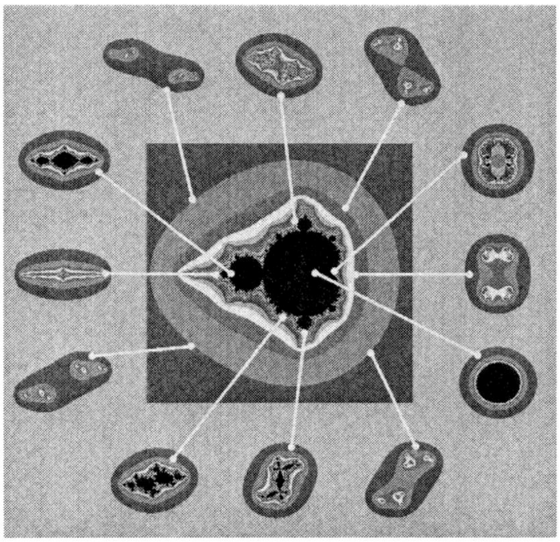

any point on the Mandelbrot set can generate a corresponding Julia set.

Of course computer programs to generate these sets focus only on a region of the complex plane between $[x_{min}, x_{max}] \times [y_{min}, y_{max}]$and usually generate only the first n_{iter} points of the sequence x. If there is a point outside of a certain bound R e.g. $|x_n| \geq R$ then the sequence is not attracted by 0. To generate these fractals we need to calculate the first n_{iter} points for each point of the region and see whether the trajectory if finite or not. If all the trajectory points are under the threshold R we can then say that the initial point x_0 is in the set. With these elements the algorithm to generate the Mandelbrot set is described in the following procedure:

```
Inputs: [xmin, ymin]*[ymin,ymax] – the region of interest
    niter – number of iteration to generate, R – the radius for infinity.
    c[0],c[1],...c[nrc-1] - a set of colors
Output: the fractal image
procedure fractal
    for each point (x,y) in [xmin, ymin]*[ymin,ymax]
    do
        construct the complex numbers c=x+j*y
        and z=0=j*0;
        for i=0 to niter do
                calculate z=f(z);
                if |z| > R then break;
        end for
        draw (x,y) with the colour c[i%nrc];
    end for
end procedure;
```

It is widely accepted that it is computationally expensive to generate fractals. The complexity of the procedure fractal depends on the number points we calculate for each iteration, as well as the number of pixels in the fractal image. The bigger these elements are, especially the number of iterations, the larger the execution time becomes.

FRACTAL GENERATION ON A MOBILE PHONE

Very little work has been done in the area of fractal image generation on mobile devices. One example of such (Kohn, 2005), generates the fractal image on the phone itself, however the image is displayed over the entire screen giving a slightly distorted look to the image as the screen width and heights differ. One recent paper (Doolan & Tabirca, 2005) dealt with the topic of using mobile devices as an interactive tool to aid in the teaching of fractal geometry. Another example (Heerman, 2002) used Mobile Phones in the teaching of Science. Many examples are available where fractal images are used as wallpaper for mobile devices.

This application has been designed to allow the user to select various options (Figure 2) for

Figure 2. Options screen, fractal image screen, results screen

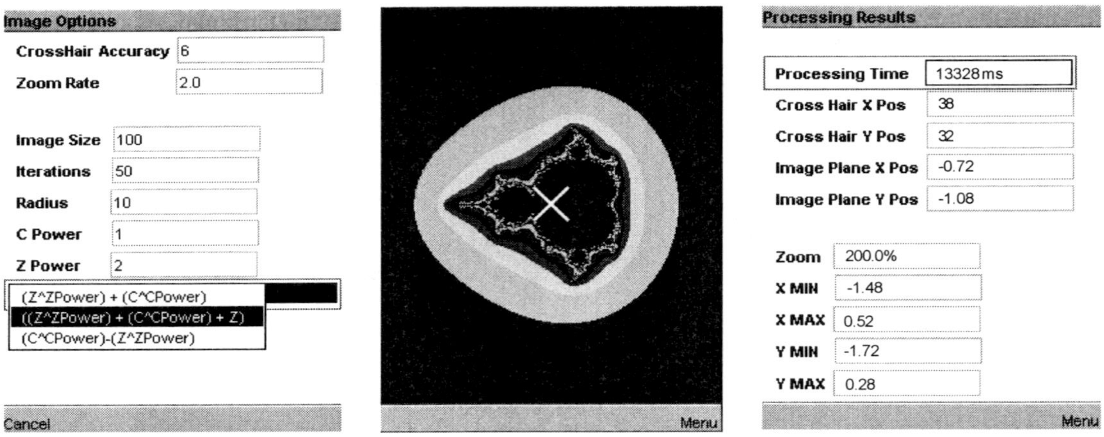

image generation such as: image size, number of iterations, radius, powers, and formula type. This allows for a rich diversity in the number of possible images the application is capable of creating. The central image of Figure 2 shows a typical example of the Mandelbrot set generated on a mobile device. The final screen shot allows the user to view the processing time of the image and some statistics, for example the zoom level of the image, the xmin, ymin, xman, ymax coordinates, and also the coordinates of the on screen cursor.

A useful addition to this application is a cursor that may be moved around the screen by the directional keys. This allows the user to select are area they may wish to zoom in on. It is also used to designate a point on the image for which the corresponding Julia set should be

Figure 3. Selection of Julia set Screen Shots

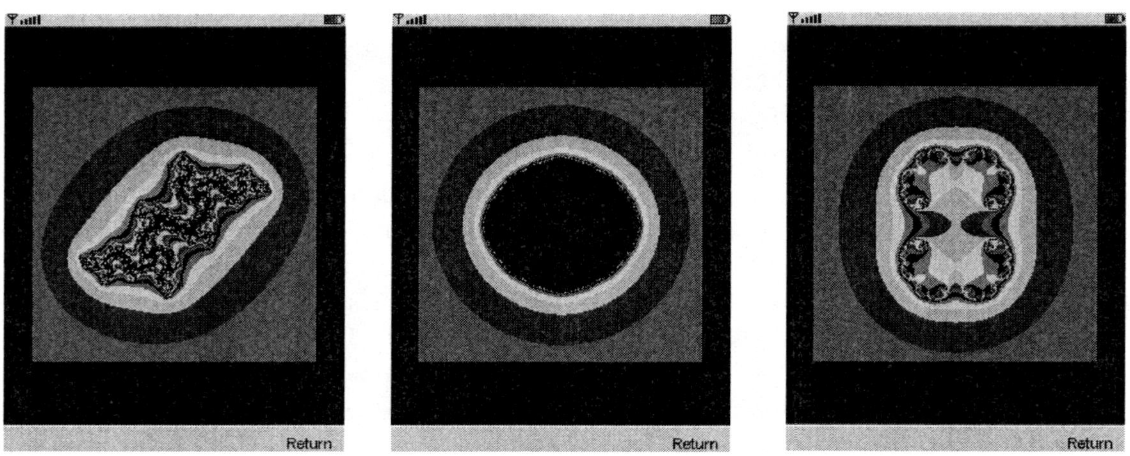

generated, Figure 3 shows some typical examples of various Julia sets the application is capable of generating.

The application is capable of generating three differing fractal images based on the formulas: $Z_{n+1} = Z^U + C^V$, $Z_{n+1} = Z^U + C^V + Z$, $Z_{n+1} = Z^U - C^V$. This results in the application being capable of generating images such as $Z^2 + C$, $Z^3 + C$, $Z^5 + C^2$, $Z^4 + C^3 + Z$, $Z^7 - C^2$. The application was designed to use a Thread for the image generation process, after a predetermined number of columns have been calculated the updated image is displayed on screen. This allows the user to see the image generation process taking place.

Table 1. Processing times for the Nokia 3320 phone

Iterations	50 Pixels	100 Pixels	150 Pixels
50	11,801 ms	56,503 ms	119,360 ms
500	77,851 ms	298,356 ms	696,075 ms

The application was tested on two differing phones: the Nokia 3320 and the Nokia 6630. The results showed a huge difference in processing times when compared with each other. The 3320 phone having a very limited heap size was unable to generate an image of 200 pixels square. To generate an image 150 pixels square at 500 iterations took in excess of 650 seconds. The generation of a 200 pixel square image on the Nokia 6630 at 500 iterations required just under 60 seconds to complete the computation task (see Figures 4, 5, and Tables 1, 2). An example of this application is available for download at the Mobile Computer Graphics Research Web site (Mobile Fractals, 2005) (see the JAD Downloads section).

SERVER SIDE COMPUTATION

The second approach uses a server to generate the fractal image, which is then returned to the mobile device and displayed. One method of carrying out server side computation is to use Servlets. The communication between the server and client (Figure 6) may be achieved by using a HttpConnection.

Figure 4. Processing times for the Nokia 3320 phone

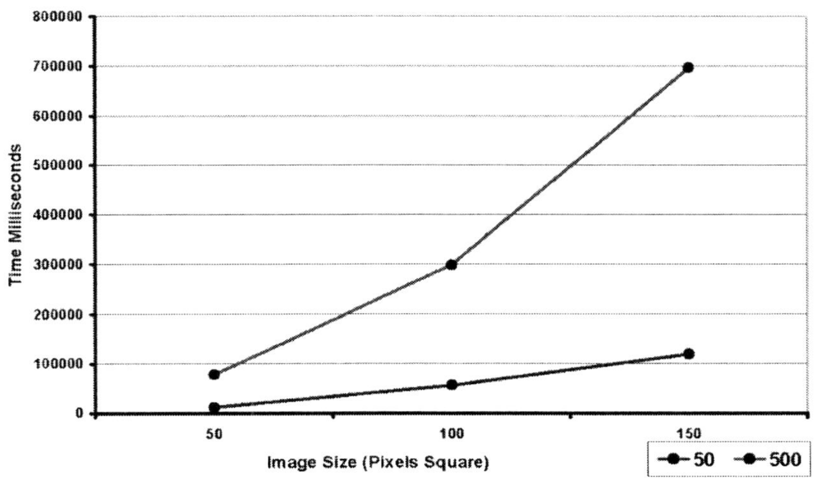

Figure 5. Processing times for the Nokia 6630 phone

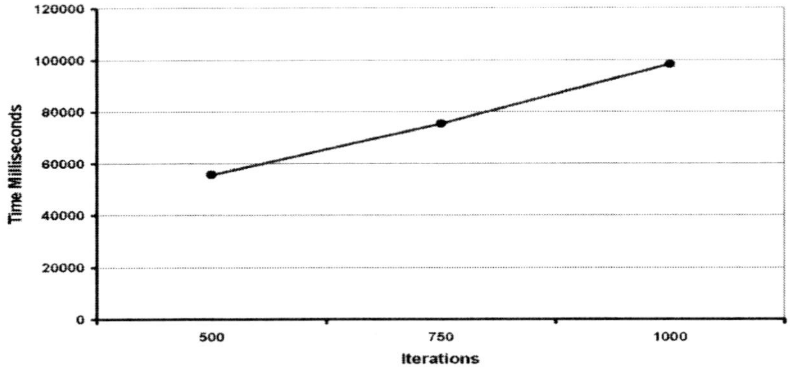

Table 2. Processing times for the Nokia 6630 phone

Iterations	500	750	1000
Processing Time	55,657 ms	75,266 ms	98,250 ms

The general methodology is that the client (mobile device) is used to enter parameters for the image type to be generated. Once the user is selected all the required parameters a message is sent to the server to generate the image corresponding to the parameters that were passed to it. The client then waits for the server to generate the image and send it back to the client. The image data is sent as a stream of integers representing the RGB values of the generated image. Once the client has received all the data, it then constructs an Image object so that it can be displayed on screen. The obvious advantage of this method of fractal generation is that the image is generated very quickly. It does however require the use of a HttpConnection which may cause the user to incur communication costs for the use of data transfer over the phone network. A successful implementation of this method was carried out with some promising results revealed (Table 3). The time to generate the image on the server is very small (1,110ms for 1000 Iterations).

Figure 6. Mobile phone to server (Servlet) communication

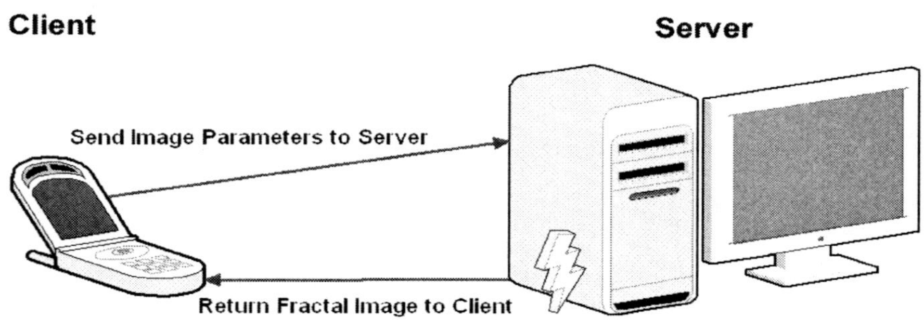

Table 3. 200 x 200 pixel Mandelbrot set, image generation using Servlets

Iterations	100	500	1000
Server Time	281 ms	594 ms	1,110 ms
Comms Time	7,828 ms	7,812 ms	7,843 ms
Total Time	8,109 ms	8,406 ms	8,953 ms

The general algorithm for this client/server communication starts with the user entering the parameters required via a GUI interface on the mobile device. Once the user issues the request to generate the image via the selection of a menu option the image parameters are sent to the server. This requires the opening of a HttpConnection object, passing the parameters to the server using a DataOutputStream.

On the server side, once a request has been received the parameters are passed to the image generation algorithm which generates the corresponding image. When the processing has been completed, the resultant data is returned to the client as an array on integer values. The actual data packet that is sent has the form of "array_size, array_data." On receipt of the complete array of RGB integers the client create a new Image object using the createImage(...) method. The image is now ready for on screen display to the user.

The Communication/Image Construction Time is composed of several distinct operations. The first stage is for the client to establish

Figure 7. Mobile phone to Servlet algorithm

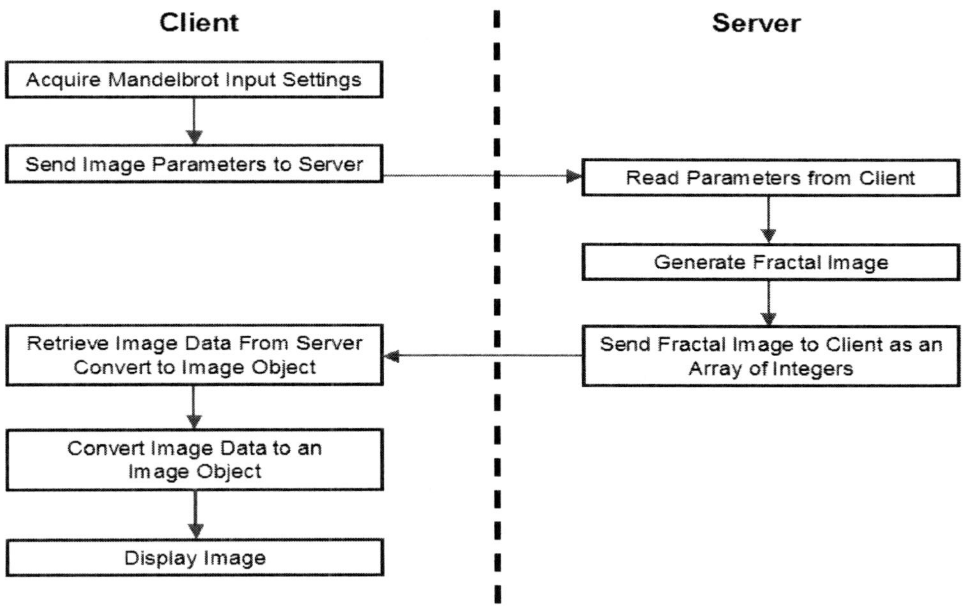

an HttpConnection to the server, once established the parameters for the fractal image are transferred to the server. The second communication stage is when the server returns the generated image to the client (see Figure 7).

For a 200 x 200 pixel image this amounts to an array of 40,000 integers being passed back to the client that requested the image. Once the client has received the pixel image array representation of the image it must generate an Image object, which takes a short period of time, the image is now ready for on screen display. It is clear from the execution results (Table 3) that the time taken to carry out these operations remains constant for the various image that were generated in the experiment. Figure 8 shows graphically the relation between server processing time vs. the total time (the time from the user requesting the image until the image is ready for display on screen).

DISTRIBUTED GENERATION WITH BLUETOOTH

The next approach splits the computation over a number of mobile devices. To achieve this Bluetooth technology is employed as the inter device communications mechanism. The system like the previous example uses a Client/Server architecture, but the method by which the architecture is used differs greatly.

Bluetooth Networking

There have been only a small number of research papers working with the Bluetooth technology so far. Some interesting work has been carried out in form of testing Bluetooth capabilities with J2ME (Klingsheim, 2004). We have also found another work (Long, 2004) which deals with the study of java games in wireless Bluetooth networks. However, both Sony-Ericsson (Sony-Ericsson, 2004) and Nokia (Nokia, 2004) have very useful developer material on how to develop with J2ME and Bluetooth technology.

Typically the first step in a Networked Bluetooth application is to discover other Bluetooth Capable devices within the catchment area (10 meters for a class 3 Bluetooth Device, 100 meters for a class 1 device). For a Bluetooth device to advertise itself as being available it must be in "discoverable mode."

Figure 8. Server time vs. total time

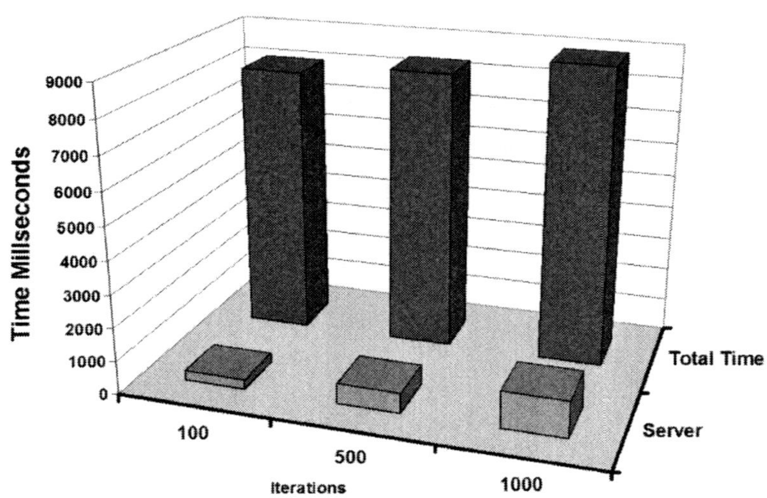

The implemented system works slightly differently to many typical client/server systems, where it is the server that carries out the processing tasks. Instead of this it is the clients that are connected to the server that carry out the actual computation task. This is akin to Seti@Home (Seti@Home, 2005) where the operation of processing data blocks is carried out by a mass of client applications. The system is designed in the fashion of a point to multi point piconet (Figure 9), this limits the number of clients that may be connected to the server at any one time to be seven. Should a larger network of client be required then it would be necessary to develop two or more networks of piconets. These would need to be connected together by a client that would act as both client (for the main piconet) and master for the secondary piconet. This interconnection of piconets is termed as a scatternet.

Client/Server Operation Mechanism

The initial stages of the process are carried out on the Server (Figure 10). Firstly it is necessary to acquire the input settings for the fractal image, a graphical user interface is provided for this. When the user issues a request to generate a fractal image the parameters are gathered from the fractal image settings GUI.

The next stage is to calculate the parameters necessary for each client (this will depend on the numbers of client currently connected). This yields a unique matrix of parameters for each client (Table 4). Several other parameters are also passed which are the same for all clients (for example: formula type, number of iterations).

There are many ways by which the matrix of image parameters can be calculated. One of the simplest methods is to divide the image in to equal sized segments based on the number of clients currently connected to the master device. The matrix of parameters can be easily calculated if the image is divided into vertical or horizontal strips (Figure 11).

Once all the parameters have been finalised the operation of sending the image parameters to each connected client can commence. The parameter data is passed in the form of a string.

Figure 9. Point to multi-point Piconet

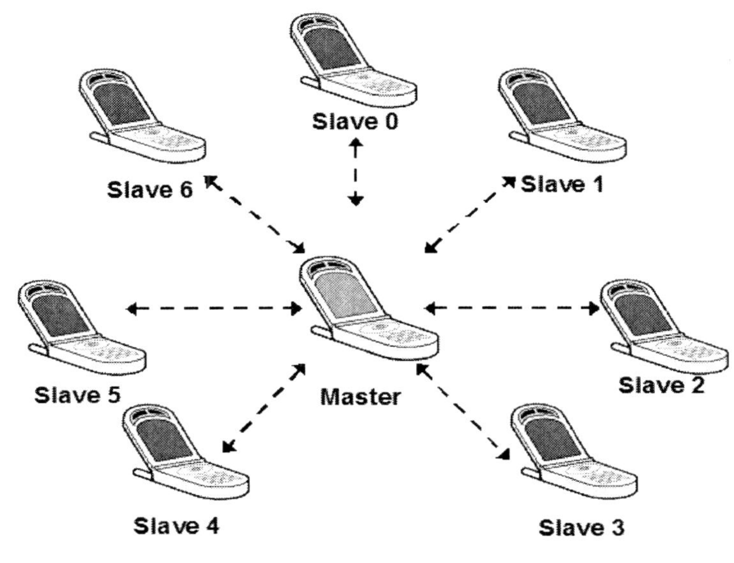

Figure 10. Server to client operating mechanism

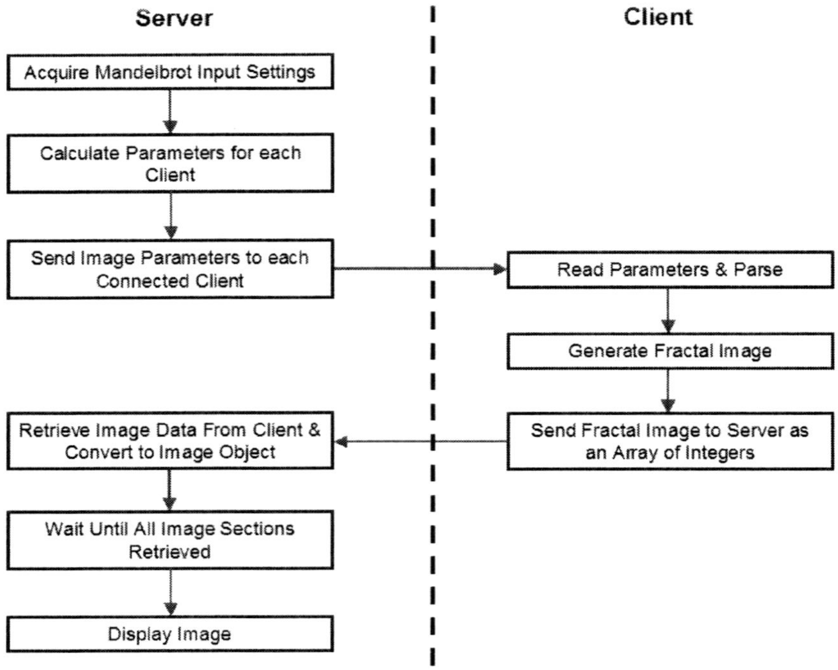

A typical example of this string has the format of "width, height, xmin, ymin, xmax, ymax, iterations, equation type, cPower, zPower, invert, image segment number." An example of the output string would be: "50, 200, -1.0, -2.0, 0.0, 2.0, 500, 0, 1, 2, 0, 1." The previous string would generate an image 50 x 200 pixels in size. The complex plane coordinates are "-1.0, -2.0, 0.0, 2.0." The client would carry out 500 iterations at each point. The generated image would be the standard non inverted Mandelbrot set Z^2+C. The final parameter "image segment number" will eventually be passed back by the client along with the generated image data so the server can place the image in its correct order.

The client has in the meantime been waiting for requests from the server. Once a request comes into the client, it must first parse the data to extract all of the required parameters necessary to generate the image. The next and most important stage is the actual generation of the

Table 4. Data matrix for a 300 pixel square image distributed to four clients

Width	Height	XMIN	YMIN	XMAX	YMAX	Slice
75	300	-2.0	-2.0	-1.0	2.0	0
75	300	-1.0	-2.0	0.0	2.0	1
75	300	0.0	-2.0	1.0	2.0	2
75	300	-2.0	-2.0	2.0	2.0	3

Figure 11. Division of fractal image into sections

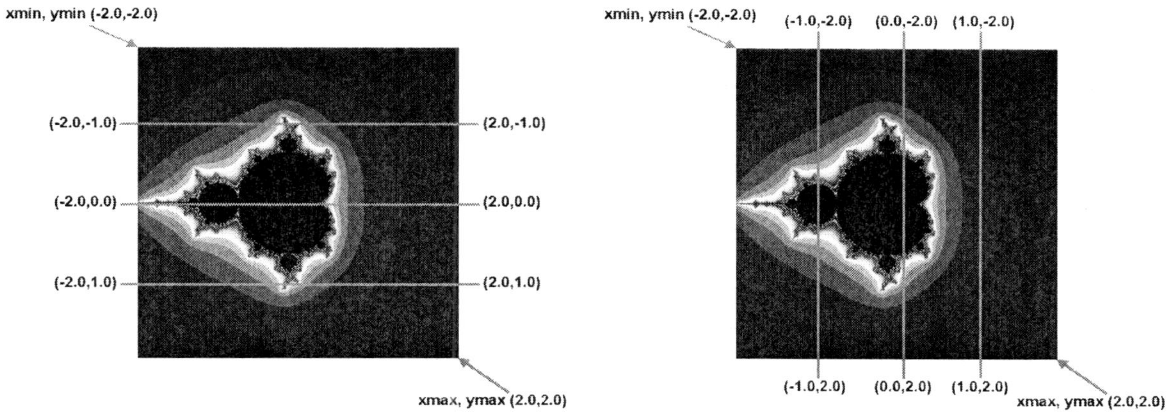

fractal image. Each client will generate a small section of the image. The image section is then sent to the server in the form of a sequence of integers. The actual format of this data can be seen in Figure 12. The image segment number is the same number that the client originally received from the server. The data size is passed to indicate to the server how much more data to expect, the final section is the actual image data itself. All this data is passed in the form of integers and is sent out to the server using a DataOutputStream object.

On the server side once it has issued its requests to all clients, it simply waits for incoming results. When a message is received from a client, the server examines the "image segment number" so the image will be placed in the correct order. Next it finds the length on the remaining incoming data, and initialises an array to be able to read all of the integer values representing the actual image. Once all the integer values have been read an Image object is created and populated into is proper location based on the "image segment number." The process of waiting for client responses continues until all image sections are retrieved. With the last image section retrieved the server displays the image segments on screen to the user.

Execution Results

Testing this system shows promising results compared to the generation of a complete image on a single phone. In the case of executing the application of four client phones, the areas of the image where more detail is present required extra processing time to areas at the extremities of the image where very little detail is to be found. Note the overall processing time is the time from the issuing of the request to generate an image to the time that the last

Figure 12. Client data output format

Segment Number	Data Size	Image Data

section of image is received by the server and converted into an Image object ready for display. In the case of the test image the difference between the longest processing time of a node and the total time averages at about 3 seconds. This difference is the time to send the initial data and the time to construct the final Image section (see Table 5 and Figure 13).

CONCLUSION

In this chapter, three methods have been explored for the generation and display of the Mandelbrot set on a mobile device. We have seen how mobile devices can be used to compute and display a fractal image. For many low-end devices, the computation time can be quite long. The latter methods examined focus on the reduction of the image generation time, by employing high-speed servers and distributed computing technologies.

The server side and distributed methods examined are not limited to just the generation of fractal images. They can be used for a wealth of computationally expensive tasks that would take a single mobile device a significant amount of time to process. Perhaps in time we

Table 5. Processing times for a 200 pixel square Mandelbrot image

Iterations	500	750	1000
Node 0	4,371 ms	4,521 ms	4,731 ms
Node 1	4,521 ms	6,559 ms	6,537 ms
Node 2	7,445 ms	7,442 ms	7,469 ms
Node 3	2,307 ms	2,538 ms	2,672 ms
Total Time	10,470 ms	10,646 ms	10,996 ms

Figure 13. Processing times for a 200 pixel square Mandelbrot image

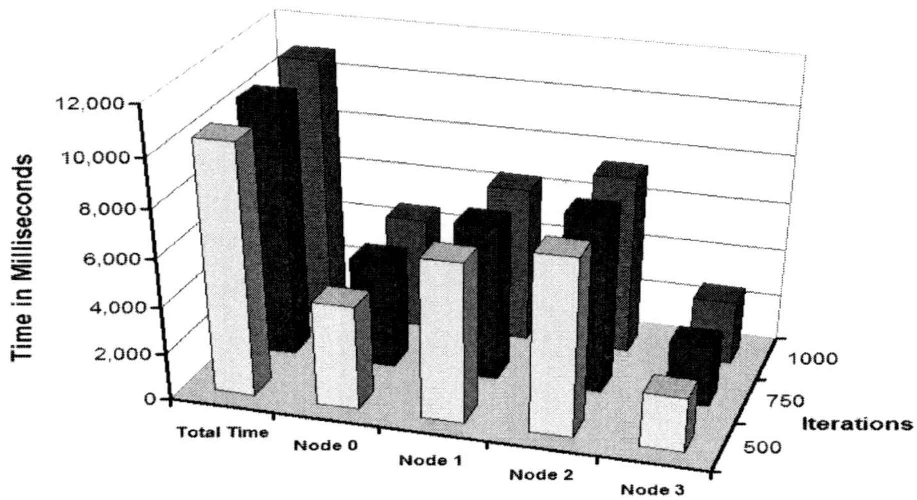

may see the Seti@Home client application running on mobile devices. As mobile devices such as phones are used by almost 100% of the population of many countries it is clear that this ubiquitous use of devices capable of carrying out complex tasks could be put to great use in the future. The combined processing power of hundreds of millions of phones is potentially immense.

The distributed fractal image generation example has shown how a small network of phones can be used together to carry out a processor intensive task. It is clear that mobile devices are here to stay and as such we should find suitable ways to employ what is potentially a massive computation resource.

REFERENCES

Doolan, D. C., & Tabirca, S. (2005). Interactive teaching tool to visualize fractals on mobile devices. *Proceedings of Eurographics Ireland Chapter Workshop*, Eurographics Ireland Chapter (7-12).

Heerman, D. W. (2002). Teaching science using a mobile phone. *International Journal of Modern Physics C*, *13*(10), 1393-1398.

Itoh, H., Seki, H., Inuzuka, N., Nakamura, T., Uenosono, Y. (1998, May). *A method to generate music using fractal coding*. Retrieved from citeseer.ist.psu.edu/77736.html

Klingsheim, A. N. (2004). *J2ME Bluetooth programming*. MSc Thesis, University of Bergen.

Kohn, M. (2005). *Mandelbrot midlet*. Retrieved from http://www.mikekohn.net/j2me.php#mandel

Long, B. (2004). *A study of Java games in Bluetooth wireless networks*. Master's Thesis, University College Cork.

Lu, N. (1997). *Fractal imaging*. San Diego, London, Boston: Academic Press.

Mandelbrot, B. (1983). *The fractal geometry of nature*. New York: Freeman.

Mobile Fractals. (2005). *Mobile computer graphics research*. Retrieved from http://www.cs.ucc.ie/~dcd1/

Musgrave, F., & Mandelbrot, B. (1991, July). The art of fractal landscapes. *IBM Journal of Research and Development*, *35*(4), 535-536, 539.

Nokia. (2004). Introduction to developing networked midlets using bluetooth. Retrieved from http://www.forum.nokia.com/info/sw.nokia.com/id/c0d95e6e-ccb7-4793-b3fc-2e88c9871bf5/Introduction To Developing Networked MIDlets Using Bluetooth v1 0.zip.html

Seti@Home. (2005). *The search for extra terrestrial intelligence at home*. Retrieved from http://setiathome.ssl.berkeley.edu/

Sony-Ericsson. (2004). *Developing applications with the Java API's for Bluetooth (JSR-82)*. Retrieved from http://developer.sonyericsson.com/getDocument.do?docId=65246, year = 2004

KEY TERMS

Bluetooth: A wireless technology that is becoming more and more widespread to allow mobile devices to communicate with each other.

Fractal: A fractal is an image that display infinite detail and self-similarity.

Julia Set: A fractal image that was discovered by French mathematician Gaston Maurice Julia.

Mandelbrot Set: A fractal image discovered in the 1970's by Benoit Mandelbrot, It acts as an index to all the possible Julia sets in existence.

Piconet: A network of Bluetooth devices, this is limited to seven devices connected to a master device.

Scatternet: A Bluetooth network of two or more interconnected piconets.

Smartphones: High end phones that typically have 3G capabilities, advanced java API's, Bluetooth technology, and much more.

Thread: Often called a "lightweight process" and is capable or executing in parallel with the main program.

Section IV

Applications and Services

The explosive growth of the Internet and the rising popularity of mobile devices have created a dynamic business environment where a wide range of mobile multimedia applications and services, such as mobile working place, mobile entertainment, mobile information retrieval, and context-based services, are emerging everyday. Section IV with its eleven chapters will clarify in a simple and self-implemented way how to implement basic applications for mobile multimedia services.

Chapter XXVIII
Mobile Multimedia Collaborative Services

Do Van Thanh
Telenor R&D, and
Norwegian University of Science and Technology, Norway

Ivar Jørstad
Norwegian University of Science and Technology, Norway

Schahram Dustdar
Vienna University of Technology, Austria

ABSTRACT

Mobile communication and Web technologies have paved the way for mobile multimedia collaborative services that allows people, team and organisation to collaborate in dynamic, flexible and efficient manner. Indeed, it should be possible to establish and terminate collaborative services with any partner anytime at anywhere on any network and any device. While severe requirements are imposed on collaborative services, their development and deployment should be simple and less time-consuming. The design, implementation, deployment and operation of collaborative services meet challenging issues that need to be resolved. The chapter starts with a study of collaboration and the different collaboration forms. An overview of existing collaborative services will be given. A generic model of mobile collaborative services is explained together with the basic collaborative services. A service oriented architecture platform supporting mobile multimedia collaborative services is described. To illustrate the development of mobile multimedia collaborative service, an example is given.

INTRODUCTION

The ultimate goal of computing is to assist human beings in their work by supporting complex, precise, and repetitive tasks. With the advent of the Internet that brought ubiquitous communication, the fundament for ubiquitous distributed computing has been laid. The next objective of computing is hence to facilitate collaboration between persons and organisations. Indeed, in the current globalisation and deregulation era, high level of dynamicity is required from the enterprises. They should be able to compete in one market as they collaborate in another one. Collaborations should be established as quickly as they are terminated. Collaborative services should be tailored according to the nature of the collaboration and to the agreement between the partners. They should be deployed quite rapidly and should function in a conformed way with the expectations of the collaborators.

With mobility, a person is able to access services anytime, at anywhere and from any device. Both, higher flexibility and efficiency can be achieved at the same time as the users' quality of life is improved considerably. Advanced collaborative services should definitely be mobile (i.e., available for the mobile users from any network and any device). While severe requirements are imposed on collaborative services, their development and deployment should simple and less time-consuming. There are many quite challenging issues that need to be resolved in the design, implementation, deployment, and operation of collaborative services. In this chapter, mobile collaborative services will be examined thoroughly. The nature of the collaboration and the different collaboration forms will be studied. Existing collaborative services will be summarized. A generic model of mobile collaborative services is explained together with the basic collaborative services. A service-oriented architecture platform supporting mobile collaborative services is described. An example of the development of mobile collaborative services is given as illustration.

BACKGROUND

Organizations constantly search for innovative applications and services to improve their business processes and to enrich the collaborative work environments of their distributed and mobile knowledge workers. It is increasingly becoming apparent that a limiting factor in the support of more flexible work practices offered by systems today lies in their inherent assumptions about (a) technical infrastructures in place (hardware, software, and communication networks), and (b) about interaction patterns of the users involved in the processes.

Emerging new ways of flexible and mobile teamwork on one hand and dynamic and highly agile (virtual business) communities on the other hand require new technical as well as organizational support, which current technologies and infrastructures do not cater for sufficiently.

Pervasiveness of collaboration services is an important means in such a context, to support new business models and encourage new ways of working. A service is a set of related functions that can be programmatically invoked from the Internet. Recent developments show a strong move towards increasingly mobile nimble and virtual project teams. Whereas traditional organizational structures relied on teams of collaborators dedicated to a specific project for a long period (Classic Teams, see Figure 1), many organizations increasingly rely on nimble teams, formed from members of possibly different branches or companies, assigned to perform short-lived tasks in an ad-hoc manner (sometimes called ad hoc teams). For team members, tasks may be small parts of their overall work activities. Such nimble collaboration styles change many of the traditional assumptions about teamwork: collaborators do not report to the same manager, they do not reside in the same location, and they do not work during the same time. As a consequence, the emerging new styles of distributed and mobile collaboration often across organizational boundaries are fostering new interaction patterns of working. Interaction patterns consist of information related to synchronous and asynchronous communication on the one hand and the coordination aspects on the other hand. So far, we have identified the following (not orthogonal) team forms:

- **Nimble teams (N-teams)** represent a short timeframe of team constellations that emerge, engage in work activities, and then dissolve again, as the circumstances require
- **Virtual project teams (V-teams)** require and enable people to collaborate across geographical distance and professional (organizational) boundaries and have a somewhat stable team configuration with roles and responsibilities assigned to team members
- **Nomadic teams (M-teams)** allow people working from home, on the move, or in flexible office environments, and any combinations thereof

Table 1 summarizes some identified emerging new forms of teamwork for a knowledge society and correlates them with relevant characteristics.

N/V/M mobile teams share the notion that they work on common goals, whereby work is being assigned to people in terms of objectives, not tasks. Whereas classic Workflow Management Systems relied on modeling a business process consisting of tasks and their control

Table 1. Characteristics of nimble/virtual/nomadic teams

	Nimble Teams	Virtual Teams	Nomadic Teams
Vision & Goals	Strongly shared	Shared	Not shared
Team coupling	Tight	Loose	None
Time span of existence	Short-lived	Project depended (short-/medium/long-lived)	Not known
Team Configuration	Flexible	Stable	Dynamic
Team Size	Compact (ca. 10)	Large (ca. 50)	Large
Example	A task force of specialists for crisis mitigation in healthcare (e.g. SARS) Scientist organizing a conference at a new location	Technical consultants for a mechanical engineering project Production team for a movie	Experts in Political conflict resolution Musicians providing composition of soundtracks Actors providing stunt or dubbing services

Figure 1. Emerging forms of mobile teams

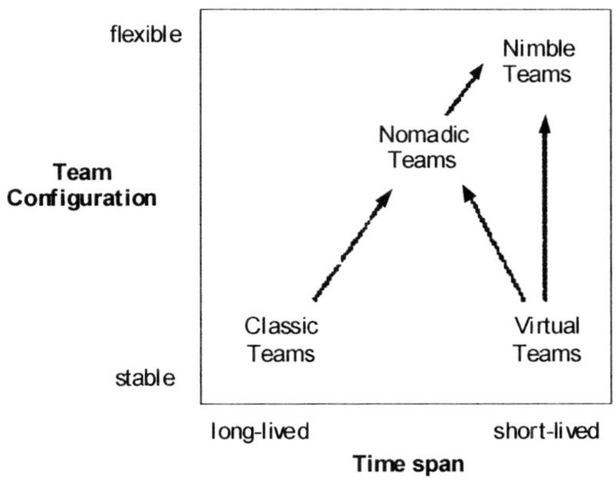

flow and data flow, emerging new forms of work in Nimble/Virtual/Nomadic teams cannot be modeled in advance. This new way of collaboration and interaction amongst activities (services) and people ultimately leads to challenging and new requirements with respect to the software infrastructure required for enabling interaction patterns to be established and managed. This is especially true in a mobile working context, where issues such as presence awareness, location-based service selection, knowledge and service sharing, etc., may imply particularly tight requirements for the underlying access network technologies and personal devices in use (e.g., PDAs, smart phones, tablet PCs, laptops, etc.). **Individuals** with specific roles and skills can establish **nimble/virtual/nomadic teams**. Multiple teams with common interests or goals establish **communities**. We distinguish between **intra-organizational communities** consisting of multiple teams within one organization and **inter-organizational communities** consisting of multiple teams residing in different organizations. Multiple communities with a goal establish a **consortium**.

REQUIREMENTS ON COLLABORATIVE SERVICES

Those team and community structures require a set of novel technological support mechanisms in order to operate efficiently and effectively. One of the main building blocks required for building service support for team processes we refer to as "context tunnelling." This concept refers to those issues when individuals from those team configurations embedded in nimble, virtual, or nomadic team settings change their "context." The view to their "world" should change accordingly. The metaphor we plan to use refers to a tunnel, connecting different work places and workspaces.

Context tunnelling deals with methods helping to transfer context information from one set of services to others. An example is the transfer and presentation of video recorded by cameras at the remote location. The impact is that people fulfilling their tasks will be able to take context information from one task (within a particular process) to other tasks. The people are, as we argued in the introduction, increasingly embedded in various emerging team forms such as nimble/virtual/nomadic teams, which provide additional challenges for our endeavour.

These team and community structures impose the following requirements on collaborative services:

- Collaborative services shall be *mobile services* that can be accessed any time at anywhere from any device
- Collaborative services shall be *pervasive* such that they support new business models and encourage new ways of working
- Collaborative services shall be *dynamic, flexible,* and *adaptable* to fit any form for collaboration

- Both *synchronous* and *asynchronous multimedia communications* shall be supported
- *Context tunnelling* shall be supported
- It shall be possible for employees of different companies to participate to collaboration team

EXISTING COLLABORATIVE SERVICES

Current collaborative services, *groupware systems*, have the potential to offer and consume services on many levels of abstraction. Consider a typical scenario of teamwork: (distributed) team members collaborate by using messaging systems for communications. In most cases, the "workspace" metaphor is used for collaboration. This means that team members have access to a joint workspace (in most cases a shared file system), where files (artifacts) and folders may be uploaded and retrieved. In many cases, (mobile) experts are part of such teams and their workspaces. One can argue that a workspace can be seen as a community of team members working on a shared project or towards a common goal. The aim of Groupware systems is to provide tool support for communication, collaboration, and to a limited extent, for coordination of joint activities (Dustdar & Gall, 2003; Dustdar, Gall, & Schmidt 2004).

Groupware systems incorporate functionalities like e-mail, voice mail, discussion forum, brainstorming, voting, audio conference, video conference, shared whiteboards, group scheduling, work flow management, etc. (Manheim, 1998). The leading products include IBM Lotus Notes, Microsoft Exchange, SharePoint, Groove, and Novell GroupWise.

The weakness of the Groupware systems lies probably on their extensive functionalities. In fact, Groupware are usually large static systems incorporating too much functionality that may not be necessary for the nimble/virtual/nomadic teams. There is no dynamicity that allow the selection of particular functionalities for a given collaboration team. Due to different work tasks in different contexts, teams and projects, it can be beneficial to dynamically extend or restrict the functionalities that are available through the collaborative system. Today, it is neither possible to remove or add new functionality during the lifetime of the collaboration. More seriously, they do not provide adequate support for the intra-organizational communities consisting of members belonging to different enterprise domains separated by firewalls. Groupware are usually centralised systems which are not adaptable to the nomadic teams that move across networks and use different devices. They lack the flexibility to replace a function with a more suitable one (e.g., change from mobile telephony to IP telephony). Quite often, personalisation of services is not allowed. The need for improved collaborative services is obvious.

Although the functionalities of the Groupware systems are numerous and vary from one system to another, they can be classified in a few types of basic collaborative services as follows:

- **Knowledge and resource sharing:** In collaboration, it is crucial to share knowledge and resource together. By sharing we mean:
  - *Presentation:* The same knowledge or resource is presented such that all the collaborators can view, experience and interpret it in the same way
  - *Generation and modification:* All the collaborators should be enabled to generate and modify knowledge and resources in such way that consistency and integrity are preserved

    - *Storage:* the knowledge or resource must be stored safely without affecting the availability
- **Communication and personal interaction:** In collaboration, communication and interaction between collaborator are crucial to avoid misunderstand and mismatch. Communications can be classified in several ways:
    - Synchronous (e.g., telephony, chat, audio conference, video conferencing, etc.) vs. asynchronous (e-mail, voice mail, newsgroup, forum, sms, voting, etc.)
    - audio, video, text, multimedia
- **Work management:** Work management services are a collection of tools designed to assist production work. They include such tools as:
    - Meeting scheduling, which assists a group in organizing and planning collective activities
    - Workflow, that supports business processes for their whole lifetime

Ideally, from the mentioned basic collaborative services, one should be able to select the needed basic services and to compose an advanced collaborative application which fits to the needs of a particular collaboration scheme. Due to the mobility of the nimble team, it is necessary to be able to use more suitable alternate basic services.

A framework allowing the construction of advanced mobile multimedia collaborative services using the basic ones will be described in later section. Let us now study the architecture of mobile multimedia collaborative services.

MOBILE MULTIMEDIA COLLABORATIVE SERVICE ARCHITECTURE

Generic Model of Mobile Multimedia Service

A collaborative service should be available to the users anywhere at any time on any network and any device and should therefore have the architecture of a mobile service.

A generic mobile service is commonly modelled as consisting of four basic components (Jørstad, Do, & Dustdar, 2005a), see Figure 2:

- **Service Logic**
- **Service Data**
- **Service Content**
- **Service Profile**

Figure 2. Composition model of MobileService

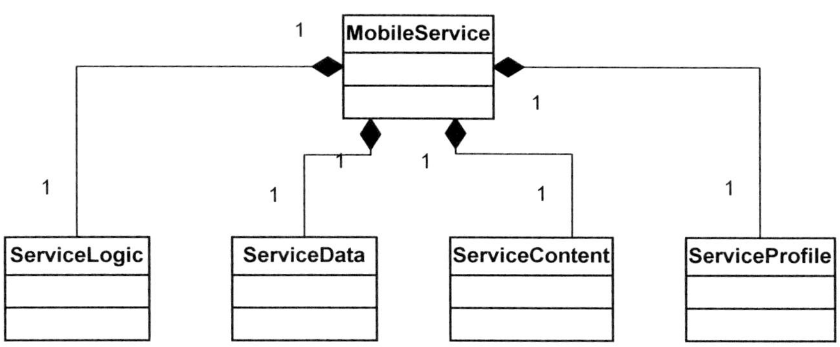

Service Logic is the program code that constitutes the dynamic behavior and provides the functions of a service. The logic of a mobile service can be subject to various distributions, as in any distributed system (ITU-T X.901 | ISO/IEC 10746-{1,2,3,4}, 1996). The most common models to describe the distribution of service logic are:

- **Standalone**
- **Client-server**
- **Peer-to-peer**
- **Multiple distributed components**

Service Data are used in the execution of the service logic and reflecting the state of it. They are for example variable values, temporal parameters, register values, stack values, counting parameters, etc.

Service Content refers to data that are the product of service usage. For example, it can be a document written in a word processor or the entries in a calendar. Service content can be produced or consumed by the user.

Service Profile contains the settings that are related to the user or/and the accessing device. It is necessary to enable personalization.

Following the mentioned model of a generic service, a collaborative service can be represented by four components: *Service Logic*, *Service Data*, *Service Content*, and *Service Profile*. Figure 3 depicts the logical architecture of a collaborative service that is used by three users.

Each user employs a *User_Interface* to collaborate with the other users via the collaboration service. The *User_Interface* can be a generic component that can be used to access several services like a browser. It could be a dedicated component that is especially built for a specific service. Each user can use different instances of the same implementation (e.g., different instances of Internet Explorer). These components can be referred to as *identical components*. They can also use different instances of different implementations. These components can be referred to as *equivalent components*.

Figure 3. Logical architecture of a collaborative service

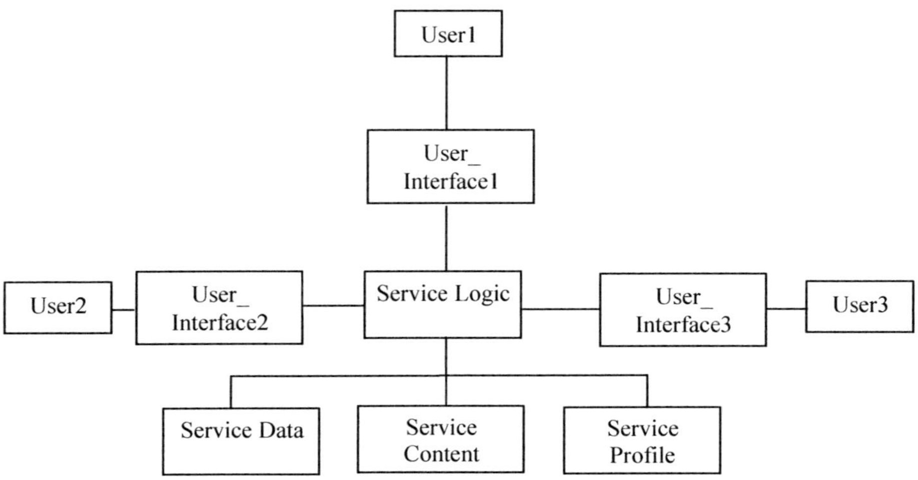

Collaborative Functions

To let several users to participate simultaneously, the *service logic* must be equipped with specific collaborative functions that we are going to examine successively.

Locking Mechanism

For knowledge and resource sharing services, it is necessary mechanisms to prevent the corruption of knowledge and resources. These mechanisms are similar to the one in database systems, locking mechanisms. There are several types of locks that can be chosen for different situations:

- **Intent:** The intent lock (I) shows the future intention of a user to acquire locks on a specific resource
- **Shared:** Shared locks (S) allow several users to view the same resources at the same time; however, no user is allowed to modify it
- **Update:** Update locks (U) are acquired just prior to modifying the resource
- **Exclusive:** Exclusive locks (X) completely lock the resource from any type of access including views

The majority of collaborative services will require a variety of locks being acquired and released on resources.

Presentation Control

Quite often, users want to experience the same resource together from their own computer (e.g., viewing the same document or the same presentation) listening to the same music song, etc. These resources are presented to the users by different applications such as word processor, presentation reader, etc. All the users may be allowed to manipulate these applications as, for example, scrolling down, going to other page, etc. Alternatively, the control can be given to only one user. In any case, it is necessary to have a presentation control component that collects all the inputs from the different users and delivers them to the respective applications according the pre-selected presentation scheme. The outputs from the applications should also be controlled by this component. This component should also support different navigation devices such as mouse, scrolling button, joystick, etc.

User Presence Management

The user belonging to a collaborating organisation should be reserved the right to decide when to participate to a collaborating activity such as viewing a multimedia documentation, editing a document, etc. It is, therefore, necessary to provide a registration or login mechanism and deregistration or logout mechanism to the different activities. It should be also possible for the user to subscribe for different log services (i.e., information about the dates and times of the different activities), information about the participants, the resources produced or modified by the activities.

Collaboration Management

There should also be a management function that allows the user in charge of the collaborative organisation to add, remove, and assign rights to the participants. The responsible user can also define different collaborative activities. Each collaborative activity may incorporate different applications and contents. For example, in activity workingGroup1, a word processing with access to folder working_group_1 is used together with chat. In activity workingGroup2, a presentation reader is used with SIP (session initiation protocol) (IETF, 2002) IP telephony.

Communication Control

In any collaboration, communication between the collaborators is decisive for the success. It should be possible to select the appropriate mean (e.g., chat, e-mail, SMS (short message service), plain old telephony, voice IP telephony, multimedia IP telephony, etc.) for each activity. To make things even easier, it should be possible to define an e-mail "notification agent" to send e-mail to a group of persons, a telephone conference to initiate telephone call to a group of persons, etc. In addition, the communication means can be used to establish context tunnelling (e.g., transfer of video recorded by cameras mounted at the communicating sites).

Generic Model of Mobile Multimedia Collaborative Service

The mentioned collaborative functions are often implemented as an integrated part of a collaborative service. Such a design is neither flexible nor efficient because it does not allow reuse or optimisation of the collaborative functions. A more optimal solution is to separate these functions into separate modules. A generic model of mobile multimedia service is shown in Figure 4. The *Collaborative Functions* are separated from the *Service Logic*. It is also placed between the different *Service Logic* and the different *User_Interface* used by the users. Indeed, a mobile multimedia collaborative service can incorporate several basic services and makes use of specific collaborative functions.

Figure 4. Generic model of mobile multimedia collaborative service

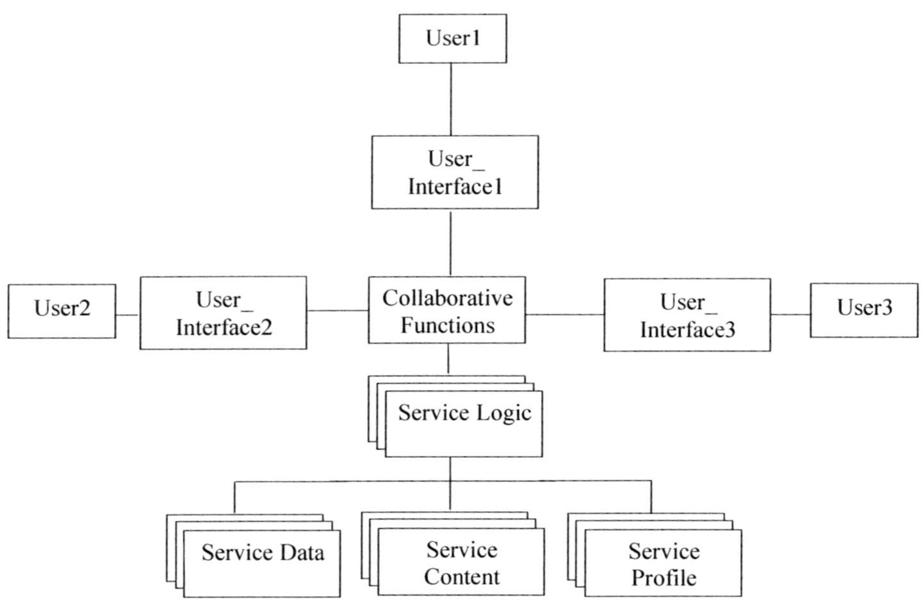

For non-collaborative services, the components service logic, service data, service content and service profile will most often exist on an individual basis (e.g., each user is associated a set of these components in conjunction with service usage). For collaborative services, however, the situation is more complicated. Some parts will be common to all participants in a collaborative session, while other parts will be individual to each user. For example, all *Service Data* will typically be pr. user, because this component contains data that is strongly associated with the user interface accessed by each user.

The *Service Content* will on the contrary be mostly shared, because this component contains work documents etc. used in projects and by all team members. The *Service Content* represents the *goal* of the collaboration; it is the result of the combined effort by all team members.

The *Service Profile*, however, must be decomposed for collaborative services. Each user in a collaborative session can choose the layout (presentation) of the service in the user interface (e.g., colors and placement of functions). However, the overall *Service Profile* (i.e., which functionalities are available and how these functionalities are tailored for the specific team, project or context, must be common to all team members). It should be possible to put restrictions on some of these functionalities due to different roles of the team members (observer, contributor, moderator etc.). The *Service Profile* shall thus describe both the overall collaborative service as well as each personal part of the collaborative service. A collaborative service can therefore also be a partially personalised service (Jørstad, Do, & Dustdar, 2004), although the main focus should be kept on sharing.

A SERVICE-ORIENTED ARCHITECTURE-BASED FRAMEWORK FOR MOBILE MULTIMEDIA COLLABORATIVE SERVICE

Service-oriented architecture (SOA) is a new paradigm in distributed systems aiming at building loosely coupled systems that are extendible, flexible and fit well with existing legacy systems. By promoting the re-use of basic components called services, SOA will be able to offer solutions that are both cost-efficient and flexible. In this paper, we propose to investigate the feasibility of using SOA in the construction of innovative and advanced collaborative services. We propose to elaborate a SOA framework for collaborative services. This section starts with an overview of SOA.

Overview of the Service Oriented Architecture

There are currently many definitions of the service oriented architecture (SOA) which are rather divergent and confusing. The World Wide Web consortium (W3C, 2004) defines as follows:

A service oriented architecture (SOA) is a form of distributed systems architecture that is typically characterized by the following properties:

- **Logical view:** The service is an abstracted, logical view of actual programs, databases, business processes, etc., defined in terms of what it does, typically **carrying out a business-level operation**
- **Message orientation:** The service is formally defined in terms of the messages exchanged between provider agents and requester agents, and not the properties of

the agents themselves. The internal structure of an agent, including features such as its implementation language, process structure and even database structure, are deliberately abstracted away in the SOA: using the SOA discipline one does not and should not need to know how an agent implementing a service is constructed. A key benefit of this concerns so-called legacy systems. By avoiding any knowledge of the internal structure of an agent, one can incorporate any software component or application that can be "wrapped" in message handling code that allows it to adhere to the formal service definition

- **Description orientation:** A service is described by machine-processable meta data. The description supports the public nature of the SOA: only those details that are exposed to the public and important for the use of the service should be included in the description. The semantics of a service should be documented, either directly or indirectly, by its description
- **Granularity:** Services tend to use a small number of operations with relatively large and complex messages
- **Network orientation:** Services tend to be oriented toward use over a network, though this is not an absolute requirement
- **Platform neutral:** Messages are sent in a platform-neutral, standardized format delivered through the interfaces. XML is the most obvious format that meets this constraint

A service is an abstract resource that represents a capability of performing tasks that form a coherent functionality from the point of view of providers entities and requesters entities. To be used, a service must be realized by a concrete provider agent.

The mentioned definition is very generic and we choose to adopt the definition inspired by Hashimi (2003):

In SOA, software applications are built on basic components called services.

A service in SOA is an exposed piece of functionality with three properties:

1. The interface contract to the service is platform-independent
2. The service can be dynamically located and invoked
3. The service is self-contained. That is, the service maintains its own state

There are basically three functions that must be supported in a service-oriented architecture:

1. Describe and publish service
2. Discover a service
3. Consume/interact with a service

A SOA Framework for Collaborative Services

In a service oriented architecture, applications are built upon the fundamental elements called services. These services can be distributed all over the Internet. This is really powerful but it might be difficult for developers to discover, understand, and use the services in a proper way. To facilitate the construction of mobile multimedia collaborative services, a SOA Framework is proposed in Figure 5.

The *Basic Service Layer* containing basic services and their descriptions constitutes the fundament of the SOA framework. These basic services are autonomous services and can operate perfectly on their own. As shown in Figure 5, the basic services are classified into three categories:

Figure 5. A SOA framework for collaborative services

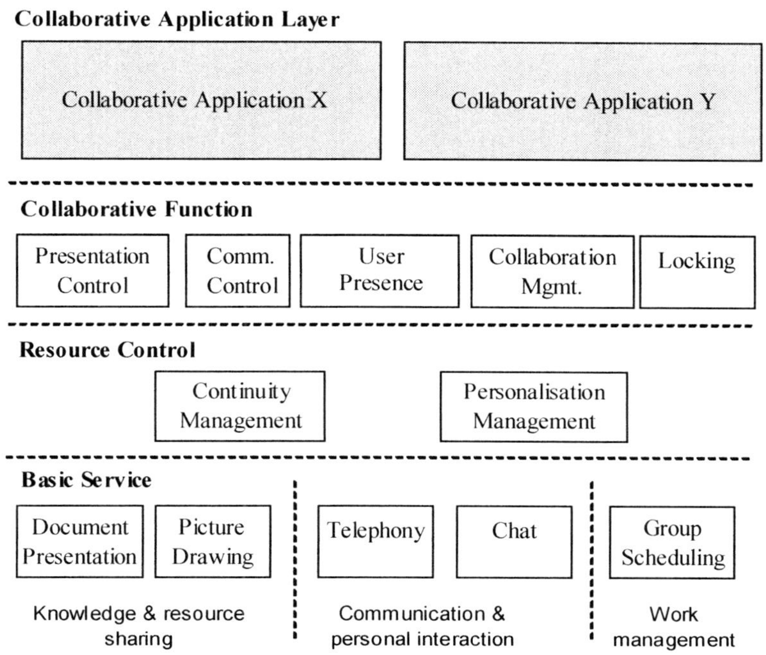

1. **Knowledge and resource sharing services:** Typical examples are Document presentation, Picture drawing, etc
2. **Communication and personal interaction services:** Typical examples are Telephony, chat, etc
3. **Work management services:** Typical examples are Group scheduling, Work flow, etc

The *Resource Control Layer* contains functions for ensuring ubiquitous access to appropriate instances in the basic service layer, as well as for providing management functionality for partial personalisation support. The functions of the Continuity Management component are summarised in (Jørstad, Do, & Dustdar, 2005a).

The *Collaborative Function Layer* contains the necessary functions for collaboration such as *locking, presentation control, user presence management, collaboration management,* and *communication control.*

On the top layer, collaborative applications can be built by utilizing the components both in the Collaborative Function Layer and the Basic Service Layer. There are two composition alternatives:

- A collaborative application can be built as a software program that invokes method calls on the service components
- It can be realised as a script that orchestrates the service components (O'Riordan, 2002; Peltz, 2003)

The service oriented architecture is realised on the World Wide Web by Web Services. A Web service is meant self-contained, modular applications that can be described, published, located, and invoked over a network (IBM, 2001). Specifically these applications use XML for data description, *SOAP* (simple object access protocol) for messaging (invocation), *WSDL* (Web Service Description Language) for service description (in terms of data types, accepted messages, etc) and *UDDI* (Universal Description, Discovery and Integration) for publishing and discovery.

The service entities in the Basic Service Layer can be distributed throughout the Word Wide Web, each entity exposed as a separate piece of functionality with the properties already discussed earlier in the section on service oriented architectures.

Based on the service oriented architecture framework for collaborative services, it is straightforward to build a service oriented architecture platform using Web services. Each SOA service is hence realised as a Web service.

Example of Building SOA Mobile Collaborative Services

To illustrate the tailoring of collaborative application to fit the needs of a specific collaboration form, one example will be considered:

- Collaborative application for nomadic teams (M-teams)

Collaborative Application for Nomadic Teams (M-teams)

For a Nomadic team, the most important requirement is the ability to work anytime anywhere and from any device in the same way as at the office. It is, therefore, crucial to have access to view documents and discuss with colleagues. For Nomadic team members, the environment is continuously changing. This means that the device used to access the collaborative functions differ over time as well as the available means for communication. This often means moving from a powerful device with high capacity network connection to a limited-resource device with limited network bandwidth and possibly an intermittent network connection.

Let us assume that an employee is participating in a collaborative session from his work place, where the other participants are at their work place, all of which are at geographically distributed locations. The basic services used in the collaborative session are telephony for communication and a white board for a shared visual display of ideas. The telephony service is realised through IP-telephony over the Internet, since it is cheaper than other telephony services. Then assume that the employee in question is required to leave this work place for some reason, but would like to keep the collaborative session active and continue to work while travelling. IP-telephony is not possible with his restricted mobile device, but the device supports ordinary GSM-telephony. The collaborative system recognises this, and the communication control mechanism together with the continuity management mechanism searches for a way to resolve this. The possible outcomes are that all participants switch to PSTN/GSM-telephony, or that the collaborative system finds a mediator (gateway) that allows routing of GSM-traffic towards the IP-telephony sessions already established within the collaborative session. For the white board basic service, only the presentation needs to be changed; the same basic service is accessed, but the view (through the presentation control service) is adapted to fit the new device. The workspace is thus extended, or retracted, due to user movements, etc. The workspace extension for the example application is illustrated in Figure 7.

For the case described in the previous paragraph, one of the most important mechanisms is the ability to search for a *replacement candi-*

Figure 6. A nomadic, collaborative application

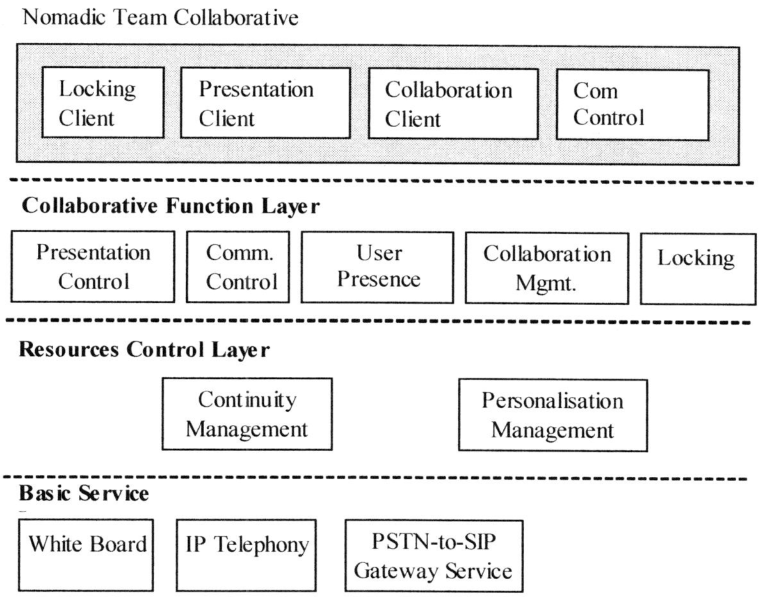

Figure 7. Extending the workspace to accommodate changes

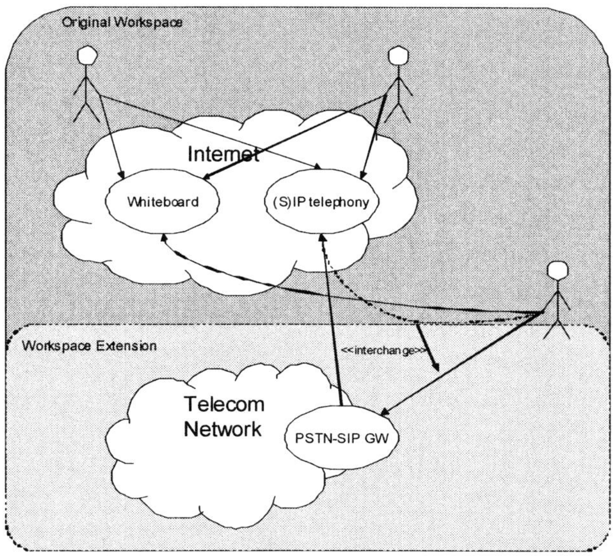

date for an existing basic service in the service architecture. Thus, the system must be able to compare the existing basic service (IP-telephony) with other basic services available in the collaborative system (e.g., a GSM service combined with a PSTN–SIP gateway). A service oriented architecture is tailored for such use, since its basic mechanisms are description (WSDL), publication, and discovery (UDDI) functions. However, there are still open issues, because there is no common framework for comparison of identicalness, equivalence, compatibility and similarity among services, which is required both on the semantic and syntactic level (Jørstad et. al., 2005b). Also, since the example case spans two different service domains (Internet and the telecom domain), the situation is even more complicated because protocol conversion and mapping are required. However, it serves as a good illustration of how a collaborative service could be supported by a service oriented architecture .

CONCLUSION

Emerging new forms of collaboration which are dynamic and agile pose severe requirements that current collaborative services do not satisfy. New architectures and technologies for mobile multimedia collaborative services are required. In this chapter, the service oriented architecture is investigated and found feasible for the construction of collaborative services. It is argued that the major benefit of using a SOA for collaborative services is the flexibility to dynamically extend or restrict the functionalities of the collaborative system in order to fit the varying requirements of Nimble, Virtual, and Nomadic teams, in mobile service environments. The generic model of collaborative service is mapped to the service oriented architecture . To alleviate the tasks of the developers, the basic collaborative services, locking, presentation control, user presence management, organisation management, and communication control are gathered into a collaborative service layer and made available to the applications. A collaborative service can be built by composing or by orchestrating the collaborative services together with other services.

REFERENCES

Andrews, T., Curbera, F., Dholakia, H., Goland, Y., Klein, J., Leymann, F. et al. (2003). *Business Process Execution Language for Web services*. Version 1.1, Copyright© 2002, 2003 BEA Systems, International Business Machines Corporation, Microsoft Corporation, SAP AG, Siebel Systems. Retrieved from http://www-106.ibm.com/developerworks/library/ws-bpel/

Dustdar, S., & Gall, H. (2003). Architectural concerns in distributed and mobile collaborative systems. *Journal of Systems Architecture, 49*(10-11), 457-473.

Dustdar, S., Gall, H., & Schmidt, R. (2004, February 11-13). Web services for groupware in distributed and mobile collaboration. *The 12th IEEE Euromicro Conference on Parallel, Distributed and Network Based Processing (PDP 2004)*. A Coruña, Spain. IEEE Computer Society Press.

Hashimi, S. (2003). *Service-oriented architecture explained*. Retrieved from http://www.ondotnet.com/pub/a/dotnet/2003/08/18/soa_explained.html

IBM, Web Services Architecture Team. (2001). *Web services architecture overview*. Retrieved December 18, 2001, from http://www106.ibm.com/developerworks/library/w-ovr/

IETF–MMUSIC RFC 3261. (2002). *Multiparty MUltimedia SessIon Control (MMUSIC) Working Group* (SIP: Session Initiation Protocol–Request For Comment

3261). Retrieved from http://www.ietf.org/rfc/rfc3261.txt?number=3261

ITU-T X.901 | ISO/IEC 10746-{1,2,3,4}. (1996). Open Distributed Processing Reference Model Part 1, 2, 3 AND 4.

Jørstad, I., Do, V. T., & Dustdar, S. (2004, October 18-21). Personalisation of future mobile services. *The 9th International Conference on Intelligence in Service Delivery Networks*, Bordeaux, France.

Jørstad, I., Do, V. T. & Dustdar, S. (2005a, March 13-17). A service continuity layer for mobile services. *IEEE Wireless Communications and Networking Conference (WCNC 2005)*, New Orleans, LA.

Jørstad, I., Do, V. T. & Dustdar, S. (2005b, June 13-14). *Service-oriented architectures and mobile services*. Ubiquitous Mobile Information and Collaboration Systems (UMICS 2005), Porto, Portugal.

Manheim, M. (1998). Beyond groupware & workflow. In *Excellence in practice: Innovation and excellence in workflow and imaging, Vol. 2. Fugure Strategies*. J. L. Kellog Graduate School of Management, Northwestern University.

O'Riordan, D. (2002). *Business process standards for Web services*. Chicago, IL: Tect.

Peltz, C. (2003, July). Web service orchestration and choreography: A look at WSCI and BPEL4WS–Feature. *Web Services Journal*. Retrieved from http://webservices.sys-con.com/read/39800.htm

W3C. (2004). Working Group Note 11 Web Services Architecture. Retrieved February 2004, from http://www.w3.org/TR/ws-arch/#stakeholder

KEY TERMS

Collaborative Service: A collaborative service is a service that supports cooperative work among people by providing shared access to common resources.

Groupware System: A groupware system is software realising one or several collaborative services.

IP Telephony: Realisation of phone calls over the Internet infrastructure, using the Internet protocol (IP) on the network layer, where the most common protocols include H.323 and session initiation protocol (SIP).

Mobile Service: A mobile service is a service that is accessible at any time and place.

Personalisation: The adaptation of services to fit the needs and preferences of a user or a group of users.

Service-Oriented Architecture (SOA): In SOA, applications are built on basic components called services. A service in SOA is an exposed piece of functionality with three properties: (1) The interface contract to the service is platform-independent. (2) The service can be dynamically located and invoked. (3) The service is self-contained. That is, the service maintains its own state (Hashimi, 2003).

Service: A service is an abstract resource that represents a capability of performing tasks that form a coherent functionality from the point of view of provider entities and requester entities. To be used, a service must be realized by a concrete provider agent.

Web Service: A self-contained, modular application that can be described, published, located and invoked over a network (IBM, 2001).

Chapter XXIX
V-Card:
Mobile Multimedia for Mobile Marketing

Holger Nösekabel
University of Passau, Germany

Wolfgang Röckelein
EMPRISE Consulting Düseldorf, Germany

ABSTRACT

This chapter presents the use of mobile multimedia for marketing purposes. Using V-Card, a service to create personalized multimedia messages, as an example, the advantages of sponsored messaging are illustrated. Benefits of employing multimedia technologies, such as mobile video streaming, include an increased perceived value of the message and the opportunity for companies to enhance their product presentation. Topics of discussion include related projects, as marketing campaigns utilizing SMS and MMS are becoming more popular, the technical infrastructure of the V-card system, and an outline of social and legal issues emerging from mobile marketing. As V-card has already been evaluated in a field test, these results can be implemented to outline future research and development aspects for this area.

INTRODUCTION

The chapter presents the use of mobile multimedia for marketing purposes, specifically focusing on the implementation of streaming technologies. Using V-card, a service for creating personalized multimedia messages, as an example, the advantages of sponsored messaging are illustrated. Topics of discussion include related projects, as marketing campaigns utilizing SMS and MMS are becoming more popular, the technical infrastructure of the V-card system, and an outline of social and legal issues emerging from mobile marketing. As V-card has already been evaluated in a field test, these results can be implemented to outline future research and development aspects for this area.

Euphoria regarding the introduction of the universal mobile telephony system (UMTS) has evaporated. Expectations about new UMTS services are rather low. A "killer application" for 3rd generation networks is not in sight. Users are primarily interested in entertainment and news, but only few of them are actually willing to spend money on mobile services beyond telephony. However, for marketing campaigns the ability to address specific users with multimedia content holds an interesting perspective.

Advertisement-driven sponsoring models will spread in this area, as they provide benefits to consumers, network providers, and sponsors. Sponsoring encompasses not only a distribution of pre-produced multimedia content (e.g., by offering wallpapers), Java games, or ringtones based on a product, but also mobile multimedia services.

Mobile multimedia poses several problems for the user. First, how can multimedia content of high quality be produced with a mobile device. Cameras in mobile telephones are getting better with each device generation; still the achievable resolutions and framerates are behind the capabilities of current digital cameras. Second, how can multimedia content be stored on or transmitted from a mobile device. Multimedia data, sophisticated compression algorithms notwithstanding, is still large, especially when compared to simple text messages. External media, such as memory cards or the Universal Media Disk (UMD), can be used to a certain degree to archive and distribute data. They do not provide a solution for spreading this data via a wireless network to other users. Third, editing multimedia content on mobile devices is nearly impossible. Tools exist for basic image manipulation, but again their functionality is reduced and handling is complex.

Kindberg, Spasojevic, Fleck, and Sellen (2005) found in their study that camera phones are primarily used to capture still images for sentimental, personal reasons. These pictures are intended to be shared, and sharing mostly takes place in face-to-face meetings. Sending a picture via e-mail or MMS to a remote phone occurred only in 20% of all taken pictures. Therefore, one possible conclusion is that users have a desire to share personal moments with others, but current cost structures prohibit remote sharing and foster transmission of pictures via Bluetooth or infrared.

V-card sets out to address these problems by providing a message-hub for sublimated multimedia messaging. With V-card, users can create personalized, high-quality multimedia messages (MMS) and send those to their friends. Memory constraints are evaded by implementing streaming audio and video where applicable. V-cards can consist of pictures, audio, video, and MIDlets (Java 2 Micro-Edition applications). Experience with mobile greetingcards show that users are interested in high-quality content and tend to forward them to friends and relatives. This viral messaging effect increases utilisation of the V-card system and spreads the information of the sponsor. Haig (2002, p. 35) lists advice for successful viral marketing campaigns, among them:

- Create of a consumer-to-consumer environment
- Surprise the consumers
- Encourage interactivity

A V-card message is sponsored, but originates from one user and is sent to another user. Sponsoring companies therefore are actually not included in the communication process, as they are neither a sender nor a receiver. V-card is thus a true consumer-to-consumer environment. It also can be expected for the near future that high quality content contains an element of surprise, as it exceeds the current state of the art of text messaging. Interactivity is fostered by interesting content, which is passed on, but also by interactive elements like MIDlet games.

Additionally, Lippert (2002) presents a "4P strategy" for mobile advertising, listing four characteristics a marketing campaign must have:

- Permitted
- Polite
- Profiled
- Paid

"Permitted" means a user must agree to receive marketing messages. With V-card, the originator of the MMS is not a marketing company but another user, therefore the communication itself is emphasized, not the marketing proposition. Legal aspects regarding permissions are discussed detailed below. Marketing messages should also be "polite," and not intrusive. Again, the enhanced multimedia communication between the sender and the receiver is in the foreground, not the message from the sponsor.

"Profiling" marketing tools enables targeted marketing and avoids losses due to non-selective advertising. Even if V-card itself is unable to match a sponsor to users, since users do not create a profile with detailed personal data, profiling is achieved by a selection process of the sender. As messages can be enhanced by V-card with media related to a specific sponsor, by choosing the desired theme the sender tailors a message to the interests of himself and the receiver. Usually, marketing messages should provide a target group with incentives to use the advertised service; the recipients need to get "paid." With V-card, sponsors "pay" both users by reducing the costs of a message and by providing high quality multimedia content.

V-CARD ARCHITECTURE

V-Card Core Architecture

Figure 1 shows the V-card core architecture and illustrates the workflow. First, the user with a mobile device requests a personalised application via the SMSC or Multimedia Messaging Service Centre (MMSC), which are part of the mobile network infrastructure. The message is passed on to the V-card core, where the connector decides which application has been called.

After the request is passed on to the appropriate application (1), it is logged in the message log. A parser receives the message (2), extracts the relevant data for customisation, and returns this data (3)–this could include the receiver's phone number, the name of the sender or a message. Then, the capabilities of the receiving phone are queried from a database which holds all relevant data (4+5) like display size, number of colours, supported video and audio codecs.

Finally, the application transmits all the data gathered to the content transformator. Here, the pre-produced content is tailored with the input delivered by the user according to the capabilities of the device (6+7). The result is then sent via the connector (8) to the receiving user. Since the personalised applications and the data are separated, new applications can be easily created.

V-Card Streaming Technology

Since video content can not be stored directly on a mobile device due to memory limitations, a streaming server supplies video data to the

Figure 1. V-Card core architecture

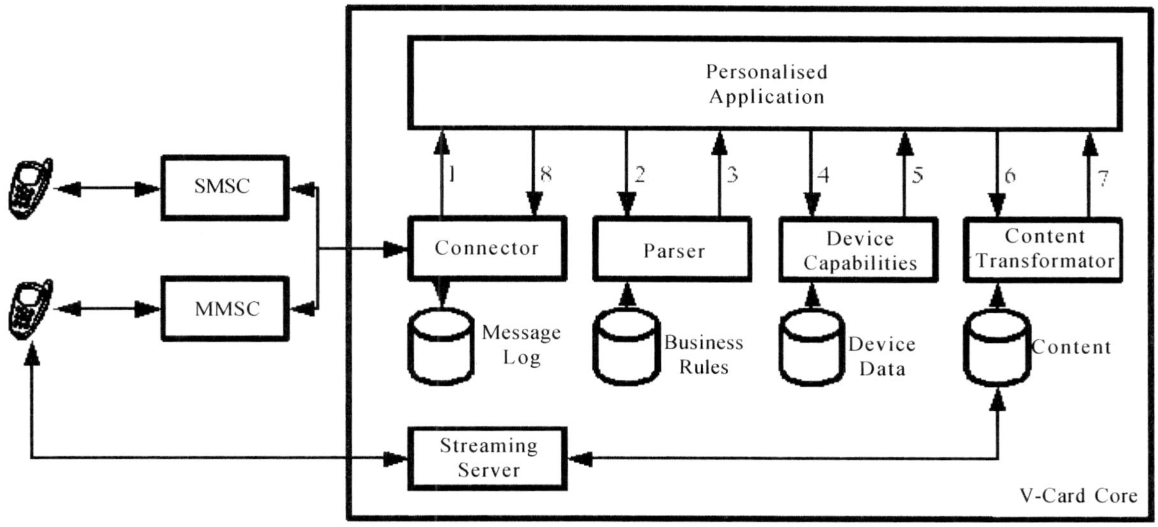

device where the video is played, but not stored with the exception of buffered data, which is stored temporarily to compensate for varying network throughput. Streaming video and audio to mobile devices can be utilized for various services (e.g., for mobile education) (Lehner, Nösekabel, & Schäfer 2003). In the case of V-card, the MMS contains a link to adapted content stored on the content server. This link can be activated by the user and is valid for a certain amount of time. After the link has expired, the content is removed from the content server to conserve memory.

Currently, there are two streaming server solutions available for mobile devices. RealNetworks offers the HELIX server based on the ReadMedia format. RealPlayer, a client capable of playing back this format, is available for Symbian OS, Palm OS 5, and PocketPC for PDAs. Additionally, it is available on selected handsets, including the Nokia 9200 Series Communicators and Nokia Series 60 phones, including the Nokia 7650 and 3650. The other solution is using a standardized 3GPP-stream based on the MPEG4 format, which can be delivered using Apples Darwin server.

An advantage of implementing streaming technology for mobile multimedia is the fact that only a portion of the entire data needs to be transmitted to the client, and content can be played during transmission. Data buffers ensure smooth playback even during short network interruptions or fluctuations in the available bandwidth. As video and audio are time critical, it is necessary that the technologies used are able to handle loss of data segments, which do not arrive in time (latency) or which are not transmitted at all (network failure). GPRS and HSCSD connections allow about 10 frames per second at a resolution of 176 by 144 pixel (quarter common intermediate format QCIF resolution) when about 10 KBit per second are used for audio. Third generation networks provide a higher bandwidth, leading to a better quality and more stable connectivity.

A drawback of streaming is the bandwidth requirement. For one, the bandwidth should be constant; otherwise the buffered data is unable to compensate irregularities. Next, the available bandwidth directly influences the quality that can be achieved — the higher the bandwidth, the better the quality. Third, a transfer of mobile data can be expensive. A comparison of German network providers in 2003 showed that 10 minutes of data transfer at the speed of 28 KBit per second (a total amount of 19 Megabyte) resulted in costs ranging from 1 Euro (time-based HSCSD tariff) up to 60 Euro (packet-based GPRS by call tariff).

V-Card Examples

Figures 2 demonstrates a picture taken with the camera of a mobile device, rendered into a video clip by the V-card core. Figure 3 combines pictures and text from the user with video and audio content from the V-card hub. Figure 4 shows how simple text messages can be upgraded when a picture and an audio clip are added to create a multimedia message.

Since sponsoring models either influence the choice of media used to enhance a message, or can be included as short trailers before and after the actual message, users and sponsors can choose from a wide variety of options best suited for their needs.

Figure 2. V-Card with picture in video

Figure 3. V-Card with picture and text in video

Figure 4. V-Card with text in picture and audio

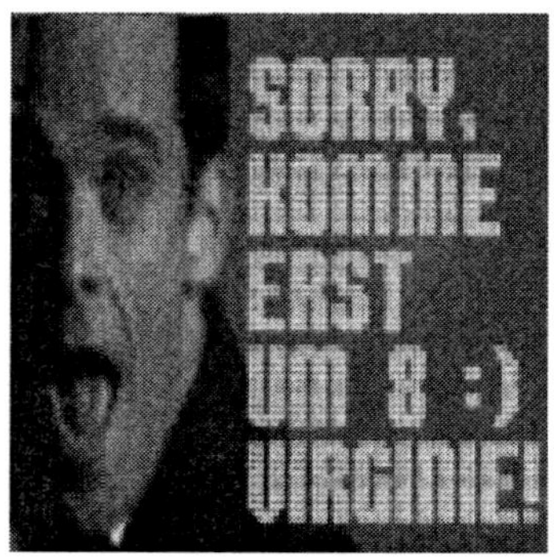

LEGAL ASPECTS

It should be noted that the following discussion focuses on an implementation in Germany and today (2005Q1)–although several EU guidelines are applicable in this area there are differences in their national law implementations and new German and EU laws in relevant areas are pending.

Legal aspects affect V-card in several areas: consumer information laws and rights of withdrawal, protection of minors, spam law, liability, and privacy. A basic topic to those subjects is the classification of V-card among "Broadcast Service" ("Mediendienst"), "Tele Service" ("Teledienst"), and "Tele Communication Service" ("Telekommunikationsdienst"). According to § 2 Abs. 2 Nr. 1 and § 2 Abs. 4 Nr. 3 Teledienstegesetz (TDG) V-card is not a "Broadcast Service" and based on a functional distinction (see e.g., Moritz/Scheffelt in Hoeren/ Sieber, 4, II, Rn. 10) V-card is presumed to be a "Tele Service."

Consumer information laws demand that the customer is informed on the identity of the vendor according to Art. 5 EU Council Decision 2000/31/EC, to § 6 TDG and to § 312c Bürgerliches Gesetzbuch (BGB) (e.g., on certain rights he has with regard to withdrawal). The fact that V-card might be free of charge for the consumer does not change applicable customer protection laws as there is still a (one-sided) contract between the costumer and the provider (see e.g., Bundesrat, 1996, p. 23). Some of these information duties have to be fulfilled before contract and some after. The post-contract information could be included in the result MMS and the general provider information and the pre-contract information could be included in the initial advertisements and/or a referenced WWW- or WAP-site. Art. 6 EU Council Decision 2000/31/EC and § 7 TDG demand a distinction between information and adverts on Web sites and can be applicable, too. A solution could be to clearly communicate the fact that the V-card message contains advert (e.g., in the subject) (analogue to Art. 7(1) EU Council Decision 2000/31/EC, although this article is not relevant in Germany). The consumer might have a withdrawal right based on § 312d (1) BGB on which he has to be informed although the exceptions from § 312c (2) 2nd sentence BGB or § 312d (3) 2 BGB could be applicable. With newest legislation the consumer has to be informed on the status of the withdrawal rights according to § 1 (1) 10 BGB-Informationspflichtenverordnung (BGB-InfoV), whether he has withdrawal rights or not.

§ 6 Abs. 5 Jugendmedienschutzstaatsvertrag (JMStV) bans advertisements for alcohol or tobacco which addresses minors, § 22 Gesetz über den Verkehr mit Lebensmitteln, Tabakerzeugnissen, kosmetischen Mitteln und sonstigen Bedarfsgegenständen (LMBG) bans certain kinds of advertisements for tobacco, Art. 3(2) EU Council Decision 2003/33/EC (still pending German national law implementation) bans advertisements for tobacco in Tele Services. Therefore a sponsor with alcohol or tobacco products will be difficult for V-card. Sponsors with erotic or extreme political content will also be difficult according to § 4, 5 and 6(3) JMStV. § 12(2) 3rd sentence Jugendschutzgesetz (JuSchG) demands a labelling with age rating for content in Tele Services in case it is identical to content available on physical media. Since V-card content will most of the time special-made and therefore not available on physical media, this is not relevant.

The e-mail spam flood has led to several EU and national laws and court decisions trying to limit spam. Some of these laws might be applicable for mobile messaging and V-card, too. In Germany a new § 7 in the Gesetz gegen den unlauteren Wettbewerb (UWG) has been introduced. The question in this area is whether it can be assumed that the sent MMS is ok with the recipient (i.e., if an implied consent can be assumed). Besides the new § 7 UWG if the implied consent cannot be assumed a competitor or a consumer rights protection group could demand to stop the service because of a "Eingriff in den eingerichteten und ausgeübte Gewerbebetrieb" resp. a "Eingriff in das Allgemeine Persönlichkeitsrecht des Empfängers" according to §§ 1004 resp. 823 BGB.

Both the new § 7 UWG and previous court decisions focus on the term of an unacceptable annoyance or damnification which goes along with the reception of the MMS. The highest German civil court has ruled in a comparable case of advert sponsored telephone calls (BGH reference I ZR 227/99) that such an implied consent can be assumed under certain conditions e.g. that the communication starts with a private part (and not with the advertisement) and that the advertisement is not a direct sales pitch putting psychological pressure on the recipient (see e.g., Lange 2002, p. 786). Therefore if a V-card message consists of a private part together with attractive and entertaining content and a logo of the sponsor the implied consent can be assumed. The bigger the advertisement content part is the likelier it is that the level of a minor annoyance is crossed and the message is not allowed according to § 7 UWG (see e.g., Harte-Bavendamm & Henning-Bodewig, 2004, § 7, Rn. 171).

If users use the V-card service to send unwelcome messages to recipients V-card could be held responsible as an alternative to the user from whom the message originated. A Munich court (OLG München reference 8 U 4223/03) ruled in this direction in a similar case of an e-mail news letter service however focusing on the fact that the service allowed the user to stay anonymously. This is not the case with the mobile telephone numbers used in V-card, which are required to be associated with an identified person in Germany. In addition to this the highest German court has in some recent decisions (BGH I ZR 304/01, p. 19 and I ZR 317/01, p. 10) narrowed the possibilities for a liability as an alternative by limiting the reasonable examination duties.

Manual filtering by the V-card service is a violation of communication secrecy and therefore not allowed (see e.g., Katernberg, 2003). Automatic filtering must not result in message suppression since this would be illegal according to German martial law § 206 (2) 2 Strafgesetzbuch.

The obligation to observe confidentiality has in Germany the primary rule that data recording is not allowed unless explicitly approved (§ 4 Bundesdatenschutzgesetz). Log files would therefore not be allowed with an exception for billing according to § 6 Gesetz über den Datenschutz bei Telediensten (TDDSG). These billing logs must not be handed over to third parties likely also including the sponsor.

As a conclusion, it can be noted that an innovative service like V-card faces numerous legal problems. During the project, however, it became clear that all these requirements can be met by an appropriate construction of the service.

EVALUATION OF V-CARD

Since V-card also has the ability to transmit personalised J2ME applications via MMS (see Figure 5 for an example), it surpasses the capabilities of pure MMS messages creating added value for the user, which normally do not have the possibility to create or modify Java programs. One example is a sliding puzzle where, after solving the puzzle, a user may use the digital camera of the mobile device to change the picture of the puzzle. After the modification, the new puzzle can then be send via V-card to other receivers.

Still, as previously mentioned, V-card requires a MMS client. It can therefore be regarded as an enhancement or improvement for MMS communication and is as such a competitor to the "normal" MMS. Hence, an evaluation framework should be usable to measure the acceptance of both "normal" MMS messaging and "enhanced" V-card messaging, creating results that can be compared with each other to determine the actual effect of the added value hoped to be achieved with V-card. While extensive research exists regarding PC-based software, mobile applications currently lack

Figure 5. V-card with MIDlet puzzle application

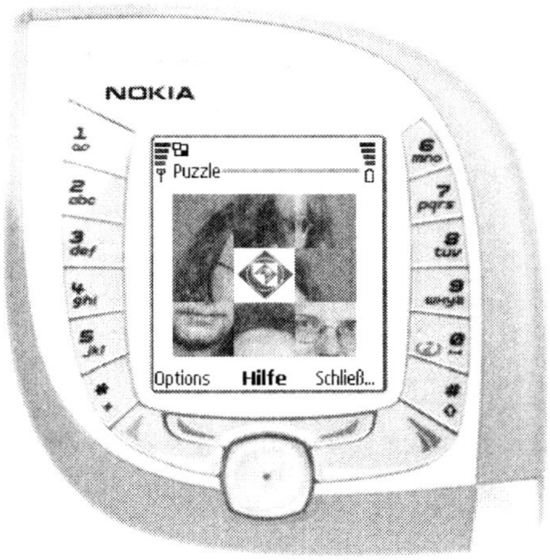

comprehensive methods for creating such evaluations. Therefore, one possible method was developed and applied in a fieldtest to evaluate V-card (Lehner, Sperger, & Nösekabel, 2004).

At the end of the project on June 3, 2004, a group of 27 students evaluated the developed V-card applications in a fieldtest. Even though the composition and size of the group does not permit to denote the results as representative, tendencies can be identified. The statistical overall probability of an error is 30%, as previously mentioned.

The questionnaire was implemented as an instrument to measure results. To verify the quality and reliability of the instrument, three values were calculated based on the statistical data. The questionnaire achieved a Cronbach alpha value of 0.89—values between 0.8 and 1.0 are regarded as acceptable (Cronbach, 1951). The split-half correlation, which measures the internal consistency of the items in the questionnaire, was calculated to be 0.77 with a theoretical maximum of 1.0. Using the Spearman-Brown formula to assess the reliability of the instrument, a value of 0.87 was achieved. Again, the theoretical maximum is 1.0. Therefore, the questionnaire can be regarded to be statistically valid and reliable.

One result of the fieldtest was that none of the students encountered difficulties in using any of the V-card applications, even though the usability of the mobile phone used in the fieldtest was regarded as less than optimal.

Overall 66% of the students thought that V-card was easy to use, 21% were undecided. It is very likely that the sample group leaned towards a negative or at least neutral rating as the usability of the end device was often criticised. This factor can not be compensated by the programmers of the mobile application. Another indicator for this rationale is the comparison with the results for the MMS client. Here, 75% of the group agreed to this statement, which is an increase of 9%. The similarity of the results suggests that also the rating for the usability of the MMS client was tainted by the usability of the device.

No uniform opinion exists regarding sponsored messages by incorporating advertising. Forty-two percent of the students would accept advertisements if that would lower the price of a message. Thirty-seven percent rejected such a method. The acceptable price for a V-card message was slightly lower compared to that of a non-sublimated MMS, which on the other hand did not contain content from a sponsor.

An important aspect for the acceptance of mobile marketing is the protection of privacy. In this area the students were rather critical. Sixty-three percent would reject to submit personal data to the provider of V-card. Since this information was not necessary to use V-card, only 17% of the sample group had privacy concerns while using V-card.

The mobile marketing component was perceived by all participants and was also accepted as a mean to reduce costs. This reduction should benefit the user, therefore a larger portion of the sample group rejected for V-card the idea for increased cost incurred by a longer or more intensive usage (88% rejected this for V-card, 67% for MMS).

As already addressed, the pre-produced content of V-card helped 50% of the users to achieve the desired results. The portion rejecting this statement for V-card was 25%, which is higher than the 8% who rejected this statement for MMS. This leads to the conclusion that if the pre-produced content is appropriate in topic and design for the intended message, it contributes to the desired message. However, it is not possible to add own content if the pre-produced content and the intention of the sender deviate. The user is therefore limited to the offered media of the service provider.

Overall, the ratings for V-card by the students were positive. Marketing messages, which were integrated into the communication during the fieldtest, were not deemed objectionable. The usability of V-card was also rated high. Main points that could be addressed during the actual implementation in the mobile market should include privacy and cost issues.

CONCLUSION

The new messaging service MMS has high potential and is being widely adopted today, although prices and availability are far from optimal. Mostly young people tend to use the fashionable messages which allow much richer content to be sent instantly to a friend's phone. This young user group is especially vulnerable to debts due to their mobile phones though, or they have prepaid subscriptions letting them only send a very limited number of messages. By incorporating a sponsor model in V-card, this user group will be able to send a larger number of messages with no additional cost and thereby offering advertising firms a possibility to market their services and goods. For those users that are not as price sensitive, the large amount of professional media and the ease of the message-composition will be an incentive to use the service. The added value of the service should be a good enough reason to accept a small amount of marketing in the messages. Since V-card offers the sender and receiver an added value, the marketing message will be more acceptable than other forms of advertising where only the sender benefits from the advertisement.

Another advantage of V-card is the fact that the system takes care of the administration and storing of professional media and the complicated formatting of whole messages, thus taking these burdens from the subscriber. At the same time, V-card offers marketers a new way to reach potential customers and to keep in dialogue with existing ones. The ease of sending such rich content messages with a professional touch to a low price or even no cost at all will convince subscribers and help push 3G networks.

Overall, it can be expected that marketing campaigns will make further use of mobile multimedia streaming, aided by available data rates and the increasing computing power of mobile devices. Continuous media (video and audio), either delivered in real-time or on demand, will possibly become the next entertainment paradigm for a mobile community.

REFERENCES

Bundesrat. (1996). *Bundesrats-Drucksache* 966/96. Köln: Bundesanzeiger Verlagsgesellschaft mbH.

Cronbach, L. J. (1951). Coefficient alpha and the internal structure of tests. *Psychometrika, 16*(3), 297-334.

Haig, H. (2002). *Mobile marketing—The message revolution*. London: Kogan Page.

Harte-Bavendamm, H., & Henning-Bodewig, F. (2004). *UWG Kommentar*. München: Beck.

Hoeren, T., & Sieber, U. (2005). *Handbuch Multimedia-Recht*. München: Beck.

Katernberg, J. (2003). *Viren-Schutz/Spam-Schutz*. Retrieved from http://www.uni-muenster.de/ZIV/Hinweise/Rechtsgrundlage VirenSpamSchutz.html

Kindberg, T., Spasojevic, M., Fleck, R., & Sellen, A. (2005). The ubiquitous camera: An in-depth study of camera phone use. *IEEE Pervasive Computing*, *4*(2), 42-50.

Lehner, F., Nösekabel, H., & Schäfer, K. J. (2003). *Szenarien und Beispiele für Mobiles Lernen*. Regensburg: Research Paper of the Chair of Business Computing III Nr. 67.

Lehner, F., Sperger, E. M., & Nösekabel, H. (2004). Evaluation framework for a mobile marketing application in 3rd generation networks. In K. Pousttchi, & K. Turowski (Eds.), *Mobile Economy—Transaktionen, Prozesse, Anwendungen und Dienste* (pp.114-126). Bonn: Köllen Druck+Verlag.

Lange, W. (2002). Werbefinanzierte Kommunikationsdienstleistungen. *Wettbewerb in Recht und Praxis*, *48*(8), 786-788.

Lippert, I. (2002). Mobile marketing. In W. Gora, & S. Röttger-Gerigk (Eds.), *Handbuch Mobile-Commerce* (pp.135-146). Berlin: Springer.

KEY TERMS

MMS: Multimedia message service: Extension to SMS. A MMS may include multimedia content (videos, pictures, audio) and formatting instructions for the text.

Multimedia: Combination of multiple media, which can be continuous (e.g., video, audio) or discontinuous (e.g., text, pictures).

SMS (Short Message Service): text messages that are sent to a mobile device. A SMS may contain up to 160 characters with 7-bit length, longer messages can be split into multiple SMS.

Streaming: Continuous transmission of data primarily used to distribute large quantities of multimedia content.

UMTS (Universal Mobile Telecommunications System): 3rd generation network, providing higher bandwidth than earlier digital networks (e.g., GSM, GPRS, or HSCSD).

Chapter XXX
Context Awareness for Pervasive Assistive Environment

Mohamed Ali Feki
Handicom Lab, INT/GET, France

Mounir mokhtari
Handicom Lab, INT/GET, France

ABSTRACT

This chapter will describe our experience concerning a model-based method for environment design in the field of smart homes dedicated to people with disabilities. An overview of related and similar works and domains will be presented in regards to our approach: adaptive user interface according to environment impact. This approach introduces two constraints in a context aware environment: the control of different types of assistive devices (environmental control system) and the presence of the user with disabilities (user profile). We have designed a service-oriented approach to make it easier the management of services life cycle, and we are designing a semantic specification language based on XML to allow dynamic generation of user interface and environment representation. With the new design of context representation, context framework, and context rule specification, we will demonstrate how changes in contexts adapts supervisor task model which in turn configure the whole system. This chapter is dedicated to researchers having strong interest in developing context aware applications based on existing framework. The application to assistive technology for dependant people is the most suitable since the demand of such pervasive environment is clearly identified.

INTRODUCTION

The **smart home** dedicated to the **dependent people** includes a whole of techniques to make home environment accessible, and provide dedicated services. In smart home concept for **people with special needs**, the design of smart system is based on the use of standard and specific devices to build an **assistive environment** in which many features are provided.

This chapter describes our experience on a model-based method for environment design in the field of smart homes dedicated to **people with disabilities**. An overview of related and similar works and domains will be presented in regards to our approach: adaptive **user interface** according to environment impact. This approach introduces two constraints in a **context aware** environment: the control of different types of **assistive devices** (environmental control system) and the presence of the user with disabilities (user profile).

The key idea of this chapter is the consideration of context awareness in order to ensure the presentation of services to end-user, to process associated features and to handle context history log file.

We have designed a service-oriented approach to improve services life cycle handling. The current development consists on designing a semantic specification language based on XML to allow dynamic generation of user interface and environment representation. Consequently, the design of a context representation, based on a context framework, and coupled with context rule specification, will demonstrate the impact on supervisor task model which in turn will configure the whole system.

In this chapter, we will focus mainly on the design of a new **context assistive framework** than on the semantic specification rules, which will be described in a future publication. This chapter is dedicated to researchers having strong interest in developing context aware applications based on existing framework. The application to **assistive technology** for dependant people is the most suitable since the demand of such **pervasive environment** is clearly identified.

WHAT IS AN ASSISTIVE ENVIRONMENT?

Dependant people, due to **disability** or **aging**, compose a significant segment of the population that would profit from usage of such technologies with the crucial condition that it is physically and economically accessible. This should be possible only if accessibility barriers are detected and considered in a global solution based on a "design for all" concept. The challenge is to consider standardization aspects from the physical low level (i.e., sensors) to application level (i.e., user interface) of any system design.

Autonomy and quality of life of people with disabilities and elderly people in daily living would benefit from smart homes designed under the "**assistive environment**" paradigm and can experience significant enhancements due to the increased support received from the environment (Sumi helal, 2003). This support includes facilities for environmental control, information access, communication, monitoring, etc., and built over various existing and emerging technologies. Nevertheless, users are usually confronted to accessibility barriers located at the level of **human-machine interface** due to heterogeneous devices, features and communication protocols involved. These problems include both, physical difficulties to handle input devices, and cognitive barriers to understand and reach suitable functionalities. Consequently, accessible unified interfaces to control all the appliances and services are needed. This is only possible if the network, devices, and mobile technologies used for smart homes are able to support interoperability and systems integration (Abascal, 2003).

FROM COMPUTING TO PERVASIVE COMPUTING

Assistive environment presented above includes smart homes technologies which are of primary importance in enhancing quality of life of people with disabilities. In such environment, the user needs to use handheld devices in order to increase his or her mobility. Besides, user would like to profit from wireless mobile technologies to ensure the availability of residential services when he or she is located even indoor (home, office, etc.) and outdoor (street, car, etc). User wishes to be served "on demand," "any time," "any where," "any system" to get commonly used services. In addition, designers should take into account the adaptation of those technologies in order to fit with end-users requirements. This situation makes the solution more complex and impose to deal with a natural extension of computer paradigm, integration of computers in people daily environment, and to manage complex environment where several heterogeneous technologies must operate together in order to provide user with new services, privacy and comfort. We can easily identify that such problematic is delimited by **pervasive** frontiers (Abowd et al., 2002) (Henning.Sc et al., 2003).

Next, we will highlight the need of adaptive **user interface** that consequently implies need of **context aware frameworks**.

THE NEED OF AWARENESS

One of the principle targets is to build generic and unified user interface (UI) to control the smart home, independent from the controlled system, or from the communication protocols, which must be flexible and personalized for each end user. However, the design of a smart environment dedicated to elderly and people with disabilities must take into account emerging technologies that may respond to user requirements and needs in term of dependence in their life.

Usability of these systems is very important and this depends on each person with disability. The ability to adapt any assistive aids according to the needs of each individual will allow the acceptance or not of the system. Besides, people with disabilities encounter static environment which allow one or many ways of communication between the user and his environment. This environment needs to be aware of some knowledge in order to provide supplementary and useful data to enrich the degree of awareness of the human machine system and the user. Context aware applications promote to respond to previous challenge. Indeed, those applications improve both mobility and communication, which are two common limitations amongst people with disabilities. User needs to manipulate intelligent systems to avoid obstacles, to make some tasks automatic and to ensure the realization of some commands at actuators level. The concept of **smart homes** permits for user to open the door of his room but if some sensors are integrated, the door could be automatically opened when it's **aware** of user presence in the proximity. A user who is using an electrical wheelchair equipped with a robotic arm is able to do some living tasks such as take a cup of water, eat, or turn his or her computer on, but there is any data that prevents user of dynamic obstacles or damage in the system. Camera or other vision sensor can contribute to assist some tasks by designing only objects, tasks and the target. Position sensor promotes to provide periodic events describing the position of the arm related to other obstacles. To summarize, in front of difficulties encountered by people with disabilities to control their environment, **adaptation** of **user interfaces** has become a necessity rather than a facility because the insufficiencies of **adapted technical aids** in one part and the increasing number of variety of devices and their use in **assistive environments** by various types of users (ordinary or handicapped) in other part. Existing systems demonstrate a lack of ability to satisfy the heterogeneous needs of **dependant people**. Those needs also vary according to the **context**, which comprises environmental conditions, device's characteristics, and the user profile. There is a need for techniques that can help the **user interface** (UI) designer and developer to deal with myriad of contextual situations. Consequently, user should be provided with the facility to have an adaptive interface that fit to changing needs.

HUMAN MACHINE INTERFACE

The user interface is the single component in such systems, upon which everything else will be judged. If the interface is confusing and badly designed, the system will be thought of in that way. Indeed, making such systems simpler is an extremely complex goal to achieve. It is, nonetheless, very important to do so.

While the implementing technologies may be similar, the interface must fit to the special needs of the user. A person with cognitive impairment may require a less complex screen, presenting him or her with limited and simpler choices at one time. The use of a greater number of menus may be necessary, as may be the use of alternative indicators such as pictures or icons. Such person may benefit from systems, which make certain choices for them or suggest actions. Artificial Intelligence is often employed in these cases (Allen, Ekberg, & Willems, 1995).

The **user interface** should be consistent with all applications the user may use from time to time and when changing environment (desktop, house, airport, station, etc). Hence, the organization of the system should be the same whether users are accessing their environmental control system, their communicator, their telephone, their local home gateway machine, or when visiting the airport, the railway station, the museum, etc.

Such situation presents a great challenge to the interface designer; requiring the involvement of various engineers, human factors specialists, ergonomists, and of course, the users themselves.

The State-of-the-Art

During our experience, we have investigated several works regarding to adaptive human machine interface Concept and experimentation, we describe briefly the most important of them:

- **TSUNAMI:** (Higel, O'Donnell, Lewis, & Wade, 2003) is an interface technology that supports a range of source's input. The system monitors users for implicit inputs, such as vague gestures or conversation, and explicit inputs, such as verbal or typed commands, and uses these to predict what assistance the user requires to fulfil their perceived goal. Predictions are also guided by **context information** such as calendars, location and biographical information
- **SEESCOA Project:** SEESCOA (Software Engineering for Embedded Systems using Component-Oriented Approach) project goals include separation of User Interface (UI) design from low level programming, and the ability to migrate UIs from one device to another while automatically adapting to new device constraints. The project seeks to adapt Component Based Development (CBD) tech-

nology. The idea was conceptualized to avoid the problem of redesigning UIs whenever new technology came into market. The experiments have used XIML as the user interface definition language (Luyten, Van Laerhoven, TConinx, & Van Reeth, 2003)

- **PALIO:** Personalized Access to Local Information and services for tourists (PALIO) proposes a framework that supports location awareness to allow the dynamic modification of information presented (according to position of user). PALIO ensures the adaptation of contents to automatically provide different presentations depending on user requirements, needs, and preferences. It provides scalability of information to different communication technologies and terminals and guarantees interoperability between different services providers in both envisaged wireless network and the World Wide Web. It is aiming to offer services through fixed terminals in public spaces and mobile personal terminals, by integrating different wireless and wired telecommunications technologies (Sousa & Garlan, 2002)
- **AVANTI Project:** AVANTI (Adaptive and Adaptable Interactions to multi-media Telecommunication applications) addresses the interaction requirements of **disabled users** using Web-based multimedia telecommunication applications and services. The project facilitates the development of **user interface** of interactive software application that adapts to individual user abilities, requirements, and preferences. The project developed a technological framework called "Unified User Interface Development Platform" for the design and implementation of user interfaces that are accessible by people with disabilities. Components of AVANTI system include a collection of multimedia databases, the AVANTI server, and the AVANTI Web browser. Databases are accessed thorough a common protocol (HTTP) and provide mobility information for disabled people. AVANTI server maintains knowledge regarding the users, retains a content model of the information system, and adapts the information to be provided, according to user characteristics (hyper-structure adaptor). AVANTI Web browser is capable of adapting itself to the abilities, requirements, and preferences of individual users (Stephanidis, Paramythis, Karagiannidis, & Savidis, 1997)

Discussion

With the ever-decreasing size and increasing power of computers, embedded processors are appearing in devices all around us. As a result, the notion of a computer as a distinct device is being replaced with an ubiquitous ambient computing presence (Dey, 2001). This proliferation of devices will present **user interface** designers with a challenge. While an average user might cope with having a different interface for their personal digital assistant (PDA), desktop PC, and mobile phone, they will certainly have difficulty if the range of devices is greatly increased. In the past, designers have suggested creating a single interface appearing on all devices; however, research has thus far not proved this to be the optimum solution. In deed, for example, developers of the Symbian OS found it was not feasible to offer the same user interface on Symbian-powered PDA's as on desktop computers. Besides, previous works are implementing one ubiquitous environment and they omit inter environment communication. The update of services presentation is done in the context of one environment discovery; however, there is no or less information of how to skip between not similar environments.

We instead propose an ambient environment **interface** within the computing environment which observes the users activities and then acts on what the user wants. The environment then handles the individual interaction. The user interface take into account also dynamic discover of services in building environment. The first step of implementation integrates only one environment. We have then included context awareness framework to ensure interspaces communications, services continuity and user interface update in real time conditions.

Design of the HMI Software and Past Implementation

The user interface has as a crucial managing role of various functionalities. Among equipment we distinguish several types of products: electrical devices (white goods), household equipment (brown goods), data-processing equipment (gray goods), and also mobile devices (mobile phones, pocket PCs, wireless devices...). The diversity of these products brings a wide range of networking protocols necessary to manage the whole smart environment (radio, infrared, Ethernet, power line communications...).

The solution consists on the design of a generic user interface with supervisor module independent of the communication protocols. This approach permits to obtain a rather acceptable time response without weighing down the task of the supervisor. Indeed, supervisor plays the central role by processing various interconnections between protocols to allow the transport the requested action to the corresponding communication object which is a specific representation of the physical devices (Feki, Abdulrazak, & Mokhtari, 2003).

Re-design of software control architecture is not sufficient to allow access to smart environment by severely disabled people. The problem is that each end user, with his or her deficiencies and his or her individual needs, is considered as a particular case that requires a typical configuration of any **assistive system**. Selecting the most adapted input device is the first step and the objective is to allow the adaptation of available functionality's according to his or her needs. For this purpose we have developed a software configuration tools, called ECS (Environment Configuration System (Abdulrazak, Mokhtari, Feki, Grandjean, Rodriguez, 2003), which allows a non expert in computer science to configure easily any selected input device with the help of different menus containing activities associated to action commands of any system. The idea is to describe equipment (TV, Robot, PC), input device (joystick, keypad, mouse), technologies (X2D, Bluetooth, IP Protocol) using XML and generate automatically all available functionalities which could be displayed in an interactive graphical user interface. According to user's needs, and to the selected input devices, the supervisor offers the mean to associate graphically the selected actions to the input device events (buttons, joystick movements...).

The ECS software is actually running and fully compatible with most home equipment. It generates an XML object as standard output which will be easily downloaded by various ways in our control system. Supervisor allows in one hand to read XML specification to create the starting display mode, and to assume the connection link with physical layers in order to recuperate changes through dynamic discover. Our implementation is mainly based on four components: (see next figures)

- **Smart home supervisor (HMI):** The smart home supervisor represents the GUI interface for all smart home compliant devices. It is able to detect the devices on the home network dynamically. It also displays the icons of the different devices. Upon clicking on a particular device by the user, the GUI will download the dynamic service discovery code and run it. The HMI supervises the whole system: it converts user events into actions according to selected output devices (Robot, TV, VCR, etc.), transmits the information to the feedback module, manages multimodal aspects, errors situations, the synchronization of modules, etc. The HMI could also be connected to the ECS for environment configuration

- **Graphic user interface (GUI):** Since household devices vary significantly in their capabilities and functionality, each device may have a different interface for configuring it. For instance, a "door" should provide an interface to open/close/lock the door. But for a VCR the interface should include controls for playback, rewind, eject etc. We would like our devices to be truly plug and play. Which means when a new smart device is employed, the

Figure 1. Smart homes concept

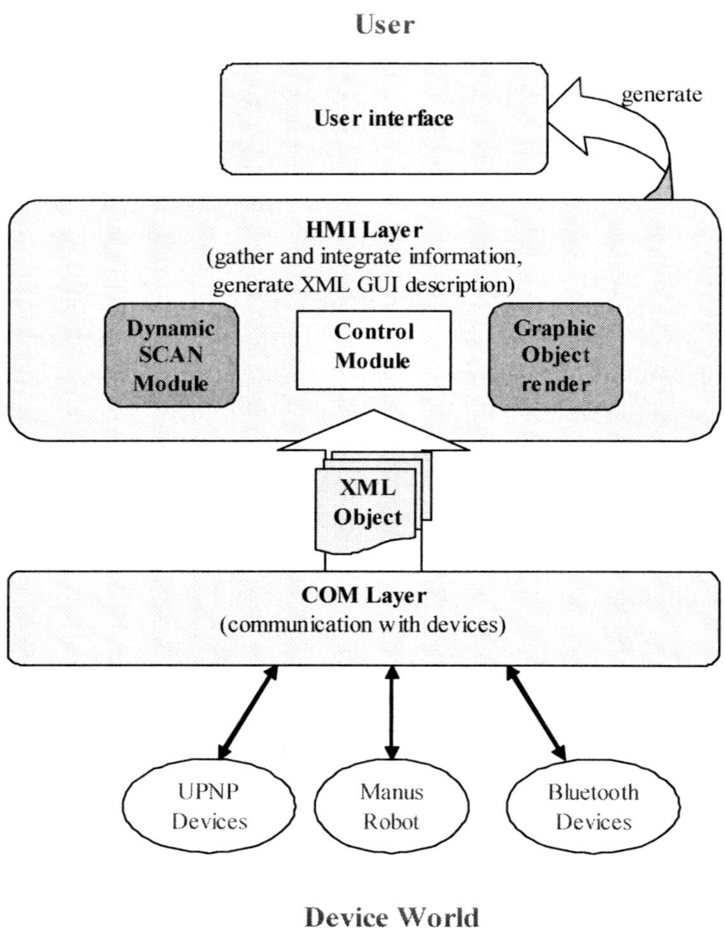

user need only to hook it up to the network after which the device is instantaneously detected by the smart home GUI without the need of loading any device drivers. We use the necessary facilities provided by UPNP coupled with JAVA THREAD technologies for creating "network plug and play" devices, which when connected to a network, they announce their presence and enable network users to remotely exploit these devices

- **Dynamic service discovery code (DSDC):** To be able to achieve the goal of being a truly plug and play devices, each of our smart device's will implement some "service discovery module" that extends Java's "JTHREAD" class and interact with JDOM Parser which is responsible for creating a standard XML Object describing all devices discovered with related services and actions (Feki et al., 2003). Here, the smart device programmer can identify what functionality the end user can control and whether features/security should be enforced or not. Once the device is detected by the GUI, the mobile code is transferred over the network using CORBA protocols and is executed on the GUI's location whenever the user desires to configure that particular device. The GUI is capable of running and detecting new smart devices without the need to add any drivers or interfaces to it. We succeed to run an effective and robust dynamic service discovery code at lower network layer which allow us to discover all devices (See Figure 2 for clarification)
- **COM Layer (CL):** Deals with specific characteristics of any output device according to its communication protocol (CAN, infrared, radio protocol, etc.). Indeed, traditional home services are proposed by home devices manufacturers by means of a proprietary control device which may be accessed either directly or from the phone network

Discussion

We presented an overview of existing works concerning human machine systems and outlined the less of plasticity and dynamicity regardless to the lack of awareness and interoperability techniques. Then we described our solution to build a human machine layer having the ability to download dynamically new services.

Our concept in its current implementation deals with myriad techniques to discover an ubiquitous system, but is still unable to make it easier inter-connection between several ubiquitous spaces. We argue that integration of **context aware** attributes should reinforce the awareness level. In next section, we present the state of the art of context awareness, an overview of similar works. We propose after that a new **framework** and describe how it affects the **human machine** layer.

CONTEXT AWARENESS: THE STATE-OF-THE-ART

While **context** has been defined in numerous ways, we present here two frequently used definitions. Dey and Abowd (Dey, 2001; Dey & Abowd, 2000) define context, context awareness, and context aware applications as: "Context is any information that can be used to characterize the situation of an entity. An entity is a person, place, or object that is considered relevant to the interaction between a user and an application, including the user and the application themselves. A system is context-aware if it uses context to provide relevant information and/or services to the user, where relevancy depends on the user's task. Context awareness is the facility to establish context. Context aware applications adapt according to location of use, collection of nearby people, hosts and accessible devices, and their changes over time. The application examines the computing environment and reacts to changes."

Chen and Kotz (2000) define context by making a distinction between what is relevant and what is critical: "Context is a set of environmental states and settings that either determines an application's behavior or in which an application event occurs and is interesting to the other user." They define the former situation as a critical case called active context and the later one as a relevant one naming it as a passive context.

CONTEXT AWARENESS: FRAMEWORKS

In order to implement definitions many **frameworks** are emerging. In next paragraphs we will try to provide an overview of the most used frameworks with a short discussion.

Context Toolkit

The main objective behind development of the context toolkit is to separate the context acquisition (process of acquiring context information) from the way it is used and delivered. (Dey and. Al, 2001) uses object-oriented approach and introduced three abstractions: widgets, servers, and interpreters. The services of context toolkit include abstraction of sensor information and context data through interpreters, access to context data through network API, sharing of context data through distributed infrastructure, storage of context data and basic access control for privacy protection. Figure 2 shows the architecture of context toolkit.

Java Context Aware Framework

Java Context Aware Framework (JCAF) (Bardram, 2003; Bardram, Bardram, Bossen, Lykke-Olesen, Madsen, & Nielsen, 2002) is the first of the kind to provide a Java-based application framework. JCAF was developed to aid development of domain specific context aware applications. One of the motivations of JCAF is to have Java API for context awareness in

Figure 2. Context toolkit architecture

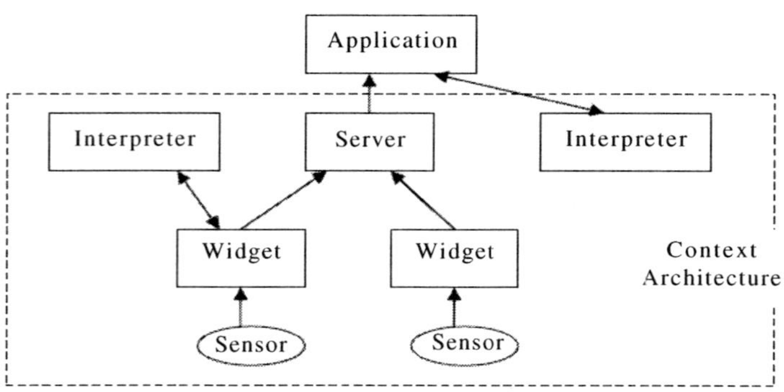

Figure 3. JCAF architecture

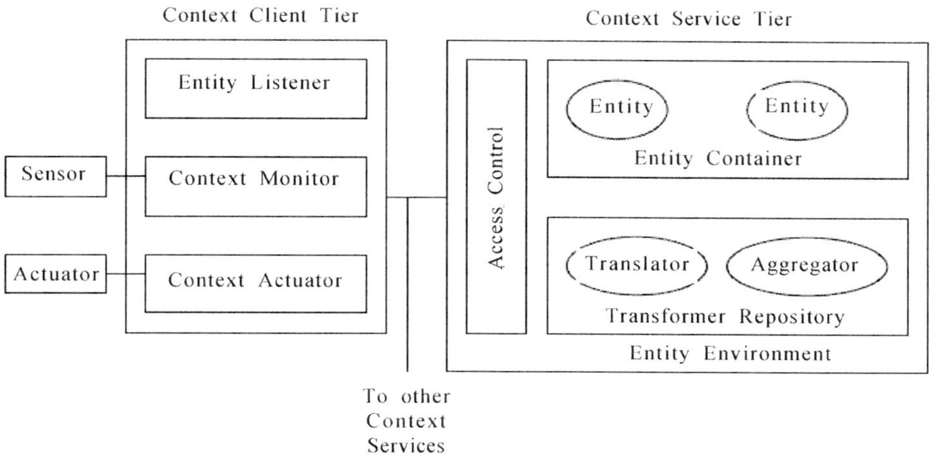

much the same way JDBC is for databases and JMS is for messaging services.

Architecture: JCAF is a distributed, loosely-coupled, service-oriented, event-based, and secure infrastructure. Components of the JCAF framework include context service, access control, remote entity listener, context client, context monitor. The architecture is based on distributed model view controller and its design principle is based on semantic free modelling abstractions.

Context Information Service

Context information service (CIS) is another object oriented framework which supports context aware applications. It is introduced by Pascoe, Ryan, and Morse's (Chen & Kotz, 2000). It is platform independent, globally scalable, and provides shared access to resources. Core features of CIS include contextual sensing, context adaptation, contextual resource discovery, and context augmentation. CIS is a layered service architecture consisting of service components that include world, world archive and sensor arrays. These components are extensible and reusable. (Pascoe, 1998).

Context Service

Context service (Brown, 2000) provides a middleware infrastructure for context collection and dissemination. The architectural components of context service include a dispatcher, configurable set of drivers and collection if utility components. Utility components include context cache, work pacer, an event engine, and privacy engine.

Two applications built using context service illustrates its use in increasing the user experience: The notification dispatcher that uses context to route messages to a device that is most appropriate to the recipient and a context aware content distribution system that uses context to envisage user's access to Web content, and uses this information to pre-process and pre-distribute content to reduce access latency.

Owl

A context-aware system aims to "gather, maintain, and supply context information to clients. It tackles various advanced issues, including access rights, historical context, quality, extensibility, and scalability." It offers a programming model that allows for both synchronous queries and asynchronous event notifications. It protects people's privacy through the use of a role-based access control (RBAC) mechanism." (Ebling, Hunt, & Lei, 2001).

Kimura

The motivation of the Kimura System (MacIntyre, Mynatt, Tullio, & Voida, 2001) is to integrate both physical and virtual context information to enrich activities of knowledge workers. It utilises a blackboard model based on tuple spaces. The four components that operate on the tuple spaces are:

1. Desktop monitoring and handling components, which uses low-level Window hooks to observe user activities and the interpreter component
2. The peripheral display and interaction components that read and display the context information so that the user can observe and utilise the context in tasks
3. Context monitoring component that writes low-level tuples, which are later interpreted
4. Interpreter component, which translates low-level tuples into tuples that can immediately be read by the whiteboard display and interaction component

MoCA

Moca (*mo*bile *c*ollaboration *a*rchitecture) is middleware architecture for developing context processing services and context-sensitive applications for mobile collaboration. Work on this architecture is part of a wider project that aims to experiment with new forms of mobile collaboration and implement a flexible and extensible service-based environment for developing collaborative applications for infrastructure mobile networks (Vagner Sacramento et al., 2004). However, Moca is designed for infrastructure wireless network. It needs adaptation to integrate cellular data networks protocols.

Discussion

From the functionalities of the **frameworks** studied above, we came up with the common set of requirements that any **context aware framework** satisfies:

1. Sensor technology to capture the contextual information: Acquire raw contextual information
2. Support for event-based programming model so as to have the ability to trigger events when certain context change is observed
3. A way to communicate the sensed contextual data to other elements in the environment and a way to interpret the collected data: provide interpreted context to application
4. Integration of a wide range of handheld devices so the context transformation should be applied to any mobile systems
5. A generic communication layer that supports heterogenic wireless and wired protocols in order to favour special needs, communication, and mobility of **people with disabilities**
6. In the case of ubiquitous environment where people having special needs are living, we include also security and pri-

vacy requirements. Hence, we need framework which is capable of adapting the content and presentation of services for use on a wide range of devices, with particular emphasis on nomadic interaction from wireless network devices. Such framework should have the capabilities in context of multiple user interfaces, and includes device and platform independence, device and platform awareness, uniformity and cross platform consistence, and user awareness

Among previous frameworks, we find that JCAF and Context toolkit are covering the most requirements described above. Moreover, Context toolkit is closely related to JCAF in the kind of features it provides since Context toolkit also provides a distributed and loosely coupled infrastructure. JCAF and the Context toolkit have similar concept which separates between sensor and data acquisition and treatment. The Context toolkit disposes more API and functionalities but JCAF is simpler to use and we can easily build application on top of it.

RESEARCH STRATEGY: JCAF AUGMENTED SOLUTION

Implementing of simple services which are utile for **people with disabilities** like automats repetitive tasks or predict user position or tasks to avoid intervention of user at interface layer is not sufficient for us. We need a complete **framework** that can answer following principal design and needs:

We notice easily that pick up information related to context allows us to use devices which are most likely not attached to the same computer running the application. In fact, sensors and actuators must be physically scattered and cannot be directly connected to the same machine. This implies that data recuperated are coming from multiple, distributed machines. Our application has to support context **distribution**.

Another problems is directly induced from previous supposition is how to support interoperability of context applications on heterogeneous platforms. The idea is to develop objects responsible for transforming data recuperated from context sources. This transformation is based on standard XML. The output is send to a smart engine for context analyze, making decision, and update the HMI layer. Moreover, most context management is specific for one environment like handling context in smart home, but occasionally it might become relevant to contact services running in other environment. Therefore, a context-awareness infrastructure should be distributed and loosely coupled, while maintaining ways of cooperating in a peer-to-peer or hierarchical communication. Besides, the core quality of context-aware applications is their ability to react to changes in their environment. Hence, applications should be able to subscribe to relevant context events and be notified when such events occur.

Based on JCAF we implemented new framework dedicated to our purpose. We started to use the entity container to build all environments. Each environment is hold under an entity container. Then, we defined entity for all containers or environments in order to precise user profile environment, and we inherit it from "person" class and so on. Next step demands to interconnect the role of supervisor for adapting user interface in order to handle communication with all entity listeners. Indeed, each entity listener is programmed to use JAVA RMI (remote method invocation) protocol to be remotely informed and updated by suitable Entity. We had to use JAVA methods to ensure interoperability between entity listeners.

This solution consists on implementing following modules based on API's provided by JCAF framework (See Figures 3 and 4 for more clarification).

1. **System entity container:** Inherits from entity container yet presented in last section, and handles modifications on system side including sensor events, actuators events, state of network traffic, etc. they represent physical devices responsible for providing data and information by different ways (signals, switch ...)
2. **Platform entity container:** Context awareness defines new decisions to adapt interface downloaded into heterogeneous pervasive computing and handheld devices (PDA, mobile phone...). Platform environment let the context module aware of related functionalities such as size of screen, memory etc
3. **User entity container:** We need to identify the user to download static preferences, capabilities, desires, and need in term of environment composition and interface display. The user profile module is responsible for enrich the awareness of the system by updating user behaviours and activities. This module also inherits from Entity Container.
4. **Sensor entity:** Each of them is associated with one or more physical sensors to recuperate raw data and make a unified data representation (standard XML). Models are available to be used by high layer applications. In order to validate functionalities of this framework (Figure 5), we coupled power of OSGI (OSGI official Web site) as an open oriented service infrastructure and our framework based on JCAF Concept. OSGI principal consists of a set of services (called bundles) that we can manage easily without interrupt the system life cycle.

We used OSCAR (OSCAR official Web site) as the OSGI frameworks and we build a new service that we called "pervasive Contextual." This service includes following bundles:

Figure 4. Context framework components

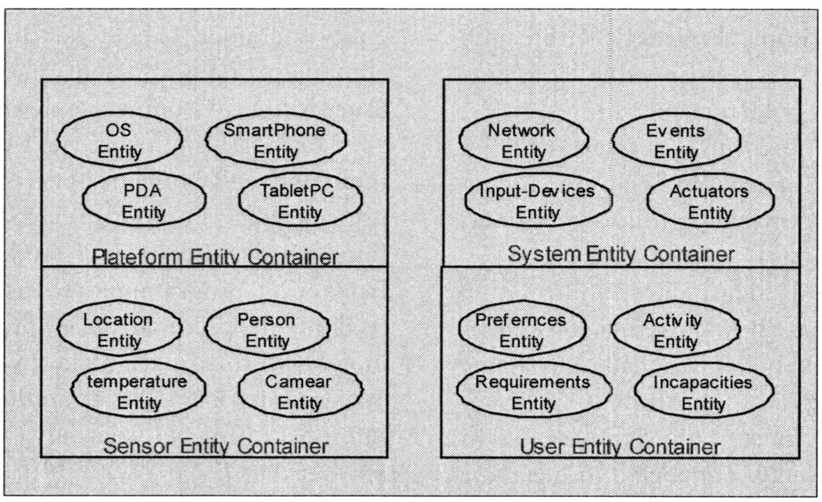

1. The principal java class that implements OSGI and JCAF APIs; it contains special methods to be conforming to OSGI specifications. It is named Activator and includes start and Stop methods
2. The JCAF bundle that provides adaptable OSGI packages
3. The Context Server bundle that interacts with the fort elements (entity listener, user entity listener, platform entity container, sensor entity) previously presented
4. The manifest file that specifies interaction with other OSGI bundles by describing import and export packages
5. The build file which is formatted as ANT (ANT official Web site) specification and has the role to organize the structure of global project by defining its resources, class's folder, jar folder, etc

Pervasive contextual service is then uploaded in OSCAR framework to allow interactions with other services in one hand and to update the human machine interface specification in the other hand. Integration of this service is ensured in both residential use and external use. In deed, RMI proposes a secure connexion between distant entities. In addition, Context Client is easily handled in smart devices such as PDA or smart phone.

CONCLUSION

We presented in this chapter, a situation of **people with disabilities** in their **assistive environment**, and we underlined the needs of awareness to enhance inter and intra interactions with such environment. We outlined problem of technologies supporting context aware applications, and we presented our approach to make the connection between existing technologies and existing assistive environments.

Facing the problem of technologies adaptation to enhance the life of people with disabilities, the increasing of the need of awareness added to systems supporting those people and the emerging of frameworks implementing context-aware applications, we proposed an OSGI/JCAF-based implementation. We aim in future

Figure 5. The context aware frameworks and its impact in smart OSGI based environments

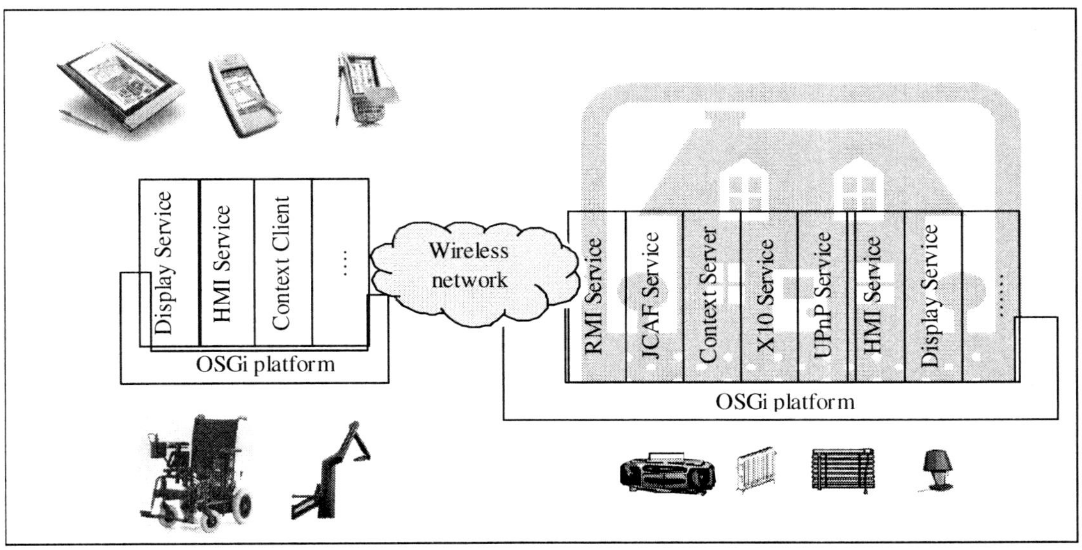

to develop a graphical builder environment (GBE) at top level in order to facilitate to non-expert user the build of context aware applications. We plan also to create task model presentation in order to make connexion between context impact and HMI update.

REFERENCES

Abascal, J. (2003, March 27-28). Threats and opportunities of rising technologies for smart houses. *Proceeding of Accessibility for all Conference*, Nice, France.

Abdulrazak, B., Mokhtari, M., Feki, M. A., Grandjean, B., & Rodriguez, R. (2003, September). Generic user interface for people with disabilities: Application to smart home concept. *Proceedings of the ICOST 2003, 1st International Conference on Smart homes and Health Telematics, "Independent living for persons with disabilities and elderly people"*, Paris (pp.45-51). IOS Press.

Abowd, G. D., Ebling, M. R., Gellersen, H.-W., Hung, G., & Lei, H. (2002, October). Context aware pervasive computing. *IEEE Wireless Communication, 9*(5), 8-9.

Allen, B., Ekberg, J., & Willems, C. (1996). Smart houses: How can they help people with disabilities? In R. Patric, & W. Roe (Eds.), *Telecommunications for all*. ECSC-EC-EAEC, Brussels*Luxembourg 1995, Printed in Belgium, CD-90-95-712-EN-C, 1995, Spanish version ISBN: 84-8112-056-1 Fundesco.

ANT official Web site, http://ant.apache.org/

Bardram, J. E. (2003, October 12). *UbiHealth 2003: The 2nd International Workshop on Ubiquitous Computing for Pervasive Healthcare Applications*, Seattle, Washington, part of the UbiComp 2003 Conference. Retrieved from http://www.healthcare.pervasive.dk/ubicomp2003/papers/

Bardram, J. E., Bossen, C., Lykke-Olesen, A., Madsen, K. H., & Nielsen, R. (2002). Virtual video prototyping of pervasive healthcare systems. *Conference Proceedings on Designing Interactive Systems: Processes, Practices, Methods, and Techniques (DIS2002)* (pp. 167-177). ACM Press.

Brown, P., Burleston, W., Lamming, M., Rahlff, O., Romano, G., Scholtz, J., & Snowdon, D. (2000, April). Context-awareness: Some compelling applications. *Proceedings the CH12000 Workshop on The What, Who, Where, When, Why, and How of Context-Awareness.*

Chen, G. & Kotz, D. (2000, November). *A survey of context-aware mobile computing research* (Tech. Rep. No. TR 2000-381). Dartmouth College, Department of Computer Science.

Dey, A. K. (2001, February). Understanding and using context. *Personal and Ubiquitous Computing, 5*(1), 4-7.

Dey, A. K., & Abowd, G. D. (2000). Towards a better understanding of context and context-awareness. *Proceedings of CHIA'00 Workshop on Context-Awareness.*

Ebling, M. R., Hunt, G. D. H., & Lei, H. (2001). *Issues for context services in pervasive computing*. Retrieved November 27, 2002, from http://www.cs.arizona.edu/mmc/13%20Ebling.pdf

Feki, M. A., Abdulrazak, B., & Mokhtari, M. (2003, Sept.). XML modelisation of smart home environment. *Proceedings of the ICOST 2003, 1st International Conference on Smart homes and Health Telematics. "Independent living for persons with disabilities and elderly people"*, Paris, September (pp.55-60). Ed. IOS Press.

Helal, A., Lee, C., Giraldo, C., Kaddoura, Y., Zabadani, H., Davenport, R. et al. (2003, September). Assistive environment for successful aging. *Proceedings of the ICOST 2003, 1st International Conference on Smart Homes and Health Telematics, "Independent living for persons with disabilities and elderly people"*, Paris (pp.55-60). Ed. IOS Press.

Higel, S., O'Donnell, T., Lewis, D., & Wade, V. (2003, November). Towards an intuitive interface for tailored service compositions. *The 4th IFIP International Conference on Distributed Applications & Interoperable Systems*, Paris.

Luyten, K., Van Laerhoven, T., & Coninx, K., Van Reeth, F. (2003). Runtime transformations for modal independent user interface migration. In *Interacting with Computers.*

MacIntyre, B., Mynatt, E. D., Tullio, J., & Voida, S. (2001). *Hypermedia in Kimura System.* Retrieved November 27, 2002, from www.cc.gatech.edu/fce/ecl/projects/kimura/pubs/kimura-hypertext2001.pdf

OSCAR official Web site, http://oscar.objectWeb.org/

OSGI official Web site, http://www.osgi.org

Pascoe, J. (1998). Adding generic contextual capabilities to wearable computers. *The 2nd International Symposium on Wearable Computers* (pp. 92-99).

Sacramento, V., Endler, M., Rubinsztejn, H. K., Lima, L. S., Goncalves, K., Nascimento, F. N. et al. (2004, October). MoCA: A middleware for developing collaborative applications for mobile users. In IEEE Distributed Systems Online 1541-4922 © 2004. *IEEE Computer Society*, *5*(10).

Schulzrinne, H., Wu, X., Sidiroglou, S., & Berger, S. (2003, November). Ubiquitous computing in home networks. *IEEE Communication Magazine, 41*(11), 128-135.

Sousa, J. P., & Garlan, D. (2002, August). Aura: An architectural framework for user mobility in ubiquitous computing environments. In *Software Architecture: System Design, Development, and Maintenance. Proceedings of the 3rd Working IEEE/IFIP Conference on Software Architecture* (pp. 29-43).

Stephanidis, C., Paramythis, A., Karagiannidis, C., & Savidis, A. (1997). Supporting interface adaptation in the AVANTI Web browser. *The 3rd ERCIM Workshop on User Interfaces for All.* Retrieved from http://www.ics.forth.gr/proj/at-hci/UI4ALL/UI4ALL-97/proceedings.html

KEY TERMS

Assistive Environment: Environment equipped with several kinds of assistive devices which interconnect and communicate together in order to give dependant user more autonomy and comfort.

Context Awarness: Any relevant information or useful data that can enrich the user interface and assist the update of environment organization and human machine interaction.

Dependant People: People having physical or cognitive incapacities (people with motor disabilities, elderly people, etc.) and suffer from less autonomy in doing their daily activities.

Pervasive Environment: Environment that include several kinds of handheld devices, wireless and wired protocols, and a set of services. The specificity of this environment is its ability to handle with any service at any time any where and any system.

Chapter XXXI
Architectural Support for Mobile Context-Aware Applications

Patrícia Dockhorn Costa
Centre for Telematics and Information Technology, University of Twente, The Netherlands

Luís Ferreira Pires
Centre for Telematics and Information Technology, University of Twente, The Netherlands

Marten van Sinderen
Centre for Telematics and Information Technology, University of Twente, The Netherlands

ABSTRACT

Context-awareness has emerged as an important and desirable feature in distributed mobile systems, since it benefits from the changes in the user's context to dynamically tailor services to the user's current situation and needs. This chapter presents our efforts on designing a flexible infrastructure to support the development of mobile context-aware applications. We discuss relevant context-awareness concepts, define architectural patterns on context-awareness and present the design of the target infrastructure. Our approach towards this infrastructure includes the definition of a service-oriented architecture in which the dynamic customization of services is specified by means of description rules at infrastructure runtime.

INTRODUCTION

Context awareness refers to the capabilities of applications that can provide relevant services to their users by sensing and exploring the user's context. Typically the user's context consists of a collection of conditions, such as the user's location, environmental aspects (temperature, light intensity, etc.), and activities (Chen, Finin, & Joshi, 2003). Context awareness has emerged as an important and desirable feature in distributed mobile systems, since it benefits from the changes in the user's context to dynamically tailor services to the user's current situation and needs (Dockhorn Costa, Ferreira Pires, & van Sinderen, 2004).

Developers of context-aware applications have to face some challenges, such as (i) bridging the gap between information sensed from the environment and information that is actually syntactically and semantically meaningful to these applications; (ii) modifying application behavior (reactively and proactively) according to pre-defined condition rules; and (iii) customizing service delivery as needed by the user and his context. These challenges require proper software abstractions and methodologies that support and ease the development process.

In this chapter, we discuss relevant concepts of context awareness and present the design of an infrastructure that supports mobile context-aware applications. Our approach tackles the challenges previously mentioned by providing a service-oriented architecture in which the dynamic customization of services is specified by means of application-specified condition rules that are interpreted and applied by the infrastructure at runtime.

In addition, we present three architectural patterns that can be applied beneficially in the development of context-aware services infrastructures, namely the event-control-action pattern, the context sources and managers hierarchy pattern and the actions pattern. These patterns present solutions for recurring problems associated with managing context information and proactively reacting upon context changes.

The remainder of this chapter is structured as follows: The section "Context Awareness" presents general aspects of context awareness, such as the definition of context, its properties and interrelationships; the section "Context-Aware Services Infrastructures" discusses the role of applications, application components and infrastructure in our approach, the section "Context-Aware Architectural Patterns" presents the architectural patterns we have identified, and the section "Services Infrastructure Architecture" introduces an infrastructure that supports the development of context-aware applications, the section "Related Work" relates our work to other current approaches, and the last section gives final remarks and conclusions.

CONTEXT AWARENESS

In the Merriam-Webster online dictionary (Merriam-Webster, 2005) the following definition of context can be found: "the interrelated conditions in which something exists or occurs." We focus on this definition as the starting point for discussing context in the scope of context-aware mobile applications. This definition makes clear that it is only meaningful to talk about context with respect to something (that exists or occurs), which we call the entity or subject of the context. Since we aim at supporting the development of context-aware applications, we should clearly identify the subject of the context in this area.

Context-aware applications have been devised as an extension to traditional distributed applications in which the context of the application users is exploited to determine how the application should behave. The services offered by these applications are called context-aware services. Furthermore, these applications have means to learn the users' context without explicit user intervention. We conclude then that in the case of context-aware applications, context should be limited to the conditions that are relevant for the purpose of these applications. The subject of the context in this case can be a user or a group of users of the context-aware services, or the service provisioning itself.

When considering context-aware applications, we should not forget that context consists of interrelated conditions in the real world, and that applications still need to quantify and capture these conditions in terms of so called "context information" in order to reason about context. This implies that context-aware applications need context models, consisting of the information on specific conditions that characterize context, its values, and relationships. The act of capturing context in terms of context information for the purpose of reasoning and/or acting on context in applications is called context modeling.

Figure 1 shows the context of a person (application user) in real world and context-aware applications that can only refer to this context through context information.

Context-aware applications strive for obtaining the most accurate and up-to-date possible evaluation of the conditions of interest in terms of context information, but the quality of the corresponding context information is strongly dependent of the mechanisms used to capture context conditions. Some context conditions may have to be measured, and the measuring mechanisms may have some level of accuracy; other context conditions may vary strongly in time, so that the measurement may quickly become obsolete. Decisions based on context information taken in context-aware applications may also take into account the quality of this information, and therefore context-applications also need meta-information about the context condition values, revealing their quality.

Figure 2 shows a simple class diagram summarizing the concepts introduced above.

Although we discuss context information previously from the point of view of condition values, context modeling can only re-used and generalized when the condition types, their semantics and relationships are clearly defined.

The following categories of context conditions have been identified in the literature (e.g., Chen et al., 2004a; Kofod-Petersen & Aamodt, 2003; Preuveneers et al., 2004):

- **Location:** The (geographical) location in which the user can be found
- **Environmental conditions:** The temperature, pressure, humidity, speed (motion), light etc. of the physical environment in which the user can be found
- **Activities:** The activities being performed by the user. These activities may be characterized in general terms (e.g., "working") or in more specific terms (e.g., "filling in an application form"), depending on the application
- **Devices:** The conditions related to the user's devices, like handheld computers, mobile phones, etc. These conditions can

Figure 1. Context in real world vs. context information in context-aware applications

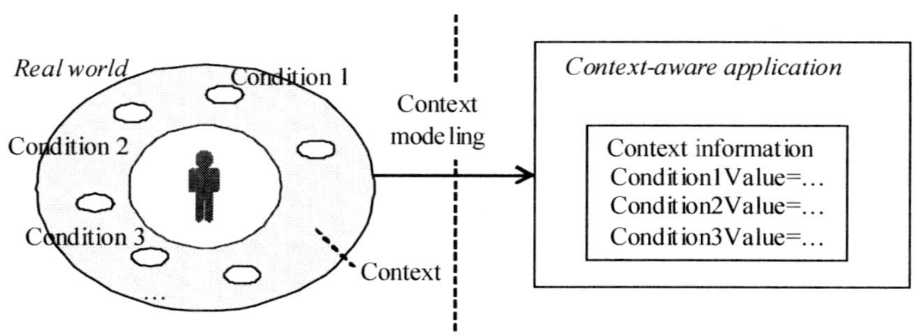

Figure 2. User diagram of the concepts related to context (real world and application)

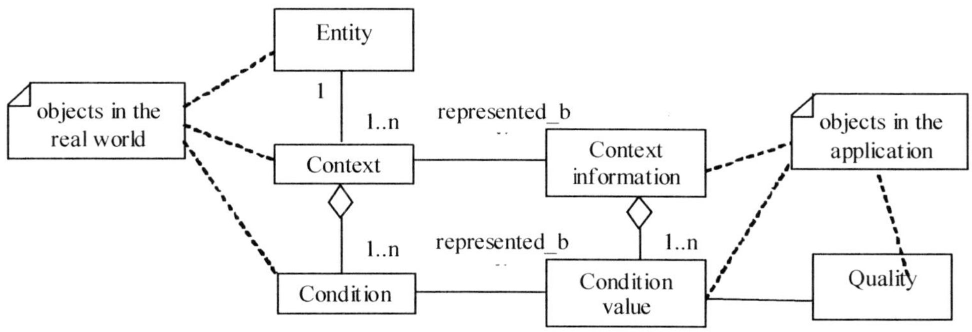

refer to configuration information (amount of memory installed, CPU speed, etc.), or available resources (memory, battery power, network connectivity, etc.)

- **Services:** The services available to the user, and possibly the state of the user in these services (e.g., pending transactions)
- **Vital signs:** The heart beat, blood pressure and even some conditions that have to be measured using more specialized medical equipment (e.g., brain activity represented in an electroencephalogram)

Some other conditions, like the user's personal information (name, gender, address, etc.) or the user's preferences concerning the use of devices and software, qualify as context according to the definition given above, but may be treated differently from the dynamic context conditions. We consider these conditions as part of the user's context just to keep the definition consistent. The same applies to histories of location, environmental conditions, activities, etc. in time.

We do not claim that the categories of context conditions mentioned above are exhaustive. Furthermore, these categories represent a specific grouping of context conditions, but many other alternative groupings can be found in the literature, and may be pursued depending on the application requirements.

Context awareness in combination with multimedia can create many interesting application opportunities. Some examples are (i) the adjustment of the quality of a real-time video streaming depending of available wireless network capabilities (e.g., user's device loses connection to a Wifi hotspot and has to reconnect using GPRS), and (ii) the delivery of multimedia services when the user enters a room with sensing capabilities (e.g., show a video clip of some products on the user's device when the user enters a shop).

CONTEXT-AWARE SERVICES INFRASTRUCTURES

In case of large-scale networked application systems, it is not feasible for each individual application to capture and process context information just for its own use. There are several reasons why a shared infrastructure should give support to context-aware applications:

- **Costs:** Sharing information derived from the same set of context sources and sharing common context processing tasks among applications potentially reduce costs
- **Complexity:** Context processing tasks may be too complex and resource inten-

sive to be run on a single application device
- **Distribution:** Information from several physically distributed context sources may be aggregated, and aggregation at the application may not be the best place for reasons of timeliness and communication resource efficiency
- **Richness:** In a ubiquitous computing world where the environment is saturated with all kinds of sensors, applications may profit from a priori unknown context sources, provided that support is provided for ad hoc networking and information exchange between such sources and context-aware applications

The support to context-aware applications from a shared infrastructure should comprise reusable context processing and management components. Such components may be based on existing mechanisms that are already deployed, but it should also be possible to dynamically add new components or mechanisms that will evolve in the future. In particular, the infrastructure may have special components that can take application-specified rules or procedures as input in order to carry out application-specific context aggregation and fusion mechanisms and control actions. This calls for a high level of flexibility of the infrastructure.

The infrastructure should also be highly scalable. The number of context sources and context-aware applications may be potentially large and will certainly grow in the near future with further developments of sensor networks and ubiquitous computing devices. At the same time, the amount of context information to be handled by the infrastructure will increase and new context-aware applications may be developed (e.g., in the gaming or healthcare domain) that require high volumes of context-related information (e.g., 3D positioning or biosignals). It should be possible to support increased numbers and volumes by adding capacity to the infrastructure without changing or interrupting the infrastructure's operation.

Context-aware applications as well as context sources may be mobile (running on a mobile device and attached to mobile objects, respectively), and therefore connections may not be pre-contemplated but ad-hoc. Mobility is an important characteristic that requires explicit consideration from the infrastructure. Different qualities for data transfer and different policies for accessing information and using resources may exist in different environments that an application or context source may experience during a single session. The infrastructure should as much as possible shield the applications from the mechanisms that are necessary to deal with such heterogeneity.

There are many technological solutions for the challenges of flexibility, scalability, and mobility. However, the following high-level guidelines are considered useful for all these solutions: (i) separate the infrastructure in a services layer and a networking layer, and (ii) enforce the use of services as the only way to interact with components in the services layer.

The networking layer is concerned with the provision of information exchange capabilities that allow components to interact, while shielding them from the intricacies of realizing such capabilities in a heterogeneous distributed environment. The services layer consists of components that provide information processing capabilities which are building blocks for the end-user applications. The services layer should comprise the context processing and management tasks, as these directly relate to the applications, not to the information exchange. Distinguishing these two layers results in a clear separation of design concerns, which facilitate maintainability in the light of requirements' and technology changes.

Each component in the services layer offers its capabilities as a service to other components, and it can make use of the capabilities of other components by invoking their services. This enforces a discipline of component composition with some important benefits. First, services do not disclose state or structure of components, and therefore components may be implemented in any way. For example, a component may consist of subcomponents, or may make use of services of other components, in order to provide the service that is associated with it. Second, a service makes no assumption as to what are its users, except that they can interact as implied by the service definition. This ensures low coupling and high flexibility. Third, services allow a hierarchical composition of components, where tasks can be delegated (a component invokes the service of another component) and coordinated (a component orchestrates the invocation of services of multiple other components). These guidelines lead to a general approach for a context-aware services infrastructure. Examples of useful patterns of component composition are presented in the section "Context-Aware Architectural Patterns", and examples of specific infrastructure components are discussed in the section "Services Intrastructure Architecture".

CONTEXT-AWARE ARCHITECTURAL PATTERNS

Architectural patterns have been proposed in many domains as means to capture recurring design problems that arise in specific design situations. They document existing, well-proven design experience, allowing reuse of knowledge gained by experienced practitioners (Buschmann, Meunier, Rohmert, Sommerlad, & Stal, 2001). For example, a software architecture pattern describes a particular recurring design problem and presents a generic scheme for its solutions. The solution scheme contains components, their responsibilities, and relationships.

In this section, we present three architectural patterns that can help the development of context-aware services infrastructures (Dockhorn Costa, Ferreira Pires, & van Sinderen, 2005), namely the Event-Control-Action pattern, the Context Sources and Managers Hierarchy pattern and the Actions pattern.

Event-Control-Action Pattern

The event-control-action (ECA) architectural pattern aims at providing a structural scheme to enable the coordination, configuration, and cooperation of distributed functionality within services infrastructures. It divides the tasks of gathering and processing context information from the tasks of triggering actions in response to context changes, under the control of an application behavior description. We assume that context-aware application behaviors can be described in terms of condition rules, such as if <condition> then <actions>.

The condition part specifies the situation under which the actions are enabled. Conditions are represented by logical combinations of events. An event models some occurrence of interest in our application or its environment. The observation of events is followed by the triggering of actions, under control of condition rules. Actions are operations that affect the application behavior. They can be a simple web services call or a SMS message delivery, or it can be a complex composition of services.

The architectural scheme proposed by the ECA pattern consists of three components, namely context processor, controller and action performer components. Figure 3 shows a component diagram of the ECA pattern scheme as

Figure 3. Event-control-action pattern

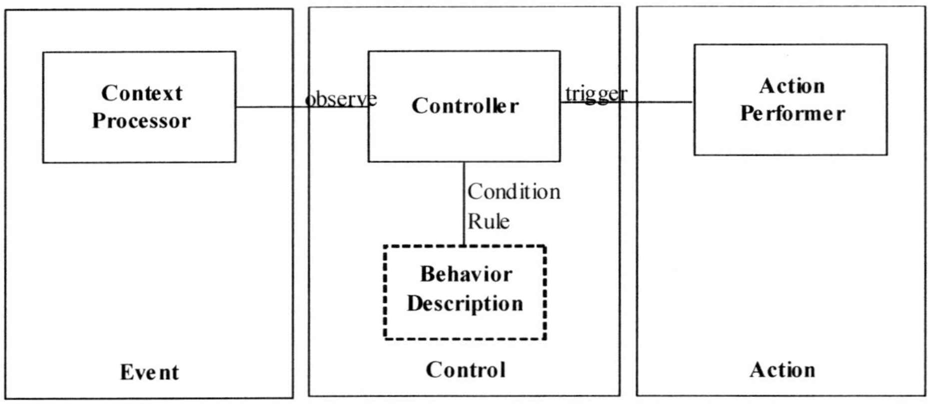

it should be applied in context-aware services infrastructures.

Context concerns are handled by the context processor component, which generates and observes events. This component depends on the definition and modeling of context information. The controller component, provided with application behavior descriptions (condition rules), observes events, monitors condition rules, and triggers actions when the condition is satisfied. Action concerns, such as decomposition and implementation binding, are addressed by the action performer component.

Consider as an example application of the ECA pattern the tele-monitoring application scenario described in Batteram et al. (2004), in which epileptic patients are monitored and provided with medical assistance moments before

Figure 4. Dynamics of the event-control-action pattern

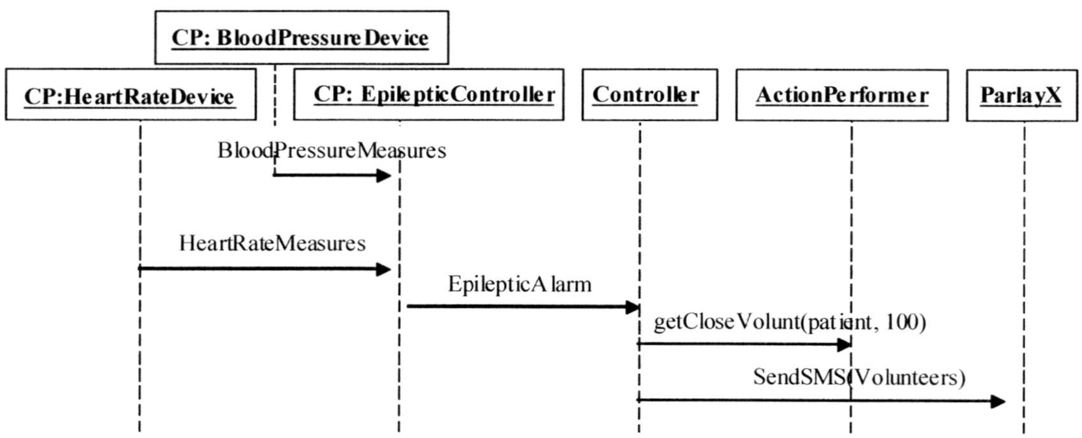

and during an epileptic seizure. Measuring heart beat variability and physical activity, this application can predict future seizures and contact volunteers or healthcare professionals automatically. We will assume here that when a possible epileptic seizure is detected, the nearest volunteers are contacted via SMS.

Figure 4 depicts the flow of information between the components of the Event-Control-Action pattern.

The condition rule defined within the Controller has the form:

```
if <event:EpilepticAlarm>
then
    <SendSMS(closeby(Volunteers, 100))>
```

The controller observes the occurrence of event EpilepticAlarm. This event is captured by the component epileptic controller, which is an instance of context processor. Blood pressure and heart beat measures are gathered from other dedicated instances of context processor. Based on these measures and a complex algorithm, the epileptic controller component is able to predict within seconds that an epileptic seizure is about to happen, and an EpilepticAlarm event is, therefore, generated.

Upon the occurrence of event EpilepticAlarm, the Controller triggers the action specified in the condition rule. The action SendSMS(closeby(volunteers, 100)) is a composed action that can be partially resolved and executed by the infrastructure. The inner action closeby (volunteers, 100) may be completely executed within the infrastructure. The execution of this action requires another cycle of context information gathering on context processors, in order to provide the current location of the patient and his volunteers, and to calculate the proximity of these persons. By invoking the operation getCloseVolunt(patient, 100) with assistance of an internal action performer, the controller is able to obtain the volunteers that are within a radius of 100 meters from the patient. Finally, the Controller remotely invokes an action provided by a third-party business provider (e.g., a Parlay X provider (Parlay, 2002)) to send SMS alarm messages to the volunteers.

Context Sources and Managers Hierarchy Pattern

The context sources and managers hierarchy architectural pattern aims at providing a structural schema to enable the distribution and composition of context information processing components. We define two types of context processor components, namely context sources and context managers. Context source components encapsulate single domain sensors, such as a blood pressure measuring device or a GPS device. Context manager components cover multiple domain context sources, such as the integration of a blood pressure and heart beat measures. Both perform context information processing activities such as, for example:

- **Sensing:** Gathering context information from sensor devices. For example, gathering location information (latitude and longitude) from a GPS device
- **Aggregating (or fusion):** Observing, collecting and composing context information from various context information processing units. For example, collecting location information from various GPS devices
- **Inferring:** Interpretation of context information in order to derive another type of context information. Interpretation may be performed based on, for example, logic rules, knowledge bases, and model-based techniques. Inference occurs, for instance, when deriving proximity information from information on multiple locations

- **Predicting:** The projection of probable context information of given situations, hence yielding contextual information with a certain degree of uncertainty. We may be able to predict in time the user's location by observing previous movements, trajectory, current location, speed, and direction of next movements

The structural schema proposed by this pattern consists of hierarchical chains of context sources and managers, in which the outcome of a context information processing unit may become input for the higher level unit in the hierarchy. The resulting structure is a directed acyclic graph, in which the initial vertexes (nodes) of the graph are always context source components and end vertexes may be either context sources or context managers. The directed edges of the graph represent the (context) information flow between the components. We assume that cooperating context source and manager developers have some kind of agreements on the semantics of the information they exchange.

Figure 5. Context sources and managers hierarchy pattern

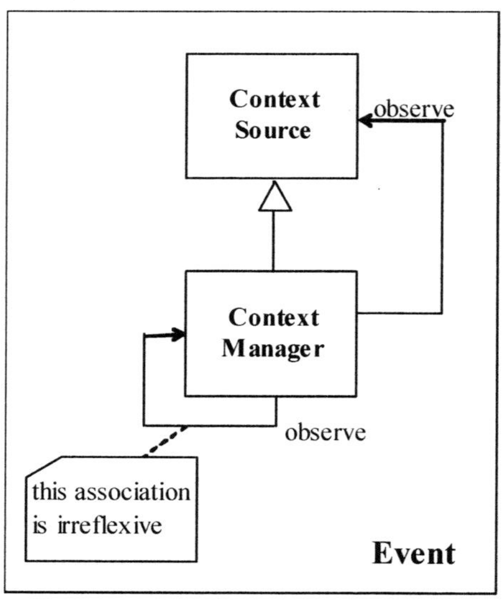

Figure 5 details in the Event part of Figure 3. It shows a class diagram of the context source and manager hierarchy pattern as it can be applied for context-aware services infrastructures.

Context managers inherit the features of context sources, and implement additional functions to handle context information gathering from various context sources and managers. A context manager observes context from one or more context sources and possibly other context managers. The association between the context manager class and itself is irreflexive. Figure 6 depicts a directed acyclic graph structure, which is an instantiation of this pattern. CS boxes represent instances of context sources and CM boxes represent instances of context managers.

Consider the tele-monitoring example again, discussed in the previous section. Figure 7 depicts the flow of information between components in the context sources and managers structure. ControllerC1 observes the occurrence of event (EpilepticAlarm ^ driving), which

Figure 6. Instance of context sources and managers hierarchy pattern

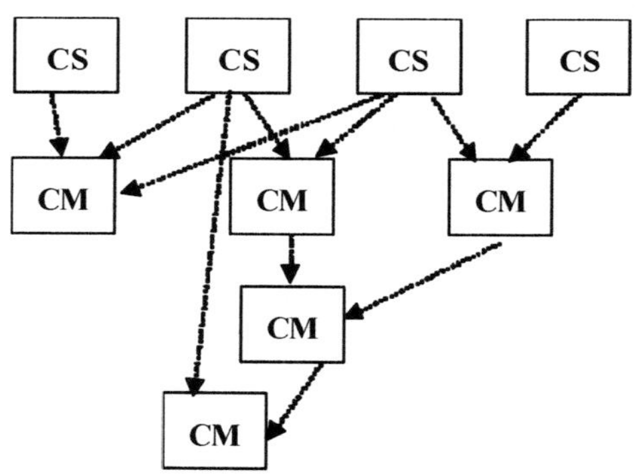

Figure 7. Dynamics of the context sources and managers pattern

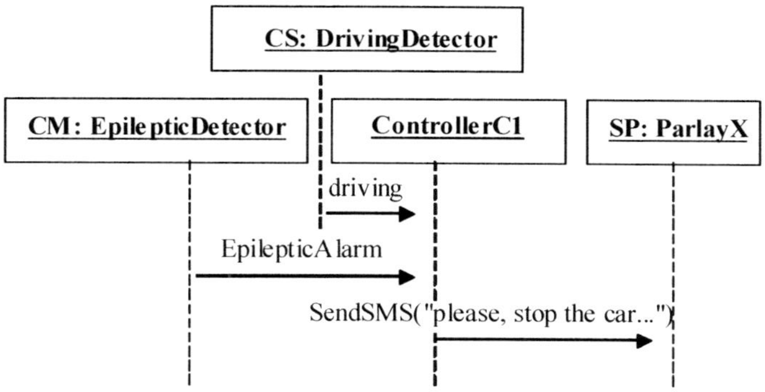

is generated from CM: EpilepticDetector and CS: DrivingDetector, respectively. When the condition turns true, (the alarm has been launched and the patient is driving), the personalized SMS message is sent to the patient.

Actions Pattern

The actions architectural pattern aims at providing a structural scheme to enable coordination of actions and decoupling of action implementations from action purposes. It involves (1) an action resolver component that performs coordination of dependent actions, (2) an action provider component that defines action purposes, and (3) an action implementor component that defines action implementations.

An action purpose describes an intention to perform an action with no indication on how and by whom these computations are implemented.

Figure 8. Actions pattern structure

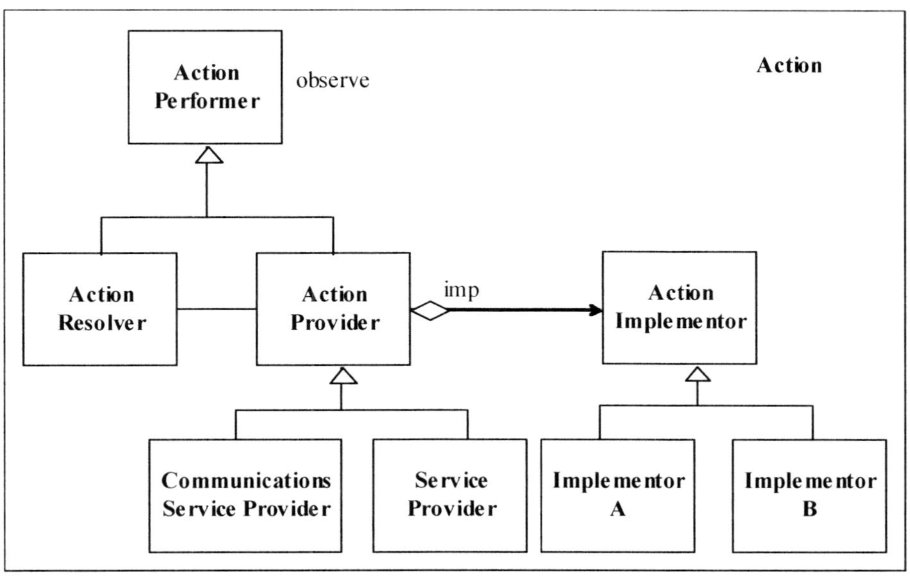

Figure 9. Dynamics of actions pattern

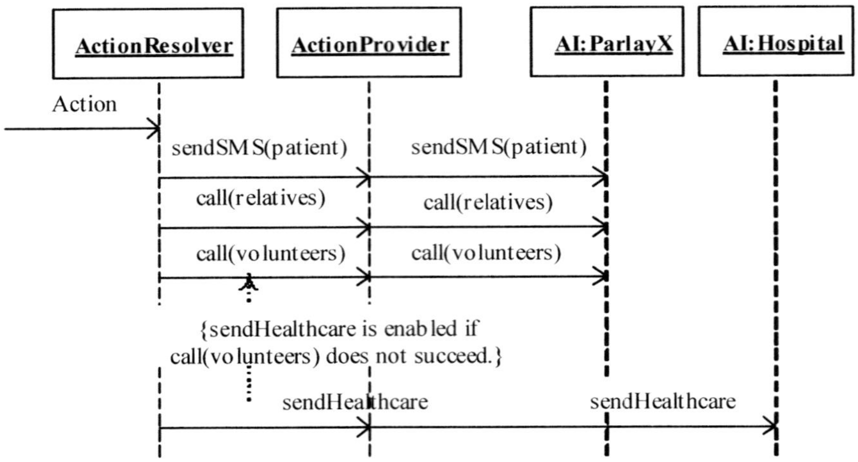

Examples of action purposes are "call relatives" or "send a message." The action implementor component defines various ways of implementing a given action purpose. For example, the action "call relatives" may have various implementations, each supported by a different telecom provider. Finally, the action resolver component applies techniques to resolve compound actions, which are decomposed into indivisible units of action purposes, from the infrastructure point of view.

Figure 8 details the action part of Figure 3. It shows a class diagram of the actions pattern as it is supposed to be applied for context-aware services infrastructures.

Both the action resolver and action provider components inherit the characteristics of the action performer component, and therefore they are both capable of performing actions. The action resolver component performs compound actions, decomposing them into indivisible action purposes, which are further performed separately by the action provider component. Action providers may be communication service providers or (application) service providers. Communication service providers perform communication services, such as a network request, while service providers perform general application-oriented services, implemented either internal or external to the infrastructure, such as an epileptic alarm generation or an SMS delivery, respectively.

An action provider may aggregate various action implementor components, which provide concrete implementations for a given action purpose. In Figure 8, two different concrete implementations are represented (Implementor A and Implementor B).

Figure 9 depicts the flow of information between components of the actions pattern for the tele-monitoring scenario.

The action resolver gets a compound action that it has to decompose so that each subaction can be executed. Provided with techniques to solve composition of services, the action resolver breaks the compound action into indivisible service units, which are then forwarded to the action provider. The action provider delegates these service units to the proper concrete action implementations. In our example, send SMS and calling actions are delegated to the ParlayX implementor and the action to send healthcare is delegated to the hospital implementor.

SERVICES INFRASTRUCTURE ARCHITECTURE

Figure 10 depicts the component-based architecture of our infrastructure. This architecture applies the event-control-action pattern, in which context concerns are decoupled from triggering actions concerns under control of an application behavior description. Context source and manager components address context specific issues, such as gathering, processing and delivering context information. The controlling component is empowered with application behavior descriptions (e.g., condition rules), which specify the conditions under which actions are to be triggered. Conditions are tested against context information observed from context source and manager components. Action performer components allow requesters to trigger actions. In our infrastructure, actions represent a system reaction to context information changes. These reactions may be the invocation of any external or internal service, such as the generation of an alarm, the delivery of a message or a web services request.

The hierarchy of context source and manager components depicted in Figure 10 illustrate the use of the context sources and managers hierarchy pattern; action performers depicted in Figure 10 illustrate the use of the Actions Pattern.

Application-specific components may directly use various components of the infrastructure, from context sources to action performers.

The components presented in this architecture offer services as in a service-oriented architecture. Therefore, services in our approach are registered and discovered in a service repository. The discovery of services is not depicted in

Figure 10. Component-based architecture

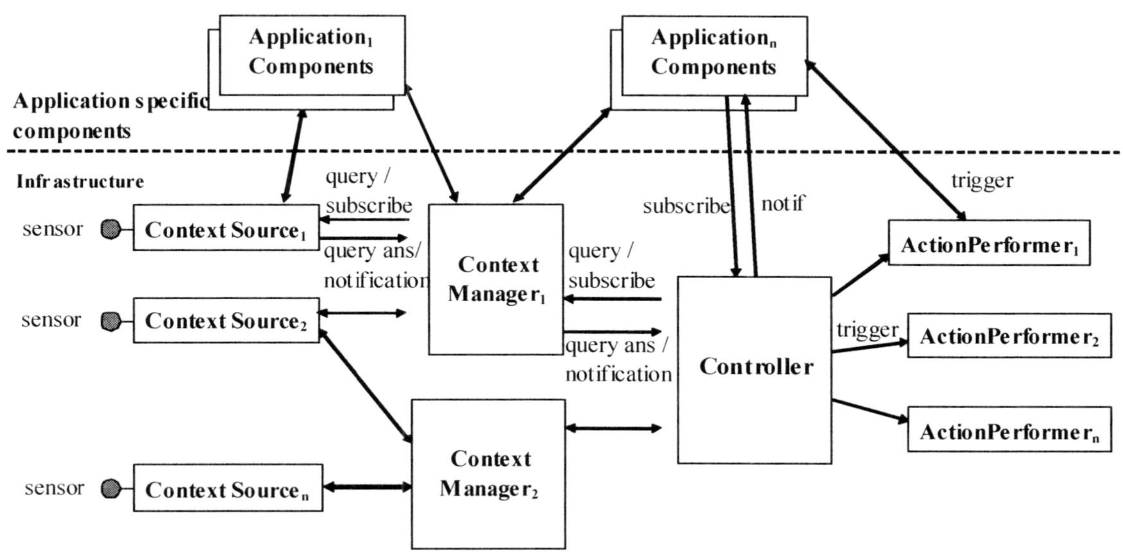

Figure 10 but it implicitly enables interactions between components in the architecture.

Discovery Services

Discovery services facilitate the offering and the discovery of instances of services of particular types. A service registry provides discovery services in our infrastructure and it can be viewed as an entity through which other entities can advertise their capabilities and match their needs against advertised capabilities. Advertising a capability or offering a service is often called "export." Matching against needs or discovering services is often called "import" (OMG, 2000).

To export or register, an entity gives the service registry a description of a service and the location of an interface where that service is available. To import or lookup, an entity asks the service registry for a service having certain characteristics. The service registry checks against the service descriptions it holds and responds to the importer with the location of the selected service's interface. The importer is then able to interact with the service. Figure 11 depicts the sequence of interactions between the service provider, service user, and the registry.

Figure 12 depicts the services that compose the discovery service, namely the register service and the lookup service.

The following data types are used in Figure 12: (1) a ServiceOffer represents a description of the service to be included in the service registered; (2) an OfferId is an identification of the

Figure 11. Interactions between a service registry and its users

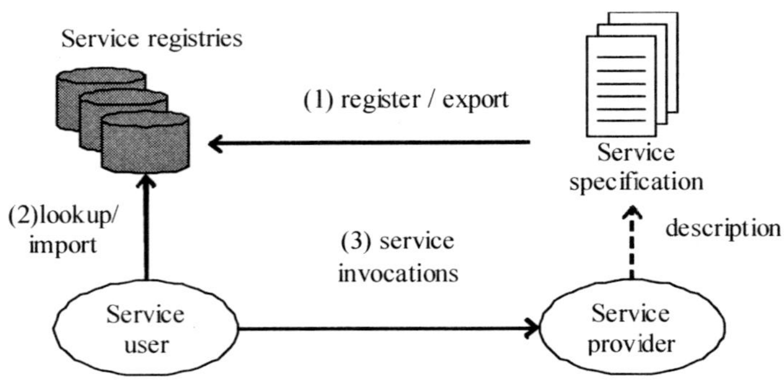

Figure 12. Discovery services

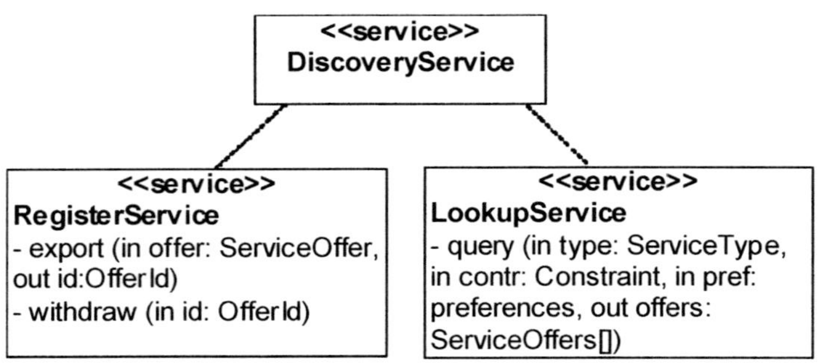

Figure 13. Difference in the interaction pattern

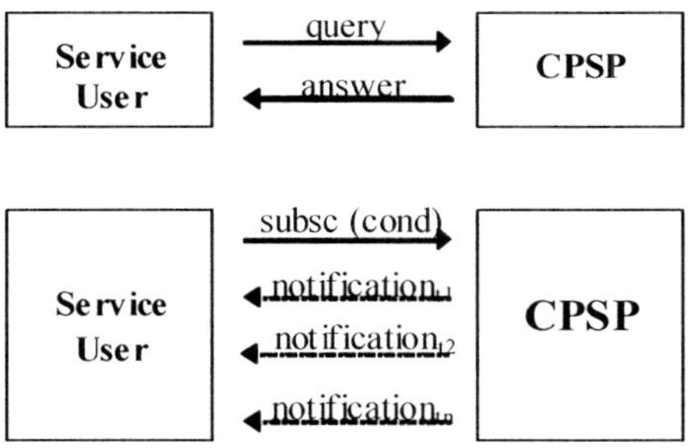

service offer; (3) Constraints define restrictions on the services offers being selected, for example, restrictions on quality of services or any other service properties defined; and (4) Preferences determine the order in which the selected services should be presented.

Context Provisioning Service

A context provisioning service facilitates the gathering of context information. This service is supported by context source and context manager components. A context provisioning service may support two types of requests: query-based or notification-based. A query-based request triggers a synchronous response while a notification-based request specifies conditions under which the response should be triggered. Example of query-based and notification-based requests are getLocation (user:John) and getLocation (user:John, condition: time=t), respectively. In the first request, the service user immediately gets the current location of user John (assuming this is available). In the second request, the service user gets John's location only when the current time is t.

Figure 13 shows the interaction pattern between a context provisioning service provider (CPSP) and its user. Query based requests trigger an immediate response, while in a subscription-based approach, the notifications are time-varying, depending on when the conditions (defined in the subscription process) are met.

Figure 14 depicts our context provisioning service.

Operation subscribe is used to register a notification request, operation unsubscribe is used to withdraw a given notification subscription and operation query is used to select specific context information instances. The specification of languages to define context subscription characterization, context query expression and context query answer is currently a topic of research.

Potential users of the context provisioning services are (1) application-specific components, (2) the controller component, and (3) other context provisioning services.

Figure 14. Context provisioning service

<<service>>
ContextProvisioningService

- **subscribe**(in characterization: ContextSubscriptionCharacterization, in subscriber: ContextSubscriptionReference, out id: ContextSubscriptionId)
- **unsubscribe** (in id:ContextSubscriptionId)
- **query** (in expression: ContextQueryExpression, out answer: ContextQueryAnswer)

Context provisioning services may be advertised and discovered using the discovery service. We may define properties of context to be used as constraints to select context provisioning services, such as quality of context properties, accuracy and freshness. The definition of such properties is highly related to the context model discussed in the section "Context Awareness".

Action Service

An action service allows users of this service to request the execution of certain actions. This service is offered by the action performer components.

Action implementers provide their action services specifications, which are wrapped into an action service supported by the infrastructure. Furthermore, action implementers should register their services in the infrastructure service registry, setting parameters and properties that should be used in the discovery process. The action performer supports a single standard operation, namely DO (action_name, parameters).

Figure 15 depicts the generation of action wrappers based on an action service specification. This action service is the SendSMS (Parlay, 2002) service offered by a telecom provider.

The SendSMSParlay service specifies two operations, SendSMS and GetSMSDeliveryStatus. This service is wrapped by a service supported by the infrastructure, containing a DO() operation. The wrapper service has pointers to the actual implementations of the operations SendSMSParlay and GetSMSDeliveryStatus. SendSMSParlay service implementers advertise this service in the infrastructure service registry, setting parameters and properties such as costs and location coverage.

Potential users of the action services are (1) specific application components, (2) the controller component, and (3) other action services. In order to find action services, action services users should first discover these services with the infrastructure service registry.

Controlling Services

The controlling service allows users of this service to (1) activate event-condition-action (ECA) rules and (2) query for specific instances of context information.

The controlling service supports the following types of operations: subscribe, unsubscribe, query and notifyApplication. Subscribe is used to activate an ECA rule within the infrastructure; unsubscribe is used to deactivate an ECA rule; query is used to select specific context information and notifyApplication is used to notify application components of the occurrence of ECA events. Figure 16 depicts the controlling service.

The definition of specification languages to define ECA subscription characterization, ECA events, context query expression, and context query answer is currently a topic of intensive research.

Figure 15. Action service

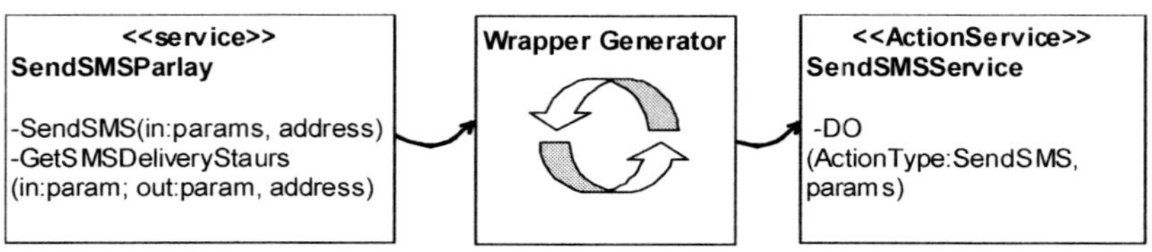

Figure 16. Controlling service

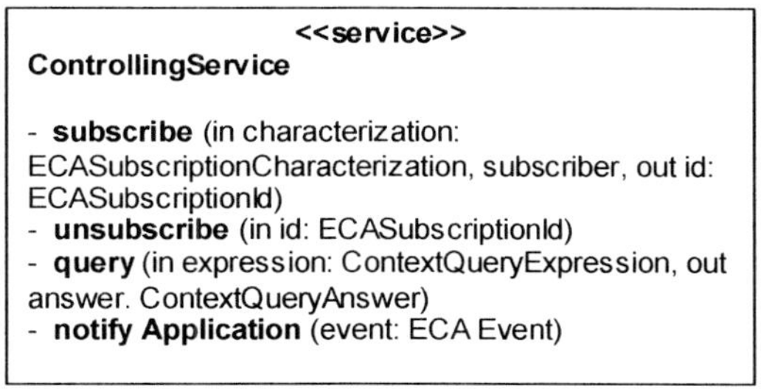

Potential users of the controlling service are application components that would like to activate ECA rules within the infrastructure. Application components may use this service to get event notifications back from the infrastructure.

The Controlling service makes extensive use of the discovery service in order to find context provisioning and action services. An ECA rule could specify, for example, a SendSMS action type with a constraint (cost < 1 Euro) and (coverage in The Netherlands).

RELATED WORK

Various frameworks for developing context-aware applications have been discussed in the literature. The approach presented in Henricksen and Indulska (2004) introduces a conceptual framework and an infrastructure for context-aware computing based on a formal, graphics oriented context-modeling technique called CML (the Context Modeling Language). CML extends object-role modeling (ORM), which uses a *fact* as the basic modeling concept. Modeling a context-aware system with CML involves the specification of fact types, entity types and their relationships. This approach is efficient to derive relational database schemas of context-aware information systems. Although this work provides an effective way to model context, it requires a centralized context repository for context reasoning, which does not satisfy our requirements on distribution and mobility.

Biegel and Cahill (2004) proposes a rule-based sentient object model to facilitate context-aware development in an ad-hoc environment. The main functionality is offered in a tool that facilitates the development process by offering graphical means to specify context aggregation services and rules. Although this approach introduces useful ideas on how to easily configure rules and aggregation services on a sentient object, it is based upon a simple model of context that is both informal and lacks expressive power.

None of the works described previously support the decoupling of context and action concerns under the supervision of a controller component, as we have discussed in our approach. In context-aware scenarios in which the collaboration of various business parties is required, the issues of separation of concerns and dynamic discovery of services need to be addressed.

A survey on context modeling has been presented in Strang and Linnhoff-Popien (2004). From this survey we noticed that many current approaches to context-aware (pervasive, ubiquitous) application development are based on the principles and technologies of the Semantic Web (Berners-Lee et al., 2001; W3C, 2005), namely the use of ontologies represented in OWL and RDF. In particular Chen et al. (2003, 2004b) report on the use of ontologies to represent context information and to provide reasoning capabilities to assert context situations in applications such as a "smart meeting room." Other developments that apply ontologies for building context-aware applications have been reported in Strang and Linnhoff-Popien (2003), Preuveneers et al. (2004), and Wang, Gu, Zhang, and Pung (2000). The main benefit of using ontologies is that general purpose reasoners can be reused for each new application, so that the design effort moves from building application-specific reasoners to defining ontologies and assertions. The potential drawbacks of using ontologies are the intensive processing required by reasoners, which may cause poor performance, and the relatively high costs of developing and validating ontologies. In order to cope with the latter, many ontologies that could be useful for context-aware applications are being made publicly available. SOUPA (Chen et al., 2004a) is possibly the most important initiative in this direction.

CONCLUSION

We have presented in this chapter, current efforts and an integrated approach towards a flexible infrastructure to support the development of context-aware applications. We have discussed (1) important aspects on context modeling, (2) architectural patterns that can be applied beneficially in the development of context-aware systems, and (3) the design of a service-oriented architecture.

Most approaches for context-aware infrastructures described in the literature do not support both context and action concerns, as discussed in this chapter. Decoupling these concerns has enabled the distribution of responsibilities in context-aware services infrastructures. Context processor components encapsulate context related concerns, allowing them to be implemented and maintained by different business parties. Actions are decoupled from control and context concerns, permitting them to be developed and operated either within or outside the services infrastructure. This approach has improved the extensibility and flexibility of the infrastructure, since context processors and action components can be developed and deployed on demand. In addition, the definition of application behaviour by means of condition rules allows the dynamic deployment of context-aware applications and permits the configuration of the infrastructure at runtime.

The hierarchical configuration of context sources and managers has enabled encapsulation and a more effective, flexible and decoupled distribution of context processing activities (sensing, aggregating, inferring, and predicting). This attempt improves collaboration among context information owners and it is an appealing invitation for new parties to join this collaborative network, since collaboration among more partners enables availability of potentially richer context information.

The use of a wrapping mechanism for action services has facilitated the integration of external actions to the infrastructure. This approach avoids permanent binding between an action purpose and its implementations, allowing the selection of different implementations by the infrastructure at runtime.

REFERENCES

Batteram, H., Meeuwissen, E., Broens, T., Dockhorn Costa, P., Eertink, H., Ferreira Pires, L., Heemstra, S., Hendriks, J., Koolwaaij, J., van Sinderen, M., Vollembroek, M., & Wegdam,

M. (2004). *AWARENESS Scope and Scenarios, AWARENESS Deliverable (D1.1)*. Retrieved June 7, 2005, from http://awareness.freeband.nl

Berners-Lee, T., Hendler, J., & Lassila, O. (2001). The semantic Web: A new form of Web content that is meaningful to computers will unleash a revolution of new possibilities. *Scientific American*. Retrieved June 10, 2005, from http://www.scientificamerican.com

Biegel, G., & Cahill, V. (2004). A framework for developing mobile, context-aware applications. *Proceedings of the 2nd IEEE Annual Conference on Pervasive Computing and Communications (PerCom2004)* (pp. 361-365). Los Alamitos, CA. IEEE Press.

Buschmann, F., Meunier, R., Rohnert, H., Sommerlad, P., & Stal, M. (2001). *Pattern-oriented software architecture: A system of patterns*. New York: John Wiley and Sons.

Chen, H., Finin, T., & Joshi, A. (2003). An ontology for context-aware pervasive computing environments. *Knowledge Engineering Review, 18*(3), 197-207.

Chen, H., Finin, T., Joshi, A., Kagal, L. Perich, F., & Chakraborty, D. (2004b). Intelligent agents meet the semantic Web in smart spaces. *IEEE Internet Computing, 8*(6), 69-79.

Chen, H., Perich, F., Finin, T., & Joshi, A. (2004a). SOUPA: Standard ontology for ubiquitous and pervasive applications. *Proceedings of the 1st Annual International Conference on Mobile and Ubiquitous Systems: Networking and Services (MobiQuitous 2004)*, Boston.

Dockhorn Costa, P., Ferreira Pires, L., & van Sinderen, M. (2004). Towards a service platform for mobile context-aware applications. In S. M. Kouadri et al. (Eds.), 1st *International Workshop on Ubiquitous Computing (IWUC 2004 at ICEIS 2004)* (pp. 48-62). Portugal: INSTICC Press.

Dockhorn Costa, P., Ferreira Pires, L., & van Sinderen, M. (2005). Architectural patterns for context-aware services platforms. In S. M. Kouadri et al. (Eds.), 2nd *International Workshop on Ubiquitous Computing (IWUC 2005 at ICEIS 2005)* (pp. 3-19). Miami, FL: INSTICC Press.

Henricksen, K., & Indulska, J. (2004). A software engineering framework for context-aware pervasive computing. *Proceedings of the 2nd IEEE Conference on Pervasive Computing and Communications (Percom2004)* (pp. 77-86). Orlando, USA: IEEE Press.

Kofod-Petersen, A., & Aamodt, A. (2003). A case-based situation assessment in a mobile context-aware system. *Proceedings of the Workshop on Artificial Intelligence for Mobile Systems (AIMS2003)*, Seattle, WA.

Merriam-Webster, Inc. (2005). *Merriam-Webster online*. Retrieved June 7, 2005, from http://www.m-w.com/

OMG Object Management Group. (2000). *Trading Object Services Specification, Version 1.0*. Retrieved June 7, 2005, from http://www.omg.org/docs/formal/00-06-27.pdf

Parlay Group. (2002). *Parlay X Web Services White Paper*. Retrieved June 7, 2005, from http://www.parlay.org/about/parlay_x/ParlayX-WhitePaper-1.0.pdf

Preuveneers, D., Van Den Bergh, J., Wagelaar, D., Georges, A., Rigole, P., Clerckx, T. et al. (2004). Towards an extensible context ontology for ambient intelligence. In P. Markopoulos, B. Eggen, E. Aarts, & J. L. Crowles (Eds.), *2nd European Symposium on Ambient Intelligence (EUSAI 2004) LNCS 3295* (pp. 148-160). Eindhoven, the Netherlands: Springer-Verlag.

Schmidt, A., Beigl, M., & Gellersen, H. W. (1999). There is more to context than location. *Computers and Graphics, 23*(6), 893-901.

Strang, T., & Linnhoff-Popien, C. (2004). A context modeling survey. *Proceedings of the 1st International Workshop on Advanced Context Modelling, Reasoning, and Management (UbiComp 2004).* Nottingham, England.

Strang, T., Linnhoff-Popien, C., & Frank, K. (2003). CoOL: A context ontology language to enable contextual interoperability. In J. B. Stefani, J. Demeure, & D. Hagimont (Eds.), *4th IFIP WG 6.1 International Conference on Distributed Applications and Interoperable Systems (DAIS2003) LNCS 2893* (pp. 236-247). Heidelberg, Germany: Springer-Verlag.

Wang, X. H., Gu, T., Zhang, D. Q., & Pung, H. K. (2004). Ontology based context modeling and reasoning using OWL. *Proceedings of the Workshop on Context Modeling and Reasoning (CoMoRea'04).* In conjunction with the *2nd IEEE International Conference on Pervasive Computing and Communications (PerCom 2004)*, Orlando, USA.

W3C. (2005). *The Semantic Web.* Retrieved June 7, 2005, from http://www.w3.org/2001/sw/

KEY TERMS

Action: A service unit that performs a computation with side-effects for one or more parties involved in the system.

Context: Collection of interrelated conditions in which something exists or occurs.

Context Awareness: Property of a system (including applications) to make use of context information.

Context-Aware Services Infrastructure: Services infrastructure that supports context-aware applications.

Context Information: Representation of context, such that it can be communicated in a system (including applications).

Context Modeling: Activity of creating context information with a representation that supports automated reasoning and/or processing.

Dynamic Customization of Services: (1) Selection of service configuration options (among a predefined set); (2) runtime composition of a predefined set of services.

Event: An occurrence of interest related to context.

Infrastructure: System that comprises common resources and services, such that it forms a shared basis for other and otherwise independent systems (including applications).

Networking Infrastructure: Infrastructure that comprises common resources and services for information exchange (or data communication).

Ontology: Formal and explicit specification of a shared conceptualization.

Rules Description (for Context-Aware Applications): Technique that allows one to specify the behavior of an application in terms of what actions should be taken if certain events occur.

Service: External perspective of a system, in terms of the behavior that can be observed or experienced by the environment (users) of the system.

Service Discovery: Process of finding relevant services according to given criteria.

Services Infrastructure: Infrastructure that comprises common resources and services for application creation, execution and management (hence excluding networking resources and services).

Service-Oriented Architecture: Architectural style, based on the concept of service.

Tele-Monitoring: Process of remotely monitoring an entity (e.g., a human being through an infrastructure).

Chapter XXXII
Middleware Support for Context-Aware Ubiquitous Multimedia Services

Zhiwen Yu
Northwestern Polytechnical University, China

Daqing Zhang
Institute for Infocomm Research, Singapore

ABSTRACT

In order to facilitate the development and proliferation of multimedia services in ubiquitous environment, a context-aware multimedia middleware is indispensable. This chapter discusses the middleware support issues for context-aware multimedia services. The enabling technologies for the middleware such as representation model, context management, and multimedia processing are described in detail. On top of our previous work, the design and implementation of a context-aware multimedia middleware, called CMM, is presented. The infrastructure integrates both functions of context middleware and multimedia middleware. This chapter also aims to give an overview of underlying technologies so that researchers in ubiquitous multimedia domain can understand the key design issues of such a middleware.

INTRODUCTION

With rapid development of wireless communication technologies like mobile data networks (e.g., GPRS and UMTS), it becomes possible to offer multimedia content to people whenever and wherever they are through personal digital assistants (PDAs) and mobile phones. The multimedia content to access can be quite overwhelming. To quickly and effectively provide the right content, in the right form, to the right person, the multimedia content need to be customized based on the user's interests and his current contextual information, such as time of day, user location, and device conditions. These services are called context-aware multimedia services.

Context-aware multimedia services have attracted much attention from researchers in recent years, and several context-aware multimedia systems have been developed. However, building context-aware multimedia systems is still complex and time-consuming due to inadequate middleware support. The application developers have to waste and duplicate their efforts to deal with context management and multimedia content processing. Software infrastructure is needed to enable context information as well as multimedia content to be handled easily and systematically so that the application developers merely need to concentrate on the application logic itself.

In this chapter, we discuss the enabling technologies for the middleware including representation model, context management, and multimedia processing. We also present the design and implementation of a context-aware multimedia middleware, called CMM.

BACKGROUND

Currently, a lot of multimedia applications have been provisioned and used through Internet, such as video conferencing, video-on-demand, and tele-learning. However, with the emergence of mobile devices, people tend to receive and enjoy multimedia content via the devices with them or around them. These trends have led to the emergence of ubiquitous multimedia. Ubiquitous multimedia refers to providing multimedia services in ubiquitous environment through various end devices connecting with heterogeneous networks. For better audio and visual experience, the provisioning of ubiquitous multimedia need to be adapted to the user's changing context involving not only the user's needs and preferences but also the conditions of the user's environment (e.g., terminal capabilities, network characteristics, the natural environment, such as the location and time, and social environment, such as companions, tasks, and activities).

Dey and Abowd (2001) state that context is any information that can be used to characterize the situation of an entity. An entity is a person, place, or object that is considered relevant to the interaction between a user and an application, including the user and the application themselves. Specifically, context in multimedia services can be user preference, location, time, activity, terminal capability, and network condition. Such context-based services are called context-aware multimedia services.

As for context-aware computing, it was first introduced by Schilit and Theimer (1994) to be software that "adapts according to its location of use, the collection of nearby people and objects, as well as changes to those objects over time." Dey and Abowd's definition (2001) states that "a system is context-aware if it uses context to provide relevant information and/or services to the user, where relevancy depends on the user's task."

Context-aware multimedia services are aware of user contexts and able to adapt to changing contexts seamlessly. In a smart-home environment, a context-aware multimedia service might, for example, record TV programs that family members are favorite of, show suitable content based on user social activities (e.g., holding a birthday party), and present content in appropriate form according to capabilities of the displaying device and network connection.

Context-based multimedia services have attracted much attention over the past decade. Traditional multimedia recommendation systems provide recommendations based on user preference, which can be classified into content-based (Yu & Zhou, 2004), collaborative (Resnick, Iacovou, Suchak, Bergstrom, & Riedl, 1994), and hybrid methods (Balabanovic & Shoham, 1997). These systems can be regarded as the early context-aware multimedia systems, though merely based on preference context. Situation context information such as location and time has recently been incorporated with preference context in multimedia recommendation systems (Adomavicius, Sankaranarayanan, Sen, & Tuzhilin, 2005), which has been proven to improve the quality of recommendation. Recently, to deliver personalized multimedia to ubiquitous devices, some researchers have considered both user preference and device/network capability context to generate appropriate presentation to terminals (Belle, Lin, & Smith, 2002). However, none of them deals with all categories of context (i.e., user preference, situation, and capability).

Although Belle et al. (2002) propose a multimedia middleware for video transcoding and summarization, they acquire context through *ad hoc* manner. QCompiler (Wichadakul, Gu, & Nahrstedt, 2002) is a programming framework to support building ubiquitous multimedia applications, which are mobile and deployable in different ubiquitous environments, and provide acceptable application-specific quality-of-service (QoS) guarantees. However, context management is not included. Other multimedia projects towards adaptation include Gamma (Lee, Chandranmenon, & Miller, 2003) and CANS (Fu, Shi, Akkerman, & Karamcheti, 2001).

Many efforts have been specifically devoted to providing generic architectural supports for context management. The Context Toolkit (Dey & Abowd, 2001) gives developers a set of programming abstractions that separate context acquisition from actual context usage and reuse sensing and processing functionality. The Context Fabric (Hong & Landy, 2001) is an open-infrastructure approach that encapsulates underlying technologies into well-established services that can be used as a foundation for building applications. The Solar project (Chen & Kotz, 2004) developed a graph-based programming abstraction for context aggregation and dissemination. Semantic Space (Wang, Dong, Chin, Hettiarachchi, & Zhang, 2004) exploits Semantic Web technologies to support explicit representation, expressive querying, and flexible reasoning of contexts in smart spaces. QoSDREAM (Naguib, Coulouris, & Mitchell, 2001) is a middleware framework providing context support for multimedia applications, which is similar to our infrastructure; however it merely handles location data.

The context-aware multimedia services proposed here take a broad spectrum of context into consideration, which includes three aspects: user preference, situation, and capability. The presented middleware covers wide range of context management functionalities from systematic perspective. Multimedia and context representation model is also described.

REPRESENTATION MODEL

Multimedia and context representation is an important part in context-aware multimedia systems. Since multimedia metadata and context information are often parsed and processed by automated systems interoperating with third-party services and applications, they need to be represented with standard-oriented, flexible, and interoperable models.

MPEG-7 is the *de facto* multimedia description standard which has been widely accepted in industrial and academic communities and popularly utilized in many applications. MPEG-7 Multimedia Description Schemes

(MDS) specify a high-level framework that allows generic description of all kinds of multimedia including audio, visual, image, and textual data.

The MPEG-7 *Creation DS* and *Classification DS* can be used to describe information about the multimedia content, such as the title, keyword, director, actor, genre, and language. This information is very useful to match user preferences and special needs. The *Variation DS* is used to specify variations of media content as well as their relationships. It plays an important role in our context-aware multimedia services by allowing the selection among the different variations of the media content in order to select the most appropriate one in adapting to the specific capabilities of the terminal devices and network conditions. A simple example of multimedia description metadata in compliance with MPEG-7 is shown in Figure 1. The title is "I Guess, Guess, Guess." A brief abstract of the content is provided, and actors or actresses of the TV show are included in the "Creator" field. The "Classification" field specifies the genre and language of the content.

The following example shows the variation description of a media item "Gone With the Wind." It comprises a source video and two variations, a WAV audio and a JPEG image.

Figure 1. A MPEG-7 based multimedia description metadata example

```
<?xml version="1.0" encoding="UTF-8"?>
< Mpeg7 xmlns="http://www.mpeg7.org/2001/MPEG-7_Schema">
<ContentDescription xsi:type="ContentEntityType">
  <MultimediaContent xsi:type="VideoType">
    <Video>
      <CreationInformation>
      <Creation>
       <Title>I Guess, Guess, Guess</Title>
       <Abstract>
         <FreeTextAnnotation>
           This is a recreational programme. The actors are Wu Zongxian, Liu Hanya, and Yang Chenglin.
         </FreeTextAnnotation>
       </Abstract>
       <Creator>
         <Role href="urn:mpeg:MPEG7RoleCS:ACTOR">
          <Name xml:lang="en">Actor</Name>
         </Role>
         <Person>
           <Name>
             <GivenName>Zongxian</GivenName>
             <FamilyName>Wu</FamilyName>
           </Name>
         </Person>
       </Creator>
       <Classification>
         <Genre><Name>AMUSEMENT</Name></Genre>
         <Language>Chinese</ Language>
       </Classification>
      </CreationInformation>
    </Video>
  </MultimediaContent>
</ContentDescription>
</Mpeg7>
```

```
<ContentDescription xsi:type="VariationDescrip
tionType">
  <VariationSet>
    <Source><Video><MediaLocator>
      <MediaUri>file://media1/GoneWithTheWind
      .mpg</MediaUri>
    </MediaLocator></Video></Source>
   <Variation><Audio><MediaLocator>
      <MediaUri>file://media1/GoneWithTheWind
     .wav</MediaUri>
   </MediaLocator></Audio></Variation>
    <Variation><Image><MediaLocator>
      <MediaUri>file://media1/GoneWithTheWind
      .jpg</MediaUri>
   </MediaLocator></Image></Variation>
 </VariationSet>
</ContentDescription>
```

As for context representation, two approaches can be adopted. One is MPEG-21 standard based and the other is user-specified ontology-based.

The MPEG-21 is defined to describe usage environment context from the perspective of the user including user profiles, terminal properties, network characteristics, and other user environments. It also includes user preferences that overlap with MPEG-7. The descriptions on terminal capabilities include the device types, display characteristics, output properties, hardware, software, and system configurations. Physical network descriptions help to adapt content dynamically to the limitation of network. An example of MPEG-21 context description is shown as Figure 2. The terminal has

Figure 2. Example of MPEG-21 context description

```
<?xml version="1.0" encoding="UTF-8"?>
<DIA xmlns="urn:mpeg:mpeg21:2003:01-DIA-NS">
        <Description>
                <UsageEnvironment xsi:type="TerminalType">
                        <TerminalCapability xsi:type="CodecCapabilitiesType">
                                <Decoding xsi:type="ImageCapabilitiesType">
                                        <Format>
                                                <mpeg7:Name>JPEG</mpeg7:Name>
                                        </Format>
                                </Decoding>
                                <Decoding xsi:type="VideoCapabilitiesType">
                                        <Format>
                                                <mpeg7:Name>MPEG</mpeg7:Name>
                                        </Format>
                                </Decoding>
                        <TerminalCapability>
                </UsageEnvironment>
                <UsageEnvironment xsi:type="NetworksType">
                        <Network xsi:type="NetworkType">
                                <NetworkCharacteristic xsi:type="NetworkCapabilityType"
                                        maxCapacity="384000" minGuaranteed="32000"/>
                                <NetworkCharacteristic xsi:type="NetworkConditionType">
                                        <AvailableBandwidth maximum="256000" average="80000" interval="330"/>
                                        <Deley packetTwoWay="330" delayVaration="66"/>
                                        <Error packetLossRate="0.05"/>
                                <NetworkCharacteristic>
                        <Network>
                </UsageEnvironment>
        </Description>
</DIA>
```

the decoding capabilities of both image (JPEG) and video (MPEG). The network capacity and condition are also specified.

Ontology is widely used for context modeling in ubiquitous computing. In the domain of knowledge representation, the term *ontology* refers to the formal and explicit description of domain concepts, which are often conceived as a set of entities, relations, instances, functions, and axioms (Gruber, 1993). Using ontology to model context offers several advantages:

- By allowing users and environments to share a common understanding of context structure, ontology enables applications to interpret contexts based on their semantics
- Ontology's hierarchical structure lets developers reuse domain ontologies (e.g., of users, devices, and activities) in describing contexts and build a practical context model without starting from scratch
- Because contexts described in ontology have explicit semantic representations, Semantic Web tools such as federated query, reasoning, and knowledge bases can support context interpretation. Incorporating these tools into context-aware multimedia services facilitates context management and interpretation

Figure 3 shows a partial context ontology delineating about: (a) user situation context, (b) user preference on media, and (c) the capability of the media terminal. The operating context of the user is captured and is evaluated in terms of location, activity, and time context. The MediaPreference class denotes a user preference on media content by indicating the <*feature*, *weight*> preference pair. Weight, ranging from -1 to 1, indicates the preference level of the corresponding feature. MediaTerminal class refers to device operating capacity in terms of its display characteristic, network communication profile, and the supported media modality.

In ontology-based modeling approach, OWL (Web Ontology Language, http://www.w3.org/TR/2004/REC-owl-features-20040210/) is usually adopted as representation language to enable expressive context description and data interoperability of context. According to aforementioned context ontology, the following OWL

Figure 3. Context ontology

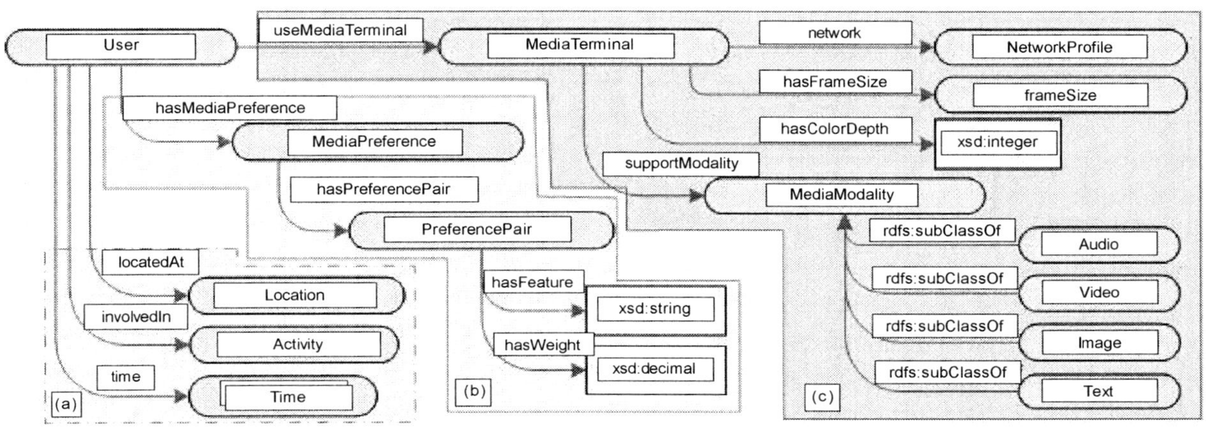

based context markup segment shows that among the many preference pairs of **David**, preference pair **PP1** has preference feature **Sci-Fi** of weight **0.81**.

```
<User rdf: ID="David">
   <hasMediaPreference   rdf:resource ="#Media
    Preference">
      <hasPreferencePair>
        <PreferencePair rdf:ID="PP1">
          <hasFeature>Sci-Fi</hasFeature>
          <hasWeight>0.81</hasWeight>
        </PreferencePair>
        ...
      </hasPreferencePair>
   </hasMediaPreference>
   ...
</User>
```

CONTEXT MANAGEMENT

While a wide variety of heterogeneous context information is ubiquitously available, a common framework is needed to manage context. Context management handles context aggregation, reasoning, learning, query, and discovery.

Context Aggregation

Context aggregation aggregates diversity of context information from an array of diverse information sources. Context aggregation helps to merge the required information which is related to a particular entity (e.g., "user"), or which is of relevance to a particular context aware system (e.g., the entire contexts needed by the smart TV service). It then asserts them into the context knowledge base for further reasoning and learning.

Context can be aggregated from a wide variety of ubiquitous context sensors. For example, the indoor location of a user can be obtained from a RFID (Radio Frequency Identification, http://www.rfid.org) location sensor system, which detects the presence of a RFID tag to conclude the location of the user wearing it, the light level is sensed by X.10 light sensors, and the noise level is obtained from noise sensors.

The acquired sensor data might not provide meaningful context that can be directly understood and utilized by context aware services, thus need to be transformed into context markups. It also provides an abstraction to separate the low-level sensing mechanism from the high-level context manipulation.

Context Reasoning

Higher-level contexts (What is the user doing? What is the activity in the room?) augment context-aware applications by providing summary descriptions about a user's state and surroundings. Although sensors cannot recognize such contexts, they provide information that lets applications infer basic contextual information (Wang et al. 2004b).

Context reasoning infers abstract, high-level contexts from basic sensed contexts, resolves context conflicts, and maintains knowledge base consistency. To support various kinds of reasoning tasks, different inference rules can be specified and preloaded into the appropriate logic reasoner. First-order logic can be adopted for reasoning about contexts. It provides forward chaining, backward chaining, and a hybrid execution model.

Gaia (Ranganathan, Al-Muhtadi, & Campbell, 2004) enables context reasoning to handle uncertainty by using mechanisms such as probabilistic logic, fuzzy logic, and Bayesian networks. Wang, Gu, Zhang, and Pung (2004) adopt logic reasoning to check the consistency of context information and to reason over low-level, explicit context to derive high-level, implicit context.

Context Learning

The context about user preference plays an important role in personalized services. User preference can be obtained through explicit input, which can reflect abrupt interest changes. However it has several disadvantages. First, the user sometimes may be bored with the questionnaires. Second, when watching media content, user rarely expresses his preferences to the system actively. Third, explicit input is "static," and can not adapt to changing user tastes. Therefore, implicit techniques should be adopted to learn user preference gradually and implicitly. Context learning implicitly deduces and updates user preference for certain object consumed in pervasive computing environment (e.g., movie, TV program, restaurant, and etc.). There are many existing machine learning algorithms for this goal (e.g., relevance feedback, genetic algorithm, probabilistic logic, expert system, decision trees, Bayesian networks, artificial neural networks, and Support Vector Machines). However, preference learning for media personalization in pervasive computing is challenging due to the following factors:

- The poor human-machine interactivity of pervasive devices causes user explicitly inputting his preference to be nearly impossible
- The limited computing power and storage make complex learning algorithm impractical
- The mobility of user and devices causes insufficient time to acquire and update user preference. In pervasive computing environments, multiple devices are attached to a user and are used at anytime anywhere. Each of these devices can independently gather user feedback information, but the information may be very fractional

Yu, Zhang, Zhou, and Li (2005) propose a user preference learning approach suitable in pervasive computing environment that uses centralized Master-Slave architecture. It applies relevance feedback and Naïve Bayes classifier approach to compiling statistical analysis on user viewing history aggregated by the context aggregator from all kinds of media playing devices (e.g., PC, television, and PDA).

Intuitively, media content is often viewed by a group of users (e.g., the family, friends in a party, etc.). Therefore, sometimes the common interest of the group users is needed for the purpose of recommendation. The context learning can also deduce the group preference by merging individual user preferences into a common one. Yu, Zhou, Hao, and Gu (2004) propose a user preference merging algorithm that is based on total distance minimization.

Context Query

Context query is responsible for handling expressive query from applications and updating applications with up-to-date query result on a continuous basis. It provides an abstract interface for applications to extract desired contexts from the context knowledge base. Unlike request-response paradigm where an application poses a query and a query engine generates a finite result set, context query allows applications to register logical specifications of interest over changing context and receive streaming results asynchronously.

Furthermore, an application may seek higher-level context (e.g., user situation, room activity) that is not directly available in context KB. In this case, context query engine interacts with context reasoner to derive higher-level context. For instance, an application seeking user's situation context needs to provide context query engine with two parameters–the query statement specifying application's contextual interests and the logic rules defining the desired way to derive user situation. Upon requested, context query engine will add the application's request in its registration table. To answer the query, the engine first triggers context reasoner to generate the inferred result, from which it then extracts user situation by query processing.

As mentioned above, the context can be represented based on MPEG-21 standards and also using ontology. If the context is MPEG-21 mark-ups, XML-oriented query languages can be deployed for context query (e.g., XPath (XML Path Language) and XQL (XML Query Language)). For RDF or OWL represented ontology-based context information, there has been several RDF query languages for the query purposes, such as RDQL (RDF Data Query Language) (Miller, Seaborne, & Reggiori, 2002), RQL (RDF Query Language) (Karvounarakis, Alexaki, Christophides, Plexousakis, & Scholl, 2002), and SPARQL Query Language for RDF (http://www.w3.org/TR/rdf-sparql-query/).

Context Discovery

Context discovery connects context providers and context consumers. It allows applications to locate context they need without knowing the underlying network address.

Service discovery protocols, for instance, UPnP (Universal Plug and Play, http://www.upnp.org), Jini (http://java.sun.com/software/jini), and SLP (Service Location Protocol, http://www.openslp.org/) have been proposed for locating appropriate services in spontaneous distributed computing environment. These protocols can be adopted for local-area context discovery.

In the Context Toolkit (Dey & Abowd, 2001), the list of components (widgets, interpreters, aggregators, and services) is maintained for context discovery. Applications may perform yellow-page looking up to search for a desired component. Solar (Chen & Kotz, 2004) enables publishing and subscribing of context information based on context-sensitive name specification (i.e., set of attributes with tag-value pair pattern). It leverages International Naming System (INS) resolver network so that context is discovered by name pattern matching in the resolver name space.

Moreover, wide-area context discovery across different local space boundaries can also be supported. Our recent work, Orion (Zhang, Chin, & Gurusamy, 2005) provides a scalable architecture for inter-space context discovery beyond smart space boundaries by using peer-to-peer techniques.

MULTIMEDIA PROCESSING

Multimedia content processing is computationally intensive. So it should be done in systematic way other than in an *ad hoc* manner. Multimedia processing framework is proposed to perform multimedia content recommendation, adaptation, and delivery.

Multimedia Recommendation

Multimedia recommendation estimates the *score* for a multimedia item and determines its presentation form in a particular context. It queries context information and multimedia description from the context knowledge base and multimedia metadata base respectively. The recommendation engine can recommend a media item with the highest score, or a list consisting of top L highest-score items for a user's choices. If the selected item has already been prepared with the variation in appropriate form (described in *Variation DS* part of the multimedia metadata), it invokes the media deliverer directly; otherwise it first calls the media adapter to make adaptation, and then triggers the media deliverer.

To date, a large number of recommendation techniques have been developed. Typical recommendation methods can be mainly classified into filtering-based, rule-based, demographic-based, knowledge-based, and utility-based techniques. Filtering-based techniques can be further divided into Boolean filtering, content-based filtering and collaborative filtering (also called social filtering).

Context-aware recommendation should take into account a broad spectrum of context information, ranging from user preference and situation to device/network capability. The recommendation result contains not only what content might be provided, but in which form it should be presented.

It is obvious that a single traditional recommendation method cannot deal with all categories of context. Hybrid recommendation approach might be a possible solution. To improve recommendation performance, traditional recommendation methods have sometimes been combined as hybrid methods (e.g., Fab [Balabanovic & Shoham, 1997]).

Multimedia Adaptation

Multimedia adaptation performs content adaptation by using two techniques: summarization and transcoding. Multimedia summarization means summarizing a content (a video or audio) into a short one by preserving important points of the source content that can allow efficient browsing and navigation. The media time of the summary content depends on the terminal's storage capacity, remainder battery power, and user time constrains. Multimedia transcoding encompasses a wide range of processing mechanisms, namely single object/data type transcoding of video or audio and cross-modal adaptation (e.g., video-to-image or speech-to-text). Being adapted, the content can be processed by a particular device or delivered by a specific network condition. For instance, most handheld computers are not capable of handling video data due to their hardware and software constraints. Transforming video into sets of images or text will enable the devices to access the information contained in the video.

Content adaptation may be performed at various locations in the delivery chain or even distributed across several nodes. Moreover, it may be performed in real time or offline providing a set of variations from which to select at delivery time (Pereira & Burnett, 2003). Advanced video compression standards, such as H.263 and MPEG, can provide the rate control mechanism by changing video frame rate, quantizer, and motion detection threshold, etc. Some other transcoding methods include video rate control, spatial resolution reduction, and heterogeneous video transcoding.

Multimedia Delivery

Multimedia delivery is responsible for streaming or downloading media content to various pervasive devices through different networks. If the recommended media modality is video or audio (i.e., continuous ones), the media deliverer streams the content to terminals. On the other hand, if the modality is image or text, i.e. discrete ones, the media deliverer merely downloads the content. The media deliverer should support a wide variety of video, audio, image formats adapting to different terminal and network conditions.

Media delivery supports transferring content over different networks (e.g., wired Internet, wireless IEEE802, Bluetooth, GPRS, and infrared communication). MPEG-4 and H.263 are widely employed as video codec for media streaming.

CONTEXT-AWARE MULTIMEDIA MIDDLEWARE ARCHITECTURE

We designed CMM, which is based on our representation model, to enable rapid prototyping of context-aware multimedia applications (see Figure 4). It consists of the following components:

- **Context aggregator** aggregates diversity of context information from various sources such as hardware sensors and software programs, converts them to OWL representations, and asserts them into the context knowledge base (CKB). We deployed various hardware sensors in our prototype system, including location sensors, lighting sensors, microphones, and video cameras. We also developed some software programs to capture context (e.g., GUIs for explicitly inputting user preferences and daily schedule); observers for capturing user feedback to a specific content; monitors for detecting terminal capabilities and network characteristics
- **Context reasoner** infers abstract, higher-level contexts (e.g., a user's activity) from basic sensed contexts, resolves

Figure 4. CMM architecture

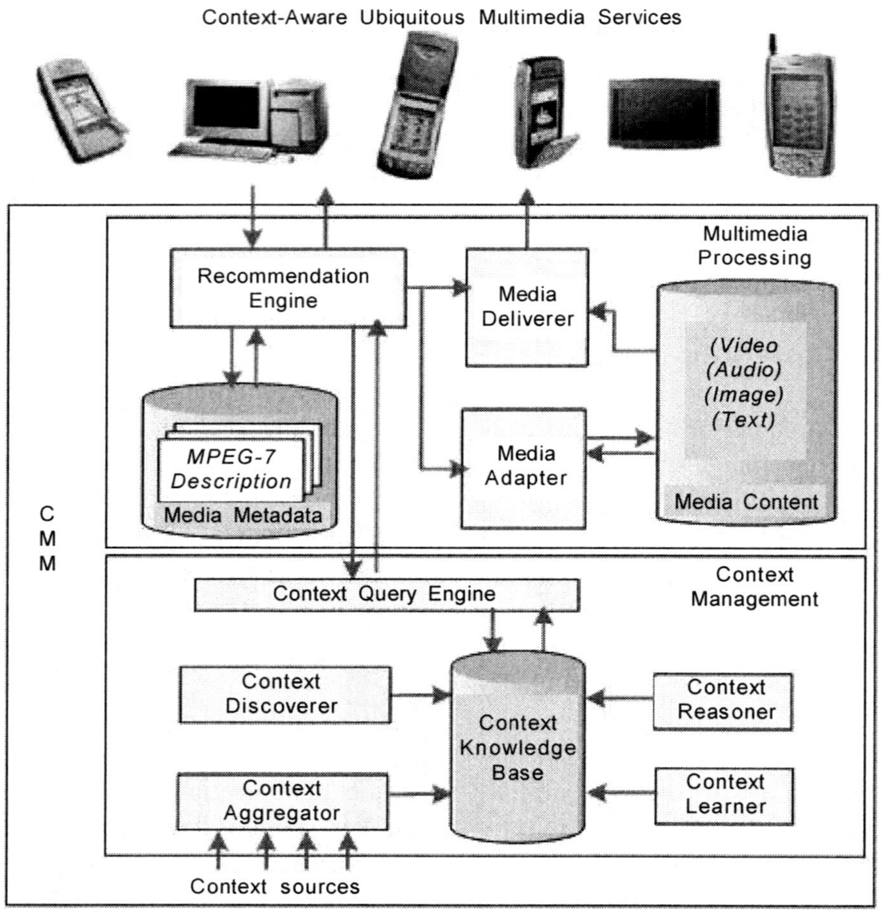

context conflicts, and maintains knowledge based consistency. Our current system applies Jena2 generic rule engine (Carroll, Dickinson, & Dollin, 2004) to support forward-chaining reasoning over the OWL represented context

- **Context learner** deduces user preference of multimedia through implicit machine learning techniques. We have designed a user preference learning approach suitable in pervasive computing environment that uses centralized Master-Slave architecture and applies relevance feedback and Naïve Bayes classifier approach (Yu, Zhang, Zhou, & Li, 2005).
- **Context query engine** handles persistent queries and allows applications to extract desired context information from the CKB. To support expressive queries, we adopt the RDF Data Query Language (Mille, Seaborne, & Reggiori, 2002) as the context query language. RDQL supports querying over semantic models based on triple (<subject, predicate, object>) patterns
- **Context discoverer** ensures the context requesters to appropriately locate the components that can provide the desired and necessary context information. The capability publishing and discovering are based on UPnP
- **Recommendation engine** estimates the score for media content and determines its appropriate form in a particular context. We propose a hybrid recommendation approach (Yu, Zhang, Zhou, Chin, Wang, & Men, 2006) to synergize the content-based, Bayesian classifier, and rule-based methods which are designed to deal with the three kinds of context, i.e. preference context, situation context, and capability context. Both desired content and appropriate presentation form are recommended
- **Media adapter** performs multimedia content adaptation enabling it to be presented in a particular terminal and transferred under a specific network condition. In our current system, the media adapter prepares content variations before delivering. The Power Video Converter (http://www.apussoft.com/) is used to convert between different video formats/frame sizes/frame rates, split video, transform video to audio, and extract images from video
- **Media deliverer** is responsible for streaming or downloading media content to various pervasive devices through different networks. Our current system integrates Java Media Framework and Windows Media Server for media streaming

The CMM supports various context-aware multimedia applications by providing developers with APIs to access its functionalities. Application developers only need to focus on application functions. They need not care how the context is managed and multimedia processing is accomplished.

Moreover, we adopt component-oriented approach in building the CMM. The algorithms or mechanisms are developed as software components. The component-based CMM architecture is pluggable and scalable. It supports efficient lifecycle management of components, ranging from install, start, remove, upgrade to stop actions.

CONCLUSION AND FUTURE WORK

This chapter described the enabling technologies for middleware to support context-aware ubiquitous multimedia services including multimedia and context representation model, context management, and multimedia processing. A context-aware multimedia middleware (CMM) is designed and demonstrated how it can be used to support development and deployment of context-aware multimedia services.

We envision several trends in future research within this topic. First, as the number of devices increases dramatically, device plug-and-play and self-integration feature becomes critical. To minimize user intervention and support device interoperability, device self-configuration, discovery and proper capability announcement are highly desirable. Self-sensing and self-integration mechanisms may fulfil this feature. Second, as the context management and multimedia processing functions increase, the middleware might be the bottleneck in the service chain. To improve performance in terms of throughput and scalability, functional components can be deployed in a network of computers. Moreover, the middleware can be distributed in different spots (e.g., the service provider side, proxy side, or consumer access gateway). It will call for the incorporation of load balancing and task/resource scheduling mechanisms. Third, applications in different domains (e.g., video-conferencing, personalized TV, and tele-learning) may differ in the requirements for context and multimedia. The functions needed among the middleware vary accordingly. Therefore, the middleware platform should support configuration in the setup phase and even reconfiguration at the running time.

REFERENCES

Adomavicius, G., Sankaranarayanan, R., Sen, S., & Tuzhilin, A. (2005). Incorporating contextual information in recommender systems using a multidimensional approach. *ACM Transactions on Information Systems, 23*(1), 103-145.

Balabanovic, M., & Shoham, Y. (1997). Fab: Content-based, collaborative recommendation. *Communications of the ACM, 40*(3), 66-72.

Belle, L. T., Lin, C. Y., & Smith, J. R. (2002). Video summarization and personalization for pervasive mobile devices. *SPIE Electronic Imaging 2002—Storage and Retrieval for Media Databases*, San Jose.

Carroll, J. J., Dickinson, I., & Dollin, C. (2004). Jena: Implementing the semantic Web recommendations. *Proceedings of the 13th International Conference on World Wide Web.*

Chen, G., & Kotz, D. (2004). Design and implementation of a large-scale context fusion network. *The 1st International Conference on Mobile and Ubiquitous Systems: Networking and Services (Mobiquitous04).*

Dey, A. K., & Abowd, G. D. (2001). A conceptual framework and a toolkit for supporting the rapid prototyping of context-aware applications. *Human-Computer Interaction Journal, 16*(2-4), 97-166.

Fu, X., Shi, W., Akkerman, A., & Karamcheti, V. (2001). CANS: Composable, adaptive network services infrastructure. *USENIX Symposium on Internet Technologies and Systems (USITS)*, March.

Gruber, T. (1993). A translation approach to portable ontology specification. *Knowledge Acquisition, 5*(2), 199-220.

Hong, J. I., & Landy, J. A. (2001). An infrastructure approach to context-aware computing. *Human-Computer Interaction Journal, 16*(2-4), 287-303.

Karvounarakis, G., Alexaki, S., Christophides, V., Plexousakis, D., & Scholl, M. (2002). RQL: A declarative query language for RDF. *WWW2002*, May 7-11, 2002, Honolulu, HI.

Lee, Y. W., Chandranmenon, G., & Miller, S. C. (2003). *Gamma: A content-adaptation server for wireless multimedia applications*. White paper, Lucent Technologies.

Miller, L., Seaborne, A., & Reggiori, A. (2002). Three implementations of SquishQL, a simple RDF query language. *Proceedings of the 1st International Semantic Web Conference (ISWC2002)* (pp. 423-435).

Naguib, H., Coulouris, G., & Mitchell, S. (2001). Middleware support for context-aware multimedia applications. *The 3rd IFIP International Conference on Distributed Applications and Interoperable Systems (DAIS2001)* (pp. 9-22).

Pereira, F., & Burnett, I. (2003). Universal multimedia experiences for tomorrow. *IEEE Signal Processing Magazine*, *20*(2), 63-73.

Ranganathan, A., Al-Muhtadi, J., & Campbell, R. H. (2004). Reasoning about uncertain contexts in pervasive computing environments. *IEEE Pervasive Computing*, *3*(2), 62-70

Resnick, P., Iacovou, N., Suchak, M., Bergstrom, P., & Riedl, J. (1994). GroupLens: An open architecture for collaborative filtering of netnews. *Proceedings of 1994 Computer Supported Cooperative Work Conference*.

Schilit, B., & Theimer, M. (1994). Disseminating active map information to mobile hosts. *IEEE Network*, *8*(5), 22-32.

Wang, X. H., Dong, J. S., Chin, C. Y., Hettiarachchi, S., & Zhang, D. Q. (2004). Semantic space: An infrastructure for smart spaces. *IEEE Pervasive Computing*, *3*(3), 32-39.

Wang, X. H., Gu, T., Zhang, D. Q., & Pung, H. K. (2004). Ontology based context modeling and reasoning using OWL. *Proceedings of Workshop on Context Modeling and Reasoning(CoMoRea04)*, In conjunction with the *2nd IEEE International Conference on Pervasive Computing and Communications (PerCom 2004)*, Orlando, FL.

Wichadakul, D., Gu, X. H., & Nahrstedt, K. (2002). A programming framework for quality-aware ubiquitous multimedia applications. *ACM Multimedia 2002* (pp. 631-640).

Yu, Z. W., Zhang, D. Q, Zhou, X. S., Chin, C. Y., Wang, X. H., & Men, J. (2006). Supporting context-aware media recommendation for smart phones. *IEEE Pervasive Computing*, *5*(3).

Yu, Z. W., Zhang, D. Q., Zhou, X. S, & Li, C. D. (2005). User preference learning for multimedia personalization in pervasive computing environment. *The 9th International Conference on Knowledge-Based Intelligent Information & Engineering Systems (KES2005)*.

Yu, Z. W., & Zhou, X. S. (2004). TV3P: An adaptive assistant for personalized TV. *IEEE Transactions on Consumer Electronics*, *50*(1), 393-399.

Yu, Z. W., Zhou, X. S., Hao, Y. B., & Gu, J. H. (2004, September 15-17). User profile merging based on total distance minimization. *The 2nd International Conference on Smart Homes and Health Telematics (ICOST2004)* (pp. 25-32). Singapore: IOS Press.

Zhang, D. Q., Chin, C. Y., & Gurusamy, M. (2005). Supporting context-aware mobile service adaptation with scalable context discovery platform. *IEEE 61st Vehicular Technology Conference (VTC2005-Spring)*.

KEY TERMS

Context: Information that can be used to characterize the situation of an entity.

Context-Aware: Property of a system to adapt its action according to context information.

Context-Aware Multimedia Middleware: Middleware that provides architectural support for context-aware multimedia applications.

Context Management: Context related functionalities, such as context storage, context acquisition, context query, and etc.

Middleware: Software that provides systematic functions for applications.

Multimedia Recommendation: Estimating the degree of a user interested in a multimedia item and determining its presentation form in a particular context.

Ubiquitous Multimedia: Provisioning multimedia services in ubiquitous environment through various end devices connecting with heterogeneous networks.

Chapter XXXIII
Mobility Prediction for Multimedia Services

Damien Charlet
INRIA-Rocquencourt (ARLES Project), France

Frédéric Lassabe
University of Franche-Comté, France

Philippe Canalda
University of Franche-Comté, France

Pascal Chatonnay
University of Franche-Comté, France

François Spies
University of Franche-Comté, France

ABSTRACT

Advances in technology have enabled a proliferation of mobile devices and a broad spectrum of novel and out breaking solutions for new applications and services. In the present, more and more people and companies are demanding mobile access to multimedia services such as real-time rich media. Today, it is necessary to be able to predict adaptation behaviour which concerns and addresses not only the mobile usage or the infrastructure availability, but the service quality especially the continuity of service. Our chapter provides insight to new challenges of mobile multimedia services and applications: Wifi indoor positioning system adapted to heterogeneous building, static and learning mobility prediction, predictive handover policy for multimedia cache management, mobile multimedia guide (such as museum), and network scalability.

INTRODUCTION

The rapid deployment and growth of multimedia applications are increasing with the appearance of new mobile services and new usages. Nowadays, taking advantage of the arrival of large bandwidth of wireless networks, it is becoming more feasible to stream numerous rich media flows towards mobile and terminal devices. However, some bottlenecks subsist when addressing, firstly, the heterogeneity of Wifi covered territories and secondly the intrinsic rich media constraints. We compare mobility to, first of all, a continuous move within a geographical space, and second a discrete space on a logical scale of the diffusion's network (from access point to access point).

This chapter deals with applications handling large size and continuous rich media communication (i.e., audio or video media). Continuous media require the installation of a specific infrastructure of diffusion guaranteeing the delivery periods. We are interested in mobiles implemented within a space provided with multiple access points, with a more or less homogeneous space cover. In such context, it is important the infrastructure reacts rapidly, or use preventive measures during the changes of access point.

Figure 1. Synthetic schema of GUINUMO's platform

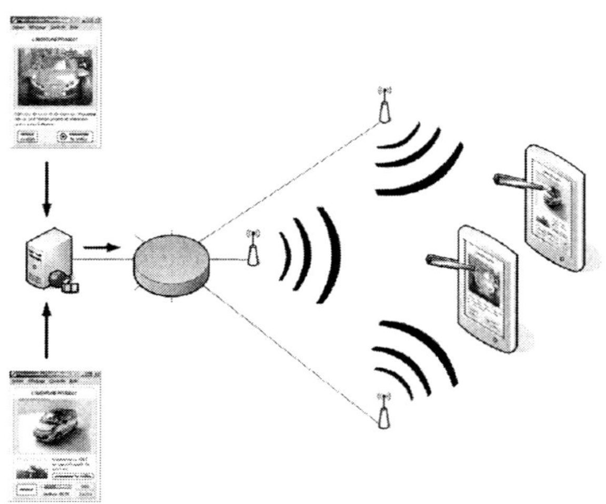

In this chapter, we do not consider the dynamic flow adaptation but rather, we consider already optimized flows dedicated to mobile devices. Thus, whatever the device nature is (pda, tablet pc, etc.), we assume there is a suitable flow adapted to each target. The reader interested in flow adaptation may refer to (Bourgeois, Mory, & Spies, 2003).

To illustrate our purpose, we use GUINUMO, a mobile numerical guide. Such guide demonstrates the accuracy and pertinence of retrieving and making use of both the visual or audio information, and the localization of the pervasive device, during the time-visit of scenarized museums. Within this framework, the media are suited to fit the specific device.

In the sequel, first of all we present the techniques of localization of the devices connected by hertzian way. We further investigate the trilateration technique and evaluate the efficiency of various methods according to several conditions of implementation.

In order to set up a preventive treatment of mobility we show that it is necessary to determine, at least statistically, the future position of a mobile. Then, we describe the methods allowing predicting, in the short run, the position of a mobile. We detail how, thanks to methods of training, it is possible to refine this prediction. In the second part of this chapter, we introduce the concept of cache, as a necessary element in the chain of continuous media diffusion. Caches make it possible to ensure the continuity and the extensibility of the diffusion's infrastructure. We start off by describing the standard methods of managements and co-operation of the caches for continuous media. We proceed by explaining the mechanisms required to manage the change of access point: *handoff*. Then we detail how preventive methods allow optimizing the continuity of flows diffusion. We also present how to integrate these mechanisms in a platform of diffusion and reception (GUINUMO). We describe use-cases of this device. Finally, we conclude with future trends about preventive treatment of mobility. We specify how the coming standards will allow optimizing the handoff and positioning determination mechanisms.

CONSIDERING MOBILITY

Positioning

Positioning is required to provide location-related services. By positioning we mean determining a mobile's geographical coordinates.

Major Positioning Techniques

Global positioning system (GPS). The GPS (Alouini, 1996) is the system allowing locating an apparatus on the surface of the planet or in the atmosphere, using reference satellites whose positions are known. This system has been essential for several years. Thanks to preliminary synchronization of the satellites with the GPS receiver, the latter can compute, based on the time of course between the satellites and the receiver, the distance separating it from each satellite. Trilateration (Fang, 1986) means the intersection of the spheres determined by the distances between the mobile device and each satellite and the receiver's position is calculated. The precision is about ±5 meters for civil applications. The main drawback is that it does not function well enough indoors, in city centres, or in raised or leafy environments.

GSM terminal positioning, standing for global system for mobile communication (GSM), is the main mobile telephony standard in use. This standard has given birth to three positioning methods. For each method, setup and accuracy vary. The methods are the *positioning by cell*, the *computation of the distance according to the signal strength,* and the *distance computation by time difference.*

The positioning by cell is basic. Interrogating the BTS (base transmitting station) is sufficient to identify to which one the mobile is connected. The mobile phone is located in the cover area of the said BTS. This method has a variable accuracy, from 100 meters to several miles (depending on the cell size, itself depending on the mobile phones density). In cases where a service like calling the mobile phone is provided, such accuracy is enough.

The positioning by measuring the signal strength is simple too. Knowing the radio wave weakening according to the distance between the transmitter and the receiver, the distances towards the BTS can be extrapolated by measuring the signal strength. With the distances towards the BTS (which positions are known), the trilateration, exposed in the paragraph concerning GPS, gives the position of the mobile phone. The accuracy obtained is between 50 and 500 meters.

Enhanced observed time difference (EOTD) is a method to estimate the time for the signal to go from the mobile phone to the BTS. This technique is usually employed to adapt the transmission timing of the mobile phones according to their distance to the BTS, thus allowing the scheduling of the packets in the time slots. Without this method, the packets would come to collision near the BTS. The adaptation of the transmission timing is called Timing Advance and requires the synchronization of the BTS and the mobile phones. The distance computation with the time difference is based on EOTD to compute the distance between the BTS and the mobile phone. Knowing the time used by the signal to go from the mobile to the

BTS and the wave speed, we can compute the distance. Then, trilateration gives the position of the mobile phone.

Wifi positioning techniques can be classified into two main categories, the one based on signal *strength cartography*, and the other which determines a relation between signal strength and distance. That makes the location computation possible using trilateration.

Within the RADAR system (Bahl & Padmanabhan, 2000), the mobile terminal positioning uses a signal strength map of the covered area. The geographic coordinates, the signal strength measurements and the mobile orientation are stored in a database. The signal strength map can either be constituted by computation or by physical measurements. The signal strength measurement from each access point is compared with the reference points stored in the database. The cartography-based positioning technique has a 2-to-3-meter precision.

Wang, Jia, and Lee (2003) present a *positioning technique based on a radio wave propagation model.* This model aims at expressing the mathematical relation between the distance from transmitter to receiver and the signal strength. The mathematical expression is obtained by polynomial regression of the third degree. The advantage of this technique is the speed of positioning. However, there is a main drawback. A lot of data are required for the regression to be accurate, which involves a high cost in measurement time. On top of that, it is possible to be confronted with singularities in the buildings where the positioning technique is implemented.

The white paper of Interlink Networks (2002) deals with security issues. Its first objective is to locate rogue mobile terminals and access points which try to infiltrate a network through its wireless part. The authors take signal strength measurements at many locations of many buildings. The results of these measurements are used to establish a radio wave propagation model. This model is based on the Friis relation. The Friis relation expresses the signal strength in function of distance, in a free space environment. The Friis-based model is adapted to fit the conditions of implementation. The precision observed is close to 2 meters. The main advantage of this technique is its setup speed. However, some singular geographic points were observed where the precision was worse than 8 meters. The main drawback of this technique is the unique exponent used in the Friis equation.

The best precision is obtained by the signal strength cartography-based technique. However, it uses lots of resources and computing time to use a signal strength map. A long setup time is also required, and it has no reactivity when topological changes occur. These drawbacks partially affect the polynomial regression-based technique because of the need for data in order to obtain the polynomial expression of the distance. Although its precision is less accurate than that of the previous technique, the technique based on an alternative to the Friis equation is very quick to setup and use. Thus, it is well adapted to topological changes.

Singular points are intrinsic to the topological heterogeneity. Buildings are composed of obstacles which interfere with radio wave propagation. The obstacles can be of various natures (Larnon, 1998) and their layout can be irregular. When facing such unfavourable cases, the signal strength cartography shows better results because it fits the building, whereas the propagation model-based techniques consider the topology uniformly.

GeoMovie

The Interlink Networks (2002) approach is chosen to implement our positioning system. It has indeed the advantages of speed and simplicity. It is interesting with mobile terminals which have little computation power. In Lassabe, Baala, Canalda, Chatonnay, and Spies (2005), we explain the drawbacks of a uniform computation in order to determine the distance according to the signal strength. We first describe the common sources of radio wave distortion and their predominance within a heterogeneous environment. Second, we highlight the radio wave distortion indoor environments with the help of our experiments and we test the model of Interlink Networks, to reveal its limits in a heterogeneous environment.

The Friis equation:

$$\frac{P_R}{P_T} = G_R G_T (\frac{\lambda}{4\pi d})^2$$

where P_R and P_T are respectively the power available at the receiving antenna and the power supplied to the source antenna;

- G_R and G_T are respectively the receiver antenna gain and the transmitter antenna gain
- λ is the carrier wavelength
- d is the transmitter-receiver distance

The Friis equation expresses the signal strength loss in function of the distance d. The radio wave absorption by obstacles is similar to the free-space loss but it is generally greater.

We use a reciprocal expression to the Friis equation to determine a value replacing the square of the distance adapted to the environment where the positioning will be achieved.

Figure 2. Geolocation based on trilateration technic

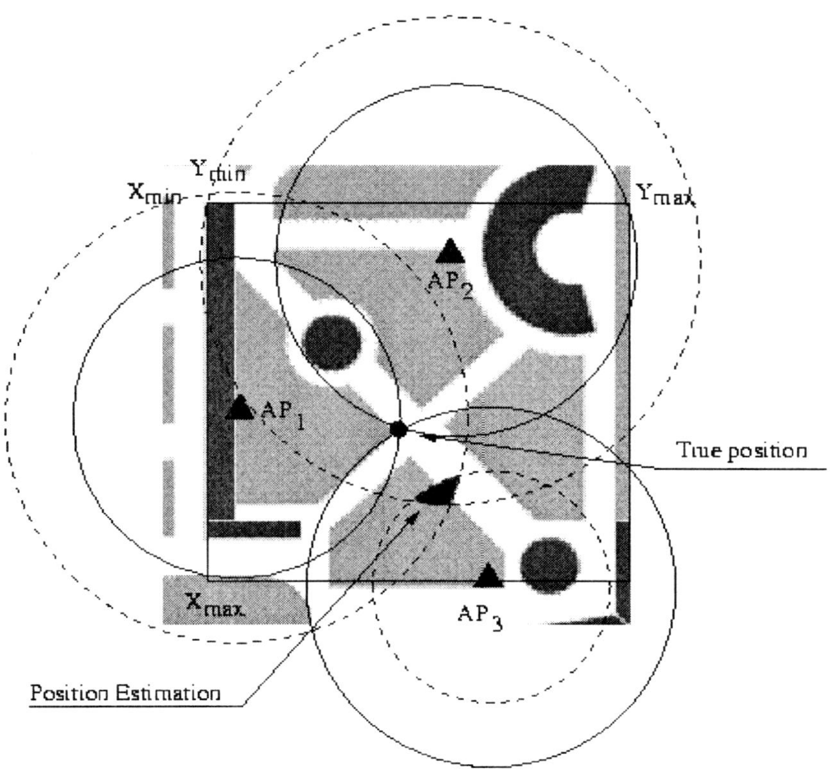

Sample measurements are used to compute the value adapted, which is different for each access point. Then, we use the expression of the distance according to the signal strength, taking into account the new coefficient in the such-modified Friis equation. Trilateration for itself is achieved by an algorithm trying to minimize the distance between the circles centred on the access points and whose radius are the distances towards the mobile terminal (see Figure 2).

The radio waves are affected by the presence of topological components altering the radio waves trajectory and therefore modify the signal strength. The phenomenon we are more likely to observe is wave reflection. The most common sources of wave trajectory distortion are metal equipment that induces huge signal reflections, preventing it from reaching areas theoretically within range. Devices functioning at frequencies close to Wifi frequencies also distort the signal by covering it with great noise.

Predictive Positioning

Knowing a mobile terminal position is not sufficient to ensure service continuity. Mobility prediction is required to do so. Predicting the mobile terminal movements requires modelling the trajectories of the mobile terminals. We present two mobility models. The first one models the *trajectories by the movement vectors* of the mobile terminals. It is a static method in which a simple trajectory equation is used. The second one is based on the *learning of the mobile terminal movements*. It is a dynamic method which can be updated when the users change their movements habits. In particular, we present the hidden Markov model (HMM) (Rabiner, 1989). These learning models can be extended by taking into account some mobility patterns.

Trajectory Calculation

In a project to increase routing efficiency in *ad hoc* networks (Lee, Su, & Gerla, 2000), mobility prediction is used. A mobile terminal trajectory at a given time is modelled by the direction and the coordinates of the speed vector of the mobile terminal trajectory. These data are sufficient to anticipate the future position of a mobile terminal. They also allow estimating the remaining connection time between two mobile terminals.

Simulations based on this model show a growth of the routing efficiency. However, the experiment hypotheses are not always realized in real conditions. The hypotheses are a positioning of the mobile terminal without error and the uniformity of its trajectory. If the mobile terminal follows a complex course, the routing loses its efficiency.

Learning from the Movements

Another mobility management method is based on the learning of the mobile terminal movements. The learning can be based on several models such as Markov model or automata. Learning the moves allow to compute transition probabilities from state to state. It is interesting because the states can be physical areas or logical states such as the dependency from a multimedia cache or the connection to a base station (in GSM, or access point in Wifi).

Use of Hidden Markov Model

In the article from the University of Liège (François, Leduc, & Martin, 2003), a mobility prediction model for mobile networks is presented. It is based on the hidden Markov model. The transition probabilities between the states are determined by a learning system. During the learning, the mobile terminals regularly send their position to the base stations. The base stations save the transitions from a state to another and compute the transition probabilities.

Simulations without positioning errors are used. The model accuracy is sufficient to predict the mobile terminal next movements. In further work (François, Leduc, & Martin, 2004), an error is added to the mobile terminal positioning. The results are less accurate but still good. When using the first two results, the prediction is 75% right with a 15% noise on the signal strength and five observations to compute the next state. These conditions are the worst tested.

Mobility Patterns

Mobility patterns are built from logs of several visits. They are used to identify classes of equivalence. Based on such identified classes of equivalence, it is possible to characterise some visitors' behaviour by attaching to visitors (Chardonnel & Van Der Knaap, 2002), and during their visit, the classes matched with behaviour logs, for example. Then, when a user is attached to one (or more) classe(s), it becomes easier to be accurate when predicting where they are going to be located and which media he is going to ask for.

Mobility patterns matching is made on the fly during visits. Thus, as the behaviour of a visitor may change during their visit, class matching must provide a mechanism for identifying new behaviour and adapt any predictions which concern either the next geolocation, or the next interaction with the mobile terminal, or else solicitation of new media. Mobility patterns are build wether online or offline, using the log file of the system.

STREAMING FOR MANY MOBILE USERS

We have seen, in the previous section, that it is possible to locate a mobile terminal in the physical space. It is achieved with accuracy depending on several criteria. The knowledge of the position also involves a possibility to predict the close future moves of the mobile terminal. The best methods are actually based on movements learning. Both the positioning and the mobility prediction allow the service continuity.

Streaming video to many mobile users brings problems which may be categorized in two classes. Firstly, the huge amount of data which has to be sent, blocks the network and the servers, and decreases the interactivity. And secondly, the tight real time requirements of video streaming cannot cope with the latency induced by the distance of clients moving away from the source.

The first class of problems may be solved by using a cache (Cao & Irani, 1997). A cache may be seen as an empty server at time 0. It may transmit the objects it possesses, like a server. Its particularity resides in the fact that it may request objects from other servers or caches and store them temporarily to serve them later. Caches are placed near the clients to reduce the length of the paths and therefore enhance the interactivity and avoid bottlenecks. Moreover, to cover wide areas and serve many users, several caches may cooperate in a set of distributed and cooperating caches.

Caches are managed by several policies: the role of the insertion policy is to decide when and which new documents should be stored on the cache, the removal policy is dedicated to the cleaning of the storage space, the admission policy considers if we can accept a new client, and the aim of the sibling (or cooperation) policy is to take care of the content of the neighbouring caches in a distributed system. Many examples of such policies may be found in the literature (Balamash & Krunz, 2004), some of them are dedicated to specific content such as video (Podlipnig & Boszormenyi, 2002; Rejaie & Kangasharju, 2001). However, none of these policies cope with the problems related to the mobility of the clients.

Mobility in Video Caches

Taking into account the mobility of the users brings new constraints to the field of caches. These not only store and serve the data according to the requests they receive any more, but they must also manage the mobility of the clients. Indeed, the usual operation of the caches is based on the observation of the requests they receive. In a traditional scheme, the topology of the served clients remains fix, and the uses similar along the time. It is thus enough for the caches to adapt their content and their operation to be effectively useful. In a mobile context, as the clients are regularly moving from one zone to another, topology is always fluctuating. A client used by a cache at time T may be out of reach at the next time step.

The integration of the mobility in the caches is a recent problem, and few solutions are exposed in the literature. Two families of solutions may be distinguished: an optimized cooperation of caches, which may imply a change of cache and a context switch where the caches become mobiles in order to follow displacements of the clients.

In Hadjiefthymiades and Merakos (2001), the authors place themselves within the framework of a cellular network in which mobile clients consult Web sites. In order to optimize the navigation of the clients as well as possible, the authors propose to insert a proxy-cache for each client within the system. This one is placed close to the user, at the level of the base stations, and stores a small set of Web pages and pictures useful for the navigation of the client. Its contents being limited, it can, at lower cost, move to follow the displacements of the client. This technique is interesting but requires that the whole end-side material (access point or base station) should be equipped with sufficient capabilities of calculation and storage. Moreover, the migration of a cache can raise the cost of the network if the stored documents are bulky, as within a video framework.

Some projects descend the level of the cache since it is placed on the client himself. Thus in Cohen, Herscovici, Petruschka, Maarek, and Soffer (2002), the authors use a cache placed on the mobile device, which is in charge of fetching a set of pages related to the navigation in progress, as soon as possible. The authors of Sailhan and Issarny (2003) present a system of caches for mobile clients functioning in *ad hoc* mode. This system is particularly useful for the users exploiting their wandering peripherals in zones where many other peers are. The example given is that of a museum offering a virtual visit thanks to personal assistants. In this architecture, the PDAs obtain their information on a server via access points disseminated through the museum. It may happen that no access point is within the range of a client. If a networked resource is to be reached, it is possible for this client to connect in *ad hoc* mode to terminals available in its neighbourhood, and to launch a strategy of exchange of cached information. This strategy is very interesting since the caches are placed on the same level as the clients, and the cooperative mechanism makes it possible to optimize the use of the bandwidth. However, it requires the use of peripherals having a significant storage capacity, and those must incorporate the cache program, which is not always possible.

Within the framework of video streaming, the mobility of the cache is hardly possible. Indeed, the volume of transmitted data being very significant, a cache migrating from one place to the other implies a very high cost of utilization. The mobility of the users have thus to be treated by optimized policies of inter-cache co-operations.

Handoff Policy to Tackle Mobility Between Caches

An optimized policy of cooperation of caches for the mobility of the users has a main function: to envisage the change of cache of the users because of their mobility. It is indeed useless to continue diffusion between a cache and a client which would have moved away too far from the former, when another more optimal cache could take over.

When a set of distributed caches is deployed, a paving of the territory in zones of diffusion can be imagined. The policy of cooperation must then supervise displacements of the client in order to detect a change of zone. This roaming from one zone to another may be compared to the change of terminal in mobile telephony or the wireless networks. One will then speak of handoff.

The first basic policy consists in carrying out no action whatsoever in any case. This corresponds to what currently occurs when no policy is implemented. The problem that arises is the lack of reactivity in the event of a handoff. Indeed, when a user looking at a film changes zones, the cache, which must take over, does not have the sequence requested and was not advised of a possible arrival of the request. The diffusion must then stop while the part being streamed is fetched in the new cache.

The second basic policy (called "broadcast") consists in contacting all the neighbouring caches to advise them to fetch the currently streamed sequence. The defect of reactivity is thus smoothed out, each cache keeping the sequence in the event a user should arrive. However, this solution is not satisfactory because it quickly causes a clogging on the network between the caches and it uses disk space of the caches for sequences which have few chances to be used.

In order to optimize the transition when a client moves, thus causing a change of zones, the caches must cooperate by exchanging information with their "neighbours" and by prefetching part or whole of one or several sequences. An optimized management should thus define a set of caches sufficient at the same time to ensure a strong probability of hit while minimizing the disk space and the lost bandwidth.

Use of the Prediction of Position in Video Caches

To succeed in treating the handoff (i.e., the change of cache in a short time), compatible with the temporal constraints of the diffusion, and also by mobilizing only the necessary resources in order not to disturb too much the diffusion towards the other users, the system cannot be only reactive. The system of caches needs to anticipate this possibility and the potentially concerned caches should be all set to stream before the user changes zones. In addition, only a minimal number of caches should be included in this preventive measure in order not to put the scalability of the system in danger.

We propose to prefetch, in a reduced number of caches, the continuation of the sequence in the course of visualization of the users who are likely to change zones. With this purpose in mind, our technique is based on an observation of the mobility of the users, on a prediction of their short-term position and on the adequacy of each cache to serve the zones having to receive new users.

Our handoff policy is based on external information to anticipate the necessity of prefetching a cache. This information is of two types: the probability for a zone to receive a client in an immediate future, and the adequacy of the connections between the caches and the zones of reception. From this information we calculate an indicator evaluating the need to fetch the sequences.

When a client is in a given zone, we need to get the set of the probabilities of presence of the client in all nearby zones, at the next moment. Our policy is based on two view points resulting from the observation of former displacements: the observation, on the one hand, of the former moves carried out by the whole set of clients and, on the other hand, of those of the particular client. In order to refine the behaviour of the system, these two approaches are balanced thanks to an exponential average. We thus obtain a probability of presence for each close zone. At the end of this calculation, we present this set of probabilities in the form of a vector called *Vp* which dimension is the number of close zones plus 1 for the zone in progress. This last value, noted P0, represents the probability that the user will remain in the zone.

The adequacy between a cache and a zone is an indicator which quantifies the effectiveness with which a cache can stream a sequence towards a zone. The cache-zone adequacy does not take the availability of a particular

Figure 3. Computation of the relevant pre fetching vector

$$M_{(adequacy)} \times \vec{V}_p = \vec{V}_e$$

$$\begin{pmatrix} Cache_0 Zone_0 & \cdots & C_0 Z_z \\ \vdots & \ddots & \vdots \\ C_c Z_0 & \cdots & C_c Z_z \end{pmatrix} \times \begin{vmatrix} P_0 \\ \vdots \\ P_z \end{vmatrix} = \begin{vmatrix} e_0 \\ \vdots \\ e_c \end{vmatrix}$$

Figure 4. Territory partitioning in streaming zones

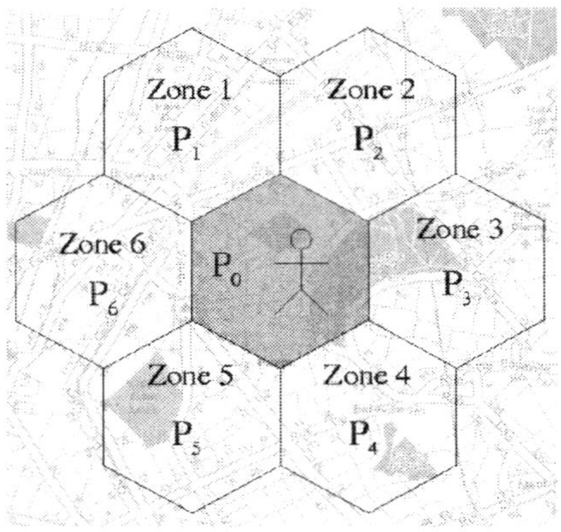

sequence in a cache into account, but qualifies the quality of the connection. A simple example of quantifier is the average Round Trip Time between a cache and a zone. It is however necessary to integrate the variations of quality in the course of time, and to recompute this value regularly. These factors of adequacy are represented in the shape of a matrix binding each cache potentially useful for the diffusion to each zone, built in a distributed way thanks to exchanges between the caches.

That still remains to quantify the relevance of caches fetching. The columns of the matrix, representing the zones, are sorted in such way as to correspond to the lines of the vector *Vp*. Then, the vector *Ve* associating to each cache a quantity, which is the sum of the products of the cache-zone adequacies by the probability of presence of the user in the zone, is defined by the product of the matrix by the vector *Vp*. The resulting values of the vector are in a range from 0 to 1. Indeed, the vector of probability of presence is a probability distribution whose sum is equal to 1. The coefficients of the matrix of adequacy are, by construction, all ranging between 0 and 1. Thus, the sum of the products of the ones by the others also lies between 0 and 1. The larger the values of the resulting vector are, the most suited for streaming to a zone having a large probability of receiving a client a cache is.

Since we have an evaluation of the interest of fetching for each cache, we still have to define the level needed to load this cache. This threshold can be modulated in order to define the level of aggressiveness of the policy. For example, in order take a multi-layers stream into account (a basic stream and a complementary stream), a first variable HS = P0 * threshold will give the value of swing beyond which the entirety of flow will be charged, and a variable $Sb = a * HS$ makes it possible to define a range within which only the basic stream will be transferred. Lastly, if the evaluation is lower than Sb, nothing will be charged.

This method makes it possible to improve the fluidity of the diffusion towards mobile clients by being based on a short-term prediction of their future positions. Thanks to the dynamic construction of a matrix of adequacy between a set of caches and a set of zones of diffusion, and with the use of the probabilities of presence, we deduce an algorithm revealing the state in which the system will have to be at moment $T+1$ will have to be. Analysing this algorithm will allow us to set up a policy of total or partial prefetching of the sequences in the neighbouring caches.

GUIDE NUMÉRIQUE MOBILE: GUINUMO

We have a demonstrator of the techniques we presented above. This product, called GuiNuMo, allows the assisted visit of the "Museum of the Peugeot Adventure." Our tool allows the geolocation of the visitor within the museum and the on-demand streaming of audio-visual content related to the environment they will come across. In Figure 4 you can see three GuiNuMo screenshots on a PDA. The picture on the left is that of the beginning of the visit; one discovers the plan of the museum and the position of the visitor, near to the entry. On the map of the museum the various zones of diffusion are shown. They are associated to themed elements of the museum (concept car, competition...). The screenshot in the middle presents the GuiNuMo interface when the user enters the zone called « concept car ». The presentation of a specific vehicle is offered and it is possible to play a video. The video, if requested, will be streamed from the cache dedicated to this zone.

During the whole visit, the position of the peripheral is computed in real time, and the available content is provided in the form of play lists that the user can choose to visualize or not. When it is possible, GuiNuMo suggests the visitor to go to a different area of the museum in order to discover objects in relation to the media he has just consulted. For example, at the end of a video depicting the "Peugeot 306 CC," GuiNuMo will suggest going to the part devoted to the "Peugeot 401 coupé transformable" which was, in 1933, the first car integrating a retractable roof. See screenshot 3 for a graphic representation.

All the actions of the visitor are recorded by the system in order to be able to analyze the visit and to feed the training algorithm. Moreover the visit can be studied by the staff of the museum to deduce some potential transformations. Indeed by analyzing the logs of the visits of multiple visitors, it is possible to note that few people visit certain sites in the museum or that a particular zone usually holds greater attention. Thus, the team of the museum may choose to rearrange spaces or to make recommendations in order to enlight under exploited spaces.

Figure 5. Three screen shots of GUINUMO application interface

The training algorithm makes it possible to determine, statistically, the probability of visualization of each media according to the actions carried out previously. Consequently, it is possible to anticipate which media will probably be visualized soon. It is thus possible to prepare the system to allow a better reactivity.

Right now, the prediction is carried out as a whole. We are currently working on the identification of the profile of the users in order to obtain a prediction of better quality. Indeed, each visitor does not visualize all the available media, but operates a themed selection set. One can make a difference between the visitors interested in technology and those interested in the history of the Peugeot Company. By analyzing the first documents requested by the user, it is possible to determine to which group of users they belong and to subsequently direct the prediction and the recommendations according to the actions carried out previously by other users of the same class of equivalence.

CONCLUSION AND FUTURE TRENDS

In this chapter, we described the interest of the predictive methods during the broadcast of multimedia flows towards mobile terminals. These methods are integrated in location services as predictions and in the architecture as caches. Caches allow the scalability of the system and the storage of data to send to the client into the future probabilistic areas.

The location predictions based on the knowledge, allows the anticipation of the possible or probable areas of location. This method is included in the cache loading strategies as well as in the establishment and continuity of the terminal connection. The union of these two mechanisms allows a fast and accurate management of the access point migration when a terminal moves from one zone to another.

The intercache communication methods, called "cache sibling," are well-known algorithms. They make it possible to manage the content deletion in the distributed caches. They require to be adapted, to take into account the specificity of the video caches. Indeed, a video cache does not maintain strong coherence. It has to communicate with its "neighbours" to facilitate the continuity of the sequence broadcast.

The localization principle used in optical or hertzian ways is rather old. It is well adapted to outdoor spaces (e.g., on seas). It is possible to use two great classes of algorithms: the triangulation or the trilateration. For clients, in Wifi-like networks, we do not have access to angular information; it is thus the trilateration which is used. In the case of the indoor localization, corrections must be made to the coefficient of the Friis equation in order to take the absorption of the wave by the building elements into account.

The prediction methods of the mobile location can be based on two types of information: a topological knowledge of the area, allowing the identification of relevant trajectories, and experiences of the previous movements deduced from a position list, the future most probable positions. The experience can be limited to previous moves of the client which receives the flow or can combine information resulting from the movements of all the users having crossed the considered area. A relevant method consists in mixing general and individual information.

The GuiNuMo1 application which was presented in this chapter to illustrate our methods is installed in a museum for demonstration purposes. This experimental platform enables us to work on real scenarios and to observe the reactivity of the system. The collected information tends to ascertain the hypothesis that mobile multimedia applications are more and more accepted by users, and drive us to consider implementing new features.

Using caches as the broadcast infrastructure of multimedia flow is a significant point. We are now thinking of installing caches into the client device and of sharing content of these caches in an *ad hoc*, peer-to-peer network. This method should allow under-dimensioning, possibly ever to remove the cache from the broadcast infrastructure, in order to rely strictly the mobile devices. In case of numerous clients, the requirement of caches is important. The available space and the density of redundant data are linked to the number of clients which will allow scalability.

The vertical handoff is a research topic currently studied by a great number of international teams. It is the opportunity to switch from a network to another, keeping a connection continuity of the current applications. Basically, the IP protocol does not offer such service, and mobileIP allows only a discrete mobility, which means without any continuity on the current connections. To achieve a goal of continuity, it is necessary to adapt or modify the TCP/IP stack of the terminal and the servers. It is also possible to modify the TCP/IP stack of a network device near the client as the access-point, in place of the servers, in order to manage this problem. The vertical handoff must integrate a policy of network selection. In fact, it is important to select the right network, which means the network with the best bandwidth. But, with this strategy, we do not have the opportunity to choose the best available throughput. In order to do so, it is necessary to dynamically choose the best network, causing additional handoff in case of network overloads. Moreover, it will be possible to connect few networks at the same time in order to have a sum of download or upload throughputs using a multi path strategy.

The step after the vertical handoff will study the management of network failures mainly due to white spots using strategies of mobile perfecting. This technique will allow to load in advance some documents during the connected times in order to deliver them latter even if a failure network occurs during the request of view.

REFERENCES

Alouini, M. (1996). Global positioning system: An overview. *Tunisian Scientific Magazine, 10*(1), 49-51.

Bahl, P., & Padmanabhan, V. N. (2000). RADAR: An in-building RF-based user location and tracking system. *Proceedings of INFOCOM, 2,* 775-784.

Balamash, A., & Krunz, M. (2004). An overview of Web caching replacement algorithms. *IEEE Communications Surveys and Tutorials*, *6*(2), 44-56.

Bourgeois, J., Mory, E., & Spies, F. (2003, November). Video transmission adaptation on mobile devices. *Journal of Systems Architecture*, *49*(10-11), 475-484.

Cao, P., & Irani, S. (1997, May). *Cost-aware WWW proxy caching algorithm* (Tech. Rep. No. CS-TR-1997-1343). Madison, WI: University of Wisconsin.

Chardonnel, S., & Van Der Knaap, W. G. M. (2002). Managing tourist time-space movements in recreational areas: A comparison between two areas with the same analysis methodology for a protected nature park in the French Alps and the Dutch National Park "De Hoge Veluwe. *Revue de Géographie Alpine, Tome, 90*(1), 37-48.

Cohen, D., Herscovici, M., Petruschka, Y., Maarek, Y. S., & Soffer, A. (2002). Personalized pocket directories for mobile devices. *Proceedings of the 11th International Conference on World Wide Web* (pp. 627-638). ACM Press.

Fang, B. T. (1986). Trilateration and extension to global positioning system navigation. *Journal of Guidance, Control, and Dynamics*, *9*(6), 715-717.

François, J. M., Leduc, G., & Martin, S. (2003). Evaluation d'une méthode de prédiction des déplacements de terminaux dans les réseaux mobiles, Réseaux mobiles et ad hoc, qualité de service, test et validation, ingénierie du traffic. *Special issue of Hermès Lavoisier* (pp. 189-202).

François, J. M., Leduc, G., & Martin, S. (2004). Learning movement patterns in mobile networks: A generic method. *European Wireless 2004* (pp. 128-134).

Hadjiefthymiades, S., & Merakos, L. (2001, May). Using proxy cache relocation to accelerate Web browsing in wireless/mobile communications. In *WWW'10 Conference* (pp. 26-35).

Interlink Networks, Inc. (2002). *A practical approach to identifying and tracking Unauthorized 802.11 Cards and Access Points.* Technical Report. Retrieved from http://www.interlinknetworks.com/graphics/news/wireless_detection_and*_*tracking.pdf

Lassabe, F., Baala, O., Canalda, P., Chatonnay, P., & Spies, F. (2005). A Friis-based calibrated model for WiFi terminals positioning. *Proceedings of IEEE WoWMoM* (pp. 382-387).

Lee, S. J., Su, W., & Gerla, M. (2000). Mobility prediction in wireless networks. *Proceedings of IEEE ICCCN* (pp. 22-25).

McLarnon, B. (1998). *VHF/UHF/microwave radio propagation: A primer for digital experimenters.* TAPR's Spread Spectrum Update, Tucson Amateur Packet Radio Corporation. Retrieved from http://www.raveontech.com/ApplicationNotes/Primer.pdf

Podlipnig, S. & Boszormenyi, L. (2002). Replacement strategies for quality based video caching. *IEEE ICME'02, 2,* 49-52.

Rabiner, L. R. (1989). A tutorial on hidden Markov models and selected applications in speech recognition. *IEEE, 77*(2), 257-286.

Rejaie, R., & Kangasharju, J. (2001). Mocha: A quality adaptative multimedia proxy cache for Internet streaming. *ACM NOSSDAV'01* (pp. 3-10).

Sailhan, F., & Issarny, V. (2003). Cooperative caching in *ad hoc* networks. *The 4th International Conference on Mobile Data Management* (Vol. LNCS 2574) (pp. 13-28).

Wang, Y., Jia, X., & Lee, H. K. (2003). An indoors wireless positioning system based on wireless local area network infrastructure. *The 6th International Symp. on Satellite Navigation Technology Including Mobile Positioning & Location Services* [CD-ROM, paper 54].

KEY TERMS

Ad Hoc Mode: Every client can talk to each other on a peer-to-peer basis.

Admission Policy: Algorithm used when a new client wants to fetch data from a cache to decide if the cache has sufficient capabilities left to serve him.

Cache: A cache gathers the functions of a server and of client. It takes place between them and can store and deliver popular documents. Being near the client, it helps resolve the problems of bottlenecks and increase reactivity.

Cooperation Policy: Used between distributed caches to cooperate and share data.

Handoff: Name of the mechanism which takes place when a user is roaming.

Insertion Policy: Algorithm of caches, decides which documents should be stored.

Mobile Network: Network in which part or all of the components are mobile.

Mobile Terminal: Every apparatus light-enough to be humanly transported and with embedded computation power, like laptops, PDA, new generation mobile phones.

Mobility: The action to move. We are interested in particular in the logical mobility (change of network, BTS, etc.) triggered by the geographical move (the action of changing physical coordinates in space).

Prefetch: Inserting documents into a cache in the hope that they are going to be requested in a near future to reduce start latency for the user. Contrarily to normal insertion, it is not triggered by clients.

Removal Policy: Management algorithm of caches, decides which stored documents should be deleted to make room for new and more popular documents.

Roaming: Action of a human moving from one zone to another.

Service Continuity: Property of a service over a mobile network. When continue, a service is not interrupted by changes in its logical position (change of AP / BTS). For example in the GSM standard, as long as you stay in covered areas, phone conversations are not interrupted when you change your BTS.

Sibling: Exchange of data between two caches.

Signal Strength: It is the power of the signal measured.

Start Latency: Time elapsed between the moment where a user requests of a document and the time it is displayed on its peripheral.

Streaming: Technique of transfer in a continuous flow to allow the display of the media while downloading.

Video Cache: A cache with specific policies, optimized for the delivery of video data.

ENDNOTE

1 GuiNuMo is a project funded by: EU, french ministry of research, Franche-Comté Council and CAPM.

Chapter XXXIV
Distribution Patterns for Mobile Internet Applications

Roland Wagner
Johannes Kepler University Linz, Austria

Franz Gruber
RISC Software GmbH, Austria

Werner Hartmann
FAW Software Engineering GmbH, Austria

ABSTRACT

After the enormous success of the internet and mobile networks, the next upcoming boost for information technology will be the combination of both. But developing applications for this domain is challenging, because first, most mobile devices provide only small memory and processor footprints, prohibiting resource intensive code at client side and second, mobile networks offer only limited bandwidth, and the probability to connection losses is relatively high compared to wired networks. Selecting the appropriate software architecture in terms of distributing the functionality of the system between server and client device is crucial. Application distribution patterns, known from conventional system development, are analysed for their applicability for the mobile environment. After the more abstract analysis of the patterns, the IP multimedia subsystem (IMS) which is part of the current specification of 3G mobile networks is introduced and its support for different application distribution patterns is examined.

OVERVIEW

The success of mobile applications strongly depends on optimal utilization of client, server, and network resources. The distribution of the application functionality between client and server has strong impact on the grade of the resource utilization. Therefore, we present a schema for application distribution patterns and analyze architectural locations where an application can be distributed and in this chapter we move our focus to the inherent problem of mobile applications to keep the data on the device and an the backend consistent.

With these distribution patterns, we will analyze several approaches for mobile Web access. In the last part of the chapter, we introduce an advanced architecture for representing mobile multimedia Web content: the IMS (IP multimedia system) with prerequisites and features as an example of a modern approach.

APPLICATION DISTRIBUTION PATTERNS

For the following we consider an application, whether it is mobile or not, consisting of three parts; The Presentation Layer, responsible for representing the visual parts of the application and doing the consummation of user input events. For distribution purposes, we divide the Presentation Layer in two sup-parts: The Dialog Representation, which is the visual painting and the reaction to events of the user and the Dialog Control which defines the sequence of the dialogs through the application.

The Business Logic Layer or the Application Kernel, responsible for the implementation of the business process, which means the origin and flow of data, which derives from the Backend Layer (Persistence Layer, Database Layer), which is responsible for retrieving and storing the data according to the requirements of the Business Logic Layer. Also for distribution analysis we divide the Backend Layer into the two subparts: The Database Access, which encapsulates the interface of the application programming language to the database (e.g., JDBC or ADO.NET) and the Database, which represents the database management system (DBMS) itself with tables, data, and stored procedures, etc.

Starting from this general architecture of an application, we want to derive a schema for making applications mobile, which means the separation of the whole application of parts of the application to a specified mobile device.

To be able to identify different software techniques to realize a device independent representation of an application, different architectural approaches have to be analyzed. First different levels where presentation and business logic can be separated are described. Finally, software techniques implementing the described design patterns are analyzed.

DISTRIBUTION LEVELS

This section describes possible distribution levels for every kind of client/server applications. The design of the client/server applications must provide functionality on the server. The client's functionality is mainly to display data ("thin client architecture"). This architectural design should ease distribution of new client versions. According to these needs, the design patterns "distributed presentation," "remote user interface" and "distributed application kernel" (taken from Renzel & Keller, 1997) are studied in more detail than "remote database" and "distributed database" pattern.

These patterns (see Figure 1) are interesting separating the representation from the business logic.

The remaining two design patterns "remote database" and "distributed database" in Figure 1 are applicable for realizing Web-based clients. A realization of these design patterns results in a "fat client" architecture, respectively the whole functionality of an application is located in the client.

Figure 1. Distribution levels (Renzel & Keller, 1997)

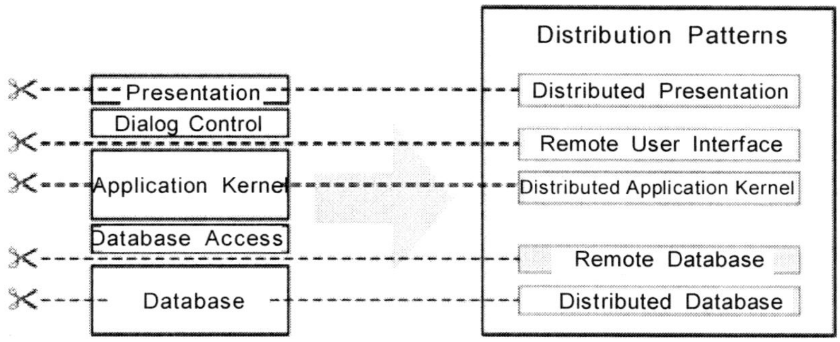

Distributed Presentation

An application design according to the "distributed presentation," design pattern means to partition the application at the presentation layer. The implementation of one part of the presentation layer acts as a client, the other part together with the rest of the layers represents server. Client/server applications designed following this design pattern result in very small clients. The whole functionality of the application as well as the user interaction handling is located at the server side. On the client side, only user interface presentation is done.

Realizing this architectural design pattern, following points have to be considered:

- This design reduces complexity of the application development, because all functionality is located in the server part of the application.
- Testing of the application is eased, because functionality of the application is not spread over various nodes in the network.
- If the client owns a Windows-based user interface (UI), network traffic can get high due to intense presentation data (the user often modifies the appearance of the UI and/or many different windows are needed) and event handling.
- Batch and transaction processing is eased, because data processing and transaction control is located at the same node in the network.
- Security mechanisms are easy to implement, although authentication, authorization, and secure communication must be realized.
- As clients do not store any data, data consistency need not be considered at client side.
- Distribution cost is limited to the system software required on the terminal nodes. The presentation data is sent to the client via network connection.

Remote User Interface

An application design according to the "remote user interface" design pattern means to partition the application between the dialog control layer and the application kernel layer (Figure 1). The client part processes all user interactions and sends, according to these actions, requests to the server part. The server part executes application functionality and sends its results back to the client part. The client program does not have any functionality; it is just responsible for user interaction logic and UI presentation and is therefore named "thin client."

Realizing this architectural design pattern, the following points have to be considered:

- Communication between client and server has to be implemented by the application programmer
- Communication effort between client and server is less than with the "Distributed Presentation" pattern as the whole user interaction and UI presentation is calculated on client side
- Security mechanisms are easy to realize because no application code is situated at client side. Nevertheless authentication, authorization and secure communication must be realized
- The client system needs an operating system, a graphical presentation system and communication software to interact with the server application
- Software distribution is more expensive than with the "distributed presentation" pattern, because clients are larger and have more functionality. Therefore; it is more difficult to distribute them at runtime

Distributed Application Kernel

An application design according to the "distributed application kernel" design pattern means to partition the application kernel level (Figure 1). In the client part, user interaction, UI presentation and parts of application functionality are implemented; the rest of the application functionality and data access is implemented in the server part. Depending on how much functionality is implemented on the client side, the client will develop rather large ("fat client").

Realizing this architectural design pattern, following points have to be considered:

This design pattern is suitable for highly interactive applications and provides good utilization of the underlying hardware.

Performance of the application is dependent on where the application kernel is cut. If done right, excellent performance can be reached.

As application functionality is located at the client and the server side, the use of batch jobs adds further complexity to the application. On the one hand you could run a batch job from the client, resulting in high network traffic; on the other hand you can run it on the server, replicating the necessary functionality for the batch job from the client (replication vs. network traffic).

Security mechanisms are more complex to realize as application data is processed at the client side.

As data is needed at the client side, consistency of the data has to be considered.

Software distribution is more complicated, as there may also be configuration issues at client side.

Remote Database

With the Remote Database application distribution pattern, one can realize applications where the full presentation logic and the business logic and part of the persistence layer but not the data itself reside on one logical tier and the storage is done on a remote location.

Note that for encapsulating the database functionality we divide the persistence layer (backend layer) into two logical partitions.

First the Database Access Layer, which consists of the connecting layer from a programming language or technology (e.g., Java or .NET) and the DBMS (database management system) itself. Well known examples of Database Access Layer software are the Java Database Connectivity (JDBC) and the Action Data Objects for .NET (ADO.NET), and second the DBMS consisting of the actual interface to the physical data stored and eventually in the DBMS located persistence logic (e.g., realized by stored procedures).

The characteristic of the remote database application distribution pattern is that the DBMS is separated from the other partitions of the application but the application programming interface (API) to the DBMS resides where the application is.

Regarding a Java application the remote database can be achieved by packaging the application and the JDBC libraries on the client and the DBMS on a server.

Realizing this architectural design pattern, following points have to be considered:

- This design pattern is suitable for highly interactive applications and provides good utilization of the underlying hardware
- Performance of the application solely depends on the client computer as mostly all the computation is done on the client tier
- Network traffic only occurs when synchronizing the persistence layer with the data displayed or manipulated on the client side
- Security mechanisms are more complex to realize as application data is processed at the client side
- As data is needed at the client side, consistency of the data has to be considered
- Software distribution is more complicated, as there may also be configuration issues at client side

Distributed Database

Additionally to the remote database application distribution pattern (see the previous section for details) the client tier gets partitions of the DBMS functionality. This pattern can be realized if the DBMS supports the division of its functionality to several clients.

This pattern is required if one needs DBMS functionality on the client tier without having the possibility to run the whole application locally.

One typical scenario for this application distribution pattern is an application realized with the lightweight additions from IBM DB2 Everyplace, which runs on a mobile device and keeps itself synchronized with a DB2 server can be located anywhere. Also Oracle, Sybase, and other major DBMS vendors have lightweight versions of their DBMSs.

Realizing this architectural design pattern, following points have to be considered:

- This design pattern is suitable for highly interactive applications and provides good utilization of the underlying hardware
- Performance of the application solely depends on the client computer as mostly all the computation is done on the client tier
- Network traffic only occurs when synchronizing the persistence layer with the data displayed or manipulated on the client side. This is fully handled by the DBMS, the application itself does no network processing (if data is not fetched from other 3rd party locations). Note that the synchronization logic can not be influenced as it is fully handled by the DBMS
- Security mechanisms are more complex to realize as application data is processed at the client side
- Software distribution is more complicated, as there may also be configuration issues at client side

Conclusion

Understanding mobile applications, one can use any of the application distribution patterns.

Which one the application developer chooses, depends on the application scenario itself. Hereby following questions can help leading the adequate application distribution patterns:

- Is there a network connection always available?
- Do I have to keep data on the client device
- Do I have enough computational resources on the client mobile device?
- Is persistent data on the client a requirement?

In answering these and other similar questions one can choose an application scenario for realization.

Up to now, we have only talked about the application distribution pattern (exactly speaking on how can one distribute and divide an application to several parts). Now we are going on to the practical implications and real life application architecture scenarios for mobile Web access.

This is done via representation of common architectures for mobile Web applications and an analysis to which application distribution pattern it belongs.

ARCHITECTURES FOR MOBILE WEB ACCESS

In this chapter we want to present several well-known architectures for mobile access to the World Wide Web. As we are talking about mobile multimedia issues, we want to put our focus especially on representing multimedia content on the Web via the introduced technologies.

The architectures we are looking at are:

- HTML/WML/ASP/JSP Web access
- Local applications and socket access
- MIDP-Java-Applications

HTML/WML/ASP/JSP Web Access

Access to the Web via Internet browser is regarding the efforts on mobile devices the simplest way to applications and data in the Internet.

Nearly the whole work (with the exception of rendering the user interface) is done on the Web server. Therefore these technologies are completely independent from the mobile device. User interfaces generated via HTML, see (Hypertext Markup Language, 2002), and (Extensible Hypertext Markup Language, 2005) or WML (Wireless Markup Language, 2002) are not as comfortable as one is used to have on a personal computer, although Active Server Pages (Keyton, 2000) and JavaServer Pages (JavaServer Pages Technology, 2005), etc. have released several advanced controls and technologies, like JavaServer Faces (JavaServer Faces, 2005) for the construction of advanced user interfaces.

However, from the basic design these technologies have no built-in mechanisms for representing multimedia content. Several extensions and plug-ins for browsers have been established; one of the most used is the Macromedia Flash Player, which is available for many platforms.

Another main disadvantage of server-based approaches is that supporting various mobile devices requires in the worst case a complete new generation of the user interface. Specialties of a mobile device cannot be regarded as in the server side technology the client is not regarded automatically. The required always-on connection to the network can be a crucial issue (e.g., on bypassing temporary network failures or availability).

The main advantage of this approach is that one does not have to redistribute software on new releases and updates. On the client side, only the browser and the required plug-ins have to be installed. Data is only stored on the server, therefore no synchronization logic is needed. Data storage is mostly realized with the XForms standard (W3C, 2005).

As application distribution pattern (see previous sections) the only realizable one is the distributed presentation pattern. Business logic and dialog control as well as backend are not accessible directly from the client device.

Applications Accessing Remote Services

Applications realized for mobile devices can access remote services provided by Web services, sockets, remote method invocation (RMI), remote procedure calls (RPC), etc, physical network access via wireless networking technology (UMTS, GSM, GPRS, etc.) is needed.

In this approach, the client runs the presentation of the application and the main dialog control. Business logic and backend need not be on the client machine, but can be located on the client. Therefore, from the architectural point of view all application distribution patterns from the previous section can be realized; only the backend storage in the DBMS has to be separated from the device, which is in fact not a restriction.

User interfaces are developed for one special device, therefore one can develop rich and full-featured user interfaces respecting the client's specialties. Multimedia content has to be integrated by the application developer and is not standardized. However, a new release requires software distribution on each client device.

Support for new versions of mobile devices with another software platform (operating system etc.) mostly causes a full new development of the application.

The architecture itself depends on the application scenario. For communication the following data transport mechanisms can be used:

- remote method invocation,
- Web services (SOAP over HTTP(S)),
- sockets,
- HTTP, and
- proprietary protocols.

A special more portable creation of user interfaces can be achieved via user interface description languages like XCC (Schmidt & Weinstein, 2002), XUL (Introduction to a XUL Document, 1999), (XWT) and UIML (User Interface Markup Language (UIML) Specification. 2002).

In this approach, a runtime is established which parses the user interface description from a server or from local file system und then renders the interface on the mobile device. Here we combine the declarative description of the user interface as we do it in HTML, etc. and the integration into a special device. Having realized such an application architecture, applications from purely local to an application with remote presentation application distribution pattern is possible, depending on the usage scenario.

MIDP Applications Accessing Remote Services

As MIDP (mobile information device profile) defines a portable way to write client managed network aware applications with respect of special concerns of mobile devices, MIDP seems to be the perfect solution for mobile multimedia applications. Technically MIDP is nothing else than a local application with a standardized runtime environment (Mobile Information Device Profile, 2002), therefore all the arguments and issues supplied in the previous section also apply here.

From an architectural point of view, MIDP supports every kind of application distribution pattern with the big advantage of being portable over many platforms.

As a more advanced and sophisticated example of a possibility for accessing multimedia data over the Web we will introduce the IP multimedia system (IMS) in the next section.

IP MULTIMEDIA SUBSYSTEM

After the introduction of wireless, mobile data connections as enabling technology for media rich, mobile applications, demand for an appropriate operation and deployment environment that eases roll out and operation of such applications arose. The IP multimedia subsystem (IMS), defined and specified by the 3GPP consortium as part of the UMTS standard, aims to provide such an infrastructure, (3GPP.TS 23.228 IP Multimedia Subsystem (IMS); Stage 2 (Release 6) 2005). 3GPP standardizes the interfaces between components, and third-party developers can build components without having to sell the entire service network. For instance, a company might specialize in mobile positioning servers and sell that service separately to operators. (Andersson, 2001) Although IMS is standardized as part of UMTS, it supports internetworking with WLAN (since release 6). Further, the CDMA2000 Multimedia Domain is based on 3GPP IMS.

IMS Architecture Overview

The IMS architecture can be subdivided into a three layered architecture, as depicted in Figure 2. The connectivity layer builds the interface to the underlying network infrastructure, while the service layer provides APIs and infrastructure to build and deploy services and applications. The control layer is responsible for session handling and user subscription management.

Connectivity Layer

The connectivity layer consist in routers and switches and is responsible for the connection to transport bearers, like 2,5G, 3G mobile networks, or WLAN. It provides an abstraction layer to enable uniform transport capacity and quality negotiation with underlying networks.

Control Layer

The control layer of IMS is responsible for session setup, modification, and release, and provides gateways to existing IP and circuit switched networks. The key elements of the control layer are the call session control functions (CSCF) and the home subscriber server (HSS). The role of CSCF can be split up into proxy CSCF (P-CSCF), interrogating CSCF (I-CSCF) and serving CSCF (S-CFCS).

The P-CSCF provides roaming functionality. If a mobile user wants to access IMS services while roaming in a foreign network, the P-CSCF is contacted. The P-CSCF detects the home network of the user and forwards the request to the home network of the user.

The I-CSCF is contacted if a user requests IMS services in his home network. The main functionality of the I-CSCF is to find out an S-CSCF which can handle the request. By routing service request to different equivalent S-CFCS servers according to their current load, it also provides the functionality of a load balancer.

The S-CSCF performs the session management of the IMS network. It handles SIP (session initialization protocol) messages, establishes sessions and negotiates with underlying transport layers to guarantee requested service qualities. SIP is an application-layer control (signalling) protocol for creating, modifying, and terminating sessions with one or more participants. (Rosenberg et al., 2002)

Figure 2. IMS architecture overview

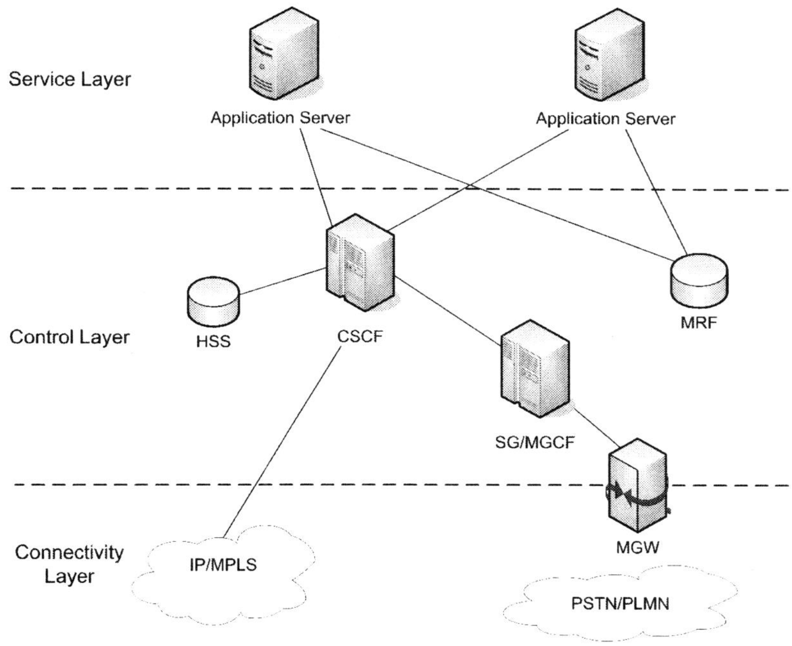

The home subscriber server (HSS) maintains a database with the unique service profile of all users. The user service profile stores all user service information and preferences in a central location. This information includes the end users current registration information, roaming information, instant messaging service information etc.

The media resource function (MRF) provides an interface to application servers and the S-CSCF to control media streams. It consists in a MRF controller which interprets incoming information, and a MRF processor which provides functionality like audio transcoding, media analysis etc. The MRFP is controlled by the MRFC.

The gateway functionality of the IMS is implemented by the media gateway control function (MGCF), the media gateway (MGW) and the signaling gateway (SG). The task of the MGCF is to control one or more MGWs, which enables scalability of the gateway system. The MGCF translates SIP messages from the CSCF, into a format that can be processed by the connected network. The MGW processes the ingoing and outgoing media streams between end users. Its primary function is to convert media from one format to another. The signalling gateway (SG) forms the signalling interface to legacy public switched telephone networks (PSTN) by transforming SS7 signalling information, which is the only format supported by PSTN, into IP and vice versa.

Service Layer

The service layer of IMS provides foundations to implement applications and services for the end user. It comprises in one or more SIP application servers that communicate with the underlying IMS functions provided by the S-CSCF. The S-CSCF is the anchor point for delivering new services since it manages the SIP sessions. (Parameshwar & Reece, 2004) application servers and S-CSCF interact via the IMS service control (ISC) interface which is based on SIP and its extension. ISC provides an event notification service, which allows Application Servers to subscribe for user specific events.

SIP application servers are programmable through scripting languages, SIP-CGI or APIs like SIP-servlets to provide the logic of value added services. The Java API for integrated networks (JAIN) and the JAIN server logic execution environment (JSLEE), which are APIs available for SIP application servers, are activities led by Sun Microsystems to create a standardized, Java-based service development API that abstracts from the underlying network architecture. See (Ferry & Lim Boon, 2004).

The architecture of the IMS service layer allows deployment of services by the operator and by 3rd party service providers through the Opens Service Access (OSA) API. The OSA API enables the secure integration of 3rd party services.

IMS Features

IMS provides a communication and service infrastructure which eases implementation and deployment of mobile multi media applications. Further, it introduces new features to the mobile domain like flexible session handling, and quality of service, which are introduced here.

Flexible Session Handling

The session initiation protocol (SIP) is the core communication protocol of the IMS. SIP is a text-based client-server protocol, similar to HTTP or SMTP that initiates session setup, routing, authentication within an IP domain. SIP enables the creation of sessions that hold different services, like voice and video. Those services can be synchronized if desired (e.g., for video telephony). Sessions can be dynamically modified, which allows adding a video component to an existing voice session. Further, IMS supports establishing multiple sessions of unrelated services, like an asynchronous textual chat session running concurrently and independent of a video conference session at one device.

Quality of Service

Real-time mobile IP communication is difficult due to fluctuating bandwidths, which severely affect the transmission of IP packets through the network. In traditional IP networks, transport quality is defined as "best effort," meaning that the network will do its best to ensure the required bandwidths, but there is no guarantee. IMS introduces a quality of service (QoS) feature to ensure that critical elements of IP transmission, such as transmission rate, gateway delay and error rates can be measured, improved and guaranteed in advance. By these means, users are able to specify the level of quality they require depending on the type of service and the user's circumstances. The policy decision function (PDF) node of the IMS negotiates service quality issues like resource authorisation and reservation, approval and removal of QoS commit with underlying network resources.

Standardized IMS Service Enablers

IMS facilitates the creation and delivery of multimedia services based on common enablers in a "write once, use many" way. These key elements in the IMS architecture are so-called service enablers. They represent generic and reusable building blocks for service creation. The service enablers developed for successful applications can become "global enablers" that are automatically included in new applications and services. Examples for these service enablers are group, list, and presence management and multi-party conferencing.

Application Integration

The architecture of the IMS server layer together with the standardized IMS service enablers fosters horizontal service/application integration. Figure 3 compares traditional vertical integrated services with the horizontal integration approach. The horizontal architecture in IMS also specifies interoperability and roaming, provides bearer control, charging and security. Horizontal integration enables the reuse of common infrastructure and service enablers, which speeds up and simplifies service creation and delivery.

Application Distribution with IMS

The mobile and wireless domain is a challenging environment for application developers. On the one hand, bandwidth and reliability of mobile networks is still not as high as desired, which could partly be antagonized by putting more application logic to the client, and facilitating caching mechanisms to enhance performance. On the other hand, the limited memory and processing resources provided by currently available mobile devices aggravates the implementation of such application architectures. To

Figure 3. Service reuse instead of replication

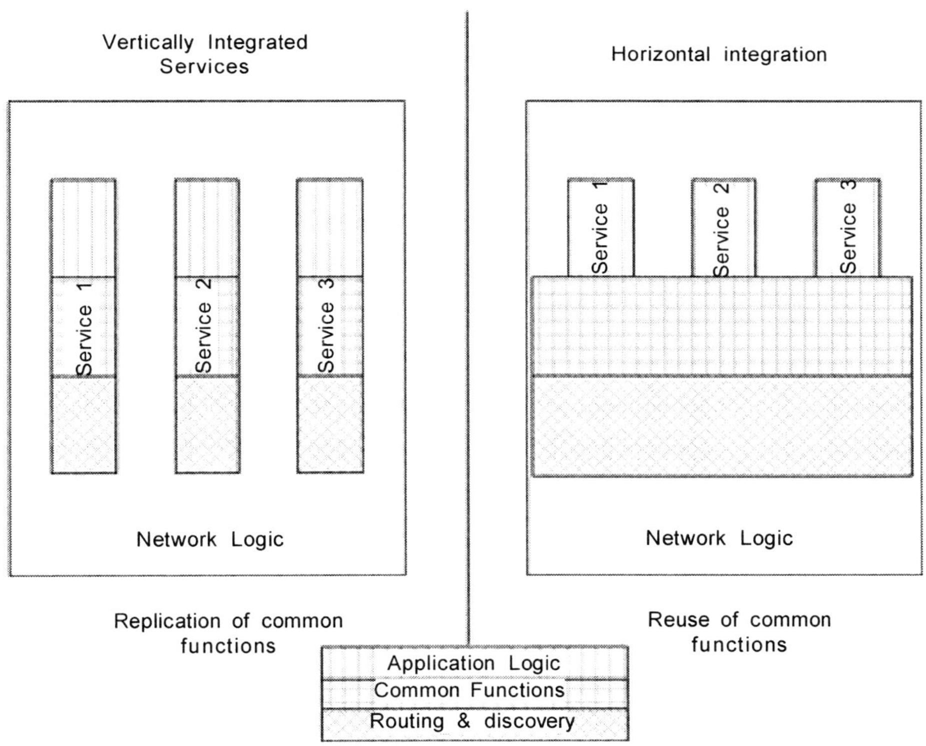

mitigate this situation, a service deployment infrastructure should show great flexibility in terms of application distribution capabilities. The SIP-based client communication of IMS enables multiple application distribution architectures. IMS even allows deployment of peer-to-peer applications, but the services provided by the IMS infrastructure, like presence, group management or billing are centralized. Thus, applications that take full advantage of the services offered by IMS are inherently client-server applications. The following sections discuss how different application distribution patterns can be implemented within IMS.

Distributed Presentation

The distributed presentation approach leaves the entire work load at the server. The client only needs to interpret and render the presentation according to the received description. No or only marginal additional software is required at the client, which eases deployment and maintenance of such applications. The centralized application logic also avoids data synchronisation and consistency problems, because data access control is located at a (logically) single server node.

The distributed presentation architecture enables applications to take full advantage of the IMS infrastructure, like service enablers, billing etc. The application server centric architecture of IMS, with the ability of horizontal application integration, fosters this distribution pattern.

The high server dependency of this application architecture makes it more vulnerable by network failures, because a lost server connection is immediately recognized by the user.

Remote User Interface

The remote user interface architecture leaves the business logic relevant tasks at the server, while moving user interface related functionality like consistency and plausibility checks etc. to the client. This includes applications which use WML-script to validate user input, but also PoC (push to talk over cellular) applications, which perform basic consistency checks like preventing double adding of users to groups. As the business logic is located at the server, this architecture allows utilisation of the IMS infrastructure.

The effort for deployment and maintenance is increased, if the application requires a dedicated client.

Distributed Application Kernel

The distributed kernel architecture splits the application logic between client and server. Applications that utilize this distribution pattern can take full advantage of IMS features like flexible session management and QoS, horizontal service integration can only be utilized by the server side part of the application. The interface between client and server part of the application need to communicate by IMS protocols like SIP or SDP (session description protocol) which may add some protocol translation overhead to the application and increase the complexity of the application.

This architecture enables to decrease the server dependency by employing a more intelligent, application specific connection management at the client. This can be used to "hide" short time disconnections from the user, and the centralized business logic avoids data synchronisation.

Remote/Distributed Database

With remote or distributed database architectures, the whole application functionality and in case of the distributed database architecture, also data is located at the client. The network infrastructure is used to synchronize distributed data bases. Although these architectures can be implemented within IMS enabled networks, taking advantage of the centralized IMS features like service enablers or billing requires considerable more effort than with server-based applications.

CONCLUSION

The IMS is the key element in the 3G architecture that makes it possible to provide ubiquitous cellular access to all the services that the internet provides (Camarillo & Garcia-Martin, 2004). The architecture of IMS clearly emphasizes server centric application architectures, which reside on application servers and take advantage of the various service enablers provided by IMS. The feature rich infrastructure of IMS fosters rapid application and service development, reuse of existing functionality and integration of different applications.

REFERENCES

3GPP.TS 23.228 IP Multimedia Subsystem (IMS); Stage 2 (Release 6). (2005). Retrieved May 16, from http://www.3gpp.org/ftp/Specs/archive/23_series/23.228/

Andersson, C. (2001). *GPRS and 3G wireless applications: Professional developer's guide*. Mississauga, Ontario, Canada: John Wiley & Sons Ltd.

Camarillo, G., & Garcia-Martin, M. (2004). *The 3G IP Multimedia Subsystem (IMS): Merging the internet and cellular worlds*. Mississauga, Ontario, Canada: John Wiley & Sons Ltd.

Extensible Hypertext Markup Language. (2005). Retrieved May 16, 2005, from http://www.w3.org/MarkUp/

Ferry, D., & Lim Boon, S. L. (2004). *JAIN SLEE 1.0 Specification, Final Release*. Sun Microsystems Inc.

Hypertext Markup Language. (2002). Retrieved May 16, 2005, from http://www.w3.org/MarkUp/

Introduction to a XUL Document. (1999). Retrieved May 16, 2005, from http://www.mozilla.org/xpfe/xptoolkit/xulintro.html

JavaServer Faces. (2005). Retrieved May 16, 2005, from http://java.sun.com/j2ee/javaserverfaces/index.jsp

JavaServer Pages Technology. (2005). Retrieved May 16, 2005, from http ://java.sun.com/products/jsp/index.html

Keyton, W. A. (2000). *ASP in a nutshell*. Sebastopol, CA: O'Reilly Associates.

Megacz, A. (2003).The XWT reference windowing toolkit. Retrieved May 16, 2005, from http://www.xwt.org/reference.html

Mobile Information Device Profile. (2002). Retrieved May 16, 2005, from http://java.sun.com/products/midp/

Parameshwar, N., & Reece, C. (2004). *Advanced SIP Series: SIP and 3GPP*. Award Solutions. Retrieved from http://www.awardsolutions.com/downloads

Renzel, K, & Keller, W. (1997). Client server distribution—A pattern language. *Proceedings of the Pattern Languages of Programming Conference (PLoP)* (Tech. Rep. No. #wucs-97-34). Monticello, WA: Washington University.

Rosenberg, J., Schulzrinne, H., Camarillo, G., Johnston, A., Peterson, J., Sparks, R. et al. (2002). *RFC 3261—SIP: Session Initiation Protocol.* Network Working Group. Retrieved May 16, 2005, from http://rfc.net/rfc3261.html

Schmidt, A., & Weinstein, T. (2002). *Design und Implementierung ultraleichter Java Clients*. NetObject Days. Retrieved May 16, 2005, from http://www.old.netobjectdays.org/pdf/01/papers/node/weinstein_schmidt.pdf

User Interface Markup Language (UIML) Specification. (2002). Retrieved May 16, 2005, from http://www.uiml.org/specs/docs/uiml30-revised-02-12-02.pdf

Wireless Markup Language. (2002). Retrieved May 16, 2005, from http://www1.wapforum.org/tech/documents/WAP-191-WML-20000219-a.pdf

W3C. (2005). *XForms—The next generation of Web forms*. Retrieved May 16, 2005, from http://www.w3.org/MarkUp/Forms/

KEY TERMS

3GPP: The 3rd Generation Partnership Project (3GPP) is a collaboration agreement that was established between European, Japanese, and North American telecommunication standardization organizations to create a globally applicable third generation (3G) mobile phone system specification.

IMS: The IP multimedia subsystem (IMS) is an open, standardized multimedia architecture for mobile and fixed services. It is based on a 3GPP variant of SIP and runs over the standard Internet protocol (IP). It enables telecom operators to offer network controlled multimedia services.

MIDP: Mobile information device profile (MIDP) is part of the J2ME framework and stands for mobile information device profile. It is specified by Sun Microsystems for the use of Java on embedded devices like cell phones or PDAs.

OMA: The Open Mobile Alliance (OMA) is an initiative of major manufacturers of end user equipment and infrastructure for mobile telecommunication networks. It aims to create interoperable services enablers to work across countries, operators, and mobile terminals. The OMA is driven by market requirements.

PoC: Push to talk over cellular (PoC) is a service that enables half duplex one to many communication over cellular networks.

SIP: Session initiation protocol (SIP) is a protocol developed by the IETF MMUSIC Working Group and proposed standard for setting up sessions between one or more clients. SIP is similar to HTTP and shares some of its design principles: it is human readable, very simple, and request-response.

Three Tier Architecture: Three-tier is a client-server architecture in which the user interface, functional process logic ("business rules"), data storage, and data access are developed and maintained as independent modules, most often on separate platforms.

Chapter XXXV
Design of an Enhanced 3G-Based Mobile Healthcare System

José Ruiz Mas
University of Zaragoza, Spain

Eduardo Antonio Viruete Navarro
University of Zaragoza, Spain

Carolina Hernández Ramos
University of Zaragoza, Spain

Álvaro Alesanco Iglesias
University of Zaragoza, Spain

Julián Fernández Navajas
University of Zaragoza, Spain

Antonio Valdovinos Bardají
University of Zaragoza, Spain

Robert S. H. Istepanian
Kingston University, UK

José García Moros
University of Zaragoza, Spain

ABSTRACT

An enhanced mobile healthcare multi-collaborative system operating over Third Generation (3G) mobile networks is presented. This chapter describes the design and use of this system in different medical and critical emergency scenarios provided with universal mobile telecommunications system (UMTS) accesses. In these environments, it is designed to communicate healthcare personnel with medical specialists in a remote hospital. The system architecture is based on advanced signalling protocols that allow multimedia multi-collaborative conferences in IPv4/IPv6 3G scenarios. The system offers real-time transmission of medical data and videoconference, together with other non real-time services. It has been optimized specifically to operate over 3G mobile networks using the most appropriate codecs. Evaluation results show a reliable performance over IPv4 UMTS accesses (64 Kbps in the uplink). In the future, advances in m-Health systems will make easier for mobile patients to interactively get the medical attention and advice they need.

INTRODUCTION

Mobile health (m-health) is an emerging area of telemedicine in which the recent development in mobile networks and telemedicine applications converge. m-health involves the exploitation of mobile telecommunication and multimedia technologies and their integration into new mobile healthcare delivery systems (Istepanian & Lacal, 2003). Wireless and mobile networks have brought about new possibilities in the field of telemedicine thanks to the wide coverage provided by cellular networks and the possibility of serving moving vehicles. One of the first wireless telemedical systems that utilized second-generation (2G) global system for mobile communications (GSM) networks addressed the electrocardiogram (ECG) transmission issues (Istepanian, 2001a). In recent years, several m-health and wireless telemedical systems based on GSM were reported (Istepanian, 2001b), allowing the accomplishment of remote diagnosis in mobile environments, as well as communication to geographic zones inaccessible by wired networks. The recent developments in digital mobile telephonic technologies (and their impact on mobility issues in different telemedical and telecare applications) are clearly reflected in the fast growing commercial domain of mobile telemedical services. A comprehensive review of wireless telemedicine applications and more recent advances on m-health systems is presented in Istepanian, Laxminarayan, and Pattichis (2005).

However, 2G-based systems lack the necessary bandwidth to transmit bandwidth-demanding medical data. The third-generation (3G) universal mobile telecommunications system (UMTS) overcomes limitations of first and second mobile network generations supporting a large variety of services with different quality of service (QoS) requirements. However, this fact makes network design and management much more complex. New applications require networks to be able to handle services with variable traffic conditions keeping the efficiency in the network resources utilization. The UMTS air interface is able to cope with variable and asymmetric bit rates, up to 2 Mbps and 384 kbps in indoor and outdoor environments, respectively, with different QoS requirements such as multimedia services with bandwidth on demand (Laiho, Wacker, & Novosad, 2000). In this kind of scenario, the emergence of 3G mobile wireless networks will permit to extend the use of m-health applications thanks to the provided higher transmission rates and flexibility over previous mobile technologies.

UMTS introduces the IP multimedia core network subsystem (IMS) (3GPP, 2005a), an IPv6 network domain designed to provide appropriate support for real-time multimedia services, independence from the access technologies and flexibility via a separation of access, transport and control. The fundamental reason for using IPv6 is the exhaustion of IPv4 addresses. Support for IPv4 is optional, but since network components require backward compatibility, it is clear that a dual stack configuration (IPv4 and IPv6) must be provided. The IMS uses the session initiation protocol (SIP) as signalling and session control protocol (Rosenberg et al., 2002). SIP allows operators to integrate real-time multimedia services over multiple access technologies such as general packet radio service (GPRS), UMTS or, ultimately, other wireless or even fixed network technologies (interworking multimedia domains). This chapter presents a 3G-based m-health system designed for different critical and emergency medical scenarios, as shown in Figure 1. Several medical specialists in the hospital take part in a multipoint conference with the ambulance personnel, receiving compressed and coded biomedical information from the patient, making it possible for them to assist in the diagnosis prior to its reception.

Figure 1. Typical medical mobility scenario

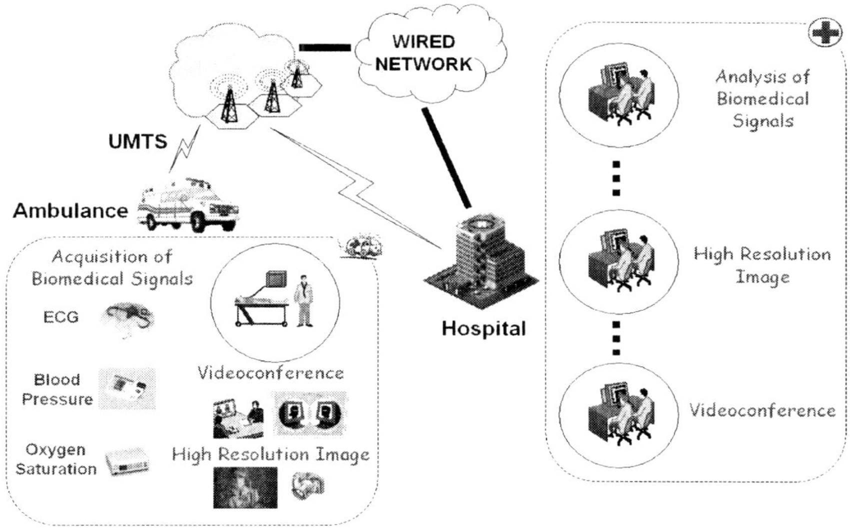

The 3G system software architecture includes intelligent modules such as information compression and coding, and QoS control to significantly improve transmission efficiency, thus optimizing the use of the scarce and variable wireless channel bandwidth compared to previous systems (Chu & Ganz, 2004; Curry & Harrop, 1998). Finally, unlike Chu and Ganz (2004), this m-health system follows a multi-collaborative design which supports IPv6/IPv4 interworking, uses SIP as the service control protocol and integrates new real-time multimedia features intended for 3G wireless networks.

Figure 2. 3G m-Health system

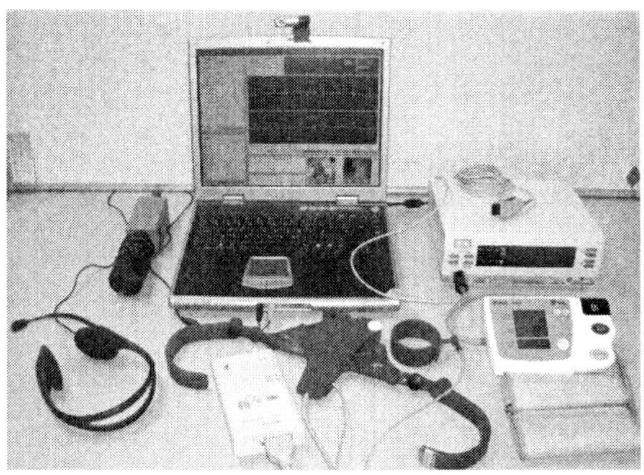

3G M-HEALTH SYSTEM ARCHITECTURE

In this section, the 3G m-health system structure is described in detail. The system (see the system components in Figure 2 and the main application graphical user interface (GUI) in Figure 3) has been built using standard off-the-shelf hardware, instead of developing propriety hardware as in Cullen, Gaasch, Gagliano, Goins, and Gunawardane (2001), uses free software and commercially available 3G wireless UMTS cellular data services. In addition, it provides simultaneous transfer of, among other services, videoconference, high-resolution still medical images and medical data, rather than only one

Figure 3. m-Health application GUI

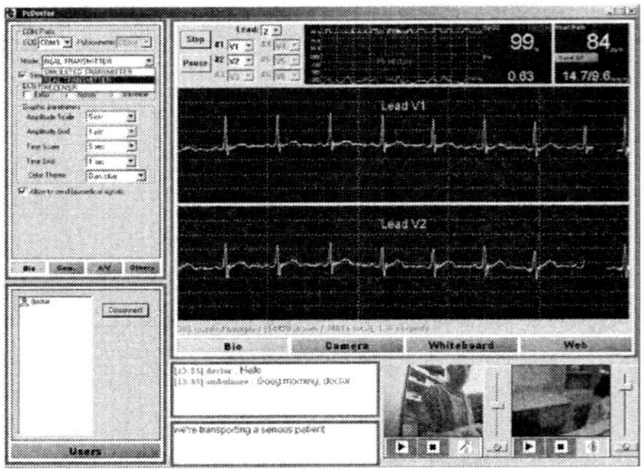

media at a time (Kyriacou, 2003; Pavlopoulos, Kyriacou, Berler, Dembeyiotis, & Koutsouris, 1998).

The 3G m-health system consists of different modules that allow the acquisition, treatment, representation and transmission of multimedia information. The modular design of each medical user service allows great flexibility. In addition, there exist other medical user services like chat and electronic whiteboard that allow data exchange in order to guide the operations performed by remote users.

The medical signals module acquires, compresses, codes, represents, and transmits medical signals in real time. The medical signals acquisition devices included are: a portable electrocardiograph that allows the acquisition of 8 real and 4 interpolated leads of the ECG signal, and that follows the standard communi-

Figure 4. 3G network scenario

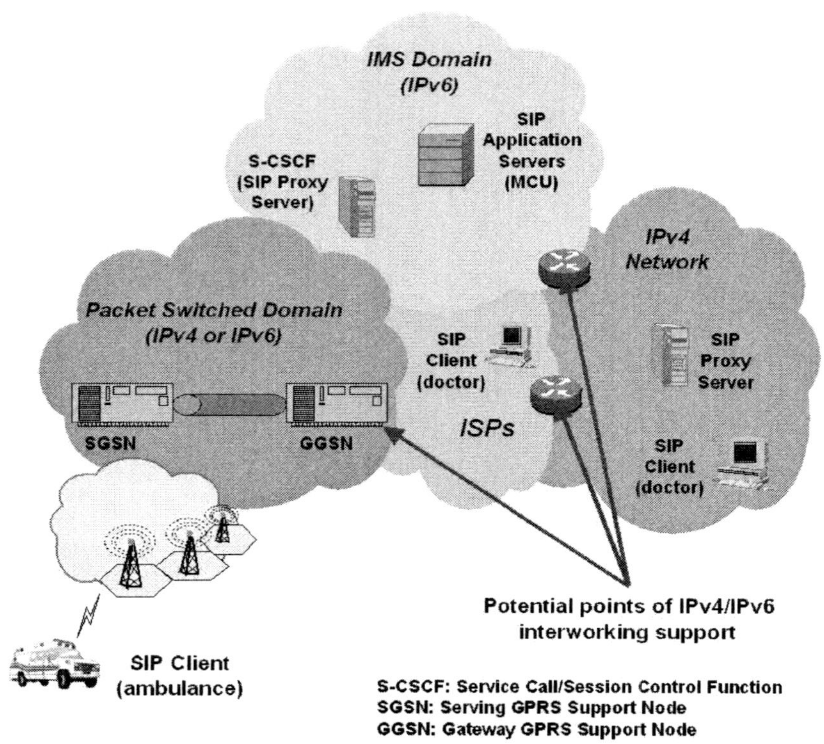

Figure 5. Block diagram of the m-Health system architecture and sub-systems

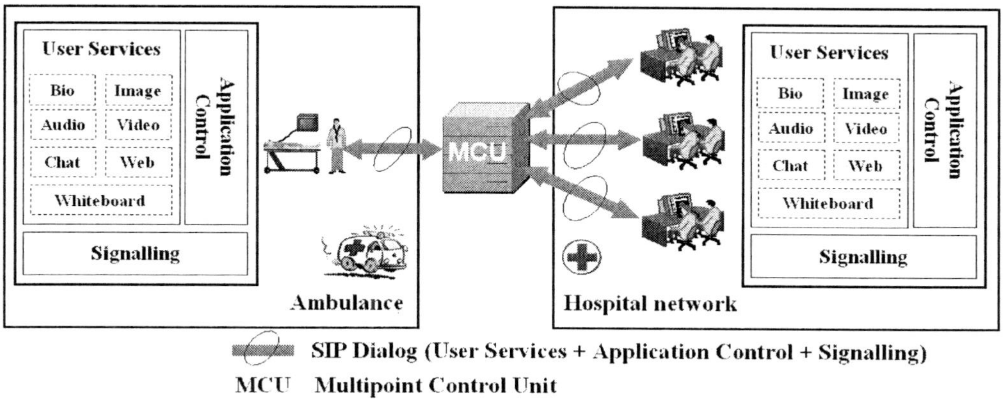

cation protocol-ECG (SCP-ECG); a tensiometer that provides systolic and diastolic blood pressure values; and a pulsioximeter that offers the blood oxygen saturation level (SpO_2) and the cardiac pulse.

The details of the 3G system architecture are shown in Figure 4 and Figure 5. As it can be seen, the system comprises of the signalling and session control, medical user services and application control sub-systems, which will be described later.

This architecture allows the 3G system to offer real-time services such as medical data transmission (ECG, blood pressure, heart rate and oxygen saturation), full-duplex videoconference, high-resolution still images, chat, electronic whiteboard, and remote database Web access.

Communication between the remote ambulance personnel and medical specialists is established by means of multipoint multi-collaborative sessions through several network environments capable of supporting the different types of multimedia traffic (Figure 4). The selected conference model (tightly coupled conference model (Rosenberg, 2004)) requires the existence of a MCU (multipoint control unit) to facilitate multipoint operation. The MCU maintains a dialog with each participant in the conference and is responsible for ensuring that the media streams which constitute the conference are available to the appropriate participants. The MCU can belong to mobile's home network (SIP application server) or to a external service platform. Furthermore, the m-health system is developed to support IPv4/IPv6 interworking (Wiljakka & Soinien, 2003) and can be integrated in a 3G network scenario (see Figure 4).

The MCU receives the information generated by each participant in the multipoint conference, processes and forwards it to the appropriate destinations. System users and the MCU exchange information associated with the different services provided (medical user services) and its presentation (application control). Moreover, they exchange information related to communication and service quality management (signalling). Next, a detailed description of the sub-systems is presented.

Signalling and Session Control

The developed signalling allows the exchange of the characteristics associated to the different information flows between system elements and is based on standard protocols that favour interoperability. Signalling tasks, performed by the SIP protocol, begin with the establishment of a SIP dialog with the MCU in which, by means of session description protocol (SDP) messages, the different services are described. In order to do that, each element in the system has a SIP user agent (UA), slightly modified in the MCU to allow the use of multiple simultaneous dialogs.

In addition to session control functions (establishment, management and termination of the multipoint conference), the SIP protocol is also useful for user mobility purposes inside the IMS environment.

Multipoint conference establishment, management and termination is performed by exchanging SIP messages between the different users. When a user connects, he creates a SIP dialog with the MCU, joining the conference. During the conference, SIP messages are exchanged between users and the MCU, varying conference characteristics and therefore allowing its management. In a similar process to that of conference joining, when a user wants to leave it, this fact must be communicated to the MCU with the necessary SIP messages. SIP messages also serve as the mean of transport of SDP messages with the description of medical user services.

The QoS in this system is mainly determined by the characteristics of the UMTS link. Mobile links are very variable, therefore a QoS-monitoring process is required in order to obtain a good system performance. This process is especially important in the MCU because it is there where the QoS-related decisions are taken. When the MCU detects that a particular conference participant needs to modify the characteristics of its multimedia session in order to improve QoS, it renegotiates the corresponding session by sending SIP/SDP messages. Hence, conference participants can modify certain upper-level protocol parameters (codecs used, transmission rates, compression ratios, etc.) in order to adapt the transmitted information to network performance.

The QoS monitoring process is possible thanks to a transport library that provides a uniform interface to send the information generated by medical user services and different QoS estimation tools developed for several types of links. This transport library offers different queuing policies and tools designed to measure the following QoS-related parameters: delay, bandwidth and packet loss rate. Due to the variable nature of wireless links, reception buffers have been properly dimensioned to minimize jitter, delay and packet loss.

Wireless Medical User Services

The medical user services included in the m-health system are associated with information shared in a multi-collaborative environment. Specifically, the system has services to share audio, video, medical data information, high-resolution still images, and graphical and textual information, as well as a Web service that allows remote access to clinical information databases. In addition to these services, there is a service designed to exchange control information (application control), which is discussed later.

Each kind of information is associated with a medical user service and uses a transport protocol and a codec according to its characteristics (see Table 1). Hence, real-time services (audio, video, and medical data information) use the real-time transport protocol (RTP), whereas the rest of the services use the transmission control protocol (TCP). Furthermore, the exchanged information can be very sensitive and requires a secure communication. The 3G m-health system uses an IP security protocol (IPSec) implementation supporting public

Table 1. Codec operation modes for 3G real-time wireless medical user services

	CODEC	CODEC RATE
Audio	AMR*	4.74 5.15 5.9 6.7 7.4 7.95 10.0 12.2 (Kbps)
Video	H.263	5 10 (Frames per second)
Biomedical Signals	WT**	5 10 20 (Kbps)

** Adaptive multi-rate, ** Wavelet transform*

key certificates in tunnel mode. This protocol ensures private communication of selected services.

The ECG signal is stored both in transmission and reception following the SCP-ECG standard. It is well known that for an efficient transmission an ECG compression technique has to be used. In our implementation, a real-time ECG compression technique based on the wavelet transform is used (Alesanco, Olmos, Istepanian, & García, 2003). This is a lossy compression technique, therefore the higher the compression ratio (lower the transmission rate), the higher the distortion at reception. It is clear that there is a trade-off between transmission rate and received ECG signal quality. From the transmission efficiency point of view, a very low transmission rate is desired but from the clinical point of view, a very distorted ECG is useless. Therefore, there exists a minimum transmission rate to be used so as the transmitted ECG is useful for clinical purposes, which was selected in collaboration with cardiologists after different evaluation tests. The minimum transmission rate used in our implementation (625 bits per second and per ECG lead) leads to a clinically acceptable received ECG signal.

Regarding blood pressure, oxygen saturation, and heart rate, these signals have low bandwidth requirements and, therefore, are not compressed.

The videoconference module captures, sends, and plays audio and video information obtained by means of a Web camera and a microphone. In order to reduce the bandwidth, these data are compressed and coded. The video signal is compressed following the H.263 standard, whereas the audio signal uses the adaptive multi-rate (AMR) codec, recommended for UMTS by the 3G Partnership Program (3GPP) (3GPP 2005b). This module provides the basic functionality for starting, pausing and stopping video signals acquisition and representation, as well as volume control for the microphone (capture) and the speakers (reproduction). Due to the fact that each participant in the conference receives a unique video signal, the system allows the user to select the particular video signal among all the users connected.

The high-resolution still image module obtains high quality images with a charge copled device (CCD) colour camera connected to the computer through an image acquisition card. This module includes options to preview the captured images and modify their main characteristics in real time: brightness, contrast, hue, etc. Captured images can be stored and transmitted in different formats, with various quali-

Figure 6. Application control (audio presentation spaces)

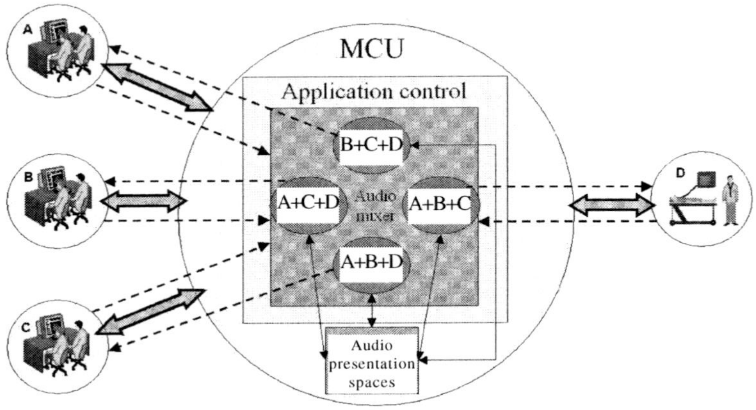

ties and compression levels. These images are sent automatically to the electronic whiteboard module of the remote users, allowing to select and mark fixed areas in a multi-collaborative fashion to facilitate a diagnostic clinical procedure.

Application Control

The MCU forwards the information generated by each medical service according to the presentation spaces defined using the control service. Each medical service has a presentation space associated with it that defines the way in which the information has to be transferred and its destination. The MCU simply forwards the information it receives, but has a special treatment for the audio, video, medical data, and control services. Regarding the Audio service, the MCU decodes the signal of each user, mixes it with the decoded signal of the other conference participants and codes the result in order to transfer a unique audio signal to each user (Figure 6). On the other hand, the MCU only forwards one video signal to each conference participant. The particular video signal forwarded to each user is selected by using the control service. Finally, the medical data service is similar to the video service: medical data are only generated in the remote location, whereas the other conference participants can only receive them.

The 3G m-health system was adapted to the critical and emergency medical scenarios. In the first stages of its design, user requirements and functional specifications were established in collaboration with medical specialists, in order to create a portable and modular m-health system that could be easily integrated in any environment, using any underlying network technology capable of supporting IP multimedia services

3G M-HEALTH SYSTEM PERFORMANCE

In order to measure the 3G m-health system performance, several tests have been carried out using the system over 64 Kbps (at IP level) UMTS accesses. Table 2 presents the results about average IP-level bandwidth used by real-time network services.

Table 2. Average IP-level bandwidth used by real-time user services

	Operation Mode		IP Bandwidth (Kbps)
Audio	Samples/packet	Codec rate (Kbps)	
	1	4.75	21.2
	2	12.2	28.8
	3	4.75	10.5
	3	12.2	18.1
Video	Frames per second		
	5		16
	10		24
Biomedical Signals	Bit Rate		
	5		5.3
	10		10.3

As it can be observed, considering more audio samples per network packet reduces the used bandwidth, since transmission efficiency (information carried by each packet to total packet size ratio) is increased. However, there is a limit in the number of audio samples per packet that can be used because more audio samples per packet yield more audio delay. For example, a more efficient transmission mode including four audio samples per packet every 80 ms causes four times the delay including only one audio sample per packet every 20 ms. Moreover, if an audio packet is lost, all the audio samples carried by it are lost and, therefore, a reduced number of audio samples per packet is more suitable to error-prone environments. Regarding the video user service, it is worth noting that the bandwidth shown in Table 2 can vary substantially with the movement of the video scene captured. Finally, the medical data service adapts well to the codec rate specified because medical data frame sizes are long enough to obtain a good transmission efficiency.

As it can be checked, the total bandwidth consumed by all real-time medical user services fits in a 64 Kbps UMTS channel, even when the most bandwidth-consuming codec rates and the lowest transmission efficiencies are used. If all medical user services are used (including non real-time services), lower codec rates should be selected. Thus, according to the previous discussions, the codec operation modes selected in this m-health system have been those highlighted in Table 2, achieving a reasonable trade-off between bandwidth, transmission efficiency, delay and loss ratio.

NEXT-GENERATION OF M-HEALTH SYSTEMS

It is evident that organizations and the delivery of health care are being underpinned by the advances in m-health technologies. In the future, home medical care and remote diagnosis will become common, check-up by specialists and prescription of drugs will be enabled at home and virtual hospitals with no resident doctors will be realized.

Hence, the deployment of emerging mobile and wireless technologies will face new challenges: inter-operability between heterogeneous networks (fourth-generation, 4G) and smart medical sensor design integrating sensing, processing, communications, computing, and networking.

With the aid of wireless intelligent medical sensor technologies, m-health can offer healthcare services far beyond what the traditional telemedical systems can possibly provide. The individual sensors can be connected wirelessly to a personal monitoring system using a wireless body area network (WBAN) and can be integrated into the user's clothing, providing wearable and ubiquitous m-health systems. A typical scenario comprises of a WBAN system that communicates with cardiac implantable devices (pacemakers, defibrillators, etc.) and that is linked to the existing 3G infrastructure, achieving "Health Anytime, Anywhere" (Figure 7).

It is expected that 4G will integrate existing wireless technologies including UMTS, wireless LAN, Bluetooth, ZigBee, Ultrawideband, and other newly developed wireless technologies into a seamless system. Some expected key features of 4G networks are: high usability, support for multimedia services at low transmission cost and facilities for integrating services. 4G advances will make easier for mobile patients to interactively get the medical attention and advice they need. When and where is required and how they want it regardless of any geographical barriers or mobility constraints.

Figure 7. Typical WBAN-3G scenario

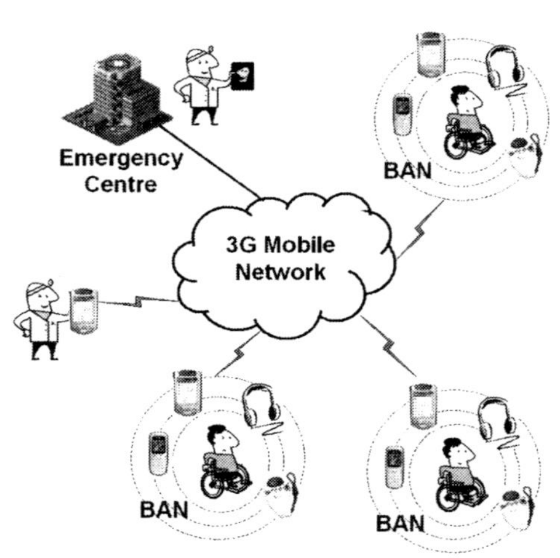

The concept of including high-speed data and other services integrated with voice services is emerging as one of the main points of future telecommunication and multimedia priorities with the relevant benefits to citizen-centered health-care systems. The new wireless technologies will allow both physicians and patients to roam freely, while maintaining access to critical patient data and medical knowledge.

CONCLUSION

This chapter has presented a feasible 3G-based m-health system targeted specifically for critical and emergency medical scenarios. The system architecture is based on 3G networks and advanced signalling protocols (SIP/SDP) that allow the integration of real-time multimedia services over multiple access channels that support IPv4 and IPv6 interworking depending on current commercial UMTS releases.

The system has the following features: simultaneous transmission of real-time clinical data (including ECG signals, blood pressure, and blood oxygen saturation), videoconference, high-resolution still image transmission, and other facilities such as multi-collaborative whiteboard, chat, and Web access to remote databases. The system has been optimized specifically to operate over 3G mobile networks using the most appropriate codecs. Evaluation results show a reliable performance over UMTS accesses (64 Kbps in the uplink).

Home telecare and chronic patient telemonitoring are other application areas in which this m-health system can be used, thus further work is currently undergone to adapt it and to evaluate its performance in each particular scenario.

REFERENCES

3GPP TS 23.228 V6.8.0. (2005). IP Multimedia Subsystem (IMS); Stage 2 (Release 6).

3GPP TS 26.235 V6.3.0. (2005). Packet switched conversational multimedia applications; Default codecs (Release 6).

Alesanco, A., Olmos, S., Istepanian, R. S. H., & García, J. (2003). A novel real-time multilead ECG compression and de-noising method based on the wavelet transform. *Proceedings of IEEE Computers Cardiology* (pp. 593-596). Los Alamitos, CA: IEEE Comput. Soc. Press.

Chu, Y., & Ganz, A. (2004). A mobile teletrauma system using 3G networks. *IEEE Transactional Information Technology in Biomedicine*, *8*(4), 456-462.

Cullen, J., Gaasch, W., Gagliano, D., Goins, J., & Gunawardane, R. (2001). *Wireless mobile telemedicine: En-route transmission with dynamic quality-of-service management.* National Library of Medicine Symposium on Telemedicine and Telecommunications: Options for the New Century.

Curry, G. R., & Harrop, N. (1998). The Lancashire telemedicine ambulance. *Journal Of Telemedicine Telecare*, *4*(4), 231-238.

Istepanian, R. S. H., Kyriacou, E., Pavlopoulos, S., & Koutsouris, D. (2001). Wavelet compression methodologies for efficient medical data transmission in wireless telemedicine system. *Journal of Telemedicine and Telecare*, *7*(1), 14-16.

Istepanian, R. S. H., Laxminarayan, S., & Pattichis, C. S. (2005). *M-Health: Emerging mobile health systems.* New York: Springer. To be published.

Istepanian, R. S. H., & Lacal, J. C. (2003). Emerging mobile communication technologies for health: Some imperative notes on m-health. *Proceedings of the 25th Silver Anniversary International Conference of the IEEE Engineering in Medicine and Biology Society 2* (pp. 1414-1416).

Istepanian, R. S. H., Woodward, B., & Richards, C. I. (2001). Advances in telemedicine using mobile communications. *Proceedings of the IEEE Engineering Medicine and Biology Society 4* (pp. 3556-3558).

Kyriacou, E., Pavlopoulos, S., Berler, A., Neophytou, M., Bourka, A., Georgoulas, A., Anagnostaki, A., Karayiannis, D., Schizas, C., Pattichis, C., Andreou, A., & Koutsuris, D. (2003). Multi-purpose healthcare telemedicine systems with mobile communication link support. *BioMedical Engineering OnLine*, *2*(7).

Laiho, J., Wacker, A., & Novosad, T. (2000). *Radio network planning and optimization for UMTS.* New York: Wiley.

Pavlopoulos, S., Kyriacou, E., Berler, A., Dembeyiotis, S., & Koutsouris, D. (1998). A novel emergency telemedicine system based on wireless communication technology–AMBULANCE. *IEEE Trans. Inform. Technol. Biomed*, *2*, 261-267.

Rosenberg, J. (2004). *A framework for conferencing with the session initiation protocol.* Internet draft. Work in progress.

Rosenberg, J., Schulzrinne, H., Camarillo, G., Johnston, A., Peterson, J., Sparks, R., Handley, M., & Schooler, E. (2002). *SIP: Session initiation protocol* (IETF RFC 3261).

Wiljakka, J., & Soinien, J. (2003). Managing IPv4-to-IPv6 transition process in cellular networks and introducing new peer-to-peer services. *Proceedings of IEEE Workshop on IP Operations and Management* (pp. 31-37).

KEY TERMS

4G: The fourth-generation (4G) is the continuation of the first, second, and third generations of mobile networks. It is a wireless access technology that provides high-speed mobile wireless access with a very high data transmission speed (2-20 Mbps) and enables users to be simultaneously connected to several wireless access technologies and seamlessly move between them in an all-IP environment. These access technologies can be any existing or future access technology. Smart antennas, low power consumption, and software-defined radio terminals will also be used to achieve even more flexibility for the user of 4G systems.

IMS: The IP multimedia subsystem (IMS) is a new open and standardized framework, basically specified for mobile networks, for providing Internet protocol (IP) telecommunication services. It offers a next generation network (NGN) multimedia architecture for mobile and fixed services, based on the session initiation protocol (SIP), and runs over the standard IP. It is used by telecom operators in NGN networks (combining voice and data in a single packet switched network), to offer network controlled multimedia services. The aim of IMS is not only to provide new services but to provide all the services, current and future, that the Internet provides. In addition, users have to be able to execute all their services when roaming as well as from their home networks. To achieve these goals the IMS uses IETF protocols. This is why the IMS merges the Internet with the cellular world; it uses cellular technologies to provide ubiquitous access and Internet technologies to provide new services.

IPv6: Internet protocol version 6 (IPv6) is the next generation protocol designed by the Internet Engineering Task Force (IETF) to replace the current version of the Internet protocol, IP version 4 (IPv4). Today's Internet has been using IPv4 for twenty years, but this protocol is beginning to become outdated. The most important problem of IPv4 is that there is a growing shortage of IPv4 addresses, which are needed by all new machines added to the Internet. IPv6 is the solution to several problems in IPv4, such as the limited number of available IPv4 addresses, and also adds many improvements to IPv4 in other areas. IPv6 is expected to gradually replace IPv4, with the two coexisting for a number of years during a transition period.

QoS: ITU-T recommendation E.800 defines the term quality of service (QoS) as "the collective effect of service performance which determines the degree of satisfaction of a user of the service." Service performance comprises of very different parts (security, operability, etc), so the meaning of this term is very broad. In telecommunications, the term QoS is commonly used in assessing whether a service satisfies the user's expectations. QoS evaluation, however, depends on functional components and is related to network performance via measurable technical parameters. A QoS-enabled network has the ability to provide better service (priority, dedicated bandwidth, controlled jitter, latency, and improved loss characteristics) to selected network traffic over various technologies.

SIP: The session initiation protocol (SIP) is a signalling protocol developed by the IETF intended for setting up multimedia communication sessions between one or multiple clients. It is currently the leading signalling protocol for voice over IP, and is one of the key components of the IMS multimedia architecture. The most important functions of SIP include name mapping and redirection (user location), capabilities negotiation during session setup, management of session participants and capabilities management during the session.

Telemedicine: Telemedicine can be defined as the rapid access to shared and remote medical expertise by means of telecommunications and information technologies, no matter where the patient or the relevant information is located. Any application of information and communications technologies which removes or mitigates the effect of distance in healthcare is telemedicine. The terms e-health and telehealth are terms often interchanged with telemedicine.

Chapter XXXVI
Securing Mobile Data Computing in Healthcare

Willy Susilo
University of Wollongong, Australia

Khin Than Win
University of Wollongong, Australia

ABSTRACT

Access to mobile data and messages is essential in healthcare environment as patients and healthcare providers are mobile. This is inline with the need of ubiquitous computing in everyday life. Mobile and wireless devices can assist in ensuring patient's safety by providing easy availability of the data at the point of care. Portability and accessibility of these devices enhances use of them in healthcare environment. However, data integrity and confidentiality of information in them need to be ensured to provide safe, effective and efficient healthcare. Mobile healthcare involves conducting healthcare related activities through using mobile devices such as a smart phone, Personal digital assistant (PDA), wireless enabled computer, iPod and so on. Mobile computing is suitable for healthcare as healthcare providers are mobile. These would be suitable for conducting patient's healthcare activities in emergencies, ward rounds, homecare, chronic disease management, conducting clinical trials, and so on. There are various projects using mobile devices to enhance patient's care. With the advancement of medical informatics, telemedicine and information technology, mobile data devices play an enormous role in healthcare system. In this chapter, we outline the need of mobile devices in healthcare, usage of these devices, underlying technology and applications, importance of security of these devices, securing mobile data communication in healthcare through different security models and case examples of applications that we have developed, in particular (1) iPathology tool on iPod, (2) securing healthcare information using Pocket PC 2003, and (3) securing information on handheld devices. There were several incidents in the past due to the insecurity of mobile devices that can leak information to anyone who does not have access to the information. In this chapter, we will illustrate several techniques that we have developed to protect these malicious activities and how these are applicable for securing mobile data computing in healthcare.

INTRODUCTION

In recent years, a wide range of handheld devices have been introduced in the market. Pocket size handheld organizers, called personal digital assistants (PDAs), have evolved from offering simple services such as calendar, address book, and e-mail, to working as a powerful mobile network client that can be used to access corporate databases and be used as part of an e-commerce platform. The healthcare industry is not an exception with healthcare providers using various handheld devices for accessing information and for telemedicine purposes. These days, the mobile devices are also equipped with communication abilities, such as GSM/GPRS ability, WiFi 802.11a/b/g Wireless LAN or Bluetooth. Handheld devices are perfect for carrying information, such as multimedia information, on the fly. Users can enjoy the information while they are on their way to work, etc.

Securing such devices includes providing protection for wireless communication, stored data and against viruses and malicious codes, together with making provisions for preventing theft of the device. The more we become dependent on handheld devices, the more threatening the attack will be. Today the U.S. military is already relying extensively on such devices. These devices can be used for GPS, viewing maps of areas, interacting with other troop units and as (remote) sensors (such as those used against bio-warfare), etc. Moreover, the U.S. military is spending large amounts of money to expand the capabilities of such devices. Other applications in critical infrastructures can be expected. For example, today many firefighters and ambulance services rely on paper maps. In regions in the world that expand rapidly, there is a significant advantage to switch to handheld devices where maps are updated more frequently. Since several manuals are primarily available via the Internet, handheld devices will be used by technicians to repair machinery in nuclear power plants, airplanes, etc.

There are several major differences between a desktop computer and a handheld device including limited input methods, small screen size, lower range processor, and limited memory for the handheld device. These differences affect the design of applications that run on these devices. Palm OS and Pocket PC handheld devices do not have a keyboard. Although a third party keyboard, such as Palm Portable keyboard or Targus Stowaway keyboard, can be used, most users enter text to the device with a *stylus* which is then recognised by a character recognition program such as Palm *Graffiti* or *Microsoft Transcriber*. Other methods of input include using the on-screen keyboard, or typing text on a desktop PC that is connected to the PDA. Communication with the device is achieved by using the HotSync/Activesync that synchronizes the data with its copy that is stored on a desktop computer, using a communication port such as the serial port, the IrDA (Infrared Data Association) port, or the USB port.

Palm 1000 was introduced by 3Com in 1996. Devices running Palm OS have dominated the handheld market and currently having around 70% of the market share (Hansmann, Merk, Nickous, & Stober, 2001). Palm OS is also adopted by companies such as HandSpring and IBM with applications ranging from Palm OS native applications including Address Book and Date Book, to other applications that are available as freeware, shareware or commercial products from third-party providers, including fax software, calculators, and software development kits for developers. The second major operating system for handheld computers is Windows CE (or pocket PC) which takes almost the rest of the market share. The most recent version is known as Windows Mobile 2005, with its support to GSM/GPRS mobile phone connection, together with its ability to connect to Wireless LAN 802.11a/b/g and Bluetooth.

Our Contribution

In this chapter, first we review the background on handheld devices, together with their known weaknesses. Then, we proceed with the use of handheld devices in healthcare environment. Finally, we provide some case studies that we have done in the effort to securing such devices.

Background of Handheld Devices

In this section, we review the necessary background and knowledge in handheld devices.

PALM OS PDA

Storage on a Palm PDA includes RAM and ROM that reside on memory cards. A memory card can store 256 MB and is divided into a dynamic heap and multiple storage heaps that are used for providing runtime environment, and storing data and applications. Newer models of Palm PDA, starting from Palm m505, m515, and Tungsten, are equipped with SD (Secure Digital) expansion slots that could be used for Bluetooth or Wireless LAN 802.11b devices.

Palm OS and built-in applications are stored in the ROM while user's data and applications are stored in RAM which is always powered to allow persistent storage. Applications and users' data can be backed-up on a desktop PC, or to the backup expansion pack (for the new models).

Development Systems

There are several development systems for building programs for Palm OS with C and C++ being the most commonly used languages for Palm Computing platform. Other available languages are Basic, Java, and Assembly.

A newly created application can be tested on a Palm OS Simulator or a Palm OS Emulator before it is installed on a Palm device. Palm OS Simulator is only available on the Macintosh OS and requires the application to be linked with a Simulator library that allows it to run as an independent application. Palm OS Emulators are available for Windows, Mac OS and Unix systems. The Emulator imitates the Palm OS device and allows an application to run without requiring additional library.

Third Party Program Installation

There are a number of ways of downloading programs to a Palm PDA. Firstly, *HotSync* program can be used to download new applications during the synchronization between the Palm and a desktop PC. A second method uses a third party program such as *Pilot Install* (envi.com, online). Applications can also be downloaded using IrDA port or a Bluetooth connection. IrDA drivers for desktop PCs can send a file to another IrDA point and so an application can be sent to a Palm PDA. An application can also be attached to a mail. This requires connection to users' mailbox on their PC via a synchronization program and so in terms of difficulty is similar to downloading the program via the HotSync program. Downloading a program is also possible through a browser (either on-line or off-line browser), such as AvantGo.

Communication and Networking

Palm OS supports two types of data communication: *serial/USB* and *infrared*. The serial/USB communication is used to synchronize the Palm via a HotSync or be used to transmit or remove data through a serial bus. The Palm is equipped with an infrared transceiver and can *beam* data (such as contact details) to another Palm using appropriate communication protocols. This ability has been used to enable a Palm to remotely control another device or to print to an infrared printer.

A TCP/IP connection can be established when a Palm connects to the Internet via a modem or a cradle with a software that supports the PPP protocol. Palm OS provides two libraries for TCP/IP networks which are known as Palm OS Net Library and Internet Library.

Wireless Portal

Many handheld devices can only access Web pages that have specially formatted data to download. Webclipping and WAP are the two commonly used technologies that meet the limitations of small displays that are not capable of displaying regular multimedia Web sites. Although Web browsers such as ProxyWeb and PalmScape, are available for Palm, the two technologies above are usually preferred due to the size limitation of the device.

Palm.Net offers mobile Internet access for Palm device with ROM version 3.5 and above. This service is available in North America and allows users to connect to the Web through a connectivity gateway, operated by Palm.

WINDOWS CE DEVICES

Windows CE was introduced in 1996 to provide a common operating system for a range of devices including handheld computers, industrial controllers, embedded appliances and wireless communication devices. Windows CE has the same API as Windows 2000 and allows the programmers to use a subset of Win32 API development tools and languages including Visual Basic and Visual C++ targeted for CE devices. There are also special versions of the compiler, known as the Embedded Visual C++ and Embedded Visual Basic, that makes it easier to create programs.

Windows CE is not an on the shelf OS. Rather, a device manufacturer needs to configure the Windows CE operating system for the processor platform of his/her choice. The operating system is usually built into the device's ROM, although sometimes it is put in flash memory.

Windows CE devices support serial and USB communication for one-to-one communication, and network connection which can use one or many devices and more sophisticated protocols such as HTTP and file exchange protocols like remote access service (RAS). To support infrared communication based on IrDA protocol, the IrSock API has been added to the operating system.

The latest version of Windows CE is known as the Windows Mobile 2005 Operating System, which is equipped with GSM/GPRS mobile phone connection together with 802.11a/b/g Wireless LAN and Bluetooth support.

VULNERABILITIES OF PALM OS DEVICES

In the following, we illustrate the known vulnerabilities of handheld devices. Without losing generality, we only illustrate the vulnerabilities against Palm OS devices. Three types of attacks have been reported on Palm.

Eavesdropping Palm Infrared Connection

The Palm OS Infrared is vulnerable to data stealing attack. Two main reported cases include constructing a program called "NotSync" which enables a Palm to *steal* the password from another Palm via its IrDA interface (@stake, online), and an eavesdropping attack (Nutheesing, online).

Denial of Service Attacks via HotSync

Palm HotSync manager is used to synchronize data between a Palm and a desktop PC. The HotSync program has a network mode that allows remote synchronization via a modem, hence adding more flexibility to Palm.

It is shown (Knight, online) that setting HotSync in the Network mode can lead to a denial of service attack by sending a large amount of data followed by a new line which results in the Palm to crash.

Palm Virus and Trojan Horse

Similar to other operating systems, Palm OS is vulnerable to viruses and Trojan Horses. The first Palm OS virus, known as the "Liberty crack" (Harris, online), is written by one of the co-authors of the Liberty program. It is a destructive Palm application, disguised as a free version of the Liberty application, that deletes all of the programs stored on a Palm and then reboots the device. A more comprehensive survey of Palm OS weaknesses can be found elsewhere (Kingpin, 2001).

PROTECTION OF HANDHELD DEVICES VIA PERSONAL FIREWALL

We have created a personal firewall for Pocket PC 2003 handheld device. Our personal firewall will allow an authorized information to pass through the device, but an unauthorized information will not be able to enter the device. The technique that we used is by rewriting the Ethernet driver for Pocket PC 2003 device. We will illustrate our work in the following section.

Background

In such a pervasive connected environment, network security is critical to protect devices and the information they contain. Central to network security is the firewall, which acts as a barrier between an untrusted network and the trusted network (or device). Due to high-profile hacker attacks and proliferation of viruses, firewalls have become a fundamental security tool. At this stage, no security system can ensure with absolute certainty that all of the information will be protected all of the time. Firewalls are one of the most effective security tools that a network administrator can deploy to limit vulnerability.

A firewall is hardware or software that monitors the transmission of packets of digital information that attempt to pass through the perimeter of a network. A firewall can be described as a system for enforcing *access control policy* between two networks. A firewall can deny unauthenticated requests or requests with potential threats, while permitting authenticated requests, thus protecting the internal network. In most cases, firewalls are used to prevent outsiders from accessing an internal network. However, firewalls can also be used to guard one highly sensitive part of a private network against its other parts. Such sensitive parts are for payroll, payment processing, R&D systems, etc.

There are essentially three types of firewalls: (1) packet-filtering firewalls, (2) stateful firewalls, and (3) application-level firewalls. A packet-filtering firewall drops packets based on their source or destination address or ports. For example, it blocks all packets from a site that we do not trust, or blocks all packets to an internal machine that should be inaccessible from an external network. Linux's *ipfwadm* and *ipchains* are examples of this type of firewall. A stateful firewall works at the session layer of the OSI model, or the TCP layer of TCP/IP. It monitors TCP handshaking between packets to determine whether a requested session is allowed. It creates a circuit then relays data between external and internal network. Stateful firewalls keep track of sessions and connections in internal state tables. It can protect against certain types of denial of service attacks. An example of this type of firewall is Checkpoint FW-1. In addition to lower levels, application-level firewalls inspect packets at the application level. A packet reaches the firewall and is passed to an application-specific proxy, which inspects the validity of the packet.

Newer types of firewalls include distributed firewalls and personal firewalls. In distributed firewalls, security policy is centrally defined but is enforced at individual network end points, such as routers, gateways, servers or user PCs. Policy can be distributed to end points in various forms. For example, policy can be pushed to end points where such policy should be enforced. Policy may also be provided to the users as credentials that will be presented when needed.

Unlike traditional firewall sitting between the external and internal networks, a personal firewall runs on a user's computers. Personal firewalls do not have problems such as end-to-end network encryption, malicious insider attacks, single point of failure, etc.

Personal Firewall for Pocket PC 2003 Handheld devices

We have designed and implemented a personal firewall for Pocket PC 2003, which filters TCP/IP traffic passing to and from the device via an Ethernet card. The concept can be extended to allow other protocols such as 802.11b to be covered. Although personal firewalls exist in the market (e.g., Norton Personal Firewall, etc.), their design and implementation are not widely available. In this chapter, we present the first open source firewall for Pocket PC 2003 operating system devices. The personal firewall is a packet-filtering type firewall, which analyzes the contents of network packets. The firewall examines, if applicable:

1. Source and destination IP address
2. Source and destination port
3. Protocol
4. HTTP domain and URL

It establishes the validity of network packets, based upon a strict set of rules. Depending on the packet validity, the firewall either accepts or rejects the network packet. If the firewall encounters a packet that is suspicious or unknown, the user will be notified, depending on the security level, and asked to respond with the appropriate action to be taken.

The design of our firewall consists of three separate modules:

1. IMDrv.dll — network device driver
2. ChkPkt.dll — dynamic link library
3. FWApp.exe — application

The relation between these three modules are illustrated in Figure 1.

We have successfully developed and implemented our personal firewall on Pocket PC 2003 handheld devices. Some of the screen captures of our implementation are shown in Figure 2.

Figure 1. Relation between the modules

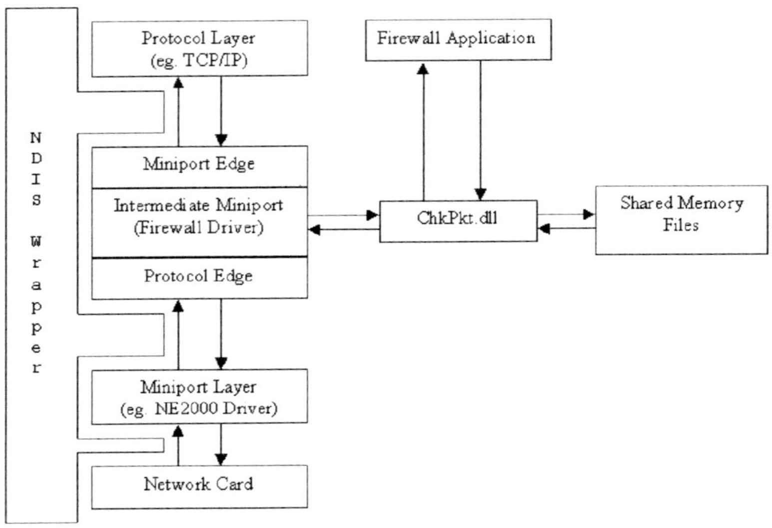

Figure 2. Screen captures of Personal Firewall for Pocket PC 2003 devices

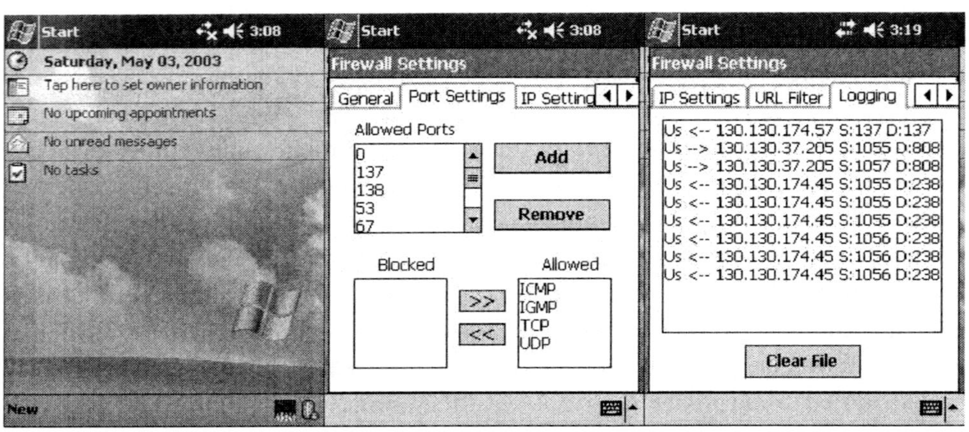

THE NEED FOR MOBILE DEVICES IN HEALTHCARE

In healthcare, mobile healthcare involves conducting healthcare related activities through using mobile devices such as a wireless phone, personal digital assistant (PDA), wireless enabled computer, iPod and so on. Portability and accessibility of these devices enhance their use of them in healthcare environment. Moreover, ease of use, affordability, and ability to read handwriting attracted use of these devices (Al-Ubaydli, 2004). The devices would be suitable for conducting patient's healthcare activities in emergencies, ward rounds, homecare, chronic disease management, conducting clinical trials, and so on.

Mobile computing is suitable for healthcare as healthcare providers are mobile. Healthcare providers would like to enter information in the patient record and process information during the ward round and at the point of care (Reuss, Menozzi, Buchi, Koller, & Krueger, 2004). Health information system located at the fixed terminal is inefficient because the necessary information will not be available at the time of query and that will interrupt the workflow (Eisenstadt et al., 1998).

USAGE OF THESE DEVICES

Mobile and wireless devices can assist in ensuring patient's safety by providing easy availability of the data at the point of care. Computerised physician order entry systems are being developed in most healthcare institutions to reduce medical errors (Ash, Stavri, & Kuperman, 2003). Mobility of mobile devices embrace usage of computerised physician order entry systems (Schuerenberg, 2005) and enhance patient safety. There are various projects using mobile devices to enhance patient's care. With the advancement of medical informatics, telemedicine, and information technology, mobile data devices play an enormous role in healthcare systems. Nursing staffs at St. Clair Hospital in Pittsburgh use PDAs equipped with peripheral scanning technology. This led to identification and prevention of 5000 errors (Anonymous, 2005). A survey and interview conducted on the usage of Personal Digital Assistants by medical residents noted that the current usage is as a single information source for personal organisation and general reference materials (Barrett, Strayer, & Schubart, 2004). However, a study conducted by Fischer, Stewart, Mehta, Wax, and Lapinsky (2003) indicated that these devices have been used for accessing medical literature, as drug information databases (pharmacopoeias), patient tracking, medical education, research data collection, business management, prescribing, speciality-specific applications such as family medicine, paediatrics, pain management, critical care, and cardiology (Fischer et al., 2003).

Two types of use of these devices can be seen in healthcare, which are download on demand (DoD) and downloading data periodically via the synchronization process (Afrin & Daniels, 2001). DoD allows the real time access of data and it is suitable for medical literature, patient tracking and various speciality specific applications. Downloading data periodically is applicable to medical educational application such as iPathology application on the iPod. A study conducted on all the American Academy of Family Physicians and American College of Osteopathic Family Practice in 2000 indicated that most commonly used operating system is the Palm on handheld computer (Criswell & Parchman, 2002). Currently, most healthcare providers use handheld devices and usage increases with emergence of smart phones and blackberry. However, only a few clinical applications are available at this stage for these devices (Al-Ubaydi, 2004).

Data integrity and confidentiality of information in these devices need to be ensured to provide safe, effective and efficient healthcare. Protecting patient confidentiality and privacy during the use of handheld device is a concern in the healthcare industry (Fischer et al. 2003). Implementing personal firewall for these devices enhances security and maintains the data integrity of health information.

CONCLUSION AND FURTHER WORK

In this chapter, we identified handheld devices as important tools for the future. We briefly reviewed the two well-known handheld devices, namely Palm OS-based and Windows CE-based handheld devices. We also illustrated our work to secure these devices. Finally, we concluded with the use of these devices in the area of healthcare. In the future, we will be developing applications that will use handheld devices for specific purpose required in the healthcare area.

REFERENCES

Afrin, L. B., & Daniels, M. S. (2001). PalMER: PalmOS-based access to the enterprise clinical data repository and clinical documentation assistant. *Proceedings of the American Medical Informatics Association Symposium 2001* Retrieved from http://www.amia.org/pubs/proceedings/symposia/start.html

Al-Ubaydli, M. (2004), Handheld computers. *British Medical Journal, 7449*(328), 1181-1184.

Anonymous. (2005). *Provider turns PDAs into scanners, mobile health data and source media*. Retrieved from http://www.mobile healthdata.com/article.cfm?articleeid=1418

Ash, J. S., Stavri, P. J., & Kuperman, G. J. (2003). A consensus statement on considerations for a successful CPOE implementation. *Journal of American Medical Informatics Association, 3*(10), 229-234.

Barrett, J. R., Strayer, S. M., & Schubart, J. R. (2004). Assessing medical residents' usage and perceived needs for personal digital assistants. *International Journal of Medical Informatics*, *1*(73), 25-34.

Criswell, D. F., & Parchman, M. L. (2002). Handheld computer use in U.S. family practice residency programs. *Journal of American Medical Informatics Association, 1*(9), 80-86.

Eisenstadt, S. A., Wagner, M. M., Hogan, W. R., Pankaskie, M. C., Tsui, F. C, & Wilbright, W. (1998), Mobile workers in healthcare and their information needs: Are 2-way pagers the answer? *Proceedings of the AMIA Symposium 1998* (pp. 135-139).

envi.com. (online). *Pilot install: Palm pilot installation*. Retrieved from http://envicon.de/e/pinstall/tentacle.html

Fischer, S., Stewart, T. E., Mehta, S., Wax, R., & Lapinsky, S. E. (2003). Handheld computing in medicine. *Journal of the American Medical Informatics Association, 2*(10), 139-149.

Hansmann, U., Merk, L., Nickous, M. S., & Stober, T. (2001). *Pervasive computing handbook*. Berlin: Springer-Verlag.

Harris, D. (online). *The "Liberty" crack: The first Palm OS Trojan Horse*. Retrieved from http://www.sans.org

Kingpin, M. (2001). Security analysis of the Palm OS and its weaknesses against malicious code threats. *Proceedings on the 10th USENIX Security Symposium* (pp. 135-151).

Knight, W. (online). *Palm Pilot open to a denial service attack*. Retrieved from http://www.zdnet.co.uk

Krause, A., Hartl, D., Theis, F., Stangl, M., Gerauer, K. E., & Mehlhorn, A. T. (2004). Mobile decision support for transplantation patient data. *International Journal of Medical Informatics*, *5*(73), 461-464.

Nutheesing, N. (online). *Padlock your Palm*. Retrieved from http://www.forbesbest.com

Reuss, E., Menozzi, M., Buchi, M., Koller, J., & Krueger, H. (2004). Information access at the point of care: What can we learn for designing a mobile CPR system? *International journal of medical informatics, 4*(73), 363-369.

Schuerenberg, B. K. (2005). *Mobility aims to boost CPOE use*. Mobile Health Data and Source Media, May 2005. Retrieved May 2005, from http://www.mobilehealthdata.com/

@Stake. (online). *Palm OS password retrieval and decoding*. Retrieved from http://www.securiteam.com/

Wikipedia, The free encyclopedia. (n.d.). Retrieved from http://en.wikipedia.org/wiki

KEYTERMS

Bluetooth: Bluetooth is an industrial specification for wireless personal area network (PANs). Bluetooth provides a way to connect and exchange information between devices like personal digital assistants (PDAs), mobile phones, laptops, PCs, printers and digital cameras via a secure, low-cost, globally available short range radio frequency (Wikpedia, http://en.wikipedia.org/wiki/).

General Packet Radio Service (GPRS): GPRS is a mobile data service available to users of GSM mobile phones. It is often described as '2.5 G', that is, a technology between the second (2G) and third (3G) generations of mobile telephony (Wikpedia, http://en.wikipedia.org/wiki/).

Global System for Mobile Communications (GSM): GSM is the most popular standard for mobile phones in the world. GSM phones are used by over a billion people across more than 200 countries. The ubiquity of the GSM standard makes international roaming very common with "roaming agreements" between mobile phone operators. GSM differs significantly from its predecessors in that both signalling and speech channels are digital, which means that it is seen as a second generation mobile phone system. This fact has also meant that data communication was built into the system from very early on (Wikpedia, http://en.wikipedia.org/wiki/).

IrDA: IrDA refers to Infrared data association, a standard for communication between devices (such as computers, PDAs and mobile phones) over short distances using infrared signals (Wikpedia, http://en.wikipedia.org/wiki/).

Wireless LAN or WLAN: WLAN is a wireless local area network that uses radiowaves as its carrier: the last link with the users is wireless, to give a network connection to all users in the surrounding area. Areas may range from a single room to an entire campus. The backbone network usually uses cables, with one or more wireless access points connecting the wireless users to the wired network (Wikpedia, http://en.wikipedia.org/wiki/).

Chapter XXXVII
Distributed Mobile Services and Interfaces for People Suffering from Cognitive Deficits

Sylvain Giroux
Université de Sherbrooke, Canada

Hélène Pigot
Université de Sherbrooke, Canada

Jean-François Moreau
Université de Sherbrooke, Canada

Jean-Pierre Savary
Division R&D CRD, France

ABSTRACT

The mobile device presented here is designed to offer several services to enhance autonomy, security, and communication for the cognitively impaired people and their caregivers. Two mobile devices are linked through a server; one is dedicated to the patient, the other one to the caregivers. The services fill three functions for patients: a simplified reminder, an assistance request service, and an ecological information gathering service. Three services are available for the caregiver: monitoring patients'ADLs, informing system and colleagues of an intervention, and planning patients' ADLs.

INTRODUCTION

The number of people suffering from **cognitive impairments** (Alzheimer disease, head injury, schizophrenia, etc.) is growing continually. For one, the aging of the society plays an important role in this trend. The always increasing needs for resources generates a huge stress on human and economical actors. Thus in Canada and United States, as in most western countries, social and medical policies try to keep people suffering from cognitive deficits at home (Hareven, 2001). Progress in cognitive rehabilitation also augments the number of semi-autonomous people that would be able to stay at home if light assistance is provided. Of course, this results in higher expectations and even more demand on resources. But most of the time, families have to take responsibility for care without access to appropriate resources. Too often, this situation then turns to an exhausting weight. Therefore, natural and professional caregivers urge for help.

Recent advances in mobile technology can provide affordable solutions to lessen the burden and anxiety put on caregivers and to collect reliable information. For many years, numerous devices have been designed to counterbalance physical and sensory deficits. Nowadays progress in technology set high hopes for cognitive orthotics that address cognitive decline (Lange, 2002; Pollack, 2004). Thanks to their small size and their versatility, **mobile devices** can offer a personalized assistance anytime anywhere. Mobile devices then become portable **cognitive prosthesis**, for instance acting as pervasive system to remind people activities of daily living (ADL) to perform when needed. They can also foster sense of security by keeping people and caregivers directly in touch. In its simplest form a direct call button may be used to request immediate assistance. **Geo referenced data** may also be used to detect crisis of schizophrenia or when a patient suffering from Alzheimer disease is lost (Médical Intelligence, 2005). **Mobile devices** can also provide non-intrusive **remote supervision** by caregivers. Besides mobile devices can gather ecological data compulsory to adapt or fine-tune diagnosis and treatments. Nevertheless, the design of devices and mobile services must put a very careful attention to user interfaces used to deliver information. Population suffering from cognitive deficits has often severe limitations and constraints.

This chapter discusses features and implementation of mobile services for cognitive assistance and remote supervision. Mobile devices provide practical solutions to the issues presented above. First, we review benefits and limitations of current reminder systems that can assist patients and/or caregivers. Then we present the needs of the population regarding the assistance necessary to stay safely at home. The multiple mobile services, designed especially for that population, are also described. We show how supervision can safely foster autonomy. Next, we describe cognitive assistance to patients and remote supervision by professionals or relatives, and then we go further and gather ecological data to foster better treatments. Finally, outdoors, safety and security of the patients is relying on geo-localization features. Implementation specific details are also sketched.

MOBILE PROSTHETIC SYSTEMS

The management of ADL is a central issue for people suffering from cognitive deficits. In the process of rehabilitation, occupational therapists provide a patient with a paper **agenda** as a tool he has to master to manage his life and autonomy. A lot of research projects and commercial applications also targeted electronic adaptation of specific functionalities of these agendas. In this section, we review the advantages and limitations of current reminder and/or agenda systems.

Some systems are device specific and are explicitly designed to be used inside the patient home and then could not help outside (Helal et al., 2003; Visions, 2005). Another category of systems is designed specifically for mobile devices in such a way that their use is not restricted to a specific location. These applications are usually running on PDAs or smart phones to either assist for needs specifically related to mobility (Patterson et al., 2004) or as general reminder/agenda systems or electronic organizers (Gorman, Dayle, Hood, & Rumrell, 2003; Haberman, Jones, & Mueller, 2005; Neuropage; Szymkowiak, Morrison, Prveen Shah, Evans, & Wilson, 2005). They are usually provided with acoustic alert or remote communication.

Activity Compass project intends to provide compensatory aid for outdoor courses for patient suffering from Alzheimer disease (Patterson et al., 2004). It consists basically of a PDA equipped with a GPS. Its role is to guide patient towards their destination by a mean of an arrow indicating the direction to take.

Neuropage is a reminder system that uses radio technology to send reminders of things to do (Neuropage). When the message arrives, the pager beeps (or vibrates), one button is pressed and the message can be read from the screen. Messages can be regular events or single-time message. Messages are added or removed by contacting Neuropage office by phone, e-mail, letter, or fax.

The ISAAC system acts as a **cognitive prosthesis** used by people experiencing dysfunctions in autonomy due to different kinds of brain injury (Gorman et al., 2003). It provides a checklist reminder to execute safely the ADL. Its hierarchical structure permits to navigate easily throughout the different pieces of advice displayed. Personalization and adaptation of pieces of advice enable ISAAC to evolve according to the patient cognitive evolution.

Providing a reminder helps the cognitively impaired patient to increase his autonomy. But the patient himself and his caregiver express needs for safety. Both want to keep in touch and be confident that activities will be performed in time. So a mean of communication must be added to the current reminders. In the next section, we review the cognitive deficits encountered and the needs that ensue.

NEEDS OF COGNITIVELY IMPAIRED PEOPLE

Cognitive impairments encountered in schizophrenia, head trauma and during the early stages of the Alzheimer disease provoke similar losses in autonomy that justify exploring a common approach to support, assistance and remediation by technology (Pigot, Savary, Metzger, Rochon, & Beaulieu, 2005). Cognitively impaired people present lacks of initiation. For instance, they remain for long periods without undertaking actions. They all call for frequent reminders to remember what to do and sometimes how to do. Alzheimer disease strikes elders and one's situation evolves from relative autonomy to an unrelenting dependence while trauma injury and schizophrenia appear most often during adulthood and one's situation evolves from dependence to relative autonomy.

Unless severely affected, cognitively impaired people are usually quite autonomous in performing basic **ADL** such as eating, dressing, and washing themselves. But they at times forget to do them and then need continuous recalls in order to initiate ADLs. Taking medication is a critical issue, as it requires a good short-term memory in order to fulfill the prescription correctly: which pills, how many and when. Without a reliable short-term memory, it can be forgotten or taken twice. At home oblivion or bad discernment in the use of domestic appliances often cause bath and sink flooding, fire, burn, cut...Beside these hazardous situations, risk of malnutrition, bad hygiene, and isolation are also real.

Cognitive impairments experienced at home have also manifest effects outside. Moreover, cognitively impaired people exhibit more difficulties in an unusual context. Oblivion leads to problems in finding one's way to a destination, in remembering the shopping list or even the reason of the trip.

Loss of autonomy requires assistance. Caregivers lavish attention to the patients days and nights long. Frequent recalls and safety are two factors especially important in the overall burden coming from constant supervision and assistance. Therefore, an electronic companion could alleviate the "threatening" recalls and improve the relationship. Safety prevents a caregiver from letting the patient alone.

MOBILE AND PERVASIVE COMPUTING

Although existing services are invaluable for patients and caregivers, overall they suffer from some drawbacks. These services are not integrated in a common portal. They must run on a specific patient device. They are either closely linked to a given location or, if they are location independent, they can not benefit from the casual presence of other devices in the environment, for instance sensors on the kitchen cupboards.

In next sections, we present the first steps of an on-going research project at DOMUS Laboratory. Mobile computing is coupled with **pervasive computing** (Weiser, 1991). Mobile computing will enable patients and caregivers to use services wherever they are. Pervasive computing will allow them to access services whatever the device is and to benefit from existing devices, sensors, and effectors present in their current environment. This approach then promotes a tighter integration of devices and services to the environment when possible while preserving mobility and independence of the patients.

Currently, a common portal and many services were implemented and are ready to be evaluated in situ. Through this portal, a patient gets access to services with a simplified adapted user interface. Caregivers also have access to their counterpart of services but the interface is more complex. For the moment, devices used are PDAs, but we intend to port them on e-mate, a platform that renders deployment of services independent of devices they are deployed on (Giroux, Carboni, Paddeu, Piras, & Sanna, 2003). So phones, laptops, TVs could be used to access services.

ASSISTANCE AND SUPERVISION SERVICES

A mobile device is designed to offer several services to enhance autonomy, security, and communication to cognitively impaired people and their caregivers. PDAs are linked through a server. Some are dedicated to patients. Others run applications presenting the caregivers point of view on information. Patient services fill four functions: a simplified reminder, a prompter to request immediate assistance, a service to gather ecological information, and a location-based health information service (Boulos & Maged, 2003) to help patient in case of crisis. On the other side, three services are available for the caregiver: monitoring patients' ADLs, informing system and colleagues of an intervention, and planning patient' ADLs. Services also enable patients and caregivers to communicate. The client component running on a patient's PDA is kept very simple. The patient benefits on an auto login procedure in order to reduce the cognitive load.

An Enhanced Agenda as a Mean for Cognitive Assistance

A basic paper agenda is designed to help remembering appointments, to fill a timetable and to prevent time conflicts. Our aim here is quite different. The agenda acts mainly as a cognitive prosthesis. Typically, the home of a patient suffering from head injury is covered with numerous notices, for instance written on a post-it. Each one indicates something to do. On the contrary, the electronic diary can show solely the activities often forgotten, and moreover just when it is necessary. By default, our prototype displays on the PDA single window the next three activities to perform in a three hours time bracket (Figure 1). Depending on the patient's cognitive abilities the number of activities displayed and the time slot can be easily reduced or increased. When an activity is performed the patient clicks on it to indicate its completion to the caregivers. The activities monitored on the **agenda** are either events occurring just one time, as appointments, or recurring events as ADLs often forgotten.

Figure 1. Patient agenda

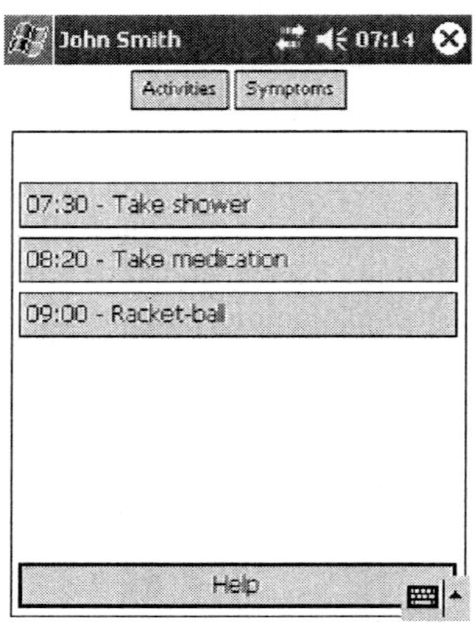

Keeping in Touch

If help is necessary, the patient clicks on the button "Help" at the bottom of the screen to warn caregivers. This button is always visible and active. This **remote supervision** ensures permanent contact between them. It provides a sense of safety for both patients and caregivers. Caregivers feel insecure to let patients alone without the opportunity to be reached. For patients, the phone is often too complicated to use. Carrying everywhere a PDA offers permanent assistance at a single click.

Monitoring ADL

Previous sections described the patient side of the system. In this section and the next two, we depict the caregiver view: how ADLs are monitored, how cooperation can be foster between caregivers, and how to manage a patient list of ADLs.

The role of the patient PDA is to foster his or her autonomy. From time to time, he or she does not perform given activities in time which may compromise his or her ability to stay alone. Equilibrium must be maintained between a patient intimacy and his or her safety. For instance, forgetting one day to take a shower does not justify an immediate intervention. However, after several days without showers, caregivers urge him or her to do it. On the other side, forgetting once to take medication could be harmful for the patient's health. Without a tool that helps to monitor ADLs, a caregiver surveys constantly the patients asking for the ADLs completion.

The current prototype supports **remote supervision** of ADLs to perform (Figure 2). Caregivers then know when an **ADL** is completed without bothering residents. On his or her **mobile device**, a caregiver can monitor every patient under his or her responsibility. On

Figure 2. Caregiver's ADL monitoring: (a) supervised patients; (b) one patient ADLs

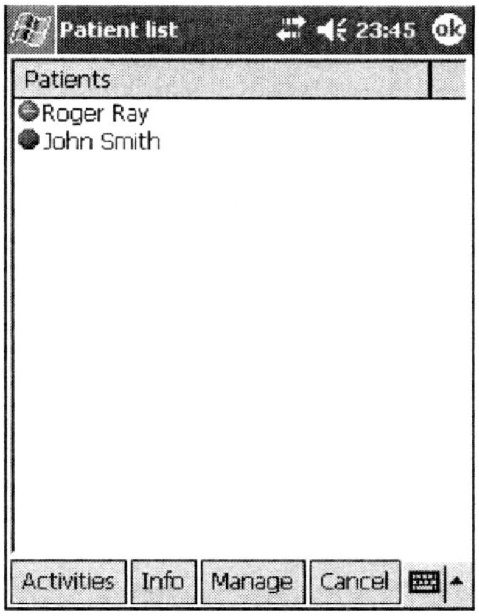

(a)

(b)

his or her patients list, colored icons indicate the status of current ADLs for each patient: green if all is OK, yellow if the patient is near to forgot to perform an ADL, and red if time planned for the ADL is over (Figure 2a). For each patient, a caregiver can know: which activities are under monitoring (Figure 2b) and the time of completion of a given ADL. The same color code prevails.

Coordination Between Caregivers

Since many caregivers can be in charge of a patient, coordination is important. Several caregivers can take care of many patients. When a patient does not perform a task in time, all responsible caregivers are informed. It is then essential to coordinate their actions, for instance in order to avoid all of them going to see the patient. As soon as a caregiver notifies his intervention to the system, this information is sent to all the **mobile devices** owned by the other caregivers. Caregivers can also exchange responsibility of patients, for instance when a working shift occurs.

Implementation section explains in more details how the information is updated on each PDA.

Planning ADL

In addition to ADL monitoring, caregivers must have also simple means to update and change rapidly the ADLs and events that have to be monitored. Like a usual organizer, the caregiver **mobile device** enables to register an activity occasionally or on a periodic basis ranging from hours to months (Figure 3a). The caregiver has to specify duration and completion time and for how long the activity has to be monitored (Figure 3b). This last information is particularly suitable during the rehabilitation period or to update a medication prescription that is regularly changed by the physician.

Figure 3. Planning activity: (a) activity period basis; (b) time activity information

(a)

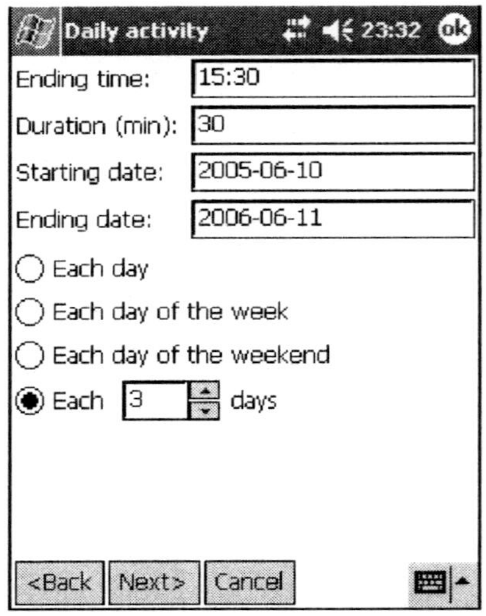

(b)

Ecological Data: Gathering Information for Better Treatments

Collecting ecological information may be of tremendous value for delivering better treatments. For instance to control the schizophrenic symptoms, a physician prescribes neuroleptics which induce severe side effects (Shriqui & Nasrallah, 1995). The physician has then to adjust medication during years to limit side effects and to avoid the recurrence of hallucinations. He or she needs valid information about the patient feelings.

On his or her **mobile device**, the patient can notify symptoms (or other information) as soon as he or she feels them (Figure 4). An intensity scale permits him or her to estimate the level of severity. The PDA is also equipped with a GPS to register automatically date, time, and position. These **geo-referenced data** are sent to the server on a batch mode. Data can be analyzed later to improve diagnosis and treatments.

Figure 4. Patient symptoms

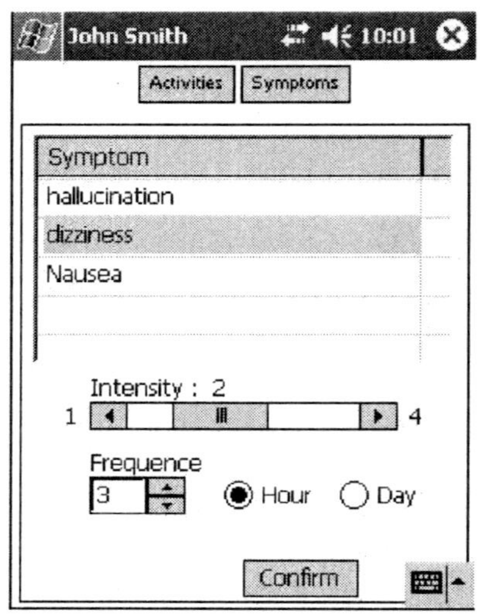

IMPLEMENTATION

The basis of the implementation is a client-server architecture (Figure 5). Patients and supervisors use iPAQs to connect to the server through wireless networks. The system is available both indoors and outdoors. Indoors, the

Figure 5. Connection infrastructure

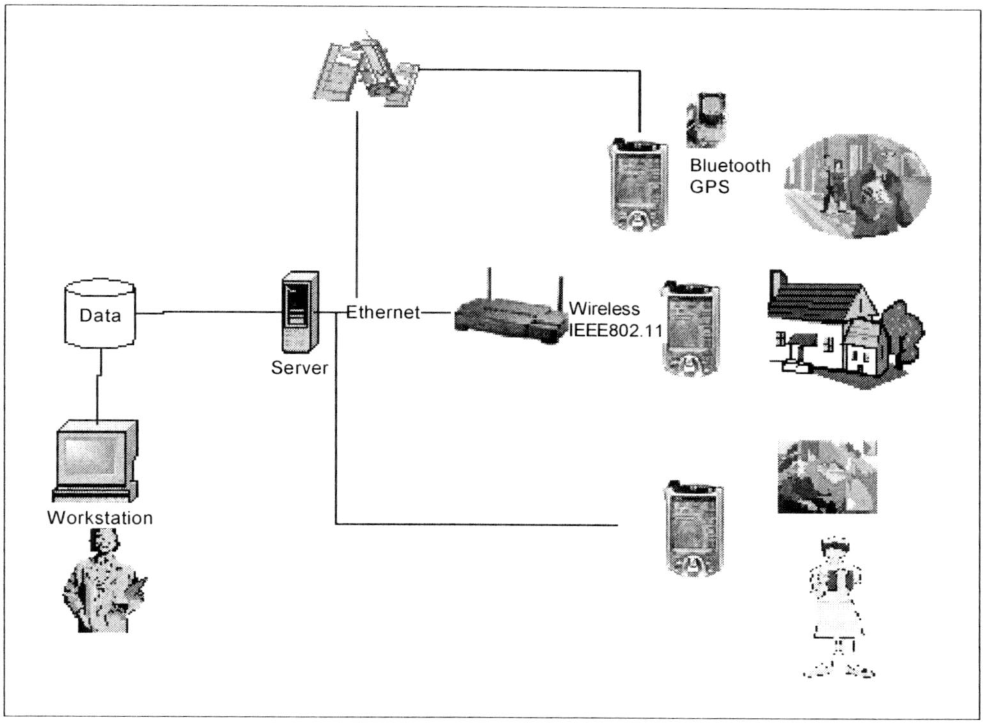

wireless connections to the server use IEEE 802.11b,g. Outdoors, the wireless connection is made through a Sierra Wireless AirCard®. Each PDA connects to the server through TCP sockets. The server allocates a different thread to each connection. If the connection is lost, the **mobile device** tries to restore it. The PDAs and the server exchange serialized objects through sockets. Outdoors, a GPS gives the patient location. To ensure rapid communication the traffic is kept low. When a patient confirms an ADL has been done or when a caregiver let know about his intervention, a message is sent to the server. The server first updates the database, and then sends a message to each PDA to inform a change happens. At that moment, some PDAs may want to display the information under change. They request all the information needed to update their screen displayed.

Different types of communication are used between the PDAs and the server. Inside, when the connection is established the **transmission** is in direct mode. Outside or when the connection is out, the communication is in batch mode. Messages are saved in a queue waiting until a connection becomes available. Finally, a configuration file is used to personalize the application: auto-login, font, font size, message strings, icons, and number of activities displayed.

The implementation is in standard Java (J2SE) for the server part. PersonalJava is used for the client part on PDAs. On the PDAs, user interfaces were implemented with SWT. On the PDAs, the virtual machine was J9 for PocketPC. The system was tested on HP iPAQ 4155 and HP iPAQ 1945 running Microsoft PocketPC OS.

DISCUSSION

Specific needs are addressed by our system for cognitively impaired people in order to foster their autonomy. Functions of an electronic agenda are redesigned to limit cognitive load, especially the burden put on memory. Presently only the patient name is displayed. We are evaluating the relevance of adding other information such as the patient address and caregiver phone number. It may be helpful when a patient is lost but it could raise the cognitive load by requiring one more screen.

The help function is germane to a panic button provided by various companies. A panic button is generally linked to a remote controller that intervenes when the button is pressed. The use of such panic systems remains generally difficult especially for elders suffering Alzheimer disease. First, we hope that providing on the PDA a help button they use frequently is more likely to be used in case of emergency. Secondly, making explicit the help button on the PDA by an icon or by writing the function is a guarantee they could understand its use.

Providing **geo-referenced data** is another critical issue. Knowing a patient position may be worthful and even vital in many situations and for a range of purposes:

- To send directly somebody at the patient location if he is in crisis or in danger
- To forbid specific zones
- To detect crisis states by analyzing paths and detecting pathologic patterns of movement of patients suffering from schizophrenia
- To predict the goal of a patient combining path history and analysis

When a patient is outdoors, a GPS connected to his PDA get his or position. This position is sent at predefined time interval to the server. Path analysis can then be performed. Currently client and server components are under development to specify and manage forbidden zones.

Mobile services are actually designed to offer the activities monitoring by professional caregivers. Families need too a similar system to be informed of difficulties. They feel insecure to let alone the patient. They could exploit geo-referenced data in the timetable information to know if it is conform to the one expected. For instance, a patient suffering from schizophrenia shows disorganization symptoms before experiencing a new crisis. Sending no information about the activity performed is a premonitory sign. Applying it to the Alzheimer disease may help to detect falls or other health problems.

CONCLUSION

In this chapter, we presented a client-server-based mobile system for patients suffering from cognitive deficits and their caregivers. Users will use mobile devices. They help especially to remember the activities a patient have to perform. We adapted traditional **agenda** to decrease the cognitive prerequisites and to provide a permanent communication between a patient and his caregiver. All along the development process, attention was paid on the real needs expressed by patients and caregivers. A sociological questionnaire and study had established previously the needs among patients suffering Alzheimer disease, head injury and schizophrenia (Pigot et al., 2005). Periodic meetings with caregivers have ensured conformity of the design of services to the needs expressed. Finally, a clinical validation is under process among people suffering from schizophrenia.

Future works will integrate mobile services to a helpful environment designed for people suffering from **cognitive impairments** in the spirit of **pervasive computing**. For instance instead of asking a patient to notify when an activity is completed, the environment will be able to recognize the activities performed by means of sensors (Pigot, 2004). The home will adapt itself to alleviate autonomy, remembering when necessary what needs to be done and how to do it. Such smart homes are an answer to the growing demand of cognitively impaired people to stay safely at home.

REFERENCES

Boulos, K., & Maged, N. (2003). Location-based health information services: A new paradigm in personalised information delivery. *International Journal of Health Geographics* 2003, *2*(2). Retrieved from http://www.ij-healthgeographics.com/content/2/1/2

Giroux, S., Carboni, D., Paddeu, G., Piras, A., & Sanna, S. (2003). Delivery of services on any device: From Java code to user interface. The 10th International Conference on Human Computer Interaction 2003, June 22-27, 2003, Crète, Grèce. In C. Stephanidis & J. Jacko (Eds.), *Human computer interaction* (Vol. 1-2). Laurence Erlbaum Associates.

Gorman, P., Dayle, R., Hood, C. A., & Rumrell, L. (2003). Effectiveness of the ISAAC cognitive prosthetic system for improving rehabilitation outcomes with neurofunctional impairment. *NeuroRehabilitation, 18,* 57-67.

Haberman, V., Jones, M., & Mueller, J. (2005). Mobile technology, compensatory aids, and usability evaluations. *Pervasive Computing*, *4*(2), 82-83.

Hareven, T. K. (2001). Historical perspectives on aging and family relations. In R. H. Bintock, & L. K. George (Eds.), *Handbook of aging and the social sciences* (pp. 141-159). New York: Academic press.

Helal, S. et al. (2003). *Smart phone based cognitive assistant*. UbiHealth 2003: The 2nd International Workshop on Ubiquitous Computing for Pervasive Healthcare Applications, Seattle, Washington, October 12, 2003.

Lange, M. L. (2002). Technology and occupation: Contemporary viewpoints. The future of electronic aids to daily living. *American Journal of Occupational Therapy*, *56*(1), 107-109.

Médical Intelligence. (2005). *A simple solution to prevent disappearance*. Retrieved from http://www.medicalintelligence.ca/en/columba.html

Neuropage. The Oliver Zangwill Centre Princess of Wales Hospital, Cambridgeshire, UK. Retrieved from http://www.neuropage.nhs.uk/

Patterson, D. J., Liao, L., Gajos, K., Collier, M., Livic, N., Olson, K., Wang, S., Fox D., & Kautz, H. (2004, October). Opportunity knocks: A system to provide cognitive assistance with transportation services. In N. Davies, E. Mynatt, & I. Siio (Eds.), *Proceedings of UBICOMP 2004: The 6th International Conference on Ubiquitous Computing* (LNCS 3205, pp. 433-450). Springer-Verlag.

Pigot, H., Lefebvre, B., Meunier, J. G., Kerhervé, B., Mayers, A., & Giroux, S. (2003, April 2-4). The role of intelligent habitats in upholding elders in residence. *The 5th International Conference on Simulations in Biomedicine,* Slovenia (pp. 497-506).

Pigot, H., Savary, J. P., Metzger, J. L., Rochon, A., & Beaulieu, M. (2005). Advanced technology guidelines to fulfill the needs of the cognitively impaired population. *The 3rd Inter-*

national Conference on Smart Homes and Health Telematic. Canada. (In press.)

Pollack, M. E. (2004). *Special committee on aging. United States Senate hearing on assistive technology for aging populations*. Retrieved from http://www.eecs.umich.edu/~pollackm/Pollack-web_files/senate-testimony.pdf

Shriqui, C. L., & Nasrallah, H. A. (1995). *Contemporary issues in the treatment of Schizophrenia*. Gilmore Academic Psychiatry.

Szymkowiak, A., Morrison, K., Prveen Shah, P. G., Evans, J. J., & Wilson, B. A. (2005). A memory aid with remote communication using distributed technology. *Personal and Ubiquitous Computing*, *9*, 1-5.

Visions. (2005). *The visions system*. Retrieved from http://www.thevisionssystem.com/

Weiser, M. (1991). The Computer for the 21st Century. *Scientific American*.

KEY TERMS

Batch Mode: The batch mode indicates the transmission mode where the data are sent to the receptor a moment after the transmittor command.

Cognitive Assistance: It is the assistance provided by the environment to compensate the cognitive deficits.

Cognitive Deficits: They are the deficits encountered by a person following a brain lesion. The memory losses and the lacks of planning, initiation, and attention are examples of cognitive deficits.

Direct Mode: The direct mode indicates the transmission mode where the data are sent to the receptor immediately after the transmitter command.

Mobile Services: Mobile services refer to the software applications available on portable devices such as PDA or smart phone.

Pervasive Computing: Pervasive computing is the next generation computing environments with information and communication technology everywhere, for everyone, at all times.

Reminder: A reminder is a prosthetics aid which displays the common agenda functions and information on memories or advices to be followed.

Remote Monitoring: See smart home.

Sensors: The sensors are electrical devices that gather information on the state of the environment and on the localization and the activities performed by the inhabitant. For instance some sensors detect the presence of the inhabited or which door is open.

Smart Home: The smart home is the home equipped with sensors and effectors that could react according to the inhabitant actions. The smart home could warn persons outside by the remote monitoring or give advices to the inhabitant.

Chapter XXXVIII
Mobile Magazines

Tom Pfeifer
Telecommunications Software & Systems Group (TSSG) and Waterford Institute of Technology (WIT), Ireland

Barry Downes
Telecommunications Software & Systems Group (TSSG) and Waterford Institute of Technology (WIT), Ireland

ABSTRACT

Mobile magazines will be magazines on a mobile computing and communication platform (in contrast to print magazines about mobile technologies), providing valuable, current multi-media content. However, there is a gap in the value chain between mobile operators and content publishers, hindering established small and medium sized publishers to enter the mobile market despite its commercial potential. A mobile operator publishing and entertainment platform enables the creation of a new category of mobile service called a mobile magazine. An m-Mag (mobile magazine) eco-system is a next generation mobile publishing service that is made available from a mobile operator's portal, that is integrated with value added mobile data services and that uses the operator's billing capabilities to charge consumers for access to the magazine. Using Parlay/OSA as an open approach, the m-Mag platform can integrate into an operator's network using standardised APIs and is portable across different operator networks. A discussion of the commercial potential analyses the route to the market.

INTRODUCTION

In his editorial of Pervasive Computing (Satyanarayanan, 2005), Satyanarayanan compares the paradigms of the "Swiss Army Knife" to the "Wallet" approach for the area of mobile computing. He finds the knife most versatile when far from civilisation, but sub-optimal in the individual functions (such as a full-size screw-driver compared to the small one on the knife). The wallet, in contrast, containing credit and identity cards, enables the user to all amenities of the civilised world, providing payment and access to plenty resources.

The mobile phone, as the first truly ubiquitous computing device (Abowd, Iftode, & Mitchell, 2004) can provide such a wallet functionality, transforming generic infrastructure into highly personalised services (Satyanarayanan, 2005). Abowd counts the current main mobile phone usage (voice and SMS) as the first two ubicomp killer applications, considering the over a billion devices deployed today (Abowd et al., 2004).

Magazines on general topics however, generating a massive revenue on the print market, and getting out of the red with their Web-based offsprings, are only beginning to discover the mobile devices.

The term "mobile magazine" in this chapter should therefore be understood in the meaning of a magazine on a mobile platform (in contrast to print magazines about mobile technologies). Besides technical marketing tools of mobile operators, only very few mobile magazines have entered the market of mobile communication since 2004.

The issues that exist already, are manually tailored for the mobile community, such as from Sendandsee and Zed/LaNetro, one of the leading providers of consumer mobile value-added services in European, Asian and American markets (Schuurmans & Poropudas, 2004; Sendandsee, 2004).

The global publishing industry, that so successfully took advantage of the Internet as a distribution channel during the 1990s, has not been able to exploit the "mobile" Internet to date. This is due to a number of technical, financial, and business issues. Recently a number of the these barriers to success have been addressed with new types of colour mobile phone and technologies such as GPRS and UMTS — but still a number of critical problems persist.

While much has been speculated about why appropriate services are not being developed and used in Europe and the U.S., in contrast, NTT DoCoMo has convincingly demonstrated the public's appetite for these services with i-mode in Japan. Takeshi Natsuno, managing director for i-mode Strategy, NTT DoCoMo (Takeshi, 2003) believes that the problem is one of market structure; that in Europe and the U.S., a value chain that supports the creation of a mobile services ecosystem is still missing, hindering the fully automated creation/conversion of time-sensitive, premium multi-media services.

Mobile operators do not have the resources or content to build a sufficiently critical mass of revenue generating 3G services to make 3G a success by themselves–they need to create an eco-system of third party service providers.

MOBILE MULTIMEDIA VALUE CHAINS

Karvonen and Warsta (2004) have analysed the value chain perspective for mobile multimedia services development within a number of companies. Based on a history of value chain and value network models, and the increasing diversification of revenue models, they derive the model as presented in Figure 1.

They clearly distinguish the infrastructure and services provision from the content. The latter needs to be created, further developed (specifically reflecting MMS content such as games and ringtones), appropriately packaged, and marketed to the consumer. Karvonen and Warsta conclude with the requirement of an eco-system for mobile multimedia development, providing room for a complex combination of separate members.

Mobile magazines are becoming part of the mobile entertainment market, which is a sub-

Figure 1. Mobile multimedia value chain by Karvonen and Warsta (2004)

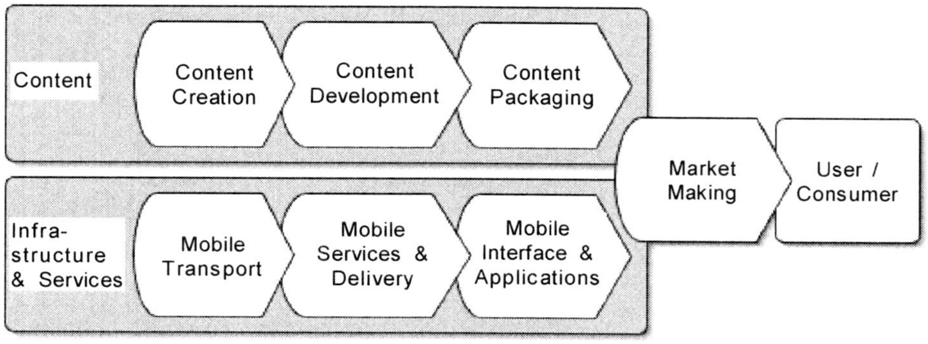

segment of the mobile data services market. This market is large and rapidly growing. According to the EITO (European Information Technology Observatory) 2002 Report (EITO, 2002), mobile entertainment services represent the largest revenue opportunity in the mobile Internet marketplace and they expect revenues to grow from EUR1.4 billion in 2001 to EUR21.8 billion in 2006 (CAGR 73%) in Western Europe alone.

The EITO believes that mobile games will drive most revenues in 2006 and will total EUR6.8 billion. This would lead to a total size of the mobile magazine market being ultimately be very similar to that of mobile games, but will lag behind games by about two years. Thus the author of this chapter believes the total market size for mobile magazines will be around EUR7 billion in 2008 in Western Europe.

A publishing platform is required that links together the publishing industry with mobile operators — but it is the mobile operators that have the need to drive mobile data revenues. Data services revenue for most Tier 1 and Tier 2 operators is currently around 15% of total revenues (even less for Tier 3), with SMS comprising on average 10-12% of total revenues. However most mobile operators are experiencing significant price erosion on voice and SMS services and need to find innovative services and content to maintain and grow ARPU (average revenue per user).

Thus the target customer of such a platform is the mobile operator. It is the operator that has the greatest need to introduce it — and it is the operator that has the commercial resources and reach necessary to promote mobile magazines both to publishers and ultimately to end consumers.

The m-Mag platform, as it will be introduced in this chapter, can broadly be viewed as part of the mobile services infrastructure industry or more narrowly as part of the Parlay value added services industry. The Parlay industry is a favourable industry to compete in for those organisations that can bridge the high barriers to market entry, which are primarily knowledge-based, contacts-based, technical and financial.

The main challenge for new entrants in the industry is the power of the customer–the mobile operator in this case. However, if the mobile operator can see the revenue potential in a particular product or service they will make a purchasing decision.

The other key issue within the industry is the long sales cycle (requiring a substantial capital reserve), which can be months to get to trial and over a year to final purchase decision. However this applies mainly to the gateway vendors as operators have recently shown great speed in buying and installing value added Parlay services where a gateway is already in place.

The broader mobile services infrastructure industry is similar, and in general this is a competitive industry, but the rewards for creating a sustainable competitive advantage are high. We believe that by embedding substantial knowledge and intellectual property into a mobile magazine platform in the areas of Parlay, billing, and multi-media services that it can form the basis of a sustainable competitive advantage.

Regarding the route to market, Parlay/OSA vendors such as AePONA use a direct sales force to sell their gateway and professional services to mobile operators. An operator buys a gateway, not for the gateway itself, but rather for the new services to be introduced. Thus Parlay vendors such as AePONA work hard to establish an eco-system of vendors that deliver services on top of their gateway.

The vision of the authors is to make the development and management of advanced programmable and content services easy for service providers with the m-Mag platform, and in doing so, reduce the barriers to entry in the mobile services market for third-party service providers, which will enable this market to explode.

RELATED WORK

To the research of the author, no comprehensive systems for supporting multi-party eco-systems for mobile content publishing have been reported about. This section therefore discusses related work providing preliminary streams into this issue.

Concepts

Read and Maurer (2003) discuss the metaphors of paper documents and Web sites, as organised in pages; in contrast to the mobile world, where small screens, restricted storage, and limited bandwidth prevent large and complex layout. Therefore, content organisation could be considered as a "deck of cards." They recognise such a metaphor underlying all current content development for mobile wireless technology.

In an approach to combine and augment print and mobile media, Ota, Shimizu, Yamazaki, and Ohashi (2004) propose a paper-based cross-media cooperation (PCMC) system. For overcoming the difficulties of the mobile user to access particular content, paper-based media (books, magazines, posters) are annotated with mark-ups that are recognised by the mobile device. Their concept (Figure 2) introduces an online electrical feature definition file relating to the paper material. A pen-like device com-

Figure 2. Paper-based cross-media cooperation (PCMC) system by Ota et al. (2004)

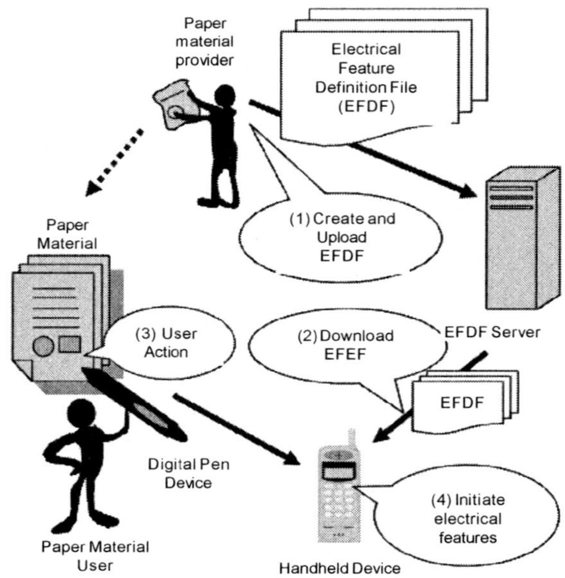

municates with the mobile device and thereby with the server while the user interacts with the paper.

General Technological Approaches

Three application-development technologies are dominant in the area of mobile wireless technology: the wireless application protocol (WAP and WAP 2.0), NTT DoCoMo's i-mode, and Sun Microsystem's Java 2 Micro Edition (J2ME) (Read et al., 2003).

While WAP, a free and open specification, initially suffered from a steep learning curve and incompatibility with existing Web content, it was endorsed by major wireless operators. WAP 2.0 adopts existing Web standards, aiming to improve the user experience, allowing a move to richer content and cheaper bandwidth.

With WAP as a protocol and i-mode as a service, Read et al. (2003) does not see them as competitors, in particular as NTT DoCoMo is a member of the WAP Forum.

The Java 2 Micro Edition is advertised by Sun Microsystems (JAVA, 2005) as a Java platform for all kinds of commodities and embedded devices, ranging from phones and PDAs to set-top boxes and vehicular telematics.

For its first launch, the mobile magazines published by Sendandsee employ their proprietary "mobiprint" publishing technology, supporting a variety of Symbian, J2ME, and WAP-enabled handsets. Different client versions for customer groups are all produced with the same production system, enabling some automation (Schuurmans et al., 2004). The publishers focus on the implementation of "new, rich media containing mobile services," not discriminating users based on their handsets. Thereby, they consider high-end handsets for development work but avoid such polarisation of user groups, which would be considered to negatively impact the commercial development of such services.

Mohomed, Chin, Cai, & de Lara (2004) address the issue of user presentation, and introduce the approach of community-driven adaptation (CDA). It focuses on the automatic content adaptation based on feedback provided by users within the same community group (sharing common characteristics and preferences), thereby priorising content and optimising screen space and bandwidth. It aims at a higher user fidelity over the total period of usage of the mobile service.

Frameworks and Architectures

The MediaBroker, introduced by Modahl, Bagrak, Wolenetz, Hutto, and Ramachandran (2004), provides a distributed framework for pervasive applications. In particular, it features a type-aware data transport mechanism that is capable of dynamically transforming data on its way from the source to the sink in complex pervasive applications.

The Gaia middleware allows the set-up of active spaces, orchestrating digital and physical environments, while the Gaia Microserver proposed by Chan, Bresler, Al-Muhtadi, and Campbell (2005) enables mobile devices to run lightweight proxy clients written in J2ME to join such active spaces. Its main motivation is to provide a collective synergy among the participants. It solves the problem of the mobile devices communicating directly with Gaia due to resource constraints by providing a level of indirection between them and Gaia. The proxy is implemented to allow the mobile devices to act as specialised input and output devices.

Joshi, Finin, and Yesha (2001) discuss problems spanning the fields of data management and pervasive computing while using multiagent systems as the enabling technology. They propose agent frameworks and service discovery and delivery protocols to be suitable to this environment. They experiment with prototypes and applications for mobile electronic commerce within dynamic communities of ad-hoc mobile services.

Berkvens, Claassen, van Gassel, and Sinitsyn (2005) analyse the distribution of media from the perspective of Philips Research. They argue, within the scope of their connected planet (CP) concept, for a new media distribution architecture. Addressing the problems of mobile devices, such as temporary and unannounced loss of network connectivity, usually short connection sessions, and scarce resources, they propose a media distribution management. It offers a single point of abstraction for applications running on the local device, handling all distribution-related operations (e.g., queries, copies, moves, deletions) on separated meta-data and content. This separation allows more efficient search than in meta-data still attached to content files.

Davidyuk, Riekki, Rautio, and Sun (2004) consider the middleware aspects for mobile multimedia application for mobile devices from the perspective of context-awareness. Their CAPNET middleware offers functionality for service discovery, asynchronous messaging, publish/subscribe event management, storing and management of context information, user interfaces and network resource handling. By handling functions such as capture and rendering, storing, retrieving and adapting of media content, it enables the development of complex, context-aware multimedia applications.

Authoring

While plenty of commercial authoring software exists for print and Web media, the development for the mobile market is in its infancy.

Figure 3. Real-world content acquisition and related information retrieval by Kidawara, Uchiyama, and Tanaka (2005)

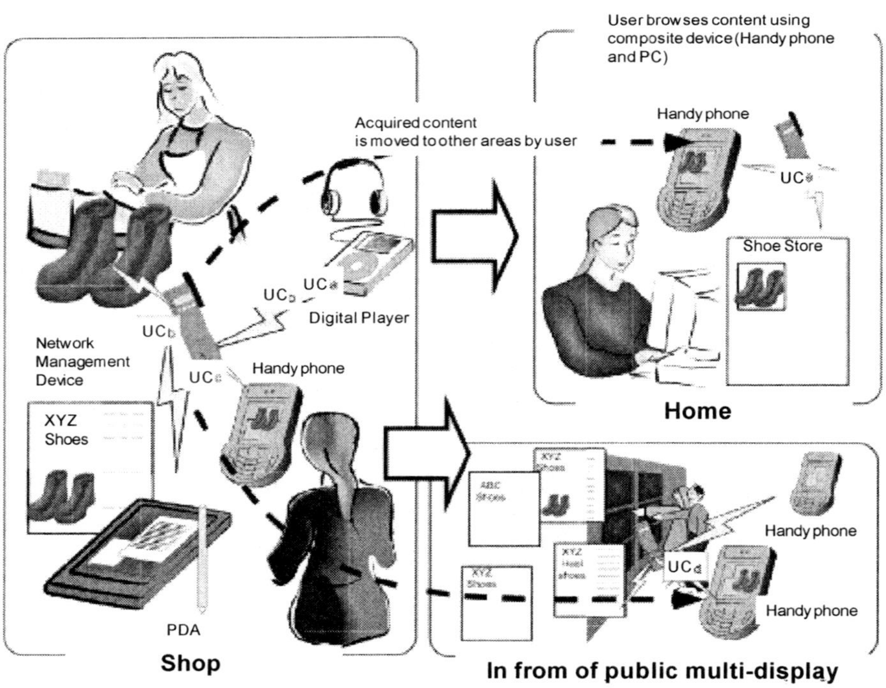

Teng, Wu, Chen, Chu, and Hsu (Teng (2004)) propose a mobile authoring tool for personal experience computing. Their "mProducer" enables the user to generate content and relate it to his personal experience. The system introduced an algorithm for Storage Constraint Uploading (SCU) of multimedia data, as well as Sensor-Assisted Automated Editing for quality-improved picture acquisition. While the authors have designed their system for personal experience only, it has capabilities to support distributed content acquisition for mobile magazines, as the image resolution of the collected data would match the requirements of the percipient.

Such collaborative content acquisition is the direct target of Kidawara, Uchiyama, and Tanaka et al. (2005), employing coordinated ubiquitous devices, as shown in Figure 3. Their content processing mechanism allows the content, augmented with the context identifier UCx, to be related to retrieved information from the internet. This information can than be compared to other items, or shared within a community group.

The multi-device authoring technology (MDAT) by Banavar, Bergman, Cardone, Chevalier, Caeremynck, Giraud, et al. (2004) from the IBM labs supports developers in building generic applications which are common to multiple devices, which can later be customised for specific devices.

As illustrated in Figure 4, MDAT is intended to reduce the complexity of creating interactive, form-based applications for heterogeneous devices. Existing (legacy) HTML pages are automatically generalised into a device-independent representation. In the next step, they are modified for specific targets (specialisation), and iteratively refined. Manual editing interaction allows for an overall semi-automatic process.

Kobayashi and Fujioka (2003) go a formal approach for optimising Web pages and link structures for mobile devices. They argue that the technique of separation of link and data, which is often used in restricted mobile devices, leads to an accumulation of redundant links, which are automatically optimised by their algorithm.

Last but not least, the content needs to be protected in accordance with the rights of the copyright owners. While this is a complete area of its own, the approach of Dusparic, Dahlem, and Dowling (2005) towards flexible rights management in pervasive environments should be exemplarily discussed. They argue that the characteristics of pervasive environments, such

Figure 4. MDAT design-time application generation (solid arrow: automatic, dashed arrow: manual) (Banavar et al. 2004)

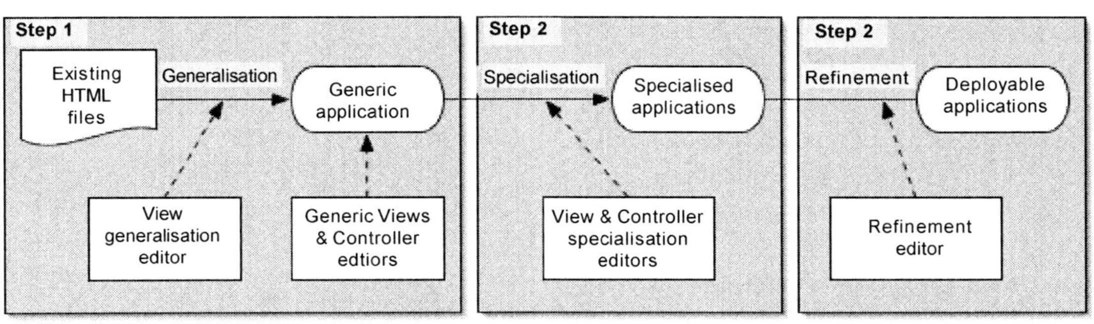

as occasional connectivity, require more flexible usage rights than traditional approaches for software and electronic content. Their pervasive applications rights management architecture (PARMA) combines features from both license management systems (LMS) and digital rights management (DRM).

THE M-MAG MOBILE MAGAZINE ECOSYSTEM

This section introduces the mobile magazine services platform, m-Mag (Downes, Pfeifer, & Ryan 2005), as developed within the TSSG.

The TSSG (Telecommunications Software & Systems Group), a special research unit at WIT (Waterford Institute of Technology), has been part of this industry since it initiated. It has managed and participated in the Opium project (OPIUM, 2004) (which recently achieved the world's first inter-network Parlay roaming) and has built a substantial knowledge base, technical expertise and set of contacts in the industry. The TSSG is currently establishing a campus company to commercialise the m-Mag platform.

The mobile magazine services platform, m-Mag, addresses two critical issues that currently prevent mobile publishing from taking-off, namely:

- enabling publishers to partner with mobile operators for the distribution of content
- the creation of mobile services by publishers that are tuned to the mobile channel

Tuning content to the mobile channel is critical to success for publishers because mobile devices have different form factors and usage patterns to the wired Internet, and publishers need to create attractive content that is integrated with premium value added mobile services such as SMS (short message service) MMS (multimedia messaging service) and LBS (location-based service).

The metaphor to support these needs is the "mobile magazine", which is publishers' content adapted for mobile devices and integrated with value added mobile services—see Table 1 for illustration.

Table 1. What is a mobile magazine?

What is a mobile magazine?
It's the mobile version of a traditional magazine such as GQ, Loaded, Cosmo, tuned for delivery over the mobile channel. It could also be the mobile version of newspapers and TV stations – in fact any multimedia publication. It is mobile Web technology integrated with value added mobile services such as messaging, video, Flash and location based services. It is integrated into an operator's portal and billing system.
What types of Value Added Services are integrated into a Mobile Magazine?
SMS – Short Message Service (Text Messaging) - Alerts - Voting - Feedback - Chat
MMS – Multimedia Message Service - Alerts - Video Clips - Slide Shows - Animations
LBS – Location Based Services - Local alerts - Local information - Positioning - Personalisation based on area
Macromedia Flash - Enables sites to use advance animated design for - Infotainment - Advertising - Delivery of content
Other Widget Services - Chat - Bulletin Boards - Dating
Download Services - Java Games - Java applications - Brew Games - Ringtones - Screensavers

Partnering with mobile operators to create mobile magazines is critical to success for publishers, because mobile Internet users do not use search tools to discover mobile content and services as they do on the wired Internet—rather they access their mobile operator's portal.

Given this, it is critical for publishers to position themselves on the portal. Furthermore publishers need to have a way of charging for their content; mobile operators can facilitate this through their billing systems. Also premium value added mobile services such as SMS, MMS and LBS can only be delivered by working with mobile operators and connecting to their low level gateways that enable these services.

Hence, without mobile operators, publishers can neither position their offerings, nor tailor them to the mobile channel. However, partnering with a mobile operator, and gaining access to the operator's mobile portal and services such as charging, messaging, and location is incredibly complex from a legal, financial, and technical perspective and is out of the reach of almost all publishers.

Operators do not want it to be so difficult for content providers to deliver content and services over their networks. However, they have complex networks that were not designed to enable a myriad of service providers to connect to them, and their business processes are not flexible enough to support them.

What is needed to advance the state of the art is an integration platform that enables publishers to create mobile content using their existing tools, integrate mobile value added services using pre-built controls that require no programming, deliver their content over mobile networks, and automatically charge subscriptions for content and per usage charges for value added services. The m-Mag platform addresses these needs by providing an open integration platform that links mobile operator and publishers together seamlessly.

Scenario-Based Analysis

The scenario-based approach (Go & Carroll, 2004; Rosson & Carroll, 2003) has been used to understand the operation between publishers,

Figure 5. Overall mobile magazine publishing scenario

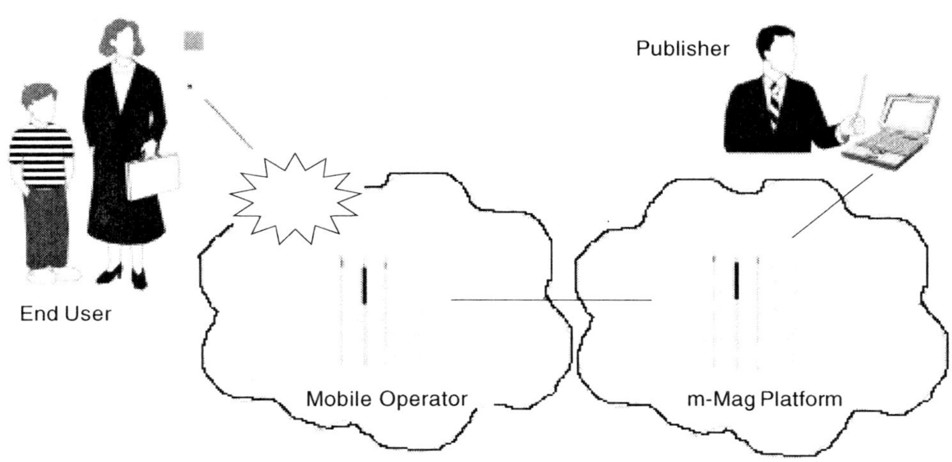

operators and end-users — and the provider of the m-Mag platform as illustrated in Figure 5. These usage scenarios have been defined as follows.

Publisher Scenario

1. A publisher decides to use the m-Mag service after being advised of it by one of their local mobile Operators.
2. They download and install the m-Mag toolbar (plugin) for their publishing environment Adobe Go Live from the m-Mag public Web portal, and use it to register their organisation and magazines with the m-Mag portal and with selected operators.
3. They create their mobile content using Adobe GoLive CS, utilising their magazine content from their Adobe print and Web magazine publishing system. They can format this content using predefined m-Mag templates or create their own.
4. They drag and drop value added services and Macromedia Flash content into the magazine from the m-Mag toolbar. For example, they drop in an alerts service into the magazine and are guided through a wizard set-up to create their alerts.
5. They create MMS slide shows and video clips in Adobe GoLive and then link them into the magazine by dragging and dropping MMS controls from the m-Mag toolbar into the magazine, whereby they are taken through an MMS creation wizard, which enables them to link the MMS documents they created, along with their end-user cost, with the magazine and the operator that transmits the MMS message.
6. They then upload the magazine to the m-Mag portal using the m-Mag toolbar site management controls, then preview it and put it live on one or more operator's networks. The mobile operator can review and approve the magazine's content before the magazine can go live for the first time.
7. Using the m-Mag toolbar they can change the content or services on the live site at any-time. Later they can view statistical reports on subscriptions and value added services revenue from the m-Mag toolbar reports controls.

Operator Scenario

1. An operator decides to launch mobile magazines using the m-Mag platform, as they already have a Parlay gateway installed. They contract with the m-Mag portal operator which specifics the terms and conditions of portal access and revenue sharing.
2. A secure connection is established between the operator and the m-Mag portal ASP (application service provider) environment.
3. The operator creates a menu link on their mobile services portal to the m-Mag portal called "mobile magazines." The operator advertises the service to publishers in their territory. The operator is asked to approve each magazine's content before it goes live for the first time.
4. Once the magazine is live the operator and m-Mag portal operator exchange revenue share reports on a monthly basis and the m-Mag portal operator shares revenue with content providers.

End-User and m-Mag Portal Scenario

1. The user selects their mobile operator's portal on their mobile phone — they establish a GPRS/UMTS context and connect to operator's WAP gateway.

2. They select the "mobile magazine" menu, which links to the m-Mag portal, browse through the list of magazines and click on the desired mobile magazine.
3. The m-Mag portal communicates with the Parlay gateway to determine the terminal capabilities and selects the most appropriate WCSS (wireless cascading style sheet). The user is delivered to the home page of the selected magazine.
4. The user can browse a few sample pages and is invited to subscribe (e.g., EUR4 per month).
5. The user accepts and the m-Mag portal communicates with the Parlay charging API, to determine if the user has EUR4 in their prepaid account (prepaid phones billed immediately). If yes an RBS (Rating Bureau Service) accounting record is written and the user is billed immediately through the Parlay gateway.
6. The m-Mag portal server then communicates with the client provisioning server to update the user's terminal settings via a WAP push over SMS. Service settings are a bookmark and possibly userid and password. The user is then free to access the full magazine.
7. If the user decides to download a Java application such as a game, the m-Mag portal communicates with the download server to download the application, and then the Parlay charging API to charge the user for this value added content.
8. Similarly if the user downloads a videoclip, the m-Mag portal communicates with the value added services server (which integrates to Parlay messaging) to deliver the MMS clip, and then the charging API to charge the user for this value added content.
9. In both cases an accounting record is written to the RBS to enable revenue share to be reported on in the customer management server and in all cases prepaid credit is checked first.

Technology

To validate that the critical issues as discussed in the introduction can be addressed, we have developed a prototype of the m-Mag service platform. The prototype is hosted as an ASP (application service provider) service for operators and publishers, enabling publishers to create mobile magazines and operators to deliver them to their consumer base.

In collaboration with industry, namely O2 Ireland Ltd. on the operator side, we have collected the requirements and validated the commercial viability. Currently, the prototype is analysed based on industry feedback, leading into the route to commercialisation. The prototype m-Mag platform has been designed to comprise of three major subsystems, as discussed below.

All operator integration is being achieved through Parlay/OSA (Lozinski, 2003). Parlay/OSA provides an open interface created by a consortium of 65 companies in the IT and telecom industries, which enables the creation of new telecommunications services. It has been defined jointly between the European Telecommunications Standards Institute (ETSI), Parlay, and the Third Generation Partnership Program (3GPP).

Parlay (2004) is implemented as a gateway that is integrated into an operator's network, which then exposes an API to third parties to enable them to access network capabilities. The m-Mag operator integration subsystem focuses on using the AePONA Parlay 3.1 APIs to enable charging, delivery of messages across an operator's network (SMS and MMS), obtaining the location of a mobile phone (terminal) and its capabilities (screen size, resolution, etc).

This subsystem integrates with Waterford Institute of Technology/WIT's Rating Bureau Service (RBS) for reporting on subscriptions and revenue shares. The RBS is a product that WIT has developed (supported by Enterprise Ireland funding) which uses a flexible XML-based tarrifing scheme to rate and report on mobile services. Using the RBS greatly simplifies the job of tracking and reporting on revenue sharing data for operators and magazine providers.

The key reason for using Parlay/OSA was to validate that the m-Mag platform can integrate into an operator's network using standardised APIs and to validate that the platform is portable across different operator networks. The key reason for using the RBS was to validate that the m-Mag platform can adequately perform rating and calculate revenue shares for all parties in the value chain.

The m-Mag portal hosts the content provider's mobile magazine and is a Web server which:

- Provides the mobile magazine portal that is linked into a mobile operator's portal
- Integrates with the m-Mag tool bar to enable a content provider to upload and maintain a mobile magazine on the m-Mag portal
- Integrates value added services into the mobile operators network via the m-Mag operator integration subsystem
- Provides reporting and statistics to the mobile operator and the mobile magazine owner on subscriptions and revenue share. This reporting module uses WIT's RBS product to report on individual magazine and publishing company revenue shares
- Provides basic customer relationship management (CRM) functionality to support churn management, customer queries, and market research

The key reason for creating the m-Mag portal was to validate that an ASP hosting and management service for mobile magazines for operators and publishers can be created and is of value.

Adobe GoLive is one of the leading Web publishing tools for the publishing industry. The latest version of the product Adobe GoLive CS is the first publishing toolset to provide support for the mobile Internet that fully integrates with the rest of its suite of products. As a result, publishers are able to design their mobile magazine using a combination of their existing publishing tool GoLive and the m-Mag toolbar. The m-Mag toolbar has been implemented as a GoLive plug-in, adding the following capabilities:

- Registration services—legal, financial, publisher, and operator details for the m-Mag platform
- m-Mag custom controls to design and integrate value added mobile services such as SMS, MMS, Macromedia Flash and LBS into the magazine (e.g., alerts, voting, personalisation)
- Subscription and accounting services for magazine content
- Magazine hosting management tools to upload and maintain the magazine on the m-Mag platform
- Circulation and revenue share data reports

The key reason for creating the m-Mag toolbar prototype was to validate that publishers can use their standard publishing tools to create and maintain mobile magazines.

DEN-ng Information Modelling

A directory-enabled network (DEN) (Strassner, 2003) integrates user profiles, applications, and network services through a common information model that stores network state and exposes network information. This information then enables bandwidth utilisation to be optimised; it strengthens security; it enables policy-based management; it provides a single point of administration of all network resources; and all this serves to lower total cost of ownership, and improves the services that end-users can rely on regardless of their physical location.

DEN-ng (DEN-new generation) is a TMF (telemanagement forum) standard which is also in the process of being adopted by the ITU (International Telecommunications Union) and ISO (International Standards Organisation). DEN-ng also provides the unique service lifecycle perspective required by the m-Mag platform.

The information model of the m-Mag platform has been developed using an eTOM-based (eTOM, 2003) DEN-ng model, based on the user requirements captured. This information model represents real world objects, not software, and is developed in a platform, language and protocol agnostic way. The model includes:

- Things of interest (entities)
- Relationships between these things (associations)
- Details/characteristics of these things (attributes)
- How these things work (methods, which represent behaviour)

The DEN-ng information model is part of the TMF SID framework and thus takes the TMF NGOSS views: business needs (what is required); system capabilities (what is possible); distributed functionality (how it works);

Figure 6. Overview and interconnection of the m-Mag hosted ASP platform by Pfeifer and Downes (2005)

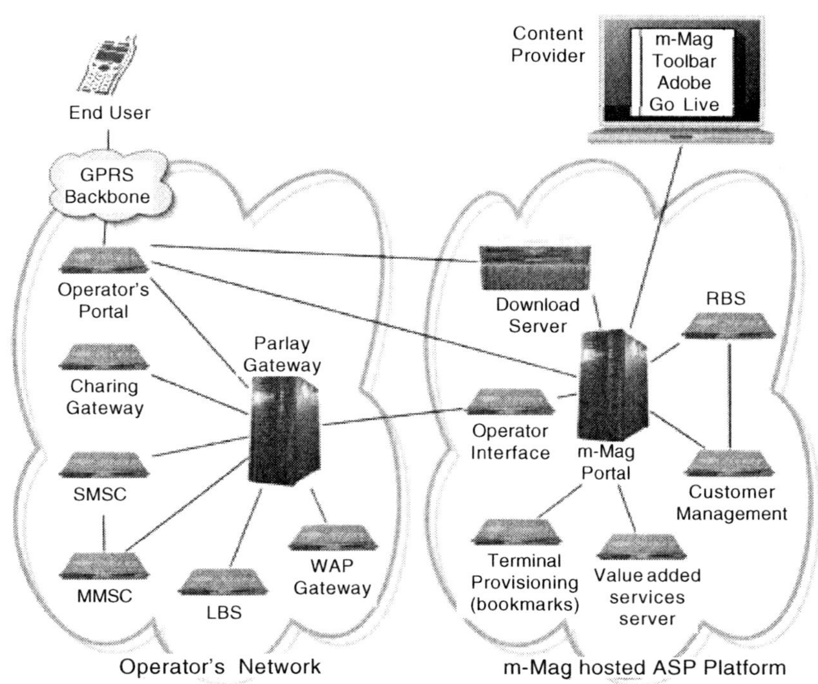

product offerings (vendor building blocks); monitoring and performance (is it working?).

m-Mag has extended the DEN-ng model for multi-media services and will engage with the TMF to standardise these extensions.

M-MAG IMPLEMENTATION

m-Mag Server Architecture

Figure 6 provides an overview of the m-Mag platform. The final commercialised version of the m-Mag prototype will be comprised of a number of servers as outlined with their functionality below:

The m-Mag Portal hosts the publisher's magazine, providing the capabilities as discussed later.

The download server:

- uses Java MIDP OTA (from the Java Community Process) to download Java applications to the user terminal (mandatory for MIDP 2.0), and
- provides ring tones, etc., via OMA OTA download mechanism.

The server for terminal provisioning:

- uses OMA Client Provisioning 1.1 to configure service settings on the user terminal for example to bookmark the magazine URL on a terminal via a WAP push over SMS.

The value added services server integrates to operator gateways and provides wizard-based service design for:

- SMS, MMS, Flash and LBS content, and
- also has a set of predefined widget services that can be dropped into the magazine: alerts, voting, chat, bulletin boards, feedback/letters, RSS feeds.

The operator interface serves the Parlay interface, which uses the following Parlay APIs:

- Charging
- Messaging (SMS, MMS)
- Terminal capabilities
- User location

The Rating Bureau Service (RBS) has been previously implemented at Waterford Institute of Technology (WIT). This is a fully functioning rating service that uses the IPDR specification for designing charging rules and schema and is used by the m-Mag platform for determining magazine revenue share.

The customer registration and management serves to:

- Registers customers
- Registers magazines
- Uses RBS to report on revenue share

These servers are logical rather than physical servers, but are built so that their functionality can be partitioned over one or more servers for security, scalability and performance as needed in the ASP environment.

Figure 6 represents a fully-grown commercial version of the m-Mag service platform, which includes a download server for Java Games, ringtones and screensavers and also a terminal-provisioning server for updating mobile phone bookmarks OTA (Over The Air).

Implementation Status

In the first iteration, a basic working version of the m-Mag platform has been built, focusing on enabling a publisher to create mobile content using the m-Mag toolbar, upload it to the m-Mag portal, maintain it, then put the content live and charge for it. The second iteration currently focuses on adding advanced features such as value added services, namely SMS, MMS and LBS integration into the m-Mag toolbar, portal and operator interface. This second iteration also focuses on RBS integration and reporting.

CONCLUSION AND FUTURE WORK

We believe in the concept of the mobile magazine, which is mobile publishing integrated with mobile data services, is a new media service category with significant revenue potential. We also believe that the creation of the m-Mag integration platform to link publishers and operators is also a key innovation that advances the state of the art in the delivery of mobile services.

The indicators that have been used to confirm the feasibility of the concept are two fold: firstly we have assessed the technical feasibility of project by building a prototype; and secondly the commercial feasibility of the concept by engaging with Industry. Both approaches gathered requirements and assessed that the final prototype does address the two critical issues of: (1) adapting content to the mobile channel and (2) providing access for publishers to mobile operators networks, in an economic way, that creates value for all participants in the value chain.

We would see it as essential to add support for different national languages in the future so that the service can be marketed across Europe. In addition, Parlay is the only open services platform of its type in production at the moment, but competing Web services initiatives are being developed by the OMA (Open Mobile Alliance) and a Microsoft/Vodafone alliance. As these initiatives come to market we believe it would be necessary to enhance the m-Mag platform to interface to them. We also see enhancements to the m-Mag tool bar such as increasing the number of publishing systems that it is developed for, including adding support for Quark and Dreamweaver. We also believe that there could be value in additional custom controls such as a control for adding Java game downloads to magazines and similar controls for ringtones and screen savers. Looking further into the future we could see initiatives that could enable the m-Mag system to become the outsourced data services platform for tier 3 operators worldwide.

As a general outlook such a magazine publishing platform will become a part on even more comprehensive service provisioning platforms.

ACKNOWLEDGMENT

This work on the m-Mag platform has been supported by Enterprise Ireland's Proof of Concept programme for new product/process development, PC/2004/340, m-Mag.

REFERENCES

3GPP. (2000). *3GPP Technical Specification Group Radio Access Network. Stage 2: Functional Specification of Location Services in UTRAN* (Technical Specification 3G TS 25.305 V3.1.0). 3rd Generation Partnership Project (3GPP/TM), March 2000. Retrieved from http://www.3gpp.org

Abowd, G., Iftode, L., & Mitchell, H. (2004). Guest editors' introduction: The smart phone—A first platform for pervasive computing. *IEEE Pervasive Computing*, *04*(2), 18-19.

Banavar, G., Bergman, L., Cardone, R., Chevalier, V., Gaeremynck, Y., Giraud, F., et al. (2004, July/September). An authoring technology for multidevice Web applications. *IEEE Pervasive Computing, 3*(3), 83-93. Retrieved from http://doi.ieeecomputer society.org/10.1109/MPRV.2004.1321033

Berkvens, H., Winfried, A., Claassen, A., van Gassel, J. P., & Sinitsyn, A. (2005). Media distribution in a pervasive computing environment. *The 3rd IEEE International Conference on Pervasive Computing and Communications Workshops (PERCOMW'05)* (pp. 204-208). Retrieved from http://doi.ieee computersociety.org/10.1109/PERCOMW.2005.47

Chan, E., Bresler, J., Al-Muhtadi, J., & Campbell, R. (2005). Gaia microserver: An extendable mobile middleware platform. *The 3rd IEEE International Conference on Pervasive Computing and Communications (PERCOM'05)* (pp. 309-313). Retrieved from http://doi.ieee computersociety.org/10.1109/PERCOM.2005.22

Davidyuk, O., Riekki, J., Rautio,V., & Sun, J. (2004). Context-aware middleware for mobile multimedia applications. *Proceedings of the 3rd International Conference on Mobile and Ubiquitous Multimedia*, ACM, College Park, MD (pp. 213-220). Retrieved from http://doi.acm.org/10.1145/1052380.1052410

Downes, B., Pfeifer, T., & Ryan, C. (2005, July 11-13). Mobile operator publishing and entertainment platform. *ICMB 2005, The 4th International IEEE Conference on Mobile Business.* Sydney, Australia.

Dusparic, I., Dahlem, D., & Dowling, J. (2005). Flexible application rights management in a pervasive environment. *2005 IEEE International Conference on e-Technology, e-Commerce and e-Service (EEE'05)* (pp. 680-685). Retrieved from http://doi.ieeecomputer society.org/10.1109/EEE.2005.76

EITO. (2002). Entering the UMTS era—mobile applications for pocket devices and services. *EITO (European Information Technology Observatory) 2002 Special Studies.* Retrieved from http://www.eito.com/previous.html

eTOM. (2003). *eTOM: enhanced Telecom Operations Map.* Retrieved from http://tmforum.org/browse.asp?catID=1647

Go, K., & Carroll, J. M. (2004, November/December). The blind men and the elephant: Views of scenario-based system design. *Interactions, 11*(6), 44-53). New York: ACM Press.

JAVA. (2005). *Java 2 Micro Edition.* Retrieved from http://java.sun.com/j2me/

Joshi, A., Finin, T., & Yesha, Y. (2001). Agents, mobility, and m-services: Creating the next generation applications and infrastructure on mobile ad-hoc networks. In B. Kig-Ries et al. (Eds.), *Developing an infrastructure for mobile and wireless systems. NSFWorkshop IMWS 2001,* Scottsdale, AZ, October 15 (LNCS Vol. 2538/2002, pp. 106-118). Springer.

Karvonen, J., & Warsta, J. (2004). Mobile multimedia services development: Value chain perspective. *Proceedings of the 3rd International Conference on Mobile and Ubiquitous Multimedia*, ACM, College Park, MD (pp. 171-178). Retrieved from http://doi.acm.org/10.1145/1052380.1052404

Kidawara, Y., Uchiyama, T., & Tanaka, K. (2005). An environment for collaborative content acquisition and editing by coordinated ubiquitous devices. Special interest tracks and posters of the *14th International Conference on World Wide Web*, Chiba, Japan (pp. 782-791). Retrieved from http://doi.acm.org/10.1145/1062745.1062748

Kobayashi, A., & Fujioka, H. (2003). Personalizing a Web site for cellular phones. IEEE/WIC 2003. *IEEE/WIC International Conference on Web Intelligence (WI'03)* (pp. 432). Retrieved from http://doi.ieeecomputersociety.org/10.1109/WI.2003.1241233

Lozinski, Z. (2003). *Parlay/OSA—A new way to create wireless services*. White paper, June 2003. Retrieved from http://www.parlay.org/specs/library/

Modahl, M., Bagrak, I., Wolenetz, M., Hutto, P., & Ramachandran, U. (2004). MediaBroker: An architecture for pervasive computing. *The 2nd IEEE International Conference on Pervasive Computing and Communications (PerCom'04)* (pp. 253). Retrieved from http://doi.ieeecomputersociety.org/10.1109/PERCOM.2004.1276863

Mohomed, I., Chin, A., Cai, J. C., & de Lara, E. (2004). Community-driven adaptation: automatic content adaptation in pervasive environments. *The 6th IEEE Workshop on Mobile Computing Systems and Applications (WMCSA'04)* (pp. 124-133). Retrieved from http://doi.ieeecomputersociety.org/10.1109/MCSA.2004.10

OPIUM. (2004). *Open platform for integration of UMTS middleware*. OPIUM. EU IST project 2001-36063, 2002-2004. Retrieved from http://www.ist-opium.org/

Ota, S., Shimizu, A., Yamazaki, S., & Ohashi, M. (2004). Proposal of paper-based cross-media cooperation system. *2004 IEEE International Conference on Mobile Data Management (MDM'04)* (pp. 152, 2004). Retrieved from http://doi.ieeecomputersociety.org/10.1109/MDM.2004.1263053

Parlay. (2004). *Parlay Specification*, Version 4.1, 2004. Retrieved from http://www.parlay.org/specs/

Pfeifer, T., & Downes, B. (2005). *m-Mag: The mobile magazine services platform*. WMCS'05: The 2nd IEEE International Workshop on Mobile Commerce and Services, Munich, Germany, July 18-19.

Pfeifer, T., Magedanz, T., & Hübener, S. (1998). Mobile guide — Location-aware applications from the lab to the market. In T. Plagemann, & V. Goebel (Eds.), *Interactive distributed multimedia systems and telecommunication services. The 5th International Workshop, IDMS'98*, Oslo, Norway, Sep 8-11 (LNCS Vol. 1483/1998, pp. 15-28). Springer.

Read, K., & Maurer, F. (2003). Developing mobile wireless applications. *IEEE Internet Computing*, *07*(1), 81-86. January/February 2003.

Rosson, M. B., & Carroll, J. M. (2003). Scenario-based design. In J. A. Jacko, & A. Sears (Eds.), *The human-computer interaction handbook: Fundamentals, evolving technologies, and emerging applications* (pp. 1032-1050). Mahwah, NJ: Lawrence Erlbaum Associates.

Satyanarayanan, M. (2005, April/June). Swiss Army Knife or wallet? *IEEE Pervasive Computing*, *04*(2), 2-3.

Schuurmans, J., & Poropudas, T. (2004). *Sendandsee and Zed/LaNetro launch mobile magazines in European markets*. Nordic Wireless Watch—November 8, 2004. Retrieved from http://www.nordicwirelesswatch.com/wireless/story.html?story_id=3903

Sendandsee, Inc. (2004). *Sendandsee Inc*. Retrieved from http://www.sendandsee.com/

Strassner, J. (2003). *Leveraging DEN-ng to model all business, system, and network properties*. White paper. Retrieved from http://www.intelliden.com/page.asp?id=Insights&subID=DEN-ng

Takeshi, N. (2003). *The i-mode wireless ecosystem*. Hoboken, NJ: John Wiley & Sons.

Teng, C., Wu, C., Chen, Y., Chu, H., & Hsu, J. Y. (2004). Design and evaluation of mProducer: a mobile authoring tool for personal experience computing. *Proceedings of the 3rd International Conference on Mobile and Ubiquitous Multimedia*, ACM, College Park, MD (pp. 141-148). Retrieved from http://doi.acm.org/10.1145/1052380.1052400

KEY TERMS

DEN-New Generation (DEN-ng): TMF (Telemanagement Forum) standard in the process of being adopted by the ITU (International Telecommunications Union) and ISO (International Standards Organisation).

Directory-Enabled Network (DEN): Integration of user profiles, applications, and network services through a common information model that stores network state and exposes network information.

Mobile Magazine: A magazine on a mobile computing and communication platform (in contrast to print magazines about mobile technologies), providing valuable, current multimedia content.

Value Added Mobile Services: Services allowing third parties (other than mobile operators) to generate revenue by selling premium mono/multimedia content of various kind over the mobile communication channel.

About the Authors

Ismail Khalil Ibrahim is a senior researcher and lecturer at the institute of Telecooperation at Johannes Kepler University Linz, Austria. He currently teaches, consults, and conducts research in mobile multimedia, norm-based computing, and ubiquitous intelligence and is also interested in the broader business, social, and policy implications associated with the emerging information technologies. Dr. Ibrahim holds a BSc in electrical engineering (1985), MSc, (1998) and PhD (2001) in computer engineering and information systems. Before joining Johannes Kepler University of Linz in October 2002, he was a research fellow at the Intelligent Systems Group at Utrecht University, Netherlands from 2001-2002 and a senior researcher and project manager at the Software Competence Center Hagenberg—Austria from 2000-2001. Dr. Ibrahim is the founder and the acting president of the international organization on information integration and Web-based applications and services, overseas collaborator in the E-Commerce Lab at the National University of Singapore and the Telematics Lab at Bandung Institute of Technology, Indonesia. He is the steering committee chair of the international (iiWAS) series of conferences on Information Integration and Web-based Applications & Services since 1999 and (MoMM) series of conferences on Advances in Mobile Multimedia since 2003. He serves as the editor-in-chief of the *International Journal of Web Information Systems*, *Journal of Mobile Multimedia*, editorial board of several journals, including, *Business Process Management Journal*, *The International Journal of Information Technology and Web Engineering*, *Journal of Ubiquitous Computing and Intelligence*, *International Journal of Wireless and Mobile Computing*, and the *International Journal of Business Data Communications and Networking*. His work has been published in various books, journals, and conferences. He is also a reviewer and PC member for several conferences and journals.

* * *

Ashraf M. A. Ahmad is currently a PhD researcher in this MIS Lab at National Chiao Tung University. His interest area includes MPEG video streaming and multimedia communication. He has published many conference and journal papers in his field of expertise. Mr. Ahmad has co-edited several conference proceedings and books. He chaired and organized many special issues in international conferences.

Marios C. Angelides is a professor of computing at Brunel University. He holds a BSc in computing and a PhD in information systems both from the London School of Economics, where he began his career as lecturer in information systems. He is a member of the ACM, the IEEE Computer Society, and the British Computer Society.

Antonio Valdovinos Bardají is a full professor in the Department of Electronics Engineering and Communications at the University of Zaragoza, Spain, and a member of Aragón Institute of Engineering Research (I3A). Research interests are in mobile communications.

Jennifer Blechar is a PhD candidate and research fellow with the Department of Informatics at the University of Oslo, Norway, and a visiting research fellow with the Department of Informatics at Copenhagen Business School, Denmark.

Christian Breiteneder is a professor for interactive media systems at the Vienna University of Technology. His current research interests include interactive media systems, media processing systems, augmented reality, content-based visual information retrieval, and e-learning.

Stéphane Bressan is a senior lecturer in the Computer Science Department of the School of Computing (SoC) of the National University of Singapore (NUS) and an adjunct associate professor at Malaysia University of Science and Technology (MUST). He joined NUS in 1998 and MUST in 2004. His research is in the area of the integration of heterogeneous and distributed information.

Philippe Canalda received his MSc and PhD degrees in computer science from the University of Orléans (France) in 1991 and 1997, respectively. He worked at INRIA Rocquencourt from 1991 to 1996 on the automatic generation of optimizing and parallel n-to-n cross-compilers. From 1996 to 1998, he worked as a research engineer in the Associated Compiler Expert start-up factory at Amsterdam, The Netherlands. Then, he worked two years at LORIA on the synchronisation of cooperative process fragment, based on workflow model, and applied to ephemeral enterprise. Since 2001, he has been an associate professor at the Computer Science Laboratory (LIFC, FRE CNRS 2661) at the University of Franche-Comté in France. His research topics deal with, on the one hand, mobility services and, on the other hand, on robust and flexible optimizing algorithms based on graph, automata, and rewriting theories.

Li-Pin Chang serves as an assistant professor in the Department of Computer and Information Science, National Chiao-Tung University, Hsin-Chu, Taiwan since 2005. His research interests include real-time and embedded systems, storage systems, and System-on-Chip (SoC).

Damien Charlet is an associate researcher at the Computer Science Laboratory of the University of Franche-Comté. He received his PhD degree of the University of Franche-Comté in 2004. Currently, he is focusing on a framework of distributed caches optimized for video streaming to mobile users.

Pascal Chatonnay received his MSc and PhD degrees in computer science from the University of Franche-Comté (France) in 1994 and 1998, respectively. He has worked at CNET Caen from 1994 to 1998 on distributed object systems. Since 1998, he is an associate professor at the Computer

Science Laboratory (LIFC, FRE CNRS 2661) at the University of Franche-Comté in France. His research topics deal with, on the one hand, mobility usage and video streaming on mobile devices and, on the second hand, on transportation systems such as on-demand transportation as an alternative to bus transportation. He is involved in two important multi-field projects with geomaticien and physicist.

Ioanna D. Constantiou is an assistant professor at Copenhagen Business School in the Department of Informatics. She holds a PhD from the Department of Management and Technology in Athens University of Economics and Business.

Patrícia Dockhorn Costa is a PhD candidate in the faculty of electrical engineering, mathematics and computer science of the University of Twente. Her research interests are context-awareness, ubiquitous computing, and service-oriented architectures.

Jan Damsgaard is a professor at the Department of Informatics, Copenhagen Business School, Denmark. He holds a PhD in information systems from Aalborg University, Denmark.

Volker Derballa is a PhD student at the Chair of Business Informatics and Systems Engineering at the University of Augsburg. Further, he is the project manager at MAN B&W Diesel working with business intelligence and knowledge management projects.

Mahieddine Djoudi received a PhD in computer science from the University of Nancy, France, in 1991. He is currently an associate professor at the University of Poitiers, France. He is a member of SIC (Signal, Images, and Communications) Research Laboratory. He is also a member of IRMA (Ingénierie des Ressources Médiatiques pour l'Apprentissage) research group. His PhD thesis research was in speech recognition; his current research interest is in e-learning, multimedia, computer supported cooperative work, and information literacy. He started and is involved in many inter-diplomacy research projects which include many researchers from different universities.

Daniel C. Doolan has spent the past decade studying computer science. He has both a BSc in computer applications, an MSc in multimedia technology, and half a dozen other qualifications in the area of business and computing. He is currently pursuing a PhD in the area of mobile computer graphics. He is a member of the British Computer Society (BCS), the Institute of Electrical Engineers (IEE), and the Institute of Electrical and Electronics Engineers (IEEE).

Barry Downes is the commercial director of the TSSG. In this role, he specializes in the management of advanced communications research and commercialization programmes. He holds a BSc from WIT and an MBA from Smurfit School of Business at University College Dublin.

Schahram Dustdar is a full professor at the Distributed Systems Group, Vienna University of Technology. While on sabbatical leave, he was a post-doctoral research scholar at the London School of Economics and a visiting research scientist at NTT Multimedia Communications Labs in Palo Alto, USA. Since 1999, he has been working as the co-founder and chief scientist of Caramba Labs

Software AG in Vienna, a software company focused on software for collaborative processes in teams. More information can be found on http://www.infosys.tuwien.ac.at/Staff/sd.

Mohamed Ali Feki is a second year PhD student at Handicom Lab Laboratory, National Institute of Telecommunication in France. He has obtained his master's degree from University of Technology in Troys France and his engineer diploma from the National School of Engineer in Sfax Tunisia. His research activity focus is on user interface design and development for pervasive software applications. He is interested in service-based approaches for context-aware application, human machine impact in case of people with disabilities.

Panayotis Fouliras is a lecturer in the Department of Applied Informatics at the University of Macedonia (Greece). His research interests include networks, multimedia streaming, security, and e-commerce technology.

Christos K. Georgiadis is a lecturer in the Department of Applied Informatics at the University of Macedonia (Greece). His research interests include e-commerce and m-commerce technology, security, and multimedia databases. He is a member of the ACM (SIGEcom).

Sylvain Giroux received a PhD in computer science from the University of Montreal, Canada. He is one of the founders of the DOMUS Laboratory at the Sherbrooke University. His main research interests are mobile computing, pervasive computing, and distributed artificial intelligence.

Franz Gruber is currently the project manager at the RISC Software GmbH, working in the field of Web engineering and mobile applications. From 2000 to 2003, he was the project manager at the Software Competence Center Hagenberg.

Stathes Hadjiefthymiades received his BSc, MSc, and PhD degrees in informatics from the Department of Informatics and Telecommunications, University of Athens (UoA). He also received a joint engineering-economics MSc from the National Technical University of Athens. In 1992, he joined the Greek consulting firm Advanced Services Group, Ltd. In 1995, he joined the Communication Networks Laboratory (CNL) of UoA. During the period 2001-2002, he served as a visiting assistant professor at the University of Aegean, Department of Information and Communication Systems Engineering. In the summer of 2002, he joined the faculty of the Hellenic Open University, Patras, Greece, as an assistant professor. Since December 2003, he has worked in the faculty of the Department of Informatics and Telecommunications, University of Athens, where he is presently an assistant professor. He has participated in numerous EU and National projects. His research interests are in the areas of Web engineering, mobile/pervasive computing, and networked multimedia. He has contributed to over 100 publications in these areas. Since 2004, he coordinates the Pervasive Computing Research Unit of CNL.

Jonna Häkkilä works as a usability expert at the Research and Development of Nokia Multimedia, Oulu, Finland. Her research interests include context-awareness and human-computer interaction, including mobile, ubiquitous, and multimodal user interfaces. She has received her MSc degree in physics, and will shortly finish her PhD at the Information Processing Laboratory, University of Oulu.

Timo Hämäläinen received his BSc in automation engineering from the Jyväskylä Institute of Technology in Finland in 1991 and his MSc and PhD degrees in telecommunication from the Tampere University of Technology and University of Jyväskylä, in 1996 and 2002, respectively. His current research interests include traffic engineering and quality of service in wired and wireless networks. Recently, he has studied different traffic engineering capabilities that can be used to dynamically control traffic flows in order to prevent congestion and optimize the availability of resources.

Saad Harous obtained his PhD in computer science from the Case Western Reserve University (CWRU), Cleveland, Ohio in 1991. He worked as a visiting assistant professor for one year at the Department of Computer Engineering and Science (CWRU). After that, he moved to Sultan Qaboos University, Oman, where he was a faculty member from 1992 until 1999 in the College of Sciences. In September 1999, he joined the Department of Computer Science at the University of Sharjah, UAE, as an associate professor. Since September 2003, he is the head of the Department of Computer Science. His teaching interests include programming, data structures, design, and analysis of algorithms, operating systems, and network. His research interests include parallel and distributed computing, distance education, computer aided learning, and e-learning. He is an IEEE senior member.

Werner Hartmann joined the Institute for Applied Knowledge Processing (FAW) in December 1999 as a research assistant. He has publications at international conferences focused on mixed reality and computer graphics (e.g., the ISMAR and the VRST).

Álvaro Alesanco Iglesias received his MSc in telecommunication engineering from the University of Zaragoza (UZ), Spain. He is an assistant professor at UZ. His research interests are mobile telemedicine and signal processing applied to telemedicine.

Robert S. H. Istepanian received his PhD from Loughborough University, UK. He has held many research academic positions in UK and Canada from 1994. He has published more than 150 scientific papers, edited three books, and contributed to several book chapters in the area of mobile communications for healthcare. He was one of the boarding editorial members of IEEE Trans. Inf. Tech. Biomedicine and other international journals in these areas.

Ivar Jørstad received his cand. scient.-degree in informatics from the Department of Informatics, University of Oslo, in December 2002. He is pursuing a PhD degree at the Norwegian University of Science and Technology in Trondheim, focusing on middleware to support future, advanced mobile services with emphasis on personalization and service continuity. Visit http://ongx.org for more information.

Panagiotis Kalnis received his diploma in computer engineering from the Computer Engineering and Informatics Department, University of Patras, Greece, and PhD from the Computer Science Department, Hong Kong UST. Currently, he is an assistant professor at the Computer Science

Department of the School of Computing (SoC), National University of Singapore. His research interests are peer-to-peer, mobile computing, OLAP, data warehouses, and spatial databases.

Ross King received his doctorate from Stanford University and is presently the head of operations of the Research Studio Digital Memory Engineering. His recent research activities concern multimedia databases, data models, and metadata.

Wolfgang Klas is a professor at the Institute for Distributed and Multimedia Systems at the University of Vienna, heading the multimedia information systems group. He is also head of the Research Studio Digital Memory Engineering at the Austrian Research Centers. His research experience includes work in the fields of object-oriented database technology, multimedia information systems, and cooperative systems.

Vassilis Kostakos holds a BSc in computer science and a PhD in human computer interaction, both from the University of Bath. He is currently working as a post-doctoral researcher at the University of Bath. His research interests include human computer interaction, mobile, pervasive, and wearable computing, usability, network and wireless security, novel interaction techniques, design theory for pervasive systems and space syntax.

Gabriele Kotsis received her master's degree (1991, honored with the Award of the Austrian Computer Society), her PhD (1995, honored with the Heinz-Zemanek Preis), and the venia docendi in computer science (2000) from the University of Vienna. She was working as a researcher and teacher at the University of Vienna (1991-2001), at the Vienna University for Economics and Business Administration (2001), and at the Copenhagen Business School (2002). Since December 2002, she is holding a full professor position at the Telecooperation Department at the Johannes Kepler University Linz. Her research interests include performance management of computer systems and networks, workgroup computing, mobile and Internet computing, telemedia, and telecooperation. Prof. Kotsis is the author of numerous publications in international conferences and journals and is co-editor of several books. Since April 2003, she is the president of the Austrian Computer Society.

Hend Koubaa is a postdoctoral researcher at the NTNU University, Trondheim, Norway. Her research interests include content location and medium access control in mobile ad hoc networks. She received her PhD in computer science in March 2003, from the Université Henri Poincaré, France. She was awarded an ERCIM fellowship for postdoctoral research in Fraunhofer IPSI Institute, Darmstadt, Germany, and in NTNU. She has been working with SINTEF, Trondheim, from March 2005 until May 2005.

Reinhard Kronsteiner is a scientific assistant at the Telecooperation Department, Johannes Kepler University Linz, Austria. Mr. Kronsteiner holds a magister degree in business-informatics from the University of Linz. He worked in the areas of software engineering, training, and strategic IT consulting in the financial and telecommunication sectors. His research focus is on business value and the application range of mobile devices.

Janne Lahti is a research scientist at VTT Electronics, Oulu, Finland. He has received his master degree from the University of Oulu. He is interested in mobile multimedia systems and mobile digital rights management.

Frédéric Lassabe received his diploma in engineering in 2003. He is preparing his PhD thesis at the Computer Science Laboratory at the University of Franche-Comté in France. Currently, he is focusing on location-based services and service continuity. Researches on efficient mobility prediction models and accurate positioning models are connected topics.

Florian Ledermann is a researcher and PhD student at the Interactive Media Systems Group, Vienna University of Technology, where he also received his master's degree in computer science and design. His current research interests include semantic systems, authoring for ubiquitous systems, and user interface design.

Say Ying Lim is currently a postgraduate student in Monash University, Australia, under the School of Computer Science and Software Engineering. She has completed her Bachelor of Business Systems and Master of of Business Systems by Projects, both also at Monash University.

Abdulhussain E. Mahdi is a senior lecturer at the University of Limerick, Ireland. He is a chartered engineer (CEng), member of the Institute of Electrical Engineers, UK (MIEE), and a member of the Engineering Council, UK. He is a graduate in electrical engineering from the University of Basrah (BSc 1st Class Hon. 1978) and earned his PhD in electronic engineering at University of Wales, Bangor, in 1990. His current research interests include: speech processing and applications in telecom and rehabilitation, and domain transformation and time-frequency analysis. He has authored and co-authored more than 75 referred journal and conference articles and has edited one book. His published work has been cited in more than 32 journal articles. He chaired one conference and co-chaired another. He is a regular reviewer and member of the Steering Committees for a number of national, European and international conferences, and has acted as a reviewer for a number of journals, books, and research funding bodies.

Jani Mäntyjärvi is a senior research scientist at Technical Research Center of Finland, VTT Electronics, in the advanced interaction technologies group. He received his PhD degree in the department of electrical engineering and information processing at the University of Oulu, 2003. His professional interests include context-aware computing for mobile devices and technologies for adaptive user interaction.

José Ruiz Mas is the associate professor in the Department of Electronics Engineering and Communications (University of Zaragoza, Spain) and a member of Aragón Institute of Engineering Research (I3A). His research interests in mobile communications and multimedia applications.

Mounir Mokhtari is an associate professor at the National Institute of Telecommunication (INT) in France. He obtained a master's in physical science from the Universite de Val de Marne France in 1992, then a master's degree in networking from INT (1994). He collaborates with the University Pierre and Marie Curie to obtain his PhD in computer science in 1997. His research activity focuses on human machine interaction, rehabilitation robotic and health thematic. He obtained his qualification of research director from the University Pierre and Marie Curie in computer sciences, November 2002.

Jean-François Moreau is a student at the Laboratory DOMUS. He works on interface designed for the cognitively-impaired people.

José García Moros is an associate professor in the Department of Electronics Engineering and Communications (University of Zaragoza, Spain) and member of Aragón Institute of Engineering Research (I3A). His research interests are in telemedicine.

Ghita Kouadri Mostéfaoui is a postdoctoral fellow at the University of Montreal (Canada) and an external scientific collaborator at the University of Fribourg (Switzerland). She holds a PhD in computer science from both the University of Fribourg and the University of Paris 6.

Julián Fernández Navajas is an associate professor in the Department of Electronics Engineering and Communications (University of Zaragoza, Spain) and a member of Aragón Institute of Engineering Research (I3A). His research interests are in telemedicine.

Eduardo Antonio Viruete Navarro received his MSc in telecommunications engineering (2003). He is a PhD student in the Department of Electronics Engineering and Communications (University of Zaragoza, Spain). His research interests are in network evaluation and telemedicine.

Holger Nösekabel is currently a research assistant at the University of Passau. His research interests include mobile computing, specifically mobile education, which was also the topic for his doctoral dissertation. Other related areas of interest encompass usability aspects and multimedia content for mobile devices.

Eamonn O'Neill is a senior lecturer in the Computer Science Department at the University of Bath. His research interests include human computer interaction, mobile, pervasive, and wearable computing, novel interaction techniques, context awareness and creativity.

Marko Palola is a research team leader at VTT focusing on mobile services, mobile service performance, and mobile video streaming. He received his master's degree from the University of Oulu.

Ai-Chun Pang serves as an assistant professor at the Graduate Institute of Networking and Multimedia, National Taiwan University, Taipei, Taiwan since 2004. Her research interests include design and analysis of personal communications services network, mobile computing, voice over IP, and performance modeling.

Johannes Peltola is a research team leader at VTT specializing on multimedia signal processing, source compression techniques, and content analysis. He has received his master's degree from the University of Oulu.

Tom Pfeifer joined the TSSG Research Group of the Waterford Institute of Technology, Ireland, as a principal investigator, director of research in ubiquitous and pervasive computing, in 2004. He has been leading consultancy projects for the Fraunhofer Institute FOKUS and TU Berlin, and teaches as a visiting professor at various universities. He received his Dr.-Ing. (PhD) in computer science at the Technical University of Berlin (2000) and holds a Dipl.-Ing. (master's) degree in electrical engineering/electronics from the Technical University of Dresden, Germany (1989). He has published numerous papers and serves in editorial boards of journals and chairs in international conferences.

Dorel Picovici is a senior researcher at the University of Limerick, Ireland. He is a member of the Institute of Electrical and Electronics Engineers (MIEEE), and a founder of the International Speech Communication Association—Student Advisory Committee (MISCA-SAC). Dr. Picovici is a graduate in electronic engineering from the Technical University of Cluj-Napoca (BSc in1999). He earned his master's of engineering and his PhD in electronic and computer engineering at the University of Limerick (2000 & 2004). His research interests are in the area of non-intrusive objective speech quality assessment. His current project investigates the development of real-time non-intrusive speech quality evaluation methods to adapt the VoIP and wireless networks to sustain a specified perception based quality of service. He has authored and co-authored more than 19 refereed journal and conference articles. To date, he acted as a reviewer for one IEEE journal and reviewer/member of the Steering Committees for three international conferences.

Hélène Pigot received a PhD in computer science from the University of Paris and a bachelor's in occupational therapy. She is a researcher of the laboratory DOMUS at the Sherbrooke University. Her main research interests are cognitive assistance and activity recognition.

Luís Ferreira Pires is an associate professor of the faculty of electrical engineering, Mathematics and computer science of the University of Twente. His research interests include design methods and architectures for telematics systems, especially for context-aware and mobile applications.

Key Pousttchi heads the Mobile Commerce Working Group at the Chair of Business Informatics and Systems Engineering at the University of Augsburg, Germany, and is a member of the visiting faculty of the University of Zurich, Switzerland.

Ioannis Priggouris received his BSc in informatics from the Department of Informatics & Telecommunications at the University of Athens, Athens, Greece in 1997 and his MSc in communication systems and data networks from the same department in 2000. Over the last few years, he has been a PhD candidate in the research area of mobile Internet. Since 1999, he has been member of the Communication Networks Laboratory (CNL) of the University of Athens. He has participated in the EUROCITI (EUROpean CITIes platform for online transaction services) and the PoLoS (Integrated Platform for Location Based Services) project implemented in the context of IST, as well as several national research projects. His research interests are in the areas of mobile computing, QoS and mobility support for IP networks, pervasive computing, and location-sensitive network management.

Jani Puttonen received his MSc degree in information technology in the field of telecommunications from the University of Jyväskylä in 2003. He recently began his PhD studies concerning IP mobility and particularly handover techniques. His current research interests include intelligent handovers in ubiquitous heterogeneous all-IP environments.

Carolina Hernández Ramos received her MSc in computer science from the University of La Laguna (2004). She is a PhD student in the Department of Electronics Engineering and Communications (University of Zaragoza, Spain). Her research interests are in telemedicine.

Wolfgang Röckelein received his master's degree (Dipl.Wirtsch.Inf.) in business informatics in 1992 from the University of Bamberg, Germany. He received his PhD (Dr. rer.pol.) in business informatics in 1998 at the University of Regensburg, Germany. Then he joined the newly-formed Telemedicine Centre at the University Hospital Regensburg and is since then actively participating in national and EU-funded research projects related to mobile computing and telemedicine. Dr. Röckelein is the project manager of the NOAH project. Since January 2002, he has also served as the managing director of the Center for Mobility and Information (ZMI) of the University of Regensburg. Dr. Röckelein's special interests are wire-based and wireless computer networks, mobile IT, inter-organisational systems, system analysis, and software design. In October 2005, he joined EMPRISE Consulting Düsseldorf, Germany.

Jean-Pierre Savary works at France Telecom, section Research and Development, in France. Ergonomist, he is responsible of the project SYNAPSE, which is aimed to provide services to the disabled people.

Christos N. Schizas is a professor of computer science at the University of Cyprus. He received a BSc degree in electronic engineering from the University of London, UK, in 1978, the MBA degree from the University of Indianapolis, Indiana, in 1988, and a PhD degree in systems theory from the University of London in 1981.

Marten van Sinderen is an associate professor of the faculty of electrical engineering, mathematics, and computer science of the University of Twente. His research interests include architectural modeling of distributed systems, development of design methodologies, and development of telematics systems and service platforms.

Anastasis Sofokleous has followed undergraduate studies in computer science at the University of Cyprus (BSc 2002), graduate studies in advance computer science at University of Cyprus (MSc, 2004), and is a PhD candidate at Brunel University.

François Spies received his PhD and the French "Accreditation to Supervise Research" degrees in 1994 and 1999, respectively. He was an associate professor at the Computer Science Laboratory at the University of Franche-Comté in France from 1996-99. Since 1999, he has held a professor position at the University of Franche-Comté. Currently, he is focusing on managing video streams on wireless and mobile architecture. Researches on cooperative video cache strategies, including mobility and video quality levels, transport, congestion control, and quality of service of video streams, are the main developed topics.

Bala Srinivasan is a professor of information technology in the School of Computer Science and Software Engineering at the Faculty of Information Technology, Monash University, Melbourne, Australia. He was formerly an academic staff member of the Department of Computer Science and Information Systems at the National University of Singapore and the Indian Institute of Technology, Kanpur, India. He has authored and jointly edited six technical books and authored and co-authored more than 150 internationally-referred publications in journals and conferences in the areas of multimedia databases, data communications, data mining, and distributed systems. He is a founding chairman of the Australasian database conference. He was awarded the Monash Vice-Chancellor medal for post-graduate supervision. He holds a Bachelor's of Engineering Honours degree in electronics and communication engineering, a master's and a PhD degree, both in computer science.

Chris Stary is currently the head of the Department of Business Information Systems – Communications Engineering at the University of Linz. His interest is in model-based and user-oriented application design. Areas of application include multimedia systems, e-learning, organisational development, and adaptive systems engineering.

Henrik Stormer studied computer science and business economics in Saarbrücken, Germany. Afterwards, he was a research assistant of the Information and Communication Group at the University of Zurich where he received his PhD. Since March 2003, he has been working in the Information Systems Group at the University of Fribourg.

Miska Sulander received his MSc degree in information technology from the field of software engineering from the University of Jyväskylä in 2003. He has extensive experience with the enterprise networking and related technologies. Currently, he is continuing his PhD studies in the field of data communications. His current research interests include intelligent network and traffic management.

Willy Susilo received a PhD in computer science from University of Wollongong, Australia. He is currently a senior lecturer at the School of Information Technology and Computer Science of the University of Wollongong. He is the coordinator of the Network Security Research Laboratory at the University of Wollongong. His research interests include cryptography, information security, computer security, and network security. His main contribution is in the area of digital signature schemes, in particular fail-stop signature schemes and short signature schemes. He has served as a program committee member in a number of international conferences. He is a member of the IACR.

Sabin Tabirca is a lecturer in the Department of Computer Science at the National University of Ireland, Cork. His main research interest is parallel computation for scientific problems with an emphasis in number theory. He has published more than 80 articles in the areas of parallel and mobile computation, number theory, and combinatorial optimization.

David Taniar received his bachelor's (Honours), master's, and PhD degrees, all in computer science, with a particular specialty in databases. His research now expands to data mining and mobile databases. He has published research papers in these fields extensively including a recent book, *Object-Oriented Oracle* . Dr. Taniar now works at Monash University, Australia. He is an editor-in-chief of international journals, including *Data Warehousing and Mining, Business Intelligence and Data Mining, Web Information Systems, Web and Grid Services, Mobile Multimedia*, and *Mobile Information Systems*. He is a fellow of the Institute for Management Information Systems (FIMIS).

Do Van Thanh obtained his MSc in electronic and computer sciences from the Norwegian University of Science and Technology and his PhD in informatics from the University of Oslo. He is now at Telenor R&D in charge of research activities focusing on SIP, XML and next generation mobile applications. He holds a professorship at the Department of Telematics, Norwegian University of Science and Technology in Trondheim.

Wee Hyong Tok received his MSc from the School of Computing (SoC), National University of Singapore (NUS). He is currently a PhD candidate with the School of Computing, NUS. His research interests include adaptive query processing, optimization issues in continuous query systems, and spatial databases.

Klaus Turowski holds the chair of business information systems and systems engineering at the University of Augsburg. Prior to assuming his current position, he was a visiting professor at the University of the Federal Armed Forces Munich and assistant professor at the University of Magdeburg. He received a diploma degree in industrial engineering and management at the University of Karlsruhe and a PhD degree in business information systems at the University of Münster. He was a visiting professor at the University of Tartu (Estonia) and had further teaching assignments at the Universities of Darmstadt and Konstanz. Besides his theoretical background, he has been working in various consulting projects.

Ari Viinikainen received his MSc degree in information technology in the field of telecommunications from the University of Jyväskylä in 1997. Since then, he has worked in a variety of teaching positions at the University of Jyväskylä while continuing his post-graduate studies. His current research interests include traffic engineering and quality of service in wired and wireless networks, optimizing handover in mobile networks and new modulation techniques for cellular networks.

Elena Vildjiounaite is a research scientist at VTT. She received her master's degree from Leningrad Polytechnic Institute. Her research interests are in the area of context-aware systems, human-technology interaction, and personalization.

Roland Wagner is a professor for information systems at the Johannes Kepler University Linz. He is director of the Institute for Applied Knowledge Processing (FAW), managing various projects in the area of information systems.

Zhou Wang is a research associate at Fraunhofer's Integrated Publication and Information Systems Institute IPSI in Germany. His research is focused on mobile distributed information systems, including networking and middleware in mobile environments, nomadic service access, and location-based applications. He received PhD in computer science from the University of Karlsruhe in 2002.

Edgar R. Weippl is an assistant professor at the Vienna University of Technology. His research focuses on applied concepts of IT-security and e-learning. He has taught several tutorials on security issues in e-learning at international conferences, including ED-MEDIA 2003 and 2004. In 2005, Springer published his book *Security in E-Learning*. After graduating with a PhD from the Vienna University of Technology, he worked for two years in a research startup. He then spent one year teaching as an assistant professor at Beloit College, WI. From 2002 to 2004, while with the software company ISIS Papyrus, he worked as a consultant for a health insurance company in New York and for Deutsche Bank in Frankfurt, Germany.

Michael Welzl works at the Institute of Computer Science, University of Innsbruck, Austria, where he leads a research team on Network Support for Grid Computing. He received his PhD degree from TU Darmstadt, Germany.

Utz Westermann is a visiting scientist at VTT. He obtained his doctoral degree from the Technical University of Vienna and his diploma degree from the University of Ulm. He is interested in context-aware multimedia systems.

M. Howard Williams has been a professor of computer science at Heriot-Watt University, Edinburgh, since 1980. He has led the Database, Information, and Knowledge Base Research Group for over 20 years and published over 200 papers.

Khin Than Win is a lecturer in the School of Information Technology and Computer Science, University of Wollongong, Australia. She is a medical doctor with a PhD in health informatics. Her research interests are in issues related to electronic health record systems and quality and safety in healthcare. She teaches health informatics subjects and supervises several honours and post graduate research students in health informatics

Laurence T. Yang is a professor in computer science at St. Francis Xavier University, Canada. His research includes high performance computing with applications, design and testing of embedded systems, ubiquitous/pervasive computing and intelligence, autonomic and trusted computing. He has published around 200 technical papers in referred journals, conference proceedings, and book chapters in these areas.

Yuping Yang received her BSc and MSc in computer science from Huazhong University of Science and Technology, PRC and her PhD from Heriot-Watt University, UK. Her research interests focus on database integration and pervasive computing.

Zhiwen Yu received his PhD in computer science from Northwestern Polytechnical University, P.R.China in 2005. His research interests include ubiquitous multimedia and context-aware systems. He is a member of the IEEE. He has joined Nagoya University, Japan as a post-doctoral researcher since Feb. 2006.

Robert Zehetmayer received his master's degree in business and computer science at the University of Vienna. His research interests concern multimedia information systems and the Semantic Web, in particular in the mobile application domain.

Evangelos Zervas received the diploma in electrical engineering from the National Technical University of Athens, Greece. He obtained the MSc and PhD degrees in 1989 and 1993, respectively, both in electrical and computer engineering, from the Northeastern University, Boston, MA. In 1995, he joined the Faculty of the Technological Education Institute of Athens, Greece, where he is presently an associate professor in the Department of Electronics. His research interests include digital communications, information theory, adaptive filtering, neural networks, and mobile networks.

Daqing Zhang is the manager of the CAS Department at the I2R, Singapore. His research interests include service-oriented computing and context-aware systems. He received his PhD from University of Rome La Sapienza and University of L'Aquila, Italy.

Baihua Zheng received her PhD in computer science from the Hong Kong University of Science and Technology. She is a member of the IEEE and the ACM. Currently, she is an assistant professor in the School of Information Systems at Singapore Management University. Her research interests include mobile and pervasive computing, and spatial databases.

Index of Key Terms

Symbols

3G 127, 191, 433, 438, 522
3G (3rd generation) network 233, 431, 439, 522
3GP format 355
3GPP (3rd Generation Partnership Project) 514
3rd generation (3G) network 233, 431, 439, 522
3rd Generation Partnership Project (3GPP) 514
4G (4th generation) 532

A

AAA (authentication, authorization, and accounting) 130
accentuation 389
access control (see also *admission control*) 24
accessibility 441
ACDI (Authoring Challenges for Device Independence) 280
action 474
activities of daily living (ADL) 545, 547, 548
ad hoc
  (infrastructureless) networks 140
  mode 505
  networks 40, 172, 496
  -environments 172
Ada 261
adaptability 267, 293, 294, 442
adaptable tool 377
adapted technical aids 443
adaptive multi-dimensional histogram (AHM) 110
adaptive system 268
ADL (activities of daily living) 545, 547, 548
admission
  control (see also *access control*) 154, 164
  policy 505
advanced mobile services 127
agenda 545, 548, 551
aggregation 59, 110, 118, 177
aging 441
AHM (adaptive multi-dimensional histogram) 110
ALF (application layer framing) 131, 137
Alzheimer disease 544, 546, 551
ambient computing 444
analysis (see also *evaluation*) 184
annotation 355
anti-join 56
anti-virus software 29
any time 455
any where 455
application 447, 449, 478
  distribution 508, 510, 511, 512, 513, 517, 518
  kernel 508
  layer framing (ALF) 131, 137
  service provider (ASP) 565
architectural support 478, 490
area (see also *district*) 392
ASP (application service provider) 565
assistive
  aid 442
  device 441
  environment 441, 443, 453, 455
  system 445
  technology 441
asynchronous service discovery 170, 171
atomic function 387, 388
audio conference 418
augmented reality 383, 385
authentication 24
authentication, authorization, and accounting (AAA) 130
Authoring Challenges for Device Independence (ACDI) 280

autonomy 551, 553
availability 24
AVANTI Project 444
aware 442

B

BACK (binding acknowledgment) 180
Backend Layer (see also *Database Layer, Persistence Layer*) 508
bandwidth 141
  skimming 42
basic ADL 546
basic language skills 382
batching 42
behavior description 461, 467
best-effort network (see also *Internet*) 367
best-effort packet networks 357
bidimensional decision 90
billing 558, 563
binding
  acknowledgment (BACK) 180
  update (BU) 180
blind authentication 30
Bluetooth 3, 27, 406, 412
  network 406, 412
broadcast 167, 177, 192
  pull 111
  push 111
browser 512
browsing 391
BU (binding update) 180
buffering 182
business
  logic 508
  Logic Layer 508
  model (see also *revenue model*) 20
    types 21

C

cache 502, 505
calculation aggregation 60
Candela 355
care-of-address (CoA) 181, 189
caregiver 545
cascading style sheets (CSS) 284, 290
CC/PP (composite capability/preference profiles) 280
CDN (content distribution network) 43, 48
CE (congestion experienced) 135
cellular networks (see also *handover, mobile networks, wireless networks*) 140
centralized service directory (see also *distributed service directory, service directory*) 167
chaining 42, 43, 46
checksum 133
cHTML (compact HTML) 283
circuit switched data (CSD) 3
CIS (context information service) 449
CKNN (continuous kNN) 108
CL (COM Layer) 447
classification 239
client 279
  -server 111
  -server model 111
cluster 173
clustering 177
CML (Context Modeling Language) 471
CMM (context-aware multimedia middleware) 476, 477, 487, 488, 490
CN (corresponding node) 181
CoA (care-of-address) 181, 189
CODEC 367
cognitive
  assistance 545, 548
  deficit 545, 554
  impairment 544, 546, 547, 553
  prosthesis (see also *compensatory aid*) 545, 546, 548
  psychology 387
collaboration 369, 376, 378, 415, 417
  service 429
COM Layer (CL) 447
common features 313
communication (see also *transmission*) 312, 323, 324, 369, 370, 376, 546, 547, 550, 554
  control 324
compact HTML (cHTML) 283
compensatory aid (see also *cognitive prosthesis*) 546
component-based architecture 467
composite capability/preference profiles (CC/PP) 280
computer security 28
computing 320
conceptual partitioning (CPM) 112
condition rule 472
configuration 458
congestion

control 130, 358, 367
experienced (CE) 135
connecting 391
connection 550, 551
consistency 292
consumer behavior 120
content 556, 557, 558, 559, 560, 572
adaptation 276, 485
distribution network (CDN) 43, 48
provider 563, 564, 566
-aware video streaming 360
-device fit 121, 122, 125
context 266, 341, 342, 345, 346, 443, 447, 456, 471, 474, 490
acquisition 490
aggregation 482
assistive framework 441
awareness (see also *context-aware*) 259, 327, 329, 441, 445, 447, 448, 455, 456, 474
data 341, 342, 345, 348
discovery 484
framework 441
information 443, 457, 467, 474
information service (CIS) 449
learning 483
management 476, 477, 478, 482, 490
modeling 458, 472, 474
Modeling Language (CML) 471
query 483
reasoning 482
tunnelling 417, 418
-aware (see also *context awareness*) 353, 441, 464, 471, 477, 481, 487, 490
framework 442, 450
multimedia middleware (CMM) 476, 477, 487, 488, 490
services infrastructure 474
-based 477
continuous
kNN (CKNN) 108
media 492
queries 105, 106
spatial queries 118
conversion 313, 314, 317, 321
cookie 280
cooperation policy 505
cooperative evaluation 85
corresponding node (CN) 181
count aggregation 59
CPM (conceptual partitioning) 112
CRM (customer relationship management) 566
cryptography 25
CSD (circuit switched data) 3
CSS (cascading style sheets) 284, 290
customer
interface 271
interfaces 268
relationship management (CRM) 566
customization (see also *personalization, tailor*) 457

D

DA (directory agent) 168
DAD (duplicate address detection) 181
Daidalos project 320
data
broadcast 111, 113
dissemination 111
database 50, 53, 55, 275
Layer (see also *Backend Layer, Persistence Layer*) 508
management system (DBMS) 508
queries 52
datagram congestion control protocol (DCCP) 130, 133, 134, 138
DBMS (database management system) 508
DCCP (datagram congestion control protocol) 130, 133, 134, 138
decision
process 86, 87
support 86, 90
DEN (directory-enabled network) 567, 572
DEN-ng (DEN-new generation) 567, 568, 572
DEN-new generation (DEN-ng) 567, 568, 572
denial-of-service (DoS) 26
dependant people (see also *elderly people, people with disabilities, people with special needs*) 440, 441, 443, 455
design 386, 387, 389
theory 387
diffusion of innovations theory (see also *Rogers*) 120
digital
rights management (DRM) (see also *licensing*) 33, 355, 356
signature 30
directional stroke recognition (DSR) 74
directory
agent (DA) 168, 178

-enabled network (DEN) 567, 572
disability 441
disabled user 444
discovery services 468
display styles (see also *rendering styles, visualization styles*) 389, 395
distance learning (see also *e-learning*) 369
distant collaboration 369
distributed service directory (see also *service directory, centralized service directory*) 169
distribution 451
Level 509
district (see also area) 392
DoS (denial-of-service) 26
DRM (digital rights management) (see also *licensing*) 33, 35, 341, 342, 351, 355
DSDC (dynamic service discovery code) 447
DSR (directional stroke recognition) 74
duplicate address detection (DAD) 181
dynamic
customization of services 474
service discovery code (DSDC) 447
Web site generation 290

E

e- (electronic)
commerce 12, 20
health (see also *telemedicine*) 533
learning (see also *distance learning*) 369
services 260
ECA rule 470
ECN (explicit congestion notification) support 135
ecological data 545, 550
EDGE 9
elderly people (see also *dependent people, people with disabilities, people with special needs*) 441
electronic commerce (e-commerce) 12, 20
entity 468
environment 350, 353
epileptic patients 462
equi-join 56
eSarine Web shop 290
evaluation (see also *analysis, formative evaluation, self evaluation*) 377, 378
event 470, 471, 474
explicit congestion notification (ECN) support 135
exploring high-resolution images 83
export 468
Extensible Stylesheet Language (XSL) 286, 290
extension (see also *plugin*) 240

F

fast
binding update (FBU) 182
broadcasting (FB) 42
FB (fast broadcasting) 42
FBU (fast binding update) 182
FDMA (frequency division multiple access) 2
FER (frame error rate) 202, 204
FFHMIPv6 (flow-based fast handover for Mobile IPv6) 180, 189
filtering 57, 58, 66
fingerprinting (see also *patterns*) 29
firewall 538
flow-based fast handover for Mobile IPv6 (FFHMIPv6) 180, 189
formative evaluation (see also *evaluation*) 378
fractal 400, 401, 412
frame error rate (FER) 202, 204
framework (see also toolkit) 441, 442, 445, 447, 448, 450, 451, 478
frequency division multiple access (FDMA) 2
functional consistency 292

G

GDSS (group decision support system) 92, 101
geo-referenced data 545, 550, 551, 552
geographic service location (see also *positioning*) 173
gesture interaction 77, 85
GIF (graphic interchange format) 32
global
positioning system (GPS) (see also *location, positioning*) 148, 164, 355
system for mobile communications (GSM) 2, 9, 522
GOP (group of pictures) 367
GPRS 9, 127, 556, 565
GPS (global positioning system) (see also *location, positioning*) 148, 164, 355
graphic
interchange format (GIF) 32
user interface (GUI) 405, 407, 446
graphical user interface 407
group

decision 86, 92, 101
decision support system (GDSS) 92, 101
of pictures (GOP) 367
groupware (see also *group decision support system (GDSS)*) 418, 429
GSM (global system for mobile communications) 2, 9, 522
GSM multiple access processes 10
GUI (graphic user interface) 405, 407, 446

H

HA (home agent) 181, 190
Hand-held Device Mark-up Language (HDML) 5
hand-off-address (HofA) 187
handheld device 329, 442, 450, 455, 535, 536, 537, 538, 540, 541, 542
handoff (see also *handover*) 164, 492, 499, 503, 505
handover (see also *cellular networks, handoff, roaming*) 155, 164, 180
hardware 33
harmonic broadcasting (HB) 42
hash 177
HB (harmonic broadcasting) 42
HCI (human-computer interaction) (see also *usability*) 71, 327
HCLP (hybrid content location protocol) 174
HDML (Hand-held Device Mark-up Language) 5
head injury 544, 552
head-of-line blocking delay 131, 138
healthcare 462, 535, 536, 541, 542
hierarchical Mobile IPv6 (HMIPv6) 181
high speed circuit switched data (HSCSD) 3
Hilbert curve 118
histogram 118
historical synopsis 110
HMI 445
HMIPv6 (hierarchical Mobile IPv6) 181
HofA (hand-off-address) 187
home agent (HA) 181, 190
HSCSD (high speed circuit switched data) 3
HTML 512
human-computer interaction (HCI) (see also *usability*) 71, 327
human-machine interface 441, 455
hybrid content location protocol (HCLP) 174

I

I-frame 367
identification 389
import 468
IMS (IP multimedia system) 514, 515, 516, 522, 532
infrared 27
Infrared Data Association 27
infrastructure 448, 449, 457, 470, 474
INR (intentional name resolver) 169
INS (intentional naming system) 169
insertion policy 505
integrity 23
intentional
name resolver (INR) 169
naming system (INS) 169
interception 26
interface 267, 375, 377, 445
design 271
Internet (see also *best-effort network*) 357, 363
protocol version 6 (IPv6) 522
interoperability 170
intrusive objective voice quality measure 231
IP (Internet protocol)
multimedia system (IMS) 514, 515, 516, 522, 532
telephony 429
-TV (television over IP) 180, 190
IPv6 522, 532
IPv6 (Internet protocol version 6) 522

J

Java
context-aware framework (JCAF) 448
Virtual Machine (JVM) 168
JCAF (Java context-aware framework) 448
join operation 57, 58, 63
JSR-82 411
Julia set 400, 401, 412
JVM (Java Virtual Machine) 168

K

k-nearest neighbor (kNN) 104, 108, 118
query 114, 118
for moving query points (k-NNMP) 108
k-NNMP (k-nearest neighbor for moving query points) 108
key
and dial text entry 80
attributes 120
kinaesthetic interaction 72, 85
kNN (k-nearest neighbor) 104, 108, 118
query 114, 118
knowledge representation 481

L

LAH (location-aided handover) 162
LAM (location-aided mobility management) 162
language learning 370, 371
LCoA (on-link CoA) 181
LEMP 46
licensing (see also *digital rights management (DRM)*) 355
lightweight process 412
linking 391
local HA 182
location (see also *positioning*) 139, 341, 342, 347
  awareness 164, 338, 444
  fingerprinting 147
  -aided handover (LAH) 161
  -aided mobility management (LAM) 162
  -aware 164, 338, 444
    query 106
  -based queries 113
  -dependent 50, 52
  -dependent information processing (see also *mobile query processing*) 50, 54, 55, 70
logbook 382
logic rule 463

M

m- (mobile)
  business 10
  commerce (see also *mobile commerce*) 1, 3, 10, 268
  health 522
  payment 4, 10
  security 4, 10
  teams 416
M3 (mobile multimedia) 267
  communication 274
  presentation 273
maintenance 251, 262
Mandelbrot set 400, 401, 402, 411, 412
manipulation 26
MAP (mobility anchor point) 181
masquerading 26
mathematical monster 399
MBMS (multimedia broadcast/multicast service) 193
MBone (multicast backbone) 43
MC (mobile commerce) (see also m-*commerce*) 11, 20
mean opinion score (MOS) 211, 227, 228, 231
media
  gateway discovery protocol (MeGaDiP) 170
  transcoding 485
mediated reality 385
MeGaDiP (media gateway discovery protocol) 170
metadata 341, 346, 355
metainformation 390
MGIS (mobile network geographic information server) 161
middleware (see also *wireless middleware*) 259, 478, 486, 488, 490
MIDP (mobile information device profile) 512, 513
MIDP 2.0 (Mobile Information Device Profile version 2.0) 355
MIPL (Mobile IPv6 for Linux) 190
MIPv6 (Mobile IPv6) 180, 190
mixed-reality 79
MMDBMS 276
MMS (multimedia message service) 3, 10, 119, 233, 326, 433, 435, 437, 438, 439
MN (mobile node) 180, 190
MobiCon 344
MobiEyes 112
MobiJoin 111
mobile (see also *wireless*)
  ad hoc network 177
  business models 12
  commerce (MC) (see also *m-commerce*) 11, 20
  communication 326, 327, 328, 329, 336, 337
    devices 336
    technology 328, 331, 337
  computing (see also *pervasive computing, ubiqutous computing*) 1, 2, 10, 71, 332
    challenges 2
  content sharing (see also *video sharing*) 345
  data services 557, 569
  database (see also *multiple remote database*) 70
  device (see also *portable device*) 1, 279, 290, 535, 541, 545, 548, 549, 550
  health (m-health) 522
  information device profile (MIDP) 512, 513
  Information Device Profile version 2.0 (MIDP 2.0) 355
  IPv6 (MIPv6) 180, 190
  IPv6 for Linux (MIPL) 190
  learner 368
  magazine 556, 565, 566, 569, 572
  marketing 431, 437, 438

multimedia (M3) 101, 267, 557, 560
collaborative services 414, 419, 425, 428
multimedia value chains 557
networks 39, 140, 522, 505
network geographic information server (MGIS) 161
node (MN) 180
operator 556, 563, 565
query processing 70
security 25
service 11, 120, 417, 429
technology 49, 71
terminal 326, 505
video capture 344
video streaming (see also *streaming, video streaming*) 433
mobility 254, 259, 267, 505
anchor point (MAP) 181
management 180
pattern 496, 497
prediction 496
prediction model 496
Model-2 281, 288
model-based user interface development 293
MOS (mean opinion score) 211, 227, 228, 231
mouse 72
MP3 382
player 379
MPEG (see also *video*) 361, 367
MPEG-7 (multimedia content description interface) 351, 356
multicast 177
backbone (MBone) 43
overlay network 43
multicasting 192
multidimensional decision 90
multihoming 132, 138
multimedia 266, 327, 331, 439
adaptation 485
broadcast/multicast service (MBMS) 193
communication 274
content 266
content description interface (MPEG-7) 351, 356
database 275
delivery 485
message service (MMS) 3, 10, 119, 233, 326, 433, 435, 437, 438, 439
middleware 233
personalization 267
processing 476, 484, 487
recommendation 484, 490
security 29
services 477, 479, 487, 529
streaming (see also *mobile streaming, streaming, video streaming*) 39
multiple remote database (see also *mobile data base*) 65
multiple servers 50

N

N-teams 416
naming 389
naR (new access router) 181
navigation 391
nCoA (new CoA) 182
nearest-neighbor (NN) query 105, 107, 118
network
reconfiguration 154
resource management 139, 164
resources 192
Simulator 2 (ns-2) 190
networking 406
infrastructure 474
new
access router (nAR) 181
CoA (nCoA) 182
news
application 233
markup 233
markup language 233
syndication 234
nimble teams 415, 416
NN (nearest-neighbor) query 105, 107, 118
nomadic teams 416
non-intrusive objective voice quality measure 231
non-repudiation 24
ns-2 (Network Simulator 2) 190

O

oaR (old AR) 182
object-oriented technologies 258
object-role modeling (ORM) 471
objective voice quality measure 211, 231
old AR (oAR) 182
OMA DRM 341, 356
on-demand model 111
on-link CoA (LCoA) 181

on-mobile location-dependent information processing 70
ontology (see also *personal ontology*) 233, 341, 343, 346, 356, 475, 480, 481
-based classification 233
orientation 391, 392
ORM (object-role modeling) 471
overlay network 48
overview 391, 392

P

P-frames 367
P2Cast 44
p2p (point-to-point) 44
packet data network (PDN) 193
PALIO 444
Palm device (see also *handheld device, personal digital assistant (PDA)*) 536, 537
paper-based cross-media cooperation (PCMC) 558
Parlay 557, 558, 564, 565, 566
patching 42, 43
patterns (see also *fingerprinting*) 457
PB (pyramid broadcasting) 41
PCMC (paper-based cross-media cooperation) 558
PDA (personal digital assistant) (see also *handheld device, Palm device*) 10, 73, 279, 476, 535, 536, 541
PDN (packet data network) 193
people (see also *dependant people, elderly people*)
with disabilities 441, 442, 450, 451, 453
with special needs 440
perceptual quality 361, 363
periodic broadcasting 40
permutation-based pyramid broadcasting (PPB) 41
Persistence Layer (see also *Backend Layer, Database Layer*) 508
personal
communication 314, 315, 320, 324
digital assistant (PDA) (see also *handheld device, Palm device*) 10, 73, 279, 476, 535, 536, 541
firewall 538, 539
ontology (see also *ontology*) 341, 350
personalization (see also *customization*) 121, 122, 124, 233, 267, 276, 418, 429
personalized redirection 312, 317, 324
pervasive 472
computing 71, 259, 320, 324, 442, 452, 547, 553, 554
environment 441, 455
technology 71
piconet 407, 412
pie menus 72
pilot study 85
PKI (public key infrastructure) 5
platform entity container 452
plugin (see also *extension*) 240
PoC (push to talk over cellular) 518
point-to-point (p2p) 44
polymorph application 294, 297
portable
device (see also *mobile device*) 12
MP3 player 382
portal 563, 564, 569
positioning (see also *location*) 146, 164, 493
aggregation 59
post-join 70
PPB (permutation-based pyramid broadcasting) 41
PRCD 317
system 312
pre-join 70
pre-reservation 164
prediction 182, 502
predictive
positioning 496
query 110
prefetch 505
presentation 292
Layer 508
proactive (see also *pull-based*) 457
scheduling 40
projection 60
protection (see also *security*) 355, 535
protocol management 157
proxy 43, 48
public key infrastructure (PKI) 5
publisher 557, 559, 562, 563, 564
pull-based scheduling (see also *proactive, reactive*) 40
push to talk over cellular (PoC) 518
push-based scheduling 40
pyramid broadcasting (PB) 41

Q

QoP (quality of perception) 272
QoS (quality of service) (see also *service quality*) 6, 48, 141, 191, 194, 211, 212, 231, 276, 522, 532
quality
of perception (QoP) 272

of service (QoS) (see also *service quality*) 6, 48, 141, 191, 194, 211, 212, 231, 276, 522, 532
scaling 362
query 52, 53, 55, 59
processing 65, 66

R

R*-tree 106
R+-tree 106
R-tree 106, 113
RA (range aggregate) 110
RA (router advertisement) 182
RAMJ (recursive and mobile join algorithm) 111
range
aggregate (RA) 110
query 105
RCoA (regional CoA) 181
RDF site summary 234
reactive (see also *pull-based*) 457
scheduling 40
real-time
embedded multimedia systems 259
transport protocol (RTP) 202
really simple syndication 234
recursive and mobile join algorithm (RAMJ) 111
regional CoA (RCoA) 181
relational algebra set operations 62, 63, 65
relationship 356
reminder 545, 546, 554
remote
database 50
monitoring 554
sensing 393
server 67
supervision (see also *tele-monitoring*) 545, 548
removal policy 505
rendering styles (see also *display styles, visualization styles*) 386
representation model 476, 477, 478, 486, 488
repudiation 26
reusability 251, 255, 261
reusable component 460
revenue 14, 21, 564, 565, 566, 568
model (see also *business model*) 14, 21
source 14
reverse nearest neighbor (RNN) 105
rich site summary 234
RNN (reverse nearest neighbor) 105
roaming (see also *handover, handoff*) 505
Rogers (see also *diffusion of innovations theory*) 120
router advertisement (RA) 182
routing 155
RSS 234
RTCP (RTP control protocol) 202
RTP control protocol (RTCP) 202
RTP (real-time transport protocol) 202
rules 441, 457
description (for context-aware applications) 475

S

SA (service agent) 168
SB (skyscraper broadcasting) 41
scalability 174, 178
scatternet 407, 412
scenarios 313, 319
schizophrenia 544, 545, 546, 551, 552
SCTP (stream control transmission protocol) 130, 136, 138
SDMA (space division multiple access) 2
SDP (session description protocol) 526
SDS (service discovery service) 170
secrecy 23
security (see also *protection*) 120, 538, 539
policy 23
SEESCOA Project 443
self-evaluation (see also *analysis, evaluation*) 378
semantic 481
pack 242
service discovery 170, 171
semiotics 386, 387
sensitive application 450
sensor entity 452
sensors 441, 451, 452
sentient 471
server side solution 290
server 55, 56, 57, 65, 67
service 178, 429, 468, 475
service
agent (SA) 168
continuity 505
directory (see also *centralized service directory, distributed service directory*) 167, 178
discovery 166, 178, 475
service (SDS) 170
location protocol (SLP) 168
-oriented architecture (SOA) 415, 423, 429, 457, 467, 475

platform 414
provider 557, 558, 563, 565
quality (see also *quality of service (QoS)*) 120, 123, 124, 125
repository 178
services 450
infrastructure 475
session
description protocol (SDP) 526
initiation protocol (SIP) 321, 516, 518, 523, 533
short message service (SMS) 2, 233, 439
sibling 505
signal strength 505
signalling protocol 530
signs 387, 390
simple
service discovery protocol (SSDP) 167
tool 377
simulation 184
SIP (session initiation protocol) 321, 516, 518, 523, 533
session timer 204
situation 461
skyscraper broadcasting (SB) 41
SLP (service location protocol) 168
smart
home 440, 442, 446, 447, 553, 554
system 440
phones 412
SMS (short message service) 2, 233, 439
SOA (service-oriented architecture) 415, 423, 429, 457, 467, 475
software 34, 256, 258, 260, 261, 293
framework 256, 258
quality 293
sorting 60
space division multiple access (SDMA) 2
spatial
awareness 392
distance join 105
join 104, 105, 118
scaling 362
-temporal
aggregate 106
databases 118
prediction (STP) tree 110
queries 106
speech quality 211
SQL (Structured Query Language) 171
SSDP (simple service discovery protocol) 167
stable tool 377
start latency 505
status 390
storage (see also *video database*) 345, 346, 351
STP (spatio-temporal prediction) tree 110
strategy 432
stream control transmission protocol (SCTP) 130
streaming (see also *mobile video streaming, video streaming*) 39, 44, 48, 277, 374, 376, 382, 431, 433, 439, 505
technology 433
video 497
stroke 85
interaction 72, 73, 85
structure 348
Structured Query Language (SQL) 171
struts 281
framework 281
subjective
listening test 211, 227
voice quality test/measure 231
summative evaluation 378
SVFTP 367
system entity container 452

T

tablet PC 77
tailor (see also *customization*) 456
TAM (technology acceptance model) 120
targeted customer segment 14
task conformance 294
taxonomy 385, 395
of database operations 54
TCP 367
-friendly 130
rate control (TFRC) 130
streaming protocol 194
TDMA (time division multiple access) 2
technology
acceptance 127
model (TAM) 120
tele-
health (see also *telemedicine, e-health*) 533
monitoring (see also *remote supervision*) 462, 464, 466, 475
telecommunications 3
telemedicine (see also *e-health*) 522, 533
television over IP (IP-TV) 190

temporal scaling 362
territories 392
testability 251
text entry 72, 73, 80, 84, 85
TFRC (TCP-friendly rate control) 130
theory of reasoned action (TRA) 120
thin client 510
  architecture 509
third generation (3G) 522
  mobile network 233
thread 412
tilt the maze 79
time
  division multiple access (TDMA) 2
  -parameterized R-tree (TPR-tree) 107
toolkit (see also *framework*) 448, 451
touch screen 72
TP spatial join 106
TPR-tree 107
TPR-tree (time-parameterized R-tree (TPR-tree) 107
TPR-tree (time-parameterized R-tree) 107
TRA (theory of reasoned action) 120
traces 393
transmission (see also *communication*) 551, 554
TSUNAMI 443
tunneling 182

U

UA (user agent) 194, 168, 526
UAC (user agent client) 194
UAS (user agent server) 194
ubiquitous 71, 525, 384, 472, 477, 487, 490
  computing 71, 252, 384
  multimedia 477, 490
    services 487
UDP (user datagram protocol) 361
  -Lite 130
UE (user equipment) 193
UI (user interface) 267, 277, 328, 333, 334, 441, 442, 443, 444
UMTS (universal mobile telecommunications system) 193, 431, 439, 522
UMTS (universal mobile telephony system) 431
unidimensional decision 90
universal
  access 324
  mobile telecommunications system (UMTS) 193, 431, 439, 522
  mobile telephony system (UMTS) 431
upload 348
usability (see also *user-perceived quality*) 75, 294, 334
  engineering 293, 295
user 468
  agent (UA) 194, 168, 526
    client (UAC) 194
    server (UAS) 194
  context 317, 321, 325
  datagram protocol (UDP) 361
  entity container 452
  equipment (UE) 193
  interface (UI) 267, 277, 328, 333, 334, 441, 442, 443, 444
  mobility 166
  model 268
  preferences 314, 317, 318, 325
  -adaptive systems 268
  -perceived quality (see also *usability*) 292

V

V-card 430, 431, 432, 433, 437
V-teams 416
value
  -added mobile services 572
  chain 556, 557, 566, 569
video
  cache 498, 499, 506
  capture 344
  conference 418
  data 360
  database (see also *storage*) 341, 354
  delivery 358
  management 341, 345
  protocol 361
  scaling 367
  sharing (see also mobile content sharing) 341, 351, 353
  streaming (see also *content-aware video streaming, mobile video streaming, streaming*) 39, 358, 367, 433
virtual
  project teams 415, 416
  reality 386
visualization styles (see also *display styles, rendering styles*) 386
voice
  clarity 211

communication system 211, 212, 231
over IP (VoIP) 180, 187, 190, 192, 194
quality 211, 231
measurement (VQM) 211, 231
VoIP (voice over IP) 180, 187, 190, 192, 194
VQM (voice quality management) (see also *subjective voice quality test/measure*) 211
VQM (voice quality measurement) (see also *subjective voice quality test/measure*) 231

W

WAN (wireless wide area network) 27
WAP (wireless access protocol) 5
watermarking 30
wayfinding 386, 391
wearable computing 71
Web
page 279
service 341, 342, 343, 348, 429
site adaptation 279, 280, 290
weight lifting trainer application 77
WEP (wired equivalency privacy) 26
wired equivalency privacy (WEP) 26
wireless 406
access protocol (WAP) 5
broadcast 113
communication 12
computing 53
connection 550
devices 2
LAN (WLAN) 26, 191
Markup Language (WML) 283
middleware (see also *middleware*) 5
networks 140
technologies and standards 2
technology 70
wide area network (WAN) 27
WLAN (wireless LAN) 26, 191
WML (Wireless Markup Language) 283

X

XML 286
XSL (Extensible Stylesheet Language) 286, 290

Z

ZigZag 44